Classification for Functional Groups

Classification test	Experiment number	Class tested for	Page
Nitrous acid	18	Amines	245
	18e	Amino acids	248
Osazones	28	Carbohydrates	263
Periodic acid oxidation	25	Carbohydrates	258
Potassium permanganate solution	36	Hydrocarbons—Alkenes	279
		Hydrocarbons—Alkynes	
		Phenols	
Schiff's reagent (fuchsin–aldehyde reagent)	15	Aldehydes	235
Schotten–Baumann reaction (ester formation)	3a	Acid anhydrides	209
		Acyl halides	
Silver nitrate (ethanolic)	33	Acyl halides	270
		Carboxylic acids	
		Halides	
		Sulfonyl chlorides	
odium detection of active hydrogen	5	Alcohols	215
		Amines	
		Hydrocarbons—Alkynes	
Sodium bicarbonate test	29	Carboxylic acids	265
Sodium bisulfite addition complex	13	Aldehydes	232
		Ketones	
Sodium hydroxide fusion	45	Sulfonamides	300
Sodium hydroxide hydrolysis	16	Amides	237
		Nitriles	
Sodium hydroxide treatment	21	Amines	252
	42	Nitro compounds	293
Sodium iodide test	34	Halides	273
		Sulfonyl chlorides	
Tollens test	14	Aldehydes	234
		Carbohydrates	
Zeisel's alkoxy method (hydroiodic acid)	32	Esters	268
		Ethers	
Zinc and ammonium chloride reduction	41	Nitro compounds	293

Classification for Functional Groups

Functional group	Name of classification test	Experiment number	Page
Acid anhydride	Anilide formation	4	211
	Ester formation (Schotten–Baumann reaction)	3a	209
	Hydrolysis	1	205
	Hydroxamic acid test	2b	206
Acyl halide	Anilide formation	4	211
	Ester formation (Schotten–Baumann reaction)	3a	209
	Hydrolysis	1	205
	Hydroxamic acid test	2b	206
	Silver nitrate (ethanolic)	33	270
Alcohol	Acetyl chloride	6	217
	Ceric ammonium nitrate	7	218
	Chromic anhydride (chromium trioxide, Jones oxidation)	8	221
	Hydrochloric acid/zinc chloride (Lucas test)	9	222
	Iodoform test	10	224
	Sodium detection of active hydrogen	5	215
Aldehyde	Benedict's solution	26	260
	Chromic anhydride (chromium trioxide, Jones oxidation)	8	221
	2,4-Dinitrophenylhydrazine	11	229
	Fehling's solution	27	262

Classification for Functional Groups

Functional group	Name of classification test	Experiment number	Page
Aldehyde (cont.)	Fuchsin–aldehyde reagent (Schiff's reagent)	15	235
	Hydroxylamine hydrochloride	12	231
	Sodium bisulfite addition complex	13	232
	Tollens test	14	234
Amide	Hydroxamic acid test	2c, 2d	207, 208
	Sodium hydroxide hydrolysis	16	237
Amine	Acetyl chloride	6	217
	Benzenesulfonyl chloride (Hinsberg's method)	17	242
	Nickel chloride, carbon disulfide, and ammonium hydroxide	19	251
	Nickel chloride and 5-nitrosalicylaldehyde	20	251
	Nitrous acid	18	245
	Sodium detection of active hydrogen	5	215
	Sodium hydroxide treatment	21	252
Amino acid	Copper complex formation	23	254
	Ninhydrin test	22	254
	Nitrous acid	18e	248
Carbohydrate	Acetyl chloride	6	217
	Benedict's solution	26	260
	Borax test	24	258
	Fehling's solution	27	262
	Osazones	28	263
	Periodic acid oxidation	25	258
	Tollens test	14	234
Carboxylic acid	Ester formation	3b	210
	Silver nitrate (ethanolic)	33	270
	Sodium bicarbonate test	29	265
Ester	Hydroiodic acid (Zeisel's alkoxy method)	32	268
	Hydroxamic acid test	2b	206
Ether	Ferrox test	30	267
	Hydroiodic acid (Zeisel's alkoxy method)	32	268
	Iodine test	31	267
Halide	Silver nitrate test (ethanolic)	33	270
	Sodium iodide test	34	273
Hydrocarbon–Alkene	Bromine test	35	276
	Iodine	31	267
	Potassium permanganate solution	36	279
Hydrocarbon–Alkyne	Bromine test	35	276
	Potassium permanganate solution	36	279
	Sodium detection of active hydrogen	5	215
Hydrocarbon–Aromatic	Azoxybenzene and aluminum chloride	38	287
	Chloroform and aluminum chloride	39	288
	Fuming sulfuric acid	37	285
Ketone	2,4-Dinitrophenylhydrazine	11	229
	Hydroxylamine hydrochloride	12	231
	Iodoform test	10	224
	Sodium bisulfite addition complex	13	232
Nitrile	Hydroxamic acid test	2c	207
	Sodium hydroxide hydrolysis	16	237
Nitro compound	Ferrous hydroxide reduction	40	292
	Sodium hydroxide treatment	42	293
	Zinc and ammonium chloride reduction	41	293
Phenol	Acetyl chloride	6	217
	Bromine water	44	298
	Ceric ammonium nitrate	7	218
	Ferric chloride–pyridine reagent	43	296
	Liebermann's test	18d	248
	Potassium permanganate solution	36	279
Sulfonamide	Sodium hydroxide fusion test	45	300
Sulfonic acid	Hydroxamic acid test	2e	208
Sulfonyl chloride	Hydroxamic acid test	2e	208
	Silver nitrate test (ethanolic)	33	270
	Sodium iodide test	34	273

The Systematic Identification of Organic Compounds

Seventh Edition

Ralph L. Shriner
Christine K.F. Hermann
Terence C. Morrill
David Y. Curtin
Reynold C. Fuson

John Wiley & Sons, Inc.
NewYork • Chichester • Weinheim • Brisbane • Singapore • Toronto

ACQUISITIONS EDITOR Nedah Rose
MARKETING MANAGER Kimberly Manzi / Catherine Beckham
PRODUCTION EDITOR Sandra Russell
DESIGNER Harold Nolan
ILLUSTRATION EDITOR Edward Starr
This book was set in 10/12 Times Roman by York Graphic Services, Inc. and printed and bound by Hamilton Printing Company. The cover was printed by Hamilton Printing Company.

This book is printed on acid-free paper. ♾

The paper in this book was manufactured by a mill whose forest management programs include sustained yield harvesting of its timberlands. Sustained yield harvesting principles ensure that the numbers of trees cut each year does not exceed the amount of new growth.

Library of Congress Cataloging-in-Publication Data:
The systematic identification of organic compounds / R.L. Shriner . . . [et al.]—7th ed.
p. cm.
Includes index.
ISBN 0-471-59748-1 (cloth : alk. paper)
1. Chemistry, Organic—Laboratory manuals. I. Shriner, Ralph Lloyd, 1899– .
QD261.S965 1997
547′.34′078—dc21 97-5545
CIP

L.C. Call no. Dewey Classification No. L.C. Card No.

Printed in the United States of America

10 9 8 7 6 5 4 3 2 1

Other books by the same authors

Robert M. Silverstein, G. Clayton Bassler, and Terence C. Morrill
Spectrometric Identification of Organic Compounds

Ralph L. Shriner and Rachel H. Shriner
Organic Syntheses. Cumulative Indices to Collective Volumes I, II, III, IV, and V.

The Systematic Identification of Organic Compounds

7th Edition

Preface

This edition is dedicated to the memory of R.L. ("Ralph") Shriner. Ralph Shriner passed away on June 7, 1994 at the age of 94 while this edition was being prepared. He was the major force in the publication of the first (1935) and other early editions of this text. Organic chemistry had long been a daunting component of the chemistry curriculum and the idea of an advanced course built on the identification of unknown organic compounds (organic qualitative analysis, or "org qual") was identified as a way of providing students with a method of organizing their thoughts about this subject in a way that would lead to improved comprehension and an appreciation of systematic methods of characterization. Shriner, along with C.S. "Speed" Marvel, then at the University of Illinois, a major center of chemistry and research, already had prepared mimeographed notes to teach the "org qual" lab course. When Marvel said he could not participate in the preparation of a book, it was up to Shriner, assisted by R.C. ("Bob") Fuson (also at Illinois), to publish the text. David Curtin was brought on board to incorporate spectroscopic methods (in the 4th edition, 1956, which was the edition used by TCM when he took org qual at Syracuse University in 1960). Interestingly, spectroscopy in the 4th edition included only infrared and ultraviolet methods. NMR was incorporated into the 5th edition (1964). One of us (TCM) was brought in in 1980 to update spectroscopic methods (6th edition, 1980), as well as to revise the book as a whole, and Chris Hermann did the heavy work necessary to get this edition (the 7th) off the ground.

Ralph Shriner received a bachelors degree in chemical engineering from Washington University (St. Louis) in 1921, followed by a PhD from the University of Illinois (Champaign-Urbana, 1925) under the direction of Roger Adams. After a two-year postdoctoral position at the SUNY–College of Agriculture, he returned to join the faculty of the University of Illinois in 1927, and he taught there (including rising to professor and department chair) until he left for the University of Indiana in 1941 to be their department chair. He left Indiana to assume the chair of the University of Iowa in 1947, and upon his "retirement" in 1963, he became a visiting professor at Southern Methodist University, retiring from that position in 1978. Professor Shriner received the 1962 James Flack Norris Award of the Northeastern Section of the American Chemical Society for his teaching. He was one of the early participants in the establishment of "Organic Synthesis", a well-known series of books that has been an excellent source of tested, and thus highly reliable, synthetic procedures. He, and his wife Rachel, published a comprehensive index for Organic Synthesis entitled "Organic Synthesis, Collective Volumes I–V". A testament to Ralph Shriner's popularity and the effect that he had upon college chemistry was the fact that when the 6th edition of "The Systematic Identification of Organic Compounds": appeared, a number of unsolicited (and very insightful) reviews came from chemistry professors in many locations, especially the midwest. Ralph Shriner was the coauthor of another book and published in excess of 120 papers. He also held a number of leadership positions in the American Chemical Society. An important niche in Professor Shriner's activities was the thoroughness with which he approached chemical or "wet" tests. He would scrutinize any new procedure and test it on literally dozens of compounds. He felt very strongly that the tests should work, and he did not want any test included in the book until it did work. A balky case had been the nitrogen test, which until the 6th edition still was somewhat of a problem child. It was Ralph Shriner's efforts at SMU which allowed the 6th edition to include a nitrogen test which has proved to be very reliable.

In the preparations of this 7th edition two major areas were addressed. First, it was clear that modernization was important, especially in areas such as spectroscopy. Secondly, reviewers and users have requested flexibility. Thus chemical and spectroscopic methods were separated in this edition. This allows users to choose how much and where wet tests vs. nuclear magnetic resonance, etc., should be used to teach this subject.

Chapter 1 continues to be an introduction including an updated description of the ever-important area of lab safety. Chapter 2 continues to be a survey of many of the following chapters; it is intended to give students an overview of how compound identification is carried out.

Preliminary examination and determination of physical properties are the main focus of Chapter 3. Preliminary examination means observing the color, appearance, etc., of the unknown. The determination of the melting and boiling points is described in this chapter. Moreover, thin-layer chromatography (TLC) and gas chromatography (GC) are introduced in this chapter. TLC and GC are here as options that may be employed to test the purity of a single unknown or the composition of a mixture. Use of chromatographic methods to separate mixtures is held until Chapter 10. Specialized methods of characterization are also covered in Chapter 3. These include specific gravity, refractive index, and optical rotation. It should be made clear that here, as well as elsewhere in this book, a number of techniques will be covered that may not be used in a given lab course. We are merely providing choices, and the instructor must pave the way by guiding students to what is required and to what is optional.

The molecular formula of a compound is an important source of information and this is the main topic of Chapter 4. The sodium fusion test has long been a way of the determining whether a compound contains nitrogen, sulfur, or halogen. Since this test brings the remnants of sodium fusion into contact with water, it clearly has hazardous aspects and requires great care and explicit guidelines from instructors. Users are challenged to think long and hard as to whether this procedure is appropriate in their teaching situation. Alternatives include simply providing students with hetero atom content and using mass spectrometric data as clues to atomic composition. We also describe combustion analysis and mass spectrometric approaches in this chapter. Actually conducting combustion analysis is an unlikely part of an organic qual course, but it is conceptually related, and we fell it is worthwhile for students to know about this research-related item. It is not uncommon for instructors to provide students with percent composition data as a source of clues to the identity of organic compounds.

Chapter 5 contains methods and theory of solubility classification. The fact that a compound is insoluble in water but soluble in a well-defined solution of base is a major clue in classification. The supposition that a compound is a carboxylic acid can be quickly checked by IR and the telltale broad O—H stretch linked to solubility classification clearly points to the carboxylic acid class of compound. Thus in this fashion solubility represents an important niche in the overall characterization of organic compounds.

As mentioned above, spectroscopic methods have been separated from chemical tests, and thus we provide spectrometric methods (IR, NMR, and UV) in Chapter 6. This of course required updating and certainly the greatest degree of development has been in the area of NMR. The expanded ability to use NMR to analyze organic compounds has been nothing short of incredible. The appearance of more common high fields (300 MHz seems to be very prevalent now), ever-expanding computer power, and the increasing ease of use of NMR software has made this area one of great activity. Special techniques such as 2D NMR are becoming more commonplace. The simple act of digitization of NMR results allows use of the "electronic superhighway" to do and teach organic chemistry. It is not uncommon now for students to be able to obtain their NMR results on a disk, and then to be able to expand signals, check integrations, etc., and finally to directly incorporate their results into a written report. IR has still has the important position of being a quick, reliable method of functional group identification. The major new development for IR is the area of Fourier transform IR (FT IR). While the basic information is essentially the same, the ability to obtain IR peak positions very accurately on very small amounts of compound, together with the power resulting from digitization of these results, is very meaningful.

Classification by chemical (or "wet") tests is covered in Chapter 7. And covered is the operative word. We have added a number of pieces of information to provide the user with the widest possible scope of information. Again we do not expect all labs to use all tests. But we are confident that an extensive menu has been provided from which to choose. An important addition here is the inclusion of cleanup procedures for every chemical test. Waste disposal has become a great source of concern and indeed an expense that is competitive with and even greater than expenses associated with the purchase of chemicals and equipment. We have endeavored to treat this area in a comprehensive fashion.

The approach to Chapter 8 (Derivatives) is very similar to that of Chapter 7. Again we provide the user with a wide array of information. We do not expect all labs to use all derivative procedures. Additionally we have cleanup procedures for every derivative procedure.

Chapter 9 contains the ever-popular "road-map" problems. The idea is to simulate a lab experience by providing the student with a wide range of clues to the structures of organic unknowns. These include simple problems, as well as those which are more challenging. Moreover we have problems that are based solely on chemical tests, problems that rely heavily on spectrometric information (IR, NMR, mass spectrometry, etc.), and problems that integrate chemical and spectrometric methods. The chapter begins with an explanation of how to do problems. Successful completion of these problems requires knowledge of the chemical tests and derivatives (information from, respectively, Chapters 7 and 8), an understanding of the value of spectrometric methods described in Chapter 6, and an appreciation of the melting point and boiling point tables for the various classes of organic compounds provided in the appendixes.

Techniques that serve to separate mixtures are described in Chapter 10. This includes extractions with aqueous acids and bases and chromatographic procedures. Chapter 11 is a description focused on literature methods useful for organic chemists. Here we have also included recent developments in electronic methods of searching, especially those provided by Chemical Abstracts.

We are grateful to the following chemists for contributing their time and ideas to this edition: Rogers Lambert (Radford University), C. F. H. Allen (Rochester Institute of Technology), William Bigler (Rochester Institute of Technology),William Closson (SUNY-Albany), Louis Freidrich (University of Rochester), Robert E. Gilman, (Rochester Institute of Technology), Jack Kampmeier (University of Rochester), P. A. S. Smith (University of Michigan), David Strack (Waters Associates), Thomas P. Wallace (Rochester Institute of Technology), Casey Swallow (Merrimack College), Charles Garner (Baylor University), Clelia Mallory (University of Pennsylvania), David Minter (Texas Christian University), Francis Knowles (University of California-Davis), William Suggs (Brown University), and Walter Zajac (Villanova University). Christine Hermann owes a special debt of gratitude to her husband, Richard Hermann, for drawing all of the new original artwork and his patience during the preperation of this manuscript.

In summary, we hope that we have provided a book which is useful. We would appreciate receiving any input from users, from such relatively mundane items as corrections of physical constants to broad recommendations we might use in future editions. Please feel free to inform us of anything you think might be of value to chemists at large.

Terence C. Morrill
Rochester, NY

Christine K.F. Hermann
Radford, Virginia

Contents

CHAPTER 1

Introduction

1.1 SYSTEMATIC IDENTIFICATION OF ORGANIC COMPOUNDS: THE NEED FOR ORGANIC QUALITATIVE ANALYSIS

The amount of information facing the student of organic chemistry is overwhelming. In roughly two dozen chapters of a standard organic text, the student encounters many chemical reactions. Literally millions of different organic compounds have been synthesized. Chemical companies sell thousands of compounds and industrial-scale production generates about 10,000 different compounds on various scales. And yet we maintain that characterization of organic compounds can be done by a handful of physical and chemical observations, if they are made in a systematic fashion. Aldrich Chemical Company lists about 9000 compounds. Many other chemical companies list only 100 to 1000 compounds. Industrially produced organic compounds number about 6000 to 10,000. Thus the list of more common and more readily available chemicals is much smaller than the millions that are possible.

In this text we have focused our attention on an even smaller list of compounds that can be used as "unknowns." The melting point-boiling point tables give a very accurate idea of the focus of this book. Instructors using this book may very well use other references (CRC reference volumes,[1] the Aldrich Company catalog, etc.) for a more extensive list of possibilities for "unknown" compounds.

Organic chemists are often confronted with either of the following extreme situations:

1. Determination of the identity of a compound that has no prior history. This is often the case for a natural-products chemist who must study a very small amount of sample isolated from a plant or animal. A similar situation applies to the forensic chemist who analyzes very small samples related to a lawsuit or crime.

[1]For example, *Handbook of Tables for Organic Compound Identification,* 3rd ed., edited by Z. Rappoport (CRC Press, Boca Raton, FL, 1967).

2. The industrial chemist or college laboratory chemist who must analyze a sample that contains a major *expected* product and minor products, all of which could be expected from a given set of reagents and conditions. It is entirely possible that such a sample with a well-documented history will allow one to have a properly preconceived notion as to how the analysis should be conducted.

The theory and technique for identifying organic compounds constitute an essential introduction to research in organic chemistry. This study organizes the accumulated knowledge concerning physical properties, structures, and reactions of thousands of carbon compounds into a systematic, logical identification scheme. Although its initial aim is the characterization of previously known compounds, the scheme of attack constitutes the first stage in the elucidation of structure of newly prepared organic compounds.

If, for example, two known compounds A and B are dissolved in a solvent C, a catalyst D is added, and the whole subjected to proper reaction conditions of temperature and pressure, a mixture of new products plus unchanged starting materials results.

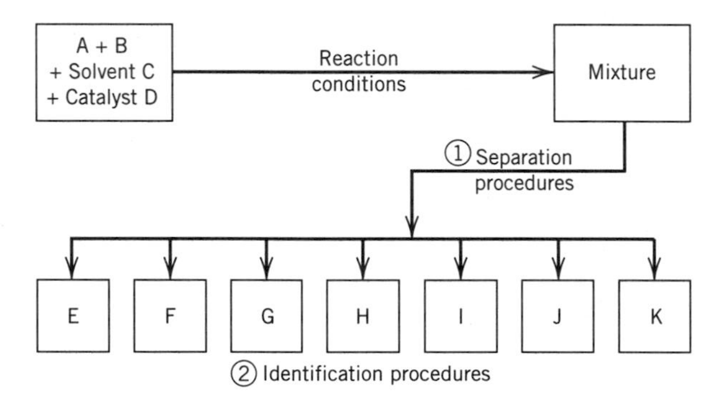

Immediately two problems arise:

1. What procedure shall be chosen to separate the mixture into its components?
2. How are the individual compounds (E through K) to be definitely characterized? Which ones are unchanged reactants? Which compounds have been described previously by other chemists? Finally, which products are new?

These two problems are intimately related. Separations of organic mixtures use both chemical and physical processes and are dependent on the structures of the constituents.

The present course of study centers attention on the systematic identification of individual compounds first. The specific steps are given in Chapter 2. Then the use of these principles for devising efficient procedures for the separation of mixtures is outlined in Chapter 10. The practical laboratory methods and discussion of principles for the steps in identification are given in Chapters 3 through 8.

In recent years the question of scale has become an issue. Scale has always been a focal point for qualitative analysis. The issue has been recognized at an even earlier point in the chemistry curriculum, and a very large number of colleges now incorporate some sort of microscale approach into their sophomore organic courses. (Here we loosely define microscale chemistry as the use of tens of milligrams of organic compound in a procedure, while macroscale reactions employ tens of grams.) Organic qualitative analysis has always been a test tube subject and thus should philosophically be in tune with the microscale revolution. We have left most of our experiments at the scale of the past editions of this text and thus many chemistry instructors may wish to scale down. We anticipate scaling down to 1/2, 1/5, or 1/10 of the cited amount should be very straightforward in most cases, and thus scale is the option of the course coordinator. The only warning is that certain reactions (for example, conversion of a carboxylic acid to an amide or of an alcohol to a 3,5-dinitrobenzoate) are notoriously sensitive to the purity of the reagents. Thus a larger scale reaction is likely desirable here.

Cleanup and Waste Disposal

A related, and in some ways bigger, issue is that of waste disposal. The trend at most colleges in recent years is to have waste disposal done by a licensed company under contract with the college. Most instructors are not qualified to dispose of waste and thus they can only provide cleanup guidelines. We have attempted to prepare this edition with that in mind. It is usually the job of the instructor to provide containers for waste disposal (it is now very rare that a chemical can be washed down the sink). Waste disposal vessels are usually labeled as to their use such as solids vs. liquids and inorganic vs. organic compounds. In some cases a special vessel is provided for especially toxic wastes such as halogenated organic compounds. Moreover, there are usually special containers for glass (especially broken glass) objects. There may be places to recycle paper, and finally there are simple trash cans for garbage. Thus there is usually a classification decision for every act of discarding material. And most importantly the students should receive instructions from their lab instructor that are in accord with local regulations.

1.2 SUGGESTIONS TO STUDENTS AND INSTRUCTORS

Schedule. An exact time schedule applicable to all schools cannot be set because of the varied use of semester, quarter, trimester, and summer session terms of instruction. However, for a semester of 15 weeks, two 3-hr laboratory periods per week, plus one "lab lecture" per week works well. Modification can be made to adapt the course to individual schools.

Lecture Material. The experiments and procedures (Chapters 7 and 8), and instructions for use of infrared and of nuclear magnetic resonance spectroscopy (Chapter 6) have been described such that students can study them as their work progresses. The first lecture should describe the course overview as outlined in Chapter 2. It is clearly not necessary to lecture on all the specific "recipes" listed in Chapters 7 and 8, although an introduction to some of the more commonly used tests is useful.

After the first one or two unknowns have been completed, it will be valuable to work some of the problems of Chapter 9 in class and discuss structure correlation with chemical reactions and spectral data.

Laboratory Work—Unknowns. By use of spectroscopic data and chemical reactions it is possible for students to work out six to eight single compounds and two mixtures (containing two or three components each) in a 15-week semester.

To get a rapid start and illustrate the systematic scheme, it may be useful to give a titratable acid to each student for a first unknown. The student is told that the substance is titratable and that he or she is to get the elemental analysis, melting or boiling point, and neutralization equivalent and to calculate the possible molecular weights.[2] Then, if

[2]Alternatively, the student can be given a compound with mass spectral data or elemental analyses (% C, H, N, O, . . .).

the unknown contains halogen or nitrogen, the student is to select and try three or four (but no more) classification tests. Next, a list of possible compounds with derivatives is prepared by consulting the table of acids (pp. 565–575). One derivative is made and turned in with the report (see pp. 22–23). This first unknown should be completed in two 3-hour laboratory periods.

Since many schools run organic qualitative analysis in a lab course connected to the second semester (or last term) of the traditional sophomore course, the decision about how to order the functional groups possible for the unknown may very well depend upon the order of coverage of these groups in the lecture course.

The other unknowns should be selected so as to provide experience with compounds containing a wide variety of functional groups.

It is often desirable to check the student's progress after the preliminary tests, solubility classification, and elemental analyses have been completed. This checking procedure is highly recommended for the first one or two unknowns for each student.

Purity of Unknowns. Although every effort is made to provide samples of compounds with a high degree of purity, students and instructors should recognize that many organic compounds decompose or react with oxygen, moisture, or carbon dioxide when stored for a considerable time. Such samples will have wide melting or boiling point ranges, frequently lower than the literature values. Hence, for each unknown the student should make a preliminary report of the observed value for melting or boiling point. The instructor should verify these data and if necessary tell the student to purify the sample by recrystallization or distillation and to repeat the determination of the physical constant in question. This avoids waste of time and frustration from conflicting data. (Read also pp. 37–50.)

Amounts of Unknowns. As a general guide, the following amounts are suggested:

Unknown No. 1, a titratable acid, 4 g of a solid or 10 mL of a liquid
Unknown No. 2, 3 g of a solid or 8 mL of a liquid
Unknown No. 3, 2 g of a solid or 5 mL of a liquid
Unknown No. 4, 1 g of a solid or 5 mL of a liquid

Mixtures should contain 4–5 g of each component. *Note:* If repurification of a sample is required, an additional amount should be furnished to the student.

The amounts listed above are essentially macroscale unknowns; use of analytical techniques and instrumentation such as thin-layer chromatography and gas chromatography may very well allow sample sizes of unknowns to be ca. 20% of that listed above. *In such cases, that is, for microscale samples, it is imperative that chemical test and derivatization procedures described in Chapters 7 and 8 be scaled down correspondingly.*

Toward the end of the term, when the student's laboratory technique has been perfected and facility in interpreting reactions has been obtained, it is possible to work with still smaller samples of compounds by using smaller amounts of reagents in the classification tests and by using a smaller scale in the derivatization procedure.

Timesaving Hints. It is important to plan laboratory work in advance. This can be done by getting the elemental analyses, physical constants, solubility behavior, and infrared and NMR spectra on several unknowns during one laboratory period. This information should be carefully recorded in the notebook and then reviewed (along with the discussion in each of these steps) the evening before the next laboratory period. A list of a few selected classification tests to be tried is made and carried out in the laboratory the next day. In some cases a preliminary list of possible compounds and desirable derivatives can be made. It is important to note that few of the 51 classification tests should be run on a given compound. It should not be necessary to make more than two derivatives; usually one derivative will prove to be unique. The object is to utilize the sequence of systematic steps outlined in Chapter 2 in the most efficient manner possible.

The instructor should guide the students so that the correct identification results by a process of logical deductive reasoning. Once the structure of the unknown is established, understanding of the test reactions and spectra becomes clear. Practice in this phase of reasoning from laboratory observations to structure is facilitated by early guidelines in Chapter 9. One method for developing this ability is for the instructor to write a structural formula on the blackboard and ask the students to predict the solubility behavior and to select the appropriate classification tests.

To tie together the identification work in this course with actual research, the instructor can select a few typical examples of naturally occurring compounds, such as nicotine, D-ribose, quinine, penicillin G, and vitamin B_1, and review the identifying reactions used to deduce these structures. The recent literature also furnishes examples of the value of infrared, ultraviolet, and NMR spectra in establishing structures. Knowledge of the mechanisms of the reactions used for classification tests and for preparing derivatives requires an understanding of the functional groups and their electronic structures.

Throughout this book, references to original articles, monographs, and reference works are given. Many of these will not be used during a one-semester course. However, the citations have been selected to furnish valuable starting sources for future work and are of great use in senior and graduate research.

The use of this manual will be greatly facilitated by the preparation of a set of index tabs for each chapter and parts of chapters. The time spent in preparing the index tabs is more than recovered in speeding up the location of experiments for functional groups, derivatization procedures, and tables of derivatives.

1.3 LABORATORY SAFETY

At all times, the instructor and students should observe safety rules. They should always wear safety glasses in the laboratory and should become familiar with emergency treatment.

Laboratories are places of great responsibility. Careful practice and mature behavior can prevent most mishaps. The following are all very important. Treating the lab with respect makes it far less dangerous.

Eye Protection. Goggles or safety glasses are to be worn at all times. These should conform to safety codes. Eyeglasses that are supposedly made from shatterproof glass are very likely inadequate. Side shields are a must for all protective eyewear.

Food and Drink. Food and drink should never be brought into a laboratory.

No Horseplay. At no time should horseplay or other antics be a part of activity in the laboratory.

No Unauthorized Experiments. At no time should unauthorized experiments be carried out. It may be common to obtain special test or derivatization procedures in texts to supplement this one. Even these experiments are not to be carried out unless approved by the instructor.

Read Labels Carefully. Most hazards are listed in detail on modern reagent bottles. Moreover, special instructions for the handling of certain reagents may be posted by the instructor. Most labs provide Manufacturers Safety Data Sheets (MSDS) for every chemical used. Students may obtain this information at any time. MSD sheets can be obtained in either paper or CD ROM form. Students are encouraged to obtain this information at any time to supplement the information obtained from their texts or instructors.

Waste Disposal. In recent years this area has become very rigidly defined. Rarely if ever are reagents merely poured down the sink. Vessels for wastes are normally provided in the lab. There may very likely be more than one type of vessel (for organics, inorganics, etc.). It may be necessary to carefully list (manifest) all reagents placed in a waste vessel. Common sense should be used; for example, it is highly likely that mixing a strong acid with a strong base will result in a violent reaction, so never allow this.

Fume Hoods. Most laboratories provide fume hood areas or bench top fume hoods. Always use these. If you think the hoods are not turned on, bring this to the attention of your instructor. Often students are provided with simple methods of testing hood efficiency and these should be used periodically. Safety regulations usually prohibit storage of toxic substances in hoods, and fume cupboards for such compounds are normally available.

Gloves. Most laboratories provide boxes of gloves. Modern gloves are quite manageable and still allow for handling of equipment with some agility. Gloves have their place and can certainly protect your hands from obnoxious odors or chemicals that can cause allergic responses. But they are not a license for sloppy technique. Moreover, they often are easily penetrated by some compounds. Due care is still required.

Compressed Gas Cylinders. Compressed gas cylinders, especially those that are nearly as tall as an adult, can be dangerous if not clamped to the bench top. Gas cylinders containing inert gases such as nitrogen or helium may well be around the lab. Cylinders containing chlorine or more toxic reagents should be stored in a fume cupboard.

Safety Equipment. The location of safety equipment should be made known to you. Moreover, you should know if and when you should use these.

Most of the following items should be readily available in the chemistry laboratory; items on this list or their description may vary due to local safety regulations:

Fire blanket
Fire extinguisher
Eye-wash fountain
Shower
First aid kit
Washes for acid or base (alkali) burns

Accident Reporting. All accidents should be reported. The manner in which they should be reported will be provided by the instructor. It is also important that someone accompany an injured person who is sent out of the laboratory for special care; if the injured person should faint, the injury could easily become compounded.

Medical treatment, except in the simplest of cases, is usually not the responsibility of the instructor. Very simple, superficial wounds can be cleaned and bandaged by the instructor. But any reasonably serious treatment is the job of a medical professional. The student should be sent to the college medical center accompanied by someone from the chemistry department. In all labs, the instructor should provide the students with instructions that are consistent with local regulations.

Explosion Hazards of Common Ethers

A number of violent explosions due to accidental detonation of peroxides, which can build up in common ether solvents, have been reported. These ethers include ethyl ether, isopropyl ether, dioxane, and tetrahydrofuran, among others. Isopropyl ether seems to be especially prone to peroxide formation. Apparently the greatest hazard exists when ethers have been exposed to air, especially for extended periods of time. The danger is enhanced when the ethers are concentrated, for example, by distillation. *Any ether solvent that displays a precipitate or that seems to be more viscous than usual may very well contain peroxides; do not handle such samples and report their condition to your instructor IMMEDIATELY.* The situation described here involves ether samples that are not acceptable for laboratory use.

There are a number of qualitative tests for the presence of peroxides in ethers; we describe two here. **Do not carry out these procedures without permission from your instructor.** Your instructor may decide that ether peroxide tests are not necessary if fresh ether is used.

Procedure A. Ferrous Thiocyanate Test for Peroxide. A fresh solution of 5 mL of 1% ferrous ammonium sulfate, 0.5 mL of 1 N sulfuric acid, and 0.5 mL of 0.1 N ammonium thiocyanate are mixed (and if necessary, decolorized with a trace of zinc dust) and shaken with an equal quantity of the solvent to be tested. If peroxides are present, a red color will develop.

Procedure B. Potassium Iodide Test for Peroxides. Add 1 mL of a freshly prepared 10% solution of potassium iodide to 10 mL of ethyl ether in a 25-mL glass-stoppered cylinder of colorless glass protected from light; when viewed transversely against a white background, no color is seen in either liquid. If any yellow color appears when 9 mL of ethyl ether is shaken with 1 mL of a saturated solution of potassium iodide, there is more than 0.005% peroxide and the ether should be purified or discarded.

Ferrous sulfate can be used to remove peroxides from ethers. Each liter of ether should be treated with 40 g of 30% aqueous ferrous sulfate. The reaction may well be vigorous and produce heat if the ethers contain appreciable amounts of peroxide; care should thus be exercised in using this procedure. The ether can then be dried (e.g., with magnesium sulfate) and distilled.

A simple method for removing peroxides from high-quality ether samples, without need for distillation apparatus or appreciable loss of ether, consists of percolating the solvent through a column of Dowex-1 ion exchange resin. A column of alumina can be used to remove peroxides and traces of water from ethyl ether, butyl ether, dioxane, and hydrocarbon solvents and to remove peroxides from tetrahydrofuran, decahydronaphthalene (decalin), 1,2,3,4-tetrahydronaphthalene (tetralin), cumene, and isopropyl ether.

CHAPTER 2

Identification of Unknowns

There are two situations in which the information outlined in this chapter can be applied. The first application is the exercise wherein a student is asked to identify a compound already described in the literature. The second use is the characterization of a new compound.

The following directions are intended as a guide in the process of identifying an unknown. Good laboratory technique dictates that students keep their own careful and systematic records of observations. The preparation of such records will, however, be greatly simplified by following the suggested sequence of operations.

We shall begin by assuming that the student has a sample, in hand, that is one compound. This compound has probably been characterized in the literature. If the sample is comprised of more than one major component, Chapter 10 on separation techniques should be consulted.

The sample would be given a preliminary examination including the ignition tests, followed by determination of the physical constants, such as melting point or boiling point. The unknown is then tested for the presence of nitrogen, sulfur, chlorine, bromine, iodine, and fluorine. Solubility tests are then used to simplify the list of possible functional groups. Infrared (IR) and ^1H nuclear magnetic resonance (NMR) spectra are obtained on the unknown. A mass spectrum of the sample may be a reasonable option. The student should consult with the instructor to confirm that the spectra are of acceptable quality. The solubility tests and the spectra are then interpreted, leading to the identification of any functional group(s) present. The student should then run at least two classification tests per sample, to confirm or deny the presence of the functional group(s) proposed for the compound. The classification tests are selected based upon the information obtained from the solubility tests and the spectra, especially the IR spectrum. The derivative tables (Appendix III) are then consulted from which the student chooses a list of possible compounds. More classification tests may be done to further restrict the choices. Preparation of one or two derivatives is the final confirmation of the identity of the unknown.

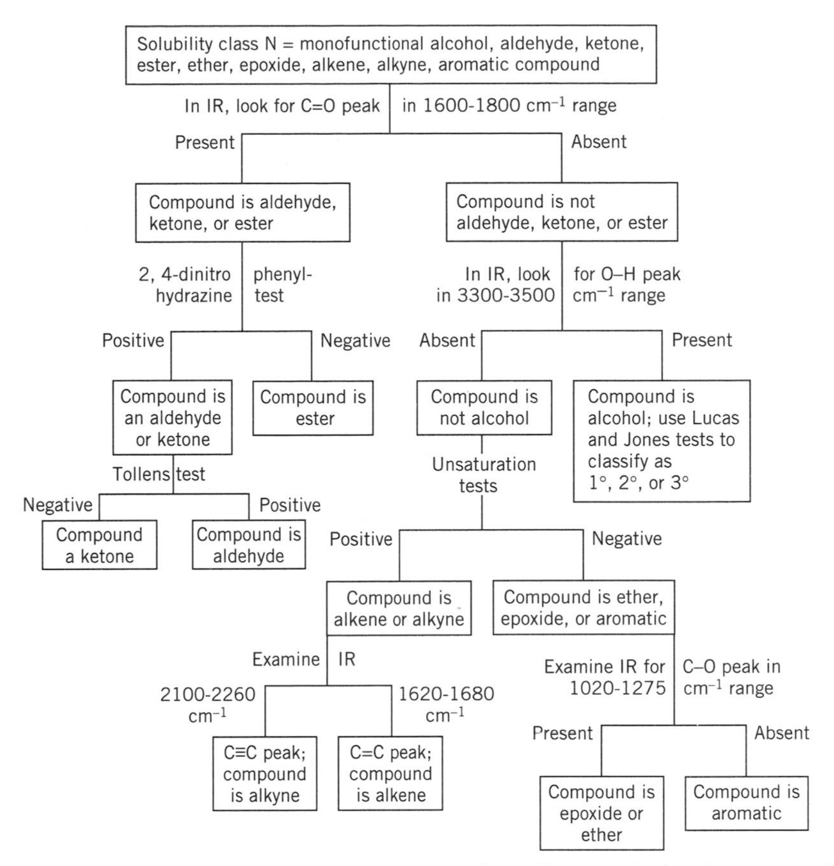

Figure 2.1 Example of a systematic approach to the identification of an unknown in solubility class N.

In Figure 2.1, a systematic approach to the identification of an unknown sample is illustrated using a flowchart format. The melting point or boiling point, the solubility class, the IR spectrum, and the NMR spectrum were determined or obtained for the unknown prior to the first instruction in this chart. In this example, its solubility class was found to be class N. The possibilities for this solubility class include alcohols, aldehydes, ketones, esters with one functional group and more than five but fewer than nine carbons, ethers with fewer than eight carbons, epoxides, alkenes, alkynes, and aromatic compounds. Then, by referring to the IR spectrum, the NMR spectrum, and the results of relevant classification tests, the functional group in the compound can be identified. Next, the melting point or boiling point of the unknown is compared with the list of compounds in Appendix III.

In Figure 2.2, an unknown with a solubility class of S_1 is analyzed in a similar manner. The S_1 solubility class includes monofunctional alcohols, aldehydes, ketones, esters, nitriles, and amides with five carbons or less. An elemental test is useful in this analysis, since it will determine the presence or absence of the nitrile or amide.

To further illustrate the concept of identifying a compound, let us apply these techniques to an actual unknown. The melting point of the unknown was determined to be 80°C. In the elemental tests with sodium fusion (pp. 83–90), the unknown did not produce a black solid with lead sulfide; thus sulfur is absent. With 2-nitrobenzaldehyde and

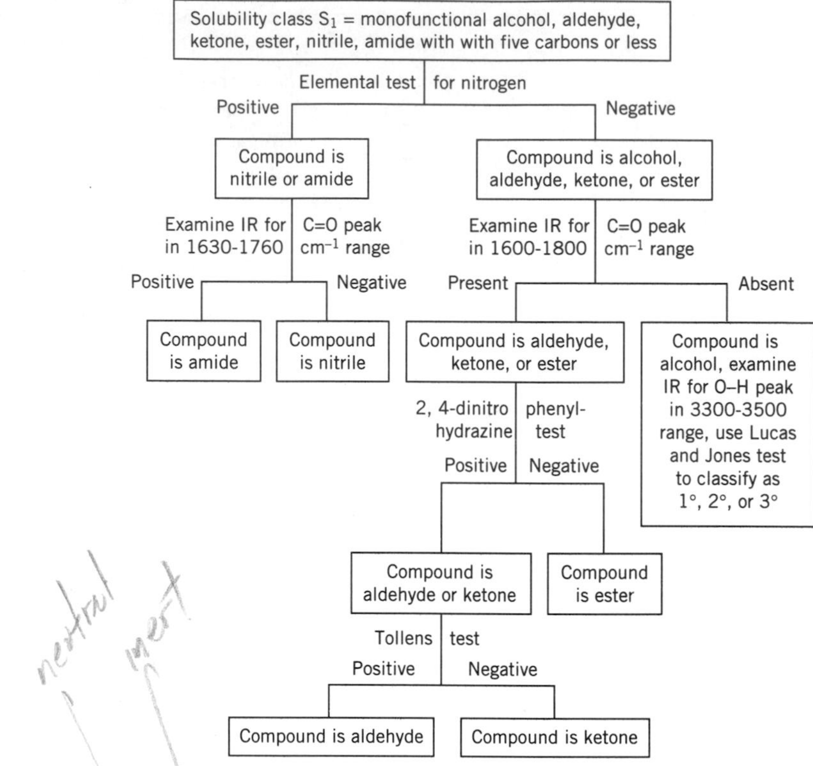

Figure 2.2 Example of a systematic approach to the identification of an unknown in solubility class S_1.

1,2-dinitrobenzene in 2-methoxyethanol, a blue-purple compound was formed, showing that nitrogen is present. No precipitate was formed upon treatment of the sodium fusion filtrate with silver nitrate. Therefore halogen is absent.

Following the solubility procedures in Chapter 5 (pp. 95–98), the unknown was found to be insoluble in water and insoluble in 5% sodium hydroxide. A definitive result was not obtained when 5% hydrochloric acid was used. Thus the unknown would be classified as B, MN, N, or I. The solubility class B includes aliphatic amines with eight or more carbons, anilines, and some ethers. The solubility class MN includes miscellaneous neutral compounds containing nitrogen or sulfur and having more than five carbon atoms. The solubility class N includes alcohols, aldehydes, ketones, and esters with one functional group and more than five but fewer than nine carbon atoms, ethers, epoxides, alkenes, alkynes, or aromatic compounds containing deactivating groups. Saturated hydrocarbons, haloalkanes, other deactivated aromatic compounds, and diaryl ethers are included in solubility class I.

In the ^{1}H NMR spectrum, a strong singlet at δ 2.77 ppm probably indicates an isolated methyl or methylene group. A multiplet in the range of δ 7.6–8.8 ppm shows that the unknown is an aromatic compound. The integration ratio of the singlet to the multiplet was 3:4, suggesting a disubstituted benzene ring and supporting the methyl group proposed above. In the IR spectrum, meta substitution on an aromatic ring is shown by peaks at 745 and 765 cm^{-1}.

Since a flowchart for solubility class N is presented in Figure 2.1, we can use it to assist us in the identification of the unknown. The presence of a strong carbonyl peak at 1670 cm^{-1} shows that the unknown is probably an aldehyde, a ketone, or an ester. The fact that this IR band is at less than 1700 cm^{-1} implies that the carbonyl group is conjugated with the benzene ring. A yellow-orange solid formed in the 2,4-dinitrophenylhydrazone test indicates that the compound is an aldehyde or a ketone. The Tollens test failed to produce a silver mirror, thus suggesting elimination of an aldehyde as a possibility. As a confirmation, the IR spectrum is examined for two peaks in the range of 2695–2830 cm^{-1} that correspond to the C—H stretch of an aldehyde. These peaks are absent in the IR spectrum of the unknown, confirming that the compound is not an aldehyde. The absence of a peak in the ^{1}H NMR spectrum in the range of δ 9.0–10.5, corresponding to the proton adjacent to the carbonyl in an aldehyde, also indicates that the unknown is not an aldehyde.

From the above information, we know that the compound is an aromatic compound containing nitrogen. In the ^{1}H NMR spectrum, a signal at δ 2.77 ppm indicates the presence of an isolated methyl group. The classification tests confirm the presence of a ketone. The IR spectrum suggests meta substitution, but if a group meta to the methyl substituent is present, this group does not contain hydrogens since no other hydrogens are seen on the NMR spectrum. The positive test for nitrogen, however, suggests that this group may be something like a nitrile group or a nitro group.

In the IR spectrum, nitriles show a C≡N stretch in the range of 2220–2260 cm^{-1} and nitro compounds show two $N(=O)_2$ bands in the ranges of 1259–1389 and 1499–1661 cm^{-1}. In the IR spectrum of the unknown, there are no peaks in the 2200 cm^{-1} range, but peaks at 1350 and 1545 cm^{-1} indicate the presence of a nitro group.

At this time, the derivative tables (Appendix III) are consulted under the headings of ketone and nitro to find any compounds with this melting point that match the criteria above. The compounds listed below are ketones and nitro compounds that have a melting point in the range of 75–85°C.

Compound	mp (°C)
1-Naphthyl phenyl ketone	76
2-Benzoylfuryl methyl ketone	76
2-Naphthoxy-2-propanone	78
4-Phenylcyclohexanone	78
4-Chlorobenzophenone	78
1,4-Cyclohexanedione	79
1,3-Diphenyl-1,3-propadione	81
3-Nitroacetophenone	81
4-Bromobenzophenone	82
Fluorenone	83

The only compound that contains both a keto group and a nitro group is 3-nitroacetophenone.

Other strategies are possible. The derivative tables can be consulted earlier in the process. Chemical tests for the nitro group can be used. Other tests for the carbonyl group, such as semicarbazide, are possible. The 2,4-dinitrophenylhydrazine result already suggests that derivatives are possible. In summary, there are often a number of reasonable ways to deduce the structure.

The following sections only briefly outline each technique. Therefore the student should consult later chapters for a more thorough discussion.

2.1 PRELIMINARY EXAMINATION

[Refer to Chapter 3, pp. 35–36.] Note whether the substance is homogeneous, and record its physical state (solid or liquid), color, and odor. The student should not directly sniff the substance, but merely note if an odor is noticeable during general laboratory operations.

Thin-Layer and Gas Chromatography. [Refer to Chapter 3, pp. 59–63 and 63–70, respectively.] Simple thin-layer chromatography (TLC) and gas chromatography (GC) are very simple and direct methods of purity determination. TLC and GC analyses are optional; consult with your instructor. Observation of only a single developed spot on a thin-layer chromatogram (after using solvents of differing polarity), a single peak on a gas chromatogram, and a sharp melting point, all lend strong support to a sample's purity. If the sample is a liquid or a solid, TLC should always be attempted. If the sample is a liquid, GC could be tried as well. Gas chromatography of reasonably volatile solids is also possible.

2.2 PHYSICAL CONSTANTS

[Refer to Chapter 3, pp. 37–58.] If the unknown is a solid, determine its melting point (pp. 37–43). If the melting point range encompasses more than 2.0°C, the compound should be recrystallized. Some pure compounds may not have a sharp melting point, especially if they undergo decomposition, such as turning dark, at or near the temperatures used for the melting point determination. If the unknown is a liquid or a very low melting solid, determine its boiling point (pp. 44–50); the range of this constant should not exceed 5.0°C except for extremely high boiling compounds. Distillation is recommended if the boiling point range indicates extensive contamination by a wide boiling point range, if the compound is heterogeneous, or if it appears to be discolored. Distillation at reduced pressure may be necessary for those compounds that show evidence of decomposition in the boiling point test.

As mentioned earlier, a sharp melting point is strong support for sample purity.[1] Narrow boiling point ranges do not, however, always indicate sample purity. Specific gravity (sp gr, pp. 50–55) was used in the past when NMR and IR were unavailable for structure determination. Occasionally specific gravity might be used for very inert compounds (e.g., certain hydrocarbons); in these situations, it might be one of the first steps in structure determination. Refractive index (pp. 56–58) values can be easily obtained and are of value in the identification of the unknown. NMR and IR spectroscopy has reduced the need for refractive index for initial structure determination.

2.3 MOLECULAR WEIGHT DETERMINATION

Molecular weight is normally very useful in determining organic structure; a reasonable estimate of the molecular formula can be postulated from the molecular weight. Mass spectrometry, discussed on pages 119–126 of Chapter 6, gives molecular weights for a

[1] Sharp melting points of mixtures can be misleading; these do not, however, occur very frequently.

wide range of organic compounds.[2] Molecular weights may also be obtained from neutralization equivalents (Procedure 1, p. 307) and saponification equivalents (Procedure 35, p. 354). These techniques apply to specific functional groups (Chapter 8, pp. 301–390).

2.4 MOLECULAR FORMULA DETERMINATION

[Consult Chapter 4, pp. 82–93] Simple "wet" or "test-tube" tests can be used to determine the presence of certain elements in the compound.

The compound should be tested for the presence of nitrogen, sulfur, chlorine, bromine, iodine, and fluorine (pp. 85–90). If a residue was noted in the ignition test, the student can identify the metal that it contains by inorganic qualitative methods.

Control Experiments. Results may be difficult to interpret, particularly if the student is unfamiliar with the procedure for decomposing the compound or with interpretation of the elemental tests. In this case, control experiments on a known compound should be carried out at the same time that the unknown is tested. The compound to be used for the control experiment should, of course, contain nitrogen, sulfur, and a halogen. A compound such as 4-bromobenzenesulfonamide is a good choice for the control experiments of this nature.

If mass spectrometry is available, an attempt should be made to determine the molecular formula of the organic compound from the cluster of peaks in the area of the molecular ion in the mass spectrum; these peaks are due to the isotopic contributions of elements in the molecular ion (p. 121). Mass spectral data can also be used to determine the presence and number of elements in the molecule that make unusually large or unusually small contributions to peaks in the molecular ion cluster (pp. 119–126).

Combustion analysis and other quantitative techniques for measuring elemental composition are useful in determining the structure of organic compounds; these procedures are generally not carried out in organic qualitative analysis labs, but the data from such procedures may be made available by the instructor.

The next stage in structural determination involves two steps. First, the student should determine the solubility (Chapter 5) to allow the placing of the unknown compound in a general structural class. Second, the student should determine the exact structure of the compound by detailed interpretation of the spectra (Chapter 6), by chemical tests (Chapter 7), and ultimately by chemical derivatization (Chapter 8).

2.5 SOLUBILITY TESTS

[Refer to Chapter 5, pp. 94–117.] Determine the solubility of the unknown following the solubility chart (Figure 5.1, p. 96) in water, ether, 5% hydrochloric acid, 5% sodium hydroxide solution, 5% sodium bicarbonate solution, and/or cold concentrated sulfuric acid (pp. 97–98). If the classification is doubtful, repeat the tests with control compounds

[2]Alternatively, instructors may feel compelled to provide the student with mass spectral data, molecular weights from colligative properties, or % C, H, N data in order to allow the student to have the experience of interpreting these data and applying them to structure determination.

that will give positive solubility tests and compounds that will give negative solubility tests. Compare the results of these tests with your unknown.

We also recommend solubility studies (pp. 115–116) in various organic solvents; results of these studies will be useful in choosing solvents for spectral analyses, for chromatographic analyses, and for purification by recrystallization.

When testing the solubility of the compound in water, the reaction to litmus (or other indicator paper) and phenolphthalein of the solution or suspension should be determined.

When the solubility behavior of the unknown has been determined, compose a list of the chemical classes to which the compound may belong. The results of these tests should agree with the information obtained from the IR and NMR spectra.

Preliminary Report. To avoid loss of time through mistaken observations, it is recommended at this point that the student consult with the instructor concerning the correct interpretation of the physical constants, elemental composition, and solubility behavior.

2.6 INFRARED, NUCLEAR MAGNETIC RESONANCE, AND MASS SPECTRA ANALYSES

Infrared (IR) and nuclear magnetic resonance (NMR) are crucial to organic structural determination. Infrared analysis (pp. 126–154) is an excellent functional group probe, which can be used in conjunction with the functional group chemical tests. Use of both IR and chemical tests *may* lead to structural diagnosis. Nuclear magnetic resonance also aids in the structure determination. NMR is essentially a method of determining the relative positions and numbers of spin-active nuclei. Both ^{1}H and ^{13}C NMR spectra can yield useful information concerning the types of protons or carbons present, such as aromatic or aliphatic; the number of adjacent protons (for ^{1}H NMR); and the number of protons attached to a particular carbon. Once some preliminary structures are chosen, mass spectrometry can be used to narrow down the choices by utilization of fragmentation patterns and molecular weight.

Interim Results. After interpretation of solubility results and IR and NMR analyses (recalling all results in the preliminary report), the student can usually propose one or more reasonable structures and subsequently proceed to the final characterization. Note that the instructor may well wish to review the student's interim results before the final characterization is attempted.

The final characterization stage involves application of the "wet" classification tests, detailed scrutiny of the NMR, IR, and perhaps the mass spectra, culminating in the derivatization of the compound; all of these steps are outlined below and discussed in detail in Chapters 6, 7, and 8.

2.7 CLASSIFICATION TESTS

[Refer to Chapter 7, pp. 200–300.] From the evidence that has been accumulated, the student must deduce what functional group or groups are most likely to be present in the unknown and test for them by means of suitable classification reagents. About 51 of the most important of these are mentioned in Chapter 7, where directions for their use may also be found. In Table 2.1 these tests are arranged according to the functional groups for which they are most useful.

Table 2.1 Classification Tests for Functional Groups (Chapter 7)

Functional group	Name of classification test	Experiment number	Page
Acid anhydride	Anilide formation	4	211
	Ester formation (Schotten–Baumann reaction)	3a	209
	Hydrolysis	1	205
	Hydroxamic acid test	2b	206
Acyl halide	Anilide formation	4	211
	Ester formation (Schotten–Baumann reaction)	3a	209
	Hydrolysis	1	205
	Hydroxamic acid test	2b	206
	Silver nitrate (ethanolic)	33	270
Alcohol	Acetyl chloride	6	217
	Ceric ammonium nitrate	7	218
	Chromic anhydride (chromium trioxide, Jones oxidation)	8	221
	Hydrochloric acid/zinc chloride (Lucas test)	9	222
	Iodoform test	10	224
	Sodium detection of active hydrogen	5	215
Aldehyde	Benedict's solution	26	260
	Chromic anhydride (chromium trioxide, Jones oxidation)	8	221
	2,4-Dinitrophenylhydrazine	11	229
	Fehling's solution	27	262
	Fuchsin–aldehyde reagent (Schiff's reagent)	15	235
	Hydroxylamine hydrochloride	12	231
	Sodium bisulfite addition complex	13	232
	Tollens test	14	234
Amide	Hydroxamic acid test	2c, 2d	207, 208
	Sodium hydroxide hydrolysis	16	237
Amine	Acetyl chloride	6	217
	Benzenesulfonyl chloride (Hinsberg's method)	17	242
	Nickel chloride, carbon disulfide, and ammonium hydroxide	19	251
	Nickel chloride and 5-nitrosalicylaldehyde	20	251
	Nitrous acid	18	245
	Sodium detection of active hydrogen	5	215
	Sodium hydroxide treatment	21	252
Amino acid	Copper complex formation	23	254
	Ninhydrin test	22	254
	Nitrous acid	18e	248
Carbohydrate	Acetyl chloride	6	217
	Benedict's solution	26	260
	Borax test	24	258
	Fehling's solution	27	262
	Osazones	28	263
	Periodic acid oxidation	25	258
	Tollens test	14	234
Carboxylic acid	Ester formation	3b	210
	Silver nitrate (ethanolic)	33	270
	Sodium bicarbonate test	29	265

Table 2.1—continued

Functional group	Name of classification test	Experiment number	Page
Ester	Hydroiodic acid (Zeisel's alkoxy method)	32	268
	Hydroxamic acid test	2b	206
Ether	Ferrox test	30	267
	Hydroiodic acid (Zeisel's alkoxy method)	32	268
	Iodine test	31	267
Halide	Silver nitrate test (ethanolic)	33	270
	Sodium iodide test	34	273
Hydrocarbon–Alkene	Bromine test	35	276
	Iodine	31	267
	Potassium permanganate solution	36	279
Hydrocarbon–Alkyne	Bromine test	35	276
	Potassium permanganate solution	36	279
	Sodium detection of active hydrogen	5	215
Hydrocarbon–Aromatic	Azoxybenzene and aluminum chloride	38	287
	Chloroform and aluminum chloride	39	288
	Fuming sulfuric acid	37	285
Ketone	2,4-Dinitrophenylhydrazine	11	229
	Hydroxylamine hydrochloride	12	231
	Iodoform test	10	224
	Sodium bisulfite addition complex	13	232
Nitrile	Hydroxamic acid test	2c	207
	Sodium hydroxide hydrolysis	16	237
Nitro compound	Ferrous hydroxide reduction	40	292
	Sodium hydroxide treatment	42	293
	Zinc and ammonium chloride reduction	41	293
Phenol	Acetyl chloride	6	217
	Bromine water	44	298
	Ceric ammonium nitrate	7	218
	Ferric chloride–pyridine reagent	43	296
	Liebermann's test	18d	248
	Potassium permanganate solution	36	279
Sulfonamide	Sodium hydroxide fusion test	45	300
Sulfonic acid	Hydroxamic acid test	2e	208
Sulfonyl chloride	Hydroxamic acid test	2e	208
	Silver nitrate test	33	270
	Sodium iodide test	34	273

The student is strongly advised against carrying out unnecessary tests, since they are not only a waste of time but also increase the possibility of error. For example, it would be pointless to begin the functional group tests of a basic nitrogen-containing compound by testing for keto or alcohol groups. On the other hand, tests that can be expected to give information about the amino group are clearly indicated.

Several of the tests for ketones and aldehydes are, in general, easier to carry out and more reliable than tests for other oxygen functions. It is advisable, therefore, in the classification of a neutral compound suspected of containing oxygen, to begin with the carbonyl tests, especially when IR analysis has indicated the presence of a carbonyl group.

In this book we have provided directions on how to interpret IR, NMR, and mass spectra and also have included sample spectra of most of the typical organic functional groups. For additional aid in interpreting the spectra of these compounds, organic and instrumental analysis texts should be consulted.

After deducing the structure of an unknown compound, or perhaps a few possible structures, derivatization should be carried out to confirm this structure. Although the melting point of the derivative may be sufficient to allow correct choice of the identity of the unknown, it may also be useful to characterize the derivative by chemical and spectral means, in a similar manner to the procedure used for the characterization of the unknown.

2.8 PREPARATION OF A SATISFACTORY DERIVATIVE

[Refer to Chapter 8, pp. 301–390.] After the solubility tests, the NMR spectrum, the IR spectrum, and perhaps the mass spectrum and the elemental tests, the student should propose a list of possible compounds for the unknown sample. These possible compounds may contain a number of structural differences. More classification tests may be needed to confirm or deny the existence of particular functional groups. Other characteristic properties, such as specific gravity, refractive index, optical rotation, or neutralization equivalent, may also be desirable. The final confirmation for the identity of the unknown can be accomplished by the preparation of derivatives. An index to derivatization procedures by functional group class is listed in Table 2.2. The melting points of these derivatives are listed in Appendix III.

Table 2.2 Index to Derivatization Procedures by Functional Group Class (Chapter 8)

Compound	Derivative	Procedure number	Page
Acid anhydride	Acid	2	308
	Amide	3b	309
	Anilide	4c	311
	4-Toluidide	4d	311
Acyl halide	Acid	2	308
	Amide	3b	309
	Anilide	4d	311
	4-Toluidide	4d	311
Alcohol	3,5-Dinitrobenzoate	10	317
	Hydrogen 3-nitrophthalate	11	318
	1-Naphthylurethane	8	315
	4-Nitrobenzoate	9	316
	Phenylurethane	8	315
Aldehyde	Dimedon derivative	16	323
	2,4-Dinitrophenylhydrazone	13	321
	4-Nitrophenylhydrazone	14	322
	Oxidation to an acid	17	324
	Oxime	15	323
	Phenylhydrazone	14	322
	Semicarbazone	12	321
Amide	Acetamide	19, then 20a	329, 333
	9-Acylamidoxanthene	18	328
	Benzamide	19, then 20b or 20c	329, 333 or 334

Table 2.2—continued

Compound	Derivative	Procedure number	Page
	4-Bromophenacyl ester	19, then 5	329, 312
	Hydrolysis to acid and amine	19	329
	4-Nitrobenzyl ester	19, then 5	329, 312
Amine—1° or 2°	Acetamide	20a	333
	Amine hydrochloride	26	338
	Benzamide	20b or 20c	333 or 334
	Benzenesulfonamide	21	335
	Phenylthiourea	22	336
	4-Toluenesulfonamide	21	335
Amine—3°	Amine hydrochloride	26	338
	Chloroplatinate	24	337
	Methyl iodide	23a	337
	Methyl 4-toluenesulfonate	23b	337
	Picrate	25	338
Amino acid	Acetamide	20d	335
	Benzamide	20d	335
	3,5-Dinitrobenzamide	29	344
	2,4-Dinitrophenyl derivative	30	344
	Phenylureido acid	28	343
	4-Toluenesulfonamide	27	343
Carbohydrate	Acetate	33	347
	4-Bromophenylhydrazone	32	347
	4-Nitrophenylhydrazone	32	347
	Phenylosazone	31	346
Carboxylic acid	Amide	3a	309
	Anilide	4a or 4b	310
	S-Benzylthiouronium salt	6	313
	4-Bromophenacyl ester	5	312
	Neutralization equivalent (NE)	1	307
	4-Nitrobenzyl ester	5	312
	Phenylhydrazide	7	313
	4-Toluidide	4a or 4b	310
Ester	Acid hydrazide	39	359
	Amide	34, then 3a	351, 309
	N-Benzylamide	38	358
	3,5-Dinitrobenzoate	37	358
	Saponification and hydrolysis	34	351
	Saponification equivalent (SE)	35	354
	4-Toluidide	36; 34, then 4a or 4b	356 351, 310
Ether—Alkyl	3,5-Dinitrobenzoate	40	359
Ether—Aromatic	Bromo derivative	44	364
	Nitro derivative	49	373
	Picrate	41	361
	Sulfonamide	42, then 43	362, 363
Halide—Alkyl	Alkylmercuric halide	45	366
	Alkyl β-naphthyl ether	47	367
	Alkyl β-naphthyl ether picrate	48	368

Table 2.2—continued

Compound	Derivative	Procedure number	Page
	S-Alkylthiouronium picrate	49	368
	Anilide	46	366
	N-Naphthylamide	46	366
Halide—Aromatic	Nitration	51	373
	Oxidation	50	370
	Sulfonamide	42, then 43	362, 363
Hydrocarbon—Aromatic	Aroylbenzoic acid	52	374
	Nitration	51	373
	Picrate	41	361
Ketone	2,4-Dinitrophenylhydrazone	13	321
	4-Nitrophenylhydrazone	14	322
	Oxime	15	323
	Phenylhydrazone	14	322
	Semicarbazone	12	321
Nitrile	Amide	53, then 3;	377, 309
		54	378
	Anilide	53, then 4a or 4b	377, 310
	Benzamide	55, then 20b or 20c	378, 333 or 334
	Benzenesulfonamide	55, then 21	378, 335
	Hydrolysis of nitrile	53	377
	α-(Imidioylthio)acetic acid hydrochloride	56	379
	Phenylthiourea	55, then 22	378, 336
	Reduction of nitrile	55	378
Nitro compound	Acetamide	57, then 20a	381, 333
	Benzamide	57, then 20b or 20c	381, 333 or 334
	Benzenesulfonamide	57, then 21	381, 335
	Reduction to amine	57	381
Phenol	Acetate	59	383
	Aryloxyacetic acid	60	384
	Benzoate	9	316
	Bromo derivative	61	384
	3,5-Dinitrobenzoate	10	317
	1-Naphthylurethane	8;	315
		58	383
	4-Nitrobenzoate	9	316
	Phenylurethane	8;	315
		58	383
Sulfonamide	Sulfanilide	64, then 62, then 63	388, 387, 387
	Sulfonic acid	64	388
	Sulfonyl chloride	64, then 62	388, 387
	N-Xanthylsulfonamide	65	389
Sulfonic acid	Benzylthiouronium sulfonate	66	389
	Sulfonamide	62, then 43	387, 363
	Sulfanilide	62, then 63	387, 387
	Sulfonyl chloride	62	387
	4-Toluidine salt	67	390
Sulfonyl chloride	Sulfonamide	43	363
	Sulfanilide	63	387
	Sulfonic acid	64	388

Properties of a Satisfactory Derivative

1. A satisfactory derivative is one that is easily and quickly made, readily purified, and gives a well-defined melting point. This generally means that the derivative must be a solid, because in the isolation and purification of small amounts of material, solids afford greater ease of manipulation. Also, melting points are more accurately and more easily determined than boiling points. The most suitable derivatives melt above 50°C but below 250°C. Most compounds that melt below 50°C are difficult to crystallize, and a melting point above 250°C is undesirable because of possible decomposition, as well as the fact that the standard melting point apparatus does not go higher than 250°C.
2. The derivative must be prepared by a reaction that results in a high yield. Procedures accompanied by rearrangements and side reactions are to be avoided.
3. The derivative should possess properties distinctly different from those of the original compound. Generally, this means that there should be a marked difference between its melting point and that of the parent substance.
4. The derivative chosen should be one that will single out one compound from among all the possibilities. Hence the melting points of the derivatives to be compared should differ from each other by at least 5°C.

For example, hexanoic anhydride (bp 257°C) and heptanoic anhydride (bp 258°C) would have very similar NMR and IR spectra. The amide derivatives, melting at 100°C and 96°C, respectively, are too similar to be useful. However, the anilide derivatives, hexananilide (mp 95°C) and heptananilide (mp 71°C), could be used to easily distinguish the, two compounds. Consult Chapter 8 and select a suitable derivative from those suggested.

When determining the physical constants for a compound, considerable latitude must be allowed for experimental error. Thus, if the boiling point is very high or the melting point is very low, the range between the observed constant and the ones listed in the book must be extended somewhat beyond 5°C. Other constants such as specific gravity (pp. 50–55), refractive index (pp. 56–58), and neutralization equivalents (p. 307) may be used, with proper allowance for experimental error, to exclude compounds from the list of possibilities. A complete list of possible compounds with all of the derivatives for each should be compiled.

Examination of the list of possibilities often suggests that additional functional group tests need to be performed. For example, if a list of possible nitro compounds contains a nitro ketone, carbonyl tests may be valuable, especially if the IR spectrum is consistent with the presence of a carbonyl group.

If this text does not describe a useful procedure for the preparation of a derivative, a literature search can be made for more procedures. The most direct way to make a thorough search for a particular compound is to look for the molecular formula in the formula indices of each of the following works in order.

Beilstein's *Handbuch der organischen Chemie,* 4th ed., Index to the 2nd to 5th Supplements covers the literature to 1979. The value of Beilstein's *Handbuch* is such that further discussion of its use is warranted. The main work (*Hauptwerk*) covers the literature through 1909 in its 27 volumes. The organization is based on structure in such a way that it is possible to find a desired compound rather easily without using the index once one is familiar with the work. For instance, acyclic hydrocarbons, alcohols, aldehydes, and ketones are contained in Volume 1; acyclic acids in Volume 2; acyclic hydroxy, aldehydo, and keto acids in Volume 3; sulfonic acids, amines, and phosphines in Volume 4. In Volume 5, cyclic hydrocarbons, including aromatic compounds, are covered, and the presentation of cyclic compounds continues along these lines until Vol. 17, where the discussion of heterocyclic compounds begins.

Once a particular compound is located in the main work, it is easily found in the supplements. The First Supplement (*Erstes Ergänzungswerk*), which covers the literature from 1910 to 1919, has the same order as the main work. Thus a compound found in Vol. 1 of the main work is in Vol. 1 of the First Supplement. Furthermore, an auxiliary set of page numbers at the top center of each page of each supplement relates the material on that page to the corresponding pages in the main work. The Second, Third, Fourth, and Fifth Supplements cover 1920–1929, 1930–1949, 1950–1959, and 1960–1979, respectively, in the same manner.

The CRC *Handbook of Tables for Organic Compound Identification*[3] and the *Dictionary of Organic Compounds*[4] list many characteristics of the individual compounds such as melting point, boiling point, refractive index, and specific rotation. The former gives references for the general procedures for the preparation of the derivatives, while the latter gives the literature references for each individual compound and derivative.

In *Chemisches Zentralblatt,* the Collective Indices cover the years, 1922–1924, 1925–1929, and 1930–1934, and the Annual Formula Indices cover the years 1935–1939.

The most reliable indexes are the *Chemical Abstracts* Subject, Author, and Formula Indices. Decennial Indices cover the literature during the years 1937–1946, and 1947–1956. At this point, the cumulative indexes cover only 5-year spans. Cumulative indexes for 1957–1961, 1962–1966, 1967–1971, 1972–1976, 1977–1981, 1982–1986, and 1987–1991 have been published. The annual indices must be consulted for later years. Although there is a Collective Formula Index to *Chemical Abstracts* for the years 1920–1946 as well as annual Formula Indices for later years, these are not complete and should not be depended upon when a thorough search is necessary.

Elsevier's *Encyclopedia of Organic Chemistry*[5] was initially intended to cover the literature of organic chemistry with the same thoroughness as Beilstein's *Handbuch.* Unfortunately, publication was discontinued after the appearance of just a few volumes. However, these volumes provide a valuable supplement to Beilstein's *Handbuch,* as they are devoted to subjects such as steroids that were omitted from the original organization of Beilstein.

Rodd's Chemistry of Carbon Compounds[6] includes a discussion of many basic techniques and lists many compounds and their derivatives.

2.9 MIXTURES

[Refer to Chapter 10, p. 000.] At some time during the course, one or more mixtures may be assigned. After obtaining the mixture from the instructor, proceed with the separation according to the methods outlined in Chapter 10. The mixture may contain a volatile component which can be removed by heating the mixture on a steam bath. This volatile com-

[3]*Handbook of Tables for Organic Compound Identification,* 3rd ed., edited by Z. Rappoport (CRC Press, Boca Raton, FL, 1967).

[4]*Dictionary of Organic Compounds* (Chapman and Hall, New York, Volumes 1–9, 1996).

[5]*Elsevier's Encyclopedia of Organic Chemistry* (Elsevier Publishing Company, Amsterdam, and New York. Josephy, Edith 1940–1964).

[6]*Rodd's Chemistry of Carbon Compounds,* edited by S. Coffey (Elsevier Publishing Company, Amsterdam and New York, 2nd edition, 1964–1986).

ponent would then be identified. In dealing with a mixture of unknown composition, it is inadvisable to attempt distillation at higher temperatures.

When the components of the mixture have been separated, identify each according to the procedure followed for simple unknowns.

2.10 REPORT FORMS

After the identification of an unknown has been completed, the results should be reported on special forms supplied by the instructor. The following reports are examples illustrating the information to be reported. In the summary of the NMR data, abbreviations such as s (singlet), d (doublet), t (triplet), q (quartet), m (multiplet), and b (broad) are used. All spectra, as well as copies of literature (fingerprint) spectra, must be included with the report. Each report should be accompanied by a vial containing the derivative. A separate report would be written up for each component in a mixture. Please note that report forms at different schools may vary.

REPORT FORM 1

Compound *1-Butanol* Name *John Smith*

Unknown Number *1* Date *February 14, 1997*

1. Physical Examination:
   - (a) Physical state *Liquid*
   - (b) Color *None*
   - (c) Odor *Choking*
   - (d) Ignition test *Burns with a bluish flame, no residue.*
   - (e) TLC ______
   - (f) GC ______
2. Physical Constants:
   - (a) mp: observed ______ ; corrected ______
   - (b) bp: observed *114–117°C* ; corrected *115–118°C*
3. Elemental Analysis:

   F – , Cl – , Br – , I – , N – , S – ,

   Metals *None*
4. Solubility Tests:

H_2O	ether	NaOH	$NaHCO_3$	HCl	H_2SO_4
+	+				

Reaction to litmus: *None*

Reaction to phenolphthalein: *None*

Solubility Class: S_1

Possible Compounds: *Monofunctional alcohols, aldehydes, ketones, esters, nitriles, and amides with five carbons or fewer*

5. Molecular Weight Determination: *None*
6. IR spectrum (attach to end of report):

   Solvent: *neat*

Significant Frequencies (cm^{-1})	Inferences
3600, 3300	*—O—H alcohol*
1025 (very broad)	*—C—O 1° alcohol*
2850	*—C—H aliphatic*

7. NMR spectrum (attach to end of report):

   Solvent: *CDCl3*

δ	Integration	Type of Peak (s, d, t, q, m)	Inferences
0.95	*3H*	*t*	$C\underline{H}_3CH_2$—
1.25–1.90	*4H*	*m*	*aliphatic*
2.15a	*1H*	*bs*	$O\underline{H}$
3.65	*2H*	*t*	—$CH_2C\underline{H}_2OH$

[a]This value is dependent upon the concentration of 1-butanol in $CDCl_3$.

8. Mass spectrum (attach to end of report):

M/Z Ratio	Inferences
15	CH_3—
29	CH_3CH_2—
74	$CH_3CH_2CH_2CH_2OH$

9. Preliminary Classification Tests:

Reagent	Results	Inference
2,4-Dinitrophenylhydrazine	*No ppt*	*No aldehyde or ketone carbonyl group*
Acetyl chloride	*Reaction; heat; fruity odor*	*Presence of hydroxyl group*
Ceric nitrate	*Red color*	*Presence of hydroxyl group*
Lucas reagent	*Dissolved in reagent—no separation of oily layer*	

Functional group indicated by these tests:

Alcohol, probably primary

10. Preliminary Examination of the Literature:

Possible Compounds	mp or bp (°C)	Suggestions for Further Tests
2-Methyl-1-propanol	*108*	
3-Methyl-2-butanol	*113*	*Contains CH_3CHOH—, should give positive iodoform test*
3-Pentanol	*116*	
1-Butanol	*117*	
2-Pentanol	*119*	*Contains CH_3CHOH—, should give positive iodoform test*
3,3-Dimethyl-2-butanol	*120*	

11. Further Classification and Special Tests:

Reagent	Results	Inference
Iodoform test	*No ppt*	*Does not contain a CH_3CHOH— unit*

12. Probable Compounds:

	Useful Derivatives and Their mp, NE, etc.			
Name	**3,5-Dinitrobenzoate (mp, °C)**	**α-Naphthylurethane (mp, °C)**	**Phenylurethane (mp, °C)**	**SP GR**
1-Butanol	*64*	*71*	*61*	*0.810*
2-Methyl-1-butanol	*86*	*104*	*86*	*0.805*
3-Pentanol	*97*	*71*	*49*	*0.820*

13. Preparation of Derivatives:

Name of Derivative	**Observed mp (°C)**	**Reported mp (°C)**
3,5-Dinitrobenzoate	*62–63*	*64*
α-Naphthylurethane	*68–69*	*71*
Phenylurethane	*57–59*	*61*

14. Special Comments:

    The identity of the OH peak in the 1H NMR spectrum was supported by the concentration dependence of its chemical shift.

15. Literature Used:

    Pouchert, Charles J. *The Aldrich Library of 1H and 13C NMR Spectra (Aldrich Chemical Company, Milwaukee, 1993).*

REPORT FORM 2

Compound *2-Amino-4-nitrotoluene* Name *John Smith*

Unknown Number *2* Date *March 17, 1997*

1. Physical Examination:

   (a) Physical state *Solid*

(b) Color *Yellow*

(c) Odor

(d) Ignition test *Yellow flame, no residue*

(e) TLC

(f) GC

2. Physical Constants:

   (a) mp: observed *107–108°C* ; corrected *109–110°C*

   (b) bp: observed ; corrected

3. Elemental Analysis:

   F – , Cl – , Br – , I – , N + , S – ,

   Metals *None*

4. Solubility Tests:

H_2O	ether	NaOH	$NaHCO_3$	HCl	H_2SO_4
–		–		+	

Reaction to litmus:

Reaction to phenolphthalein:

Solubility Class: *B*

Possible Compounds: *Aliphatic amines with eight or more carbons, anilines (one phenyl group attached to nitrogen), some ethers*

5. Molecular Weight Determination: *150 ± 4 (freezing point depression, in camphor)*

6. IR spectrum (attach to end of report):

   Solvent: *KBr pellet*

Significant Frequencies (cm^{-1})	Inferences
3300–3400 (2 bands)	*—NH_2 primary*
1500, 1600	*C=C aromatic*
1260, 1550	*—NO2*
2870	*C—H aliphatic*

7. NMR spectrum (attach to end of report):

   Solvent: *$CDCl_3$/DMSO–d_6*

δ	Integration	Type of Peak (s, d, t, q, m)	Inferences
2.20	*3H*	*s*	*Ar$C\underline{H}_3$*
4.70	*2H*	*bs*	*—$N\underline{H}_2$*
6.9–7.6	*3H*	*m*	*aromatic*

8. Mass spectrum (attach to end of report):

M/Z Ratio	Inferences

9. Preliminary Classification Tests:

Reagent	Results	Inference
Hinsberg	*NaOH: clear solution; HCl: ppt*	*Primary amine*
Nitrous acid	*Orange ppt with 2–naphthol*	*Primary aromatic amine*

Functional group indicated by these tests:

Primary aromatic amine

10. Preliminary Examination of the Literature:

Possible Compounds	mp or bp (°C)	Suggestions for Further Tests
4-Aminoacetophenone	*106*	*Test for methyl ketone needed*
2-Amino-4-nitrotoluene	*107*	*Test for nitro group needed*
2-Naphthylamine	*112*	
3-Nitroaniline	*114*	*Test for nitro group needed*
4-Amino-3-nitrotoluene	*116*	*Test for nitro group needed*
		Run UV spectrum

11. Further Classification and Special Tests:

Reagent	Results	Inference
2,4-Dinitrophenylhydrazine	*No ppt*	*Not 4-aminoacetophenone*
Iodine and sodium hydroxide	*No iodoform*	*Not a methyl ketone*
Zinc and ammonium chloride; followed by Tollens reagent	*Silver mirror*	*Nitro group present*

12. Probable Compounds:

	Useful Derivatives and Their mp, NE, etc.		
Name	Benzenesulfonamide (mp °C)	Acetamide (mp °C)	Phenol (mp °C)
2-Amino-4-nitrotoluene	*172*	*150*	*118*
3-Nitroaniline	*136*	*155*	*97*
4-Amino-3-nitrotoluene	*102*	*96*	*32*

13. Preparation of Derivatives:

Name of Derivative	Observed mp (°C)	Reported mp (°C)
Benzenesulfonamide	*170–171*	*172*
2-Hydroxy-4-nitrotoluene	*116–117*	*118*

14. Special Comments:

4-Amino-3-nitrotoluene has been reported to be hydrolyzed to 4-hydroxy-3-nitrotoluene [Neville and Winther, *Ber.* 1882, *15,* 2893]. *The unknown gave only starting material under these conditions.The unknown was converted to the phenol by the method reported by* Ullmann and Fitzenkam, *Ber.,* 1905, *38,* 3790.

15. Literature Used:

Additional References:

Pouchert, Charles J. *The Aldrich Library of 1H and 13C NMR Spectra* (Aldrich Chemical Company, Milwaukee, 1993).

REPORT FORM 3

Compound *2-Naphthol* Name *John Smith*

Unknown Number *3* Date *March 30, 1997*

1. Physical Examination:
   - (a) Physical state *Solid*
   - (b) Color *White*
   - (c) Odor *Suggests moth balls*
   - (d) Ignition test *Smoky flame; no residue; suggests aromatic compound*
   - (e) TLC ______
   - (f) GC ______
2. Physical Constants:
   - (a) mp: observed *121–122.5°C* ; corrected ______
   - (b) bp: observed *284–286°C* ; corrected ______
3. Elemental Analysis:

   F – , Cl – , Br – , I – , S – ,

   Metals *None*

4. Solubility Tests:

H_2O	ether	NaOH	$NaHCO_3$	HCl	H_2SO_4
−		+	−		

Reaction to litmus: ______

Reaction to phenolphthalein: ______

Solubility Class: *A2*

Possible Compounds: *Weak organic acids: phenols, enols, oximes, imides, sulfonamides, thiophenols, all with more than five carbons, 1, 3-diketones, nitro compounds with α-hydrogens, sulfonamides*

5. Molecular Weight Determination: *144 (mass spectrometry; unit mass resolution). Molecular formula from high-resolution spectrometry: $C_{10}H_8O$*

6. IR spectrum (attach to end of report):

   Solvent: *$CHCl_3$*

Significant Frequencies (cm^{-1})	Inferences
3300 (broad)	*—OH alcohol*
3600 (sharp)	*—OH alcohol*
1605	*C=C aromatic*
1200	*C—O aromatic alcohol*
3020	*C—H aromatic*

7. NMR spectrum (attach to end of report):

   Solvent: *acetone-d_6*

δ	Integration	Type of Peak (s, d, t, q, m)	Inferences
7.0–7.8	*7H*	*m*	*aromatic*
8.35	*1H*	*bs*	*—O<u>H</u>*

8. Mass spectrum (attach to end of report):

M/Z Ratio	Inferences

9. Preliminary Classification Tests:

Reagent	Results	Inference
Bromine water	*Precipitate*	*Phenol*
Ferric chloride	*Green solution*	*Phenol*

Functional group indicated by these tests:

Phenol

10. Preliminary Examination of the Literature:

Possible Compounds	mp or bp (°C)	Suggestions for Further Tests
4-Hydroxy-benzaldehyde	*115*	*Should give carbonyl test*
Hydroquinone monobenzyl ether	*122*	*Should be cleaved by acid*
2,4,6-Trinitrophenol	*122*	*Test for nitro group; unlikely*
2-Naphthol	*122*	
1,3-Dihydroxy-4-nitrobenzene	*122*	*Test for nitro group*
2-Methyl-1,4-benzoquinone	*124*	*Should be readily oxidized to quinone*

11. Further Classification and Special Tests:

Reagent	Results	Inference
2,4-Dinitro-phenylhydrazine	*No ppt*	*Not an aldehyde or a ketone*
Zinc and ammonium chloride; followed by Tollens reagent	*No silver mirror*	*Nitro group absent*

12. Probable Compounds:

	Useful Derivatives and Their mp, NE, etc.	
Name	**Acetate (mp, °C)**	**4-Nitrobenzoate (mp, °C)**
Picric acid	*76*	*143*
2-Naphthol	*72*	*169*
1,3-Dihydroxy-4-nitrobenzene	*91 (di)*	

13. Preparation of Derivatives:

Name of Derivative	Observed mp (°C)	Reported mp (°C)
Acetate	*73–74*	*72*
4-Nitrobenzoate	*166–167*	*169*

14. Special Comments:

2-Methyl-1,4-benzoquinone is unlikely on several grounds. Reported to be readily soluble in water. Ferric chloride in concentrated aqueous solution gives a brownish red color and in dilute solution a yellow color (Beil., VI, 874), while the unknown gave a green color.

15. Literature Used:

Additional references:

Simons, William H. *The Sadtler Guide to NMR Spectra* (Sadtler, Philadelphia, 1978).

CHAPTER 3

Preliminary Examination, Physical Properties, and Purity Determination

The investigator begins at this point when he or she has in hand a sample that is believed to be primarily one compound; if the investigator believes that the sample contains more than one component, Chapter 10 on separations should be consulted. The assumption that predominantly one component is present may be based on (1) instructor's guidance, (2) the method of synthesis, (3) the method of isolation, and/or (4) chromatographic or other analytical results.

This chapter contains two major portions. The first (Sections 3.1 and 3.2) deals with the usual simple physical properties and the second (Sections 3.2–3.6) with more detailed and specialized methods of characterization. The simple properties include physical state, color, odor, and ignition tests (Section 3.1), and the simple physical constants (Section 3.2: melting point, boiling point, and less frequently, specific gravity and index of refraction). In most teaching environments those two sections represent the bottom line of chemical characterization.

Often, however, it is useful to do more: for example, if a melting point is unusually broad, and the appearance of the unknown shows more than one physical form, it might be very useful to run thin-layer chromatography analysis (Section 3.3.1) on this sample. The remainder of the chapter discusses useful chromatographic methods and other analytical techniques that may be most helpful for working up (recrystallization, Section 3.5) and characterizing (gas chromatography, Section 3.3.2; optical rotation, Section 3.4) organic compounds.

3.1 PRELIMINARY EXAMINATION

3.1.1 Physical State

Note whether the unknown substance is a liquid or solid; tables of compounds (Appendix III) are subdivided on the basis of phase. In addition, insofar as the phase relates to solubility and volatility, an aid to choice of purification method is provided: liquids are usually purified by distillation or by gas chromatography (below and Chapter 10 pp. 474–476), and solids are purified by recrystallization (or sublimation).

3.1.2 Color

The color of the original sample is noted, as well as any change in color that may occur during the determination of the boiling point (Section 3.2.2), distillation, or after chromatographic separation.

The color of some compounds is due to impurities; frequently these are produced by slow oxidation of the compound by oxygen in the air. Aniline, for example, is usually reddish brown, but a freshly distilled sample is nearly colorless.

Many liquids and solids are definitely colored because of the presence of chromophoric groups in the molecule. Many nitro compounds, quinones, azo compounds, stable carbocations and carbanions, and compounds with extended conjugated systems are colored. If an unknown compound is a stable, colorless liquid or a white crystalline solid, this information is valuable because it excludes chromophoric functional groups as well as many groups that by oxidation would become chromophores.

3.1.3 Odor

We cannot in good conscience recommend that you examine the odor of an organic compound by direct inhalation. Frequently organic compounds have at least some degree of toxicity and their uses are often regulated. The odor of many organic compounds will, however, quickly become evident during the course of normal handling, and when that happens you should make note of it.

It is not possible to describe odor in a precise manner, but some basic facts are well known. Amines often have a distinctly fishy smell and thus they frequently are easily identified. Some amines have common or trivial names that suggest odors: for example, cadavarine and putrescine. Thiols (or mercaptans) and organic sulfides (thioethers) are easily detected by their rotten egg smell, an odor that you may have encountered when dealing with hydrogen sulfide. Carboxylic acids of low molecular weight have distinct and noxious odors: acetic acid yields the bad smell in vinegar, butanoic (or butyric) acid has the smell of unwashed gym socks. Esters usually have pleasant smells that are often characterized as fruity. For example, 3-methylbutyl ethanoate (isopentyl or isoamyl acetate) is often referred to as "banana oil". Hydrocarbons can have very different smells: naphthalene has been used as mothballs and thus the odor should be recognizable; pinenes are components of turpentine and therefore they have the odor of paint thinner. Benzaldehyde, nitrobenzene, and benzonitrile all have odors that have been described as "cherry-like" or the odor of "bitter almonds". The origins of some organic compounds suggest distinct smells: eugenol (from cloves), coumarin (from lavender oil and sweet clover), and methyl salicylate (oil of wintergreen). Other compound classes have distinguishable but less pronounced odors. Thus aldehydes are different from ketones, and both are different from alcohols. Phenols also have unique odors and isonitriles have very disagreeable odors.

The theory of odor certainly is dependent upon stereochemistry. A pertinent case is that of carvone: the (+), or dextrorotatory, stereoisomer has an odor quite consistent with the fact that it can be isolated from caraway or dill seeds. On the other hand, the (−), or levorotatory, form is a major component of spearmint.

carvone

Toxicity information is usually available on the bottle label and in catalogs such as that available from Aldrich Chemical Co. Most labs must provide *MSD* sheets describing the toxicity of any organic compound used in that laboratory. The *Merck Index* can also be consulted for more information.

3.1.4 Ignition Test

Procedure. A sample of 10 mg of the substance is placed in a porcelain crucible cover (or any piece of porcelain) and brought to the edge of a flame to determine flammability. It is then heated gently over a low flame and behind a safety shield and eventually heated strongly to accomplish thorough ignition. A note is made of (1) the flammability and nature of the flame (is the compound explosive?); (2) if the compound is a solid, whether it melts and the manner of its melting; (3) the odor of gases or vapors evolved (*caution!*); (4) the residue left after ignition. Will it fuse? If a residue is left, the lid is allowed to cool, a drop of distilled water is added, and the solution tested with litmus. A drop of 10% hydrochloric acid is added. Is a gas evolved? A flame test with a platinum wire is made on the hydrochloric acid solution to determine the metal present.

When using this test, as well as any other test, control runs should be made on standard compounds of known composition. When possible, a known compound of composition similar to that suspected for the unknown should be chosen. Control compounds, which can be used to indicate the range of results possible for this ignition test, are ethanol, toluene, barium benzoate, copper acetate, sodium potassium tartrate, and sucrose.

Discussion. Many liquids burn with a characteristic flame that assists in determining the nature of the compound. Thus, an aromatic hydrocarbon (which has a relatively high carbon content) burns with a yellow, sooty flame. Aliphatic hydrocarbons burn with flames that are yellow but much less sooty. As the oxygen content of the compound increases, the flame becomes more and more clear (blue). If the substance is flammable, the usual precautions must be taken in subsequent manipulation of the compound. This test also shows whether a melting point of a solid should be taken and indicates whether the solid is explosive.

If an inorganic residue[1] is left after ignition, it should be examined for metallic elements. A few simple tests will often determine the nature of the metal present.[2] If the

[1]A "residue" becomes identifiable with a little experience. A residue is more than a small streak of blackened remains; the amount should correspond to a reasonable percentage of the original sample. The control samples cited earlier containing metal (sodium, barium) ions should be ignited as a reference.

[2]Consult a book on inorganic qualitative analysis.

flame test indicates sodium, a sample of the compound should be ignited on a platinum foil instead of a porcelain crucible cover. (Why?)

3.1.5 Summary and Applications

The tests of this section are extremely useful for decisions as to whether further purification is necessary and as to what type of purification procedures should be used. If various tests in this section indicate that the compound is very impure, recrystallization (Section 3.5) or chromatography is almost certainly required. Although liquids are very often easily analyzed by gas chromatography (Section 3.3), those that leave residues upon ignition should *not* be injected into the gas chromatograph.

3.2 DETERMINATION OF PHYSICAL PROPERTIES

3.2.1 Melting Points

The melting point of a compound is the range of temperature at which the solid phase changes to liquid. Since this process is frequently accompanied by decomposition, the value may be not an equilibrium temperature but a temperature of transition from solid to liquid only. If the ignition test indicates that a solid melts easily (25–300°C), the melting point should be determined by Procedure A. For higher melting point ranges (300–500°C), use special equipment (see, e.g., Figure 3.10). If a melting point determination by Procedure A indicates definite decomposition or transition from one crystalline state to another, Procedure B is recommended. Compounds melting between 0 and 25°C may be analyzed by the freezing point method described on pp. 42–43.

Melting points for a large number of compounds and their derivatives are listed in this book. Frequently a small amount of impurity will cause a depression (and broadening) of the observed melting point. Thus the procedure of determining melting points of mixtures described below is strongly recommended. If, for any of a number of reasons, one has a compound that is contaminated by minor amounts of impurities, the section on recrystallization should be consulted (see Section 3.5).

Mixture Melting Points. The "mixed melting-point" method provides a means of testing for the identity of two solids (which should, of course, have identical melting points) by examination of the melting point behavior of a mechanical mixture of the two. In general, a mixture of samples of nonidentical compounds shows a melting point depression. Although the use of mixed melting points is valuable at certain points of the identification procedure, a mixed melting point of an unknown with a known sample from the side shelf will not be accepted in this course as proof of the structure.

A few pairs of substances when mixed show no melting point depression, but more frequently the failure to depress may be observed only at certain compositions. It requires little additional effort to measure the melting points of mixtures of several compositions if the following method is used.

Make small piles of approximately equal sizes of the two components (A and B) being examined. Mix one-half of pile A with one-half of pile B. Now separate the mixture of A and B into three equal parts. To the first add the remainder of component A, and to the third, the remainder of component B. It is seen that three mixtures with the compositions 80% A, 20% B; 50% A, 50% B; and 20% A, 80% B are obtained. The melting points of all three mixtures may be measured at the same time by any of the following procedures.

Procedure A. For many melting point determinations, 6-cm-long melting point tubes of capillary size (ca. 1 mm width) are used; if such capillary tubes are not available, they may be made by melting and drawing out (clean) soft glass tubing (e.g., 15-mm tubing) or drawing soft glass test tubes. A modest amount of the compound (powdered, if necessary) is placed on a hard, clean surface; a small amount is picked up by tapping the mouth of the capillary tube on the compound (Figure 3.1*a*). The capillary tube is then held vertically (Figure 3.1*b*, open end up) and rubbed with a file or a coin with a milled edge, or tapped on the table top or dropped through a glass tube (Figure 3.1*c*) to pack the compound at the bottom. The capillary tube should contain a 2–3-mm column of sample. This charged tube can now be used in any of the devices shown in Figures 3.2, 3.3, and 3.4 (consult these figures). Use of the Fisher–Johns stage apparatus (Figure 3.6) requires microscope cover glasses rather than capillary tubes.

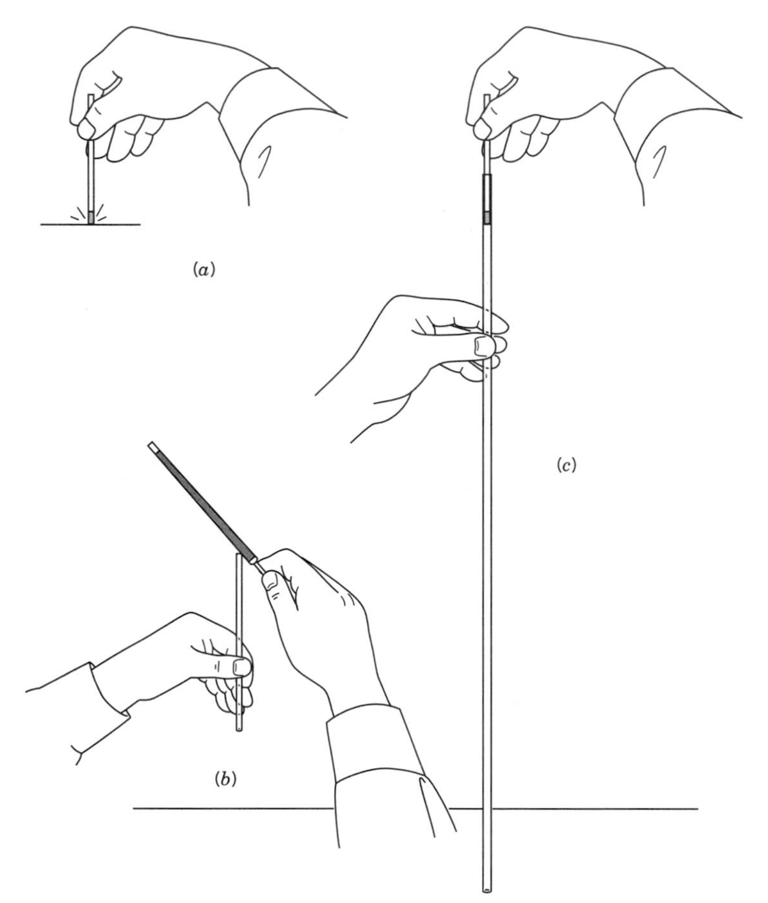

Figure 3.1 Charging (*a*) and packing (*b, c*) capillary melting point tubes.

It is often timesaving to run a preliminary melting point determination, raising the temperature of the bath very rapidly. After the approximate melting point is known, a second determination is carried out by raising the temperature rapidly until within 5°C of the approximate value and then proceed slowly as described above. A fresh sample of the compound should always be used for each determination.

Failure to use clean capillary melting tubes is one of the causes of low melting points and wide melting point ranges. These are due to alkali on the surface of the glass, which catalyzes the aldol condensation of aldehydes or ketones, or the mutarotation of sugars

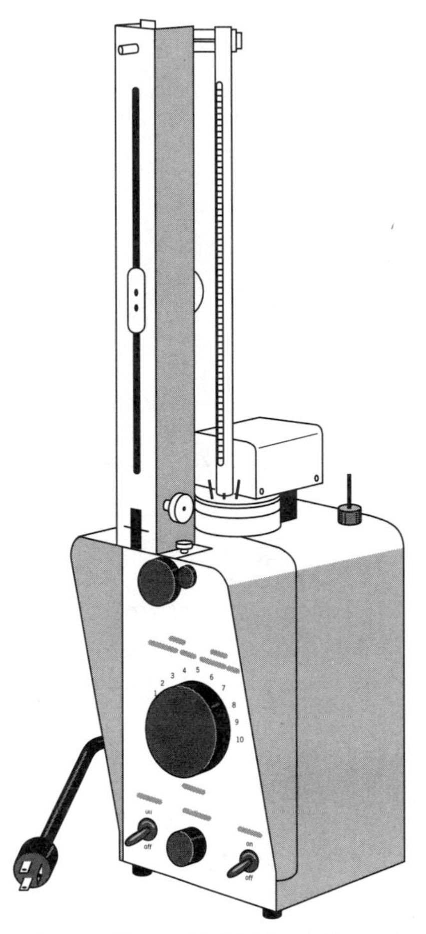

Figure 3.2 Thomas–Hoover Uni-Melt melting point apparatus.

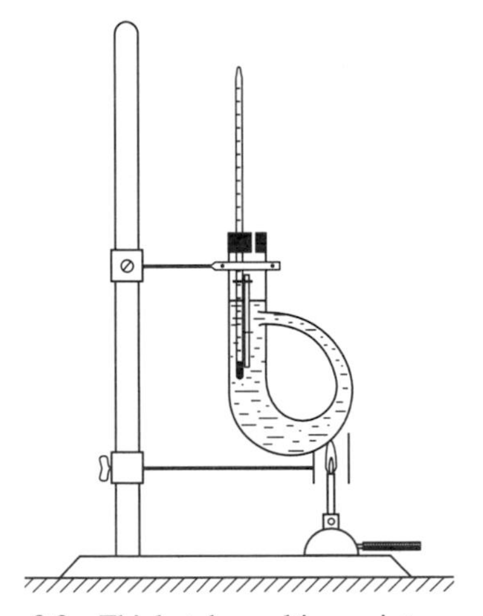

Figure 3.3 Thiele tube melting point apparatus.

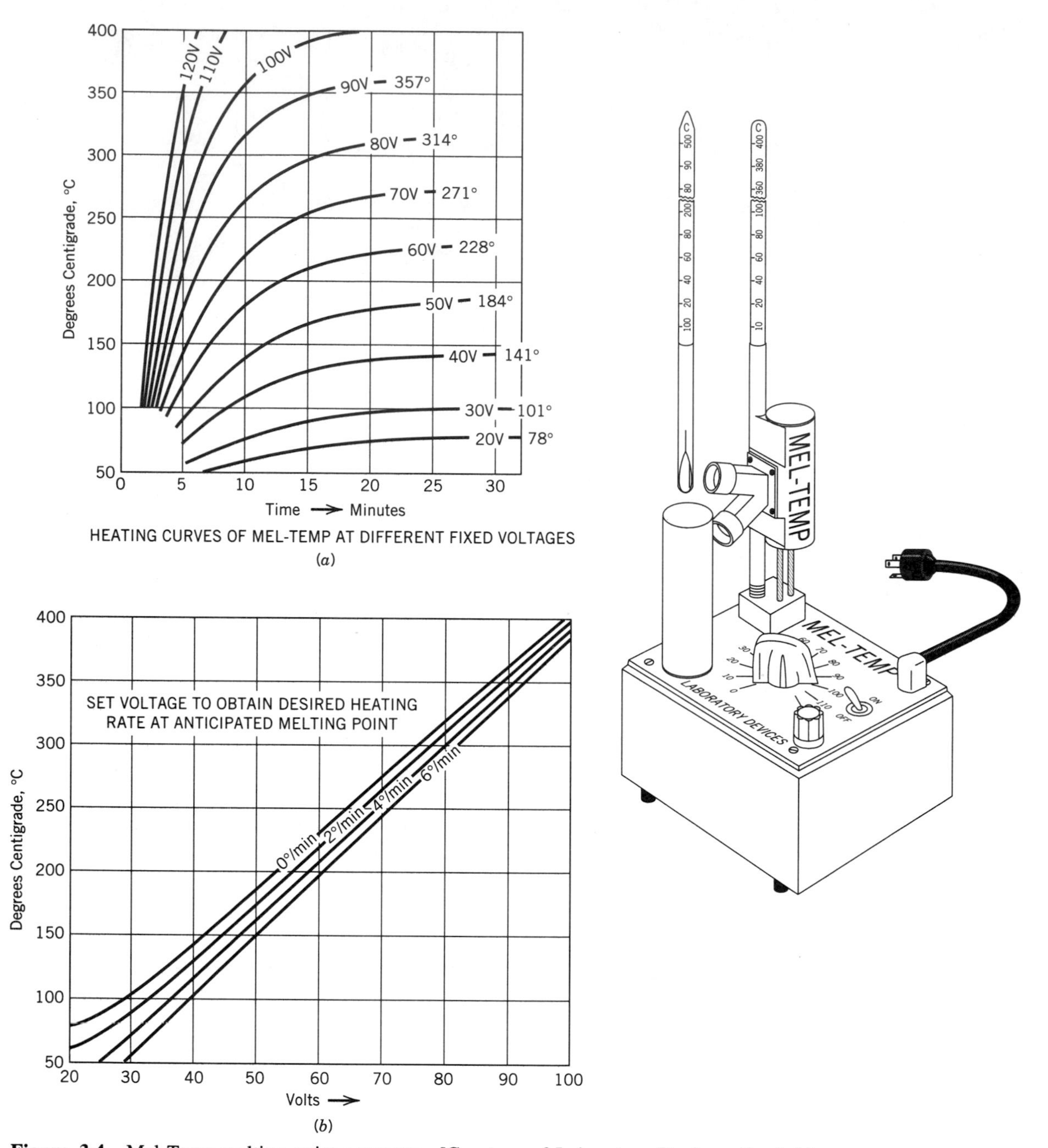

Figure 3.4 Mel-Temp melting point apparatus. [Courtesy of Laboratory Devices, Cambridge, MA.] Graph *b* suggests that a setting of 50 volts will result in a heating rate of 2°/min in the vicinity of 170°.

and their derivatives, and so on. For example, in an uncleaned tube α-D-glucose softened at 133°C and melted at 143–146°C, whereas in a dealkalinized tube the softening occurs at 142°C and the melting at 146–147°C. The use of clean tubes made from Pyrex glass avoids these problems.

Corrected Melting Points. The thermometer should always be calibrated by observing the melting points of several pure compounds (Table 3.1). If care is taken to use the same apparatus and thermometer in all melting point determinations, it is convenient and timesaving to prepare a calibration curve such as that shown in Figure 3.5. The observed melting point of the standard compound is plotted against the corrected value, and a curve, *DA*, is drawn through these points. In subsequent determinations the observed value, *B*, is projected horizontally to the curve and then vertically down to give the corrected value, *C*. Such a calibration curve includes corrections for inaccuracies in the thermometer and stem correction. The thermometer should be calibrated by observing the melting points of several pure compounds.

Table 3.1 Melting Point Standards

mp (corr.) (°C)		mp (corr.) (°C)	
0	Ice	187	Hippuric acid
53	*p*-Dichlorobenzene	200	Isatin
90	*m*-Dinitrobenzene	216	Anthracene
114	Acetanilide	238	1,3-Diphenylurea
121	Benzoic acid	257	Oxanilide
132	Urea	286	Anthraquinone
157	Salicylic acid	332	*N,N'*-Diacetylbenzidine

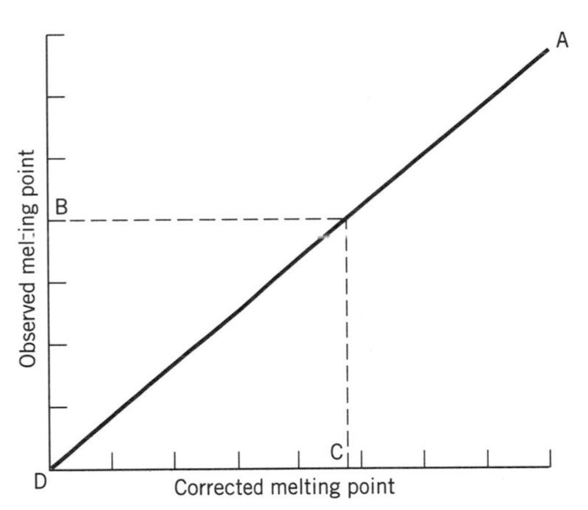

Figure 3.5 Melting point calibration curve.

It is important to record the melting point range of an unknown compound, because this is an important index of purity. A large majority of pure organic compounds melt within a range of 0.5°C or melt with decomposition over a narrow range of temperature (about 1°C). If the melting point range or decomposition range is wide, the compound should be recrystallized from a suitable solvent and the melting or decomposition point determined again. Some organic compounds, such as amino acids, salts of acids or amines, and carbohydrates, melt with decomposition over a considerable range of temperature.

The Fisher–Johns apparatus, shown in Figure 3.6, has an electrically heated aluminum block fitted with a thermometer reading to 300°C. The sample is placed between two 18-mm microscope cover glasses which are put in the depression of the aluminum

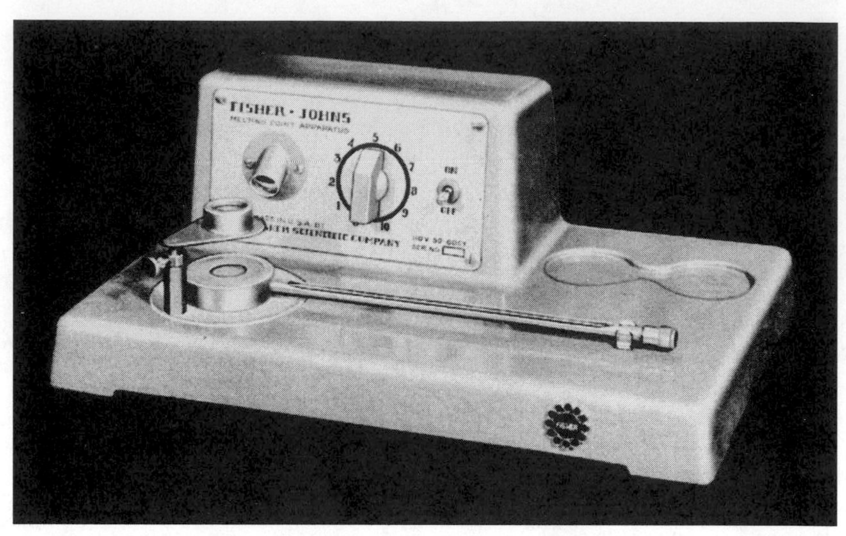

Figure 3.6 Fisher–Johns melting point apparatus. [Courtesy of Fisher Scientific Co., Pittsburgh, PA.]

block. The temperature is regulated by means of a variable transformer, and the melting point is observed with the aid of the illuminator and magnifying glass. A calibration curve must be prepared for the instrument by reference to known compounds as described above.

Procedure B. The Nalge–Axelrod instrument (Figure 3.7) consists of a 25-power microscope with Polaroid inserts, a light source, an electrically heated aluminum block controlled by a variable transformer, and a thermometer reading to 400°C. The sample is placed between two 18-mm microscope cover glasses in the depression of the heating block. The cover, which has a small aperture, is put in place, the light turned on, and the microscope focused. The tube is rotated so as almost to cross the Polaroids, and the crystals are observed as the temperature is raised. The melting points of anisotropic crystals are easily noted, because the polarization colors disappear when melting occurs. Also, the transition from one allotropic crystal modification to another is easily observed. A calibration curve must be prepared (above).

In the usual experiment (Procedure A), the melting point is defined as the temperature range between the first appearance of a liquid phase and the disappearance of the last crystals of the solid phase. With the microscope slide techniques, there are difficulties with the application of the definition. Since the crystals are largely separated in the microscope slide procedure, the melted crystals are normally not in contact with the unmelted crystals. Thus, there is often a very wide range of temperature over which the melting can occur. This procedure is therefore to be avoided; the melting point determined in a capillary tube is a more accurate and more meaningful criterion of purity.

Freezing Points

About 5–10 milliliters of the liquid is placed in an ordinary test tube fitted with a thermometer and a wire stirrer (made of copper, nickel, or platinum). The tube is fastened in a slightly larger test tube by means of a cork and cooled in an ice or ice–salt bath or an acetone–dry ice mixture, and the liquid is stirred vigorously (Figure 3.8). As soon as crystals start to form, the tube is removed from the bath, and vigorous stirring is continued while the temperature on the thermometer is being read. The freezing point is the tem-

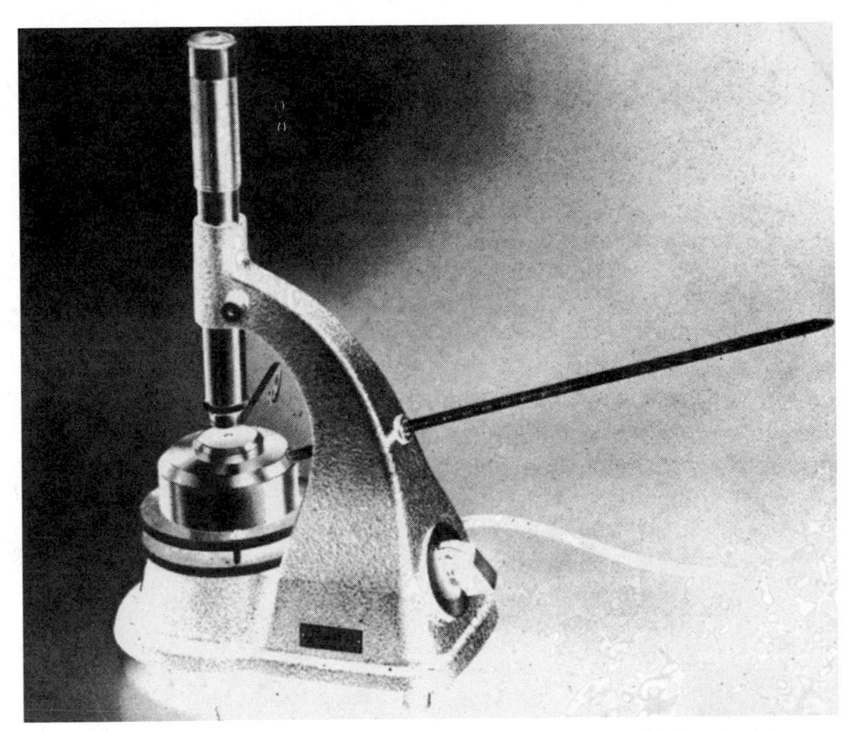

Figure 3.7 Nalge–Axelrod melting point apparatus. [Courtesy of the Nalge Co., Rochester, NY.]

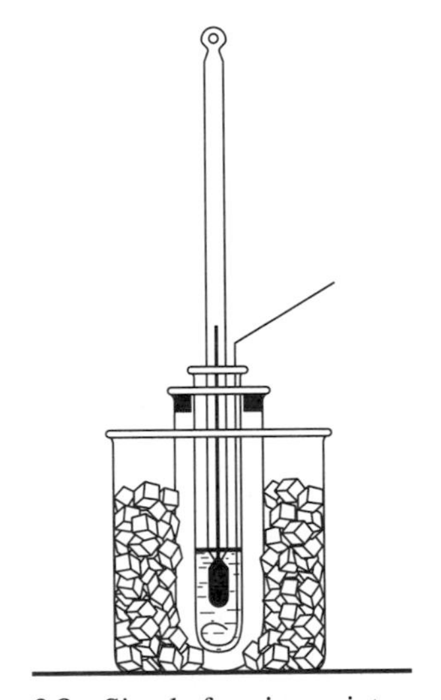

Figure 3.8 Simple freezing point apparatus.

perature reached after the initial supercooling effect has disappeared. The temperature of the cooling bath should not be too far below the freezing point of the compound. Freezing points of most organic liquids as ordinarily determined are only approximate (note that the determination requires a relatively large sample).

A more elaborate apparatus for determining freezing points (down to −65°C) has been described.[3]

[3]R. J. Curtis and A. Dolenga, *J. Chem. Educ.*, *52*, 201 (1975).

3.2.2 Boiling Points

The use of boiling points (bp) for compound identification has been introduced in Chapter 2 (p. 12).

Procedure A. A small-scale distillation apparatus similar to that shown in Figure 3.9 is set up. Test tubes immersed in a beaker of ice are used to condense the vapors and act as receivers. A few boiling chips and 10 mL of the liquid whose boiling point is to be determined are added. The thermometer is inserted so that the top of the mercury bulb is just *below* the side arm. The liquid is heated to boiling.[4] The liquid is distilled at as uniform a rate as possible. After the first 2–3 mL of distillate has collected, the receiver is changed without interruption of the distillation, and the next 5–6 mL is collected in a clean, dry test tube. There will be a considerable lag of the thermometer reading, but usually the boiling point range can be determined during the collection of the second portion of distillate. This boiling point range should be recorded.

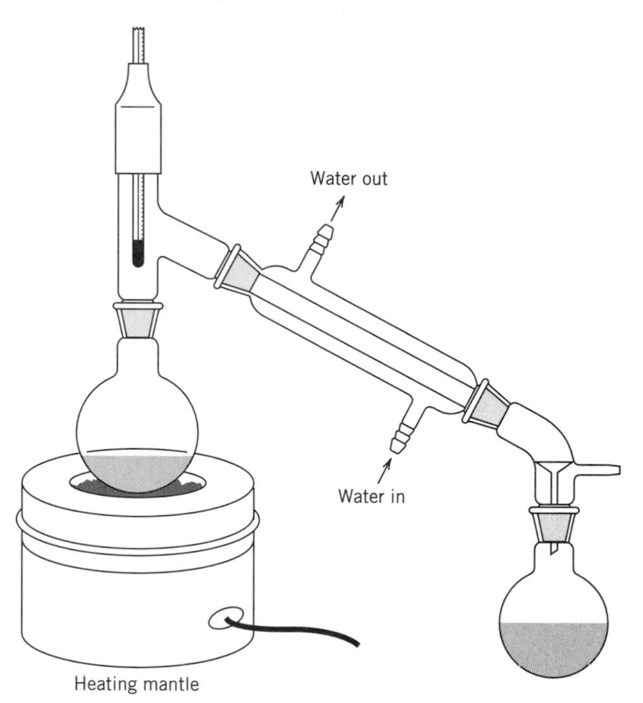

Figure 3.9 A small-scale simple distillation apparatus. Sand has been used to fill in the well.

Large-scale distillations can be carried out using an apparatus like that shown in Figure 3.10.

Great care should be exercised against overrelying upon boiling point as a criterion of purity or a basis for identity. Atmospheric pressure variations have a significant effect upon boiling point. Many organic liquids are hygroscopic, and some decompose on standing. Generally the first few milliliters of distillate will contain any water or more volatile impurities, and the second fraction will consist of the pure substance. If the boiling point range is large, the liquid should be refractionated through a suitable column (see Chapter 10 pp. 444–449).

The boiling point determined by the distillation of a small amount of liquid as described above is frequently in error. Unless special care is taken, the vapor may be su-

[4]An electrical heating method, employing a mantle (if macroscopic) or a sand bath or metal block (if microscopic), is preferred.

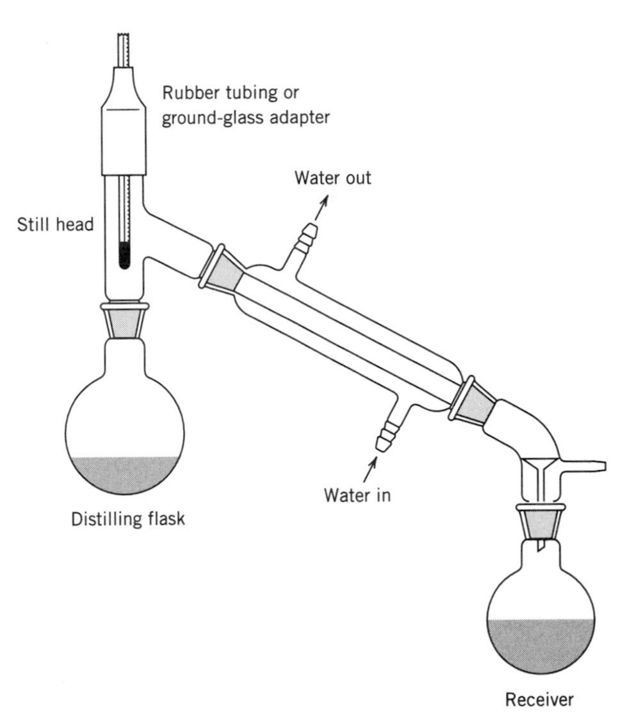

Figure 3.10 Standard taper apparatus for simple distillation. Heating and cooling devices and clamps are not shown.

perheated; also, the boiling points observed for high-boiling liquids may be too low because of the time required for the mercury in the thermometer bulb to reach the temperature of the vapor. The second fraction collected above should be used for a more accurate boiling point determination by Procedure B below. *Portions of the main fraction should also be used for the determination, as far as possible, of all subsequent chemical, spectral, and physical tests.*

Procedure B. Boiling points may be determined by placing a thermometer approximately 1–1.5 cm above a small amount (~0.5 mL) of the liquid in a clamped test tube (Figure 3.11). Then liquid is slowly heated to boiling such that the boiling liquid condenses on the thermometer while making certain that the entire mercury bulb is immersed in vapor. This technique is also useful for determining the boiling point of some low melting solids provided they are thermally stable. When this temperature holds at a constant value, the value is recorded as the boiling point of the liquid.

Procedure C. A *micro* boiling point set-up is prepared as shown in Figure 3.12. The outer tube is a small (5-mm) test tube; a (sealed) capillary melting point tube is inverted and placed in the test tube. Two drops of the liquid whose boiling point is to be determined are added, and the test tube is fastened to the thermometer in the apparatus used for the determination of melting points (Figures 3.3 and 3.4). The temperature is raised until a rapid and continuous stream of bubbles comes out of the small capillary and passes through the liquid. The heat is then removed from the melting-point apparatus and the heated medium is allowed to cool while being stirred continuously. The temperature is noted at the instant bubbles cease to come out of the capillary and just before the liquid enters it. This temperature is the boiling point; such a result is usually much more accurate than that determined by Procedure A.

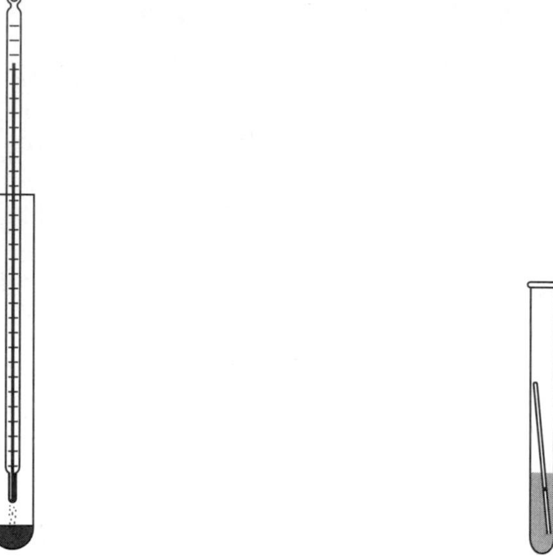

Figure 3.11 Boiling point determination: Procedure B.

Figure 3.12 Micro boiling point tube (ca. 5 cm tall).

Procedure D. In the ultramicro boiling point determination procedure, a standard Pyrex glass melting point capillary is charged with both the liquid of interest and a fused-glass bell (Figure 3.13*a* and *b*). The entire unit is placed in a standard melting point determination apparatus (either a Thomas–Hoover or Mel-Temp apparatus can be used, but the more sensitive the temperature control, the better) and the boiling point is measured.

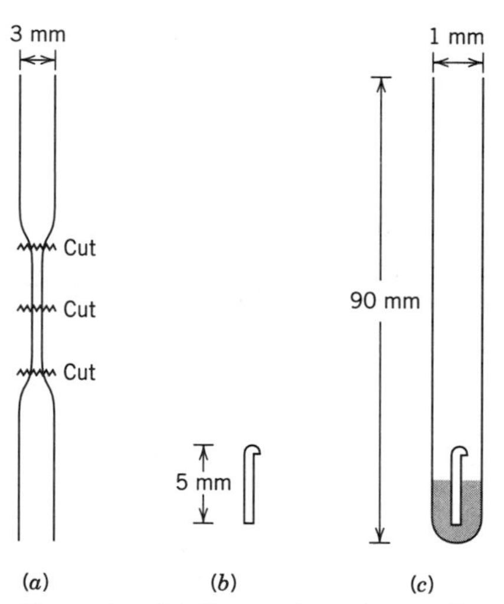

Figure 3.13 Ultramicro boiling point. (*a*) Preparation of a small glass bell; (*b*) the completed glass bell with a fused tip; (*c*) the glass bell in a capillary melting point tube. [From D. W. Mayo, R. M. Pike, S. S. Butcher, and M. L. Meredith, *J. Chem. Educ., 62,* 1114 (1985).]

Glass bells are purchased commercially or can be prepared by heating (Bunsen burner) 3-mm (O.D.) Pyrex glass tubing and drawing it (Figure 3.13*a*) to a thin enough diameter that the bell is easily fitted into the melting point capillary. The drawn tube is amply fused at one end in order to give the bell sufficient weight. The capillary is injected with 3–4 μL of the liquid (a 10-μL GC syringe can be used for this injection) and the glass bell is inserted with its open end down (fused end up). Both the liquid and the bell can be centrifuged to the bottom, if necessary. The boiling point is measured by rapidly raising the temperature to 15–20°C below the boiling point of the liquid (estimated by a preliminary run on an unknown). Then the heating rate is slowed to an increase of 2°/min and this rate is continued until a fine stream of bubbles is emitted from the end of the bell. Then the heating is adjusted such that the temperature drops and the boiling point is taken at the point where the last bubble collapses. This procedure may be redone for repeat determinations on the same or new sample.

Effect of Pressure on Boiling Point. At the time the boiling point is being determined, the atmospheric pressure should be recorded. Table 3.2 illustrates the magnitude of such barometric "corrections" of boiling point for pressures that do not differ from 760 mm by more than about 30 mm.

Table 3.2 Boiling Point Changes per Slight Pressure Change

		Correction in °C for 10-mm Difference in Pressure	
b.p. (°C)	b.p. (K)	Nonassociated[a] liquids	Associated[a] liquids
50	323	0.38	0.32
100	373	0.44	0.37
150	423	0.50	0.42
200	473	0.56	0.46
300	573	0.68	0.56
400	673	0.79	0.66
500	773	0.91	0.76

[a]Associated liquids are those liquids that have substantial intermolecular associations due to hydrogen bonding; an example is methanol.

It is evident that small deviations in pressure from 760 mm, such as 5 mm, may be neglected in ordinary work.

Investigators working in laboratories at high altitudes[5] and low barometric pressures have found it convenient to determine a set of empirical corrections to be added to observed boiling points in order to get boiling points at 760 mm. The corrections are obtained by distilling a number of different types of compounds with different boiling points. The difference between the boiling point recorded in the literature and the observed boiling point gives the correction.

Nomographs for boiling point versus pressure data of organic compounds have been devised; these charts are useful for vacuum distillations. An example is provided in Appendix I in this text.

In order to give an idea of the change in boiling point with pressure, the data on three pairs of nonassociated and associated compounds are given in Table 3.3. The temperatures are given to the nearest whole degree. The data indicate that, as the pressure is reduced, the boiling point of an associated compound does not fall off as much as the boiling point of a nonassociated liquid.

[5]At the top of Mt. Evans in Colorado, water boils at 81°C (average pressure 460–470 mm; altitude 14,200 ft). Water boiling at the University of Colorado (ca. 5000 ft) will have a temperature of ca. 90°C.

Table 3.3 Boiling Points (°C) at Reduced Pressures

	Pressure in millimeters of mercury (torr)					
Compound	**760**	**700**	**650**	**600**	**550**	ΔT^a
Heptane	98	96	94	91	88	10
1-Propanol	97	95	93	91	89	8
Iodobenzene	188	185	182	179	175	13
Pentanoic acid	186	183	180	178	175	11
Fluorene	298	294	290	286	282	16
2-Naphthol (β-naphthol)	295	292	288	284	280	15

[a] $\Delta T = bp_{760} - bp_{550}$.

Correlations of Boiling Point with Structure. The boiling points of the members of a given homologous series increase as the series is ascended. The boiling points rise in a uniform manner, as shown in Figure 3.14, but the increment per CH_2 group is not constant, being greater at the beginning of the series than for the higher members (Table 3.4).

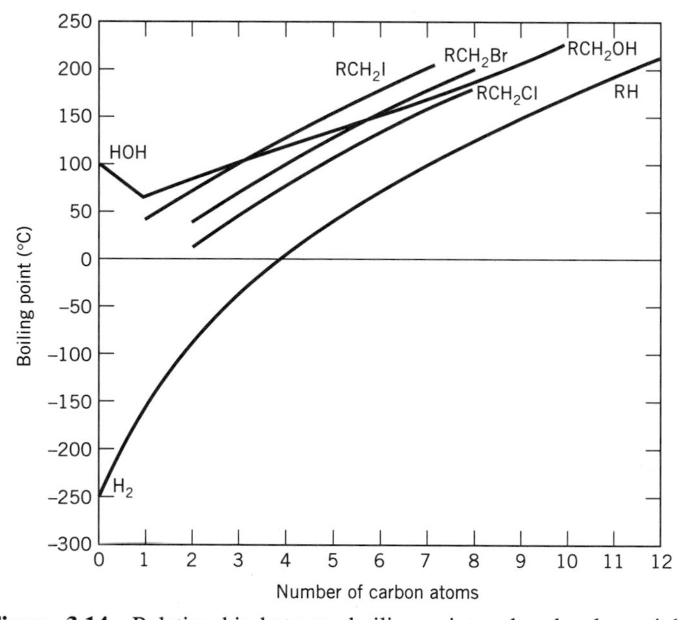

Figure 3.14 Relationship between boiling point and molecular weight.

If a hydrogen atom of a saturated hydrocarbon (alkane) is replaced by another atom or group, an elevation of the boiling point results. Thus alkyl halides, alcohols, aldehydes, ketones, acids, and so on, boil higher than the hydrocarbons with the same carbon skeleton.

If the group introduced is of such a nature that it promotes association, a very marked rise in boiling point occurs. This effect is especially pronounced in the alcohols (Figure 3.14) and acids because of hydrogen bonding. For example, the difference in boiling point between propane (nonassociated) and 1-propanol (associated) is 142°C—a difference far greater than the change in molecular weight would indicate. As more hydroxyl groups are introduced, the boiling point rises, but the change is not as great as that caused by the first hydroxyl group. Nevertheless, the increment per hydroxyl group is much greater than the increment per methylene group (Tables 3.4 and 3.5).

Table 3.4 Boiling Point and Chain Length for Straight-Chain Alkanes

	bp (°C)	Δ^a
Pentane	36	
		32
Hexane	68	
		30
Heptane	98	
		27
Octane	125	
		24
Nonane	149	
		24
Decane	173	
		21
Undecane	194	
		21
Dodecane	215	

[a] Δ=change in boiling point for addition of one methylene group.

If the hydroxyl groups are converted to ether linkages, the association due to hydrogen bonds is prevented and the boiling point drops. The following series illustrates this effect.

	CH_2OH	$CH_2OC_2H_5$	$CH_2OC_2H_5$	$CH_2OC_2H_5$
	CHOH	CHOH	CHOH	$CHOC_2H_5$
	CH_2OH	CH_2OH	$CH_2OC_2H_5$	$CH_2OC_2H_5$
bp(°C):	+290	+230	+191	+185

A comparison of oxygen derivatives with their sulfur analogs also shows that association is a more potent factor than molecular weight. The thiol (RSH) compounds are as-

Table 3.5 Boiling Point and Hydroxyl Group Substitution

	CH_3		CH_3		CH_2OH		CH_2OH
	CH_2		CH_2		CH_2		CHOH
	CH_3		CH_2OH		CH_2OH		CH_2OH
bp(°C)	−45		+97		+216		+290
Δ/OH(°C)		142		119		74	

sociated only slightly and hence boil lower than their oxygen analogs, even though the former have higher molecular weights than the latter.

	bp (°C)		bp (°C)
HOH	100	HSH	−62
CH_3OH	66	CH_3SH	+6
CH_3COOH	119	CH_3COSH	93

Ethers and thio ethers are not associated, and hence the alkyl sulfides boil higher than the ethers because they have higher molecular weights:

	bp (°C)		bp (°C)
$(CH_3)_2O$	−24	$(CH_3)_2S$	+38
$(C_2H_5)_2O$	+35	$(C_2H_5)_2S$	+92

These data on sulfur and oxygen compounds, and on hydrocarbons, alkyl chlorides, bromides, and iodides illustrate the general rule that replacement of an atom by an atom of higher atomic weight causes a rise in the boiling point, provided that no increase or decrease in the extent of association takes place as a result of this substitution.

Just as with solubility relationships (Chapter 5 pp. 102–103), branching of the chain and position of the functional group influence the boiling point. The saturated aliphatic alcohols (Table 3.6) serve to illustrate the following generalizations.

1. Among isomeric alcohols, the straight-chain isomer has the highest boiling point.
2. If comparisons are made of alcohols of the same type, the greater the branching of the chain the lower the boiling point.
3. A comparison of the boiling points of isomeric primary, secondary, and tertiary alcohols shows that primary alcohols boil higher than secondary alcohols, which, in turn, boil higher than tertiary alcohols provided that isomeric alcohols with the same maximum chain length are compared.

A knowledge of the boiling points of some simple compounds is frequently of value in excluding certain types of compounds. The following simple generalizations are helpful.

1. An organic chloro compound that boils below 132°C must be aliphatic. If it boils above 132°C, it may be either aliphatic or aromatic. This follows from the fact that the simplest of aryl halide, chlorobenzene, boils at 132°C.
2. Similarly, an organic bromo compound that boils below 157°C or an iodo compound that boils below 188°C must be aliphatic. Other bromo and iodo compounds may be either aliphatic or aromatic.

3.2.3 Specific Gravity

The use of specific gravity in compound identification can be a useful fingerprint. Recall that specific gravity (sp gr), for substance 2, is defined as

$$\text{sp gr}\,^{T_2}_{T_1} = \frac{w_2}{w_1}$$

Table 3.6 Alcohol Boiling Point and Branching

Primary alcohols		Secondary alcohols		Tertiary alcohols	
Structure	**bp (°C)**	**Structure**	**bp (°C)**	**Structure**	**bp(°C)**
CH_3OH	66				
CH_3CH_2OH	78				
$CH_3CH_2CH_2OH$	97	$CH_3CH(OH)CH_3$	82		
$CH_3CH_2CH_2CH_2OH$	116	$CH_3CH_2CH(OH)CH_3$	99	$CH_3{-}C(CH_3)(OH){-}CH_3$	83
$CH_3CH(CH_3)CH_2OH$	108				
$CH_3CH_2CH_2CH_2CH_2OH$	138	$CH_3CH_2CH_2CH(OH)CH_3$	119		
$CH_3CH(CH_3)CH_2CH_2OH$	131	$CH_3CH_2CH(OH)CH_2CH_3$	115		
$CH_3CH_2CH(OH)CH_2CH_3$	129				
$CH_3{-}C(CH_3)_2{-}CH_2OH$	114	$CH_3CH(CH_3){-}CH(OH)CH_3$	111	$CH_3CH_2{-}C(CH_3)(OH){-}CH_3$	102

where w_2=weight of a precise volume of substance 2 (the unknown), w_1=weight of precisely the same volume of substance 1 (usually water), and T_2, T_1 are the temperatures of these substances. The density (d) of substance 2 can be obtained from

$$d_2=\left(\text{sp gr } \tfrac{T_2}{T_1}\right)_2 (d_1)_{T_1}$$

where $(d_1)_{T_1}$=the density of water (or other reference substance) at temperature T_1, and d_2 is the density of substance 2 at temperature T_2. Such densities are available from standard chemistry handbooks.

Specific gravity may be determined by means of a small pycnometer.

Procedures. If a small pycnometer with a capacity of 1–2 mL is not available, either of the two forms shown in Figures 3.15 and 3.16 may be used.

The pycnometer in Figure 3.15 is made from a piece of capillary tubing (1–2 mm). It is bent into the shape shown, a small bulb blown in the middle, and one end drawn out to a fine capillary. A scratch is made on the other arm at the same height as the tip of the capillary. The pycnometer is suspended by means of a fine Nichrome, aluminum, or plat-

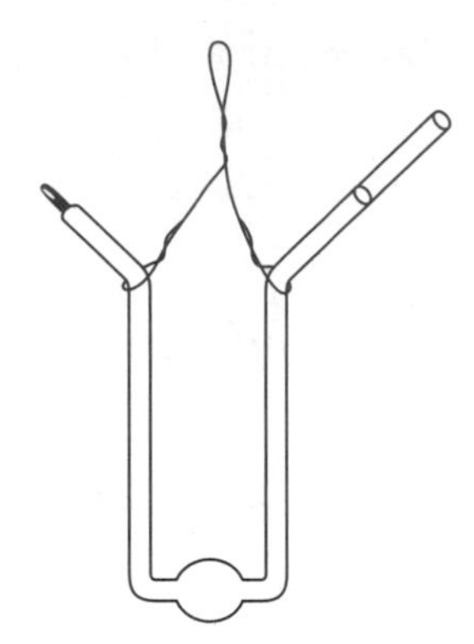

Figure 3.15 Micropycnometer.

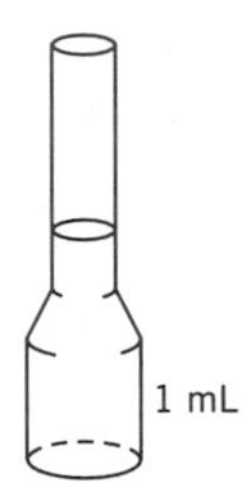

Figure 3.16 Specific gravity bulb (small volumetric flask).

inum wire, and its weight is determined.[6] The pycnometer is then filled with distilled water to a point beyond the mark and suspended in a constant-temperature bath (e.g., at 20°C). After about 10 min, the amount of liquid in the tube is adjusted by holding a piece of filter paper to the capillary tip until the meniscus in the open arm coincides with the mark. The pycnometer is then removed from the beaker, dried, and weighed.

Figure 3.16 shows another specific gravity device. Commercial 1.00-mL (or somewhat larger) volumetric flasks may be used. The weight of the empty bulb is determined. It is then filled with distilled water and suspended by a wire in a constant-temperature bath (e.g., at 20°C). The level of the water in the bulb is adjusted to the mark by means of a disposable pipet. The bulb is then removed from the beaker, dried, and weighed.

The weight of the empty pycnometer and its weight when filled with distilled water are recorded and kept permanently. To determine the specific gravity of a liquid the bulb or pycnometer is filled with the liquid at 20°C and its weight determined.

$$\text{sp gr}_{20}^{20} = \text{weight of sample}$$

The apparatus must be carefully cleaned and dried immediately after use.

Care must be taken that the sample used for this determination is pure. It is best to use a portion of the center fraction collected from distillation or a gas chromatographic collection corresponding to a single peak (Chapter 10 pp. 474–476). Sometimes it is necessary to determine the density with reference to that of water at 4°C. This may be done by means of the factor 0.99823:

$$\text{sp gr}_{4}^{20} = \text{weight of sample} \times 0.99823$$

Another micropycnometer has been described.[7]

Discussion. The specific gravity of a liquid may often be used to exclude certain compounds from the list of possibilities. It varies with the composition as well as the structure of the compound.

Hydrocarbons are usually lighter than water. As a given homologous series of hydrocarbons is ascended, the specific gravity of the members increases, but the increment per methylene radical gradually diminishes. Curves I, II, and III in Figure 3.17 show the

[6] Analytical balance (±0.1 mg).

[7] M. M. Singh, Z. Szafran, and R. M., Pike, *J. Chem. Educ., 70,* A36 (1993).

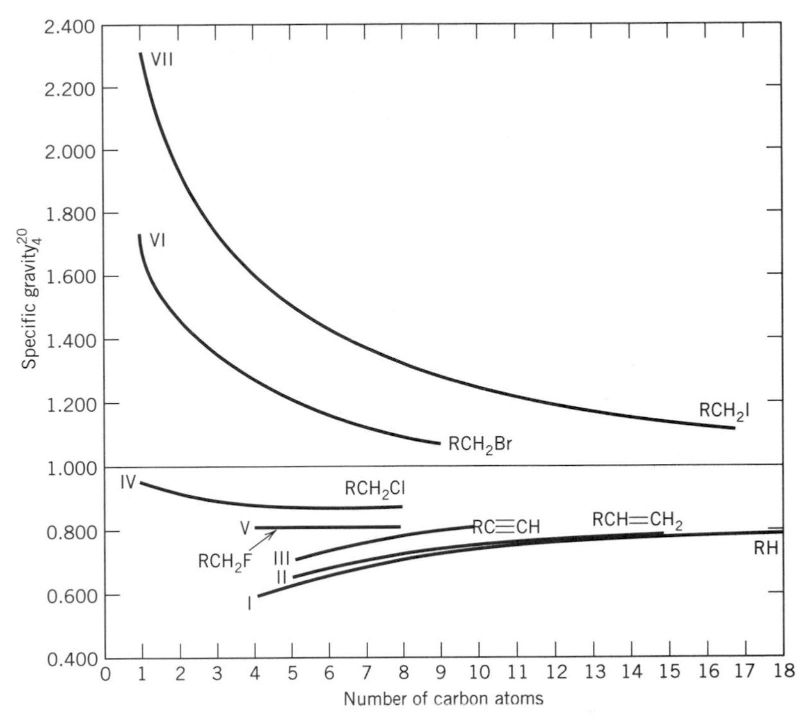

Figure 3.17 Relationship between specific gravity and molecular weight.

change in density for the alkanes, 1-alkenes, and 1-alkynes. It will be noted that the specific gravity of the acetylenic hydrocarbon is greater than that of the corresponding olefin, which in turn is more dense than the alkane hydrocarbon with the same number of carbon atoms. The position the unsaturated linkage occupies also influences the density. Moving the double bond nearer the middle of the molecule causes an increase in the specific gravity. The data in Table 3.7 illustrate this change.

Table 3.7 Specific Gravity and Double Bond Position

Name	Compound	sp gr
1-Pentene	$CH_2{=}CHCH_2CH_2CH_3$	0.645_4^{25}
2-Pentene	$CH_3CH{=}CHCH_2CH_3$	0.651_4^{25}
1,4-Pentadiene	$CH_2{=}CHCH_2CH{=}CH_2$	0.659_4^{20}
1,3-Pentadiene	$CH_2{=}CHCH{=}CHCH_3$	0.696_4^{20}
2,3-Pentadiene	$CH_3CH{=}C{=}CHCH_3$	0.702_4^{20}
1-Hexene	$CH_2{=}CHCH_2CH_2CH_2CH_3$	0.673_4^{20}
2-Hexene	$CH_3CH{=}CHCH_2CH_2CH_3$	0.681_4^{20}
3-Hexene	$CH_3CH_2CH{=}CHCH_2CH_3$	0.722_4^{20}

The replacement of one atom by another of higher atomic weight usually increases the density. Thus curve IV, which represents the specific gravities of the normal alkyl chlorides, lies above the curves of the hydrocarbons. It will be noted that the alkyl chlorides are lighter than water and that the specific gravities *decrease* as the number of carbon atoms is increased.

The rather limited data on the alkyl fluorides are shown by curve V. The graph is interesting because it reveals only a very slight change in density as the number of carbon atoms is increased.

Curves VI and VII show that the specific gravities of the primary alkyl bromides and iodides are greater than 1.0 and that in these homologous series the specific gravity decreases as the number of carbon atoms is increased. The slopes of curves IV, VI, and VII are decreasing because the halogen atom constitutes a smaller and smaller percentage of the molecule as the molecular weight is increased by increments of methylene units. The relative positions of the curves show that the specific gravity increases in the order

$$RH < RF < RCl < RBr < RI$$

provided that comparisons are made on alkyl halides with the same carbon skeleton and of the same class. Similar relationships are exhibited by secondary and tertiary chlorides, bromides, and iodides.

The specific gravities of aryl halides also arrange themselves in order of increasing weight of the substituent (Table 3.8).

Table 3.8 Boiling Point and Specific Gravity of Aryl Halides

Compound	bp (°C)	sp gr$_4^{20}$
Benzene	79.6	0.878
Fluorobenzene	86	1.024
Chlorobenzene	132	1.107
Bromobenzene	156	1.497
Iodobenzene	188	1.832

An increase in the number of halogen atoms present in the molecule increases the specific gravity. Compounds containing two or more chlorine atoms or one chlorine atom together with an oxygen atom or an aryl group will generally have a specific gravity greater than 1.000 (Table 3.9).

Table 3.9 Specific Gravity Change per Number of Chlorine or Oxygen Atoms

Compound	sp gr	Compound	sp gr
Benzyl chloride	1.1026_4^{15}	Carbon tetrachloride	1.595_4^{20}
Benzal chloride[a]	1.2557_4^{14}	Ethylene chlorohydrin	1.213_4^{20}
Benzotrichloride	1.3800_{20}^{20}	Chloroacetone	1.162_4^{16}
Methylene chloride	1.336_4^{20}	Methyl chloroacetate	1.235_{20}^{20}
Chloroform	1.4984_4^{15}		

[a]Benzylidene chloride, $C_6H_5CHCl_2$.

The introduction of functional groups containing oxygen causes an increase in the specific gravity. The curves in Figure 3.18 represent the change in specific gravity of some of the common types of compounds. The ethers (curve VIII) are the lightest of all the organic oxygen compounds. The aliphatic alcohols (curve IX) are heavier than the ethers but lighter than water. The specific gravity of the alcohols becomes greater than 1.0 if a chlorine atom (ethylene chlorohydrin), a second hydroxyl (ethylene glycol), or an aromatic nucleus (benzyl alcohol) is introduced. The dip in curve IX is due to the fact that methanol is more highly associated than ethanol. The amines (curve X) are not as dense as the alcohols and are less associated. Association also causes the specific gravity of formic acid and acetic acid to be greater than 1.000; the higher liquid fatty acids are lighter than water (curve XI).

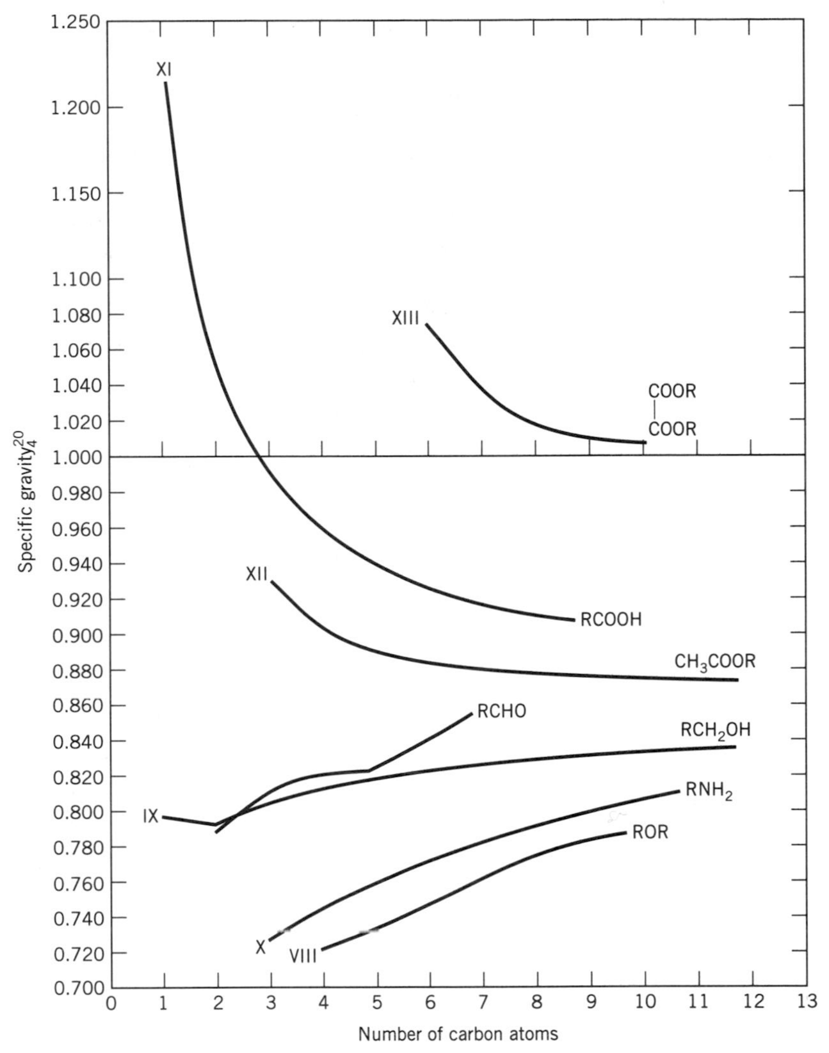

Figure 3.18 Relationship between specific gravity and molecular weight (linear compounds).

The simple esters (curve XII) and aldehydes (RCHO) are lighter than water, whereas esters of polybasic acids (curve X3), halogenated, keto, or hydroxy esters are heavier than water. Introduction of the aromatic ring also may cause esters to be heavier than water. Examples of esters of these types that are heavier than water are phenyl acetate, methyl benzoate, benzyl acetate, ethyl salicylate, butyl oxalate, triacetin, isopropyl tartrate, and ethyl citrate. Since the hydrocarbons are lighter than water, it is to be expected that esters containing long hydrocarbon chains will show a correspondingly diminished specific gravity.

In general, compounds containing several functional groups—especially those groups that promote association—will have a specific gravity greater than 1.0. Merely noting whether a compound is lighter or heavier than water gives some idea of its complexity. This is of considerable value in the case of neutral liquids. If the compound contains no halogen and has a specific gravity less than 1.0, it probably does not contain more than a single functional group in addition to the hydrocarbon or ether portion. If the compound is heavier than water, it is probably polyfunctional.

3.2.4 Index of Refraction of Liquids

The refractive index of a liquid is equal to the ratio of the sine of the angle of incidence of a ray of light in air to the sine of the angle of refraction in the liquid (Figure 3.19). The ray of light undergoes changes in wave velocity ($v_{air} \longrightarrow v_{liquid}$) and in direction at the boundary interface, and these changes are dependent on temperature (T) and wavelength (λ) of light. Direct measurements of the angles of incidence and refraction are not feasible; hence optical systems have been devised that are dependent on the critical angle of reflection at the boundary of the liquid with a glass prism of known refractive index. The Abbe type of refractometer operates on this principle, and a number of instruments are available commercially.[8] The advantages of the Abbe type of refractometer are that (1) a white light source of illumination may be used, but the prism system gives indices of refraction for the sodium D line; (2) only a few drops of liquid are needed; (3) provision for temperature control of prisms and sample is incorporated; and (4) the compensating Amici prisms permit the determination of specific dispersion.

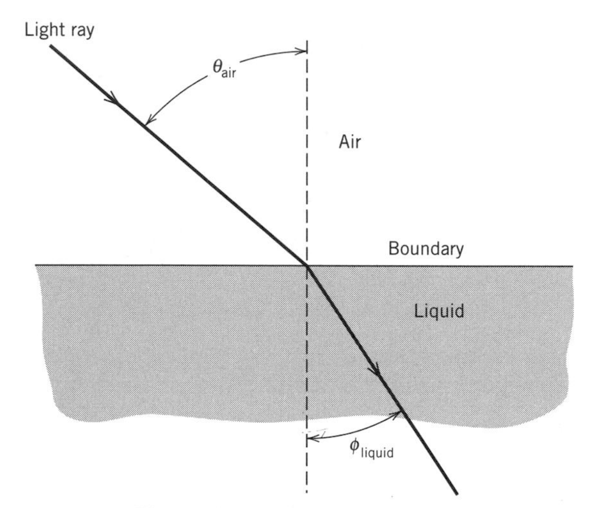

Figure 3.19 Refraction of light:

$$n_\lambda^T = \frac{\sin \theta_{air}}{\sin \phi_{liquid}} = \frac{v_{air}}{v_{liquid}}$$

n = index of refraction

A schematic drawing of the optical system of the Bausch & Lomb Abbe 3L refractometer is shown in Figure 3.20, and the instrument in Figure 3.21. Water at 20°C is allowed to flow through the jackets (J) surrounding the prisms (P_1, P_2). If the liquid sample is free flowing, it is introduced by means of a pipet through one of the channels beside the prisms (D). If the sample is viscous, the upper prism is lifted and a few drops spread on prism P_2 with a wooden applicator. The prisms are then closed slowly, any excess liquid being squeezed out. The lamp (L) is turned on. While one looks in the eyepiece (E), the coarse adjusting wheel (A) is turned and the position of the lamp (L) is also adjusted so as to obtain a uniformly lighted field. Then the eyepiece (E) is focused on the cross hairs (H) and the wheel (A) is turned so that the dividing line between the light and dark halves coincides with the center of the cross hairs. Usually the borderline is colored and is achromatized by turning the milled screw (Z) until a sharp black-to-white dividing line

[8]Bausch & Lomb Optical, Co., One Bausch & Lomb Place, Rochester, N.Y., 14604; consult the annual instrument guides in the journals *Science* or *Analytical Chemistry* for additional sources.

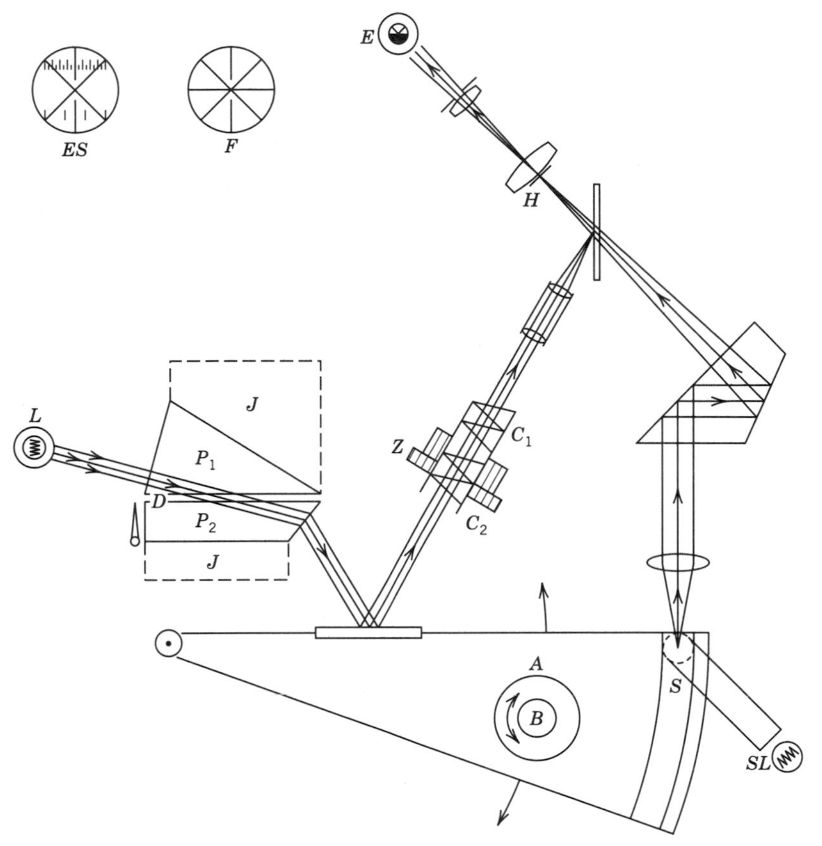

Figure 3.20 Schematic diagram of the optical system of the Abbe 3L refractometer. [Redrawn from the operating manual by courtesy of Bausch & Lomb Optical Co., Rochester, NY.]

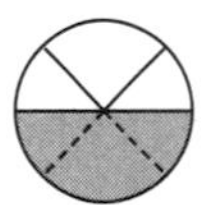

Figure 3.21 Bausch & Lomb Abbe 3L refractometer. [Courtesy Bausch & Lomb Optical Co., Rochester, NY.]

is obtained at the center.[9] This adjusting screw (*Z*) rotates the two Amici prisms (C_1, C_2), which compensate for differences in the degree of refraction of light of different wavelengths present in white light. After the dividing line is made as sharp as possible, the micrometer screw (*B*) is turned so that the dividing line is exactly at the center of the cross hairs, as shown at F. Then the switch on the left side of the instrument is depressed to cause illumination of the scale (*S*). The index of refraction for the sodium D line is read to three decimal places in the eyepiece (*ES*) and the fourth place estimated. The result is recorded in the following form:

$$n_{\mathrm{D}}^{20} = 1.4357$$

[9]Certain liquids with unusual dispersions do not give a sharp line of demarcation when an electric light is used as a source of illumination. In such cases, the lamp should be swung down and a sheet of white cardboard placed at such an angle to the illuminating prism that daylight from a window will be reflected into the instrument. This usually provides a sharp division line with careful adjustment of the dispersion screw (Figure 3.20).

At the same time that the refractive index is read, the reading on the drum (*D*) should be noted. The prisms should then be cleaned by means of a cotton swab dipped in toluene or petroleum ether for water insoluble compounds.[10] Distilled water is used to remove water soluble compounds. Extreme care must be taken not to scratch the prisms. Metal or glass applicators should be avoided, and only clean absorbent cotton (free from dust) used to clean the prisms. Manuals for operating various instruments should be consulted for variations in operating procedures. A complete discussion of the principles of refractometry and refractometers is available.[11]

Discussion. The values for density and refractive index are useful in excluding certain compounds from consideration in the identification of an unknown. Care must be taken, however, that the sample is pure. It is best to determine these physical constants on a center cut from distillation (bp determination) or a gas chromatography collection.

These two constants can also serve as a final check on an unknown after its identity and structure have been established. They are of value in research work for checking structures. This checking is accomplished by comparing the observed index to the literature value (see tables in Appendix 3).

3.3 INTRODUCTION TO CHROMATOGRAPHY

Chromatography may be defined[12] as the science of separation techniques involving a mobile phase (e.g., the solvent in column chromatography) passing by a stationary phase (e.g., the alumina in column chromatography). The ability to separate various components of a mixture of organic compounds is based on selective and preferential absorption of these components in the mobile phase by the stationary phase.

Organic chemists are interested in two major classes of chromatography: gas chromatography (GC) and liquid chromatography (LC). Gas chromatography is useful for relatively volatile and thermally stable organic compounds; this method involves a gaseous mobile phase (helium or, less frequently, nitrogen) and a liquid stationary phase. The stationary phase is usually spread evenly over a solid support, such as crushed brick. Gas chromatography is introduced in some detail below (Section 3.3.2).

Liquid chromatography (LC) involves a liquid mobile phase (usually common organic solvents, Section 3.3.2) and either solid stationary phases (e.g., the alumina or silica gel of column and thin-layer chromatography) or stationary phases of liquids spread over solids (e.g., as used in high-pressure liquid chromatography). Thin-layer chromatography (an example of liquid chromatography) is introduced below (Section 3.3.1).

Most organic chemists are concerned with subdividing a chromatographic method into analytical and preparative classes. Techniques developed on an analytical scale usually involve handling quite small amounts of material; care must be exercised in extrapolating the methods to the different equipment and larger samples used in preparative-

[10] Do *not* use acetone to clean prisms.

[11] N. Bauer, K. Fajans, and S. Z. Lewin, "Refractometry," Chap. 18, pp. 1140–1281, in *Physical Methods of Organic Chemistry,* 3rd ed., Vol. I, Part 2, edited by A. Weissberger (Interscience, New York, 1960).

[12] The linguistic origin of the word chromatography is based on color (from the Greek word *chroma* meaning color and *graphy* meaning written.); early chromatography was carried out on paper using colored derivatives of naturally occurring compounds.

scale work. Preparative separations require working samples large enough that a number of chemical and spectral analyses and chemical reactions can be carried out.

In choosing between GL and LC the following facts should be considered before making that choice.

Gas chromatography (GC):

1. The sample should be at least moderately volatile and reasonably stable to heat. Specifically, the compound must be stable enough to survive the conditions necessary to convert it to the gas phase.
2. Simple GC instruments are inexpensive, easy to operate, and usually give results rapidly; instruments are now available for very modest prices.

Liquid chromatography (LC):

1. Procedures are usually time-consuming, especially for classical gravity-flow conditions and preparative-scale work; rapid analyses can, however, be carried out with dry column chromatography (Chapter 10) and high-performance (high-speed) liquid chromatography (Chapter 10).
2. Slow procedures mean that careful consideration must be made of the possible chemical reactivity of the sample of the column (e.g., a sample of alcohol $+ Al_2O_3 \longrightarrow ?$). Proper choice of conditions, however, allows virtually any organic compound (other than, e.g., salts) to be analyzed by LC.
3. High-performance liquid chromatography (HPLC, Chapter 10) is more expensive in terms of initial cost because of the high-quality pumps and column packings that are necessary.

These chromatography possibilities are routinely used by most organic chemists. Purity checks by GC and LC are routinely used.

3.3.1 Thin-Layer Chromatography

Thin-layer chromatography (TLC) is perhaps the most rapid, easiest, and most often applied method for assessing the purity (complexity, and often the nature) of organic compounds. Chromatography may be defined as a separation by differential partitioning of chemical components by distribution between two phases, one phase being immobile (e.g., a surface) and the other being the transport medium (e.g., the solvent or eluant). The general subject of chromatography will be discussed thoroughly in Chapter 10 (pp. 461–476). In TLC, the immobile phase is a thin layer of adsorbent spread over a sheet of glass or plastic. Calcium sulfate or an organic polymer serves to bind the adsorbent to the sheet. A small amount of sample is placed at the bottom of the slide and this spotted slide is placed on end in a container with a shallow layer of solvent (see Figure 3.23); the distance to which the solvent moves the compound up the chromatogram sheet is dependent on the ability of the compound to adhere to this adsorbent system, as well as many other factors. More often than not, adsorbent–solvent systems can be found to separate most components of a given mixture. This procedure is especially useful for compounds that are heat sensitive or nonvolatile, that is, those compounds that are not amenable to boiling point or gas chromatographic determination.

Compounds can be detected on TLC sheets in various ways. The simplest is to use a low-power hand-held UV light. This procedure requires the use of TLC adsorbent that has been mixed with a fluorescent indicator. In such cases the eluted compounds will appear as dark spots because they block out the fluorescent indicator. Other procedures are described below, including methods involving *p*-anisaldehyde and involving phosphomolybdic acid, both of which are much more sensitive than fluorescent indicators.

Procedure. Commercial sheets, precoated with alumina (Al_2O_3) or silica gel ($SiO_2 \cdot x\ H_2O$), often containing fluorescent material, are available.[13] A preliminary aid to choice of solvent is the following: prepare fine capillary tubes (e.g., drawn mp tubes) containing solutions of the sample in each of a series of solvents by capillary action. Table 3.10 lists those solvents that can be screened for possible use. Merely touch a chromatographic sheet with the tube (Figure 3.22). When the solvent circle has reached its final position, a useful solvent will have moved the sample ring to a position about 0.3 to 0.5 that of the solvent circle radius. If no other solvent information (e.g., from recrystallization attempts) is available, chloroform and solvents of similar polarity may be tried.

Table 3.10 Chromatographic Solvents[a]

Solvent	
Petroleum ether	Increasing polarity[b] ↓
Cyclohexane	
Carbon tetrachloride	
Benzene	
Methylene chloride	
Chloroform (alcohol free)	
Ethyl ether	
Ethyl acetate	
Pyridine	
Acetone	
1-Propanol	
Ethanol	
Methanol	
Water	
Acetic acid	

[a]Mixtures of two or more solvents can be used as developing solvents in chromatographic separations.

[b]Polarity, in this context, is meaningful only for chromatography and is not necessarily the same as polarity as measured by, for example, dielectric constant.

A development chamber (Figure 3.23) that can be capped is prepared by filling it to a depth of ca. 3 mm with the solvent of choice. A solution (1–10%) of the sample of interest is prepared in any volatile solvent. Fine capillary tubing is prepared by drawing out melting point capillary tubing; touching the fine tubing to the surface of the solution causes some of the solution to rise up into the tubing (capillary action). Draw a *pencil* line across

[13]Consult the annual Instrument (or Equipment) Guides of *Science* or of the journal *Analytical Chemistry* for guides to sources of chromatography equipment. One source is Eastman Kodak Co., Phone: 800-242-2424.

the TLC sheet about 5 mm from the bottom. Spot the sheet with sample by touching the solution-charged tubing to a position on the pencil line (Figure 3.24). This sheet is now placed, spotted-end down, upright in the development chamber. The sheet *must* be so placed as to position the spot above the solvent at the beginning of the analysis. Solvent is allowed to climb up the plate until the solvent front is ca. 1 cm from the upper end of the adsorbent-covered region (Figure 3.23); during this development, the system should be enclosed to ensure that the atmosphere is saturated with solvent vapor. Saturation may be further ensured by fitting a filter paper (wick) to the inside wall of the chamber. After development (ca. 5 min for sheets of microscope-slide size and 0.5–1.0 hr for large plates), the position to which the solvent front has moved is marked by scratching a line along this solvent front. Any obvious sample spots are scratch-marked (circled) also. Allow the chromatogram to dry (usually for a few minutes) before visualization.

Special procedures must be used to visualize colorless compounds. If the chromatogram originally contained fluorescent material, the developed chromatogram may be observed under ultraviolet light (***caution!***);[14] dark spots are observed where sample spots block out the fluorescence. These spots (and any detected by alternative methods below) should be marked by circling. Alternatively, spots may be brought out by placing the developed chromatogram in an encloseable chamber containing iodine crystals

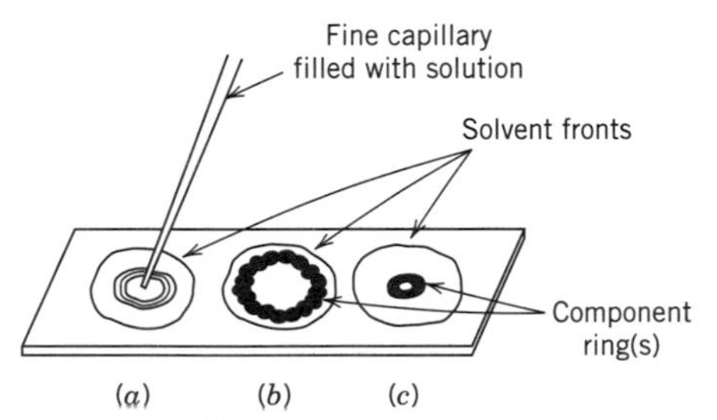

Figure 3.22 Thin-layer chromatography: determination of solvent choice. (*a*) Good development. (*b*) Overdeveloped. (*c*) Underdeveloped.

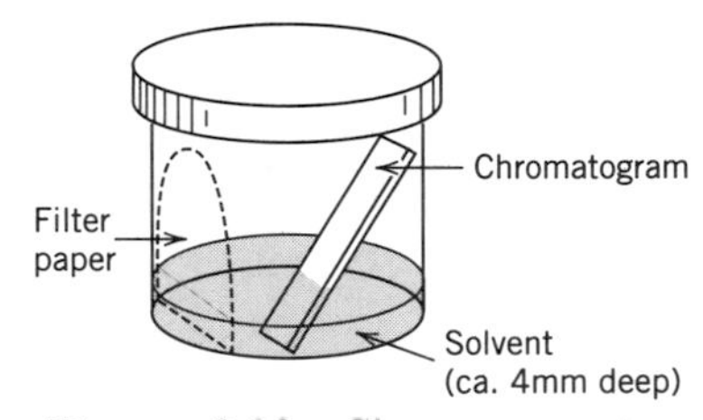

Figure 3.23 Chromatographic development unit.

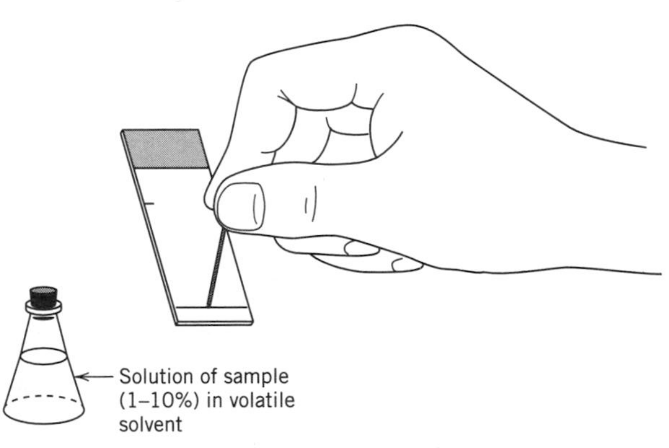

Figure 3.24 Preparation for TLC.

[14]Do not look directly at the UV light as it can cause eye damage.

(Figure 3.25). Nearly all compounds (except alkanes and aliphatic halides) form iodine charge-transfer complexes; the formation of these complexes is (often rapidly) reversible, so the position of the dark spot should be marked quickly. Another visualization technique involves spraying the chromatogram with sulfuric acid to cause the colorless compounds to darken by charring; this is, of course, a sample-destructive analytical technique.

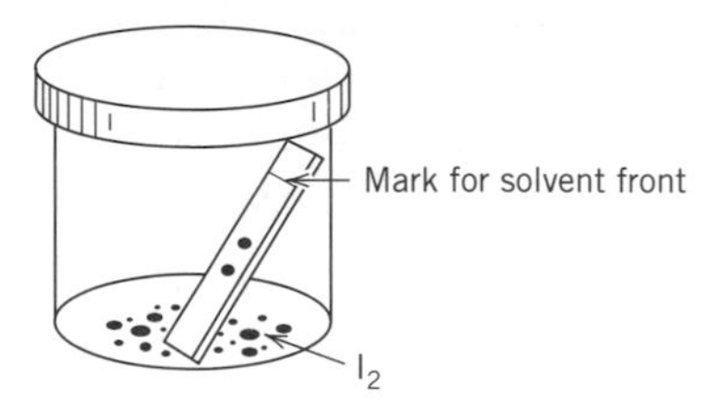

Figure 3.25 Iodine TLC spot-marking chamber.

A very sensitive method for detecting compounds involves use of a spray containing *p*-anisaldehyde. TLC plates should be made of glass (if plastic sheets are to be used, a test should be run to see if the sheets can survive the heating conditions) and the adsorbent should not be mixed with fluorescent indicator. This procedure requires the preparation of a solution (combine all reagents with cooling) of 0.5 mL of *p*-anisaldehyde (*p*-methoxybenzaldehyde) in 9 mL of 95% ethanol containing 0.5 mL of concentrated sulfuric acid and a few drops of glacial acetic acid. This solution is sprayed on the developed plates, and these are then baked to dryness on a hot plate until the dark spots develop.

Another procedure using 5% phosphomolybdic acid has become popular. The spots can be marked by dipping the developed silica sheets into the 5% phosphomolybdic acid. This method does not suffer from interference from indicators mixed into silica gel.

If commercial chromatographic sheets are not available, glass plates are easily prepared. Prepare a slurry of adsorbent (35 g of silica gel G in 100 mL of chloroform or 60 g of alumina in 100 mL of 67 vol%/33 vol% chloroform/methanol). Stir the slurry thoroughly; dip a pair of back-to-back plates (e.g., standard microscope slides) into the slurry and slowly withdraw them. Wipe excess adsorbent from the edges and separate the plates. Wipe adsorbent from the back of the plates and allow them to dry (ca. 5 min). Such plates may be used immediately or stored in a desiccator.

Discussion. Development of different chromatograms (alumina, silica gel, silica acid) of the same sample that clearly yield only a single spot in a variety of solvents is a very good indication of purity. In addition, the identity of an unknown can be supported (but not completely proved; why not?) by comparison of spot positions of known and unknown materials (Figure 3.26). Figure 3.26 also implies how the progress of a reaction can be monitored. The most common method of reporting TLC results is by the R_f value (Figure 3.27) in a specific solvent on a specific type of chromatographic sheet:

$$R_f = \frac{\text{distance spot of interest has moved from origin } (b)}{\text{distance solvent front has moved from origin } (a)}$$

Alternatively, the results may be reported as R_x, where

$$R_x = \frac{\text{distance spot of interest has moved from the origin}}{\text{distance spot of reference compound has moved from the origin}}$$

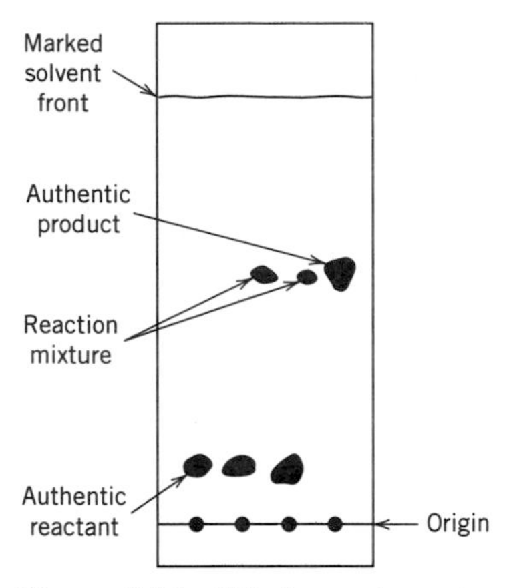

Figure 3.26 Thin-layer chromatogram; reaction mixture composition analysis.

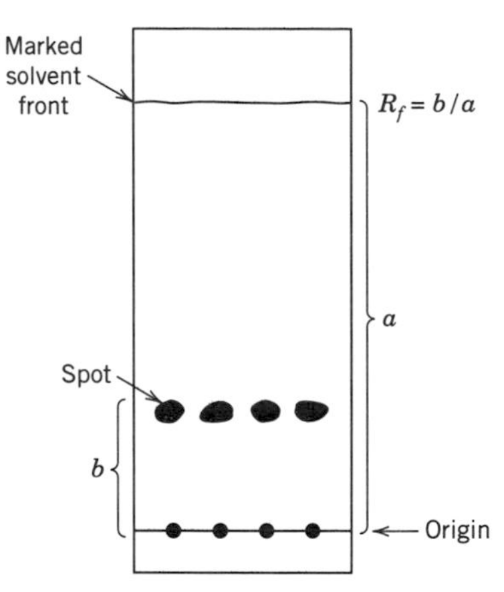

Figure 3.27 Determination of R_f on TLC chromatogram.

Information obtained from TLC is useful for solvent and adsorbent choices in (preparative-scale) column chromatography (see Chapter 10).

3.3.2 Gas Chromatography

Gas chromatography (GC) is the science of separating compounds that have been converted to the gas phase on the basis of their boiling points or upon their polarity differences. Liquid samples can be directly injected onto the GC instrument; solids must be dissolved in a solvent and they can be analyzed only if they are sufficiently volatile. Liquid samples are frequently injected as solutions. Samples upon injection are subjected to a very hot injection port which immediately converts them to the gas phase. A carrier gas (usually helium) transports the gases into a heated column which is internally coated with a stationary phase. A primary driving force for separation of organic components by GC is simple boiling point differences: higher boiling compounds are eluted more slowly. Boiling point differences are of substantial importance even for the so-called polarity-based columns described below. There are two major classifications of columns: nonpolar and polar. Nonpolar columns separate components almost completely by boiling point differences; that is, the higher boiling compounds are eluted more slowly (have longer retention times). Polar columns separate components at least to some degree on the basis of differences in dipole–dipole interactions between each component and the stationary phase. The stationary phase is thus a material that separates components by selective attraction (selective adsorption). More polar compounds are held back longer (have longer retention times) and less polar compounds elute more rapidly (have shorter retention times). Separated peaks are passed through a detector and the detector responses are converted to peak shapes on an electronic recorder. Part of this process involves computer-based conversion of the detector's analog signal to a digitized format. Retention times of the components can be used to identify them (knowns can be injected to measure standard retention times), and relative peak areas are used to measure the amount of each compound.

There are two major categories of GC instrument based upon the sizes of columns employed. One type uses "packed" columns, which are typically 3–6 mm in width and 1–5 m long. The other type uses open tubular or capillary columns 0.10–0.70 mm in width and 15–100 m long. Commonly the widths are in the 0.20–0.53-mm range. The latter, capillary technique is much more efficient and is the method of choice for analytical work. The former, packed-column method is used when preparative-scale collections are required.

Gas chromatography is useful for purity determinations and component analysis of sufficiently volatile organic compounds. Observation of a single, large peak in a variety of GC determinations (various columns, temperatures, etc.) for a given sample is a strong indication of its purity. Samples must thus be thermally stable to volatilization conditions for GC analysis.

A simplified schematic diagram of a gas chromatograph is shown in Figure 3.28. A specific gas chromatograph is shown in Figure 3.29. The instrument consists of the following:

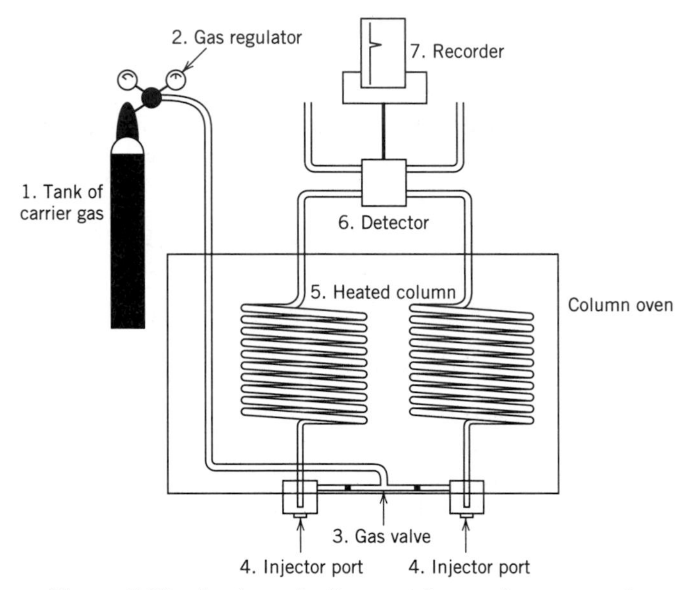

Figure 3.28 A schematic diagram of a gas chromatograph.

1. A tank of carrier gas (see point 1), which is usually helium (or, less frequently, nitrogen). Helium is used because it is inert and has a high thermal conductivity.
2. A method of controlling the gas flow. This usually involves a valve on the tank (points 1 and 2) and a regulator on the instrument (point 3).
3. An injection port (see point 4). This is a metal cap, with a hole over a piece of rubber or plastic material that can be pierced with a syringe.
4. A heated column (see point 5). This is a metal or glass tube that contains the solid support and stationary phase. This "column" is coiled to fit in the heated compartment and is connected from the injection port to the detector.
5. A detector (see point 6). The two most common detectors are of the thermal conductivity and flame ionization types. The thermal conductivity detector measures changes, at a filament, in the conductivity of the carrier gas as a function of sample content; this change is electrically passed on to the recorder (see point 7). Thermal conductivity (TC) detectors are used on instruments that employ packed columns. Flame ionization detection (FID) measures sample content by burning the eluted sam-

Figure 3.29 Gas chromatograph and recorder: GOW-MAC series 350.

ple in a small hydrogen flame, followed by determining the amount of ions so produced. Detection by flame ionization detection is frequently used on instruments that are equipped with capillary columns.

6. Recorder (see point 7). The analog signal from the detector is converted to digital output on an electronic recorder in the form of peaks (Figure 3.30). As described earlier, the chromatogram occurs as a series of these peaks. The retention times are used to identify compounds, and the peak areas are used to measure the amounts of the various compounds in the sample. The retention time of a component is taken as the time at which the maximum for the peak occurs relative to the time of injection (or relative to the retention time of some volatile standard).

Procedure. The first procedure is for the operation of GCs equipped with packed columns and thermal conductivity detectors. The detector and injector blocks should be heated continuously; these points come to temperature very slowly (1–2 hr) and are usually several tens of degrees warmer than the column. The column temperature is usually set such that it is 10–20°C above the boiling point of the sample of interest;[15] it should,

[15]If the sample is a solid and is being introduced in solution, an educated guess as to these temperature settings is in order.

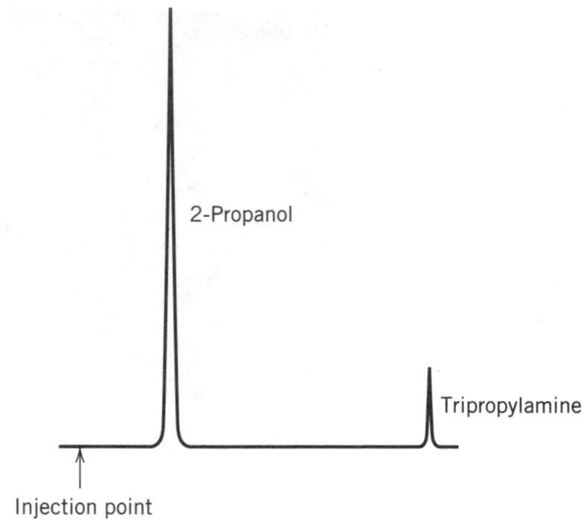

Figure 3.30 Gas chromatograph trace recording.

however, *always* be below the upper limit recommended for the stationary phase (Figure 3.31). Typical column temperatures are 50–200°C, and the corresponding detector and injector temperatures are 80–250°C; clearly, sample stability should influence temperatures chosen. To begin instrument operation, one should be *certain* that there is helium flow[16] before turning on the filament (bridge) current in a thermal conductivity instrument; flow is assured by checking the tank and instrument gas gauges *and* by checking the detector exit port with a bubble meter. Allow about 30 min for the flow rate and column temperatures to reach a steady state and measure the flow rate with the instrument gauge or with a bubble meter. The current is then turned on and set typically between 100 and 200 mA. The chart pen should be zeroed and the recorder started. A sample is injected (e.g., 1 μL if neat or 10–100 μL if a 1–10% solution in a volatile solvent) by means of a microsyringe (Figure 3.32) into one port or the other. The chart should be marked at the time of injection; a record should also be made of the column size and type used, the column temperature, gas flow rate, and other details. As the peaks are traced on the recorder, one should be ready to make attenuation (detector response) changes such that the top of the peak can be observed. Retention time is taken as the time from injection until the time this peak maximum is obtained. After all of the components have been eluted, the chromatograph is shut down by turning off the filament current and only then reducing the carrier flow to a trickle.

The operation of a modern capillary GC is generally similar to that described above for packed columns. The main difference is that capillary instruments are electronically operated. The following instructions are very general and it is very important that you consult your laboratory instructor for specific instructions for the instrument in your laboratory.

The column temperature is entered and usually this is about 20°C higher than the boiling point of the least volatile sample to be analyzed.

A "ramp" choice must be made. "Ramping" refers to the practice of using regular increases in column temperatures over time to allow GC analysis in a minimum length of time. A typical ramp might mean that the column is increased in temperature from, say, 100°C to 150°C over a given period of time. This could be done so that the main components of interest, if more volatile, could be analyzed at 100°C before ramping, and then

[16]Failure to have an inert gas around the detector filament will very likely result in destruction of the filament of a thermal conductivity detector.

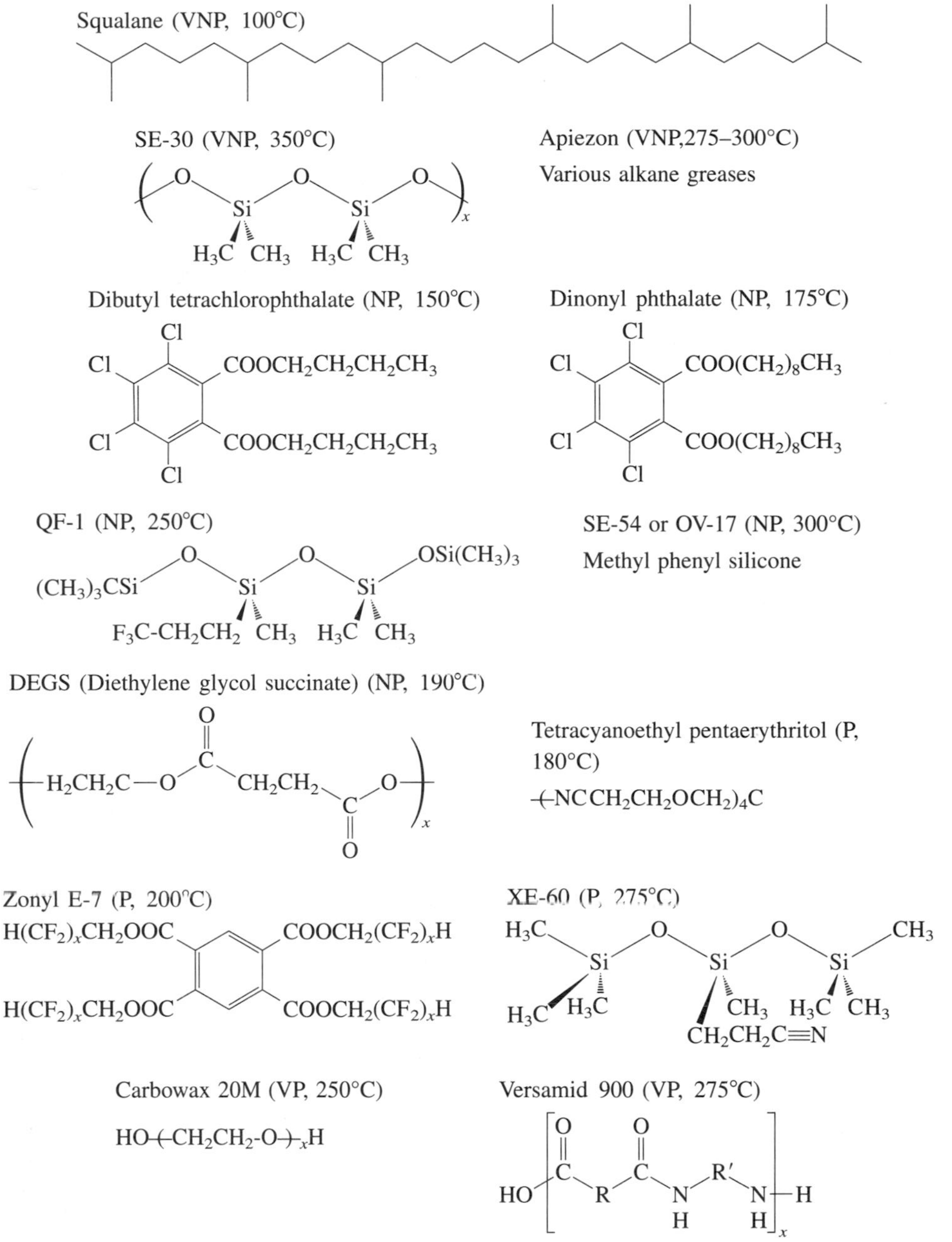

Squalane (VNP, 100°C)

SE-30 (VNP, 350°C)

$(\!-O-Si(CH_3)_2-O-Si(CH_3)_2-O-)_x$

Apiezon (VNP, 275–300°C)
Various alkane greases

Dibutyl tetrachlorophthalate (NP, 150°C)

$C_6Cl_4(COOCH_2CH_2CH_2CH_3)_2$

Dinonyl phthalate (NP, 175°C)

$C_6Cl_4(COO(CH_2)_8CH_3)_2$

QF-1 (NP, 250°C)

$(CH_3)_3CSi-O-Si(CH_3)(CH_2CH_2CF_3)-O-Si(CH_3)_2-OSi(CH_3)_3$

SE-54 or OV-17 (NP, 300°C)
Methyl phenyl silicone

DEGS (Diethylene glycol succinate) (NP, 190°C)

$(-H_2CH_2C-O-C(=O)-CH_2CH_2-C(=O)-O-)_x$

Tetracyanoethyl pentaerythritol (P, 180°C)

$(NCCH_2CH_2OCH_2)_4C$

Zonyl E-7 (P, 200°C)

$H(CF_2)_xCH_2OOC$, $COOCH_2(CF_2)_xH$, $H(CF_2)_xCH_2OOC$, $COOCH_2(CF_2)_xH$

XE-60 (P, 275°C)

$H_3C-Si(CH_3)(CH_3)-O-Si(CH_3)(CH_2CH_2C\equiv N)-O-Si(CH_3)(CH_3)-CH_3$

Carbowax 20M (VP, 250°C)

$HO(-CH_2CH_2-O-)_xH$

Versamid 900 (VP, 275°C)

$HO[-C(=O)-R-C(=O)-NH-R'-NH-]_xH$

Figure 3.31 Common stationary phases.* See polarity classifications in Table 3.11.

the elevated temperature would drive off less volatile side products, clearing the column of all injected materials. Simple samples do not require ramped temperature changes. If a ramping procedure is to be used, the initial and final temperatures, the start and stop times of the ramp, as well as the heating rate must be programmed.

The temperature and pressure of the inlet system must be entered. Usually the temperature of the inlet system is 20–30°C higher than that of the column. A typical inlet temperature that works for a wide range of organic compounds is 250°C.

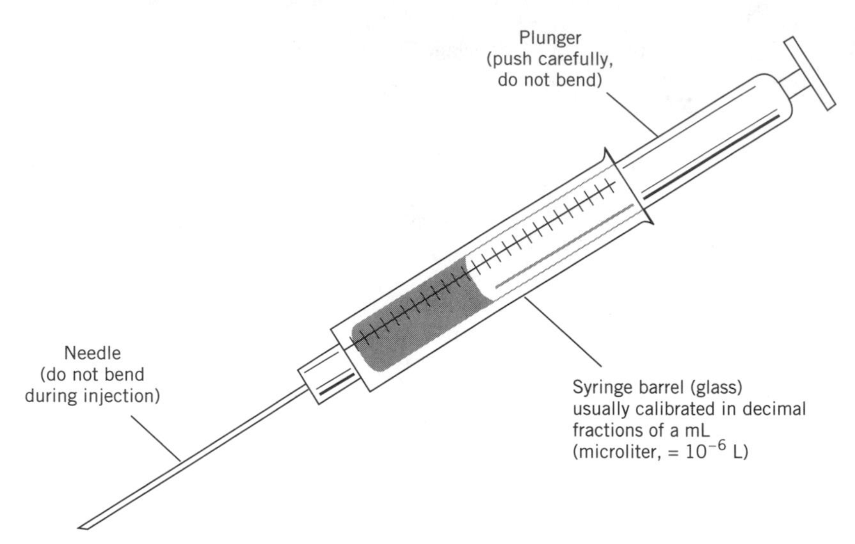

Figure 3.32 Syringe.

The detector conditions must be set. The detector temperature must be chosen and usually this is 20–30°C higher than that of the column. A typical detector temperature that works for many organic compounds is 250°C.

Since capillary GC instruments are so sensitive, they are often fitted with a splitter which functions to make the sample concentration much more lean when it is introduced onto the column. For example if 1 μL (1×10^{-6} L) of a 1% solution of compound is used, a typical splitter setting might be such that only 1/40th of the vaporized sample be allowed to pass onto the column.

The electronic signal (base line) must be zeroed.

Capillary GC involves the use of open tubular columns (Figure 3.33) whose walls are made of fused silica (SiO_2) of 15–100 m length. These columns have inside diameters of 0.10–0.70 mm (typical columns are in the 0.20–0.53-mm range), and they are frequently coated internally with a stationary phase adsorbent that is 0.1–5 μm ($\mu m = 10^{-6}$ m) thick. There are three categories of stationary phase coatings (Figure 3.33*c*): simple wall coatings (WCOT), coatings dispersed on porous solid support (SCOT) to increase adsorbent surface area, and a dispersion of porous solid particles on the inside wall (PLOT; e.g., Zeolite particles are used to separate gases). Stationary adsorbent phases may be modified by chemical cross-linking or by chemical bonding to the solid support to reduce loss by bleeding. Detectors for capillary GC are either FID or electron-capture detectors (ECD) since these are more sensitive, reliable, and reproducible than TC detection—especially for the small sample sizes that can be used on these columns. Since FID and ECD detection is destructive and since the sample size is much smaller than with packed columns, capillary GC is restricted to analytical procedures. Preparative-scale operations are normally carried out on packed columns.

There are several advantages to capillary GC (compared to packed-column GC):

1. Very accurate analyses are quickly attained on very small samples.
2. Since capillary columns contain so little stationary phase, column "bleed" is normally not a serious problem. Thus the loss of packing materials as columns age is not nearly as serious as with packed columns which contain much larger amounts of stationary

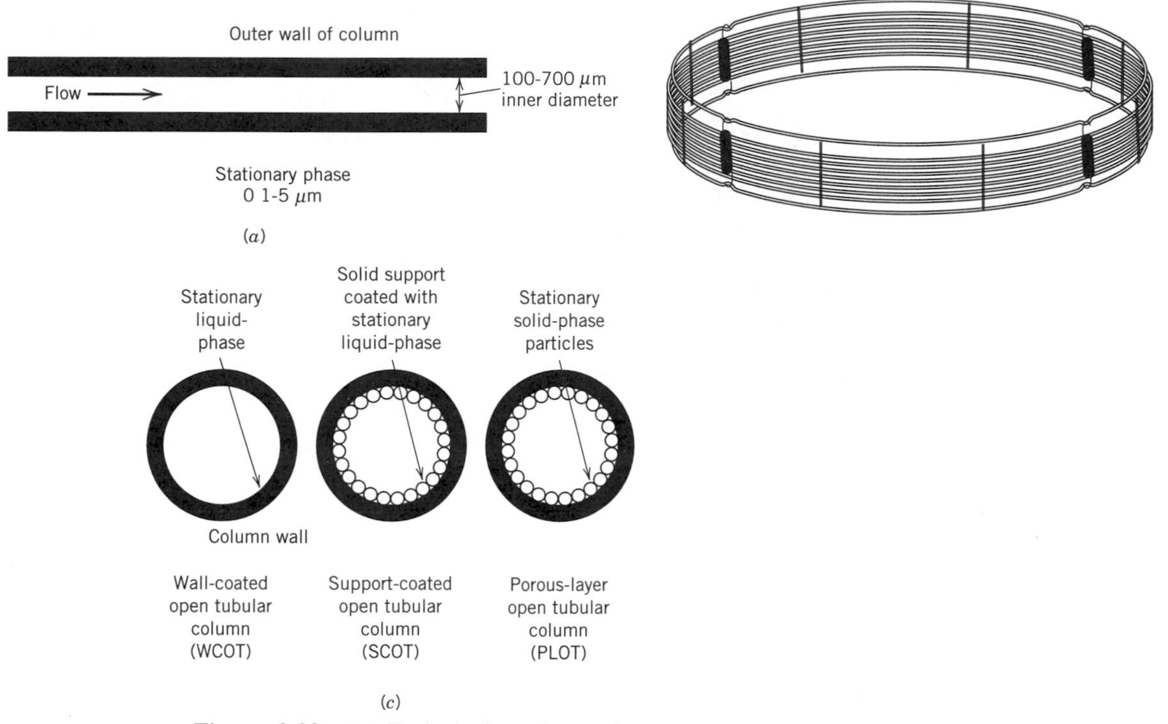

Figure 3.33 (*a*) Typical dimensions of open tubular column for gas chromatography. (*b*) Aluminum-clad fused silica chromatography column. (*c*) Cross-sectional views of wall-coated, support-coated, and porous-layer columns.

phase. As volatile materials bleed off, they can be confused with samples of compounds of interest with similar retention times.

A wide array of stationary phases are available for GC and a number of these have been listed in Figure 3.31. The choice of stationary phase used is based upon the polarity of the compound being analyzed (Table 3.11) and on the temperature that is required. The polarity of compounds classified on Table 3.11 ranges from polar to nonpolar in four stages. For example, if the analysis is to be conducted on alkanes of low molecular weight (and thus high volatility), squalene can be used. Polar organic compounds such as carboxylic acids which frequently have very low volatility require a stationary phase such as XE-60.

Capillary GC has become very common and its great sensitivity and reliability make it the method of choice for analytical work. Packed-column GC is very useful where it is desirable to collect larger, preparative-scale samples (see also Chapter 10) and it can be used for analytical work keeping in mind its limited sensitivity.

Discussion. Retention times (or retention volumes = flow rate × retention time) are characteristic of the compound of interest. One should avoid precise, quantitative comparisons to literature retention values; too many parameters have to be reproduced to justify such correspondence. However, injection of an unknown mixed with an authentic sample of the suspected compound resulting in a single, sharp peak is strong support for identity of the two components of this mixture. This mixture should be analyzed on both polar and nonpolar GC columns. Reaction progress (reactant, product, side product quantities, etc.), as well as purity, can be monitored by gas chromatography.

If a peak is due to only one compound, the area under a peak is proportional to the number of moles of compound causing that peak. In Figure 3.30, for example, areas could

Table 3.11 Polarity Classifications of Organic Compounds for GC Analysis[a]

Very Nonpolar (VNP)
Saturated hydrocarbons
Olefins
Aromatic hydrocarbons
Alkyl halides
Thioalcohols (mercaptans)
Thioethers
Nonpolar (NP)
Ethers
Aldehydes, ketones
Esters
Tertiary amines
Nitro compounds with no α-hydrogens
Nitriles with no α-hydrogens
Polar (P)
Alcohols
Carboxylic acids
Primary (1°) and secondary (2°) amines,
Oximes
Nitro compounds with α-hydrogens
Nitriles with α-hydrogens
Very Polar (VP)
Polyhydroxy compounds (alcohols, carbohydrates)
Amino alcohols
Hydroxy acids
Polyprotic acids
Polyphenols

[a]More details can be found in D. C. Harris, *Quantitative Chemical Analysis,* 3rd ed. (Freeman, New York, 1991).

be determined by measuring the heights the alcohol and amine peaks, and multiplying these heights by the widths of the corresponding peaks at half-height.[17] The area for any peak should be adjusted if it was measured at a different recorder sensitivity.

A concept that is important for quantitative gas chromatography is the molar response. Molar response (m.r.) is simply the area of a peak per mole of compound that is producing that peak. The molar response can be measured by use of the following simple formula:

$$\text{m.r.} = \frac{\text{area of peak}}{\text{moles of compound producing the peak}} \qquad (\text{cm}^2/\text{mole} = \text{common units})$$

Where possible, the molar response of all components in a gas chromatogram should be determined. This is done by chromatographing known volumes of standard solutions of each component. These molar responses are then used to convert peak areas of all components in a mixture to moles of compound and thus mole percentages of all components in the mixture can be calculated. Calculations of peak areas and molar composition are illustrated in Chapter 10 (pp. 473–474).

[17]This triangulation method for measuring peak areas should be used only if the peaks are symmetrical.

3.4 OPTICAL ROTATION

Stereochemistry is chemistry caused by structural differences due to variations in spatial arrangements of atoms and groups in a pair of molecules. A molecular structure that does not possess an internal mirror plane is called chiral (possesses "handedness"). A chiral compound has the potential of being optically active. The compound 2-bromobutane is a simple example of a chiral compound. Its structure reveals one stereocenter (a carbon center bearing four different substituents) at C-2. Thus 2-bromobutane can exist in two different forms called enantiomers:

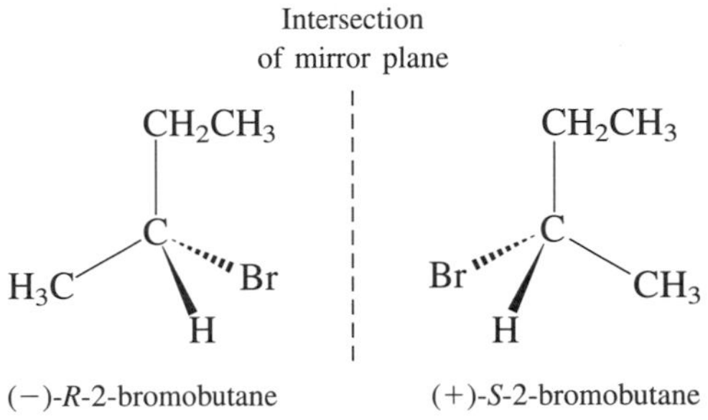

Intersection
of mirror plane

CH_2CH_3 CH_2CH_3

H_3C C Br Br C CH_3

H H

(−)-*R*-2-bromobutane (+)-*S*-2-bromobutane

Enantiomers are stereoisomers (isomers that have the same structural connectivity) that are nonsuperimposable mirror images. The *R* and *S* forms of 2-bromobutane have identical IR spectra, NMR spectra, boiling points, densities, and chromatographic retention times. They differ only in their abilities to rotate plane-polarized light (and thus only in their optical rotations, or optical activity). The *R* enantiomer rotates plane-polarized light 12° in the negative (or levorotatory) direction and the *S* enantiomer rotates plane-polarized light 12° in the positive (or dextrorotatory) direction. The terms *R* and *S* refer to the three-dimensional absolute configurations of the two forms of 2-bromobutane as defined by the Cahn–Ingold–Prelog notation (this is described in detail in organic lecture texts). Thus for 2 bromobutane the *R* isomer is also the (−) isomer and the other is the *S*-(+)-isomer. These two forms of 2-bromobutane thus have optical activity that is measurable on a polarimeter as described below. Configuration (*R* or *S*) does not necessarily correspond to a particular rotation. For example, a molecule with the *R* configuration can have either a positive (+) or negative (−) rotation.

The optical rotation is determined only if the list of possible compounds contains optically active substances.

3.4.1 Preparation of the Solution

Procedure. An accurately weighed sample (about 0.1–0.5 g) of the compound is dissolved in 25 mL of solvent in a volumetric flask. The solvents commonly used are water, ethanol, and chloroform. The solution should be clear; it should contain no suspended particles of dust or filter paper. It should also be colorless if possible. If the solution is not clear, either the original compound should be recrystallized or 50 mL of the solution made up and filtered through a small dry filter paper. The first 25 mL of filtrate is discarded; the last 25 mL is used for the determination.

3.4.2 Filling the Polarimeter Tube

Procedure. The cap is screwed on the small end of the polarimeter tube (T, Figure 3.34), the tube is held vertically, and the solution is poured in until the tube is full and the rounded meniscus extends above the top end of the tube. The glass plate is caused to slide over the end of the tube so that no air bubbles are imprisoned. The brass cap is then screwed on. Some sample tubes are loaded through side arms rather than end caps.

Precautions

1. A rubber washer should be placed between the glass plate and the brass cap. There is no washer between the glass end plate and the glass tube. This is a glass-to-glass contact.
2. The ends should not be screwed on too tightly. They should be turned up enough to make a firm, leakproof joint. If the ends are screwed on too tightly, the glass end plates will be strained and a rotation will be observed with nothing in the tube at all. For substances with very low rotations it is advisable to loosen the caps and tighten them again between readings.
3. If the brass ends come off the glass tube, they may be cemented in place again by means of litharge-glycerol cement. In putting an end back on, care must be taken that the glass part extends 1 mm beyond the brass end.

3.4.3 The Use of the Polarimeter

One form of the polarimeter is an instrument of the Lippich double-field type. Schematic diagrams of the working parts are shown in Figures 3.34 and 3.35. Modern instruments automatically read rotations to within $\pm 0.003°$.

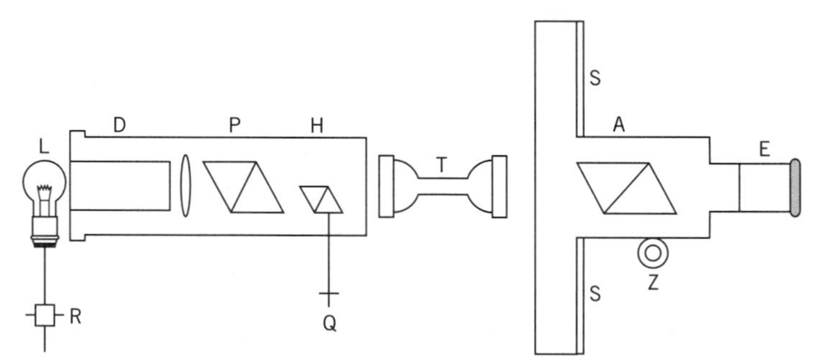

Figure 3.34 Schematic diagram of polarimeter (longitudinal section). L, light (may be replaced by sodium flame or electric sodium lamp); D, dichromate filter; P, polarizing Nicol prism; Q, half-shadow adjustment; H, half-shadow Nicol prism; T, tube containing solution; A, analyzing Nicol prism; E, eyepiece; S, scale; R, switch.

Procedure. The light (L) is turned on by means of snap switch R. The eyepiece (E) is focused by turning to right or left until the line dividing the two fields is sharp.

To determine the *zero reading,* the locknut (N) is loosened and the two fields are matched approximately by means of the coarse adjustment lever (C). To use C it is necessary first to loosen N and then to tighten C by turning the white knob on the end of C. After approximate equality of the fields is obtained, N is tightened and the fields are matched exactly by turning the micrometer adjustment (M) to the right or to the left. The lamp (X) is turned on, and the scale (S) is read. The main circle is divided into degrees and 0.25 degree. The vernier or outside scale is divided into 25 divisions, enabling the

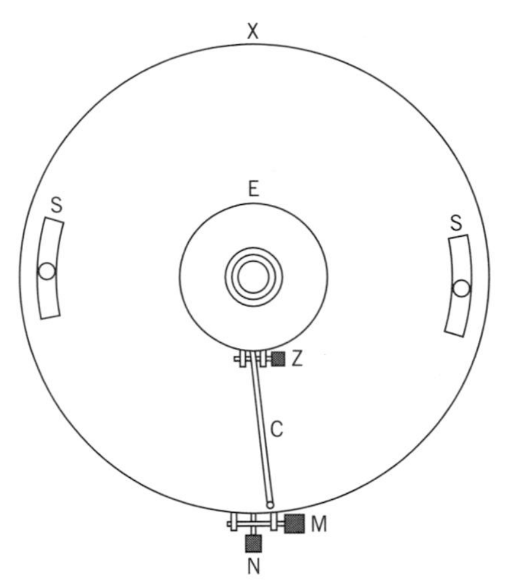

Figure 3.35 Polarimeter (end view). E, eyepiece; Z, zero adjustment; S, scale; C, coarse adjustment; N, locking nut; M, micrometer adjustment; X, lamp.

reading to be made to 0.01 degree. At least five readings should be made and the values averaged. For very exact work, both scales (S) should be read and the readings averaged in order to correct for any lack of centering of the scale with reference to the eyepiece and Nicols.

To get the *rotation* of the solution, the polarimeter tube is placed in the trough. The cover is closed, and the procedure that was followed for the zero reading is repeated. The average of at least five readings should be taken. The observed rotation is the difference between this value and the zero reading.

Notes and Precautions

1. The instrument is usually set up for use with yellow light corresponding to the sodium D line. Light of this wavelength is most readily obtained by means of electrically heated sodium lamps, which produce a brilliant yellow glow. A sodium flame may also be used, but it is not as brilliant. A white light with the following solution in a 3-cm filter cell gives good results for compounds possessing low observed rotations. The filter solution is made of 8.9 g of hydrated copper sulfate, 9.4 g of potassium dichromate, and 300 g of water. The solution is filtered and allowed to stand in order to permit dust particles to settle.

   A mercury arc makes it possible for the green mercury line to be used. To use the green mercury line, the light (L) is removed and the dichromate filter (D) is unscrewed. The arc is substituted for L, and the above procedure is repeated.

2. The instrument is set up and adjusted for all ordinary work. The half-shadow lever (Q) or the screw (Z) should not be changed.
3. In making readings, it is best to start with the scale (S) adjusted nearly to zero. If the arm (C) is moved very much, reversal of the fields will occur.
4. The lever (C) and locknut (N) should not be screwed up too tightly. It is only necessary to turn them firmly.

3.4.4 Expression of Results

The specific rotation of a substance is calculated by one of the following formulas:

For pure liquids: $$[\alpha]_{\mathrm{D}}^{25^\circ} = \frac{\alpha}{ld}$$

For solutions: $$[\alpha]_{\mathrm{D}}^{25^\circ} = \frac{100\alpha}{lc}$$

where

$[\alpha]_{\mathrm{D}}^{25^\circ}$ = specific rotation at 25°C (using the D line of sodium)
α = observed rotation
l = length of tube (decimeters)
d = density in g/mL
c = grams in 100 mL of solution

It should be noted that the specific rotation may be quite sensitive to the nature of the solvent and, in certain cases, even to the concentration of the substance being examined. The wavelength of the light used for measurement can also affect not only the magnitude but also the sign of rotation. Attention should be paid, therefore, to the exact conditions under which a rotation reported in the literature was measured.

The following is the correct way to report specific rotation.[18]

$$[\alpha]_{546}^{25^\circ} = -40 \pm 0.3^\circ \qquad (c = 5.44\ \mathrm{g/100\ mL\ water})$$

The preceding relationship refers to a specific rotation determined at 25°C, in water, with light of a wavelength of 546 nm, at a concentration of 5.44 g/100 mL of solution. It is necessary to determine the observed rotation, α, at two different concentrations. In the simplest cases, observed rotations will be decreased by the same factor as the concentration decrease; for example,

α	**Concentration**
−50°	x
−5.0°	$0.1x$
0.50°	$0.01x$

Thus in such a case the $[\alpha]$ determined from all three experiments will be the same, and one has in hand a value of $[\alpha]$ that can be safely compared to literature values of $[\alpha]$ determined at other concentrations.[19] In this way the value of $[\alpha]$ in such simple cases can be used to confirm the identity of the compound of interest. If the value of $[\alpha]$ has been determined upon a liquid sample using no solvent, the specific rotation should be reported as follows:

$$[\alpha]_{\mathrm{D}}^{25^\circ} = +40^\circ\ (\text{neat})$$

As described earlier, the existence of a single stereocenter (also called chiral or asymmetric center) in a compound such as 2-bromobutane can cause optical activity. Other commonly encountered organic compounds with one stereocenter include

[18]All of these items should be reported. It will be common, however, to find data reported that omit the concentration and/or the solvent.

[19]In a few special cases the magnitude of the specific rotation will depend on concentration. For example, concentration-dependent hydrogen bonding of an alcoholic solvent to a polar chiral substrate could cause this dependence. Thus, it is important to include the concentration of the sample.

R-(−)-lactic acid   L-alanine

The L designation used above for alanine refers to its absolute configuration (here the L form has the *S* configuration). The enantiomer of L-alanine would be D-alanine (the D and L notation for absolute configuration is frequently encountered for amino acids and carbohydrates, and is less general than *R* and *S* notation). The L notation is not to be confused with *l*, which stands for levorotatory, or (−), rotation of plane-polarized light. In like manner, D absolute configuration is not the same as *d* rotation. In fact it is not uncommon to find that a D stereoisomer has *l*, or (−), rotation!

Chiral compounds with more than one stereocenter are common. Glucose, for example, has four stereocenters when acyclic and five when cyclic.

D-glucose   α- and β-D-glucose

Tartaric acid exists in three stereoisomeric forms, (+), (−), and *meso*.

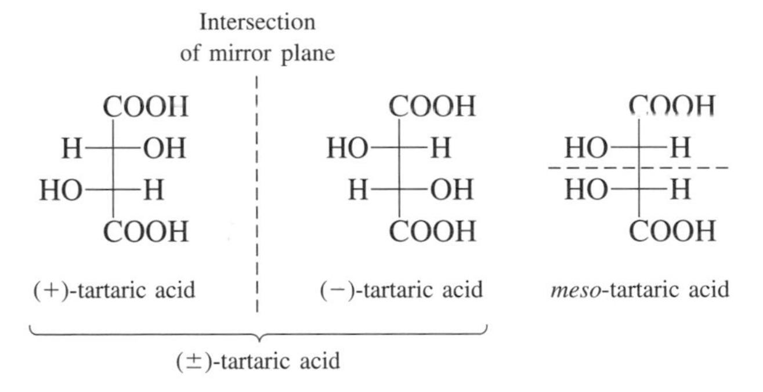

The *meso* form shows no optical activity; although it has two stereocenters, the effects of these two centers cancel resulting in an optically inactive (or achiral) compound. The lack of activity is not surprising in view of the fact that this compound bears a *meso* plane (the intersection of which is illustrated by the dotted line). Since this *meso* plane is an internal mirror plane, this compound by definition is achiral and thus optically inactive.

3.4.5 Enantiomeric Excess (Optical Purity) and Resolution

In introducing optically active compounds above, we have discussed compounds that are one enantiomer (e.g., *S*-2-bromobutane) or the other (*R*-2-bromobutane). Clearly, mixtures are possible, and one of the more well known cases is that of a racemic mixture (50% one enantiomer, 50% the other). Mixtures other than 50/50 are, of course, also possible. A criterion of optical purity, called enantiomeric excess (ee, also called optical purity), is useful; it is defined as

$$\% \text{ ee} = \frac{[\alpha]_{obs}}{[\alpha]} \times 100$$

Here $[\alpha]_{obs}$ is the specific rotation observed for the mixture of interest, $[\alpha]$ is the specific rotation for a sample of just one enantiomer. The specific rotations should be carried out at the same temperature, concentration, and in the same solvent. The resulting enantiomeric excess can be related to the solution composition as follows: A 2-bromobutane sample of $+12°$ is clearly 100% (*S*)-(+) form. If a sample of $-6°$ rotation is isolated, the enantiomeric excess is

$$\% \text{ ee} = \frac{[\alpha]_{obs}}{[\alpha]} \times 100 = \frac{-6°}{-12°} \times 100 = 50\% \text{ ee}$$

The composition of the mixture of $-6°$ rotation can be calculated on a 100 molecule basis: 50 molecules (50%) give rise to the $-6°$ rotation and have the *R* configuration. Since the remaining 50 molecules have no net optical activity, they must have the composition of a racemic mixture, and thus there must be 25 *S* and 25 *R* molecules in this mixture. Thus the entire mixture has 25 *S* molecules and 50 + 25 = 75 *R* molecules.

As you might expect, since enantiomers have only subtly different properties, their separation (resolution) can be challenging. Chromatographic separation and resolution by formation of diastereoisomers are the two main methods of separation. Both require formation of diastereomeric derivatives of enantiomers for separation. Diastereomers have different properties and thus would be expected to have different melting points and chromatographic retention times. Thus, if an enantiomeric pair (represented by *R/S*) is treated with a chiral reagent (abbreviated Rgt*), a separable diastereomeric pair can be formed:

$$\left.\begin{matrix} R \\ S \end{matrix}\right\} \xrightarrow{\text{Rgt*}} \left\{\begin{matrix} R\text{-Rgt*} \\ S\text{-Rgt*} \end{matrix}\right\} \xrightarrow{\text{separation}} \begin{matrix} R\text{-Rgt*} \xrightarrow[\text{reaction}]{\text{simple chemical}} R \\ S\text{-Rgt*} \xrightarrow[\text{reaction}]{\text{simple chemical}} S \end{matrix}$$

Essentially the same concept applies when the *R/S* pair is separated on a chiral chromatographic adsorbent (see below). This chromatographic technique has the advantage that it does not require chemical formation of diastereomeric derivatives. We can separate the enantiomers of α-phenylethylamine using (+)-tartaric acid as the chiral reagent. The enantiomeric amines form a separable pair of diastereomeric pair of salts.

$$\left.\begin{matrix} R \\ S \end{matrix}\right\} \xrightarrow[\text{(Rgt*)}]{\text{(+)-tartaric acid}} \left\{\begin{matrix} R\text{-Rgt(+)*} \\ S\text{-Rgt(+)*} \end{matrix}\right\} \xrightarrow{\text{separation}} \begin{matrix} R\text{-Rgt(+)*} \xrightarrow{\text{base}} R \\ S\text{-Rgt*(+)} \xrightarrow{\text{base}} S \end{matrix}$$

(±)-α-phenyl-
ethylamine

Upon treatment with base the separated salts yield the two enantiomeric amines.

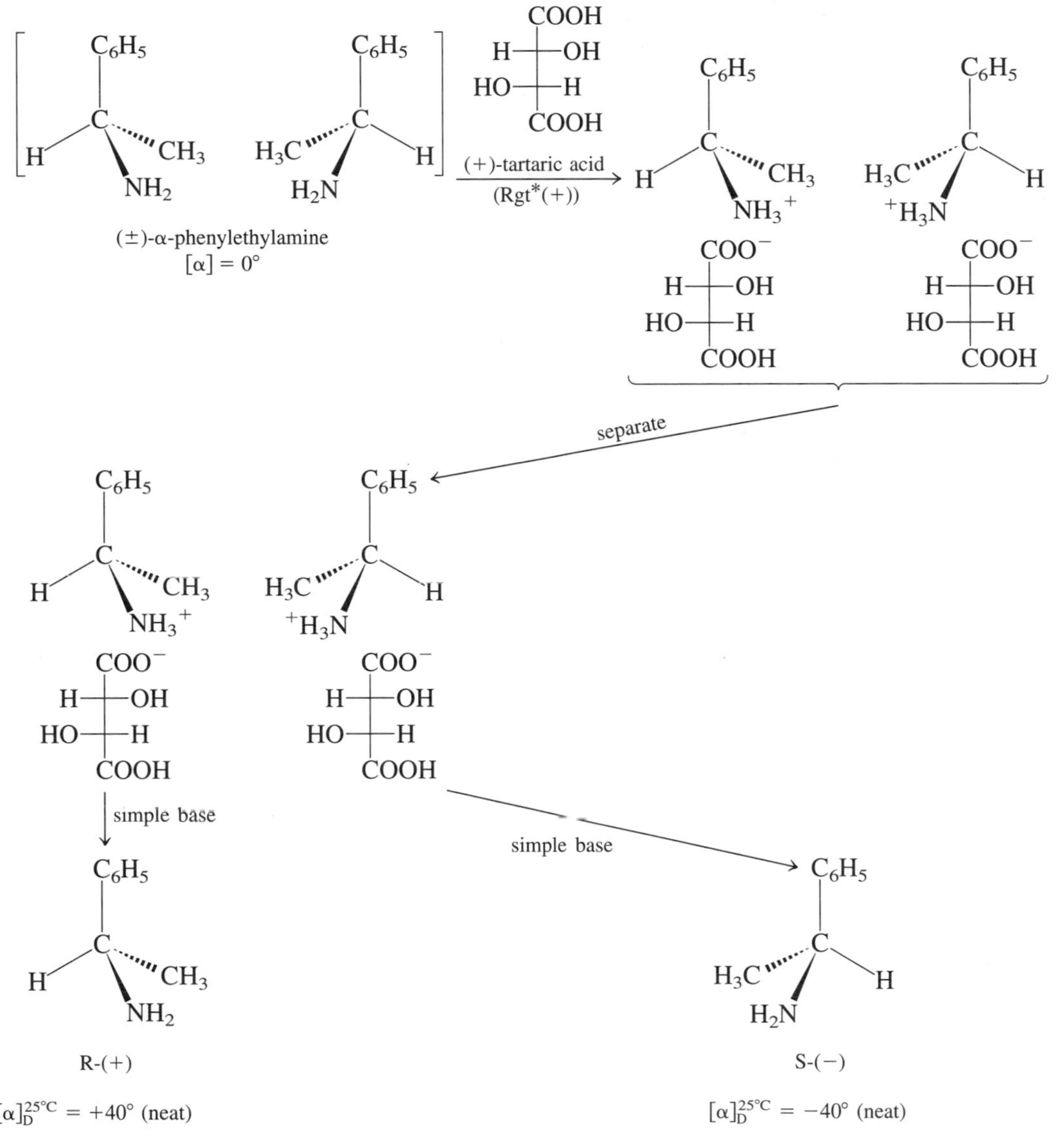

(±)-α-phenylethylamine
$[\alpha] = 0°$

(+)-tartaric acid
(Rgt*(+))

separate

simple base

simple base

R-(+)

$[\alpha]_D^{25°C} = +40°$ (neat)

S-(−)

$[\alpha]_D^{25°C} = -40°$ (neat)

Table 3.12 provides references for the separation of α-phenylethylamine and, as well, for separations of a variety of types of organic compounds.

Chiral chromatographic adsorbents[20] play the role of Rgt* in the equations above. Chiral adsorbents have been successfully used in liquid chromatography (LC), gas chromatography (GC), thin-layer chromatography (TLC), and high-performance liquid chromatography (HPLC).

[20]Carbon tetrachloride, and sometimes chloroform is prohibited from labs for toxicity reasons. In such cases methylene chloride (dichloromethane) is often a reasonable substitute.

Table 3.12 Reagents and References to Procedures for Optical Resolution

Functional group	Reagent used	Reference[a]
Alcohols	Amines, via phthalate esters of alcohols	Boyle[b]
Aldehydes, ketones	Amine tartrate and amine sulfate complexes	Ault[c]
	Methyl *N*-aminocarbonate	Boyle
Alkenes	Pt complexes	Boyle
Amines	(+)-Tartaric acid	Ault
Amino acids	α-Phenylethylamine	Ault
	Paper chromatography	Boyle
Aromatic (biphenyls)	Column chromatography (on lactose or potato starch)	Boyle
Carboxylic acids	α-Phenylethylamine	Ault
Esters of amino acids	GC (peptide adsorbents)	Boyle
Salts of acids	Ion exchange chromatography	Boyle

[a]Boyle is not a primary reference but rather a lead to primary references. Ault describes a detailed procedure for the resolution of α-phenylethylamine.

[b]P. H. Boyle, *Quart. Rev.* (London), 323 (1971).

[c]A. Ault, *Org. Syn., Coll. Vol. V,* edited by H. E. Baumgarten (Wiley, New York, 1973), p. 932. (A procedure that appears in *Org. Syn. has been verified independently.* Most procedures that appear in standard journals, however, have not been checked.) See also S. H. Wilen, *Resolving Agents and Resolutions in Organic Chemistry* (University of Notre Dame Press, Notre Dame, IN, 1971).

3.5 RECRYSTALLIZATION

Recrystallization depends on the decreased solubility of a solid in a solvent, or mixture of solvents, at lower temperature. Thus, one should be familiar with the theory of solubility (pp. 98–106) in order to understand better the theory of recrystallization.

In the simplest cases, recrystallization is accomplished by dissolving a solid, oil, or semisolid material in a solvent; the procedure is often most effective when heating is necessary to dissolve the material completely. The warm solution is then allowed to cool slowly down to room temperature (or lower). In the "ideal" case, uniform crystals slowly appear; frequently, an overnight wait or longer is required for complete recrystallization. After the crystals have been isolated (by filtering and air drying on, e.g., a Büchner funnel), they are checked for purity (mp, GC, NMR, or TLC, etc.).

Actual recrystallization procedures can be quite complex; standard laboratory manuals for organic chemistry normally have exercises to introduce these procedures. Impure acetanilide is often recrystallized as a standard laboratory exercise. In this section we shall deal with some principles that may be useful in solving recrystallization problems in the more unpredictable situations involved in organic qualitative analysis.

The first consideration in planning a recrystallization is the choice of solvent. A rule of thumb suggests that the solid sample should be five times as soluble in the hot solvent as in the cold solvent. It may be very convenient to test the solubility by the use of gas chromatography described in Section 3.3.2. Table 3.13 lists a variety of recrystallization solvents. Table 3.14 lists solvent mixtures that can be employed. Even if the precise solute–solvent pair cannot be found on Table 3.13 or 3.14, the tables should at least give an idea as to what general class of solute–solvent pairs are appropriate. For example, if a phenylurethane derivative does not recrystallize from petroleum ether, then petroleum

Table 3.13 Common Solvents for Recrystallization of Standard Functional Classes

Sample to be recrystallized	Solvent[a]	Solvent bp (°C)	Co-solvent possibilities
1. Acid anhydride	Methylene chloride	40	Ether, toluene, hydrocarbons
2. Acid chloride	Methylene chloride	40	See line 1
3. Acid chloride	Chloroform[b]	61.7	Hydrocarbons
4. Amide	Acetic acid	118	Water
5. Amide	Dioxane	102	Water, toluene, hydrocarbons
6. Amide	Water	100	Acetone, alcohols, dioxane, acetonitrile
7. Aromatic	Toluene	111	Ether, ethyl acetate, hydrocarbons
8. Bromo compound	Acetone	56	Water, ether, hydrocarbons
9. Bromo compound	Ethyl alcohol	78	Water, hydrocarbons, ethyl acetate
10. Carboxylic acid	Acetic acid	118	See line 4
11. Carboxylic acid	Water	100	See line 6
12. Complex	Toluene	111	See line 7
13. Ester	Ethyl acetate	77	Ether, hydrocarbons, toluene
14. Ester	Ethyl alcohol	78	See line 9
15. General	Acetone	56	See line 8
16. General	Chloroform[b]	61.7	See line 3; also ethyl alcohol
17. General	Ethyl acetate	77	See line 13
18. General	(Ethyl) ether	34.5	Acetone, hydrocarbons, ethyl acetate, toluene, methylene chloride
19. General	Methylene chloride	40	Ethyl alcohol, hydrocarbons
20. General	Ethyl alcohol	78	See line 9; bromo compounds
21. Hydrocarbon	Toluene	111	See line 7
22. Hydrocarbon	Hexane	69	Any but acetronitrile, acetic acid, water
23. Low-melting compound	Ether	34.5	See line 18
23. Low-melting compound	Methylene chloride	40	See line 19
25. Nitro compound	Acetone	56	See line 8
26. Nitro compound	Ethyl alcohol	78	See line 9
27. Nonpolar compound	Methylene chloride	40	See line 1
28. Osazone	Acetone	56	See line 8
29. Polar compound	Acetonitrile	81.6	Water, ether, toluene
30. Salt	Acetic acid	118	See line 4
31. Salt	Water	100	See line 6
32. Sugar	Methyl cellosolve		Water, toluene, ether

[a]More details on these and other solvents, especially with regard to solvent toxicity, flammability, and practical handling comments, may be found in A. J. Gordon and R. A. Ford. *The Chemists' Companion* (Wiley, New York, 1972), pp. 442–443. **Caution: Remember that many of these solvents of which benzene is an important example, have very significant toxicity characteristics.**

[b]In revising earlier forms of this table, toluene has been substituted for benzene and methylene chloride (dichloromethane) for carbon tetrachloride. Methylene chloride is also a possible alternative for chloroform.

ether/toluene might work. Or, if you have a *m*-nitrobenzyl ester, the characterization of which is not published, you may very likely find that it recrystallizes from methanol–water (why?).

A common technique for crystallization from a solvent pair involves dissolving the sample with warming in the solvent in which the sample is more soluble, then slowly

Table 3.14 Solvents and Solvent Pairs for Recrystallization of Common Derivatives

Derivative	Solvent or solvent system[a]
Acetate	Methanol; ethanol
Amide	Methanol; ethanol
Anilide	Methanol–water; ethanol
Benzoate	Methanol; ethanol
Benzyl ester	Methanol–water; ethanol
Bromo compound	Acetone–alcohol; methanol; ethanol
3,5-Dinitrobenzoate	Methanol; ethanol
3,5-Dinitrophenylurethane	Petroleum ether–toluene
Ester	Ethyl acetate; methanol; ethanol
Hydrazone	Methanol–water; ethanol
α-Naphthylurethane	Petroleum ether
p-Nitrobenzyl ester	Methanol–water; ethanol
Nitro compound	Methanol; ethanol; acetone–alcohol
p-Nitrophenylurethane	Petroleum ether–toluene
Osazone	Acetone–alcohol
Phenylurethane	Petroleum ether
Picrate	Toluene; ethanol; methanol–water
Quaternary ammonium salt	Ethyl acetate; isopropyl ether
Semicarbazone	Ethanol; methanol–water
Sulfonamide	Methanol–water; ethanol
Sulfonyl chloride	Chloroform; methylene chloride
p-Toluidide	Methanol; ethanol
Xanthylamide	Dioxane–water

[a]See footnote a on Table 3.13.

adding the second solvent until the point at which it is expected that cooling the solution should cause crystallization of the sample. When addition of the second solvent results in a clouding of the solution that disappears *only slowly* with stirring, the proper solvent combination may be in hand. Any approach will normally involve a number of trial-and-error sequences.

All too frequently recrystallization attempts will result in oil formation rather than the desired solids. The formation of an oil may very possibly be due to the fact that the sample is impure. If there is reason to believe this, the sample may be dissolved in a solvent in which it is readily soluble—for example, ether—and this solution can be treated with decolorizing carbon, followed by drying agent (e.g., magnesium sulfate). The solvent is removed and recrystallization attempts are repeated, if necessary.

Oils may persist, even after repeated purifications. This may be due to the fact that the sample is inherently difficult to crystallize or to the fact that last traces of impurity must be removed by recrystallization. The following techniques may be tried:

1. During a recrystallization, add a small *seed crystal* of the pure sample desired. This should be added after the solution has been supersaturated: for example, add a seed crystal to a solution that has been allowed to cool to room temperature but has as yet not produced crystals.
2. A site for crystal nucleation can initiate the crystallization process. A glass rod, a scratched surface on the inside of the flask, a wooden stick, or a boiling stone may provide the surface necessary to initiate solid formation.

3. Lower temperatures may be necessary. The purpose is to decrease the solubility of the sample, *not* to freeze the solvent or sample. An increase in the solvent volume by 20–30% can be used effectively to require a lower temperature for crystal formation. The following conditions produce increasingly lower temperatures (see Appendix I for more details):

   Room temperature
   Refrigeration
   Ice–water bath
   Refrigerator freezer or salted ice bath
   Dry ice–acetone
   Liquid nitrogen

Be careful not to confuse frozen solvent or frozen amorphous oils with crystals; frozen oils will melt and form oils at room temperature.

Sometimes it is necessary to mash the neat oil sample, for example, with a stirring rod, to induce crystallization. This mashing can also be done in contact with the mother liquor. The other techniques mentioned above, such as seeding can be used with this. In any case, it may take a very long period of mashing and grinding to induce crystallization.

CHAPTER 4

Determination of Molecular Formulas

Once the chemist has established the purity, the next step is to determine the elements present in the sample. If the compound's empirical formula is determined (e.g., by combustion analysis as described in Section 4.2) and the molecular weight is known, then the molecular formula can be determined. Chemical tests (Chapter 7), IR (Chapter 6), NMR (Chapter 6), and derivatization (Chapter 8) can then lead to complete characterization and structure determination of the compound.

We will find, however, that in an educational setting, and especially when using this book, the sequence of characterization and identification will likely proceed as follows. After the melting or boiling point of the unknown has been determined, possibly followed up by TLC, the student should have the basis for making an educated guess as to the purity of the unknown compound. Next, qualitative elemental analysis might be carried out for the presence of nitrogen, sulfur, and halogen (Section 4.1). IR and NMR spectra of the sample are obtained next. Classification tests are performed to confirm or deny the existence of particular functional groups. As an option, the molecular weight might next be determined by neutralization equivalent (NE, Chapter 7), saponification equivalent (SE, Chapter 7), or by mass spectrometry (MS, Chapter 6). After the interpretation of the accumulated data and consulting the derivative tables of Appendix III, the student can select compounds which have similar characteristics to the unknown. Additional classification tests may be needed to narrow down the choices. A final confirmation of the identity of the unknown is achieved by the preparation of a derivative (Chapter 8).

4.1 QUALITATIVE ELEMENTAL ANALYSIS

Organic chemists do not usually employ chemical tests for carbon, hydrogen, and oxygen. It is often valuable, however, to determine the existence of other elements such as nitrogen, sulfur, fluorine, chlorine, bromine, and iodine. The detection of these elements, by means of chemical tests, is usually straightforward. Many of these chemical tests are quite sensitive, so all aqueous solutions should be *carefully* prepared using only distilled or deionized water. Samples which show indications of explosive character in the ignition test (1) should not be analyzed by the sodium fusion procedure or (2) should be analyzed by a smaller scale procedure than that described below. Compounds that are known to react explosively with molten sodium are nitroalkanes, organic azides, diazo esters, diazonium salts, and some aliphatic polyhalides such as chloroform or carbon tetrachloride. *Safety glasses, with side shields, must be worn at all times when any procedure or experiment is being conducted. The safety of other members of the class must be taken into account and therefore care must be taken so that reaction flasks are not pointed toward others in the lab.*

Controls should be run for all tests which leave the slightest doubt about the decisiveness of the results; for example, if you are unsure about the validity of the observations associated with a positive nitrogen test run on an unknown, the same test should be carried out on a compound that is known to contain nitrogen. A sample of 4-bromobenzenesulfonamide is recommended as a control. This sample can be decomposed by sodium and analyzed for nitrogen, bromine, and sulfur by the procedures described below. It may even be advantageous to run a control on a compound that is known *not* to contain the element of interest; observations associated with this control allow one to draw conclusions about tests yielding negative results or about the purity of the reagents involved.

Knowledge of the elemental composition of an organic compound being studied is essential for the following reasons. Such knowledge aids in the selection of the appropriate classification experiments, which serve as tests for functional groups (Chapter 7), and in the selection of procedures for the preparation of derivatives (Chapter 8). Additionally, mass, IR, and NMR spectra (Chapter 6) can be interpreted accurately for the deduction of molecular formula (Chapter 4), functional groups (Chapters 5 and 7), and structure (Chapters 7 and 8).

Almost all of the elements listed in the periodic table can be a part of an organic compound. In this text we will, however, be concerned with detection of only a few of the elements more commonly found in organic compounds; detection of the other elements is a subject for another course such as instrumental analysis, which covers atomic absorption and other instrumental procedures. For an introduction to the identification of "unknown" organic compounds, it is recommended that the possible elements be limited to C, H, P, N, S, F, Cl, Br, and I. A few of the most common salts, such as those containing Na, K, and Ca, might also be included.

4.1.1 Fusion of Organic Compounds with Sodium

$$\text{C, H, O, N, S, X} \xrightarrow{\text{Na}} \begin{cases} \text{NaX} \\ \text{NaCN} \\ \text{Na}_2\text{S} \end{cases}$$

X = halogen

The unknowns can be treated with sodium to form the sodium fusion filtrate according to one of the following three procedures. The filtrate is then used for determination of the presence of sulfur, nitrogen, and halogen. A compound such as 4-bromobenzenesulfonamide can be used as a control for a positive test for bromine, sulfur, and nitrogen. *The first time that the student prepares the sodium fusion filtrate, it must be done under supervision. It is imperative that the preparation of the sodium fusion filtrate be done in the hood.*

Procedure A.[1] A small test tube (approximately 100 × 13 mm) made of *soft* glass (not Pyrex® or Kimax®) is supported in a vertical position with a clamp from which the rubber has been removed. With a knife, a cube piece of *clean* sodium metal about 4 mm on an edge is isolated and placed in the test tube. *Prior to placing the sodium in the test tube, the sodium metal is placed in a beaker containing a small amount of hexane to remove all traces of kerosene, the liquid in the container of sodium. The knife and any other utensils that were in contact with the sodium are also rinsed with hexane. Ethanol is then added dropwise to the hexane washings to consume any traces of sodium.* The lower part of the tube is heated until the sodium melts and sodium vapors begin to rise in the tube. Next, the heat is removed and one-third of a mixture of 100 mg of the compound with 50 mg of powdered sucrose[2] is added and heat is reapplied. The addition and heating are repeated a second and third time, and then the bottom of the tube is heated to a dull red. The tube is allowed to cool, and 1 mL of ethanol is added to dissolve any unchanged sodium. The tube is heated to a dull red again and, while still hot, is dropped into a small beaker containing 20 mL of distilled water (***caution***!). During the heating, alcohol vapors may ignite at the mouth of the tube; this should not affect the analysis. The tube is broken up with a stirring rod, and the solution is heated to boiling and filtered. The filtrate, which should be colorless, is used for the specific tests for various elements described below after Procedure C.

Procedure B. Into a small (100 × 13 mm) Pyrex® or Kimax® glass test tube is placed 10 mg or 10 μL of the unknown and a freshly cut, pea-size (about 50 mg) piece of sodium metal (***caution***!), following the precautions used in Procedure A. The test tube is heated until the bottom of the test tube is a glowing red; the glowing and charred residue is allowed to cool to room temperature. A few drops of ethanol are added, with stirring, to ensure the reaction of all remaining elemental sodium; this is repeated until no further bubbles of hydrogen gas are evolved. Two milliliters of water is added to this solution, and the solution which is then boiled and filtered. The filtrate, which should be colorless, is used for the specific element tests outlined after Procedure C. Any indication of incomplete fusion (e.g., the presence of color) at this point requires that the entire procedure be repeated on a fresh sample.

Procedure C.[3] A small Pyrex® or Kimax® glass test tube (100 × 13 mm) containing 500 mg of sodium–lead alloy is placed in a vertical position. The test tube is heated with a flame until sodium vapors rise in the tube. *Do not heat the test tube or its contents to*

[1] K. N. Campbell and B. K. Campbell, *J. Chem. Educ., 27,* 261 (1950).

[2] Powdered sugar (sucrose) is sold in supermarkets as "confectioner's sugar". It typically contains 97% sucrose and 3% starch. The mixture of the unknown and powdered sugar provides a charring and reducing action so that compounds containing nitrogen such as amides, nitrosos, nitros, azos, hydrazos, and heterocyclic rings produce sodium cyanide. Sulfur compounds such as sulfides, sulfoxides, sulfones, sulfonamides, and heterocyclic sulfur compounds produce sodium sulfide.

[3] J. A. Vinson and W. T. Grabowski, *J. Chem. Educ., 54,* 187 (1977).

redness. Four to six drops or about 10 mg of the unknown is added directly onto the sodium without getting any of the sample on the walls of the test tube. If the unknown is a volatile liquid with a boiling point of less than 100°C, then approximately 50 mg of powdered sucrose should be mixed with the unknown prior to its addition into the test tube. The reaction mixture is heated gently to initiate the reaction. The flame is removed until the reaction ceases, then the test tube and its contents are heated to redness for 2 min. The test tube is allowed to cool to room temperature. Three milliliters of water is added and the test tube is heated gently for 2 min. The reaction mixture is filtered while still warm. The test tube is rinsed with 2 mL of water and the rinsings are filtered. These filtrates are combined together and used for the specific tests described below.

Note: If a sharp explosion occurs when the initial portion of the unknown is heated with the sodium, the procedure is stopped. About 0.5 g of fresh unknown is reduced by boiling gently with 5 mL of acetic acid and 0.5 g of zinc dust. *Do not heat too strongly, as compounds such as low molecular weight amine acetates may be lost by evaporation.* After most of the zinc has dissolved, the mixture is evaporated to dryness and the entire residue is then decomposed by Procedure A, B, or C.

Cleaning Up. Any of the sodium fusion solutions from Procedure A, B, or C which remain after the elemental tests are completed are placed in the aqueous solution container. The hexane washings, after they have been in the hood for an hour with sufficient ethanol to consume any residual sodium, are placed in the organic solvent container.

Specific Tests for Elements The sodium fusion filtrate, obtained by Procedure A, B, or C, is used in the experiments below to test for the presence of sulfur, nitrogen, or halogen.

Sulfur. Either of the following procedures may be used to test for the presence of sulfur.

Procedure (a) for Sulfur. One milliliter of the sodium fusion filtrate is acidified using acetic acid, after which a few drops of 1% lead acetate solution are added. A black precipitate of lead sulfide indicates the presence of sulfur in the unknown.

$$\underset{}{Na_2S} + \underset{\text{lead acetate}}{Pb(O\overset{\overset{\displaystyle O}{\|}}{C}CH_3)_2} \longrightarrow \underset{\text{(black precipitate)}}{PbS} + 2CH_3COONa$$

Procedure (b) for Sulfur. To 1-mL sample of the sodium fusion filtrate are added two drops of 2% sodium nitroprusside; a deep blue-violet color indicates the presence of sulfur.

$$Na_2S + \underset{\text{sodium nitroprusside}}{Na_2Fe(CN)_5NO} \longrightarrow \underset{\text{(blue-violet)}}{Na_4[Fe(CN)_5NOS]} + 2NaOH$$

Cleaning Up. The solutions from both tests for sulfur are placed in the aqueous solution container.

Nitrogen. The presence of nitrogen can be detected using any of the procedures below. The use of these elemental tests, in combination with data obtained from the classification tests, can give definite information as to the type of nitrogen-containing functional group. Such classification tests include the treatment of the unknown with sodium hydroxide which can determine the presence of an amide (Experiment 16), amine (Experiment 22), nitrile (Experiment 16), or nitro compound (Experiment 46).

Procedure (a) for Nitrogen.[4] In a small test tube, 1 mL of 1.5% 4-nitrobenzaldehyde in 2-methoxyethanol solution, 1 mL of 1.7% 1,2-dinitrobenzene in 2-methoxyethanol solution, and 2 drops of 2% sodium hydroxide solution are mixed. Two drops of the sodium fusion filtrate are added. A positive test for nitrogen is the appearance of a deep blue-purple compound. The deep-purple compound is due to a dianion produced when sodium cyanide, which is formed from nitrogen in the original compound, undergoes reaction with 4-nitrobenzaldehyde and 1,2-dinitrobenzene.

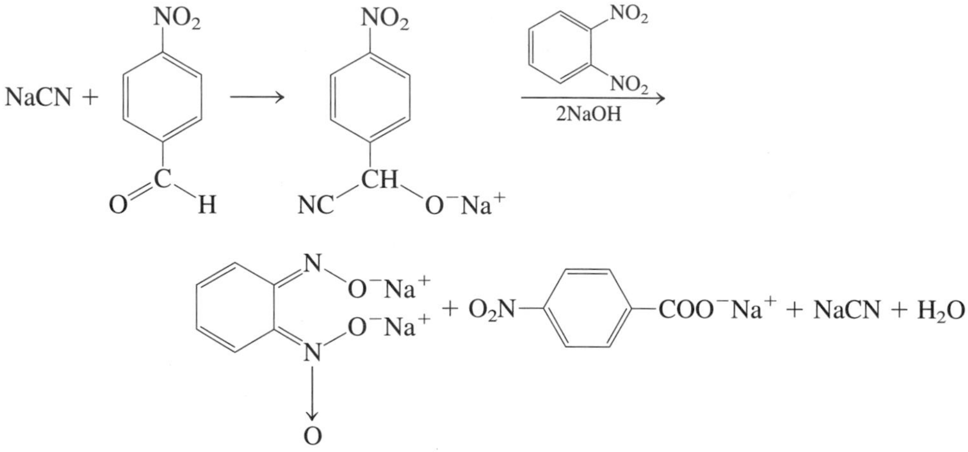

NaCN + 4-O_2N–C_6H_4–CHO ⟶ 4-O_2N–C_6H_4–CH(CN)O^-Na^+ $\xrightarrow[2NaOH]{1,2\text{-}C_6H_4(NO_2)_2}$

blue-violet dianion + O_2N–C_6H_4–COO^-Na^+ + NaCN + H_2O

blue-violet dianion

A yellow or tan solution is a negative test. This test for nitrogen is more sensitive than the Prussian blue test described in the 5th or earlier editions of this text.

This test is valid in the presence of NaX (X=halogen) or Na_2S; in other words it is reliable even if the original unknown also contains halogen or sulfur. The products of the above reaction provide the explanation as to why this test is especially sensitive. Acidification of the solution of the purple dianion results in a yellow solution of 2-nitrophenylhydroxylamine (an acid–base indicator).

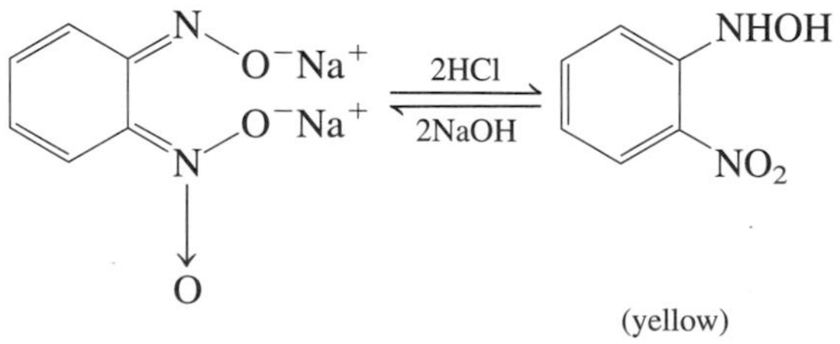

(blue-violet) $\underset{2NaOH}{\overset{2HCl}{\rightleftharpoons}}$ 2-NHOH–C_6H_4–NO_2 (yellow)

Procedure (b) for Nitrogen. Two drops of a 10% ammonium polysulfide solution are added to 2 mL of the sodium fusion filtrate, and the mixture is evaporated to dryness

[4]G. G. Guilbault and D. N. Kramer, *Anal. Chem., 38,* 834 (1966); *J. Org. Chem., 31,* 1103 (1966).

on a steam bath. Five milliliters of a 5% hydrochloric acid solution is added, and the solution is then warmed and filtered. A few drops of 5% ferric chloride solution are added to the filtrate. The presence of a red color indicates nitrogen was present in the original unknown.

$$\mathrm{NaCN} + \underset{\text{ammonium polysulfide}}{(\mathrm{NH_4})_2\mathrm{S}_x} \longrightarrow \mathrm{NaSCN} + (\mathrm{NH_4})_2\mathrm{S}_{x-1}$$

$$6\mathrm{NaSCN} + \mathrm{FeCl_3} \longrightarrow \underset{\text{(red)}}{\mathrm{Na_3Fe(SCN)_6}} + 3\mathrm{NaCl}$$

Cleaning Up. All solutions used in the test for the presence of nitrogen are placed in the aqueous solution container.

Halogens. The tests listed below are used to check for the presence and identity of halogens. Data from these tests, in conjunction with the classification tests of ethanolic silver nitrate (Experiment 37) and sodium iodide (Experiment 38), can be used to determine the specific halogen and whether the halogen is primary, secondary, tertiary, or aromatic.

Unless otherwise stated, determine the acidity of a solution by placing a drop, with a stirring rod, of the solution on red or blue litmus paper. The solution is acidic when the litmus paper turns red and basic when the litmus paper turns blue.

Procedure (a) for Presence of a Halogen. Beilstein's test is a very general test to see if any halogen is present. For this test, a small loop is made in the end of a copper wire which is then heated with the Bunsen burner until the flame is no longer green. The wire is cooled; the loop is dipped in a little of the *original* unknown compound and heated in the edge of the flame. A green flame indicates halogen and is not sustained for very long.

This test is extremely sensitive *but should always be cross-checked* by the silver nitrate test described below because minute traces of impurities containing halogen may produce a green flame. Another drawback of this test is the possibility of highly volatile liquids evaporating completely prior to the wire becoming sufficiently hot to cause decomposition, thus resulting in a possible false negative result.

Also, certain nitrogen compounds not containing halogen cause a green color to be imparted to the flame; among them are quinoline and pyridine derivatives, organic acids containing nitrogen, urea, and copper cyanide. Some inorganic compounds also give green flames.

Procedure (b) for One Halogen—Chlorine, Bromine, or Iodine. *In the hood,* about 2 mL of the sodium fusion filtrate is acidified with 5% nitric acid in a small test tube and boiled gently for a few minutes to expel any hydrogen cyanide or hydrogen sulfide that might form if the original compound contained nitrogen or sulfur.

$$\mathrm{NaCN} + \mathrm{HNO_3} \longrightarrow \mathrm{HCN}\uparrow + \mathrm{NaNO_3}$$
$$\mathrm{Na_2S} + 2\,\mathrm{HNO_3} \longrightarrow \mathrm{H_2S}\uparrow + 2\,\mathrm{NaNO_3}$$

The solution is allowed to cool. If any precipitation occurs at this point, filter the solution. A few drops of 0.1 M silver nitrate solution are added to the liquid. An immediate heavy formation of a solid indicates the presence of chlorine, bromine, or iodine. Silver chloride is white, silver bromide is pale yellow, and silver iodide is yellow. Since silver fluoride is soluble in water, it cannot be detected by this test. If only a faint turbidity is produced, it is probably due to the presence of impurities in the reagents or in the test tube.

$$NaCl + AgNO_3 \longrightarrow \underset{\text{(white)}}{AgCl(s)} + NaNO_3$$

$$NaBr + AgNO_3 \longrightarrow \underset{\text{(pale yellow)}}{AgBr(s)} + NaNO_3$$

$$NaI + AgNO_3 \longrightarrow \underset{\text{(yellow)}}{AgI(s)} + NaNO_3$$

If a silver halide is present, then continue to add sufficient 0.1 M silver nitrate to completely precipitate all of the halogens as silver halides. The precipitate is isolated by filtration.

Silver chloride, silver bromide, and silver iodide have different solubilities in 5% ammonium hydroxide. Exactly 2 mL of 5% ammonium hydroxide is added to the solid. Silver chloride is soluble in ammonium hydroxide due to the formation of $Ag(NH_3)_2Cl$; silver bromide is slightly soluble because it only partially forms its salt; and silver iodide does not undergo reaction with the ammonium hydroxide and thus remains insoluble.

$$AgCl + 2NH_4OH \longrightarrow Ag(NH_3)_2^+\ Cl^-(aq) + 2H_2O$$

Procedure (c) for One Halogen—Bromine, Iodine, or Chlorine.[5] About 2 mL of the sodium fusion filtrate is acidified using 5% nitric acid. Ten drops of a 1.0% aqueous potassium permanganate solution are added and the test tube is shaken for 1–2 min. About 20–30 mg of oxalic acid is added to barely decolorize the solution. Add 1 mL of methylene chloride and again shake the mixture. Observe the color of the bottom methylene chloride layer against a white background. A brown color indicates that bromine is present. A purple color indicates the presence of iodine. If the methylene chloride layer is colorless, then chlorine is the halogen present.

Procedure (d) for Bromine and/or Iodine; or Chlorine. About 3 mL of the sodium fusion filtrate is acidified with 10% sulfuric acid and boiled for a few minutes. The solution is cooled after which 1 mL of methylene chloride is introduced, followed by a drop of 5.25% sodium hypochlorite (household bleach). The color of the solution is viewed against a white background. The production of a purple color in the bottom methylene chloride layer indicates the presence of iodine.

$$2NaI + Cl_2(H_2O) \longrightarrow 2NaCl + \underset{\text{(purple)}}{I_2(CH_2Cl_2)}$$

The addition of 5.25% sodium hypochlorite is continued, drop by drop, with the solution being shaken after each addition. The purple will gradually disappear and will be replaced by a reddish brown color if bromine is present.

[5] D. W. Mayo, R. M. Pike, and P. K. Trumper, *Microscale Organic Laboratory,* 3rd ed. (Wiley, New York, 1994), p. 701.

$$I_2(CH_2Cl_2) + Cl_2(H_2O) \longrightarrow 2ICl$$

(purple) (colorless)

$$2NaBr + Cl_2(H_2O) \longrightarrow 2NaCl + Br_2(CH_2Cl_2)$$

(reddish brown)

If a positive test was obtained in Procedure *b*, but neither bromine or iodine are present according to the results obtained in this procedure, then chlorine is the halogen present. The chlorine does not produce a color when in solution in methylene chloride. One of the procedures listed below should be used to confirm the presence of chlorine.

Procedure (e) for More than One Halogen—Chlorine, Bromine, and Iodine. About 10 mL of the sodium fusion filtrate is acidified with 10% sulfuric acid and boiled for a few minutes. To 1 mL of the cooled solution, 0.5 mL of methylene chloride is added, followed by a few drops of a 20% sodium nitrite solution. A purple color indicates iodine. If iodine is present, then the solution is treated with 20% sodium nitrite and the iodine is extracted with methylene chloride. Next, the solution is boiled for a minute, then cooled. To 1 mL of this solution, 0.5 mL of methylene chloride and 2 drops of 5.25% sodium hypochlorite (household bleach) are added. A brown color indicates the presence of bromine. The remaining 9 mL of the acidified solution is diluted to 60 mL, and 2 mL of concentrated sulfuric acid is added, followed by 0.5 g of potassium persulfate ($K_2S_2O_8$). The solution is boiled for 5 min. After the mixture has been cooled, a few drops of a 0.1 M silver nitrate solution are added; a white precipitate indicates chlorine.

Procedure (f) for Bromine. To 3 mL of the sodium fusion filtrate in a test tube are added 3 mL of acetic acid and 0.1 g of lead dioxide. A piece of filter paper, moistened with a 1% solution of fluorescein, is placed over the mouth of the test tube, and the contents of the tube are heated to boiling. If bromide is present in the solution, brown vapors cause the yellow fluorescein to turn pink owing to the formation of eosin. Chlorides and cyanides do not interfere with this test. Iodides produce a brown color.

Procedure (g) for Iodine. Two milliliters of the sodium fusion filtrate is acidified with 2 M nitric acid, followed by the addition of 1 mL of a 5% mercury(II) chloride solution. The formation of a yellow solid, which changes to orange-red upon standing for a few minutes, indicates the presence of iodine.

Procedure (h) for Chlorine in the Presence of Nitrogen, Sulfur, Bromine, and Iodine. *In a hood,* about 2.0 mL of the sodium fusion filtrate is acidified with 5% nitric acid and boiled to expel any hydrogen cyanide and hydrogen sulfide which might be present. Sufficient 0.1 M silver nitrate is added to precipitate completely all of the halogens as silver halides, and the precipitate is isolated by filtration. If both nitrogen and sulfur have been determined to be present, the precipitate is boiled for 10 min with 6.0 mL of concentrated nitric acid to undergo reaction with any silver thiocyanate which may be present. The resulting mixture is diluted with 6.0 mL of distilled water and filtered. The precipitate of silver halides is then boiled with 4.0 mL of 0.1% sodium hydroxide for 2 min. The solution is again filtered, the filtrate is acidified with 5% nitric acid, and a few drops of 0.1 M silver nitrate are added. A white precipitate indicates the presence of chlorine.

Procedure (i) for Fluorine. About 2 mL of the sodium fusion solution is acidified with glacial acetic acid, and the solution is boiled and cooled. One drop of the solution is placed on a piece of zirconium-alizarin test paper. A yellow color on the red paper indicates the presence of fluorine. The test paper is prepared by dipping a piece of filter paper into 3 mL of 1% ethanolic alizarin solution, and then dipped into 2 mL of a 0.4% solution of zirconium chloride (or nitrate). The red filter paper is dried and, immediately before use, is moistened with a drop of 50% acetic acid.

Cleaning Up. The silver, mercury, and lead salts are filtered off and placed in a hazardous solid waste container. The bottom methylene chloride layers are placed in the halogenated organic solvent container. The aqueous layers are neutralized with sodium bicarbonate, if acidic, or 5% hydrochloric acid, if basic, and placed in the aqueous solution container.

4.2 QUANTITATIVE ELEMENTAL ANALYSIS

4.2.1 Combustion and Related Analyses

Quantitative analytical data are routinely reported for confirmation of structure of new organic compounds; these data are extremely useful for structure determination of unknown compounds. Such microanalyses are usually determined by commercial firms[6] equipped with combustion or other appropriate analytical equipment. Unknown samples must be checked for purity by thin-layer chromatography (Chapter 3 pp. 59–63) and/or gas chromatography (Chapter 10 pp. 474–476) after they have been recrystallized (Chapter 3 pp. 78–81) or distilled (Chapter 3 pp. 44–45). The unknown can be dried to remove residual solvents in an Abderhalden drying pistol (Figure 4.1). A small amount of material is spread thinly in a porcelain boat or on glazed weighing paper. The sample can be also be placed in a small vial with a high-quality laboratory tissue held around the mouth of the container with a rubber band. The container is then placed in the horizontal portion of the drying pistol.

The drying bulb of the drying pistol is charged with fresh, anhydrous drying agent. For removal of water, the drying agent can be phosphorus pentoxide or, less efficiently, calcium sulfate or calcium chloride. The entire system is then evacuated with a vacuum pump. The speed at which the sample loses the solvent may be incrcased by allowing toluene or xylene to reflux up from the lower flask; the sample must be stable and have a melting point higher than the boiling point of the solvent. If it is believed that hydrocarbon solvents are present in the sample, wax shavings are used instead of the desiccant. Samples can be examined under a magnifying lens for filter paper fibers or other extraneous material.

Normally 5 mg of the sample is needed for carbon and hydrogen analysis and another 5 mg for each additional analysis for other atoms such as sulfur, halogen, or deuterium. Oxygen analyses are not generally obtained; percent of oxygen is normally obtained by the difference in percentages. Confirmation of a molecular formula is satisfactory when the calculated and experimentally determined percentages agree within $\pm 0.4\%$; for

[6]Quantitative analyses of C, H, N, S, and X are determined by Galbraith Analytical Laboratories, Knoxville, TN; Baron Consulting, Orange, CT; Huffman Microanalytical Laboratories, Wheatridge, CO; and Atlantic Microlabs, Norcross, GA.

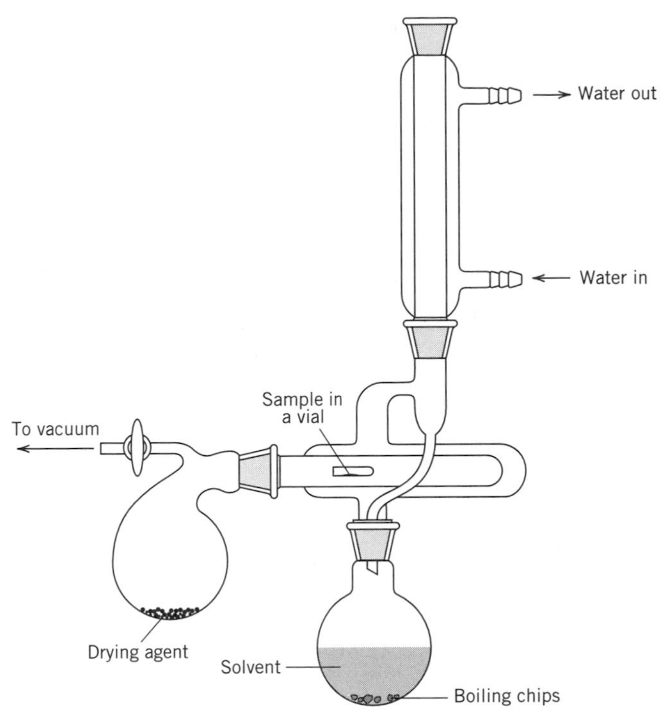

Figure 4.1 Abderhalden drying pistol.

Calculated for $C_{13}H_{16}O$:	C, 82.93; H, 8.57
Observed:	C, 82.87; H, 8.67

When the molecular formula is unknown prior to the analysis, the empirical formula can be determined by combustion. The molecular formula can then be calculated if the molecular weight is known.

Two methods[7] are used in order to determine the amounts of carbon, hydrogen, and nitrogen in a sample. In the first method,[8] the instrument combusts the sample at approximately 900°C with oxygen introduced in a helium stream. The products from the combustion are carbon dioxide (from carbon), water (from hydrogen), and nitrogen oxides (from nitrogen). The combustion products are passed over copper at 600°C, which reduces the nitrogen oxides to nitrogen. The resulting product mixture is swept over three thermal conductivity detectors. Water is removed between the first and second detectors by an appropriate trap material. Between the second and third detectors, the carbon dioxide is also removed by trapping. The amount of the element is determined by the differences of the readings by the three detectors and by calibration with other known compounds.

Another manufacturer[9] separates the combustion products on a gas chromatograph connected to a thermal conductivity detector. These instruments can also be adapted for the determination of sulfur and oxygen.

[7]Private communication with Paul Rosenberg, Fisons Pharmaceuticals, Rochester, NY.

[8]The first method is based on an elemental analyzer manufactured for Perkin-Elmer by Exeter Analytical, North Chelmsford, MA.

[9]The second method is based on an elemental analyzer manufactured by Fisons Instruments, Danvers, MA.

To determine the amount of halogen present, the sample is combusted in a Schöniger micro combustion flask. The combustion products are titrated with mercuric nitrate to a diphenylcarbazone colorimetric end point.

For example, 11.55 mg of a compound produced 16.57 mg of carbon dioxide and 5.09 mg of water from combustion. Another 5.12 mg of the same compound was found to contain 1.97 mg of chlorine. The molecular weight had been previously determined to be 368.084 g/mole by another method.

The mg of C is determined by multiplying the weight of CO_2 by the ratio of the atomic weight of carbon to the molecular weight of carbon dioxide (C/CO_2). The percentage of C is then calculated by the ratio of mg of C to the total weight of the sample.

$$\text{mg C} = 16.57 \text{ mg } CO_2 \times \frac{12.011 \text{ (C)}}{44.010 \text{ (}CO_2\text{)}}$$

$$= 4.52 \text{ mg C in original sample}$$

$$\% \text{ C} = \frac{4.52}{11.55} \times 100 = 39.13\% \text{ C}$$

The mg of H is determined by multiplying the weight of H_2O by the ratio of the atomic weight of two hydrogens to the molecular weight of water ($2H/H_2O$). The percentage of H is determined by the ratio of mg of H to the total weight of the sample.

$$\text{mg H} = 5.09 \text{ mg } H_2O \times \frac{2.016 \text{ (2H)}}{18.015 \text{ (}H_2O\text{)}}$$

$$= 0.57 \text{ mg H in original sample}$$

$$\% \text{ H} = \frac{0.57}{11.55} \times 100 = 4.943\% \text{ H}$$

Since 1.97 mg of Cl was found in 5.12 mg of sample, then the percentage of chlorine can be determined by the ratio of mg of Cl to the weight of the sample used for the analysis.

$$\% \text{ Cl} = \frac{1.97}{5.12} \times 100 = 38.46\% \text{ Cl}$$

The percentage of oxygen is determined by the difference.

$$\% \text{ O} = 100 - (39.13 + 4.94 + 38.48)$$

$$= 17.45\% \text{ O}$$

Each percentage is then divided by the atomic weight of that element to obtain the ratio of the elements.

$$\text{C} = \frac{39.13}{12.011} = 3.258$$

$$\text{H} = \frac{4.94}{1.008} = 4.901$$

$$\text{Cl} = \frac{48.46}{35.453} = 1.085$$

$$\text{O} = \frac{17.48}{16.000} = 1.093$$

The ratios are then divided by the lowest ratio (in this example, 1.085). If a fraction is obtained which is not close to a whole number, then all of the ratios are multiplied by whatever integer is necessary to obtain whole number ratios. For example, if one of the ratio numbers contains a 0.2, then all of the ratios are multiplied by 5. With a 0.25, all are multiplied by 4; with a 0.33, all are multiplied by 3; and with a 0.5, all ratios are multiplied by 2.

$$C = \frac{3.258}{1.085} = 3.028;\ 3.028 \times 2 \approx 6$$

$$H = \frac{4.901}{1.085} = 4.517;\ 4.517 \times 2 \approx 9$$

$$Cl = \frac{1.085}{1.085} = 1.000;\ 1.000 \times 2 \approx 2$$

$$O = \frac{1.093}{1.085} = 1.007;\ 1.007 \times 2 \approx 2$$

Thus the empirical formula is $C_6H_9Cl_2O_2$, with an empirical weight of 184.042 g/mole. The formula weight, which had been previously determined to be 368.084 g/mole, is then divided by the empirical weight, to obtain the number of empirical units, *n*. Then multiply the subscripts of the empirical formula by *n* to obtain the molecular formula.

$$\frac{368.084}{184.0542} = n \approx 2$$

$$C_{6\times2}H_{9\times2}Cl_{2\times2}O_{2\times2} = C_{12}H_{18}Cl_4O_4 = \text{molecular formula}$$

The empirical formula could not possibly be the molecular formula because all compounds containing only carbons, hydrogens, oxygens, and halogens must have an even number of hydrogens plus halogens.

From the molecular formula, the number of sites of unsaturations can be determined. Each double bond is one site, and each ring is one site. Triple bonds are counted as two sites. The unsaturation number is calculated by the following formula.

$$\text{Unsaturation Number} = C - \tfrac{1}{2}X + \tfrac{1}{2}Y + 1$$

C is the number of carbons or other tetravalent atoms such as silicon. X is the number of monovalent elements such as hydrogen and the halogens. Y is the number of trivalent elements such as nitrogen and phosphorus. Divalent elements such as oxygen and sulfur do not change the number of unsaturations, and thus do not appear in the formula.

For example, for the formula $C_9H_{11}NO$ the number of unsaturations is $9 - \tfrac{1}{2}(11) + \tfrac{1}{2}(1) + 1 = 9 - 5.5 + 0.5 + 1 = 5$.

4.2.2 Formula Determination by Mass Spectrometry

A sample of sufficient thermal stability and volatility to result in a measurable molecular ion should yield the molecular weight by mass spectrometry. Tables[10] have been published that correlate molecular weights to four decimal places with molecular formulas. Also, certain elements such as bromine, chlorine, and sulfur have distinctive patterns in a mass spectrum. This information is discussed in detail in Chapter 6.

[10] R. M. Silverstein, G. C. Bassler, and T. C. Morrill, *Spectrometric Identification of Organic Compounds,* 5th ed. (Wiley, New York, 1991), pp. 44–84.

CHAPTER 5

Classification of Organic Compounds by Solubility

In this chapter we begin the process of determining the structural composition of organic compounds. Elemental composition information obtained in Chapter 4 can be of great use here. Both solubility and spectrometric analyses (Chapter 6) often lead to the same kinds of structural deduction. Deductions based upon interpretation of simple solubility tests can be extremely useful in organic structure determination.

The solubility of organic compounds can be divided into two major categories: solubility in which a chemical reaction is the driving force, for example, the following acid–base reaction,

$$4\text{-}NO_2C_6H_4COOH \xrightarrow[H_2O]{OH^-} 4\text{-}NO_2C_6H_4COO^-Na^+$$

4-nitrobenzoic acid
water insoluble

sodium 4-nitrobenzoate
water soluble

and solubility in which simple miscibility is the only mechanism involved, such as dissolving ethyl ether in carbon tetrachloride. Although the two solubility sections below are interrelated, the first section deals primarily with the identification of functional groups and the second with the determination of solvents to be used in recrystallizations, spectral analyses, and chemical reactions.

5.1 SOLUBILITY IN WATER, AQUEOUS ACIDS AND BASES, AND ETHER

Three kinds of information can often be obtained about an unknown substance by a study of its solubilities in water, 5% sodium hydroxide solution, 5% sodium bicarbonate solution, 5% hydrochloric acid solution, and cold concentrated sulfuric acid. First, the presence of a functional group is often indicated. For instance, because hydrocarbons are insoluble in water, the mere fact that an unknown is partially soluble in water indicates that a polar functional group is present. Second, solubility in certain solvents often leads to more specific information about the functional group. For example, benzoic acid is insoluble in a polar solvent, water, but is converted by 5% sodium hydroxide solution to a salt, sodium benzoate, which is readily water soluble. In this case, then, the solubility in 5% sodium hydroxide solution of a water insoluble unknown is a strong indication of an acidic functional group. Finally, certain deductions about molecular size and composition may sometimes be made. For example, in many homologous series of monofunctional compounds, the members with fewer than about five carbon atoms are water soluble, whereas the higher homologs are insoluble.

Compounds are first tested for solubility in water. In considering solubility in water, a substance is arbitrarily said to be "soluble" if it dissolves to the extent of 3.3 g/100 mL of solvent. This standard is dictated by the limitations inherent in the method employed, which depends on rough semiquantitative visual observations, as will be seen. Care is needed in interpreting the classifications of "soluble" and "insoluble" in other references because different standards for solubility may have been followed.

When solubility in 5% acid or base is being considered, the significant observation to be made is not whether the unknown is soluble to the extent of 3% or to any arbitrary extent, but, rather, whether it is significantly more soluble in aqueous acid or base than in water. This increased solubility is a positive test for a basic or an acidic functional group.

Acidic compounds are identified by their solubility in 5% sodium hydroxide solution. Strong and weak acids (classes A_1 and A_2, respectively; see Table 5.1 and Figure 5.1) are differentiated by the solubility of the former but not the latter in the weakly basic solvent, 5% sodium bicarbonate solution. Compounds that behave as bases in aqueous solution are detected by their solubility in 5% hydrochloric acid solution (class B); no attempt is made here to differentiate among the various strengths of bases in the class.

Many compounds which are neutral toward 5% hydrochloric acid solution behave as bases in more acidic solvents such as concentrated sulfuric acid. In general, compounds containing sulfur or nitrogen have an atom with a unshared pair of electrons and would be expected to dissolve in a strong acid. No additional information would be gained, therefore, by determining such solubility, and for this reason, when the elemental analysis has shown the presence of sulfur or nitrogen, no solubility tests beyond those for acidity and basicity in aqueous solution are carried out. Compounds that contain nitrogen or sulfur and are neutral in aqueous acid or base are placed in solubility category MN.

Most compounds which are neutral in water and contain oxygen in any form are reasonably strong bases in concentrated sulfuric acid. Solubility in or any other evidence of a reaction with sulfuric acid indicates the presence of an oxygen atom or of a reactive hydrocarbon function such as an olefinic bond or easily sulfonated aromatic ring. These compounds are in the solubility class N. Compounds that are too weakly basic to dissolve in sulfuric acid are placed in class I (inert compounds).

Since the solubility behavior of water soluble compounds gives no information about the presence of acidic or basic functional groups, their aqueous solutions are tested with litmus or pH paper. The behavior of an acid in 5% hydrochloric acid solution and of a

Table 5.1 Organic Compounds Comprising the Solubility Classes of Figure 5.1[a]

Class	Compounds
S_2	Salts of organic acids (RCO_2Na, RSO_3Na); amine hydrochlorides (RNH_3Cl); amino acids ($R{-}CH(NH_3^+){-}CO_2^-$); polyfunctional compounds with hydrophilic functional groups: carbohydrates (sugars), polyhydroxy compounds, polybasic acids, etc.
S_A	Monofunctional carboxylic acids with five carbons or fewer; arylsulfonic acids.
S_B	Monofunctional amines with six carbons or fewer.
S_1	Monofunctional alcohols, aldehydes, ketones, esters, nitriles, and amides with five carbons or fewer.
A_1	Strong organic acids: carboxylic acids with more than six carbons; phenols with electron-withdrawing groups in the *ortho* and/or *para* position(s); β-diketones (1,3-diketones).
A_2	Weak organic acids: phenols, enols, oximes, imides, sulfonamides, thiophenols, all with more than five carbons; β-diketones (1,3-diketones); nitro compounds with α-hydrogens.
B	Aliphatic amines with eight or more carbons; anilines (only one phenyl group attached to nitrogen); some ethers.
MN	Miscellaneous neutral compounds containing nitrogen or sulfur and having more than five carbon atoms.
N	Alcohols, aldehydes, ketones, esters with one functional group and more than five but fewer than nine carbons, ethers, epoxides, alkenes, alkynes, some aromatic compounds (especially those with activating groups).
I	Saturated hydrocarbons, haloalkanes, aryl halides, other deactivated aromatic compounds, diaryl ethers.

[a]Acyl halides and carboxylic acid anhydrides have not been classified because of their high reactivity.

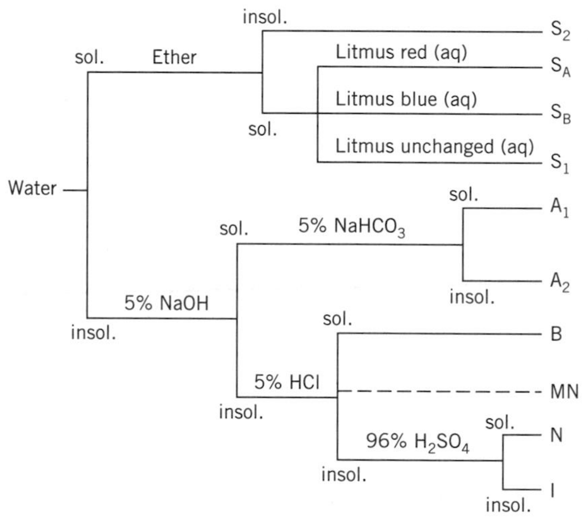

Figure 5.1 Classification of organic compounds by solubility: determination in water, acids, bases, and ethers (see Table 5.1 for compounds comprising each class). sol. = soluble, insol. = insoluble; litmus is red at pHs below 4.5 and blue above 8.3.

base in 5% sodium hydroxide solution should be examined routinely, since the molecule may have both acidic *and* basic functional groups (class A_1–B or A_2–B).

Directions for determining the solubility class of an unknown compound are given below, followed by an explanation of the solubility of various types of compounds.

5.1.1 Determination of Solubilities

Procedure for Water Solubility. Place 0.05 mL (approximately 1 drop) or 25 mg of the compound in a small test tube, and add 0.75 mL of water in small portions. Shake vigorously after the addition of each portion of solvent, being careful to keep the mixture at room temperature. If the compound dissolves completely, record it as soluble.

Solids should be finely powdered to increase the rate of dissolving of the solid. If the solid appears to be insoluble in water or ether, it is sometimes advisable to heat the mixture gently. If the solid dissolves with heating, the solution is cooled to room temperature and is shaken to prevent supersaturation. The cooled solution is then "seeded" with a crystal of the solid. Care should be taken in weighing the sample; it should weigh 25 ± 1 mg.

Liquids are handled most conveniently by means of a graduated pipet which permits the accurate measurement of the liquid. When two colorless liquid phases lie one above the other, it is often possible to overlook the boundary between them and thus to see only one phase. This mistake can generally be avoided by shaking the test tube vigorously when a liquid unknown seems to have dissolved in the solvent. If two phases are present the solution will become cloudy. In the rare cases where two colorless phases have the same refractive index, the presence of a second phase will escape detection even if this precaution is taken.

Acid–base properties of water soluble compounds should be determined with litmus paper; compounds with a $pK_a < 8$ will fall in class S_A (see Table 5.1 and Figure 5.1). Compounds with a $pK_b < 9$ will fall in class S_B. Consequently, phenols with pK_a of about 10 give aqueous solutions too weakly acidic (pH about 5) to turn litmus paper red. Litmus is red at pHs below 4.5 and blue above 8.3. For similar reasons, an aromatic amine such as aniline is too weak a base (pK_b 9.4) to turn litmus blue in aqueous solution. Although more refined procedures can be developed using a pH-indicating paper, it is preferable to rely more on the tests discussed in Chapter 7.

Cleaning Up. Place the test solutions in the aqueous solution container.

Procedure for Ether Solubility. Place 0.05 mL (approximately 1 drop) or 25 mg of the compound in a small test tube, and add 0.75 mL of diethyl ether in small portions. Shake vigorously after the addition of each portion of solvent, being careful to keep the mixture at room temperature. If the compound dissolves completely, record it as soluble.

Cleaning Up. Place the test solution in the organic solvent container.

Procedure for Solubility in Aqueous Acid or Base. To test for solubility in aqueous acid or base, shake a mixture of 0.05 mL (approximately 1 drop) or 25 mg of the unknown compound and 0.75 mL of 5% sodium hydroxide solution, 5% sodium bicarbonate solution, or 5% hydrochloric acid solution thoroughly. Separate (filter if necessary) the aqueous solution from any undissolved unknown, and *neutralize* with acid and base. Examine the solution very carefully for any sign of separation of the original unknown. Even a cloudy appearance of the neutralized filtrate is a positive test.

When solubility in acid or alkali is being determined, heat should *not* be applied because it might cause hydrolysis to occur. If the mixture is shaken thoroughly, the time required for the unknown to dissolve should not be more than 1–2 min.

Often it is possible to utilize a single portion of unknown for tests with several different solvents. Thus if the compound is found to be insoluble in water, a fairly accurate measure of its solubility in 5% sodium hydroxide solution can be obtained by adding about 0.25 mL of a 20% solution of sodium hydroxide to a mixture of the compound in 0.75 mL of water. The resulting 1.0 mL of solvent will then contain a 5% solution of sodium hydroxide. If the substance is very insoluble, it often may be recovered and used subsequently for the hydrochloric acid test. Although it is possible to conserve the unknown in this manner, it is recommended that a fresh unknown sample be used for each test.

Cleaning Up. Place the test solutions in the aqueous solution container.

Procedure for Solubility in Concentrated Acid. With sulfuric acid, it is more convenient to place 0.6 mL of solvent in the test tube and then add 0.05 mL (approximately 1 drop) or 25 mg of the unknown compound. For purposes of solubility classification, unknowns that react with sulfuric acid to produce heat and/or color changes should be classified as soluble, even if the sample does not appear to dissolve.

Cleaning Up. Carefully neutralize the test solution with 10% sodium hydroxide solution and place the test solution in the aqueous solution container.

5.1.2 Theory of Solubility

Polarity and Solubility. When a solute dissolves, its molecules or ions become distributed more or less randomly among those of the solvent. In crystalline sodium chloride, for example, the average distance between sodium and chloride ions is 2.8 Å. In a 1 M solution the solvent has interspersed itself in such a way that sodium and chloride ions are about 10 Å apart. The difficulty of separating such ions is indicated by the high melting point (800°C) and boiling point (1413°C) of pure sodium chloride. Another indication of the importance of solvent is the fact that sodium chloride readily forms ions in water, while it takes several hundred kilocalories per mole to form ions from sodium chloride in the solid state.

The work required to separate two oppositely charged plates is reduced by the introduction of matter between them by a factor called the *dielectric constant.* It is not surprising that water, with a high dielectric constant of 80, facilitates the separation of sodium and chloride ions and dissolves sodium chloride readily, whereas both hexane (dielectric constant 1.9) and ether (dielectric constant 4.4) are extremely poor solvents for ionic salts. Water molecules positioned between two ions (or the charged plates of a condenser) are actually small dipoles, which orient themselves end to end in such a way as to partially neutralize the ionic charges and thus stabilize the system. An assumption might be made that the solvating ability and dielectric constant are related. However, this is not entirely the case. A high dielectric constant is required but is not the only characteristic for an effective ion solvent. For example, hydrogen cyanide, with a dielectric constant of 116, is a very poor solvent for salts such as sodium chloride. Although the situation is quite complex, one major factor responsible for the efficiency of water and other hydroxylic solvents is their ability to form hydrogen bonds with the solute.

The high dielectric constant and hydrogen-bonding ability of water, which combine to make it a good solvent for salts, also make it a poor solvent for nonpolar substances. In pure water, molecules are oriented in such a way that positive and negative centers are adjacent. Attempting to dissolve a nonpolar substance such as hexane in a solvent such as water is analogous to separating unlike charges in a medium of low dielectric constant. As a general rule, a polar solvent may be expected to readily dissolve only polar solutes, and nonpolar solvent only nonpolar solutes. This generalization has been summarized more succinctly as "like dissolves like".

Table 5.1, related to Figure 5.1, lists the solubility class of various types of compounds. A discussion of the solubility trends of compounds is given below.

Since most organic molecules have both a polar and a nonpolar entity, it can be deduced that its solubility would depend on the balance between the two parts. As the percentage of the hydrocarbon portion increases, the properties of the compounds approach those of the parent hydrocarbons. As a result, water solubility decreases and ether solubility increases. A similar change in solubility occurs as the number of aromatic hydrocarbon residues in the molecule increases. Thus 1-naphthol and 4-hydroxybiphenyl are less soluble in water than phenol:

Water Solubility

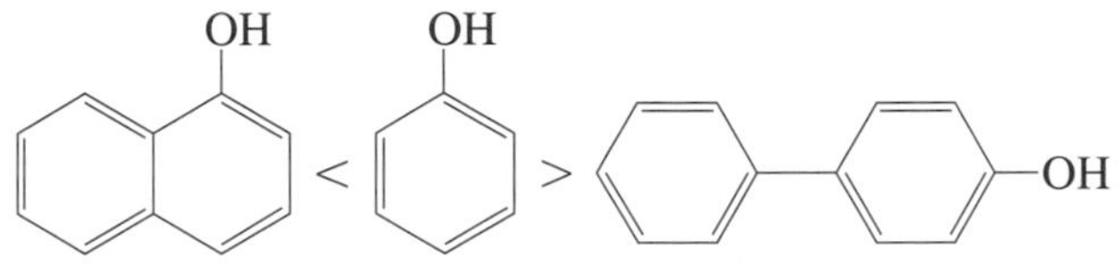

1-naphthol < phenol > 4-hydroxybiphenyl

The phenyl group, when present in aliphatic acids, alcohols, aldehydes, and similar compounds, has an effect on water solubility approximately equivalent to a four-carbon aliphatic unit. Benzyl alcohol, for example, is about as soluble in water as 1-pentanol, and 3-phenylpropanoic (hydrocinnamic) acid exhibits a solubility similar to that of heptanoic acid:

Water Solubility

$$C_6H_5CH_2OH \approx CH_3CH_2CH_2CH_2CH_2OH$$

benzyl alcohol 1-pentanol

$$C_6H_5CH_2CH_2COOH \approx CH_3CH_2CH_2CH_2CH_2CH_2COOH$$

3-phenylpropanoic acid (dihydrocinnamic acid) heptanoic acid

The solubility of a substance is a measure of the equilibrium between the pure substance and its solution. Such an equilibrium is affected not only by the solvent–solute interactions previously discussed but also by the intermolecular forces present in the pure solute. These forces are independent of the polarity or other properties of the solvent, and their relative strengths may be estimated by a comparison of melting and boiling points, since these processes involve a separation of molecules that is somewhat related to the separation which occurs on solution.

The dicarboxylic acids illustrate the inverse relationship of melting point and solubility. The data in Table 5.2 show that each member with an even number of carbon atoms melts at a higher temperature than either the immediately preceding or following acid containing an odd number of carbon atoms. The intracrystalline forces in the members with an even number of carbon atoms evidently are greater than in those with an odd number of carbons. Since the solubility limit for solids is generally set at 3.3 g/100 mL of water, it is evident that hexanedioic acid (adipic acid, six carbons) is water insoluble but heptanedioic acid (pimelic acid, seven carbons) is water soluble.

Table 5.2 Water Solubility of Dicarboxylic Acids, HOOC—$(CH_2)_{x-2}$—COOH

Number (x) of carbon atoms	mp (°C)	Solubility (g/100 g of water at 20°C)
Even number of carbon atoms		
Ethanedioic acid (2) (oxalic acid)	189	9.5
Butanedioic acid (4) (succinic acid)	185	6.8
Hexanedioic acid (6) (adipic acid)	153	2
Octanedioic acid (8) (suberic acid)	140	0.16
Decanedioic acid (10) (sebacic acid)	133	0.10
Odd number of carbon atoms		
Propanedioic acid (3) (malonic acid)	135	73.5
Pentanedioic acid (5) (glutaric acid)	97	64
Heptanedioic acid (7) (pimelic acid)	103	5
Nonanedioic acid (9) (azelaic acid)	106	0.24

The relationship of high melting point and low solubility is further illustrated by the isomers *cis*- and *trans*-2-butenedioic acid (maleic and fumaric acids). *trans*-2-Butenedioic acid sublimes at 200°C and is insoluble in water. *cis*-2-Butenedioic acid melts at 130°C and is soluble in water. Among *cis-trans* isomers, the *cis* form generally is the more soluble. Similarly, with polymorphous substances such as benzophenone,[1] the lower melting forms possess the higher solubilities.

HOOC COOH / H H

cis-2-butenedioic acid
(maleic acid)
Z-isomer

HOOC H / H COOH

trans-2-butenedioic acid
(fumaric acid)
E-isomer

The diamides of dicarboxylic acids constitute another group of compounds for which the melting point is a valuable index of the forces present in the crystals. Urea (mp 132°C) is water soluble. On the other hand, ethanediamide (oxalic acid diamide) has a quite high melting point of 420°C and a low solubility in water. Substitution of methyl groups for the hydrogen atoms of the amide group lowers the melting point by reducing intermolecular hydrogen bonding and increases the solubility in water; *N,N'*-dimethylethanediamide and *N,N,N',N'*-tetramethylethanediamide are water soluble. Hexanediamide is water insoluble, whereas its *N,N,N',N'*-tetramethyl derivative is water soluble.

[1]Benzophenone occurs in at least four crystalline forms.

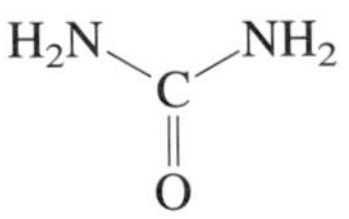

urea
mp 132°C

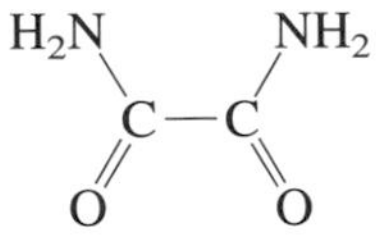

ethanediamide
mp 420°C

N,N'-dimethyl-
ethanediamide
mp 217°C

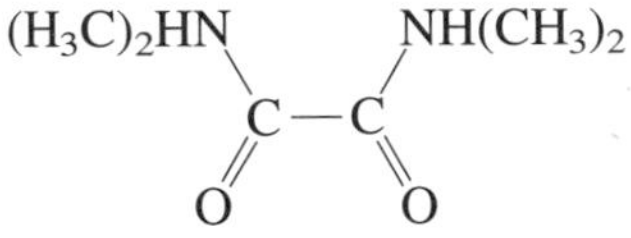

N,N,N',N'-tetra-
methylethanediamide
mp 80°C

Regarding water solubility, amides of the type $RCONH_2$ and RCONHR agree with the general rule that the borderline compounds contain about five carbon atoms (above). However, *N,N*-dialkylamides ($RCONR_2$), which have lower melting points than the corresponding unsubstituted amides, are much more soluble in water. The water solubility limit for the *N,N*-dialkylamides is in the nine to ten carbon atom range. Amides having the group $—CONH_2$ may act both as acceptors and as donors in forming hydrogen bonds. Intermolecular hydrogen bonding cannot occur with the *N,N*-disubstituted amides ($RCONR_2$), and hence their state of molecular aggregation is low, as indicated by their lower melting points and higher solubilities.

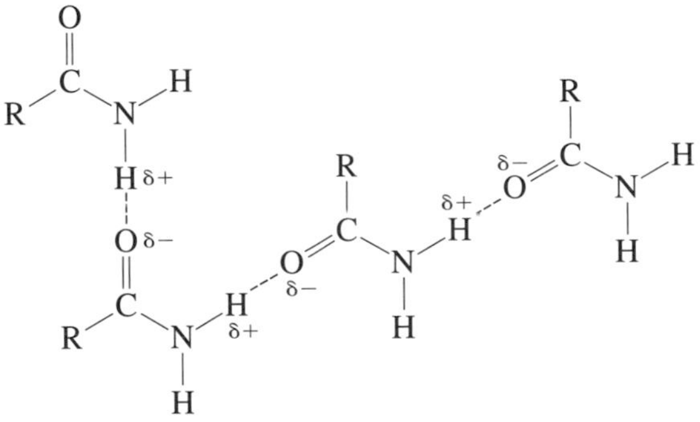

In general, an increase in *molecular weight* leads to an increase in intermolecular forces in a solid and decreased solubility. Therefore, polymers and other compounds of high molecular weight generally exhibit low solubilities in both water and ether. Thus formaldehyde is readily soluble in water, whereas paraformaldehyde is insoluble.

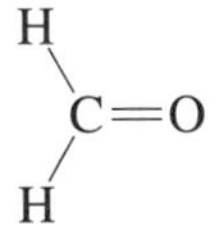

formaldehyde
(methanal)
water soluble

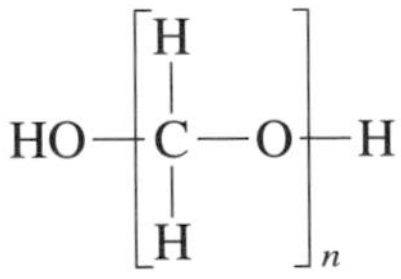

paraformaldehyde
water insoluble

Methyl acrylate is soluble in water, but its polymer is insoluble. The fact that a pi bond is lost upon polymerization may also affect solubility.

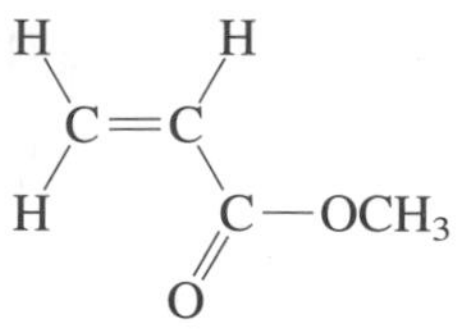

methyl acrylate

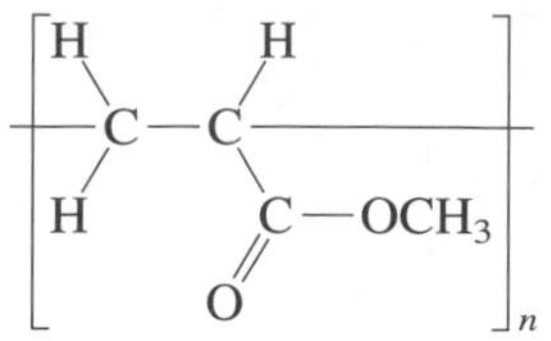

polymer of methyl acrylate

Glucose is soluble in water, but its polymers (starch, glycogen, and cellulose) are insoluble. Fibrous proteins are insoluble in water, but globular proteins are soluble in water. Many amino acids are soluble in water, but their condensation polymers, the proteins, are usually insoluble. The tendency of certain proteins, dextrins, and starches to form colloidal dispersions may lead to inaccurate conclusions regarding their solubility.

Another method of increasing the molecular weight of a molecule is by the substitution of halogens for hydrogens. The result is a lower water solubility of the halogenated compound.

The five-carbon upper limit for water solubility follows from a very general principle, that increased structural similarity of the solute and the solvent is accompanied by an increase in solubility. Since water is a polar solvent, compounds that are water soluble contain polar functional groups. As a homologous series is ascended, the nonpolar hydrocarbon part of the molecule increases while the polar function remains essentially unchanged, thus resulting in a decrease in solubility in polar solvents such as water.

The tendency of certain oxygen-containing compounds to form hydrates also contributes to water solubility. The stability of these hydrates is therefore a factor in determining water and ether solubility. Compounds such as 2,2,2-trichloroethanal probably owe their solubility in water to hydrate formation.

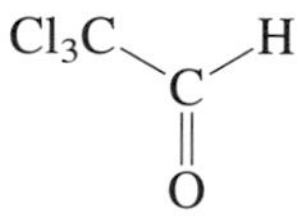

2,2,2-trichloroethanal
(chloral)

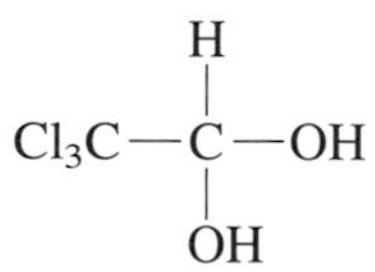

2,2,2-trichloroethanalhydrate
(chloral hydrate)

Low molecular weight esters of methanoic (formic) and 2-oxopropanoic (pyruvic) acids are hydrolyzed by water at room temperature, as indicated by the fact that the aqueous layer becomes distinctly acid to litmus.

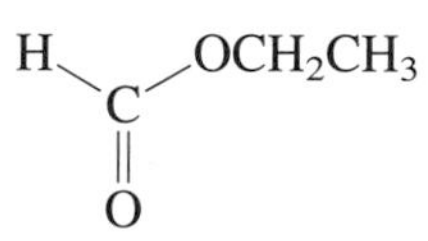

ethyl methanoate
(ethyl formate)

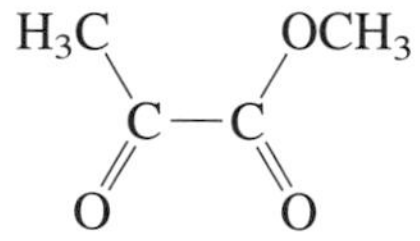

methyl 2-oxopropanoate
(methyl pyruvate)

Effect of Chain Branching on Solubility. Since branching of the hydrocarbon chain lowers the boiling points of the lower homologous series, such as the hydrocarbons and alcohols, an assumption can be made that branching also lowers intermolecular forces and decreases intermolecular attraction. Therefore, a compound having a branched chain is more soluble than the corresponding straight-chain compound. This is a general rule and is particularly useful in connection with simple aliphatic compounds. For example, the solubility of an *iso* compound differs widely from that of its normal isomer and is close to that of the next lower normal member of the same homologous series. The effects of chain branching are shown in Table 5.3.

Table 5.3 Water Solubility of Various Organic Compounds

Type of compound	Soluble	Borderline	Insoluble
Acid	2,2-Dimethylpropanoic acid (C_5) (pivalic acid)	3-Methylbutanoic acid (C_5) (isovaleric acid)	Penatonic acid (C_5) (valeric acid)
Acid chloride	2-Methylpropanoyl chloride (C_4) (isobutyryl chloride)	Butanoyl chloride (C_4) (butyryl chloride)	
Alcohol	2,2-Dimethyl-1-propanol (C_5) (neopentyl alcohol)	3-Methyl-2-butanol (C_5)	1-Pentanol (C_5) (amyl alcohol)
Amide	2-Methylpropanamide (C_4) (isobutyramide)	Butanamide (C_4) (butyramide)	
Ester	1-Methylethyl ethanoate (C_5) (isopropyl acetate)	Propyl ethanoate (C_5) (propyl acetate)	
Ketone	3-Methyl-2-butanone (C_5) (isopropyl methyl ketone)	2-Pentanone (C_5) (methyl propyl ketone)	
Nitrile		2-Methylpropanenitrile (C_4) (isobutyronitrile)	Butanenitrile (C_4) (butyronitrile)

The position of the functional group in the carbon chain also affects solubility. For example, 3-pentanol is more soluble than 2-pentanol, which in turn is more soluble than 1-pentanol. When the branching effect is combined with the position of the functional group toward the center of the molecule, as in the case of 2-methyl-2-butanol, a very marked increase in solubility is noted. *Normally, the more compact the structure, the greater the solubility, provided that comparisons are made on compounds containing the same functional group(s) and molecular weight.*

5.1.3 Theory of Acid–Base Solubility

Effect of Structure on Acidity and Basicity. In general, the problem of deciding whether a water insoluble unknown should dissolve in dilute acid or base is primarily a matter of estimating its approximate acid or base strength. In doing this, we must be concerned with the structural features that will stabilize the organic anion, A^-, and position the following equilibrium[2] farther to the right:

$$HA \rightleftharpoons H^+ + A^-$$

that is, we increase the magnitude of K_a, where

$$K_a = \frac{[H^+][A^-]}{[HA]}$$

or decrease the magnitude of $pK_a = \log(1/K_a) = -\log K_a$.

[2] H^+ is used interchangeable with H_3O^+.

The two principal effects influencing structural control of acid–base strength are electronic and steric.

Electronic Effects on Acidity and Basicity. Extensive studies have been done on the correlation of structure with acid or base strength of substituted organic compounds. These effects have been rationalized[3] on an electronic basis; thus the *para* and *ortho* positions of aromatic compounds are more sensitive to electronic influences than are the *meta* positions:

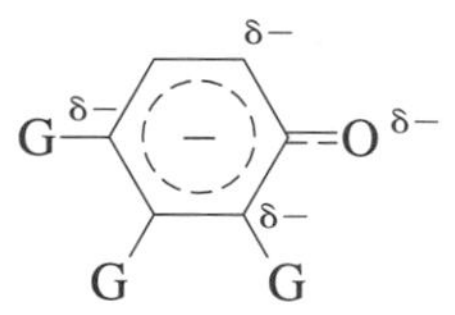

The *ortho* and *para* positions are centers of partial negative charge and thus these centers respond to substitution by polar groups; *ortho*-substitution considerations are sometimes hindered by steric factors.

Most carboxylic acids have dissociation constants in water, at 25°C, of 1×10^{-6} or greater and therefore are readily soluble in 5% sodium hydroxide solution. Phenols, with a dissociation constant of 1×10^{-10}, are less acidic. Phenols are soluble in strongly basic sodium hydroxide solution, but insoluble in 5% sodium bicarbonate solution.[4] However, the substitution of certain functional groups onto the ring may have a profound effect on their acidity. Thus 2- and 4-nitrophenol have dissociation constants of about 6×10^{-8}. The introduction of an *ortho* or *para* nitro group onto the ring increases the acidity of phenol by a factor of about 600. Therefore, the addition of two nitro groups, such as in 2,4-dinitrophenol, increases the acidity to such an extent that the compound is soluble in 5% sodium bicarbonate solution. The acidity-increasing effect of the nitro group is due to stabilization of the phenoxide anion by further distribution of the negative charge on the nitro group.

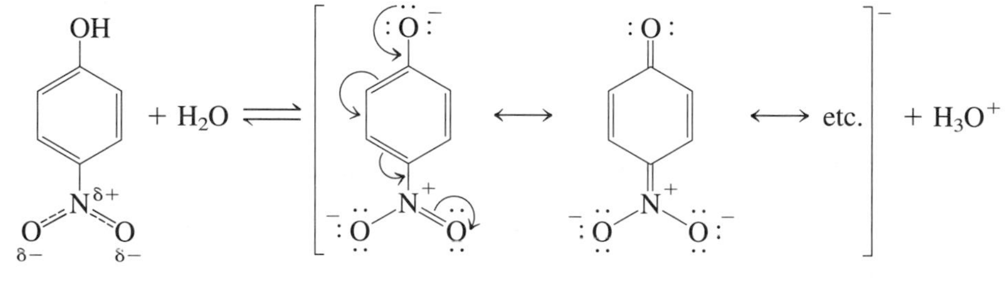

A similar increase in acidity is observed when a halogen is substituted onto phenol. Thus an *ortho* bromine atom increases the acidity of phenol by a factor of about 30 and a *para* bromine atom by a factor of about 5. The presence of more halogens increases solubility so that a compound such as 2,4,6-tribromophenol is a sufficiently strong acid to dissolve in 5% sodium bicarbonate solution. The increase in acidity can be explained by inductive effects.

Similar electronic influences affect the basicity of amines. Aliphatic amines in aqueous solution have basicity constants, K_b, of about 1×10^{-3} or 1×10^{-4}, which is ap-

[3]Consult any of the books on theoretical organic chemistry such as J. March, *Advanced Organic Chemistry,* 4th ed. (Wiley-Interscience, New York, 1992).

[4]The K_a of 2,4-dinitrophenol is about 10^{-4}.

proximately the basicity constant of ammonia, 1×10^{-5}. Introduction of a conjugated phenyl group[5] lowers the basicity by about 6 orders of magnitude, for example, aniline has a K_b of 5×10^{-10}. The phenyl ring stabilizes the free amine by resonance and also decreases the basicity of nitrogen inductively.

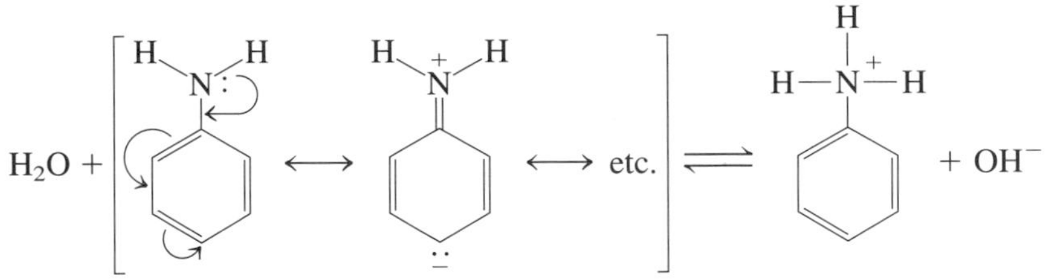

H₂O + [resonance structures of aniline ⟷ ⟷ etc.] ⇌ anilinium ion + OH^-

A second phenyl substituent decreases the basicity to such an extent that the amine is no longer measurably basic in water. For example, diphenylamine is insoluble in 5% hydrochloric acid solution. Substitution of a nitro group on the phenyl ring of aniline lowers the base strength because this electron-withdrawing group destabilizes the anilinium ion, the conjugate acid, while stabilizing the free base.

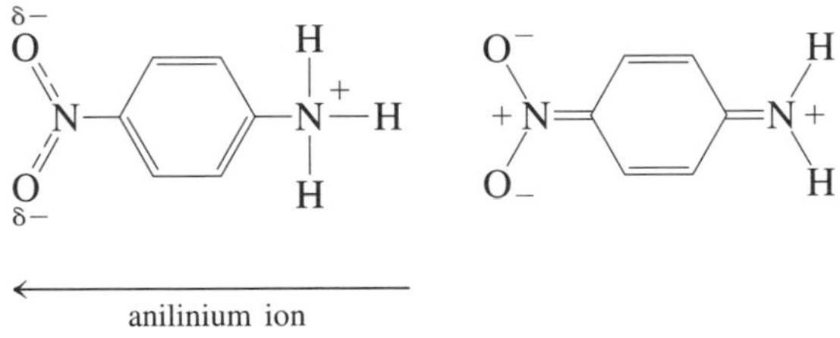

anilinium ion

Steric Effects on Acidity and Basicity. The *ortho*-disubstituted phenols have reduced solubility in aqueous alkali, and the term "cryptophenol" has been used to emphasize this characteristic. Claisen's alkali (35% potassium hydroxide in methanol–water) has been used to dissolve such hindered phenols. An extreme example is 2,4,6-tri-*t*-butylphenol, which fails to dissolve in either aqueous sodium hydroxide or Claisen's alkali. It can be converted to a sodium salt only by treatment with sodium in liquid ammonia or sodium amide. 2,4,6-Tri-*t*-butylaniline shows similar behavior. It is such a weak base that the pK_a of the conjugate acid is too low to be measured in aqueous solution. 2,6-Di-*t*-butylpyridine is a significantly weaker base than is dimethylpyridine. It has been suggested that the weakening of the base strengths of the *ortho*-disubstituted amines is due to the steric strain created about the protonated nitrogen atom. Thus, the instability of the 2,6-di-*t*-butylphenoxide ion and of the other hindered ammonium ions mentioned is probably due primarily to the steric interference with the solvation of ions.

Steric strain may either increase or decrease the acidity of carboxylic acids. For example, substitution of alkyl groups on the *a*-carbon atom of acetic acid tends to decrease acidity by destabilizing the conjugate base through steric inhibition by solvation. *Ortho*-disubstituted benzoic acids, on the other hand, are appreciably stronger than the corresponding *para* isomers. The *ortho* substituents cause the carboxyl group to be out of the

[5]The nonbonded electron pair on nitrogen is said to be conjugated with formal "double" bonds of the benzene ring; i.e., it possesses an —N̈—C═C unit with overlapping orbitals.

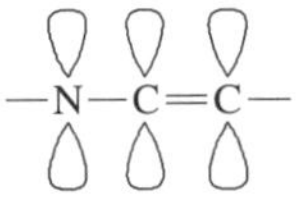

—N—C═C—

plane of the ring, disrupting resonance overlap between the carboxylate group and the ring, thus resulting in greater destabilization of the acid than of the anion.[6] This latter case is an example of a second kind of steric effect, which is commonly referred to as *steric inhibition of resonance.* Although 4-nitrophenol is about 2.8 pK_a units stronger than phenol, 3,5-dimethyl-4-nitrophenol is only about 1.0 pK_a units stronger.

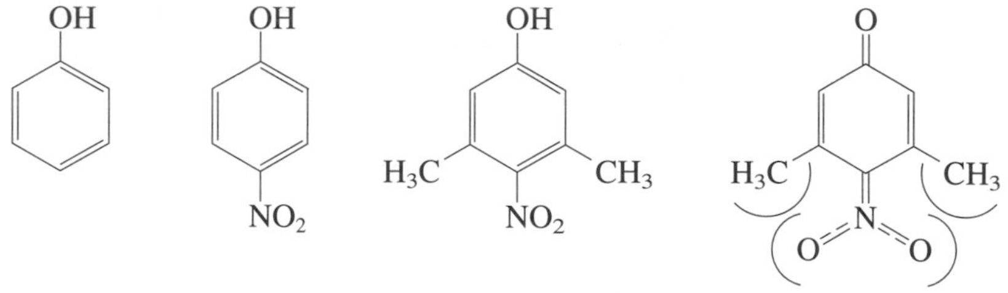

pK_a at 25°C: 9.99 7.21 8.24 A

Part of the effect of the two methyl groups, in reducing the acidity of the nitrophenol, seems to be due to steric inhibition of resonance in the anion. Thus, structure A requires coplanarity or near coplanarity of all of the nitro group's atoms and the aromatic ring. Such coplanarity is inhibited by the presence of the methyl groups in the ion.

5.1.4 Solubility in Water

Water, being a polar solvent, is a poor solvent for hydrocarbons. The presence of double bonds, triple bonds, or aromatic rings does not affect the polarity greatly and such substances are not appreciably different from alkanes in their water solubility. The introduction of halogen atoms changes the water solubility. As a halogen is substituted for a hydrogen, the water solubility decreases. Salts are extremely polar and are usually water soluble (class S_2).

As might be expected, acids and amines generally are more soluble than nonpolar compounds. The amines are highly soluble owing to their tendency to form hydrogen bonds with water molecules. This is consistent with the fact that the solubility of amines decreases as the basicity decreases. It also explains the observation that many tertiary amines are more soluble in cold than in hot water. At lower temperatures, the solubility of the hydrate is involved, whereas at higher temperatures, the hydrate is unstable and the solubility measured is that of the free amine.

Monofunctional ethers, esters, ketones, aldehydes, alcohols, nitriles, amides, acids, and amines may be considered together with respect to water solubility. *In most homologous series of this type, the longest chain with appreciable water solubility will be reached at about five carbons.*

5.1.5 Solubility in 5% Hydrochloric Acid Solution

Primary, secondary, and tertiary aliphatic amines form polar ionic salts with hydrochloric acid. Aliphatic amines are readily soluble in 5% hydrochloric acid solution and are placed in class B, if water insoluble. The presence of conjugated aryl groups decreases the basicity of the nitrogen atom. For example, primary aromatic amines, although more weakly

[6]More details on this can be found in a physical organic book such as T. H. Lowry and K. B. Richardson, *Mechanism and Theory in Organic Chemistry,* 3rd ed. (Harper and Row, New York, 1987). (Also see footnote 3.)

basic than primary aliphatic amines, are soluble in 5% hydrochloric acid solution. However, diphenylamine, triphenylamine, and carbazole are insoluble in 5% hydrochloric acid solution. Arylalkylamines, such as benzylamines, containing not more than one aryl group, are soluble in 5% hydrochloric acid solution.

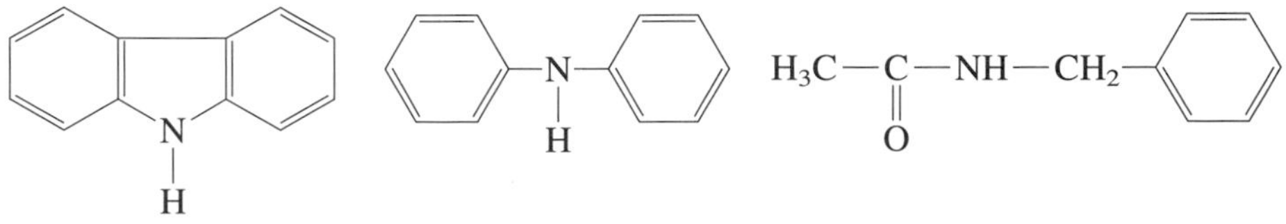

carbazole   diphenylamine   *N*-benzylacetamide

Disubstituted amides ($RCONR_2$) of sufficiently high molecular weight to be water insoluble are soluble in 5% hydrochloric acid solution. Simple amides ($RCONH_2$)[7] and most monosubstituted amides (RCONHR) are neutral compounds. *N*-Benzylacetamide, however, is a basic compound.

Amines may undergo reaction with 5% hydrochloric acid solution to form *insoluble* hydrochlorides, which may lead to errors in classification. For example, certain arylamines, such as 1-aminonaphthalene, form hydrochlorides that are sparingly soluble in 5% hydrochloric acid solution. By warming the mixture slightly and diluting it with water, it may make the compound soluble. The appearance of a solid will indicate if the amine has undergone a change. In order to decide doubtful cases, the solid should be separated and its melting point compared with that of the original compound. A positive halogen test with alcoholic silver nitrate would indicate formation of a hydrochloride.

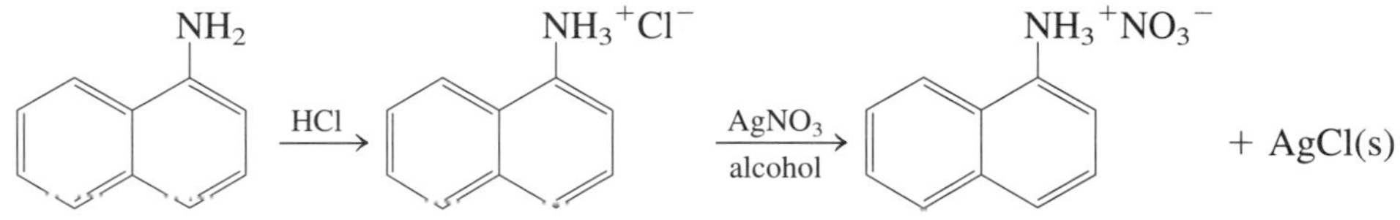

NH_2 —HCl→ $NH_3^+Cl^-$ —$AgNO_3$, alcohol→ $NH_3^+NO_3^-$ + AgCl(s)

1-aminonaphthalene (α-naphthylamine)   1-aminonaphthalene hydrochloride

Another useful technique is to dissolve the suspected base in ether and treat that with 5% hydrochloric acid solution with shaking. Formation of a solid at the interface of the two layers indicates the presence of a basic amine.

A few types of oxygen-containing compounds that form oxonium salts upon treatment with 5% hydrochloric acid solution also are basic.

5.1.6 Solubility in 5% Sodium Hydroxide and 5% Sodium Bicarbonate Solutions

A list of the various types of organic acids is given in Table 5.4. The reasoning behind most of these classifications can be understood in terms of the stability due to structural features of the conjugate base anion.

Aldehydes and ketones are sufficiently acidic to react with aqueous alkali to yield anions which serve as reaction intermediates in such reactions as the aldol condensation.

[7]Amides are generally comparable to water in basicity and small structural changes, such as alkylation, need change their K_b by only $\sim 10^2$ to move them into the "basic" category.

Table 5.4 Solubility Classes of Various Organic Acids

Name	General structure	Solubility class[a]
Carboxylic acids	RCO_2H	A_1
Sulfonic acids	RSO_3H	A_1
Sulfinic acids	RSO_2H	A_1
Enols	—C=C—OH	A_2
Imides	—C(=O)—NH—C(=O)—	A_2
Nitro[b]	>CH—NO_2	A_2
Arenesulfonamides[c]	$ArSO_2NHR$	A_2
β-Dicarbonyl compounds[d] (1,3-diketones)	—C(=O)—CH—C(=O)—	A_2[d]
Oximes	>C=N—OH	A_2

[a]Borderline cases are named in Table 5.5.

[b]Primary (RCH_2NO_2) and secondary (R_2CHNO_2) nitroalkanes only.

[c]The acidity of the N—H proton is utilized in the Hinsberg test (Experiment 17). This category also includes sulfonamides of ammonia and other sulfonamides of primary amines.

[d]Highly electronegative groups, e.g., trifluoromethyl, on the carbonyl group can move these compounds into class A_1.

They are far too weakly acidic, however, to dissolve to any measurable extent in 5% sodium hydroxide solution. When two carbonyl groups are attached to the same carbon atom, as they are in acetoacetic esters, malonic esters, and 1,3-diketones, the acidity increases sharply because of the added stabilization of the anion, since the negative charge is distributed over the two oxygen atoms as well as the central carbon atom.

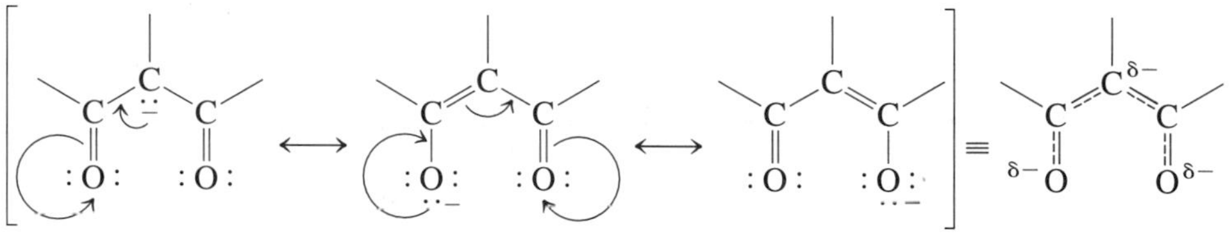

Although 1,3-dicarbonyl compounds are approximately as acidic as the phenols, the rate of proton removal from carbon may be relatively slow and as a result the rate of solution of such compounds may be so slow that they appear to be, at first, insoluble in 5% base.

Nitro compounds have a tautomeric form, the *aci* form, which is approximately as acidic as the carboxylic acids. The *aci* form of nitroethane has a K_a of 3.6×10^{-5}.

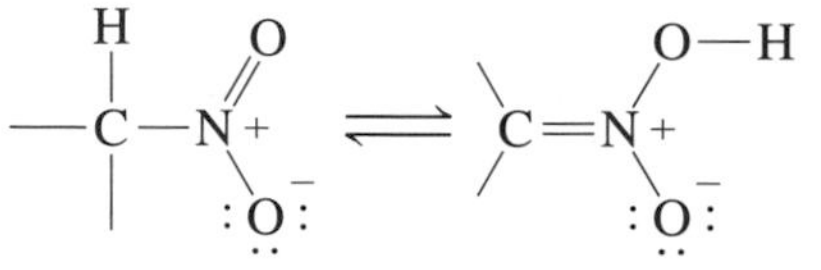

aci form

The presence of one nitro group confers sufficient acidity on a substance to make the compound soluble in 5% sodium hydroxide solution. For example, nitroethane has a K_a

of about 3.5×10^{-9}. This value should be compared to the K_a values for the following 1,3-dicarbonyl compounds:

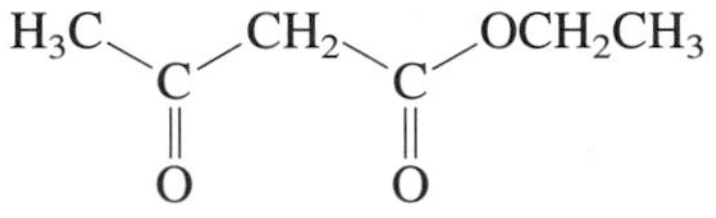

$H_3C{-}C({=}O){-}CH_2{-}C({=}O){-}OCH_2CH_3$

$K_a = 2 \times 10^{-11}$

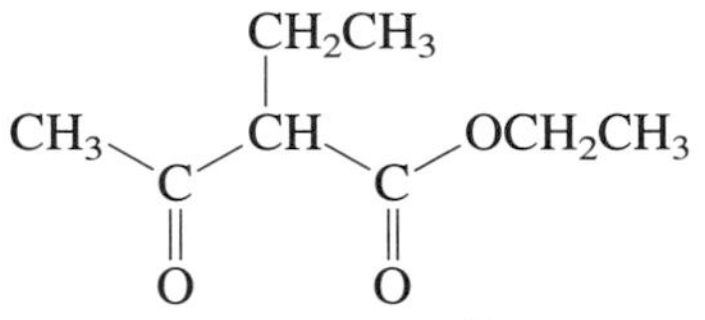

$CH_3{-}C({=}O){-}CH(CH_2CH_3){-}C({=}O){-}OCH_2CH_3$

$K_a = 2 \times 10^{-13}$

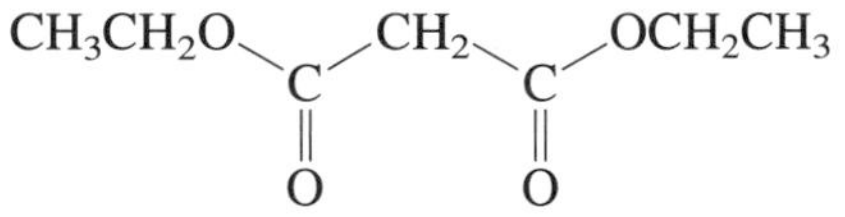

$CH_3CH_2O{-}C({=}O){-}CH_2{-}C({=}O){-}OCH_2CH_3$

$K_a = 5 \times 10^{-14}$

$CH_3{-}C({=}O){-}CH_2{-}C({=}O){-}CH_3$

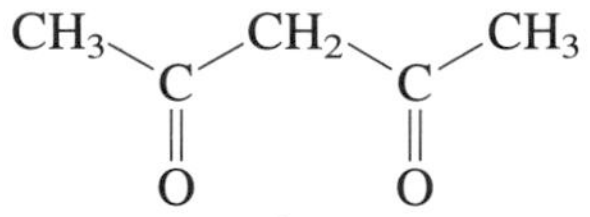

$K_a = 1 \times 10^{-9}$

Just as the grouping

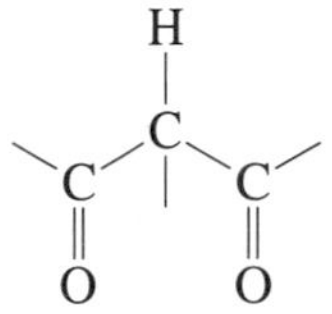

$-C({=}O){-}C(H)<{-}C({=}O){-}$

is acidic, so is the imide grouping,

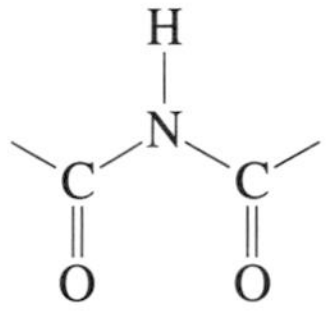

$-C({=}O){-}N(H){-}C({=}O){-}$

Imides are soluble in 5% sodium hydroxide solution but not in 5% sodium bicarbonate solution. A 4-nitrophenyl group makes the —CONH— function weakly acidic in aqueous solution. Thus 4-nitroacetanilide[8] dissolves in 5% sodium hydroxide solution but not in 5% sodium bicarbonate solution. Sulfonamides show the same solubility trends in base as 4-nitroacetanilide. Oximes, which have a hydroxyl group attached to a nitrogen atom, display similar solubility behavior.

Esters with five or six carbon atoms that are almost completely soluble in water may be hydrolyzed by continued shaking with 5% sodium hydroxide solution.[9] The alkali should not be heated and the solubility or insolubility should be recorded after 1–2 min.

Monoesters of dicarboxylic acids are soluble in 5% sodium bicarbonate solution. These esters are rapidly hydrolyzed, even with weak aqueous bases such as 5% sodium bicarbonate solution.

[8]Compounds of this type may also form adducts (Meisenheimer complexes) by bonding hydroxide to the carbon bearing the amide group:

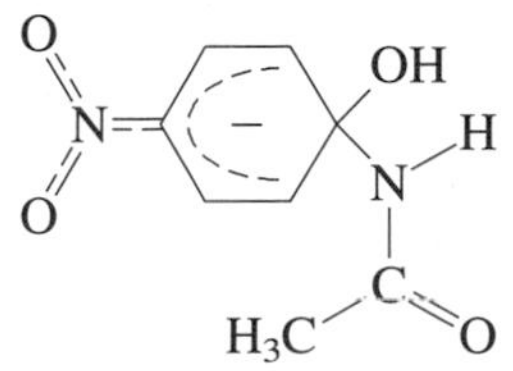

$O_2N{=}C_6H_4^{-}(OH)(N(H){-}C({=}O){-}H_3C)$

[9]Use of lithium hydroxide in place of sodium hydroxide will often yield water soluble salts.

Fatty acids containing 12 or more carbon atoms react with the alkali slowly, forming salts which are commonly referred to as soaps. The mixture is not homogeneous but, instead, consists of an opalescent colloidal dispersion which foams when shaken. Once this behavior has been observed, it is easily recognized.

Certain of the sodium salts of highly substituted phenols are insoluble in 5% sodium hydroxide solution. Certain phenols which are very insoluble in water may precipitate due to hydrolysis and, hence, appear to be insoluble in alkali.

5.1.7 Solubility of Amphoteric Compounds

Compounds containing both an acidic and a basic group are referred to as amphoteric. Low molecular weight amino acids exist, largely, as dipolar salts. They are soluble in water and may yield solutions which produce a neutral litmus in aqueous solutions (class S_2).

$$R{-}CH(NH_2){-}C(=O){-}OH \rightleftharpoons R{-}CH(^{+}NH_3){-}C(=O){-}O^{-}$$

an amino acid — dipolar or zwitterion form

The water insoluble amphoteric compounds act both as bases and as strong or weak acids, depending on the relative basicity of the amino group, since its basicity determines the extent to which the acidic group will be neutralized by salt formation. If the α-amino group contains only aliphatic substituents with no hydrogens directly attached to the nitrogen, the compounds will dissolve in 5% hydrochloric acid and 5% sodium hydroxide solutions, but not in 5% sodium bicarbonate solution (class A_2 or B):

$$R{-}CH(NR_2){-}C(=O){-}OH$$

The presence of an aryl group on the nitrogen atom, however, diminishes its basicity, so that such compounds are soluble even in 5% sodium bicarbonate solution. This is illustrated by the following compounds (class A_1 and B):

$$C_6H_5{-}NH{-}CH_2{-}C(=O){-}OH \qquad H_2N{-}C_6H_4{-}C(=O){-}OH \qquad R{-}CH(N(CH_3)C_6H_5){-}C(=O){-}OH$$

If two aryl groups are attached to the nitrogen atom, the compound is not basic. Its solubility classification is that of a strong acid (class A_1):

$$(C_6H_5)_2NCH_2COOH$$

5.1.8 Solubility in Cold, Concentrated Sulfuric Acid

Cold, concentrated sulfuric acid is used with neutral, water insoluble compounds containing no elements other than carbon, hydrogen, and oxygen. If the compound is unsaturated, is readily sulfonated, or possesses a functional group containing oxygen, it will dissolve in cold, concentrated sulfuric acid. This is frequently accompanied by a reaction such as sulfonation, polymerization, dehydration, or addition of the sulfuric acid to olefinic or acetylenic linkages. In many cases, however, the solute may be recovered by dilution with ice water. The following examples illustrate some of the more common reactions:

$$RCH{=}CHR + H_2SO_4 \longrightarrow [RCH_2{-}\overset{+}{C}HR] \xrightarrow{HSO_4^-} RCH_2\underset{\displaystyle OSO_3H}{\underset{|}{C}}HR$$

$$RCH_2OH + 2H_2SO_4 \longrightarrow RCH_2OSO_3H + H_3O^+ + HSO_4^-$$

The water arising from sulfate ester formation is converted to the hydronium ion by concentrated sulfuric acid.

$$RCH_2\overset{\displaystyle RCH_2}{\overset{|}{C}}HOH + 2H_2SO_4 \longrightarrow RCH_2\overset{\displaystyle RCH_2}{\overset{|}{C}}HOSO_3H + H_3O^+ + HSO_4^-$$

$$RCH_2\underset{\displaystyle RCH_2}{\underset{|}{\overset{\displaystyle RCH_2}{\overset{|}{C}}}}OH + H_2SO_4 \longrightarrow (RCH_2)_2C{=}C(H)R + H_3O^+ + HSO_4^-$$

$$CH_3O{-}C_6H_5 \xrightarrow{2H_2SO_4} CH_3O{-}C_6H_4{-}SO_3H + H_3O^+ + HSO_4^-$$

$$1,3,5\text{-}(CH_3)_3C_6H_3 \xrightarrow{2H_2SO_4} (CH_3)_3C_6H_2{-}SO_3H + H_3O^+ + HSO_4^-$$

$$(C_6H_5)_3C{-}OH \xrightarrow{2H_2SO_4} (C_6H_5)_3C^+ + H_3O^+ + HSO_4^-$$

$$(C_6H_5)_2C{=}O \xrightarrow{H_2SO_4} (C_6H_5)_2\overset{+}{C}{-}OH + HSO_4^-$$

Alkanes, cycloalkanes, and their halogen derivatives are insoluble in sulfuric acid. Simple aromatic hydrocarbons and their halogen derivatives do not undergo sulfonation under these conditions and are also insoluble in sulfuric acid. However, the presence of two or more alkyl groups[10] on the aromatic ring permits the compound to be sulfonated quite easily. Therefore, polyalkylbenzenes such as 1,2,3,5-tetramethylbenzene (isodurene) and 1,3,5-trimethylbenzene (mesitylene) dissolve rather readily in sulfuric acid. Occasionally the solute may react in such a manner as to yield an insoluble product. A few high molecular weight ethers, such as phenyl ether, undergo sulfonation so slowly at room temperature that they may not dissolve.

Many secondary and tertiary alcohols are dehydrated readily by concentrated sulfuric acid to give olefins which subsequently undergo polymerization. The resulting polymers are insoluble in cold, concentrated sulfuric acid and will form a distinct layer on top of the acid. Benzyl alcohol and other similar alcohols react with concentrated sulfuric acid resulting in a colored precipitate.

In summary, a student should not conclude that formation of a black tarry substance means that a compound is insoluble in sulfuric acid. The original compound must have dissolved to induce a reaction and the precipitate arises from formation of a new compound.

5.1.9 Borderlines Between Solubility Classes

In Table 5.5 are listed a number of compounds selected in such a way as to show the position of the most important of the various borderlines between solubility classes with re-

[10]Other activating groups often facilitate sulfonation.

Table 5.5 Borderlines Between Solubility Classes

Compound	Solubility class(es)
Alcohols	
1-Butanol (butyl alcohol)	S_1
2-Methyl-2-propanol (*t*-pentyl alcohol)	S_1
3-Methyl-2-butanol	S_1–N
3-Methyl-1-butanol (isopentyl alcohol)	S_1–N
Benzyl alcohol	N
Cyclopentanol (cyclopentyl alcohol)	N
Aldehydes	
2-Methylpropanal (isobutyraldehyde)	S_1
Butanal (butyraldehyde)	S_1–N
3-Methylbutanal (isovaleraldehyde)	N
Amides	
Methanamide (formamide)	S_1–S_2
Ethanamide (acetamide)	S_1–S_2
Propanamide (propionamide)	S_1–S_2
2-Methylpropanamide (isobutyramide)	S_1–S_2
Butanamide (butyramide)	S_1–MN
Methananilide (formanilide)	S_1–MN
Ethananilide (acetanilide)	MN
Amines	
Diethylamine	S_B
3-Methylbutylamine (isopentylamine)	S_B
Pentylamine	S_B
Benzylamine	S_B
Piperidine	S_B
Cyclohexylamine	S_B
Dipropylamine	S_B–B
Dibutylamine	B
Aniline	B
Tripropylamine	B
Carboxylic acids	
Chloroethanoic acid (chloroacetic acid)	S_A
Butanoic acid (butyric acid)	S_A
2-Chloropropanoic acid (α-chloropropionic acid)	S_A
trans-2-Butenoic acid (crotonic acid)	S_A
3-Methylbutanoic acid (isovaleric acid)	S_A–A_1
Pentanoic acid (valeric acid)	A_1
Esters	
Ethyl ethanoate (ethyl acetate)	S_1–N
Methyl propanoate (methyl propionate)	S_1
Propyl methanoate (propyl formate)	S_1
2-Methylethyl ethanoate (isopropyl acetate)	S_1
Propyl ethanoate (propyl acetate)	S_1–N
Methyl 2-methylpropanoate (methyl isobutyrate)	S_1–N
Butyl methanoate (butyl formate)	S_1–N
Methyl 3-methylbutanoate (methyl isovalerate)	N
1-Methylpropyl ethanoate (*sec*-butyl acetate)	N
Butyl ethanoate (butyl acetate)	N
Benzyl ethanoate (benzyl acetate)	N
Ethyl octanoate (ethyl caprylate)	N
Ethyl benzoate	N

Dimethyl carbonate	S_1–N
Diethyl ethanedioate (ethyl oxalate)	S_1–N
Dimethyl propanedioate (methyl malonate)	S_1–N
Diethyl carbonate	S_1–N
Diethyl butanedioate (ethyl succinate)	N
Diethyl 1,2-benzenedicarboxylate (ethyl phthalate)	N
Diethyl propanedioate (ethyl malonate)	N
Dibutyl carbonate	N
Dibutyl ethanedioate (butyl oxalate)	N
Ethers	
Ethyl methyl ether	S_1
Diethyl ether	S_1–N
Ethyl 1-methylethyl ether (ethyl isopropyl ether)	S_1–N
Di-1-methylethyl ether (isopropyl ether)	N
Dibutyl ether	N
Hydrocarbons (aromatic)	
1,3,5-Trimethylbenzene (mesitylene)	N
1,2,3,5-Tetramethylbenzene (isodurene)	N
(1-Methylethyl) toluene (cymene)	I
1,4-Dimethylbenzene (*p*-xylene)	N–I
Diphenylmethane	N–I
1,3-Dimethylbenzene (*m*-xylene)	N–I
1,2-Dimethylbenzene (*o*-xylene)	N–I
Naphthalene	I
Ketones	
Butanone (ethyl methyl ketone)	S_1
3-Methyl-2-butanone (isopropyl methyl ketone)	S_1
2-Pentanone (methyl propyl ketone)	S_1–N
3,3-Dimethyl-2-butanone (pinacolone)	S_1–N
3-Pentanone (diethyl ketone)	S_1–N
Cyclopentanone	S_1
Cyclohexanone	S_1–N
Acetophenone	N
5-Nonanone (dibutyl ketone)	N
Benzil	N
Benzophenone	N
Nitriles	
Propanenitrile (propionitrile)	S_1
2-Methylpropanenitrile (isobutyronitrile)	S_1–MN
Butanedinitrile (succinonitrile)	S_1–S_2–MN
Pentanedinitrile (glutaronitrile)	S_2–MN
Butanenitrile (butyronitrile)	MN
Nitro compounds	
Nitromethane	S_1A_2
Nitroethane	A_2
Nitrobenzene	MN
Phenols	
Hydroquinone	S_A
Chlorohydroquinone	S_A–A_2
1,3,5-Trihydroxybenzene (phloroglucinol)	S_2–A_2
Phenol	S_A–A_2

spect to the number of carbons present in the compound. These compounds have been grouped, so far as is possible, according to their chemical nature. Within each group an attempt has been made to include the borderline members together with one or more members on either side of their respective borderlines. Thus, the table shows 1-butanol to be in class S_1; it follows that the other butanols and all lower homologs are also in this class. Similarly, since 3-methyl-1-butanol is in class N, it follows that 1-pentanol and all higher alcohols are in this class as well.

Although it is often possible to predict the solubility class of a particular compound, solely by reference to its structural formula, there are many cases where this would result in an incorrect prediction. Sometimes, it is difficult to classify a compound by solubility, since many compounds, as shown in Table 5.5, occupy borderline positions.

5.2 SOLUBILITY IN ORGANIC SOLVENTS

The solubility of organic compounds in organic solvents should be determined in order to plan a variety of laboratory operations. A range of solvents, useful to the organic chemist, is tabulated in Table 5.6. These solvents are useful for running organic reactions, dissolving substrates for spectral analyses (Chapter 6), and for standard laboratory maintenance such as cleaning glassware. Virtually all of the solvents listed, as well as mixtures of these and other solvents, are useful for column chromatography (Chapter 3 pp. 78–81) and thin-layer chromatography (Chapter 10 pp. 463–474), for recrystallizations (Chapter 3 pp. 59–63), and for extractions during the workup of reaction products. The sample can be readily recovered by evaporation from virtually all of these solvents except *N,N*-dimethylformamide (DMF) and dimethyl sulf-oxide (DMSO). Since all of the solvents listed in Table 5.6 are used for a variety of purposes, they are often encountered as impurities in samples of interest.

Procedure. Carry out solubility tests in organic solvents using the simple procedure described for water solubilities on page 97. Use 10 mg or 10 μL of compound per 0.5 mL of solvent for an NMR sample and 50 mg or 15 μL of compound per 1 mL of solvent for an IR sample. For FTIR, the lower concentration levels can be used.

Example

Structure and name	**Expected solubility behavior and class**
$H_2N-C_6H_4-Cl$ 4-chloroaniline	Insoluble in water because it has six carbon atoms and a chlorine atom. It is basic, only one aryl group being attached to the amino group, and hence is insoluble in 5% sodium hydroxide solution, but soluble in 5% hydrochloric acid solution, thus in solubility class B.
$H_3C-C(=O)-(CH_2)_4CH_3$ 2-heptanone	Insoluble in water because it has more than five carbon atoms. Since it is a neutral compound, containing only carbon, hydrogen, and oxygen, it is soluble in cold concentrated sulfuric acid, thus in solubility class N.

Table 5.6 Common Organic Solvents[a]

Name	Structure	Common use (Code)[b]	Dielectric constant (25°C)
Acetone[a]	$CH_3—C(=O)—CH_3$	C, R, NMR[d]	21
Acetonitrile	CH_3CN	R, UV, NMR[d]	36
Benzene[e]	C_6H_6	R, NMR[d]	4.2
Carbon disulfide	CS_2	R, IR, NMR	2.6
Carbon tetrachloride[c]	CCl_4	R, IR, UV, NMR	2.2
Chloroform[c]	$CHCl_3$	All five[d]	4.7
N,N-Dimethylformamide	$H—C(=O)—N(CH_3)_2$	R, NMR[d], R	37
Dimethyl sulfoxide (DMSO)	$CH_3—S(=O)—CH_3$	NMR[d], R	49
Ethanol[c]	CH_3CH_2OH	R, UV, C, NMR[d]	24
(Ethyl) ether[c]	$(CH_3CH_2)_2O$	R, C	4
Hexane	$CH_3(CH_2)_4CH_3$	UV, C, R	2.0
Methanol	CH_3OH	R, UV, NMR[d], C	33
Methylene chloride	CH_2Cl_2	R, C, IR, NMR[d]	9
Pyridine	(pyridine ring, N)	R, NMR[d]	12
Tetrahydrofuran	(five-membered ring, O)	R, NMR[d]	7.3
(Water)	(H_2O)	(All five)[d]	(78.5)

[a]The IR and NMR (proton) spectra of many of these compounds may be found in the Sadtler collection and in R. M. Silverstein, G. C. Bassler, and T. C. Morrill, *Spectrometric Identification of Organic Compounds,* 5th ed. (Wiley, New York, 1991).

[b]C = glassware cleaning; R = reaction medium; solvents to dissolve samples for spectral analysis are denoted IR, NMR, or UV.

[c]Preliminary solubility analysis should employ these solvents.

[d]Deuterated solvents, e.g., acetone-d_6 = CD_3COCD_3, are available for determination of proton magnetic resonance spectra.[e]Toluene can be substituted to reduce toxicity problems.

Questions

1. Tabulate the structure, name, and solubility behavior of the following compounds.

a. 1-chlorobutane
b. 4-methylaniline
c. 1-nitroethane
d. alanine
e. benzophenone
f. benzoic acid
g. hexane
h. 4-methylbenzyl alcohol
i. ethylmethylamine
j. propoxybenzene
k. propanal
l. 1,3-dibromobenzene
m. propanoic acid
n. benzenesulfonamide
o. 1-butanol
p. methyl propanoate

q. 4-phenylcyclohexanone
r. 4-aminobiphenyl
s. 4-methylacetophenone
t. naphthalene
u. phenylalanine
v. benzoin
w. 4-hydroxybenzenesulfonic acid

2. Arrange the following compounds in the approximate order of their basicity toward 5% hydrochloric acid solution.

a. benzanilide
b. pentylamine
c. diphenylamine
d. benzylamine
e. 4-methylaniline
f. butanamide

3. List the solubility class(es) for each compound in Question 2.

4. Arrange each group of compounds in the order of increasing solubility in water.

a. methanol, isopropyl alcohol, ethanol, 1-butanol
b. butane, 1,4-butanediol, 1-butanol
c. 1-butanol, 1,1-dimethylpropanol, 2-butanol
d. ammonium butanoate, 3-pentanone, benzaldehyde, trimethylamine

5. List the solubility class(es) for each compound in Question 4.

6. Arrange the following compounds in the approximate order of decreasing activity.

a. *meso*-tartaric acid
b. 2-naphthol
c. benzohydroxamic acid
d. 4-toluenesulfonic acid
e. 4-tolenesulfonamide
f. 2-bromo-6-nitrophenol
g. octadecamide
h. saccharin
i. benzyl phenyl ketone (deoxybenzoin)
j. 1-naphthoic acid

CHAPTER 6

Spectrometric Methods

Complete and rapid characterization of organic compounds continues to require spectrometric methods. It is now possible to quickly characterize organic compounds by these methods since great advances have recently been made in the sensitivity and speed with which these analyses can be carried out. Thus the amount of compound which is needed for analysis is so small that synthetic chemists can very rapidly[1] prepare sufficient compound for complete structure determination. It is now routine for pharmaceutically important compounds to be synthesized much more rapidly than they can be tested. Mass spectrometry needs only a few milligrams of sample and this is the only destructive spectrometric technique. Thus the same sample of only a few milligrams (or even micrograms) can be used for each of IR, NMR, and UV spectrometry. This has been brought about by the advent of Fourier Transform (FT) methods in the areas of both IR and NMR, as well as by the development of a number of other computer-based procedures. The role of relatively inexpensive but extensive computer memory cannot be overstated.

Despite the more recent developments in spectrometric methods, their fundamental roles remain essentially the same:

Mass Spectrometry (Section 6.1): Source of molecular weight and molecular formula.

Infrared (IR) Spectrometry (Section 6.2): Method for functional group identification.

Nuclear Magnetic Resonance (NMR) Spectrometry (Section 6.3): Method for structure determination by analysis of the relative positions of carbons and hydrogens.

Ultraviolet (UV) Spectrometry (Section 6.4): Method of detection and characterization of conjugated π-systems and conjugated nonbonded electron pairs.

Although it is clear that this list is oversimplified (for example, we can readily obtain complete structures by computer-match-augmented analysis of mass spectra), it is a

[1]Combinational methods facilitate synthesis of a very large number of compounds.

good introduction, pedagogically sound, and finally it represents a reasonable order (that is use mass spectrometry first, then IR, etc.) for characterizing organic compounds.

6.1 MASS SPECTROMETRY

In recent years mass spectrometry (MS) has become available to a very large number of chemistry laboratories. It has been common for these laboratories to acquire mass spectrometry in the form of a GC–MS instrument, which contains a gas chromatograph (GC) interfaced with a mass spectrometer (MS). The GC, commonly containing capillary size columns in modern instruments, is a superb separator of organic compounds, and the mass spectrometer quickly analyzes the components of a mixture of organic compounds as they elute from the GC column. Comparison of the mass spectrum of a sample to the spectrum stored in the computer file often quickly identifies the sample. It is not uncommon to have computer files that contain spectra for 40,000 compounds, or more. Thus compound identification is carried out by comparing the often numerous peaks of an experimentally-observed mass spectrum to those peaks for the compounds in the computer file. Even when there is no match, reasonably "close" structures are proposed by the computer.

Mass spectrometry provides the most accurate method for determining molecular weights of organic compounds, provided that the sample is of sufficient thermal stability at the inlet temperature and is structurally amenable to providing a substantial molecular (or parent) ion peak. If evidence of decomposition is shown in boiling point or gas chromatographic analyses, alternative methods of molecular weight determination (including colligative properties such as freezing point depression or vapor phase osmometry) should be considered. The alternative methods are, however, far inferior to mass spectrometry. The usefulness of the exact (nominal) molecular weight determined by mass spectrometry cannot be overstated.

It is instructive to examine the mass spectrometric characteristics of the very simple compound methane (CH_4). First let us consider electron impact to form the molecular ion:

$$\underset{MW = 16}{CH_4} + e^- \longrightarrow \underset{M,\ m/z\ 16}{CH_4\cdot^+} + 2e^-$$

This occurs in the ionizing region of the instrument shown in Figure 6.1. When the molecule loses an electron to form the molecular ion ($M\cdot^+$, a cation radical), a negligible mass change occurs, but the chemistry of the molecular ion is very different. Thus $CH_4\cdot^+$ can undergo covalent bond cleavage, and a number of fragmentation routes are possible:

$$\underset{m/z\ 16}{CH_4\cdot^+} \xrightarrow{-H\cdot} \underset{15}{CH_3^+} \xrightarrow{-H\cdot} \underset{14}{CH_2\cdot^+} \xrightarrow{-H\cdot} \underset{13}{CH^+} \longrightarrow \text{etc.}$$

$$CH_4\cdot^+ \xrightarrow{-H_2} CH_2\cdot^+$$

Charged species can be detected by the mass spectrometer. The charge allows the ion to be accelerated by a potential difference and this accelated ion is then subjected to a magnetic field (Figure 6.1). The radius of curvature of the charged particle is proportional to m/z (mass/charge) and since the charge (z) is usually $+1$, a direct measure of the mass of the particle is obtained. In Figure 6.2 we see peaks at m/z 16, 15, 14, and 13 corresponding to the four species in the equation above.

Complex reaction networks for fragmentation are possible; for example the $CH_4\cdot^+$ might give rise to the $CH_2\cdot^+$ species directly by simultaneous loss of two H· atoms. In

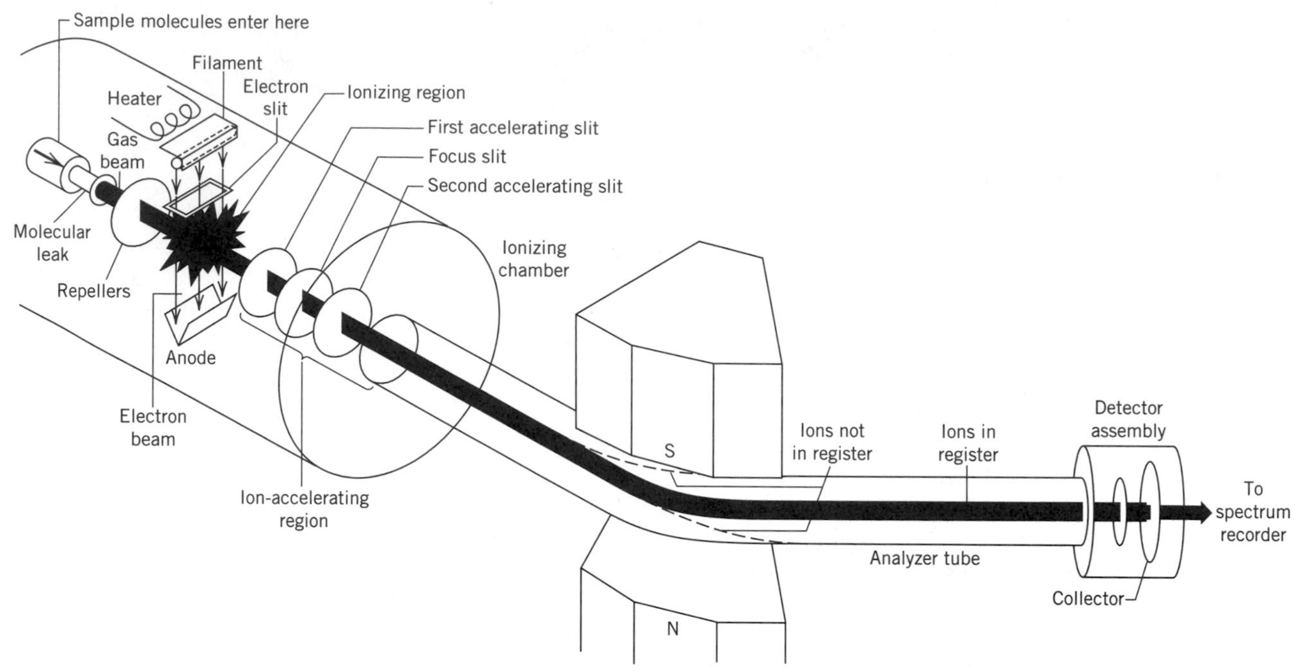

Figure 6.1 Mass spectrometer. Schematic diagram of CEC model 21-103. The magnetic horizontal field that brings ions of varying mass/charge (*m/z*) ratios into register is perpendicular to the horizontal planes of the magnets and to the flight path of the ions. [From John R. Holum, *Organic Chemistry: A Brief Course* (Wiley, New York, 1975). Used with permission.]

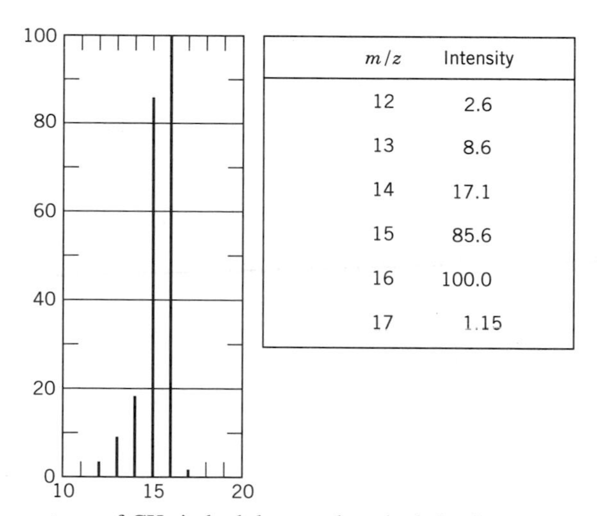

m/z	Intensity
12	2.6
13	8.6
14	17.1
15	85.6
16	100.0
17	1.15

Figure 6.2 Mass spectrum of CH_4 in both bar graph and tabular form.

practice the formidable complexity suggested here is frequently not the case. A number of fundamental cleavage rules have evolved and they can be used to deduce the structure of a compound. These rules are introduced below. Suffice it to say here that the propensity for an organic compound to fragment is based upon the stability of this compound in radical cation form, the strength of the bonds in that form, the rate at which covalent bonds cleave, and the relative stability of the fragments that form compared to the stability of the molecular ion or other fragment precursor.

In unit-resolution spectrometry (mass to nearest whole number), the cluster of peaks of successive mass numbers usually observed in the region of the molecular ion is due to the significant abundance of more than one isotope of one or more of the elements in a compound (Table 6.1). The molecular ion is composed of atoms of the most abundant isotopes (^{12}C, ^{1}H for methane); these correspond to the isotopes of lowest mass. Contributions are made to the smaller peaks (of higher mass) in this cluster by molecules composed of less abundant isotopes (e.g., ^{13}C, ^{2}H).

Table 6.1 Isotope Abundances Based on the Common Isotope Set at 100%

Element	Abundance (%)					
Carbon	^{12}C	100	^{13}C	1.11		
Hydrogen	^{1}H	100	^{2}H	0.016		
Nitrogen	^{14}N	100	^{15}N	0.38		
Oxygen	^{16}O	100	^{17}O	0.04	^{18}O	0.20
Fluorine	^{19}F	100				
Silicon	^{28}Si	100	^{29}Si	5.10	^{30}Si	3.35
Phosphorus	^{31}P	100				
Sulfur	^{32}S	100	^{33}S	0.78	^{34}S	4.40
Chlorine	^{35}Cl	100			^{37}Cl	32.5
Bromine	^{79}Br	100			^{81}Br	98.0
Iodine	^{127}I	100				

In order to appreciate the significance of isotopes of carbon and hydrogen (Table 6.1), we might write $^{12}C_1{}^{1}H_4$, rather than simply CH_4, for the molecular formula of methane. It is clear that methane could also be found in the form $^{13}C_1{}^{1}H_4$, which would contribute to the *m/z* 17 peak in Figure 6.2. Moreover, contributions to the *m/z* 17 peak would arise from substituting a deuterium atom for each of the four H atoms in methane. Table 6.2 summarizes in detail the relative abundances of all species in the methane molecular ion cluster. Peaks from species containing minor isotopes of more than two elements are usually too weak to make significant contributions to the mass spectrum. In principle the intensities of the M + 1 and M + 2 peaks compared to that of the M peak set at 100 can

Table 6.2 Isotope Abundance Aspects of Methane (CH_4)

Possible molecular species	Nominal mass	Nuclide	Abundance relative to the most abundant isotope
$^{12}C^{1}H_4$	16	^{12}C	(100)[a]
$^{13}C^{1}H_4$	17	^{13}C	1.11
$^{14}C^{1}H_4$	18	^{14}C	[b]
$^{12}C^{1}H_3{}^{2}H_1$	17	^{2}H	0.016
$^{12}C^{1}H_2{}^{2}H_2$	18	—	—
$^{12}C^{1}H_1{}^{2}H_3$	19	—	—
$^{12}C^{2}H_4$	20	—	—
$^{12}C^{1}H_3{}^{3}H_1$	18	^{3}H	[b]
$^{13}C^{2}H_4$	Highly improbable species		

[a] Arbitrary standard; set at 100%.

[b] Radioactive isotope; relative abundance varies with time and is negligible for mass spectral considerations.

be used to determine the molecular formula of a compound. This is conceptually possible since the relative intensities of the satellite peaks should be based on both the molecular formula (after all, this formula provides the ratio of the number of atoms of each element) and on the known natural abundances of isotopes of each element. Thus in theory tables of M + 1 and M + 2 peak intensities for all possible formulas of a given molecular weight could be used to determine the molecular formula for a compound. Table 6.3 suggests how this procedure could be used to assign the molecular formulas to two compounds of the same nominal molecular weight (28): carbon monoxide and molecular nitrogen. Thus in theory the differing M + 1 values for CO and N_2 would allow them to each be identified. Although success is sometimes achieved using this method, it often requires time-consuming, repetitive runs to obtain average intensities. Moreover, often ion–molecule reactions cause erroneously high M + 1 results.

$$\mathrm{M}\cdot^{+} + \mathrm{M} \longrightarrow \underset{\text{“M + 1”}}{(\mathrm{M}+\mathrm{H})^{+}} + \underset{\mathrm{M}-1}{(\mathrm{M}-\mathrm{H})\cdot}$$

Table 6.3 Masses and Isotope Abundance Ratios for All Combinations of C, H, N, and O Corresponding to Mass 16

Mass 16	M + 1[a]	M + 2[a]
CO	1.12	0.2
N_2	0.76	—

[a]Based on molecular ion (M, *m/z* 16) = 100% intensity.

When it can be assumed that a molecule contains only C, H, N, O, F, P, and/or I, the intensities of the M + 1 and M + 2 peaks (with the M again set arbitrarily at 100%) can be calculated by the following formulas:

$$\%\,[\mathrm{M}+1] = 100\{[\mathrm{M}+1]/[\mathrm{M}]\} = 1.1\text{ (no. of C atoms)} + 0.38\text{ (no. of N atoms)}$$

$$\%\,[\mathrm{M}+2] = 100\{[\mathrm{M}+2]/[\mathrm{M}]\} = [1.1\text{ (no. of C atoms)}]^2/200 + 0.20\text{ (no. of O atoms)}$$

Thus for nitrogen (N_2): % [M + 1] = 1.1(0) + 0.38(2) = 0.72%, and % [M + 2] = $[1.1(0)]^2/200$ + 0.20(0) = 0%, while the results for CO are % [M + 1] = 1.1(1) + 0.36(0) = 1.1% and % [M + 2] = $[1.1(1)]^2/200$ + 0.20(1) = 0.21%. Comparison to Table 6.3 reveals a good fit.

Far superior to molecular formula determination by unit (or nominal) mass resolution is high-resolution mass spectrometry. High-resolution spectra are obtained when both an electric field and a magnetic field are used to obtain the spectrum. An application of this is displayed in the last column of Table 6.3. Thus CO has an exact mass =27.9949 obtained from the sum of 12.0000 (the exact mass of the nuclide ^{12}C) plus 15.9949 (the exact mass of the nuclide ^{16}O). In like fashion the exact masses of two N_2 atoms (mass 14.0031 each) add up to 28.0062. Thus if an experimental value of 27.9938 is obtained, this is clearly in much better agreement with CO than with N_2. High-resolution results with at least three or four significant figures beyond the decimal point are quite common. It must be emphasized that the exact mass of the most abundant isotope of the atom must be used here, and not the elemental mass from the periodic table that reflects the mass average for all naturally occurring isotopes of that element. Table 6.4 displays masses for the commonly involved atoms; the second column lists masses that are the average of all naturally occurring isotopes, the fourth column lists the exact mass of the indicated isotope.

Table 6.1 provides data which allow use of M + 2 peak intensities to reveal the presence of S, Cl, and Br atoms. Since Table 6.1 reveals the fact that ^{34}S has 4.40% the abundance of ^{32}S (the most abundant isotope), this heavier isotope makes a major contribution to the M + 2 peak. Thus, for example, if a compound contains two sulfur atoms, the M + 2 peak will have an intensity of 2(4.4%) = 8.8% (plus a very small contribution due to C, H, N, and/or O) of the M peak (set at 100%). In like manner the (M)/(M + 2) ratios of monochloride and monobromides are 100/33 and 100/98 from the ratios, respectively, of ^{35}Cl to ^{37}Cl and of ^{79}Br to ^{81}Br (see Table 6.4). Polyhalogen compounds (Br and/or Cl) have more complex patterns as seen in the bar graphs in Figure 6.3. Finally, with some experience it is possible to detect F, I, and/or P in an organic compound since these elements are monoisotopic and thus make no contribution to the M + 1, M + 2, . . . peaks, as shown in the unusually low intensity M + 1 and M + 2 peaks which are obtained. Note that in all of these cases, the most abundant isotope is the lightest and is the isotope that contributes to the molecular ion (M) mass.

Table 6.4 Exact Masses of Isotopes

Element	Atomic weight	Nuclide	Mass
Hydrogen	1.00794	^{1}H	1.00783
		$D(^{2}H)$	2.01410
Carbon	12.01115	^{12}C	12.00000 (std)
		^{13}C	13.00336
Nitrogen	14.0067	^{14}N	14.0031
		^{15}N	15.0001
Oxygen	15.9994	^{16}O	15.9949
		^{17}O	16.9991
		^{18}O	17.9992
Fluorine	18.9984	^{19}F	18.9984
Silicon	28.0855	^{28}Si	27.9769
		^{29}Si	28.9765
		^{30}Si	29.9738
Phosphorus	30.9738	^{31}P	30.9738
Sulfur	32.066	^{32}S	31.9721
		^{33}S	32.9715
		^{34}S	33.9679
Chlorine	35.4527	^{35}Cl	34.9689
		^{37}Cl	36.9659
Bromine	79.9094	^{79}Br	78.9183
		^{81}Br	80.9163
Iodine	126.9045	^{127}I	126.9045

The mass spectrum of 2-hexanone (Figure 6.4) has a weak cluster of peaks in the molecular ion region and thus the M + 1 peak is of little value in determining the molecular formula. The mechanism for the cleavage process is shown below. Since the nonbonded electron pairs of the acylium ion (*m/z* 58) are involved in resonance, this species is reasonably stable and the *m/z* 43 peak is the largest peak (called the base peak) of the spectrum. It is also possible that simple fragmentation to yield a $C_3H_7^+$ cation may make some contribution to the *m/z* 43 peak. The peak at *m/z* 58 is distinctive because its mass is an even number. This requires the cleavage of two sigma bonds and the transfer of a hydrogen atom; the six-membered ring (A in the equation below) for this rearrangement is the crucial point in this process, called a McLafferty rearrangement. Broader generalizations regarding fragmentation are summarized below.

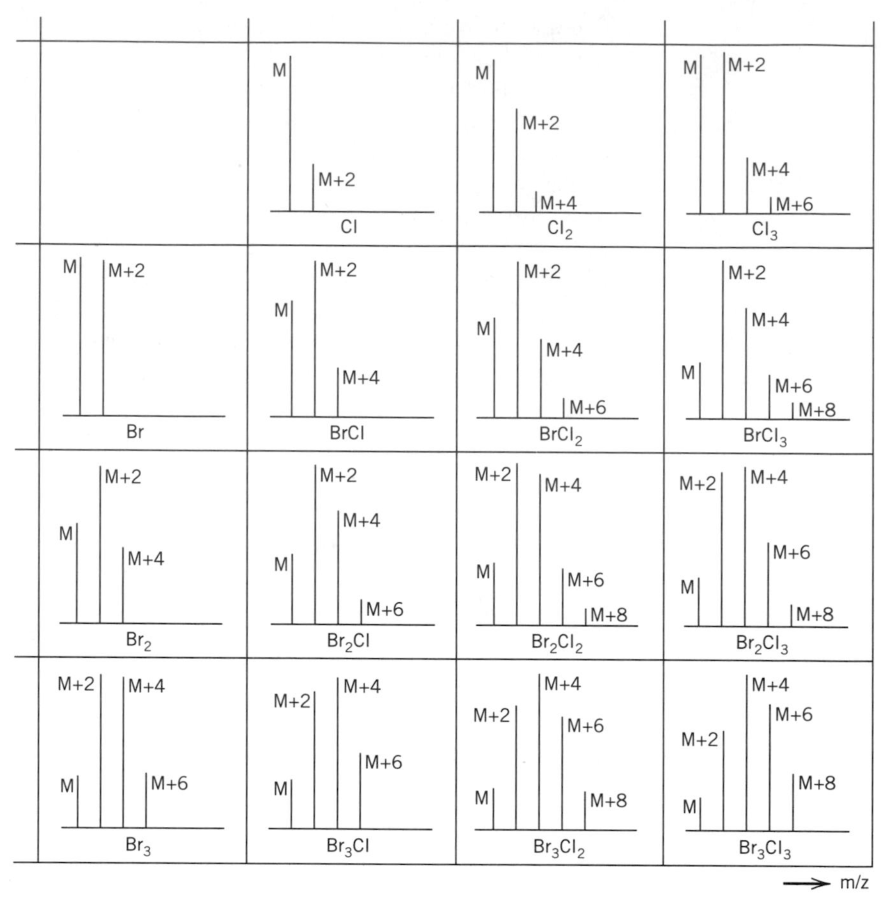

Figure 6.3 Peaks in molecular ion region of bromo and chloro compounds. Contributions due to C, H, N, and O are usually small compared to those for Br and Cl.

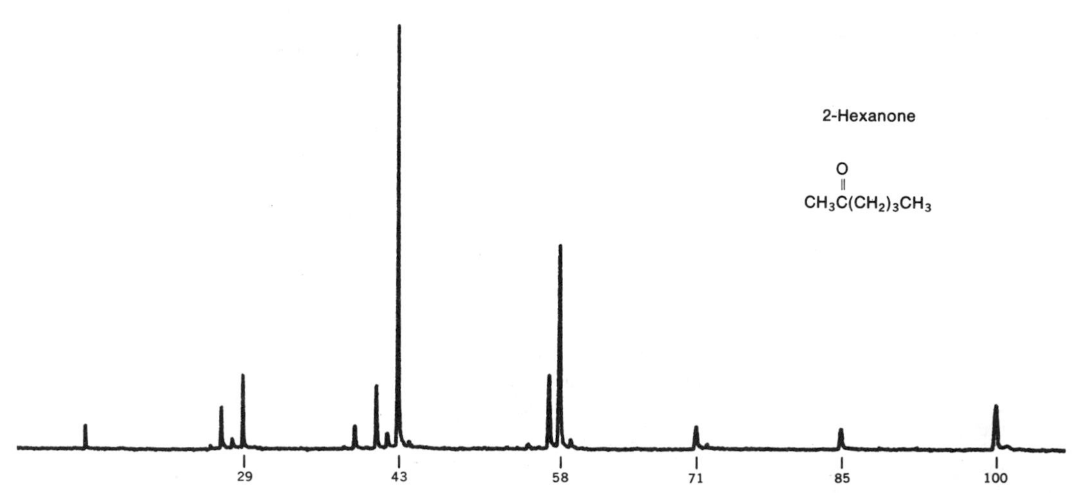

Figure 6.4 Mass spectrum of 2-hexanone. [Courtesy of Varian Instruments Division, Palo Alto, CA.]

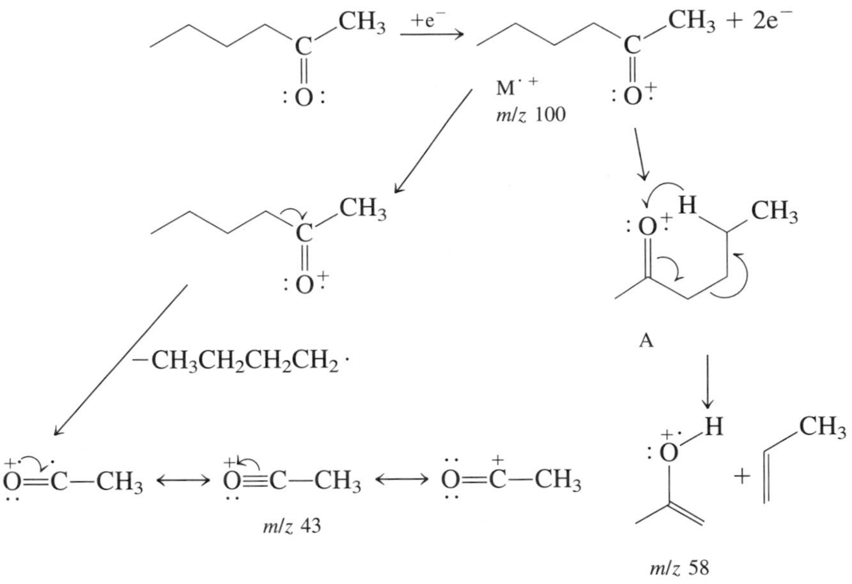

$M^{\cdot +}$ m/z 100

$-CH_3CH_2CH_2CH_2\cdot$

m/z 43

A

m/z 58

Cleavage Rules

1. Cleavage is favored at branch sites as this leads to more substituted carbocations and radicals, and it is well known that the order of stability of both carbon radicals and carbocations is tertiary > secondary > primary. Thus:

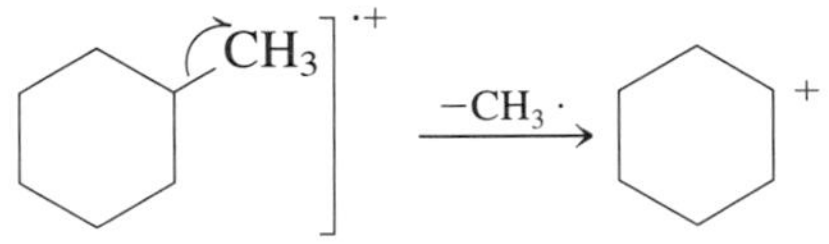

$\xrightarrow{-CH_3\cdot}$

2. Cleavage is favored at allylic and benzylic sites:

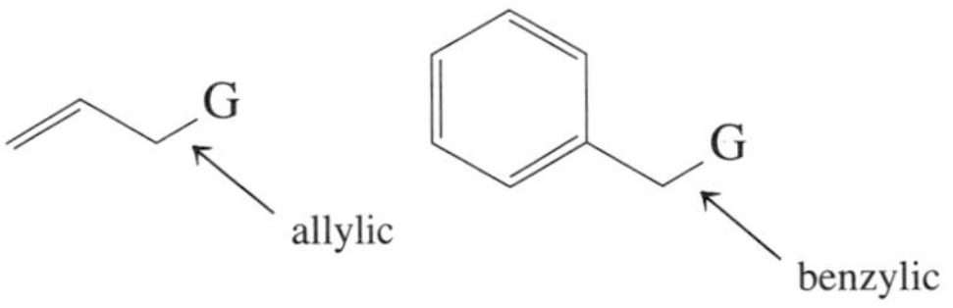

allylic

benzylic

Thus:

$\xrightarrow{-CH_3\cdot} C_7H_7^+$

m/z 91

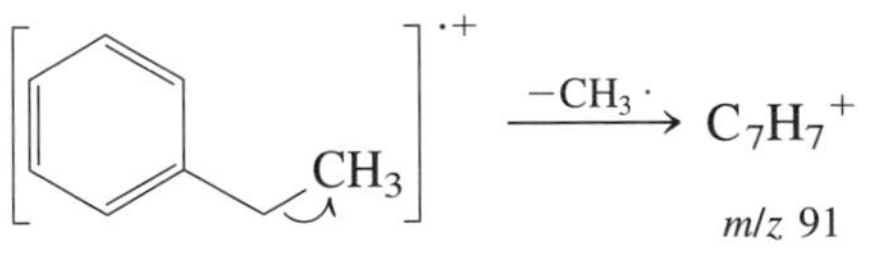

3. Cleavage is facilitated by hydroxyl and carbonyl groups:

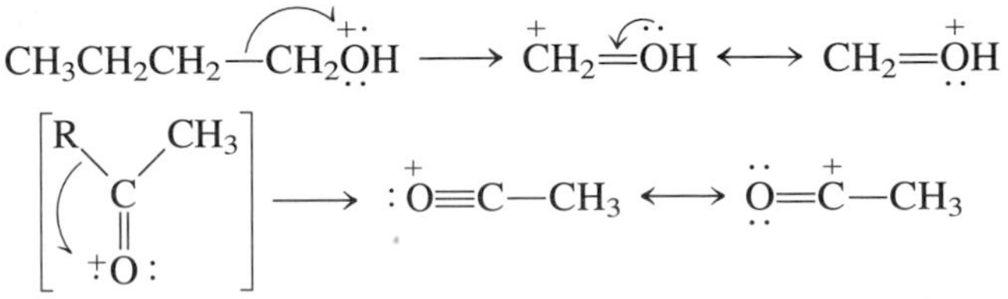

$$CH_3CH_2CH_2{-}CH_2\overset{+\cdot}{O}H \longrightarrow \overset{+}{C}H_2{-}\ddot{O}H \longleftrightarrow CH_2{=}\overset{+}{O}H$$

$$\longrightarrow :\overset{+}{O}{\equiv}C{-}CH_3 \longleftrightarrow O{=}\overset{+}{C}{-}CH_3$$

4. Fragmentation often results in the elimination of small, common, neutral molecules including H_2O, H_2S, NH_3, HCN, CO, and CO_2. In some cases, such as the dehydration of alcohols, the processes are straightforward and are not surprising in view of similar results obtained for the corresponding solution state reactions. In other cases, however, such as the elimination of HCN from nitrogen-containing heterocycles, the mechanism for the process is complex.

6.2 INFRARED (IR) SPECTROMETRY

Infrared (IR) analysis of organic compounds is nearly synonymous with functional group determination, since most simple organic compounds reveal their functional groups in the infrared spectrum. Specifically a beam of electromagnetic radiation of wavelengths somewhat shorter than that of visible light when passed through an organic compound will be absorbed by the organic compound when this beam has an energy corresponding to that of the vibrating organic bond. Thus, not surprisingly, IR analysis has been called vibrational analysis. Below is shown the equation that relates energy (ΔE) and frequency (ν) of the electromagnetic radiation, through Planck's constant, h (6.6242×10^{-27} erg sec):

$$\Delta E = h\nu$$

It is common to use the wavenumber, $\bar{\nu}$, rather than frequency, ν, to describe the magnitude of IR radiation, and this is outlined by the following equation:

$$\Delta E = hc\bar{\nu}$$

Although it is now routine to supply IR band positions in wavenumbers, some years ago wavelength (λ) in μm (micrometers, 10^{-6} meter) or earlier μ, microns (also 10^{-6} m) were used. Thus an equation for wavelength and energy is

$$\Delta E = hc/\lambda$$
$$hc/\lambda = hc\bar{\nu}$$

The equation relating wavelength (in μm) and wavenumbers (cm^{-1}) is

$$10{,}000/\lambda = \bar{\nu} \text{ (in cm}^{-1}\text{)}$$

Typically the standard IR scan runs from 4000 to 600 or 400 cm^{-1} (2.5 to 16.7 or 25 μm). Examples of typical IR spectra are shown in Figure 6.5. Here we see the IR spectra of the same compound (polystyrene, a compound which has been used to calibrate spectrometers) run on three different instruments: an FT IR spectrometer (Figure 6.5*a*), and scanning instruments with grating (*b*) and prism (*c*) optics. It is clear that these spectra have quite different appearances, despite having been run on the same compound. The FT IR and grating spectra are linear in cm^{-1} and the spectrometer with prism optics is linear in wavelength; this is one reason why the spectra have different appearances. We should find that the spectra show all of their maxima at exactly the same positions, but we will find that the appearance (breadth, etc.) of the peaks is very different from one type of spectrometer to another.

(a)

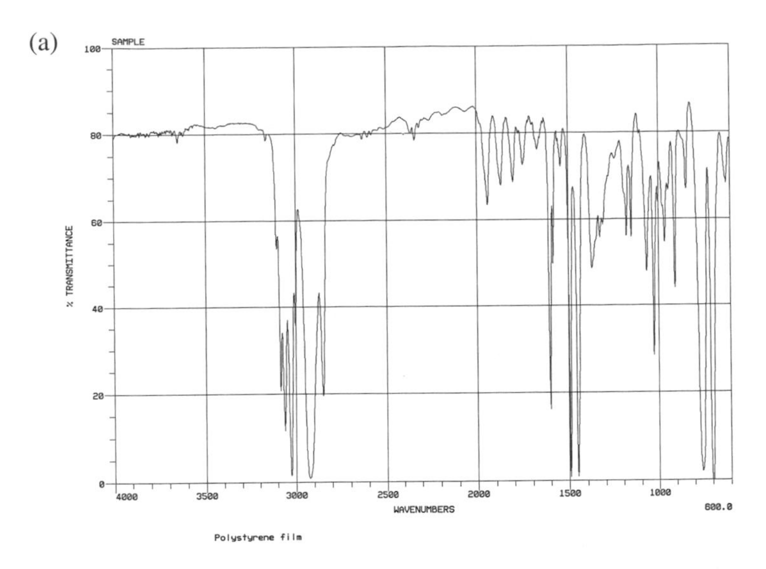

(b)

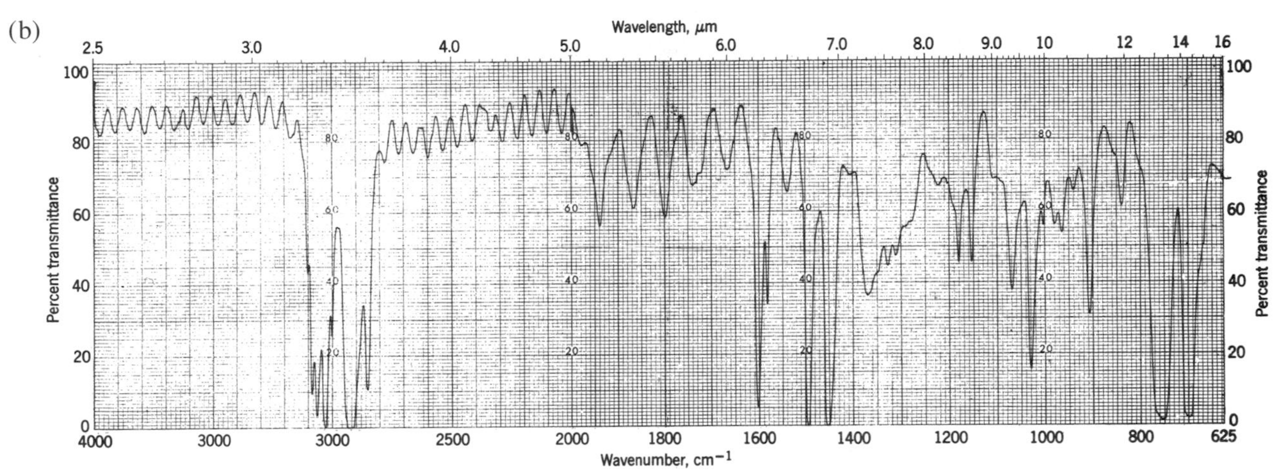

(c)

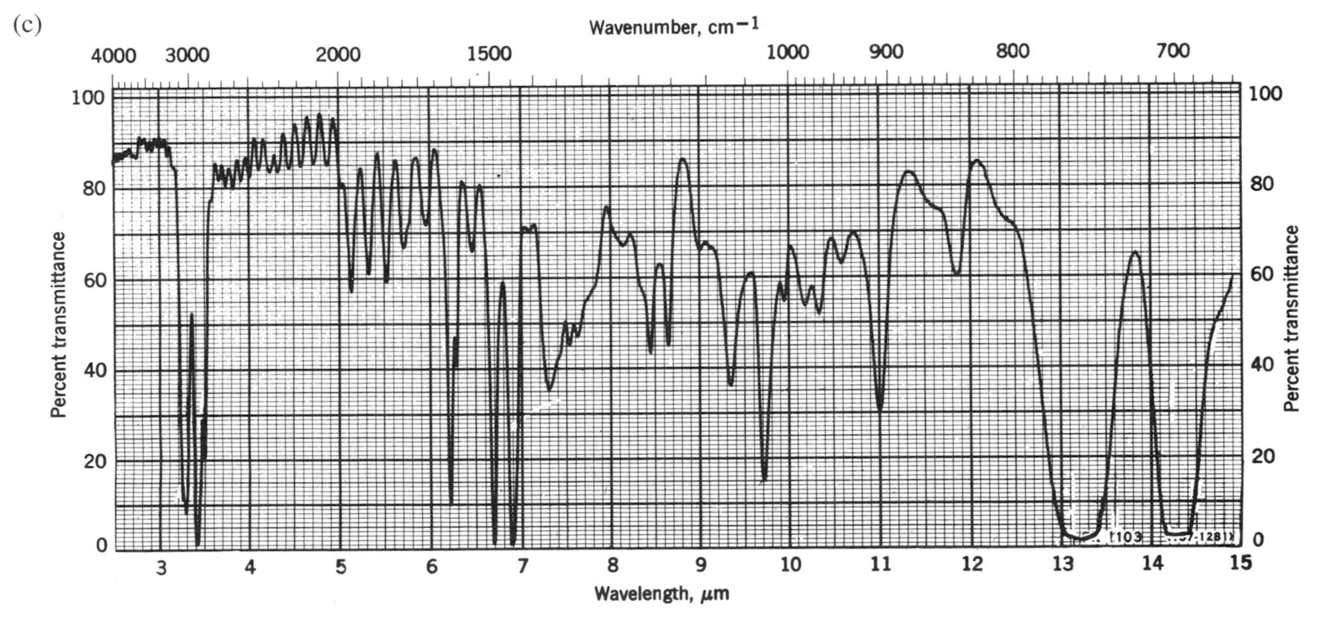

Figure 6.5 Polystyrene: same sample for a, b, and c. (*a*) FT IR spectrum, linear in cm^{-1}; (*b*) grating spectrum, linear in cm^{-1}; (*c*) prism optics, linear in wavelength (μm). The band at 1600.6 cm^{-1} is often used as a calibration mark. Perkin Elmer (Norwalk, CT.) provides such calibration windows.

Scanning IR instruments involve a simple optical device which routes the path of IR radiation through a sample while incrementally varying the wavelength, or wavenumber, of the impinging radiation. This type of spectrometer is displayed in Figure 6.6. We shall now trace the path of IR radiation through this spectrometer in some detail. A source, for example, a heated filament, provides radiation which is directed, via mirrors, through the sample and reference cells. These two beams of radiation are directed, via a series of additional mirrors, to a chopper. These mirrors and the chopper effectively produce a single beam which possesses alternate pulses of each of the sample and the reference beams. Mirrors focus this "chopped" beam into the slit entrance to the monochromator. Prior to reaching the monochromator, this beam is composed of the various energies emitted by the source. The beam's energy is dispersed by gratings (or prisms) and slits within the monochromator area such that at any one instant (grating setting) a *specific* magnitude of radiational energy is being directed toward the exit slits and on the detector. When a beam contains radiation which has been partly absorbed by the sample, this absorption is perceived by the detector as an off-null perturbation. Such perturbations are transmitted by the detector, as an electrical impulse, to the servo motor. The servo motor causes an attenuator to be pushed into the reference beam to cause the sample and reference beams to be rebalanced. The detector thus, by detecting infrared absorptions and electrically creating a reference beam compensation, maintains the combined beam at an optical null. The recorder receives two dimensions of input. One is the monitoring of changes in position of the grating (or prism) which is connected electrically to changes in position on the abscissa (i.e., it records the energy or wave-

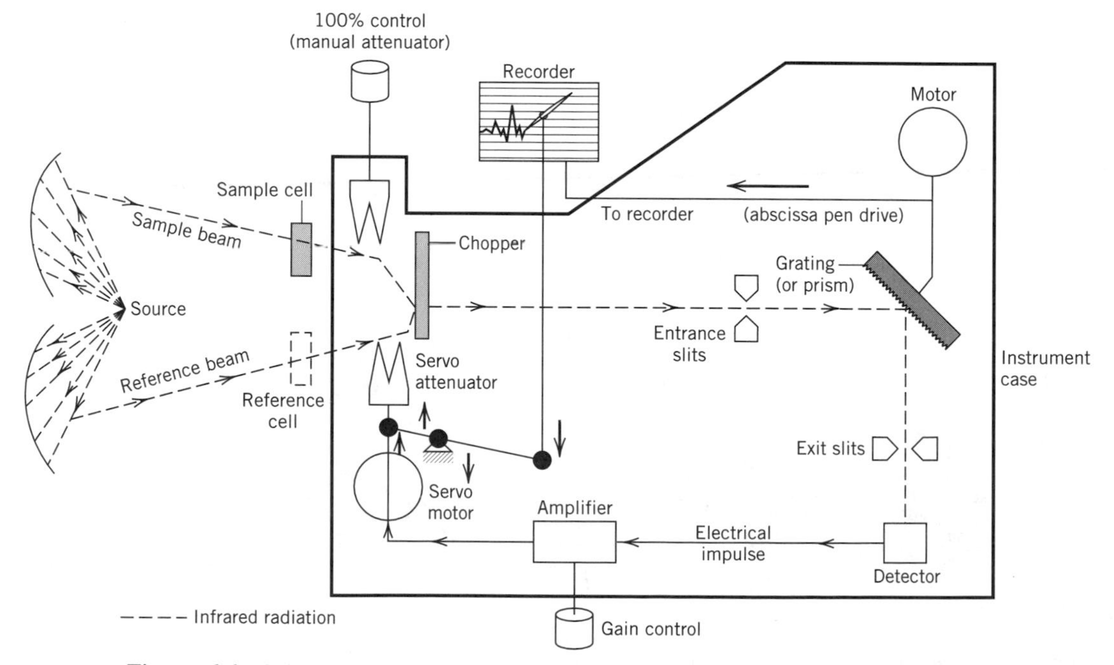

Figure 6.6 Schematic diagram of a double-beam infrared spectrometer.

length under consideration). The other is a recording of the insertion of the servo attenuator; that is, this is a measure of the degree of absorption (measured on the ordinate).

In more recent times the Fourier Transform (FT) IR spectrometer (Figure 6.7) has become the instrument of choice for the organic chemist and this type of spectrometer is rapidly replacing the types described above. FT IR spectrometers offer a number of advantages. Since the data are collected and stored in a computer in digitized form, the data can easily be manipulated, transported, and displayed. Moreover, spectra can be added or subtracted, and this technique is useful for subtracting the solvent spectrum from a spectrum of a sample in that solvent. Moreover FT IR spectrometers are sensitive,[2] fast, and accurate, especially when compared to the older spectrometers.

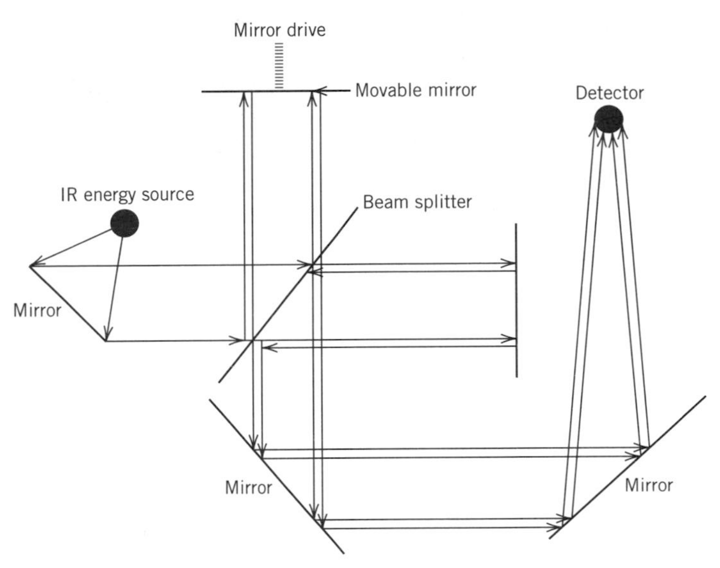

Figure 6.7 Schematic diagram of a Fourier Transform (FT) infrared (IR) spectrometer.

Let us trace the path of radiation through the FT IR spectrometer schematic depicted in Figure 6.7. The source provides the energy of the proper magnitude, which is directed by the mirror on the left toward the beam splitter. This splitter can be a partially coated mirror which can be manipulated to alternately let the radiation pass either toward the fixed mirror or up toward the movable mirror. The movable mirror is carefully incrementally adjusted so that the path length is regularly varied. The difference between path length to the fixed mirror compared to the variable path length mirrors controls the difference in wavelengths of energy (the phase) coming from the two paths. These differences produce an interferogram and when the difference between the two wavelengths is an integer multiple of the invariant length beam, constructive interference is the result. If the difference between the two is an odd integer multiple of one quarter of the invariant beam wavelength, destructive interference is the case. Fourier Transform (FT) mathematics is used to convert this interferogram to a frequency domain spectrum (such as the one shown in Figure 6.5*a*).

We shall be most concerned here with the sampling area of the IR instrument (Figure 6.8). As in any double-beam absorption technique, we are interested in the absorption due to the sample cell relative to a reference cell (solvent or air).

[2]Using dispersive-reflectance FT IR, it is possible to obtain a spectrum on one mg of solid merely sprinkled on a bed of dry KBr.

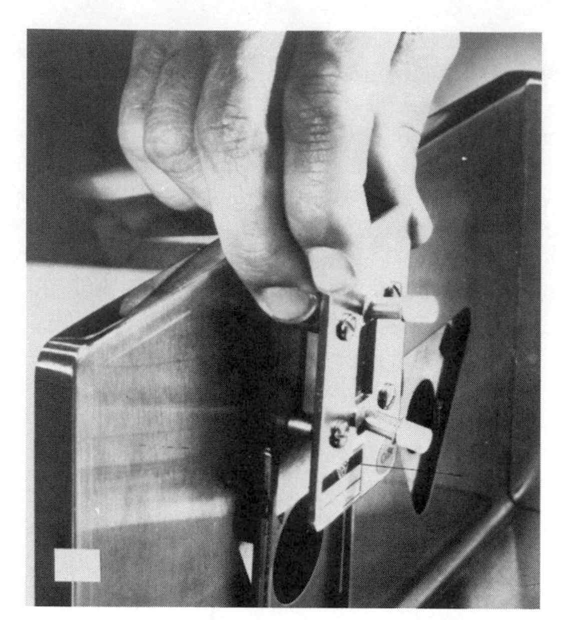

Figure 6.8 (*a*) Sampling area of an IR spectrometer (double-beam); placing the sample cell in its holder. (*b*) Matched cells for liquids are placed in the spectrometer.

Compounds which are to be subjected to IR (or other spectral analysis) should be pure. Solids should be samples from recrystallization (resulting in sharp melting points) or collected from chromatography. Liquids should be samples from distillation (or from gas chromatographic collections).

Samples can be prepared for analysis by any one of the following methods. Liquids can be examined directly as a thin film ("neat") between plates (Figure 6.9)[3]; in principle, however, the spectrum of a compound dissolved in a nonpolar organic solvent, for example, carbon tetrachloride, provides a spectrum that is least distorted due to associations caused by solvent–solute or sample–sample aggregates. Positions of significant absorption of standard IR solvents and mulls are given in Figure 6.11. Mulls do not involve substantial miscibility of the substrate in the solvent; intermolecular associations (e.g., hydrogen bonding) persist in such samples. From the standpoint of lack of absorption by the supporting medium, potassium bromide mulls ("pellets") are the most attractive; preparation of such pellets, however, demands the most careful technique of any method (Figure 6.12).

Procedure. Be sure that the infrared instrument is running to ensure sufficient pre-analysis warm-up. Prepare the sample for a thin-film cell (Figure 6.9), for a solution (ca. 10% for 0.1-mm cells; see Figures 6.10 and 6.11) cell, or in a mull. Mulls are prepared by blending roughly 1% of the sample in the mulling medium by thoroughly mashing the pair with a pestle in a smooth agate mortar. Potassium bromide mulls are analyzed as in Figure 6.12; Nujol, Fluorolube, and hexachlorobutadiene mulls are analyzed by the cells in Figure 6.9. Make sure that the chart paper and initial wavelength setting are correctly zeroed and interrelated on the instrument; exact operation guidelines should be obtained from the instructor owing to variations in instrument characteristics.

[3] Alternatively sampling devices in which a piece of paper acts a wick to a liquid are available (IR cards from 3M in St. Paul, MN.).

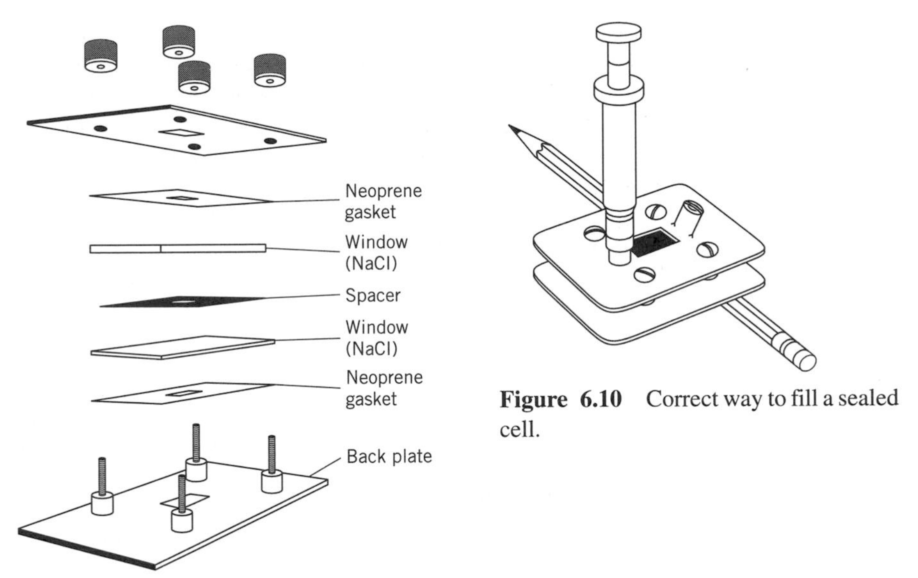

Figure 6.10 Correct way to fill a sealed cell.

Figure 6.9 Demountable cells for IR analysis of liquids. Assembly procedure: (1) Place bottom gasket and lower NaCl plates in nest of bolts on back (lower) plate. Note: NaCl plates should not be scratched or wetted on surfaces; handle them carefully by their edges. (2) Place spacer on lower window and add a few drops of sample; spacer may be omitted for sufficiently nonvolatile or viscous liquids. Extremely volatile liquids should be analyzed in enclosed solution cells (Fig. 6.10). (3) Place upper plate, top gasket, and front plate on the cell; carefully tighten nuts until the sample is evenly dispersed between the plates. *Do not overtighten the nuts, as this may break the salt plates.*

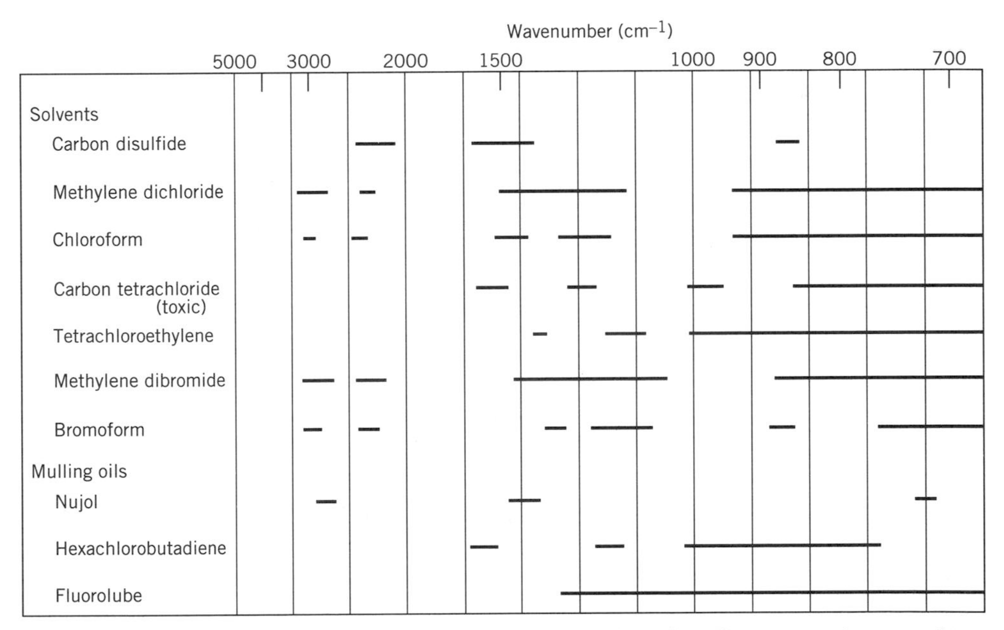

Figure 6.11 Transparent regions of IR solvents and mulling oils. The open regions are those in which the solvent transmits more than 25% of the incident light at 1 mm thickness. The open regions (no horizontal bar) for mulling oils indicate transparency of thin films.

Figure 6.11 displays the absorption characteristics of common IR solvents and mulling compounds. Although KBr mulls might seem very desirable since this inorganic compound does not absorb in the 4000–400 cm^{-1} region, it is often difficult to obtain KBr mulls which are completely dry. Moreover, mulls (KBr, Nujol, fluorolube) are not true solutions and thus aggregates of sample lead to features due to strong intermolecular hydrogen bonding (e.g., in alcohols). Therefore if IR spectra of unassociated compounds are desired, spectra of dilute solutions of compounds in CCl_4 (*very toxic*) are useful. Carbon disulfide (*flammable!*) and tetrachloroethylene represent a solvent pair which can be used to obtain a complete spectrum of a compound. We would use carbon disulfide for the 1400–700 cm^{-1} region, and tetrachloroethylene for the 4000–1400 cm^{-1} region.

Before we examine the appearance of the IR spectra associated with major functional groups, let us review the basic principles that correlate functional groups with the nature of their absorptions. The IR spectrum may be broken down into three regions:

1. The functional group region (4000–1600 cm^{-1}).
2. The fingerprint region (1600–1000 cm^{-1}).
3. The aromatic region (1000–400 cm^{-1}).

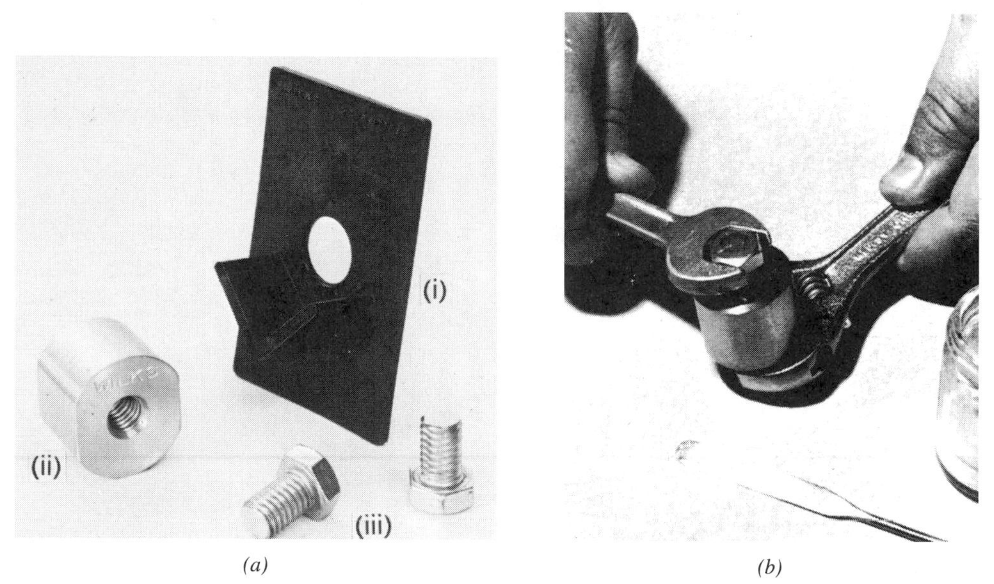

Figure 6.12 Using a Mini-Press for preparation of potassium bromide pellets.

1. Prepare a mull of ca. 1.0 mg of sample[a] in 100 mg of oven-dried, anhydrous KBr by thoroughly grinding in an agate mortar and pestle.
2. Place one of the bolts in the cell (part ii in (*a*)) and insert the mixture into the cavity of the cell (see *c*).
3. Assemble the press apparatus by inserting the other bolt and tightening the bolts as shown in (*b*).
4. The KBr disk so formed can be scanned in the holder (see part i of (*a*)). Note that only the most careful techniques (e.g., use of evacuated die presses, oven-dried equipment, etc.) will minimize the appearance of water absorption (O—H stretch at 3800–3000 cm^{-1} and O—H bend at 1750–1520 cm^{-1}).

[a]These amounts serve only as rough guidelines; trial and error may suggest changes.

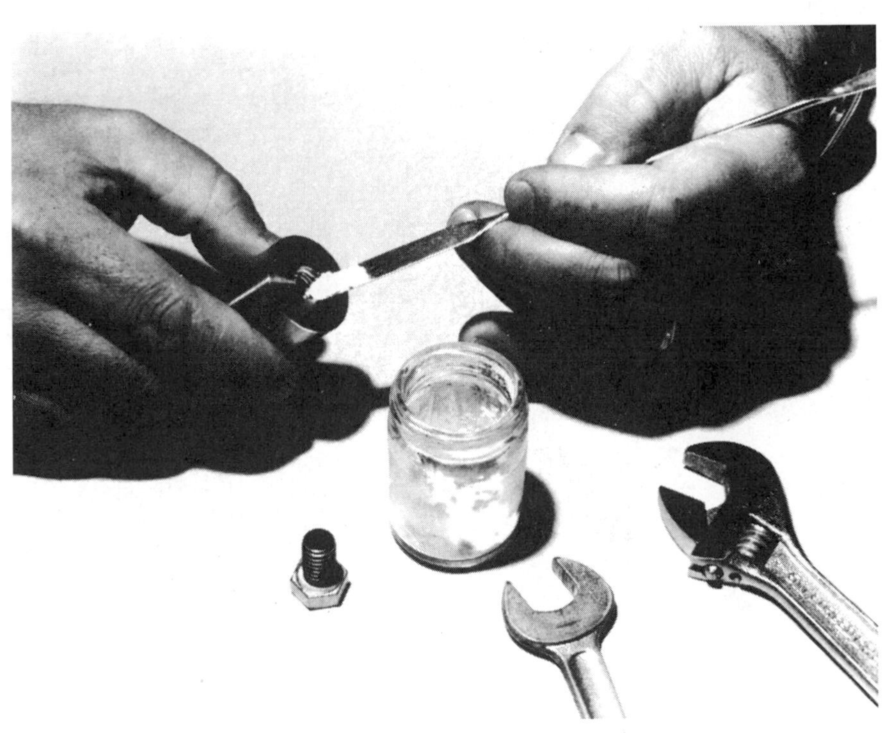

Figure 6.12 *(Continued)*

A few comments are in order for each region.

1. The functional group region (4000–1600 cm^{-1}) is the region in which most organic functional groups absorb. Most of these absorptions are at least of moderate intensity, and many are quite strong (organic chemists rarely if ever report absorbance or transmission of IR spectra quantitatively; bands are merely stated to be s = strong, m = medium, or w = weak). Moreover, the functional group region is relatively free from overlap or other interferences.
2. The fingerprint region (1600–1000 cm^{-1}) is often quite complex. Thus it is often not good for anything more than fingerprinting: that is, a band by band comparison of the spectrum of a known compound to the spectrum of the unknown compound in order to identify the compound. We are unlikely, however, to be able to assign the vibrations leading to these bands. Exceptions are the strong bands associated with the C—O stretching of alcohols, esters, and other oxygen-containing molecules. These are often strong and easily assigned.
3. The aromatic region (1000–400 cm^{-1}) is useful for identifying aromatic (especially benzenoid) compounds. The out-of-plane bending vibrations of both the ring C⎓C and C—H bonds occur in this region. They are usually reasonably strong and in principle can be used to determine the number and relative positions of groups on a benzene ring. This procedure often fails when the benzene ring is substituted by polar substituents.

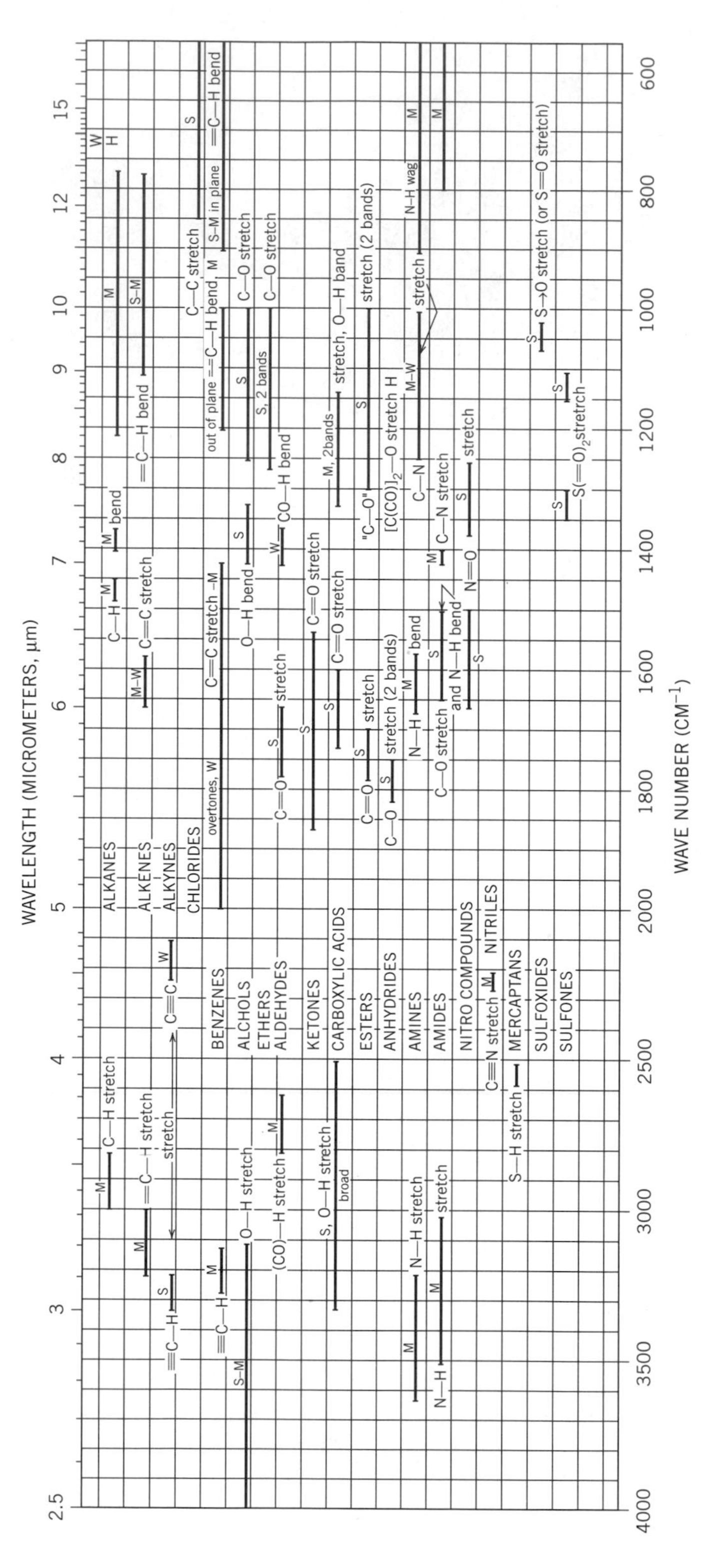

Figure 6.13 Correlation of infrared absorption with organic functional groups. Rows below the first row show only unique bands for the new functional group (e.g., the alkene row shows *only* bands due to the double bond and does not show C—H absorption bands for *saturated* side chains).

Figure 6.13 can be used to make a preliminary identification of a compound's functional groups. The Colthup table (Figure 6.14) provides much more detail.

The position of a band depends upon a number of characteristics of a bond, only one of which will be mentioned here. The higher the bond order, the higher the wavenumber for the stretching vibration for a bond. For example, for carbon–carbon bonds, triple bond stretching occurs at approximately 2200 cm^{-1}, double bonds stretch at about 1600–1650 cm^{-1}, and single bonds have weak stretching at lower wavenumbers (these are usually lost in the fingerprint region). Carbonyl groups stretch at 1700–1750 cm^{-1} and C—O stretching is usually between 1000 and 1100 cm^{-1}.

Another factor that is important to predict is band intensity. It is clear that fundamental vibrations give more intense bands when the vibration causing the band results in a significant dipole moment change for the molecule. This accounts for the weakness of absorptions for the nonpolar carbon–carbon bonds of alkanes. Moreover, we can understand why the highly polar carbonyl group gives rise to a strong absorption (in the 1700–1750 cm^{-1} region).

Since we are concerned with the correlation between structure and infrared spectra, it is useful to be aware of coupled vibrations. A number of these are depicted below. If a pair of vibrations are of the same symmetry species, they can couple, and this phenomenon is made clear by looking at one example: consider the primary amine, octyl amine. The two hydrogens move both away from the nitrogen (symmetrical stretch), or one toward and one away (asymmetrical stretch).

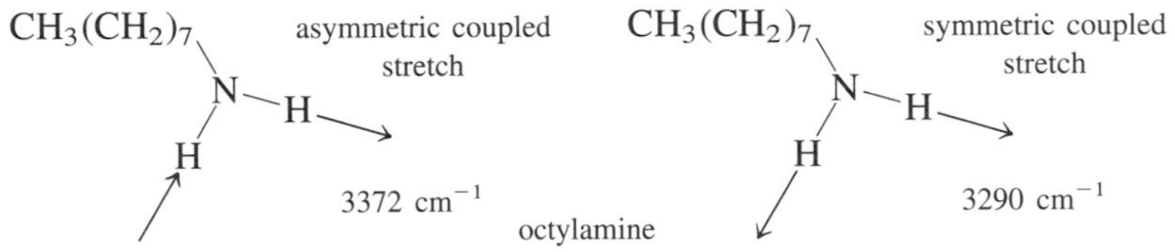

$CH_3(CH_2)_7$—NH—H: asymmetric coupled stretch, 3372 cm^{-1}

$CH_3(CH_2)_7$—NH—H: symmetric coupled stretch, 3290 cm^{-1}

octylamine

Coupled pairs are also possible for anhydrides, primary amides, and nitro compounds:

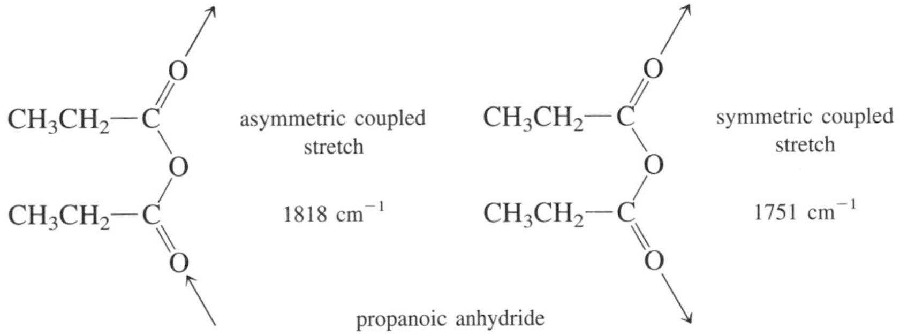

CH_3CH_2—C(=O)—O—C(=O)—CH_2CH_3: asymmetric coupled stretch, 1818 cm^{-1}

CH_3CH_2—C(=O)—O—C(=O)—CH_2CH_3: symmetric coupled stretch, 1751 cm^{-1}

propanoic anhydride

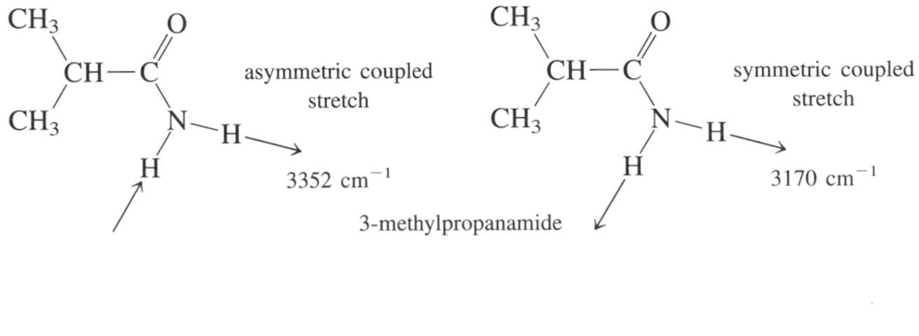

$(CH_3)_2CH$—C(=O)—NH—H: asymmetric coupled stretch, 3352 cm^{-1}

$(CH_3)_2CH$—C(=O)—NH—H: symmetric coupled stretch, 3170 cm^{-1}

3-methylpropanamide

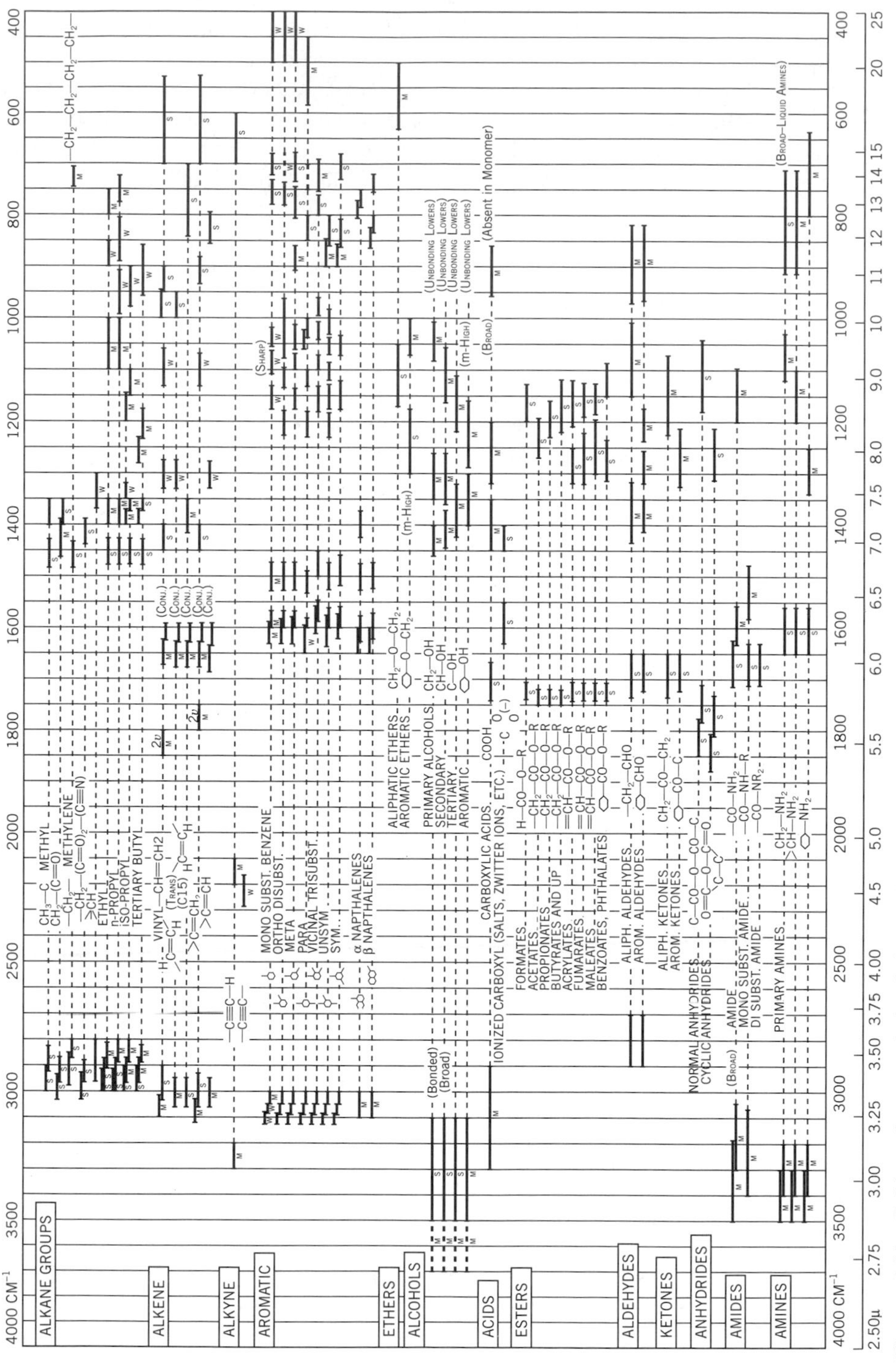

Figure 6.14 Colthup chart correlating infrared absorption with organic functional groups.

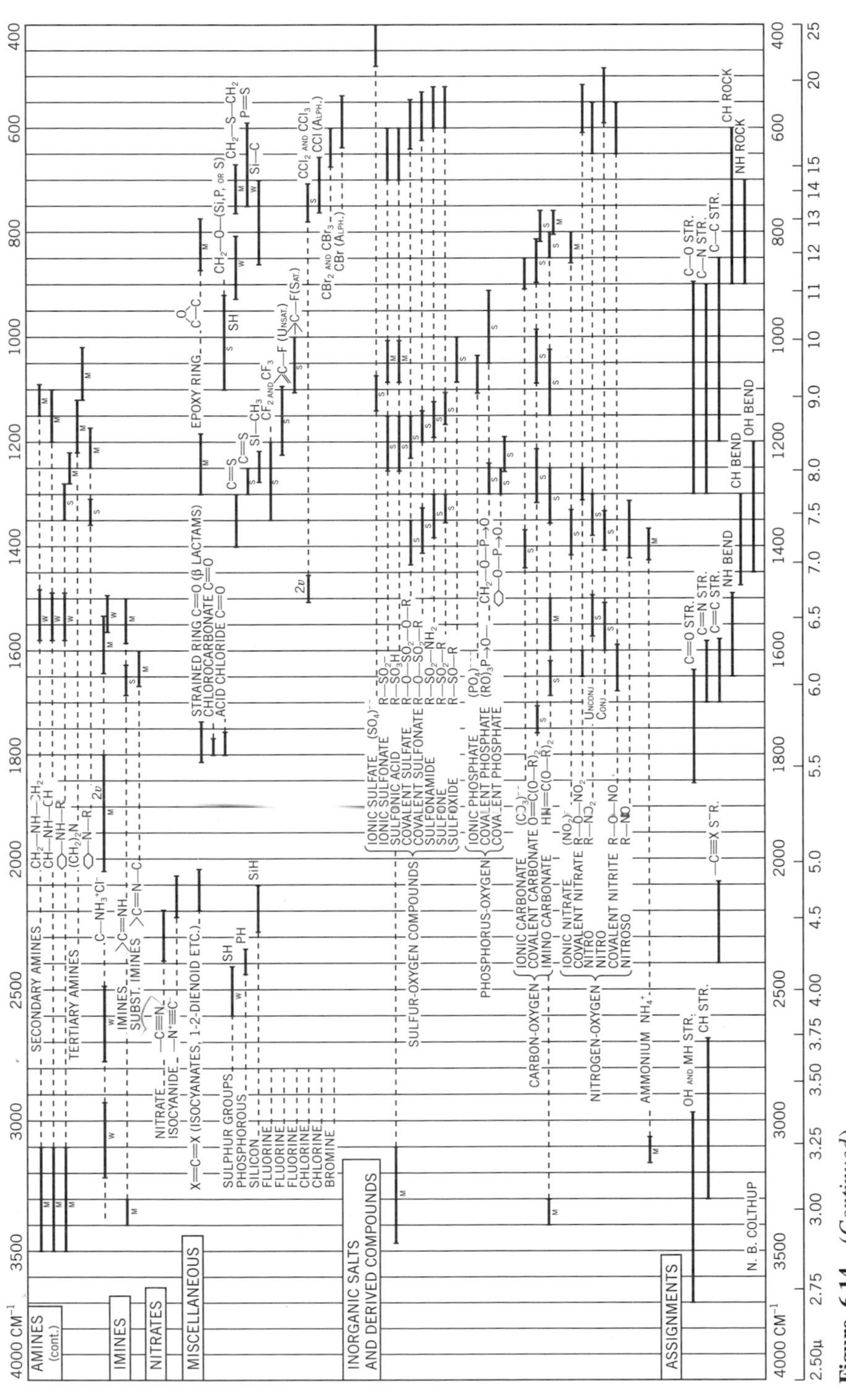

Figure 6.14 *(Continued)*

asymmetric coupled stretch

1532 cm^{-1}

symmetric coupled stretch

1347 cm^{-1}

nitrobenzene

Let us now take a tour through the IR spectra of several common organic compounds. During this tour we will see IR spectra of compounds containing most standard functional groups, and as we go from functional group to functional group, we should pay special attention to the new bands that arise as a result of those groups.

Nujol, a commercially available compound, is a viscous oil composed of long-chained saturated hydrocarbons. Nujol is used to prepare infrared mulls. The IR spectrum of Nujol (Figure 6.15) shows typical C—H stretching (just under 3000 cm^{-1}) and bending modes (near 1400 cm^{-1}) of hydrocarbons, and we should expect to see something like this whenever we are dealing with other compounds containing substantial aliphatic character. Thus the structure of cyclohexene should provide many of the same IR features as does a hydrocarbon and as well have bands due to the carbon-carbon double bond. We find that this is the case in the spectrum of cyclohexene (Figure 6.16). Cyclohexene shows a weak C═C stretch at 1653 cm^{-1}. This weakness is due to the symmetry of cyclohexene: stretching the C═C bond does not result in a substantial change in the molecular dipole. We would expect that a terminal alkene, like 1-hexene, would show a much stronger C═C stretch. Also, a band indicative of the presence of the double bond is found at 3020 cm^{-1} due to ═C—H stretch. Most of the remaining cyclohexene bands are due to the aliphatic absorptions of hydrogens on sp^3-hybridized carbons.

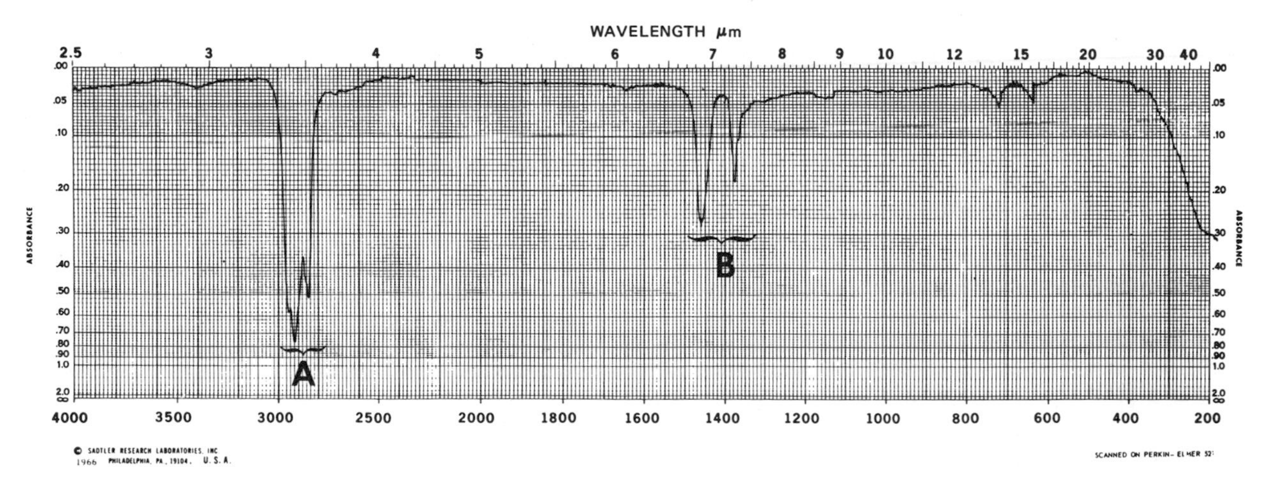

Figure 6.15 Infrared spectrum of Nujol: neat. Nujol is composed of saturated hydrocarbons. C—H stretch: **A,** 2950, 2920, 2850 cm^{-1}, ν_{C-H} (alkane). C—H bend: **B,** 1460, 1375 cm^{-1}, ν_{C-H} (alkane).

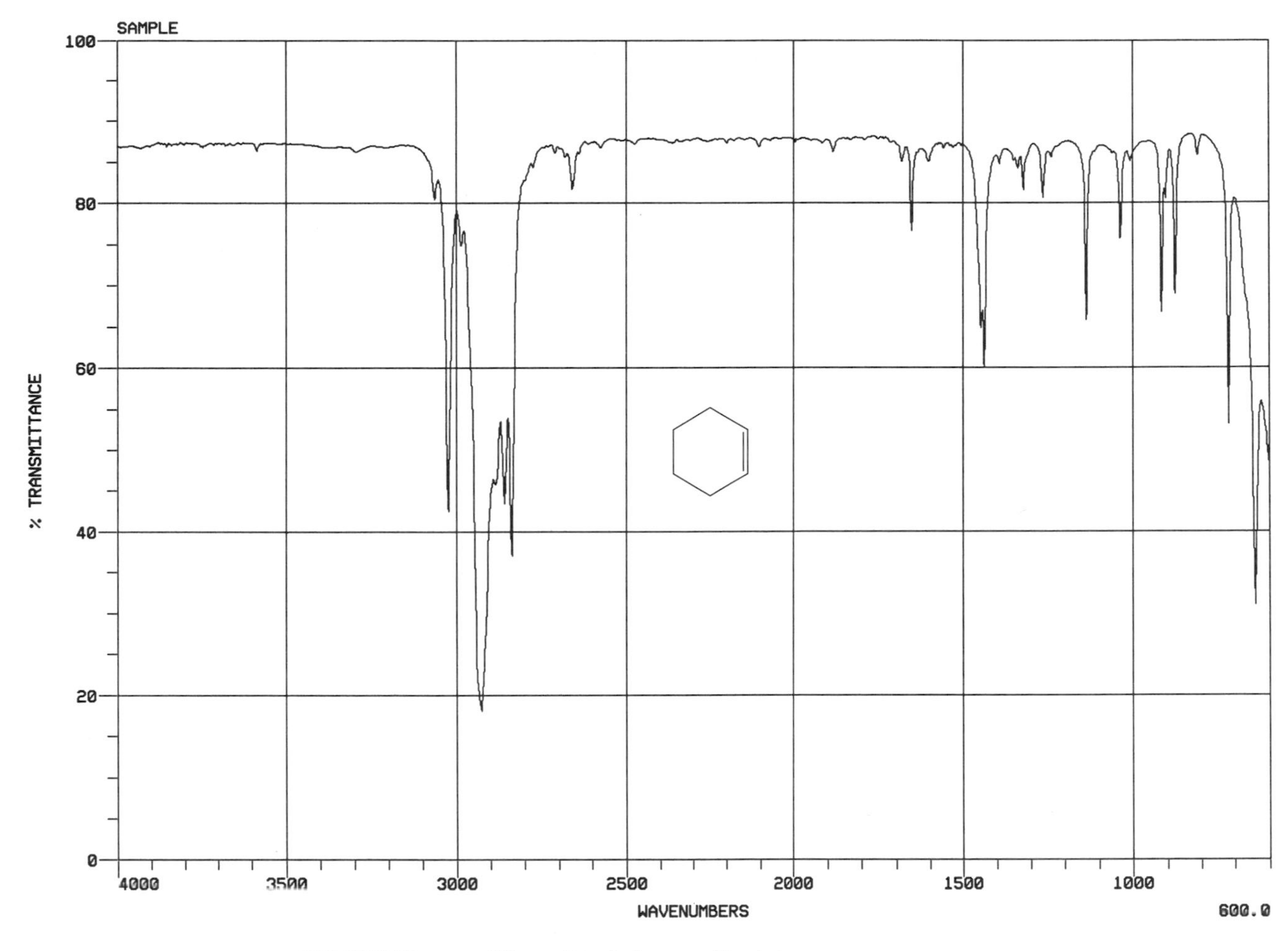

Figure 6.16 Infrared spectrum of cyclohexene: neat. C—H stretch: 3022 cm^{-1}, =C—H stretch; 2926–2837 cm^{-1}, aliphatic C—H stretch. C=C stretch: 1653 cm^{-1}, $\nu_{C=C}$ (unconjugated). C—H bond: 1437 cm^{-1}, olefinic C—H bend (*cis* double bond). Spectrum courtesy of R. Tomasi, Sunbelt R&T, Inc., Tulsa, Oklahoma.

The spectrum of phenylacetylene (phenylethyne, Figure 6.17) should reveal the features of a triple bond and the benzene ring. A weak C≡C stretch does occur at 2115 cm^{-1}. Moreover, a strong ≡C—H stretch occurs at 3290 cm^{-1}. The presence of the benzene ring is revealed by C⩦C stretching vibrations in the 1667–1429 cm^{-1} region, and aromatic bands in the low wavenumber (800–625 cm^{-1}) region. The weak bands between 2000 and 1667 cm^{-1} are due to overtones of aromatic bands, and the fact this compound is a monosubstituted benzene ring is suggested by comparison to Figure 6.18. Such use of the 2000–1667 cm^{-1} region to identify substitution patterns on benzene rings can be difficult since these bands are weak, and strong bands (such as carbonyl bands) in this region obscure the overtone patterns.

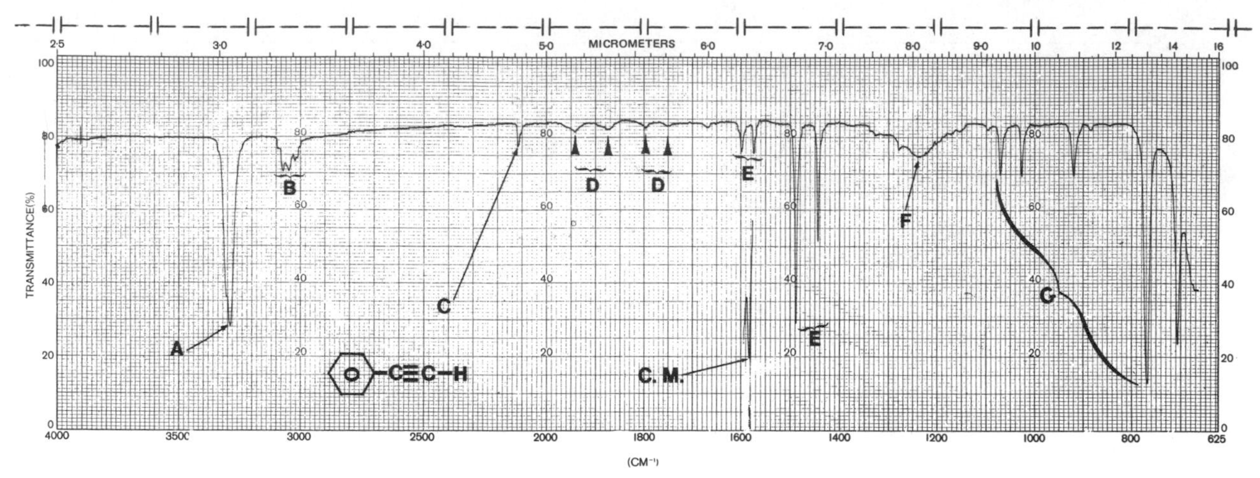

Figure 6.17 Infrared spectrum of phenylacetylene: neat. C—H stretch: **A,** 3290 cm^{-1}, ≡C—H stretch; **B,** 3100–3000 cm^{-1}, aromatic C—H stretch. C≡C stretch: **C,** 2115 cm^{-1}, $\nu_{C\equiv C}$. Aromatic: **D,** 2000–1667 cm^{-1}, aromatic overtones; **E,** 1667–1429 cm^{-1}, C═C stretch. Acetylenic: **F,** 1237 cm^{-1}, overtone of ≡C—H bend. Aromatic: **G,** 800–625 cm^{-1}, ═C—H out-of-plane bend and/or C═C ring breathing. C.M. = calibration band (1583 cm^{-1} band of a polystyrene window). Since the calibration marker, within the experimental error of observation of its position on the chart paper, is in the correct position, no corrections were necessary for tabulations of the band positions.

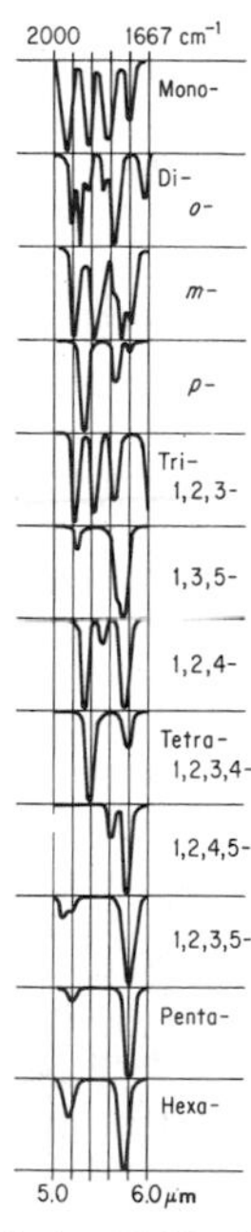

Figure 6.18 Schematic representation of the 2000–1667 cm^{-1} IR region for benzenoid compounds of all substitution types [From J. R. Dyer, *Applications of Absorption Spectroscopy of Organic Compounds,* © 1965, p. 52. Reprinted by permission of Prentice-Hall, Inc., Upper Saddle River, NJ.]

The spectrum of a simple aromatic hydrocarbon, 1,3-dimethylbenzene (*m*-xylene), is shown in Figure 6.19. The C—H stretch region centers at 3000 cm^{-1} and is broken up into two regions: the absorptions above 3000 cm^{-1} are due to C—H stretch on sp^2-hybridized carbons, and those below 3000 cm^{-1} to C—H stretch on the sp^3 carbons. The weak overtone bands of this aromatic compound are, as expected in the 2000–1667 cm^{-1} region (compare Figure 6.18), and coupled C═C stretch occurs as a set of four bands between 1650 and 1450 cm^{-1}. There are also aromatic bands at 789 and 690 cm^{-1}.

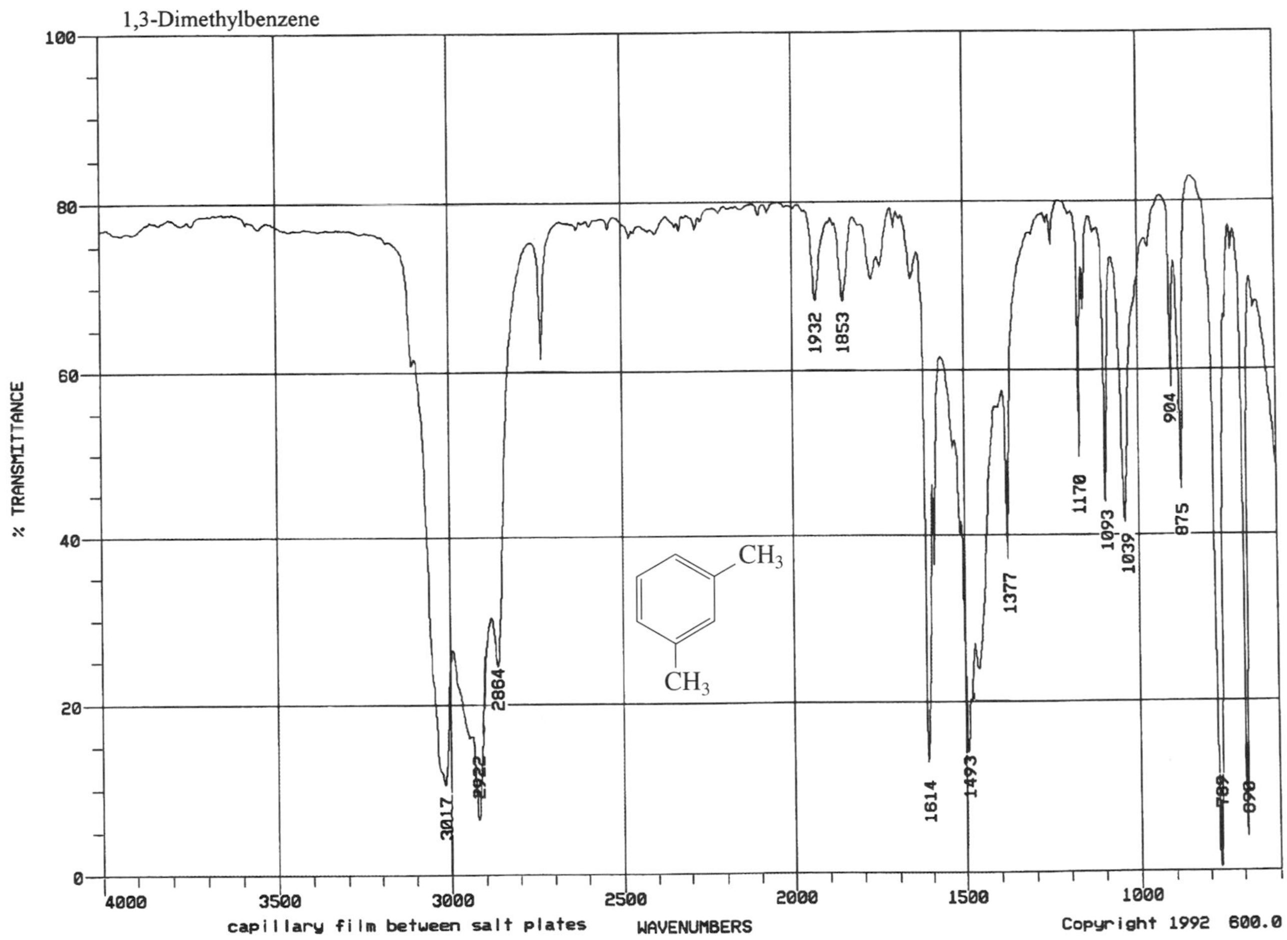

Figure 6.19 Infrared spectrum of *m*-xylene; neat. C—H stretch: 3017 cm^{-1} aromatic $\nu_{C—H}$; 2922 cm^{-1}. ν_{asym} CH_3 (aliphatic); 2864 cm^{-1} ν_{sym} CH_3 (aliphatic). Overtone bands: 2000–1667 cm^{-1} (m), indicative of *meta* substitution (see Figure 6.18). C═C stretch: 1614, 1493, cm^{-1}, $\nu_{C—C}$. C—H bend: 1170, 1093, 1039 cm^{-1}, δ_{CH_3}; 789, 690 cm^{-1} out-of-plane ═C—H bend. Spectrum courtesy of R. Tomasi, Sunbelt R&T, Inc., Tulsa, Oklahoma.

The spectrum of a simple alcohol, methanol (methyl alcohol), is shown in Figure 6.20. The O—H group is readily revealed: a broad O—H stretch is centered at 3352 cm^{-1}. A sharp, strong C—O stretch is seen at 1030 cm^{-1}.[4]

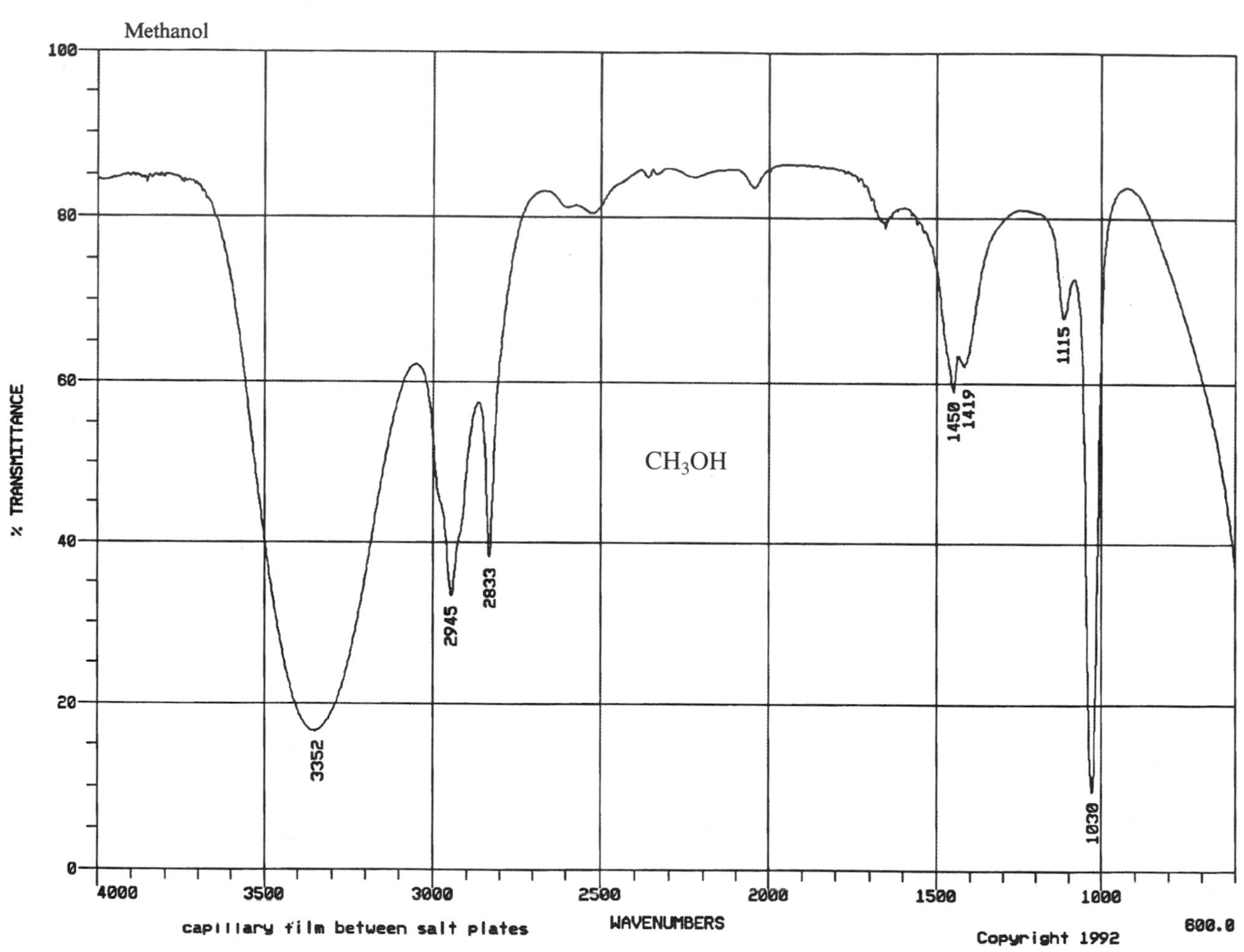

Figure 6.20 Infrared spectrum of methanol: thin film. O—H stretch: 3352 cm^{-1}, ν_{O-H}, associated, C—H stretch: 2945 cm^{-1} $\nu_{asym}CH_3$; 2833 cm^{-1}, $\nu_{sym}CH_3$. Bending vibrations: 1450 cm^{-1}, $\delta_{asym}CH_3$ of CH_3O; 1419 cm^{-1}, in-plane O—H bend. C—O stretch: 1030 cm^{-1}, ν_{C-O}. Spectrum courtesy of R. Tomasi, Sunbelt R&T, Inc., Tulsa, Oklahoma.

[4] The C—O bonds to sp^2 carbons show their stretching vibrations in the 1100–1250 cm^{-1} region; the C—O bonds to the sp^3 alkyl carbons show their stretching in the 1000–1150 cm^{-1} region.

The IR spectrum of 2-methylphenol (*o*-cresol) is shown in Figure 6.21 and it reveals a broad, strong O—H stretch centered at 3335 cm^{-1} and a strong C—O stretch at 1235 cm^{-1}.[4]

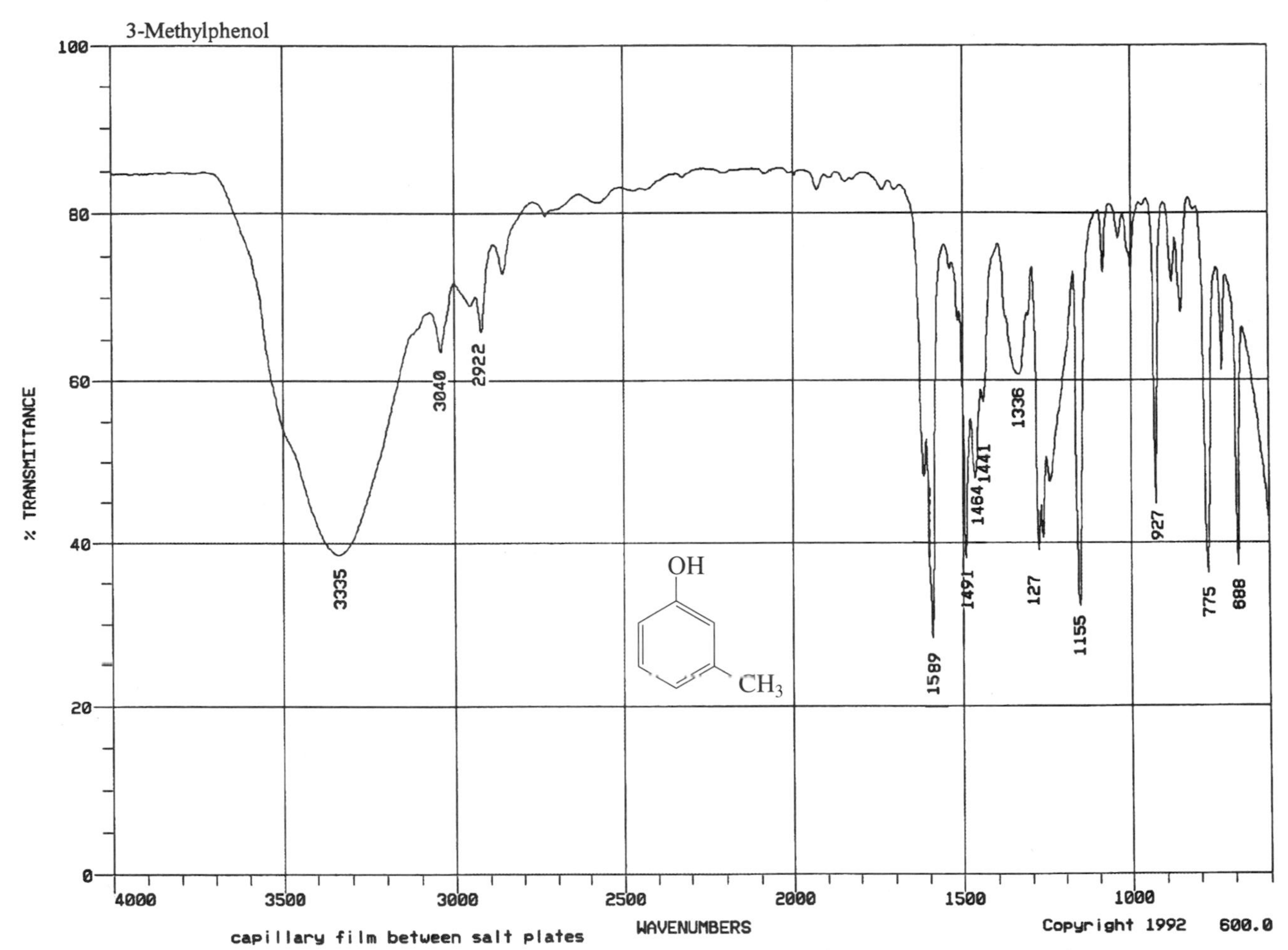

Figure 6.21 Infrared spectrum of *o*-cresol: neat. O—H stretch: 3335 cm^{-1}, ν_{O-H} (associated). C—H stretch: 3040 cm^{-1}, ν_{asym} CH_3; 2922 cm^{-1}, ν_{sym} CH_3. C$\cdots$C stretch: 1589, 1491, 1464 cm^{-1}, $\nu_{C\cdots C}$. O—H bend: 1336 cm^{-1}, δ_{O-H} (out-of-plane). C—O stretch: 1235 cm^{-1}, ν_{C-O} (aryl-oxygen). C$\cdots$C—H bend (in-plane): 1155 cm^{-1}. $\cdots$C—H bend (out-of-plane): 775, 688 cm^{-1}. Spectrum courtesy of R. Tomasi, Sunbelt R&T, Inc., Tulsa, Oklahoma.

We would expect and we find that the IR spectrum of ethyl ether (Figure 6.22) would show no O—H band (no band centered near 3300 cm^{-1}). A strong C—O stretch is expected for the asymmetric C—O—C vibration and this is seen at 1124 cm^{-1}.[1]

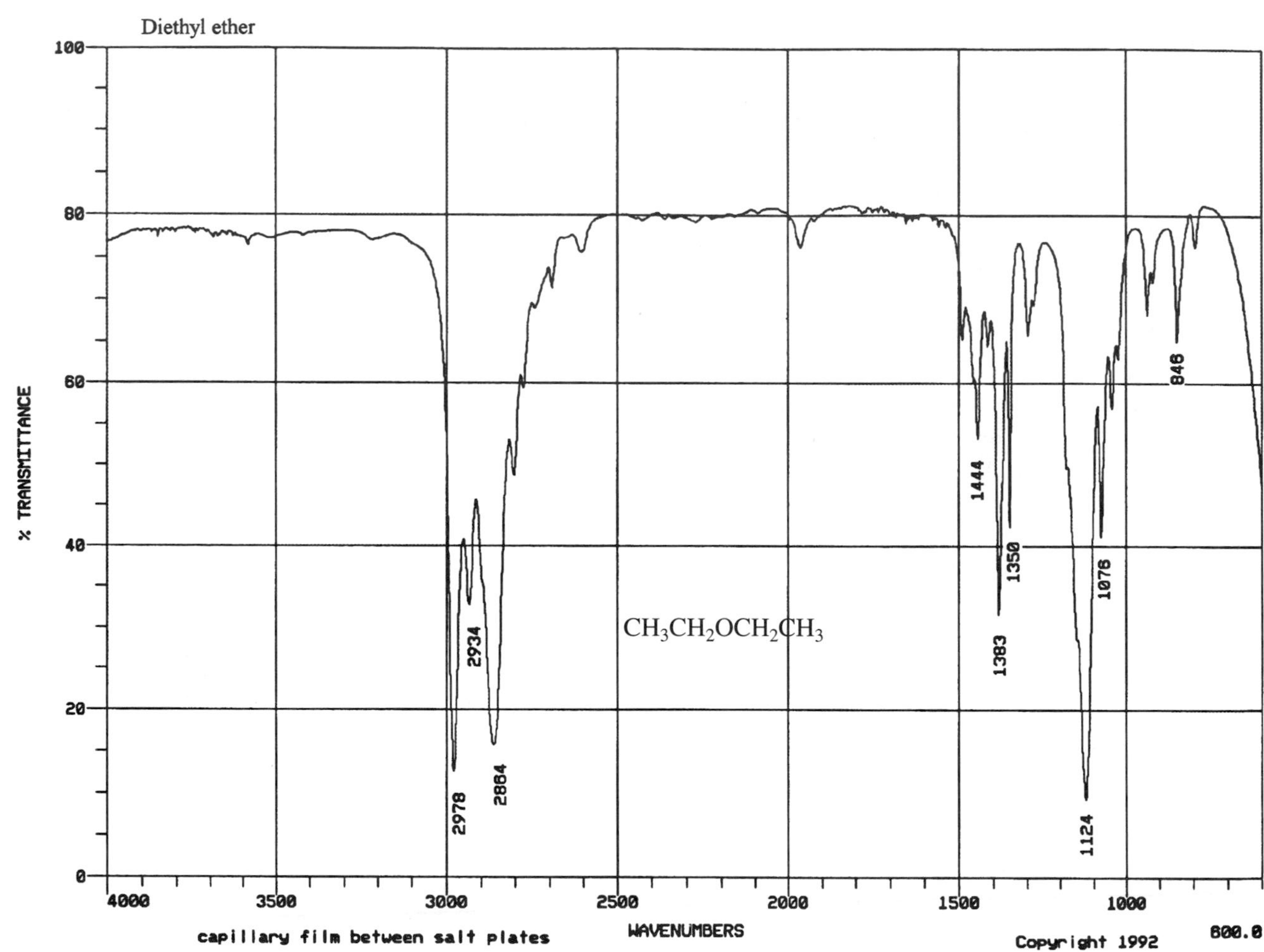

Figure 6.22 Infrared spectrum of ethyl ether: neat. C—H stretch; 2978 cm^{-1}, ν_{asym} CH_3; 2934 cm^{-1}, ν_{asym} CH_2; 2864 cm^{-1}, ν_{sym} CH_3. C—H bend: 1383 cm^{-1}, δ_{sym} CH_3 (bend). "C—O" stretch: 1124 cm^{-1}, ν_{asym} C—O—C (stretch). Spectrum courtesy of R. Tomasi, Sunbelt R&T, Inc., Tulsa, Oklahoma.

With the spectrum of acetone (2-propanone, Figure 6.23) we begin the analysis of carbonyl compounds. The prominent C=O stretch at 1714 cm^{-1} is typical of simple ketones. Ring strain in cyclic ketones moves this band to a higher wavenumber; conjugation with a C=C or aromatic ring unit moves the band to lower wavenumbers.

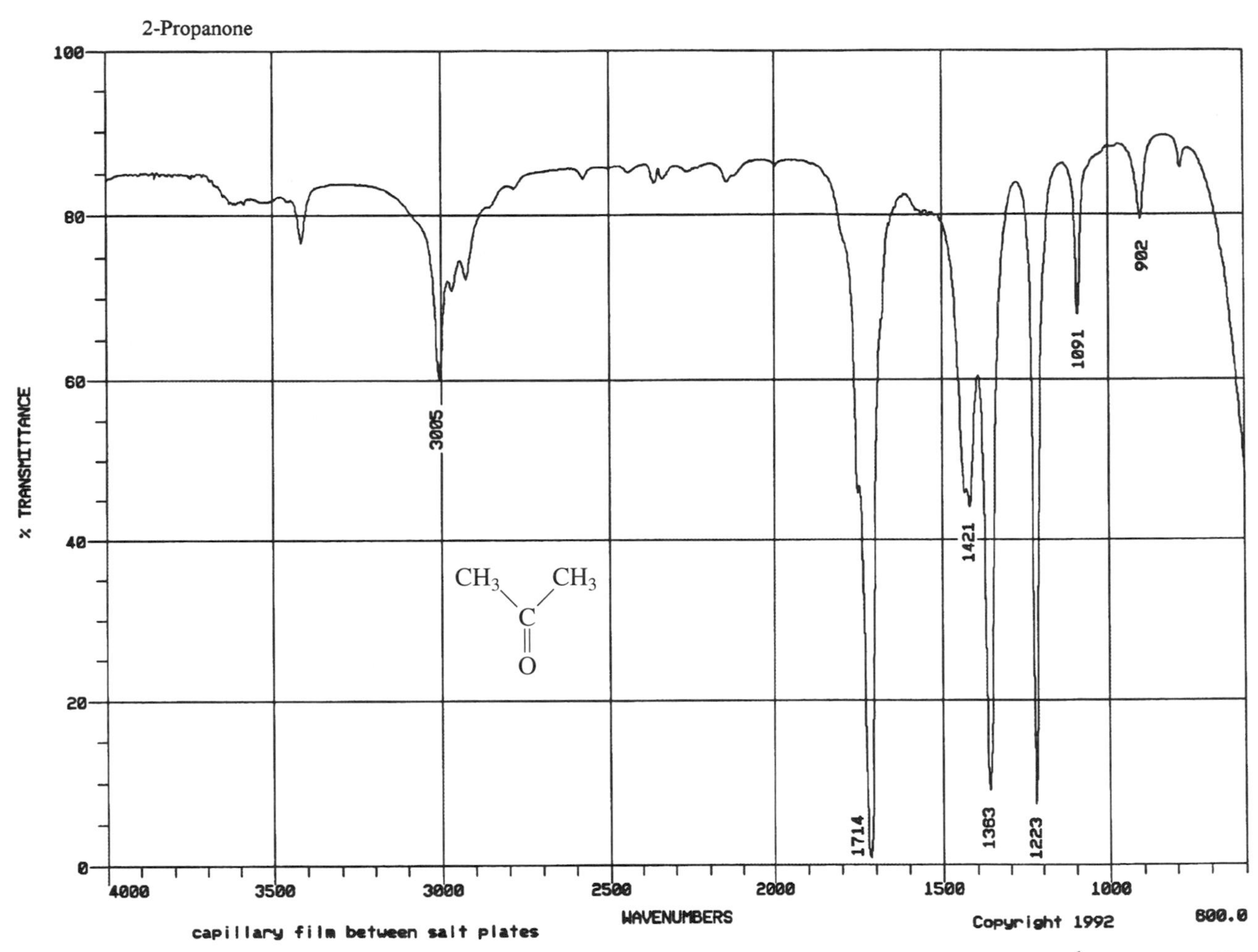

Figure 6.23 Infrared spectrum of acetone: neat. C—H stretch: 3005, 2960 cm^{-1}, ν_{asym}, ν_{sym} CH_3 overtone of δ_{sym} (in order). C=O stretch: 1714 cm^{-1}, $\nu_{C—O}$. C—H bend: 1421 cm^{-1}, δ_{asym} CH_3 of CH_3CO; 1363 cm^{-1}, δ_{sym} CH_3 of CH_3CO; 1223 cm^{-1}, C—C(=O)—C stretch and bend. Spectrum courtesy of R. Tomasi, Sunbelt R&T, Inc., Tulsa, Oklahoma.

Benzophenone (diphenyl ketone) yields the IR spectrum of Figure 6.24. The strong C=O stretch at 1660 cm^{-1}, as predicted above, is at a lower wavenumber than for simple aliphatic ketones.

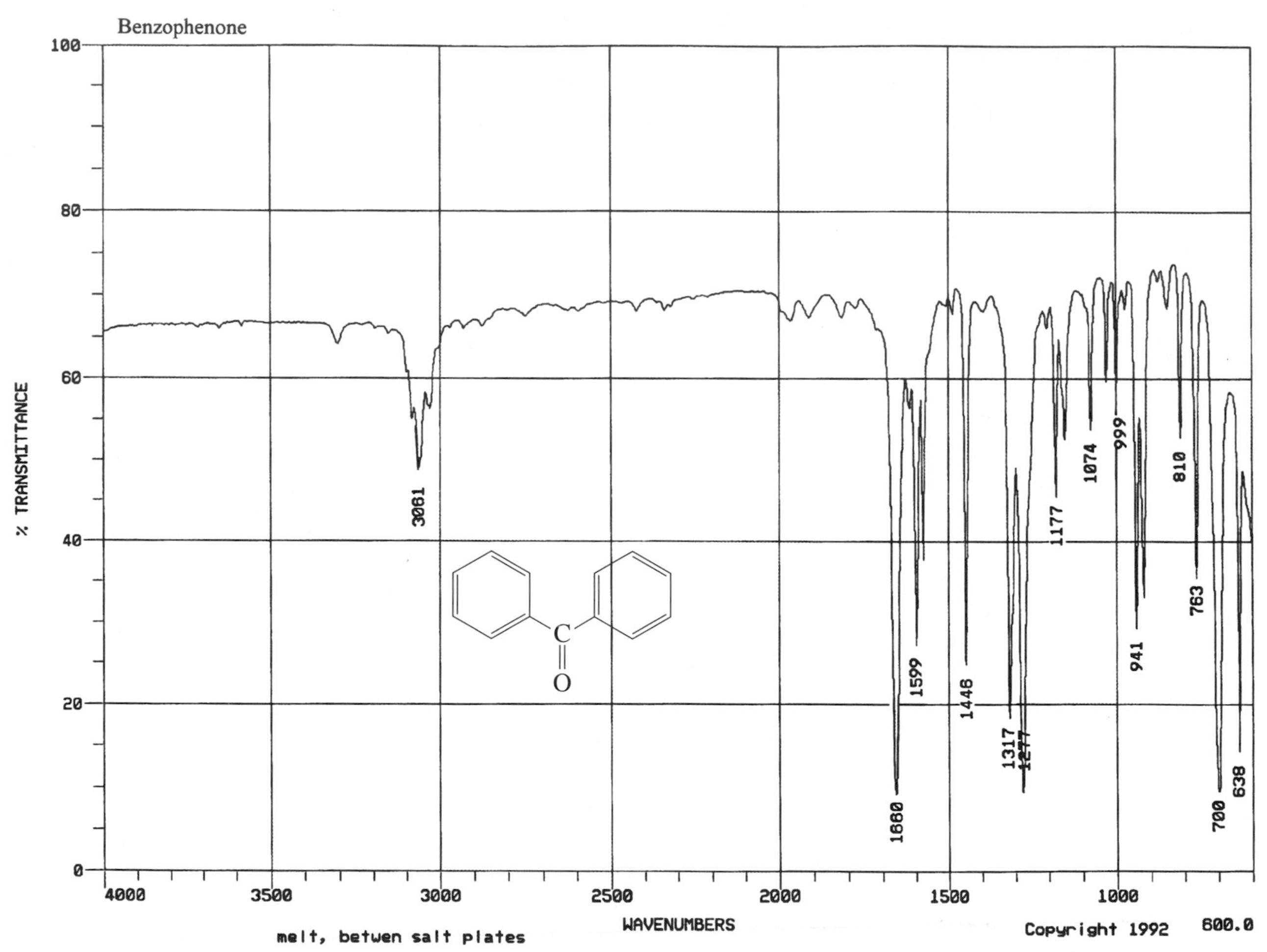

Figure 6.24 Infrared spectrum of benzophenone. C—H stretch: 3061 cm^{-1}. ν_{C-H} (aromatic). C=O stretch: 1660 cm^{-1}, $\nu_{C=O}$ (conjugated). C$\equiv$C stretch: 1599 cm^{-1}, 1446 cm^{-1}, $\nu_{C\equiv C}$. C—(C=O)—C stretch and bend: 1277 cm^{-1}. C$\equiv$C bend: 763, 700 cm^{-1}. Spectrum courtesy of R. Tomasi, Sunbelt R&T, Inc., Tulsa, Oklahoma.

Benzaldehyde yields the IR spectrum of Figure 6.25. The strong C═O stretch is now at 1703 cm^{-1}, supporting conjugation, and the aldehydic C—H stretch (for the CHO group) is observed as a pair of bands at 2820 and 2737 cm^{-1}.

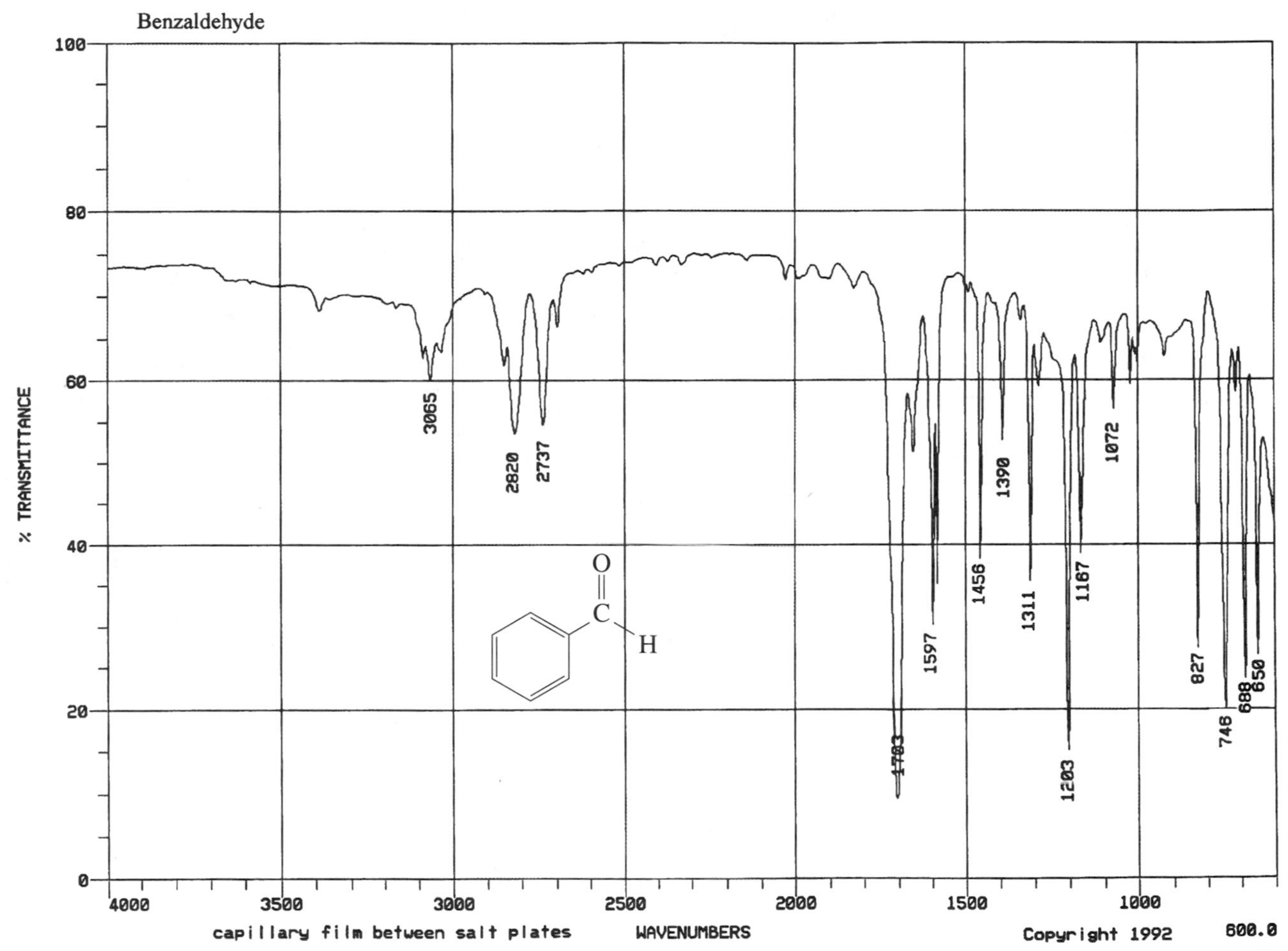

Figure 6.25 FT IR Spectrum of benzaldehyde, neat. Aldehydic (C═O)—H stretch: 2820, cm^{-1}, 2737 cm^{-1}. C═O stretch: 1703 cm^{-1}, $\nu_{C=O}$. Spectrum courtesy of R. Tomasi, Sunbelt R&T, Inc., Tulsa, Oklahoma.

When we examine the structure of a carboxylic acid, we expect IR bands for both a carbonyl and an O—H group. These are seen in the spectrum of *p*-chlorocinnamic acid (4′-chloro-3-phenyl-2-propenoic acid, Figure 6.26). The carbonyl stretch at 1695 cm^{-1} is broadened because the OH group is hydrogen-bonded to the hydroxyl group of another molecule. The O—H stretch band is very broad owing to the same hydrogen bonding and occurs over a wide range (3226–2326 cm^{-1}).

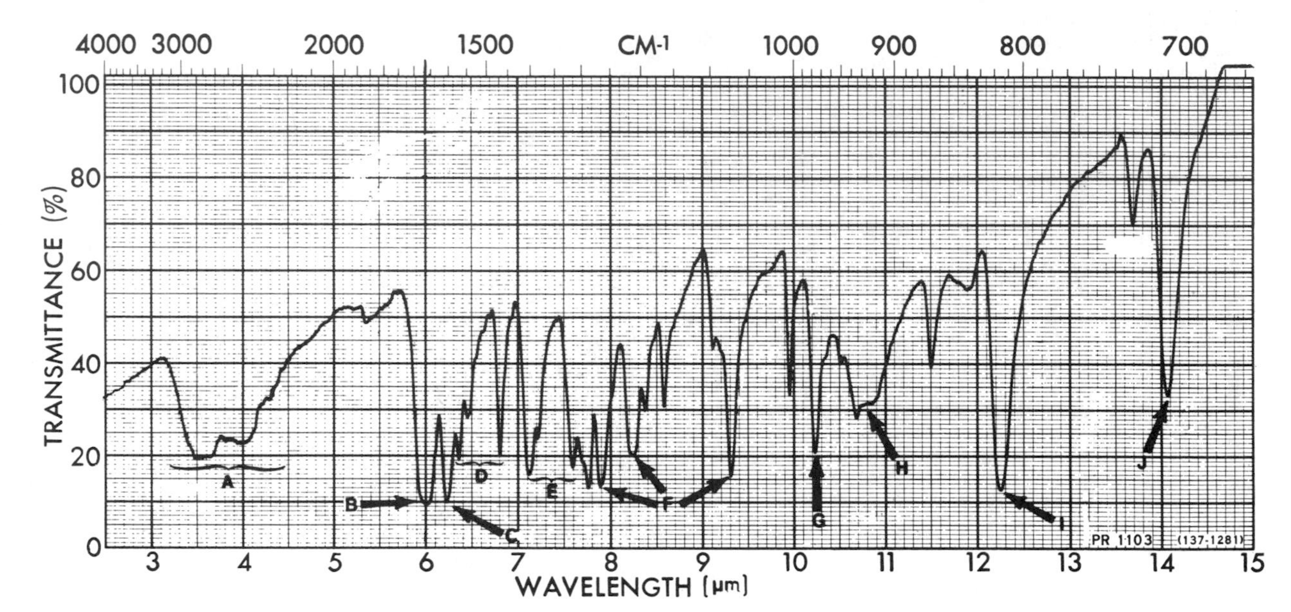

Figure 6.26 Infrared spectrum *p*-chlorocinnamic acid; KBr pellet.[a]

Cl—C_6H_4—CH=CH—CO_2H

O—H stretch: **A,** 3226–2326 cm^{-1},[b] $\nu_{\mathrm{O-H}}$ (associated). C=O stretch: **B,** 1695 cm^{-1}, $\nu_{\mathrm{C=O}}$ (conjugated). C=C stretch: **C,** 1634 cm^{-1}, $\nu_{\mathrm{C=C}}$ (conjugated). C==C stretch; **D,** 1600, 1577, 1493 cm^{-1}, $\nu_{\mathrm{C==C}}$. C—O stretch, O—H bend (in-plane): **E,** 1425, 1307 cm^{-1}, $\nu_{\mathrm{C-O}}$, $\delta_{\mathrm{O-H}}$, coupled. ==C—H bend (in-plane): **F,** 1282, 1227, 1087 cm^{-1}, $\delta_{\mathrm{==C-H}}$ (in-plane). =C—H bend (out-of-plane, olefinic): **G,** 988 cm^{-1}, $\delta_{\mathrm{=C-H}}$ (out-of-plane). O—H bend (out-of-plane): **H,** ca. 930 cm^{-1}, $\alpha_{\mathrm{O-H}}$ (out-of-plane). ==C—H bend (out-of-plane): **I,** 823 cm^{-1}, $\delta_{\mathrm{==C-H}}$ (out-of-plane): **J,** 715 cm^{-1}, $\delta_{\mathrm{==C-H}}$ (out-of-plane).

[a]Calibration: since the tracing is skewed to slightly lower wavenumbers (cm^{-1}), all band positions on spectrum increased on tabulation.

[b]Typical broad band corresponding to carboxylic acid dimer.

Upon formation of a carboxylic acid derivative such as an ester, the O—H group is lost and thus the corresponding IR band (near 3300 cm^{-1}) should also disappear, and this is the case for ethyl acetate (ethyl ethanoate, Figure 6.27). We do see a strong carbonyl stretch at 1741 cm^{-1} as well as C—O stretching due to the "acid" and "alcohol" portions of the molecule at, respectively, 1240 and 1047 cm^{-1}.[5]

[5]The stretching of an "acid" C—O bond to the sp^2 carbonyl carbon is expected to be in the 1100–1250 cm^{-1} region; the stretching of an "alcohol" C—O bond to the sp^3 alkyl carbon is expected to be in the 1000–1150 cm^{-1} region.

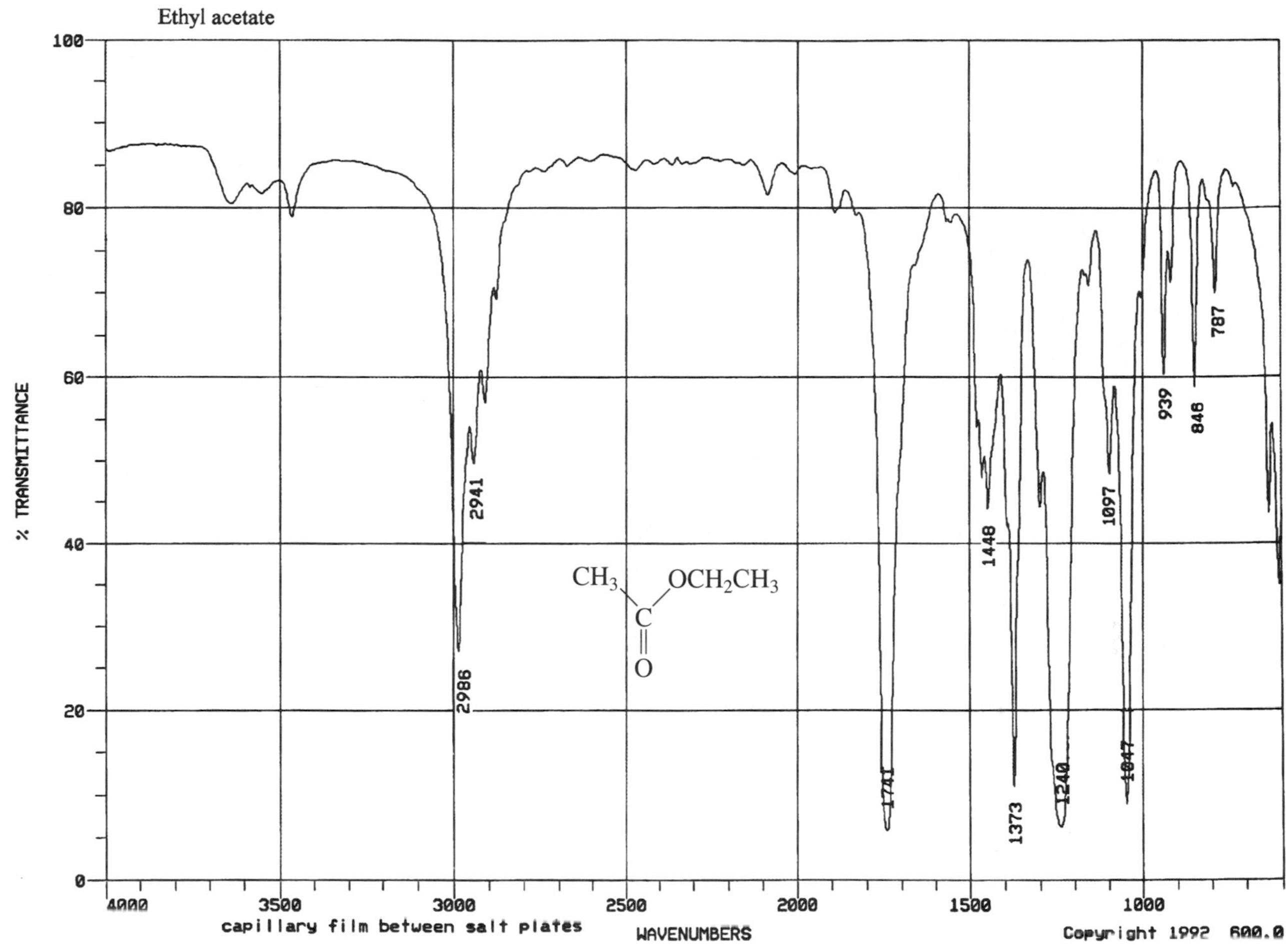

Figure 6.27 Infrared spectrum of ethyl acetate: thin film. C—H stretch: 2986 cm^{-1}, ν_{asym} CH_3; 2941 cm^{-1}, ν_{asym} CH_2. 2910 cm^{-1}, ν_{sym} CH_3. C=O stretch: 1741 cm^{-1}, $\nu_{C=O}$. C—H bend: 1373 cm^{-1}, δ_{sym} CH_3. C—O stretch: 1240 cm^{-1}, C(C=O)—O stretch: 1047 cm^{-1}, O—C—C stretch. Spectrum courtesy of R. Tomasi, Sunbelt R&T, Inc., Tulsa, Oklahoma.

The IR spectrum of another carboxylic acid derivative, benzoyl chloride, is shown in Figure 6.28. Fermi resonance[6] causes the carbonyl stretch to appear as two peaks: one at 1770 cm^{-1} and the other at 1727 cm^{-1}. A C—Cl stretching band occurs at 872 cm^{-1}, and most of the remaining bands are due to the benzene ring.

[6]Fermi resonance occurs as the result of an interaction between two fundamental bands, or of a fundamental band with a combination or overtone band.

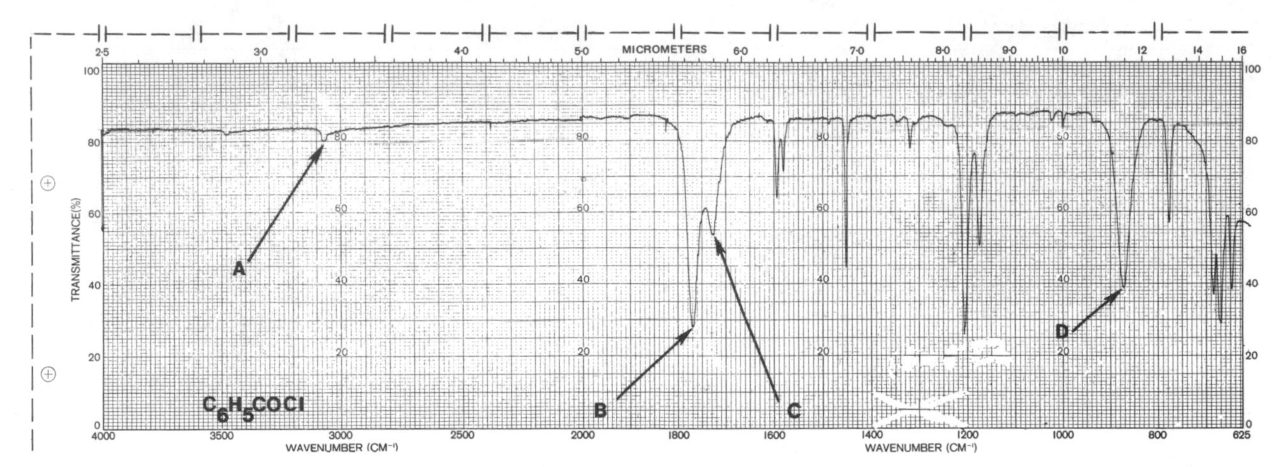

Figure 6.28 Infrared spectrum of benzoyl chloride; neat. C—H stretch: **A,** 3070 cm^{-1}, $\nu_{C—H}$ (aromatic). C=O stretch: **B,** 1770 cm^{-1}, $\nu_{C=O}$; **C,** 1727 cm^{-1}, [Fermi resonance band for resonance between B and overtone of 872 cm^{-1} band]. C—Cl stretch: **D,** 872 cm^{-1}, $\nu_{C—Cl}$.

The structure of an anhydride provides a symmetry relationship between carbonyl groups which permits asymmetric and symmetric coupling (see above) and the resulting pair of bands is revealed in the IR spectrum of the cyclic anhydride in Figure 6.29. The C=O stretch appears as both an asymmetric band at 1850 cm^{-1} and a symmetric band at 1775 cm^{-1}.

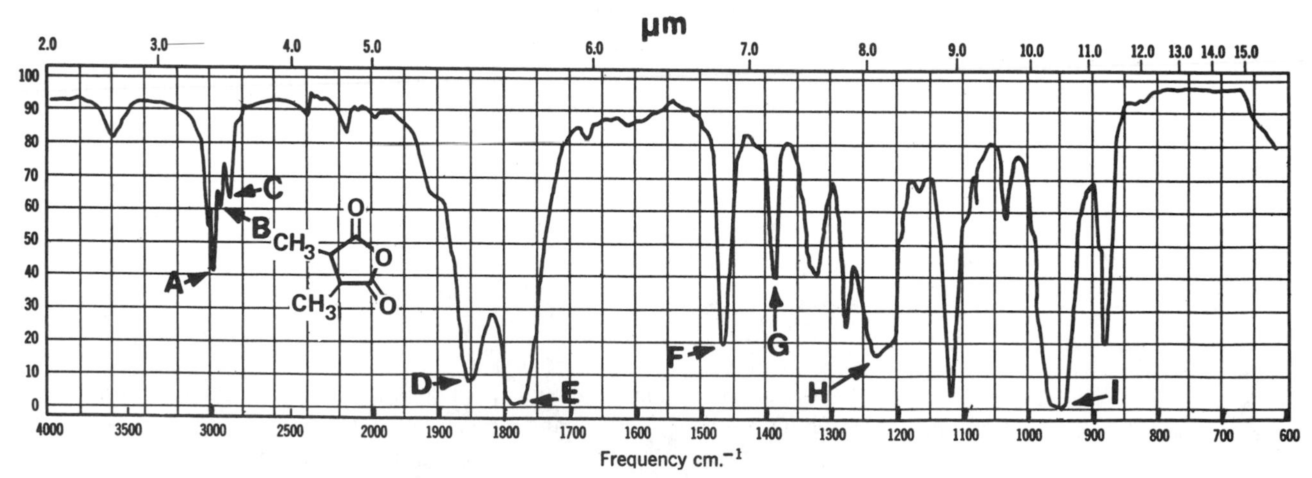

Figure 6.29 Infrared spectrum of α, α'-dimethylsuccinic anhydride. C—H stretch: **A,** 2980 cm^{-1}, ν_{asym} CH_3; **B,** 2910 cm^{-1}, $\nu_{C—H}$ (methine characteristically weak); **C,** 2860 cm^{-1}, ν_{sym} CH_3. C=O stretch: **D,** 1850 cm^{-1}, ν_{asym} C=O; **E,** 1775 cm^{-1}, ν_{sym} C=O, coupled. C—H bend: **F,** 1470 cm^{-1}, δ_{asym} CH_3; **G,** 1380 cm^{-1}, δ_{sym} CH_3; C—(C=O)—O—(C=O)—C stretch (coupled): **H,** 1230 cm^{-1}, $\nu_{C—O}$; **I,** 950 cm^{-1}, $\nu_{C—O}$.

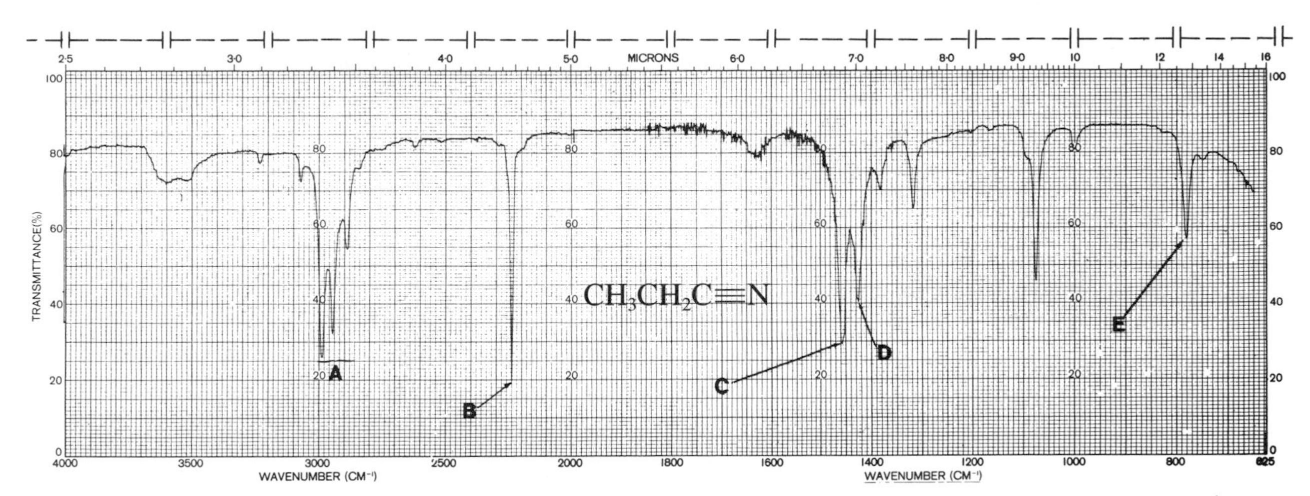

Figure 6.30 Infrared spectrum of propionitrile: neat. C—H stretch: **A,** 3010–2890 cm^{-1}, $\nu_{C—H}$ (aliphatic). C≡N stretch: **B,** 2240 cm^{-1}, $\nu_{C\equiv N}$ (aliphatic). C—H bend: **C,** 1458 cm^{-1}, δ_{asym}. CH_3C (aliphatic); **D,** 1427 cm^{-1}. δ_{sym} CH_2CN: **E,** 781 cm^{-1}, methylene rock.

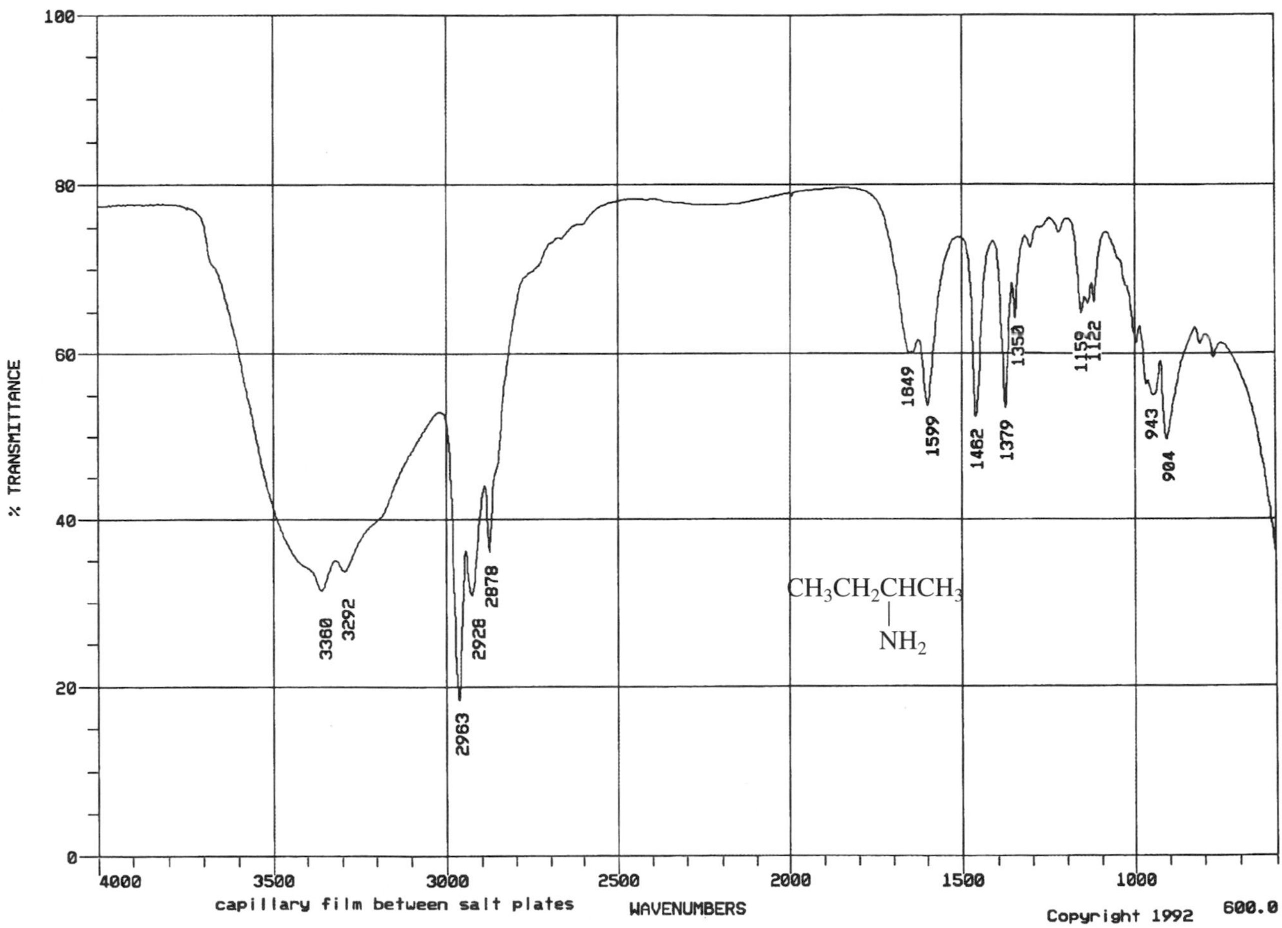

Figure 6.31 Infrared spectrum of *sec*-butylamine: neat. N—H stretch: 3360 cm^{-1}, ν_{asym}. NH_2: 2963 cm^{-1}, ν_{sym}. NH_2. C—H stretch: 3000–2850 cm^{-1}, ν_{CH}, ν_{CH_2}. N—H bend: 1599 cm^{-1}, δ_{NH_2} (scissoring). C—H bend: 1462 cm^{-1}. δ_{CH_3} (scissoring): 1379, 1350 cm^{-1}, δ_{CH_2}, δ_{CH}. Nitrogen: 1200–800 cm^{-1}, C—N stretch and N—H wag (neat sample), mostly N—H wag. Spectrum courtesy of R. Tomasi, Sunbelt R&T, Inc., Tulsa, Oklahoma.

A strong C≡N stretch is observed for propanonitrile (Figure 6.30) at 2240 cm^{-1}. The remainder of the spectrum shows simple aliphatic characteristics.

The IR spectrum of 2-aminobutane (*sec*-butylamine) in Figure 6.31 also reveals a pair of bands due to coupled vibrations. Specifically, primary amines should show coupled N—H stretching (see above) and we see it here at 3360 cm^{-1} (asymmetric band) and 3292 cm^{-1} (symmetric band). The C—N stretch of amines is difficult to find and is rarely, if ever, used for structure determination.

We find that the IR spectrum of another primary amine, 3-methylaniline (*m*-toluidine), in Figure 6.32 again reveals a pair of bands due to coupled vibrations: the coupled N—H stretching bands are seen here at 3433 cm^{-1} and 3354 cm^{-1}. The remainder of the spectrum is largely due to the aromatic ring.

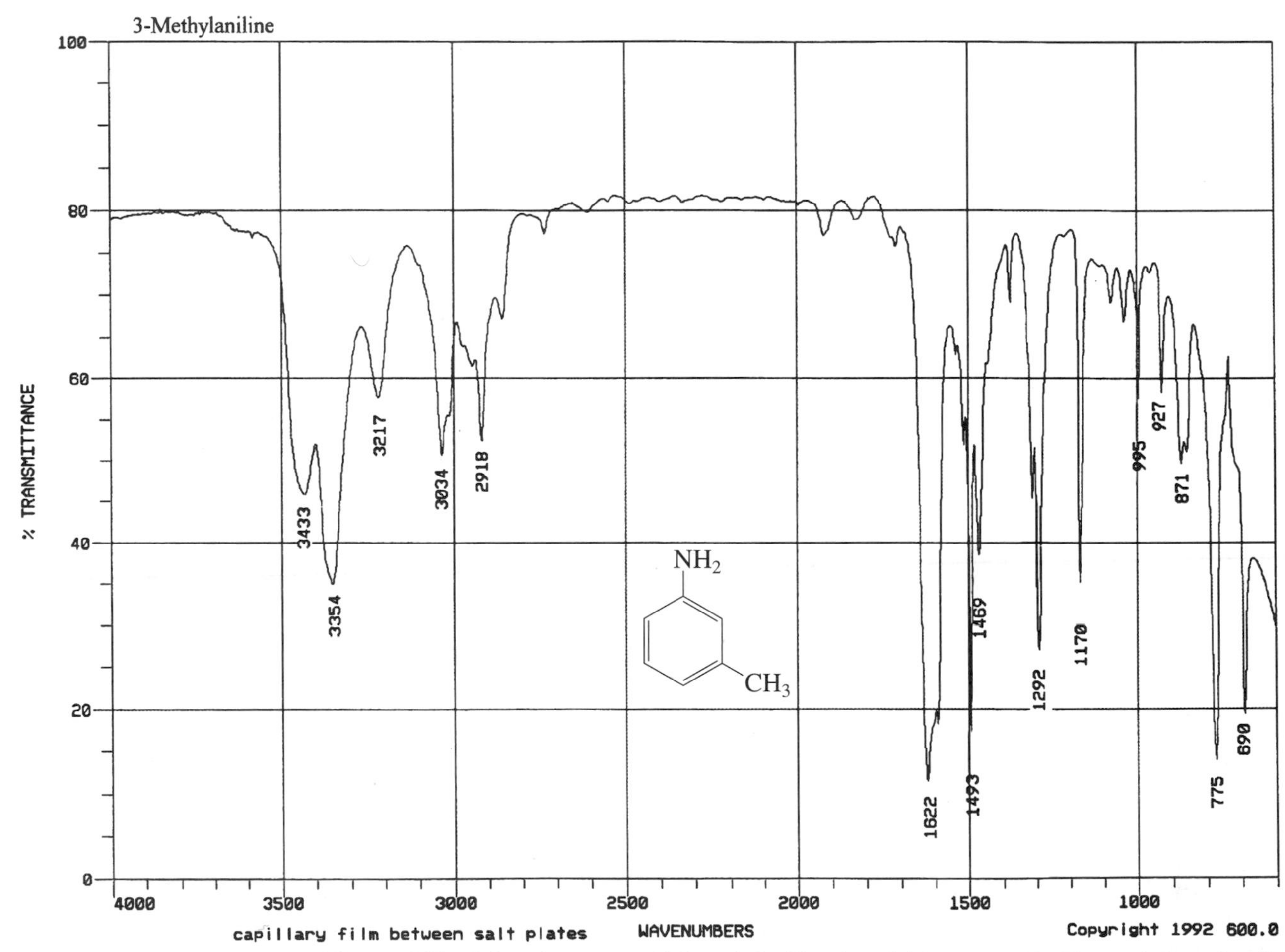

Figure 6.32 Infrared spectrum of 3-methylaniline (*m*-toluidine) neat. N—H stretch (associated): 3354, 3433 cm^{-1}, ν_{NH_2} (coupled pair). C—H stretch: 3034 cm^{-1}, $\nu_{C{=}C}$ (aromatic); 2918 cm^{-1}, ν_{C-H} (aliphatic). C=C stretch: 1622 cm^{-1}, $\nu_{C{=}C}$ (aromatic). N—H bend; 1622 cm^{-1}, ν_{N-H} (scissoring). C=C stretch: 1493, 1469 cm^{-1}, $\nu_{C{=}C}$ (aromatic). N—H wag: 871 cm^{-1}. =C—H bend (out of plane): 775 cm^{-1}. C=C bend (out-of-plane): 690 cm^{-1}. Spectrum courtesy of R. Tomasi, Sunbelt R&T, Inc., Tulsa, Oklahoma.

A word is in order about the use of N—H stretching patterns to classify amines (or amides) as primary, secondary, or tertiary. In principle, only primary amines (or amides) should show a coupled pair of bands, secondary amines and amides should show one N—H band, and tertiary compounds (and quaternary salts) should show no N—H band. In practice, poor instrument resolution can cause the pair for primary compounds to be unresolved. Moreover, a variety of hydrogen-bonding environments can cause the N—H band for secondary compounds to have multiple maxima, and tertiary or quaternary compounds are often hygroscopic and thus the O—H stretch of a water peak may be a misleading part of the IR spectrum.

Figure 6.33 displays the IR spectrum of DMF (*N,N*-dimethylformamide). The carbonyl peak is at 1676 cm^{-1} and is a part of the complex Amide I and II bands.

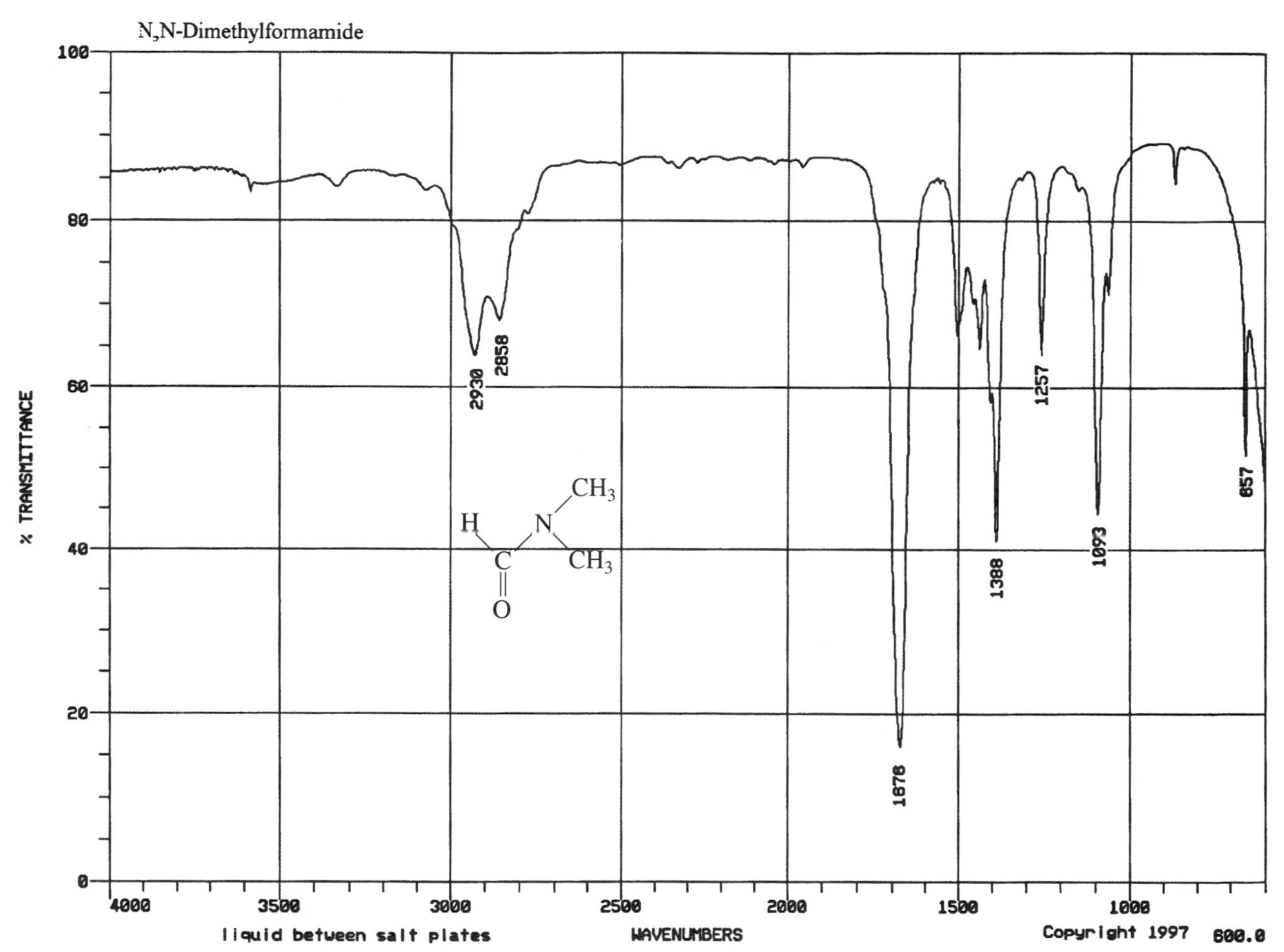

Figure 6.33 Infrared spectrum of *N,N*-dimethylformamide (DMF): neat. C—H stretch: 2930 cm^{-1}; $\nu_{C—H}$ (aliphatic); 2858 cm^{-1}; $\nu_{C—H}$ of CHO group. Amide I and II bands: 1676 cm^{-1} $\nu_{C=O}$ overlap of $\nu_{C=O}$ (Amide I) with N—H bend (Amide II). C—N band: 1388 cm^{-1}; $\nu_{C—N}$. Spectrum courtesy of R. Tomasi, Sunbelt R&T, Inc., Tulsa, Oklahoma.

Nitrobenzene contains a functional group which also should show a coupled pair of IR bands, and indeed the asymmetric and symmetric coupled NO bands are seen in Figure 6.34 at 1520 and 1345 cm^{-1}. The remainder of the spectrum is largely due to the benzene ring.

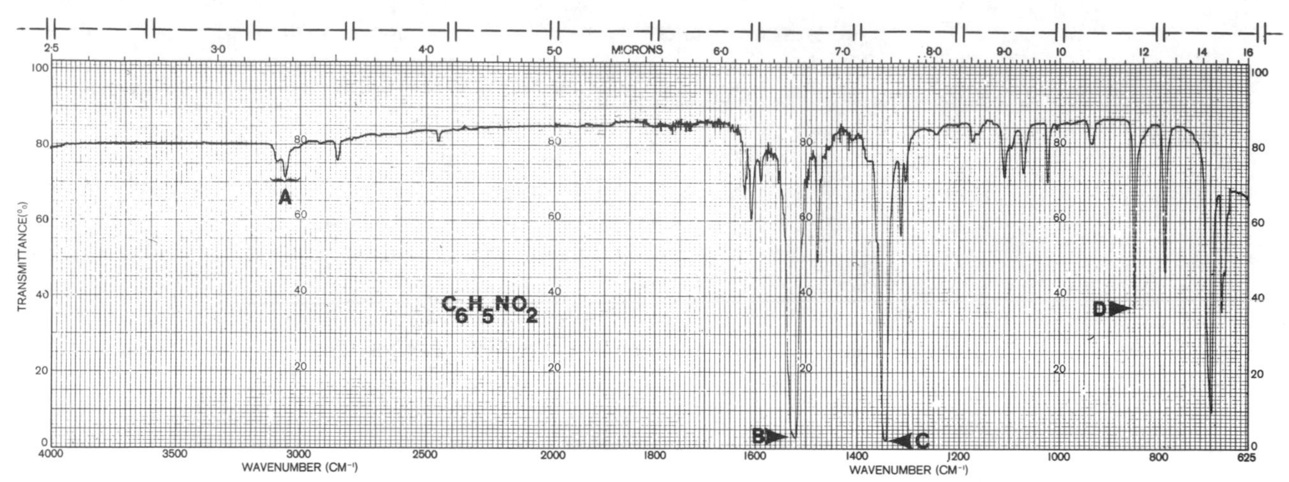

Figure 6.34 Infrared spectrum of nitrobenzene: neat. C—H stretch: **A,** 3100, 3080 cm^{-1}, $\nu_{\mathrm{C-H}}$ (aromatic). N(—O)$_2$ stretch: **B,** 1520 cm^1, $\nu_{\mathrm{N(\cdots O)_2}}$, asymmetric; **C,** 1345 cm^{-1}, $\nu_{\mathrm{N(\cdots O)_2}}$, symmetric. C—N stretch: **D,** 850 cm^{-1}, $\nu_{\mathrm{C-N}}$ (aromatic). The "l" low-frequency (<800 cm^{-1}) bands are of little use here for determination of the number and relative positions of ring substituents. This is because many of these bands are due to the interaction of the polar nitro group vibrations with the out-of-plane bending vibrations of the C—H groups.

The IR spectrum of chloroform (Figure 6.35) shows the presence of chlorine and can be used for reference when solution spectra are run using this compound as a solvent.

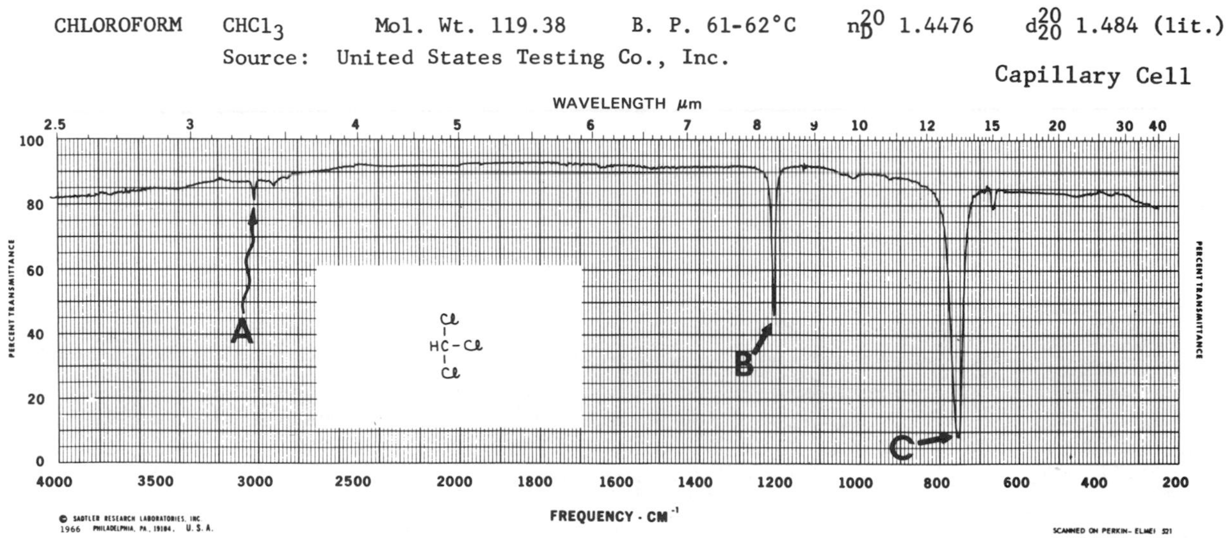

Figure 6.35 Infrared spectrum of chloroform: neat. C—H stretch: **A,** 3010 cm^{-1}, $\nu_{\mathrm{C-H}}$ (methine C—H stretch is characteristically weak). C—H bend: **B,** 1215 cm^{-1}, $\delta_{\mathrm{C-H}}$. C—Cl stretch: **C,** 755 cm^{-1}, $\nu_{\mathrm{C-Cl}}$.

6.3 NUCLEAR MAGNETIC RESONANCE

Over the past 35 years nuclear magnetic resonance (NMR) has gone from a useful tool for the organic chemist to a tool which is indispensable: the value of NMR cannot be overstated. This has been made possible by the development of higher magnetic fields and as well the expanding capabilities of computer technology.

The basis for NMR is the fact that in the presence of a magnetic field ($\mathbf{B}_0$, earlier $\mathbf{H}_0$) certain nuclei are distributed between two states of different energies. For organic compounds, the two most important are ^{1}H and ^{13}C. The discussion of these two different states begins with the Pauli exclusion principle. This principle allows two electrons in a given orbital or energy level only if these two electrons have different, or opposite spins: either $+\frac{1}{2}$ or $-\frac{1}{2}$:

⥮

In similar fashion, nuclei can have spins and certain nuclei (e.g., ^{1}H, ^{13}C, ^{19}F, and ^{31}P) have two spin states, again $+\frac{1}{2}$ and $-\frac{1}{2}$. It is the fact that the magnetic field, $\mathbf{B}_0$, causes these two states (also called α and β) to be at different energy levels and the ability to induce transitions between these two states give rise to NMR spectra. Since nuclei in different environments give rise to NMR transitions of various energies, measuring these transition energies yields the variety of information that makes NMR spectra useful.

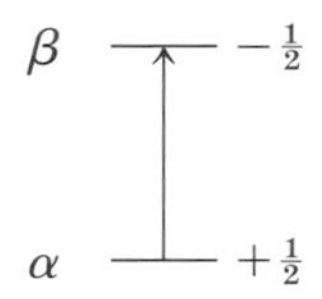

The basic principles of NMR can be applied to the simple molecule propane. Propane has two different kinds of carbons, the center carbon, and two identical carbons that are terminal. Similarly, propane has two kinds of protons, and thus this compound shows two different fundamental absorptions in both the ^{1}H and the ^{13}C NMR spectra.

$$CH_3CH_2CH_3$$
propane

In Figure 6.36 an analogy is made between a rotating nucleus (such as a proton) and a bar magnet. The rotation of the nucleus (Figure 6.36*a*) induces a magnetic field much as that caused by a bar magnet (*b*). In the absence of an instrument magnetic field, the fields of the individual nuclei are randomly oriented (Figure 6.37*a*). When the instrument magnetic field is applied, two different orientations of the "bar magnet" relative to the applied magnetic field ($\mathbf{B}_0$) can be seen (*b*), either with the field (α), or against the field (β). At room temperature in the presence of the applied (or instrument) magnetic field, $\mathbf{B}_0$, there is a very slight excess (on the order of parts per million) of nuclei in the lower energy state. This distribution between states is known as a Boltzmann distribution, and it is a type of equilibrium distribution.

Let us now consider the details of the NMR experiment. When the "spin active" nuclei (say, protons) are distributed between the two energy states, the energy required to excite the nuclei from the lower (α) to the upper (β) state can be measured. It is the energy of this transition, ΔE (which can be converted to frequency, ν) that is characteristic of a given nucleus. Thus since

$$\Delta E = h\nu$$

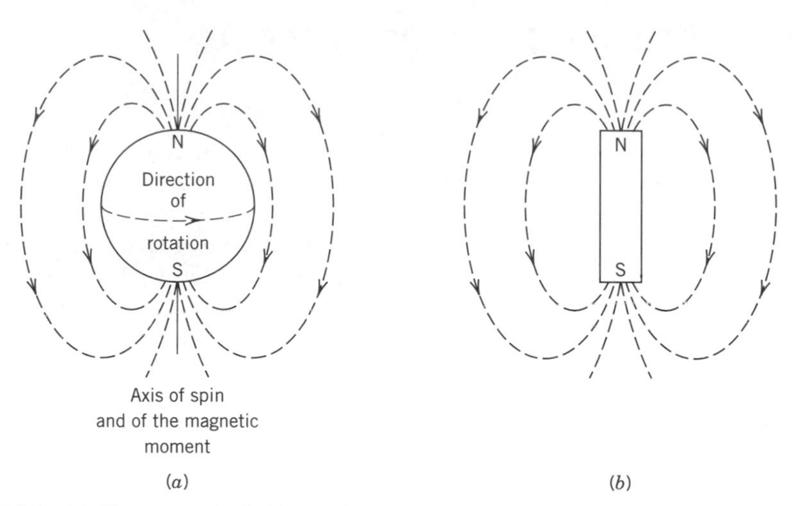

Figure 6.36 (*a*) The magnetic field associated with a spinning proton. (*b*) The spinning proton resembles a tiny bar magnet.

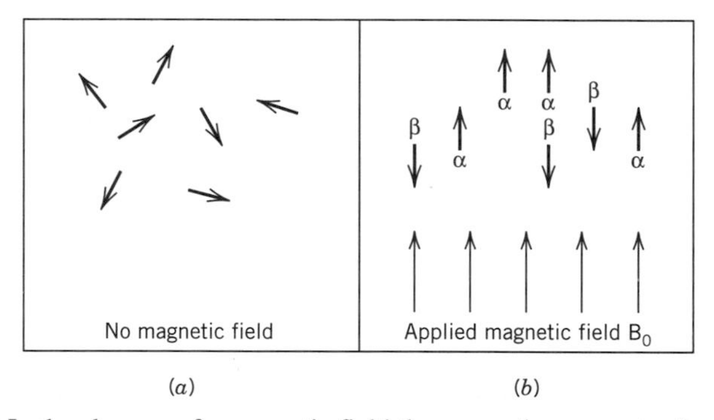

Figure 6.37 (*a*) In the absence of a magnetic field the magnetic moments of protons (represented by arrows) are randomly oriented. (*b*) When an external magnetic field ($\mathbf{B}_0$) is applied, the protons orient themselves. Some are aligned with the applied field (α spin state) and some against it (β spin state).

and since Planck's constant (h) is known, we can carry out the E to ν conversion.

A continuously changing wave of energy (or changing field) can be applied to these oriented nuclei, and when this applied energy (or frequency) matches the energy gap between the two spin states, resonance occurs and a signal is obtained. It is important to realize that the magnitude of the split between the states increases if the applied (or instrument) field increases (this is shown in Figure 6.38). This analysis method is called the continuous wave (CW) method.

In modern instruments, the CW approach has been replaced by the Fourier Transform NMR (or FT NMR) approach. This method involves using an intense, broad pulse of radiation which unselectively activates all of the ^{1}H (or ^{13}C, or other) nuclei in a molecule to the upper (β) state. These activated nuclei are then allowed to decay back to the ground (α) state, and this decay signal (called a free induction decay, or time domain signal) is mathematically converted to a frequency domain spectrum. Frequency domain spectra are the most common and several examples are provided below. This method readily lends itself to computer technology. For example, the computer can be programmed to do many pulses, and repetitive storage of large numbers of signals on the computer in digital form

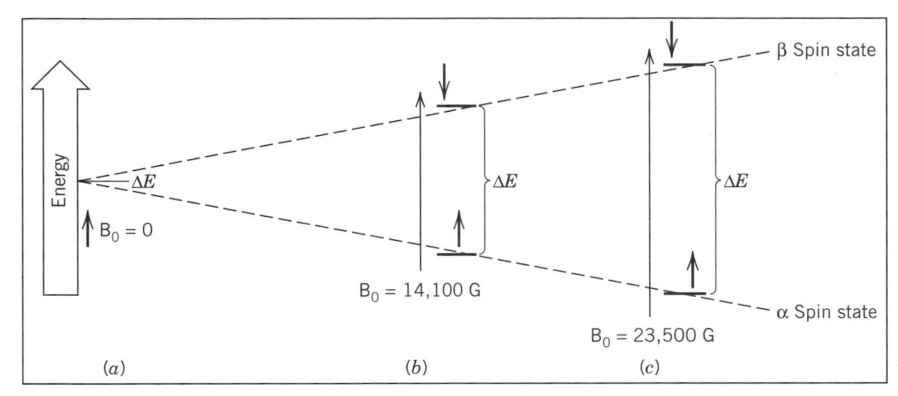

Figure 6.38 The energy difference between the two spin states of a proton depends on the strength of the applied external magnetic field, $\mathbf{B}_0$. (*a*) If there is no applied field ($\mathbf{B}_0$=0), there is no energy difference between the two states. (*b*) If $\mathbf{B}_0 \approx$14,100 G, the energy difference corresponds to that of electromagnetic radiation of 60×10^6 Hz (60 MHz). (*c*) In a magnetic field of approximately 23,500 G, the energy difference corresponds to electromagnetic radiation of 100×10^6 Hz (100 MHz). Instruments are available that operate at these and even higher frequencies (as high as 800 MHz).

allows signal reinforcement, which produces a more intense spectrum (higher signal/noise ratio) and thus a more readily interpreted spectrum. The signal/noise (S/N) ratio increases by the square root of the number of acquisitions (e.g., 64 acquisitions would give rise to an eight-fold increase in S/N). It is now common to use very small samples (say 10–20 mg) and these often require only eight pulses; multiple acquisitions (often in the thousands, or more) can be collected for compounds that display weak signals.

Experimental methods require that the compound (usually at least 10 mg) dissolved in an appropriate solvent (Tables 6.5 and 6.6) be placed in an NMR tube (see Figure 6.39) The solution should be homogeneous and care should be exercised to exclude iron filings (which cause signal broadening), moisture, or residual wash acetone. Acetone should be driven off tubes overnight in a hot oven since FT NMR instruments can reveal residual acetone that cannot be seen by the naked eye. Moreover, FT NMR instruments can pick up two different kinds of water in $CDCl_3$. The ^{1}H NMR spectra of $CDCl_3$ will reveal bulk or suspended water at $\approx\delta$ 4.7 and dissolved (monomeric) water at $\approx\delta$ 1.5. Use of C_6D_6 (here H_2O is at $\approx\delta$ 0.4) avoids this problem.

The ^{1}H and ^{13}C NMR spectra for pentane are shown in Figure 6.40. We might ask why ^{13}C, rather than ^{12}C, is the nucleus that is analyzed. After all ^{12}C is much more plentiful. This is because ^{13}C has a spin quantum number, *I*, of $\frac{1}{2}$ and is thus magnetically active, while ^{12}C with a spin number of zero is not. The low abundance (^{13}C is $\approx$1.1% as abundant as ^{12}C) of the magnetically active nucleus suggests the need for multiple acquisitions and computer storage. The proton-decoupled[7] spectrum in Figure 6.40 reveals three singlets at about δ 14, 22.4, and 34.4 (ppm). These signal positions are called chemical shifts and are calibrated relative to an internal standard, here the carbon signals of

[7]Proton decoupling is the removal of splitting of carbon signals due to the coupling of attached protons by the irradiation of all protons with a special magnetic field.

Table 6.5 Shift Positions of Residual Protons in Commercially Available Deuterated NMR Solvents[a]

Solvent	Isotopic purity of atom (%D)	Positions of residual protons (δ values)					
		Group	δ	Group	δ	Group	δ
Acetic acid-d_4	99.5	Methyl	2.05	Hydroxyl	11.53[b]		
Acetone-d_6	99.5	Methyl	2.05				
Acetonitrile-d_3	98	Methyl	1.95				
Benzene-d_6	99.5	Methine	7.20				
Chloroform-d	99.8	Methine	7.25				
Cyclohexane-d_{12}	99	Methylene	1.40				
Deuterium oxide	99.8	Hydroxyl	4.75[b]				
1,2-Dichloroethane-d_4	99	Methylene	3.69				
Diethyl-d_{10} ether	98	Methyl	1.16	Methylene	3.36		
Dimethylformamide-d_7	98	Methyl	2.76	Methyl	2.94	Formyl	8.05
Dimethyl sulfoxide-d_6	99.5	Methyl	2.50				
p-Dioxane-d_8	98	Methylene	3.55				
Ethyl alcohol-d_6 (anh.)	98	Methyl	1.17	Methylene	3.59	Hydroxyl	2.60[b]
Hexafluoroacetone deuterate	99.5	Hydroxyl	9.00[b]				
Methyl alcohol-d_4	99	Methyl	3.35	Hydroxyl	4.84[b]		
Methylcyclohexane-d_{14}	99	Methyl	0.92	Methylene	1.54	Methine	1.65
Methylene chloride-d_2	99	Methylene	5.35				
Pyridine-d_5	99	Alpha	8.70	Beta	7.20	Gamma	7.58
Silanar-C ($CDCl_3$ + 1% TMS)	99.8	Methyl (TMS)	0.00[c]	Methine	7.25		
Tetrahydrofuran-d_8	98	α-Methylene	3.60	β-Methylene	1.75		
Tetramethylene-d_8 sulfone	98	α-Methylene	2.92	β-Methylene	2.16		

[a]Data furnished by Merck Sharp and Dohme of Canada, Ltd.

[b]This value may vary considerably, depending on the solute.

[c]By definition.

tetramethylsilane, $Si(CH_3)_4$, set at $\delta = 0.00$. Chemical shifts (δ) are calculated by measuring the frequency (ν = cycles per second, or hertz = Hz) of interest relative to the frequency of the internal standard (ν_{TMS}) divided by the frequency of the instrument (ν_0, in megahertz, MHz = 10^6 Hz):

$$\delta = [(\nu - \nu_{TMS})10^6]/\nu_0$$

Thus if a signal occurs at 60 Hz on an 60-MHz instrument, the chemical shift is

$$\delta = [(60 \text{ Hz} - 0)10^6]/60 \text{ MHz} = 1.00$$

If this signal is a singlet or a well-defined multiplet, it will occur at exactly δ 1.00 on instruments of field strengths of 100, 300, 500, or 600 MHz. This is because the chemical shift measured in δ (a unitless number[8]) has been adjusted (by the division) for instrument field strength. While this signal would also be at δ 1.00 on an instrument of field strength 300 MHz, the shift in frequency units would be 300 Hz. On a 100-MHz instrument the signal would be at 100 Hz.

Now let us examine another simple NMR spectrum, in this case the ^{1}H NMR spectrum of chloroethane (ethyl chloride, Figure 6.41). By symmetry we see two types of protons (methyl and methylene) for this alkyl chloride, we thus expect two signals. There are

[8]We often see δ values listed as ppm (parts per million) since signals are usually measured in Hz, and the instrument field strength is given in MHz (megahertz or Hz $\times$ 10^6).

Table 6.6 The ^{13}C Chemical Shifts, Couplings, and Multiplicities of Common NMR Solvents[a]

Structure	Name	(ppm)	J_{C-D}(Hz)	Multiplicity[b]
$CDCl_3$	Chloroform-d_1	77.0	32	Triplet
CD_3OD	Methanol-d_4	49.0	21.5	Septet
CD_3SOCD_3	DMSO-d_6	39.7	21	Septet
$D\overset{O}{\overset{\parallel}{C}}N(CD_3)_2$	DMF-d_7	30.1	21	Septet
		35.2	21	Septet
		167.7	30	Triplet
C_6D_6	Benzene-d_6	128.0	24	Triplet
$D_2C—CD_2$ D_2C CD_2 O	THF-d_8	25.2	20.5	Quintet
		67.4	22	Quintet
O D_2C CD_2 D_2C CD_2 O	Dioxane-d_8	66.5	22	Quintet
D D D D N D	Pyridine-d_5	123.5 (C-3, 5)	25	Triplet
		135.5 (C-4)	24.5	Triplet
		149.2 (C-2, 6)	27.5	Triplet
$CD_3\overset{O}{\overset{\parallel}{C}}CD_3$	Acetone-d_6	29.8 (methyl)	20	Septet
		206.5 (carbonyl)	<1	Septet[c]
CD_3CN	Acetonitrile-d	1.3 (methyl)	32	Septet
		118.2 (CN)	<1	Septet[c]
CD_3NO_2	Nitromethane-d_3	60.5	23.5	Septet
CD_3CD_2OD	Ethanol-d_6	15.8 (C-2)	19.5	Septet
		55.4 (C-1)	22	Quintet
$(CD_3CD_2)_2O$	Ether-d_{10}	13.4 (C-2)	19	Septet
		64.3 (C-1)	21	Quintet
$[(CD_3)_2N]_3P{=}O$	HMPA-d_{18}	35.8	21	Septet
CD_3CO_2D	Acetic acid-d_4	20.0 (C-2)	20	Septet
		178.4 (C-1)	<1	Septet[c]
CD_2Cl_2	Dichloromethane-d_2 (methylene chloride-d_2)	53.1	29	Quintet

[a]From E. Breitmaier and W. Voelter, *Carbon-13 NMR Spectroscopy,* 3rd ed. (VCH, New York, 1987), p. 109; with permission. Also Merck & Co., Inc.

[b]Triplet intensities = 1:1:1, quintet = 1:2:3:2:1, septet = 1:3:6:7:6:3:1.

[c]Unresolved, long-range coupling.

indeed two large absorptions, one centered at δ 1.48 and the other at δ 3.57. Closer observation quickly reveals that the signal at δ 1.48 is split into three lines (called a triplet, t), and the signal centered at δ 3.57 into four lines (quartet, q). The reason for the splitting is the so-called spin-spin coupling of the protons on one carbon with the vicinal protons on the adjacent carbon.

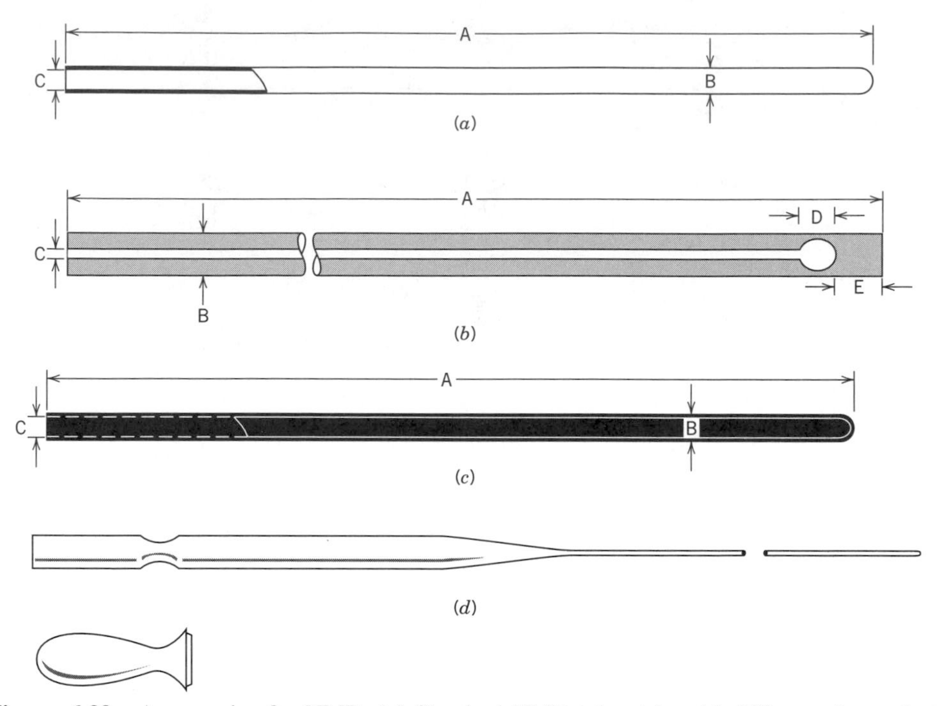

Figure 6.39 Accessories for NMR. (*a*) Standard NMR tube, takes 10–100 mg of sample in 0.4–0.5 mL of solution. Dimensions: **A,** 6–8 in.; **B,** 4 mm; **C,** ca. 3.2 mm. (*b*) NMR Microtube for NMR (requires 25μL minimally). Dimensions: **A,** 7 in.; **B,** 5 mm; **C,** 1.5 or 2.0 mm; **D,** 25μL; E, 10 mm. Tubes with cavity depths that can be adjusted are superior. (It is also suggested that rectangular, rather than spherical, cavities offer a better chance for regularity of shape and, therefore, better spinning properties.) (*c*) Tube of amber glass (to protect light-sensitive samples); dimensions similar to (*a*). (*d*) Disposable pipette (70–100 mm length) and bulb used for NMR analysis; the constriction near the top can be used for cotton plugs (allowing sample to be filtered as introduced). [Illustrations courtesy of Wilmad Glass Co., Buena, NJ].

The result of the coupled interactions is illustrated by Figure 6.42. Consider two nonequivalent protons, H_A and H_X. Assume that they couple only to each other (and to no other nuclei). We define J as the coupling constant. Assume that the magnitude of coupling here (J_{AX}, in Hz) is much smaller than the chemical shift difference (ν_{AB}, in Hz) between A and X. Let us examine the signal of H_A. Assume that the length of the arrow labeled $\mathbf{B}_{applied}$ (the applied field detects the protons and is not to be confused with the instrument magnetic field) is indicative of the field (or energy) necessary to cause H_A to resonate in the absence of coupling. Coupling means that proton H_A feels the effect of the field of H_X. The field of X can be aligned against the instrument field, and in such a case, a higher applied field of magnitude ($\mathbf{B'}_{applied}$) is needed to accomplish resonance for proton H_A, and the signal thus occurs at a somewhat higher field position ($\mathbf{B'}$). Alternatively, the field of H_X may be aligned with the applied field and here we only require an applied field magnitude equal to $\mathbf{B''}_{applied}$ to gain resonance, resulting in a lower field signal, $\mathbf{B''}$. The amount of increase or decrease in field is exactly the same, so the two peaks (called a doublet) obtained for proton H_A are centered about the single absorption that would arise in the absence of coupling. Moreover, since it is equally probable that the coupling can be in each of these two directions, the two peaks occur in an intensity ratio of 1:1. Finally, the signal for proton H_X would have exactly the same splitting, but would occur at a different chemical shift.

Pentane

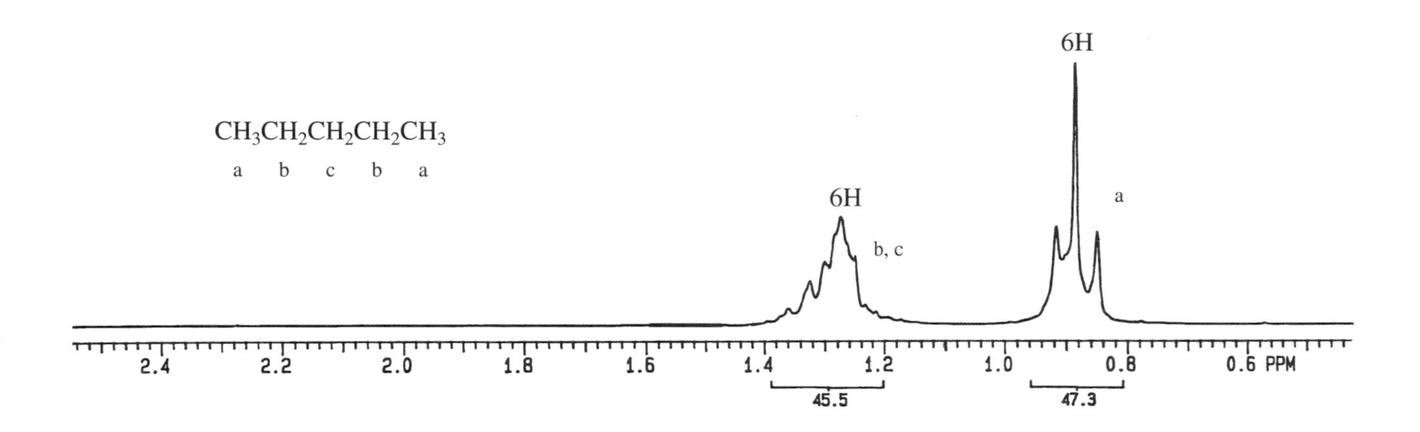

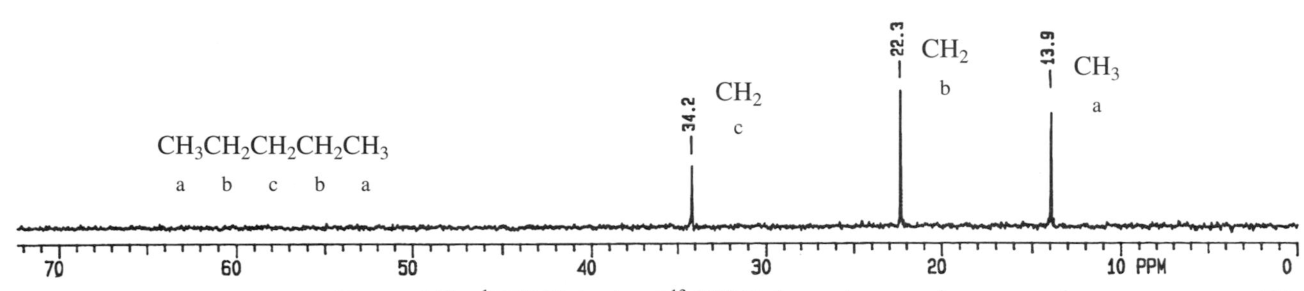

Figure 6.40 ^{1}H NMR (top) and ^{13}C NMR (bottom) spectra for pentane. Spectrum courtesy of R. Tomasi, Sunbelt R&T, Inc., Tulsa, Oklahoma.

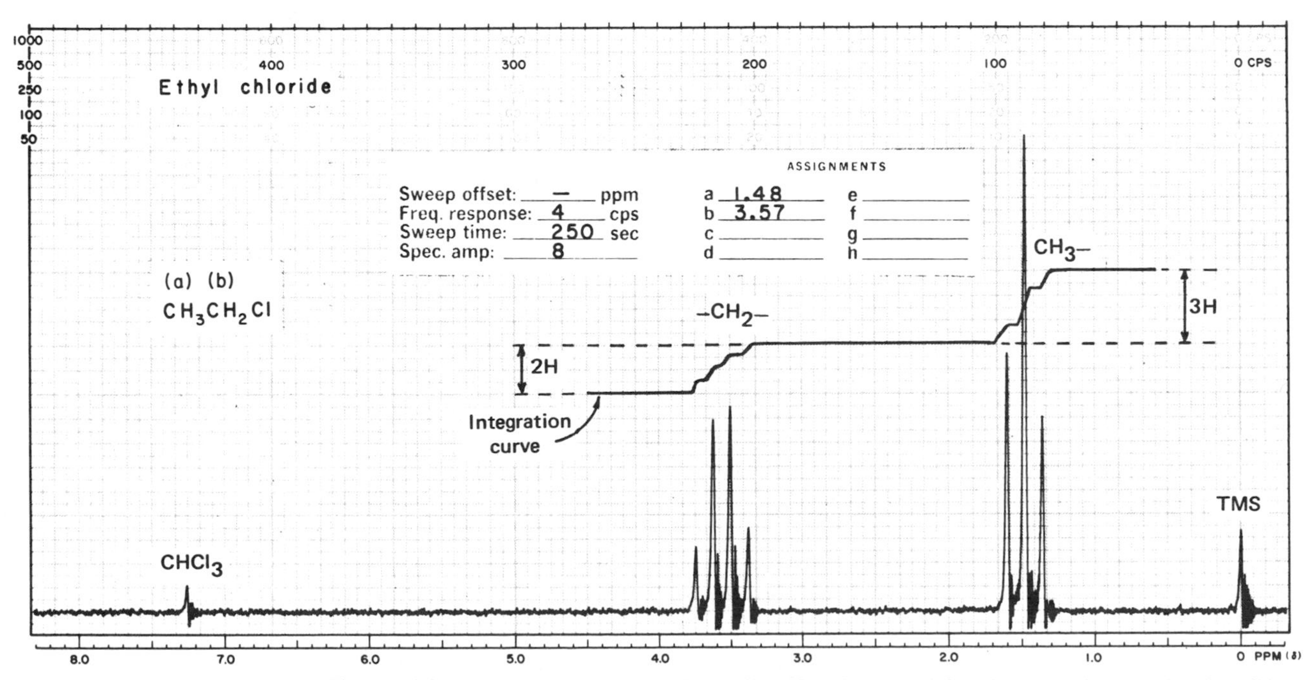

Figure 6.41 Ethyl chloride in $CDCl_3$ at 60 MHz. [Courtesy of Varian Associates, Palo Alto, CA]

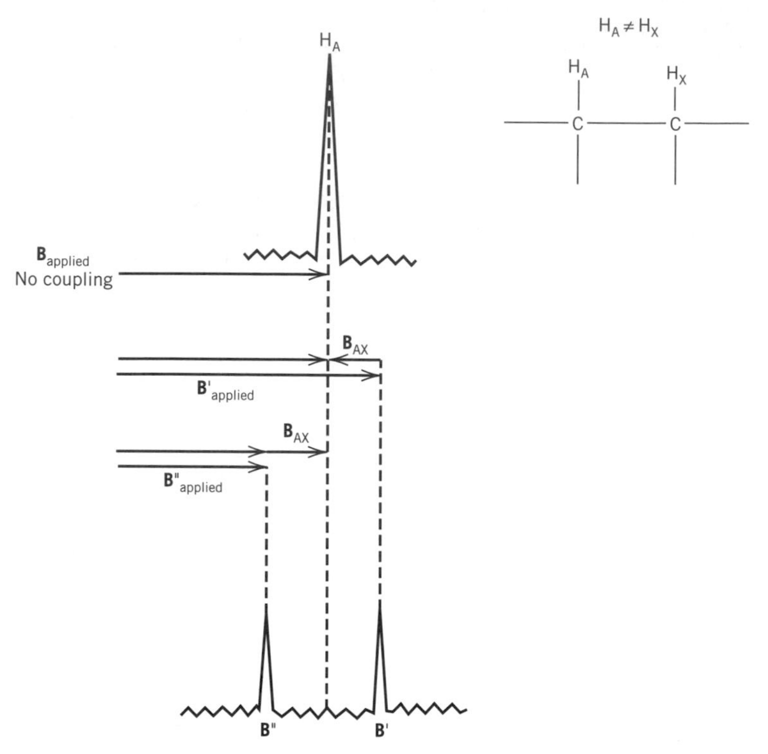

Figure 6.42 Spin-spin (J_{AX}) coupling splitting a singlet into a doublet ($J_{AX} << \nu_{AX}$).

The coupling argument for the doublet above can be extrapolated to more complex cases. Figure 6.43 shows the result when two equivalent protons (H_p, H_q) couple with a third (H_A). The H_A proton will feel the equivalent effects of the H_p and H_q fields. The fields of H_p and H_q can be aligned with or against the applied sweep field. If they are both aligned against the applied field, the required applied field is somewhat greater than needed for the uncoupled case, and the resulting peak is slightly upfield of the uncoupled signal. If both H_p and H_q are aligned with the applied field, the required applied field is somewhat less than needed for the uncoupled case, and the resulting peak is slightly downfield of the uncoupled signal. Again, the amounts of upfield and downfield offsets are equal, giving rise to the same splitting ($= J_{Ap} = J_{Aq}$). It is also possible for the two fields of protons p and q to cancel and this can occur in two ways (see Figure 6.43), giving rise to a large component at exactly the same shift as the unsplit signal. Simple probability then allows us to understand the relative intensities (1:2:1) of the signals.

Spin-spin splitting in simple cases can be predicted on a straight-forward basis using "first-order" rules. The multiplicity expected for a signal can be predicted from the n + 1 rule,[9] a simple first-order rule. That is, a given magnetically active nucleus coupled by n equivalent neighbors should appear as an $n + 1$ multiplet. Thus 1, 2, 3, 4, 5, 6, ... neighbors give rise to a signal containing $n + 1 = 2$ (doublet, d), 3 (triplet, t), 4 (quartet, q), 5, 6, 7, ... lines. Moreover, the relative intensities within the multiplet are predictable. We have already discussed the fact that doublets have 1:1 intensities and triplets 1:2:1. The rigorous theoretical measure of line composition in multiplets arises from the coefficients of an expanded polynomial. We do, however, have a shortcut to the details of these mul-

[9]The $n + 1$ rule applies to nuclei of spin $\frac{1}{2}$, and thus to ^{1}H, ^{13}C, and ^{19}F.

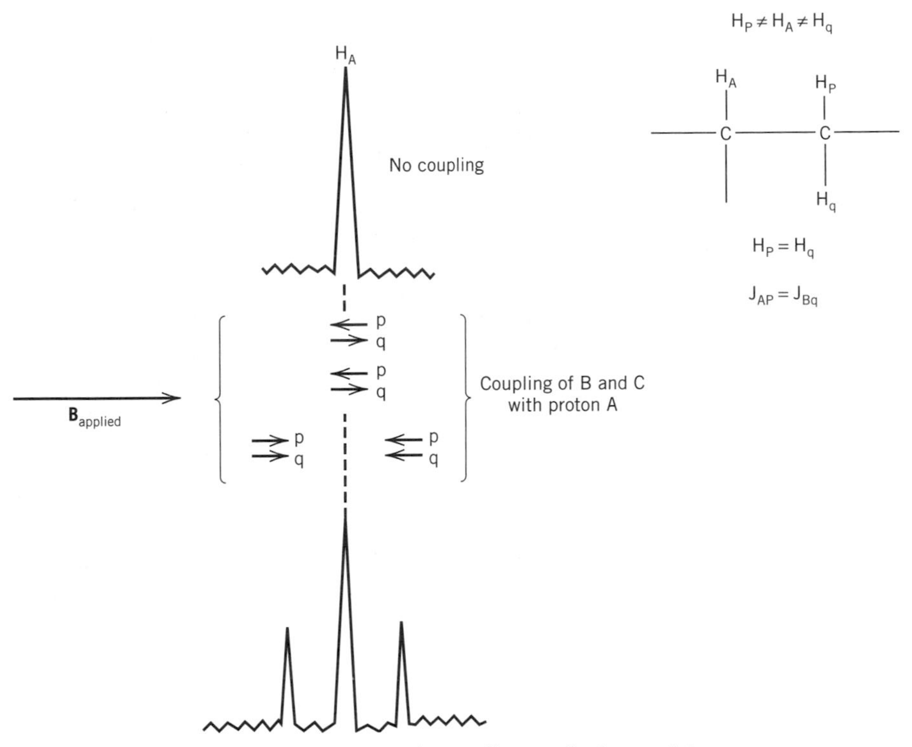

Figure 6.43 Spin-spin coupling producing a triplet.

tiplets in the form of Pascal's triangle (Figure 6.44). This triangle can be extended beyond that which is presented in Figure 6.44.

When the ^{1}H NMR spectra of certain organic compounds (alcohols, phenols, carboxylic acids, amines) are examined, the concept of coupling to the exchangeable OH (or NH) proton has a special twist. In general, we might expect that the hydroxylic proton of,

n	Relative Intensity
0	1
1	1 1
2	1 2 1
3	1 3 3 1
4	1 4 5 4 1
5	1 5 10 10 5 1
6	1 6 15 20 15 6 1

Figure 6.44 Pascal's triangle. Relative intensities of first-order multiplets; n = no. of equivalent coupling nuclei of spin $\frac{1}{2}$ (e.g., protons).

for example, ethanol, would couple with the α-protons (i.e., the protons on the adjacent methylene group). This is in fact the case when the NMR analysis of an alcohol is carried out in DMSO-d_6 or acetone-d_6 solvent. However, it is much more common to analyze organic compounds using $CDCl_3$ (deuterochloroform) solvent and in such cases even traces of HCl (or DCl) in the solvent promote a rapid exchange of hydroxylic protons:

$$R_aOH_a + R_bOH_b = R_aOH_b + R_bOH_a$$

In fact, this exchange is rapid enough that the hydroxylic proton cannot experience coupling with nearby protons on carbon and thus the OH proton occurs as a broadened singlet. Moreover, nearby protons (e.g., the protons on C-1 of ethanol) do not experience coupling by the OH proton. This allows the classification of alcohols: we should expect a singlet for the OH proton of 2-methyl-2-propanol (*t*-butyl alcohol), a doublet for the OH of 2-propanol (isopropyl alcohol), a triplet for the OH of ethanol (ethyl alcohol), or a quartet for the OH of methanol (methyl alcohol) when the analysis is conducted in DMSO-d_6 or acetone d_6. A number of factors which strongly affect the ^{1}H NMR spectra of exchangeable hydrogens are listed here:

1. Concentration (dilution shifts the OH peak upfield).
2. Acid catalysis (changes the positions and shapes of OH peaks).
3. Temperature (an increase enhances the rate of exchange).
4. D_2O Exchange (causes the organic OH peak to disappear and replaces it with a HOD peak at $\delta \approx 4.5$)

Let us discuss the ^{1}H and ^{13}C NMR spectra of some typical organic compounds. Table 6.7 displays the chemical shifts for both protons and carbons in a variety of organic

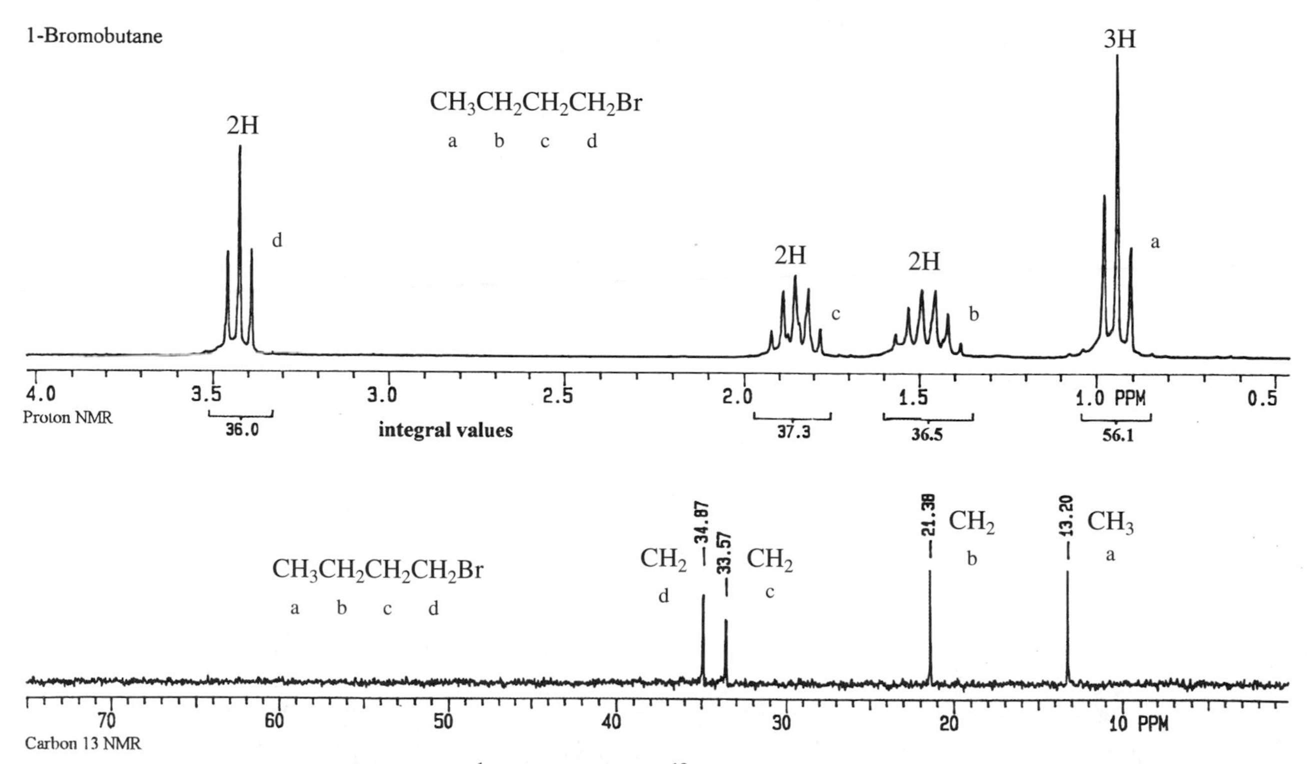

Figure 6.45 ^{1}H NMR (top) and ^{13}C NMR (bottom) spectra of 1-bromobutane. Spectra courtesy of R. Tomasi, Sunbelt R&T, Inc., Tulsa, Oklahoma.

Table 6.7 Chemical Shift Ranges for Protons[a] and Carbons of Organic Compounds

Class	δ H	δ C
Alkane	δ 0.0–2.0	δ 0.0–30.0
Alkyne (≡C—H)	δ 2.0–3.0	δ 75–95
Monosubstituted alkane[a,b]	δ 2.0–5.0	δ −6.0–70.0
Alkene (=C—H)	δ 4.3–7.3	δ 100–170
Aromatic (=C—H)	δ 6.0–8.8	δ 110–150
Aldehydic (—CH=O)	δ 9.0–10.0	δ 195–205
Alcohol (ROH)	δ 1.0–6.0[c]	
Carboxylic acid (RCOOH)	δ 10.0–13.0[c]	
Phenol (ArOH)	δ 4.0–7.5[c]	
Amine (—NH)	δ 0.5–5.0[c]	
Amide [R(C=O)NHR]	δ 5.0–9.0	δ 160–170
Nitrile (R—C≡N)		δ 105–115
Ester [R(C=O)OR′]		δ 167–174
Ketone [R(C=O)R′]		δ 200–220
Anhydride [(RC=O)$_2$O]		δ 167–174
α-protons		
Fluoride (HC—F)	δ 4.0–4.5	δ 70–80[b]
Chloride (HC—Cl)	δ 3.0–4.0	δ 40–50[b]
Bromide (HC—Br)	δ 2.5–4.0	δ 30–40[b]
Iodide (HC—I)	δ 2.0–4.0	δ 5–15[b]
Alcohol (HC—OH)	δ 3.4–4.0	δ 55–70[b]
Ether (HC—OR)	δ 3.3–4.0	δ 60–80[b]
Ester [HC—O(C=O)R]	δ 3.7–4.1	δ 60–70[b]
Carbonyl compound (HC—C=O)	δ 2.0–2.7	δ 30–50[b]

[a]Since chemical shifts depend upon solvent (and these data have been determined in a wide range of solvents) and other conditions, these ranges should be treated as being approximations. The carbon shifts of this table are supplemented by the carbon shifts provided on Table 6.9.

[b]The broad range of proton shifts for monosubstituted alkanes (δ 2.0–5.0) is subdivided at the bottom of the table (α-protons). Some α-carbon data are presented in this table, but a superior approach is to use Table 6.9.

[c]Proton shifts for such exchangeable hydrogens vary greatly since they depend greatly on concentration, solvent, and other factors which affect the exchange. Thus these ranges may not be totally reliable.

compounds. Figure 6.45 provides both the ^{1}H NMR and ^{13}C NMR spectra of 1-bromobutane. The ^{1}H NMR spectrum is easily accounted for by first-order rules. From low to high field the proton signals are assigned: the δ 3.45 (±0.1) triplet (integrates for 2H)[10] is due to the CH_2Br protons and is a triplet in view of coupling by the neighboring (C-2) methylene protons. Moreover the δ 3.45 signal is most downfield since the C-1 protons are closest to the electronegative, and thus deshielding, bromine atom. The next signal is due to the somewhat less deshielded protons at C-2 and is a quintet because of the equal coupling from the methylene groups at C-1 and at C-3. Note that the couplings from C-1 and C-3 are equal; the protons in each of these two methylene groups are not equivalent and thus the signal could have arisen from unequal coupling and in such a case

[10]The area of the peak is proportional to the number of protons causing that peak.

it would have been much more complex. The sextet at δ 1.48 (2H) is due to the $CH_3\underline{CH_2}CH_2$ protons, which are equally coupled by the neighboring C-2 methylene and C-4 methyl protons. Finally, the terminal methyl at δ 0.95 (3H) is a triplet in view of coupling by the neighboring (C-3) methylene protons. Figure 6.45 also contains the (proton-decoupled) ^{13}C NMR spectrum. Since proton coupling has been completely removed (and since the low abundance of ^{13}C precludes adjacent carbons which couple), the signals are all singlets: δ 34.87, C-1; δ 33.57, C-2; δ 21.38, C-3; δ 13.20, C-4. We find here (as in general) that deshielding may be used in the same sort of way to predict carbon chemical shifts as it is to predict proton shifts. Table 6.8 provides the chemical shifts of carbons for some alkanes and Table 6.9 has chemical shifts for the carbons of wide variety of substituted alkanes.

Table 6.8 Chemical Shifts of Carbons in Straight- and Branched-Chain Alkanes[a]

Compound	C-1	C-2	C-3	C-4	C-5
Straight chain					
Methane	−2.1				
Ethane	5.9	5.9			
Propane	15.6	16.1	15.6		
Butane	13.2	25.0	25.0	13.2	
Pentane	13.7	22.6	34.5	22.6	13.7
Hexane	13.9	22.9	32.0	32.0	22.9
Heptane	13.9	23.0	32.4	29.5	32.4
Octane	14.0	23.0	32.4	29.7	29.7
Nonane	14.0	23.1	32.4	29.8	30.1
Decane	14.1	23.0	32.4	29.9	30.3
Branched chain					
Isobutane	24.3	25.2			
Isopentane	22.0	29.9	31.8	11.5	
Isohexane	22.5	27.8	41.8	20.7	14.1
Neopentane	31.5	27.9			
Neohexane	28.9	30.4	36.7	8.7	
3-Methylpentane	11.3	29.3	36.7 18.6[b]		
2,3-Dimethylbutane	19.3	34.1			
2,2,3-Trimethylbutane	27.2	32.9	38.1	15.9	
3,3-Dimethylpentane	6.8 4.4[b]	25.1	36.1		

[a]From D. G. Grant and E. G. Paul, *J. Amer. Chem. Soc.*, *86*, 2984 (1964); J. D. Roberts et al., *ibid.*, *92*, 1338 (1970); H. Spiescke and W. G. Scheider, *J. Chem. Phys.*, *36*, 722 (1961).

[b]Branch methyl carbon. Chemical shifts in ppm downfield from TMS. Since these are obtained from spectra using various internal standards (TMS, benzene, CS_2), the error in shift position is at least 0.2 ppm.

Normally we shall use proton-decoupled ^{13}C NMR spectra for our purposes. However, it is useful to obtain proton coupling information and this can be done in a variety of direct or indirect ways. The simple fact that a methyl group (CH_3) is expected to be a quartet, a methylene (CH_2) a triplet, a methine (CH) a singlet, and quaternary carbon a singlet is clearly very useful for structure determination. All of this will be discussed below when we review Figure 6.48.

The ^1H NMR and ^{13}C NMR spectra of the aromatic hydrocarbon styrene are displayed in Figure 6.46. In general the ring current of π-systems is expected to deshield nearby protons. Thus both the olefinic and aromatic protons of styrene are downfield

Table 6.9 Shift Effect Due to Replacement of H by Functional Groups (R) in Alkanes[a]

$n = \cdots$ (chain with positions γ β α ending in R) $br = \cdots$ (chain with R on central carbon; positions γ β α β γ)

R	α		β		
	n	*br*	*n*	*br*	λ^b
CH_3	+9	+6	+10	+8	−2
COOH	+21	+16	+3	+2	−2
COO^-	+25	+20	+5	+3	−2
COOR	+20	+17	+3	+2	−6
COCl	+33	+28		+2	
COR	+30	+24	+1	+1	−2
CHO	+31		+0		−2
Phenyl	+23	+17	+9	+7	−2
OH	+48	+41	+10	+8	−5
OR	+58	+51	+8	+5	−4
OCOR	+51	+45	+6	+5	−3
NH_2	+29	+24	+11	+10	−5
NH_3^+	+26	+24	+8	+6	−5
NHR	+37	+31	+8	+6	−4
NR_2	+42		+6		−3
NO_2	+63	+57	+4	+4	
CN	+4	+1	+3	+3	−3
SH	+11	+11	+12	+11	−4
SR	+20		+7		−3
F	+68	+63	+9	+6	−4
Cl	+31	+32	+11	+10	−4
Br	+20	+25	+11	+10	−3
I	−6	+4	+11	+12	−1

n = straight-chain compounds, *br* = branched-chain compounds.

[a]From F. W. Wehrli and T. Wirthin, *Interpretation of Carbon-13 NMR Spectra* (Hayden, New York, 1976), with permission.

[b]The effect of the λ position is virtually the same for straight- and branched-chain compounds.

(^{1}H NMR, δ 5.0–8.0). The aromatic ring has three nonequivalent protons (*o, m, p*) and thus the signal is not simple. We cannot interpret this signal because the chemical shift differences are so small ($\Delta\nu/J$ is small). The olefinic protons are, however, well resolved. The signal for the proton geminal to the phenyl group is at δ 6.73, that *cis* to the ring at δ 5.75 and that *trans* to the ring at δ 5.25. This is as would be expected as the deshielding effect is expected to be weaker (and thus the protons would occur at higher field) as they get farther from the benzene ring. Moreover, the three signals are all doublets of doublets (d of d). Approximate splittings may be read from the spectrum, keeping in mind that since a 200-MHz instrument was used, $\Delta\delta = 1.00$ corresponds to $\Delta\nu = 200$ Hz and $\Delta\delta = 0.10$, 20 Hz. The geminal hydrogen at δ 6.73 shows splitting of 16 and 12 Hz (due to coupling by, respectively, the protons *trans* and *cis* to it). Thus we would expect and do see the larger coupling (*trans* coupling = 20 Hz) in the signal at δ 5.75,[11] and the smaller (J_{cis}) in the δ 5.25 signal.[8] Thus both chemical shifts and coupling tell the same story. Tables 6.10 and 6.11 provide chemical shifts for, respectively, protons and carbons of aromatic compounds.

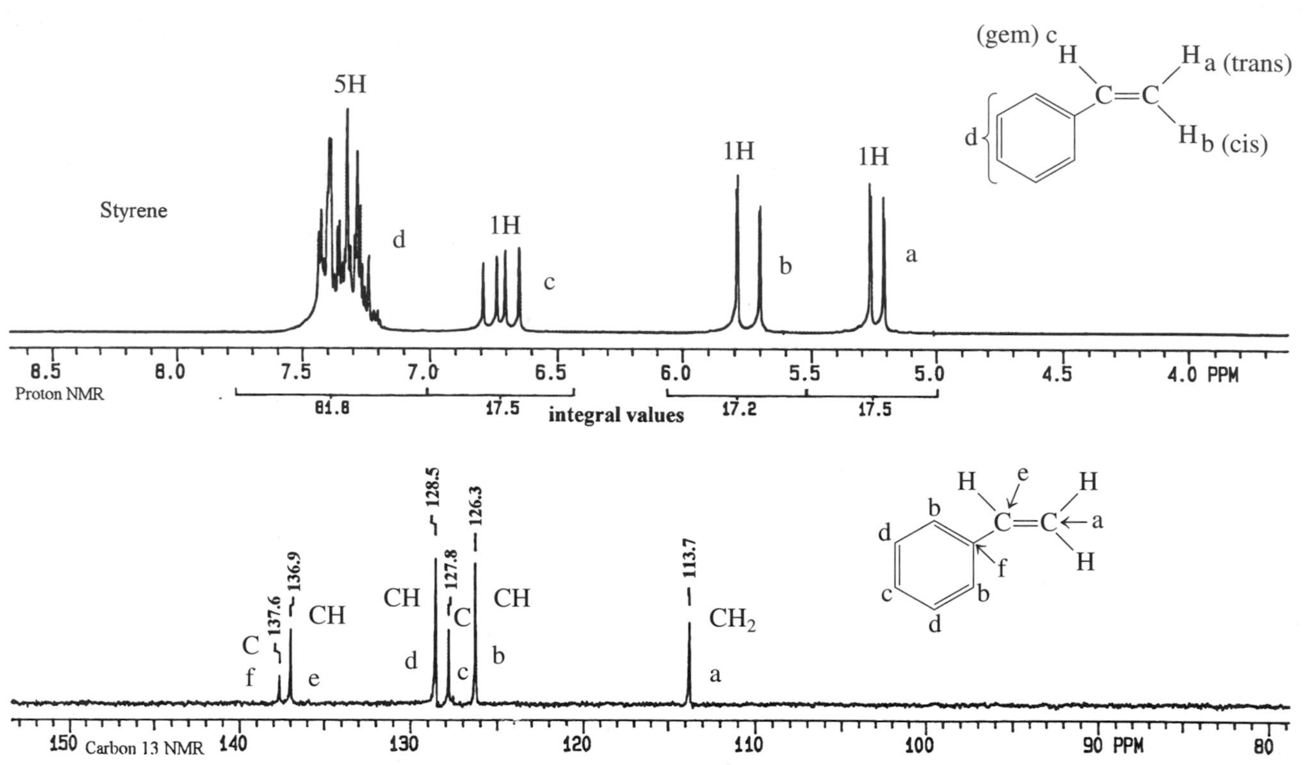

Figure 6.46 ^{1}H NMR (top) and ^{13}C NMR (bottom) spectra of styrene. Spectra courtesy of R. Tomasi, Sunbelt R&T, Inc., Tulsa, Oklahoma.

The ^{13}C NMR spectrum for styrene is also shown in Figure 6.46 and we can relate it to the structure of this molecule. The symmetry of the structure calls for six nonequivalent carbons, which is consistent with the fact that our proton-decoupled spectrum shows six singlets. The signal at δ 137.6 is assigned to the ring carbon bearing the vinyl group. It is identifiable by its chemical shift (most downfield due to the deshielding by the attached vinyl group) and its weak intensity. This weakness is partly due to the fact that carbons bearing no attached hydrogens usually show weak carbon signals (unless special NMR procedures are used). It is tempting to associate the weakness with the fact that there is only one such carbon (while there are two *ortho* and *para* carbons), but assignments based on intensities ascribed to the number of equivalent carbons are dangerous since this is only one contributing factor (and often a minor factor) affecting signal size. The olefinic carbons are at δ 136.9 and δ 113.7 and since their chemical shifts mingle with the shifts for aromatic carbons, it is not likely that shifts can be used to differentiate aromatic and olefinic carbons. The signal at δ 136.9 is the olefinic carbon attached to the ring and its more downfield position is due to deshielding by the nearby ring. The ring C-1 carbon is a weak signal at δ 137.6 since this carbon bears no protons. This signal is downfield of the other ring carbons since it bears the deshielding ethenyl (or vinyl) group.

[11] Actually the signals at δ 5.25 and δ 5.75 are split further (J_{ab}) due to geminal coupling. This splitting is so small (1–2 Hz) that it is essentially impossible to see in Figure 6.46.

Table 6.10 Chemical Shifts of Protons of Monosubstituted Benzenes (Benzene, Singlet, δ = 7.27)

	Compound	Ortho	Meta	Para
	$C_6H_5CH_3$[a]	7.10	7.18	7.10
	$C_6H_5—C_6H_5$	7.45	7.27	7.35
	$C_6H_5—SR$	7.25	7.3	7.3
Electron-withdrawing substituents	$C_6H_5—NO_2$	8.22	7.44	7.60
	$C_6H_5—CHO$	7.85	7.48	7.50
	$C_6H_5—(C{=}O)R$	7.9	7.45	7.55
	$C_6H_5—CO_2H$	8.17	7.41	7.47
	$C_6H_5—CO_2R$	8.2	7.35	7.5
	$C_6H_5—(C{=}O)Cl$	8.10	7.43	7.6
	$C_6H_5—CN$	7.54	7.38	7.57
	$C_6H_5—NH_3^+$	7.62	—	—
Electron-donating substituents	$C_6H_5—NH_2$	6.52	7.03	6.64
	$C_6H_5—OH$	6.77	7.13	6.87
	$C_6H_5—OR$	6.75	7.2	6.75
	$C_6H_5—O(C{=}O)CH_3$	7.15	7.26	7.26
	$C_6H_5—OSO_2C_6H_4CH_3$-*p*	7.01	7.22	7.22
	$C_6H_5—NH(C{=}O)CH_3$	7.52	—	—
Halides	$C_6H_5—R$	6.97	7.25	7.04
	$C_6H_5—Cl$	7.29	7.21	7.21
	$C_6H_5—Br$	7.49	7.14	7.24
	$C_6H_5—I$	7.67	7.00	7.24

[a]Toluene will show only a slightly broadened singlet for the aromatic protons at 60 MHz, 600 Hz sweep width. A separation of greater than 0.1–0.2 Hz is necessary for splitting of the aromatic signals under these conditions.

The *ortho*, *meta*, and *para* carbons have been assigned (respectively, δ 126.3, 127.8, and 128.5); these assignments require information that will not be covered here. If proton coupling information were available, we would expect the *ortho, meta*, and *para* carbons to be doublets, the ring C-1 carbon to be a singlet, the δ 136.9 signal to be a doublet, and the δ 113.7 signal to be either a triplet (if both geminally attached protons couple equally) or a doublet of doublets.

Spectra for acetanilide are provided in both Figures 6.47 and 6.48. Let us quickly deal with the ^{13}C NMR spectrum (Figure 6.47, bottom). The total of six singlets is consistent with the number of nonequivalent carbons. A weak signal for the amide carbonyl occurs at δ 169.5. The most downfield aromatic carbon is that at δ 138.2 (deshielded by the attached substituent and weak as this carbon bears no protons). The methyl carbon is upfield (δ 24.3) and the remaining ring carbons have been assigned (*o, m, p* = δ 120.4, 128.8, 124.3). Comparing the two ^1H NMR (Figure 6.47, top, and Figure 6.48) spectra makes clear the increased resolution provided by the higher field (200 MHz, Figure 6.47). The 60-MHz spectrum (Figure 6.48) reveals the singlet for the methyl protons and broadened singlets for the NH proton (δ 9.0), but the aromatic region at this lower field strength is too complex to analyze. If we go to the 200-MHz spectrum (Figure 6.47), we find the three signals for the *o, m,* and *p* protons are well resolved. Assuming that only the large, *ortho* coupling is visible, the doublet (δ 7.5) is assigned to the *ortho* protons, and peak areas are used to differentiate the triplets due to the *meta* (δ 7.25) and *para* protons

Table 6.11 Effect of Substituents on the C-13 Shift of Benzene Ring Carbons[a]

	Position			
Substituent	**C-1**	**Ortho**	**Meta**	**Para**
—Br	−5.5	+3.4	+1.7	−1.6
—CF_3	−9.0	−2.2	+0.3	+3.2
—CH_3	+8.9	+0.7	−0.1	−2.9
—CN	−15.4	+3.6	+0.6	+3.9
—C≡C—H	−6.1	+3.8	+0.4	−0.2
1,4-di—C≡C—H	−5.6	+3.8	—	—
—$COCF_3$	−5.6	+1.8	+0.7	+6.7
—$COCH_3$	+9.1	+0.1	0.0	+4.2
—COCl	+4.6	+2.4	0.0	+6.2
—CHO	+8.6	+1.3	+0.6	+5.5
—COOH	+2.1	+1.5	0.0	+5.1
—CO_2CH_3	+2 0	+1.2	−0.1	+4.8
—COC_6H_5	+9.4	+1.7	−0.2	+3.6
—Cl	+6.2	+0.4	+1.3	−1.9
—F	+34.8	−12.9	+1.4	−4.5
—H	0.0	—	—	—
—NCO	+5.7	−3.6	+1.2	−2.8
—NH_2	+18.0	−13.3	+0.9	−9.8
—NO_2	+20.0	−4.8	+0.9	+5.8
—OCH_3	+31.4	−14.4	+1.0	−7.7
—OH	+26.9	−12.7	+1.4	−7.3
—C_6H_5	+13.1	−1.1	+0.4	−1.2
—SH	+2.3	+1.1	+1.1	−3.1
—SCH_3	+10.2	−1.8	+0.4	−3.6
—SO_2NH_2	+15.3	−2.9	+0.4	+3.3
—O (oxyanion)	+39.6	−8.2	+1.9	−13.6
—OC_6H_5	+29.2	−9.4	+1.6	−5.1
—$OCOCH_3$ (acetoxy)	+23.0	−6.4	+1.3	−2.3
—$N(CH_3)_2$	+22.6	−15.6	+1.0	−11.5
—$N(CH_2CH_3)_2$	+19.9	−15.3	+1.4	−12.2
—NH(C=O)CH_3	+11.1	−9.9	+0.2	−5.6
—CH_2OH	+12.3	−1.4	−1.4	−1.4
—I	−32.0	+10.2	+2.9	+1.0
—$Si(CH_3)_3$	+13.4	+4.4	−1.1	−1.1
—CH=CH_2	+9.5	−2.0	+0.2	−0.5
—CO_2CH_3	+1.3	−0.5	−0.5	+3.5
—COCl	+5.8	+2.6	+1.2	+7.4
—CHO	+9.0	+1.2	+1.2	+6.0
—$COCH_2CH_3$	+7.6	−1.5	−1.5	+2.4
—$COCH(CH_3)_2$	+7.4	−0.5	−0.5	+4.0
—$COC(CH_3)_3$	+9.4	−1.1	−1.1	+1.7

[a]Data obtained in various solvents (CCl_4, neat, DMF = *N,N*-dimethylformamide) with various internal standards (TMS, CS_2). A positive value means a downfield shift, negative means an upfield shift. Estimated error is ± 0.5 ppm. Chemical shift of unsubstituted benzene is 128.7 ppm.

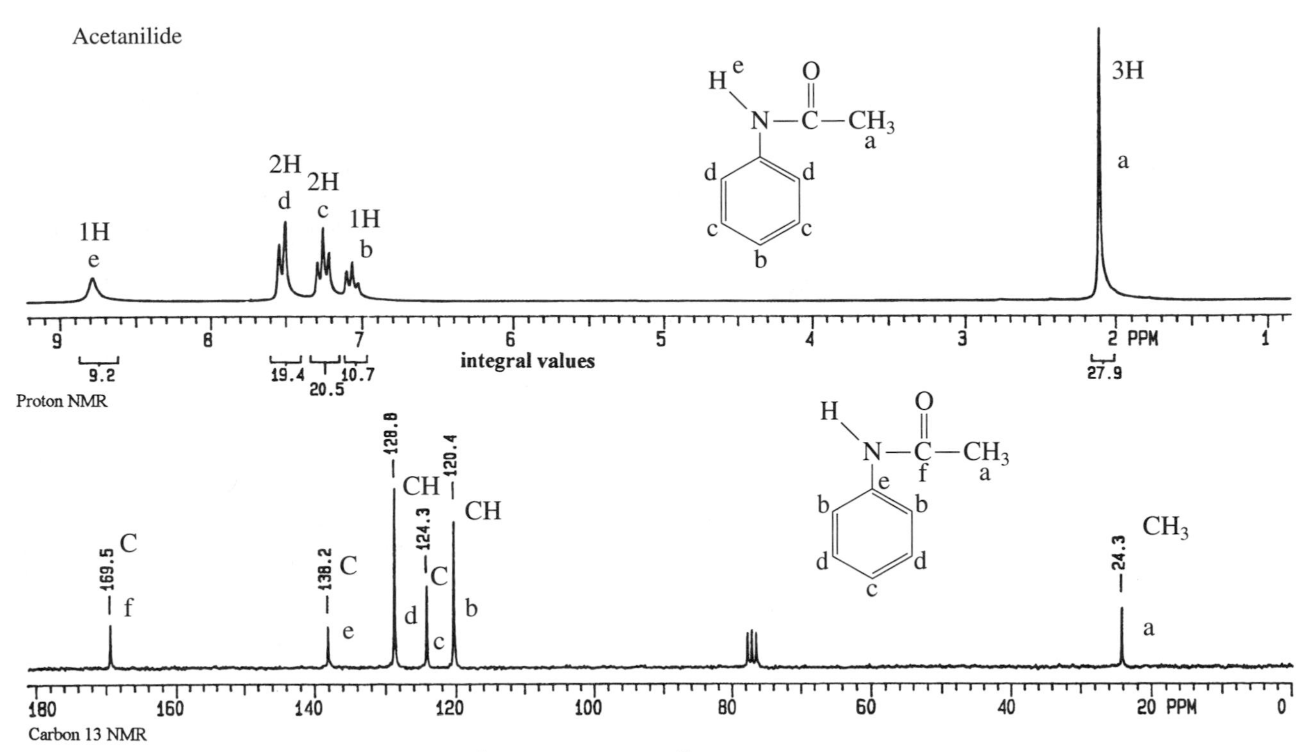

Figure 6.47 ^1H NMR (top) and ^{13}C NMR (bottom) spectra of acetanilide. Spectra courtesy of R. Tomasi, Sunbelt R&T, Inc., Tulsa, Oklahoma.

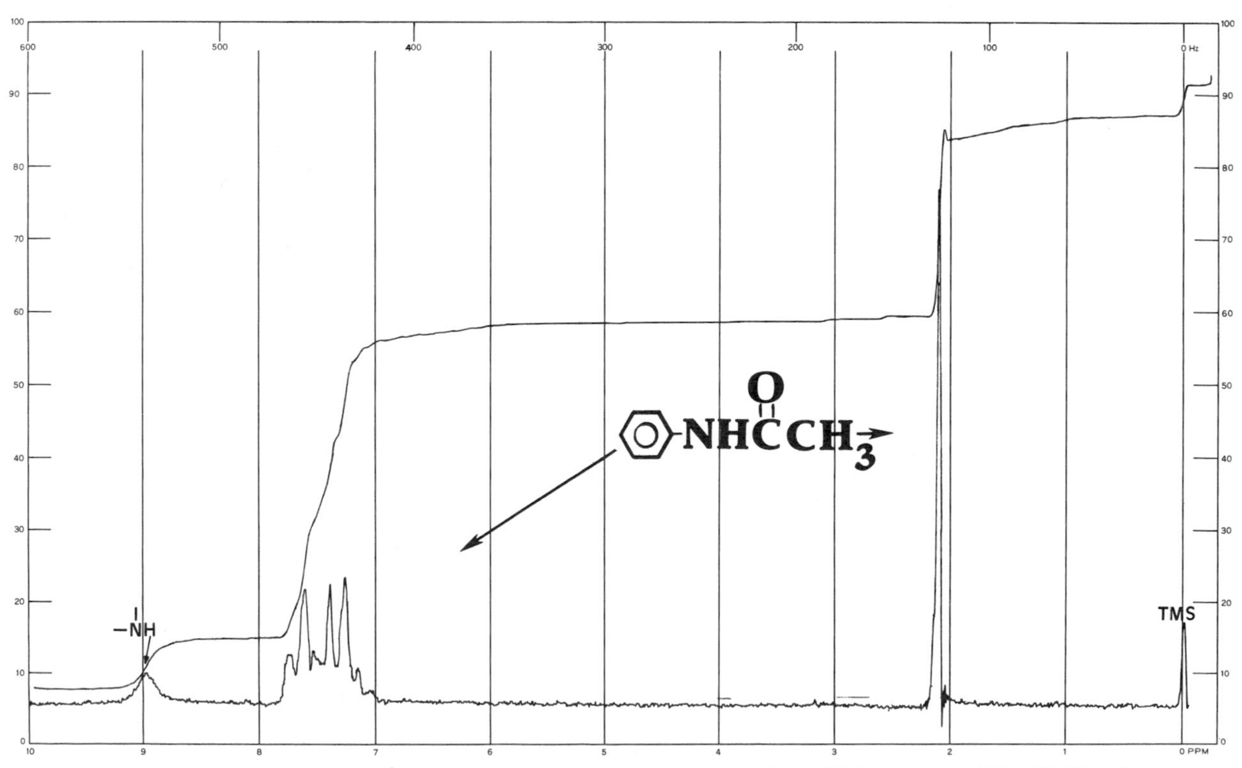

Figure 6.48 ^1H NMR spectrum of acetanilide: 60 MHz, 600-Hz sweep width, $CDCl_3$ solvent.

(δ 7.1). This is a reasonable assumption since usually $J_o \approx 9$ Hz, $J_m \approx 3$ Hz, and $J_p \approx 0$ Hz.

When piecing together an organic structure, clearly it would be of interest to be able to know the number of protons attached to a carbon. Since both carbons and protons have the same spin multiplicity, they can couple to each other in a similar fashion and the resulting split of the carbon signals is shown in Figure 6.49. Since C—H coupling constants are normally large (100–300 Hz), it is not uncommon to see overlap in the spectra of even simple compounds (such as 3-methylpentane, Figure 6.49). Thus it is the usual practice to decouple the protons completely for the first examination in ^{13}C NMR. Figure 6.49 shows an old approach to simplifying ^{13}C spectra. This so-called off-resonance-decoupled approach is done by reducing the proton-decoupling field and offsetting it somewhat. Thus we arrive at an intermediate position where some of the decoupling has "leaked" back in. This then gives rise to spectra with multiplets indicative of methyl (CH_3, shows as a quartet, q), methylene (CH_2, shows as a triplet, t), methine (CH, shows as a doublet, d), or quaternary (C, shows as a singlet, s) carbons. Thus, for example, the signal near δ 30 in Figure 6.49*b* is due to a methylene carbon since it is a triplet (t). These offset signals are not true multiplets, but they resemble the true multiplets sufficiently so they can be used to determine organic structures. Moreover, since these multiplets are more tightly spaced than the true multiplets, overlap is far less likely to occur. This technique has been largely replaced by 2D-related techniques such as the

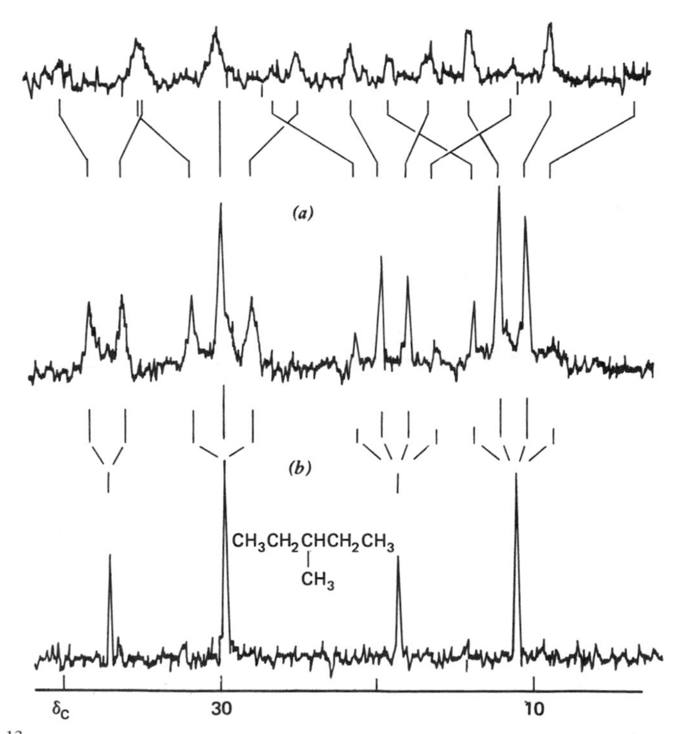

Figure 6.49 ^{13}C NMR spectra of 3-methylpentane. (*a*) No proton decoupling ("coupled" spectrum). (*b*) Protons decoupled at an off-response position ("off-resonance-decoupled" spectrum). (*c*) Protons completely decoupled ("completely decoupled" spectrum). Spectra determined at 25.15 MHz. [From *Carbon-13 NMR Spectroscopy,* by J. B. Stothers, Academic Press, New York, copyright © 1972; with permission.]

APT (attached proton test). APT-type approaches in theory never lead to overlapping signals.

Now let us bring a problem-solving format to NMR analysis. Figure 6.50 provides the ^{1}H and ^{13}C NMR spectra for an unknown ester of molecular formula $C_9H_{10}O_2$. It is straightforward to arrive at the conclusion that this compound contains an ethyl group (δ 4.4, q, 2H would be the methylene protons connected to a methyl group, and δ 1.4, t, 3H, would be the methyl protons connected to a methylene group). Moreover, the 5H integration of the aromatic protons (δ 7.4–7.6, 8.05) indicates a monosubstituted benzene ring. Since the structure must be that of an ester, this compound must be either ethyl benzoate, or phenyl propanoate. That ethyl benzoate is the case is determined by using the shifts of α-protons on Table 6.7; clearly the δ 4.4 chemical shift suggests that the methylene of an ethyl group is attached to a highly deshielding oxygen atom. As an add-on to this analysis, we will use this simple compound to illustrate the use of Table 6.11: the shifts of the ring carbons can be calculated using this table; for example, we would calculate a shift of $\delta_C = 128.7 + 2.0 = 130.7$ for C-1 (the carbon bearing the ester group). The fact that the signal is seen at δ 130.6 constitutes excellent agreement. The calculated shifts (and parenthetically, the observed shifts) for all of the ring carbons are shown here:

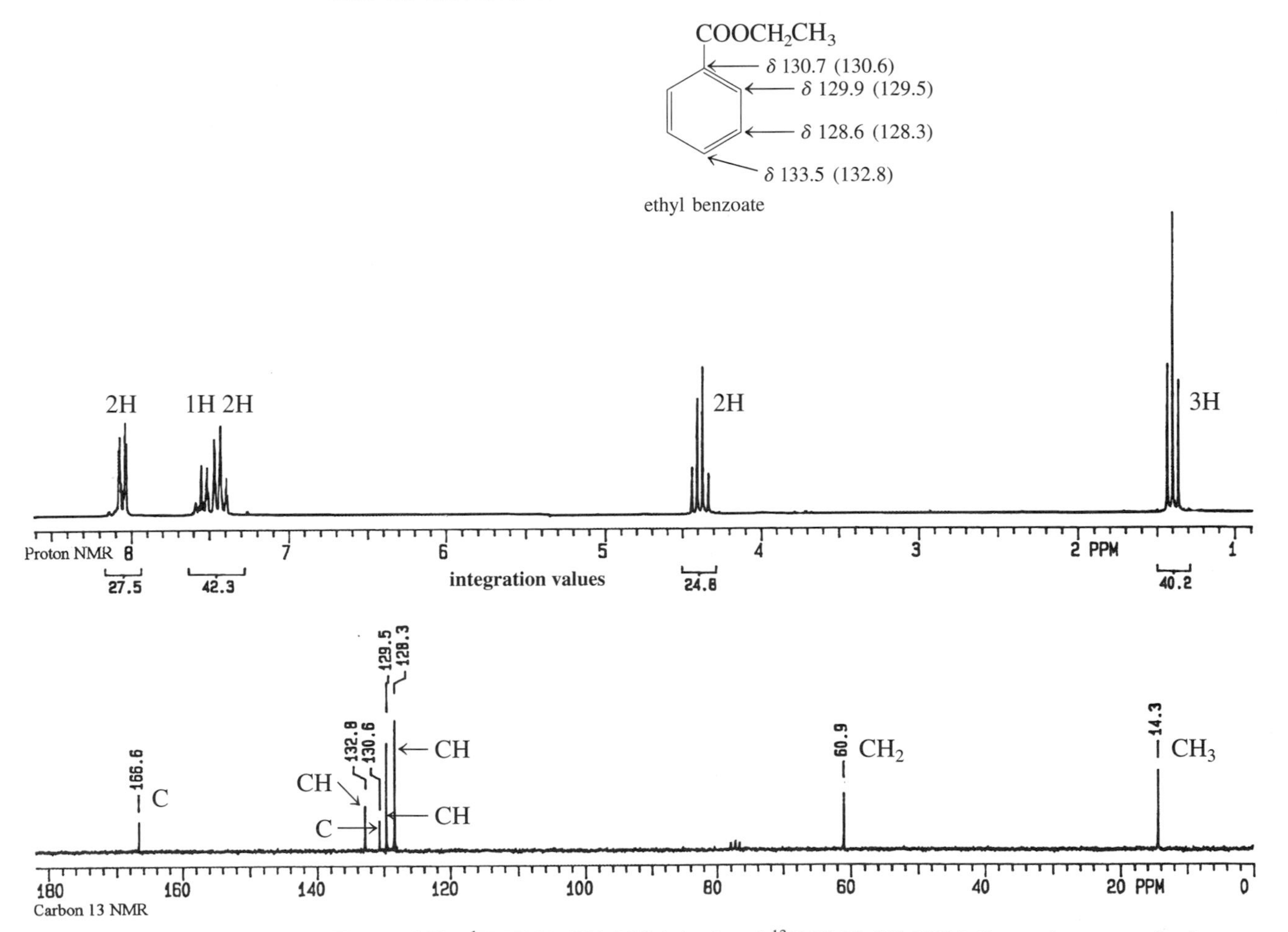

Figure 6.50 ^{1}H NMR (200 MHz) (top) and ^{13}C NMR (50 MHz) (bottom) spectra of unknown ester, $C_9H_{10}O_2$. Spectra courtesy of R. Tomasi, Sunbelt R&T, Inc., Tulsa, Oklahoma.

The problem solving presented by the spectra in Figure 6.51 provides a greater challenge. Again we have an unknown ester, here of molecular formula $C_8H_8O_3$. The ^{1}H NMR spectrum reveals that this compound contains a methyl group (δ 3.92, s, 3H), whose chemical shift suggests attachment to oxygen and thus a methyl ester (Table 6.7). Moreover, it is straightforward to use the 4H integration of the aromatic protons (δ 6.80–7.83) to determine that we have a disubstituted benzene ring. If we assume that we have a substituted methyl benzoate, the only possibility for the third oxygen atom is in a phenolic OH group. The δ 10.78 chemical shift supports a phenolic OH group, and we have either the *ortho*, *meta*, or *para* isomer of methyl hydroxybenzoate.

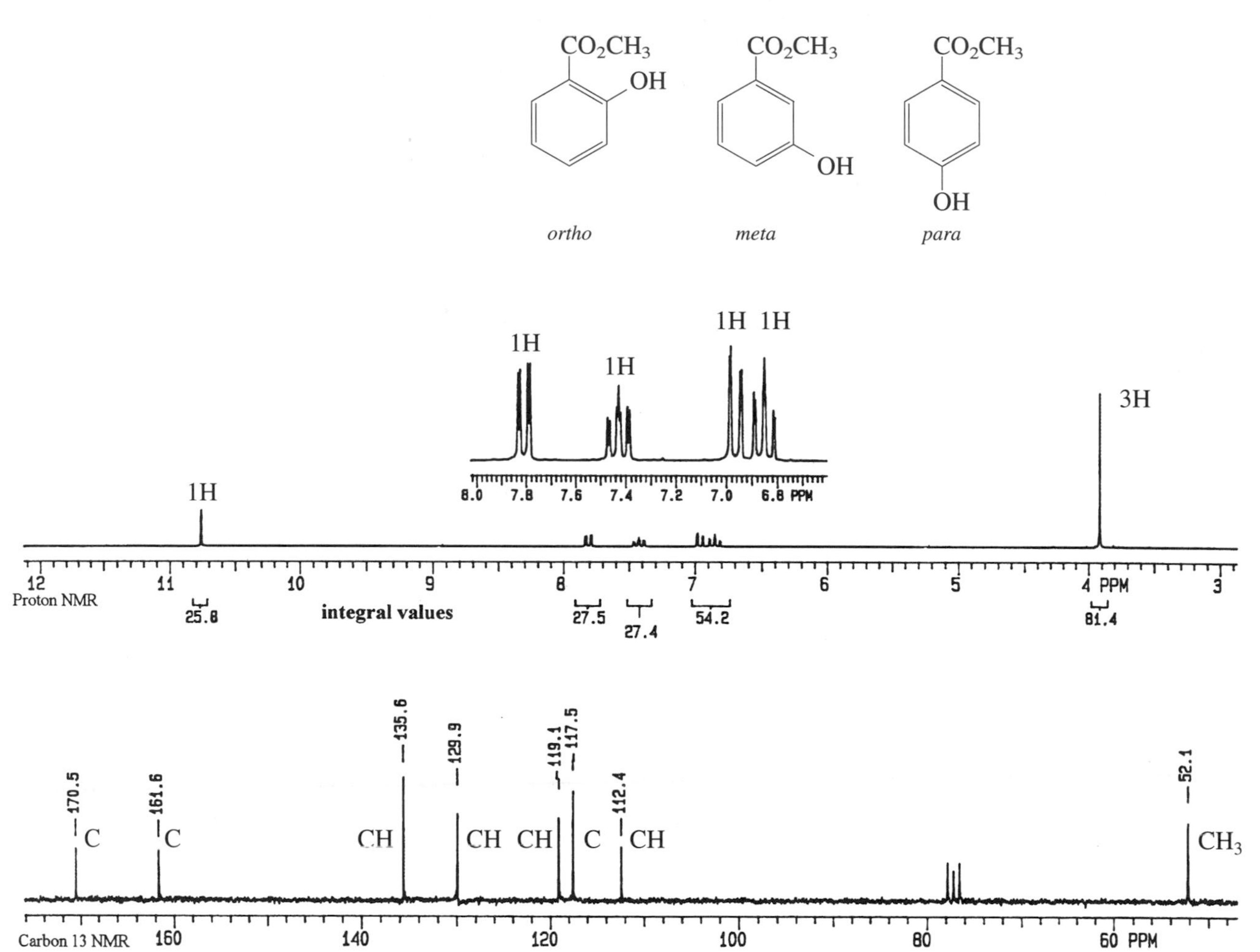

Figure 6.51 ^{1}H NMR (200 MHz) (top) and ^{13}C NMR (50 MHz) (bottom) spectra of unknown ester, $C_8H_8O_3$. Spectra courtesy of R. Tomasi, Sunbelt R&T, Inc., Tulsa, Oklahoma.

We can deduce whether the structure is that of the *ortho*, *meta*, or *para* isomer by use of the ^{1}H and ^{13}C NMR spectra. First consider the ^{1}H NMR spectrum of Figure 6.51. We see that the blowup of the aromatic protons shows essentially two doublets and two triplets (smaller coupling is just barely discernible). If we make the reasonable assumption that only the *ortho* coupling is readily discernible, we may arrive at the correct isomer choice. The *para* isomer should show only two signals (both doublets). The *meta* isomer is expected to show one of the aromatic signals (due to the proton isolated between the two functional groups) as essentially a singlet. Thus the isomer of choice is the *ortho* isomer, and we can finally choose the structure methyl 2-hydroxybenzoate (methyl salicylate, oil of wintergreen):

^{1}H NMR

Assigned signals

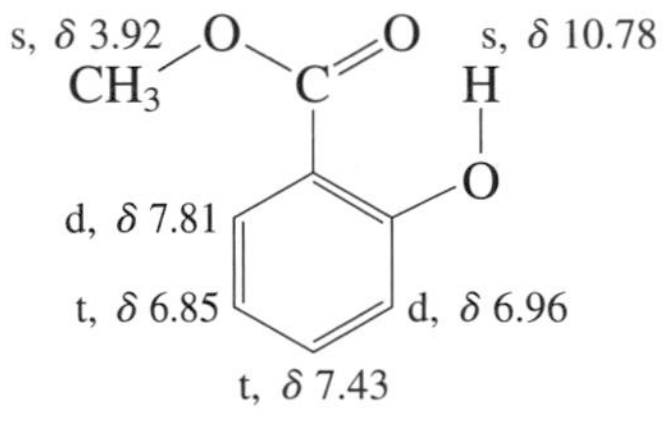

s, δ 3.92 CH_3–O–C(=O); s, δ 10.78 H–O

d, δ 7.81; d, δ 6.96

t, δ 6.85; t, δ 7.43

methyl salicylate

The carbon spectrum is also consistent with the *ortho* isomer; using data from Table 6.11 the calculated carbon shifts shown below are obtained (and they can be compared to the parenthetically listed observed signals).

^{13}C NMR

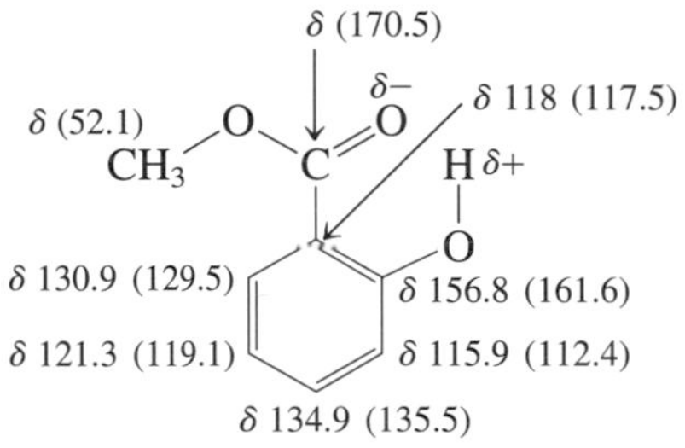

δ (170.5)

δ (52.1) CH_3–O–C=O ($\delta-$); H $\delta+$–O

δ 118 (117.5)

δ 130.9 (129.5); δ 156.8 (161.6)

δ 121.3 (119.1); δ 115.9 (112.4)

δ 134.9 (135.5)

6.4 ULTRAVIOLET SPECTROMETRY

With the advent of IR and especially NMR spectrometry, ultraviolet (UV) spectrometry has been relegated to a lesser position in the overall picture of structural characterization by spectrometric methods. It is, however, still important for the characterization of certain classes of organic compounds, especially conjugated π-systems. Moreover, it is clear that photochemistry, an important area of organic chemistry, has its foundation in the domain of ultraviolet (UV) spectrometry.

Ultraviolet (UV) spectrometry is the spectrometry of electronic transitions. A simple molecular orbital approach gives rise to the energy levels displayed in Figure 6.52. The promotion of an electron from either the π or n (nonbonding) energy level to the π^* level (thus, respectively, $\pi \longrightarrow \pi^*$ and n $\longrightarrow \pi^*$ transitions) gives rise to the majority of UV bands of concern for simple characterization of organic compounds.

What kind of organic structure gives rise to UV absorptions? Based on Figure 6.52 it should be no surprise that the molecule should contain a structural unit with nonbonded (n) electrons or π-electrons. Such a unit giving rise to a UV absorption band is called a chromophore. Although chromophores containing nonbonded electrons (such as on the oxygen atom of ethers or carbonyl groups) are of some importance, here we are more frequently concerned with the π-chromophore, especially when it is part of a conjugated π-system.

σ^*

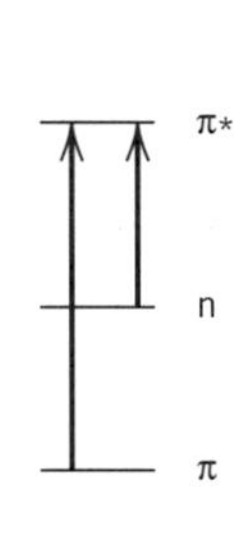

σ

Figure 6.52 Energy levels of molecular orbitals. Illustrated are the $\pi \longrightarrow \pi^*$ and n$\longrightarrow \pi^*$ electronic transitions.

The fundamental nature of a π-chromophore is made clear by realizing the same chromophoric unit (a C=C bond conjugated with a C=O bond) is the source of the UV bands for both mesityl oxide and cholest-4-en-3-one:

mesityl oxide
(4-methyl-3-penten-2-one)

cholest-4-en-3-one

The UV spectra of the two conjugated compounds mentioned above are virtually identical.

In order to place UV in perspective, we should examine the small portion of the electromagnetic spectrum for UV absorptions shown in Figure 6.53. The ultraviolet region of

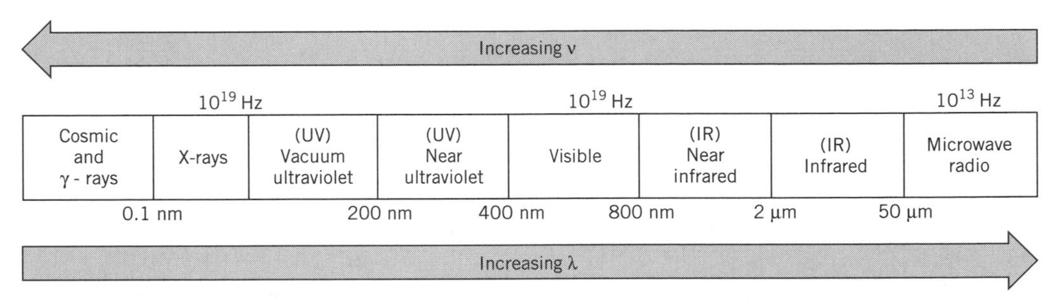

Figure 6.53 The electromagnetic spectrum.

interest involves wavelengths (λ) from 200 to 400 nm (nm = nanometer = 10^{-9} m). This means UV wavelengths are just shorter than those for the visible region (400–800 nm). The region called the vacuum ultraviolet (wavelengths under 200 nm) is not accessible under standard instrument conditions since atmospheric oxygen absorbs in this region.

A fundamental spectrometric equation should be examined here:

$$\Delta E = h\nu$$

where ΔE is the energy change for the transition of interest, h is Planck's constant (6.63 $\times 10^{-34}$ J-s), and ν = frequency. Moreover frequency can be described as

$$\nu = c/\lambda$$

where c is the speed of light (3.00×10^8 m/s) and λ is the wavelength. Thus wavelength is inversely proportional to the size of the energy gap (ΔE), and therefore UV transitions are high energy, short wavelength transitions compared to the lower energy, longer wavelength vibrations in the IR (infrared). Specifically, the wavelength of maximum intensity (λ_{max}) for ethene (ethylene) is 171 nm, which corresponds to an energy of 695 kJ/mol (166 kcal/mol), and this energy is somewhat in excess of the C—C and C—H bond energies found in most organic compounds. IR bands, in contrast, have much lower energies (ca. 20.9 kJ/mol, or 5 kcal/mol) associated with them.

Another aspect of a UV absorption is the degree to which a compound absorbs, and this is measured in terms of its molar absorptivity (ϵ, sometimes called the molar extinction coefficient), where

$$\epsilon_{max} = A_{max}/bC$$

A_{max} is the maximum absorbance (measured at λ_{max}), b is the cell length (in cm), and C = the molar concentration. Thus isoprene (2-methyl-1,3-butadiene) can be characterized by the following observations:

isoprene
colorless, λ_{max} = 222.5 nm
ϵ_{max} = 10,800

A very important aspect of UV bands is how they move with conjugation. An increase in conjugation of a π-system normally results in an increase (bathochromic, or red, shift) in wavelength (λ_{max}) of the principal absorption. A survey of the following compounds reveals that indeed the greater the degree of conjugation, the longer the wavelength of the λ_{max}:

λ_{max}: 171 nm 217 268 330

222.5 178

Sufficiently extensive conjugated systems such as those which exist in vitamin A and β-carotene result in wavelengths long enough to place these compounds in the visible region of the electromagnetic spectrum. Thus these compounds are highly colored.

O—H

vitamin A
yellow, $\lambda_{max} = 325$ nm
$\epsilon_{max} = 50{,}100$

β-carotene
yellow-orange, $\lambda_{max} = 455$ nm
margarine food coloring

The molecular orbitals (MOs) which describe the $\pi \longrightarrow \pi^*$ transition are shown in Figure 6.54. Clearly this concept corresponds to a literal breaking of the π-bond of ethylene, and in fact chemical reactions due to the breaking of that π-bond can be induced in such a fashion.

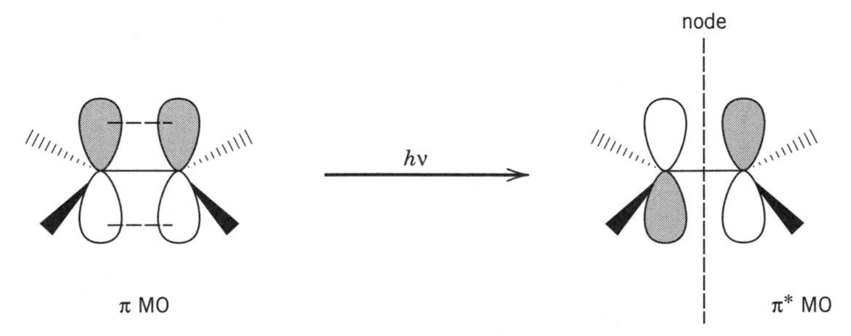

Figure 6.54 Molecular orbital diagrams for ethylene. Irradiation at an energy ($h\nu$) corresponding to 171 nm causes excitation of an electron from the π MO to the π^* MO.

We can extend this MO picture to the 1,3-butadiene system (Figure 6.55) and it is of interest to note that the extended conjugation results in a splitting of the π energy levels such that the highest occupied molecular orbital (HOMO) is now much closer to the lowest unoccupied molecular orbital (LUMO) in 1,3-butadiene than was the case in ethylene. Thus the corresponding $\pi \longrightarrow \pi^*$ transition occurs at a longer wavelength (λ_{max} = 217 nm) for the conjugated butadiene compound (compare the λ_{max} of 171 nm for ethylene).

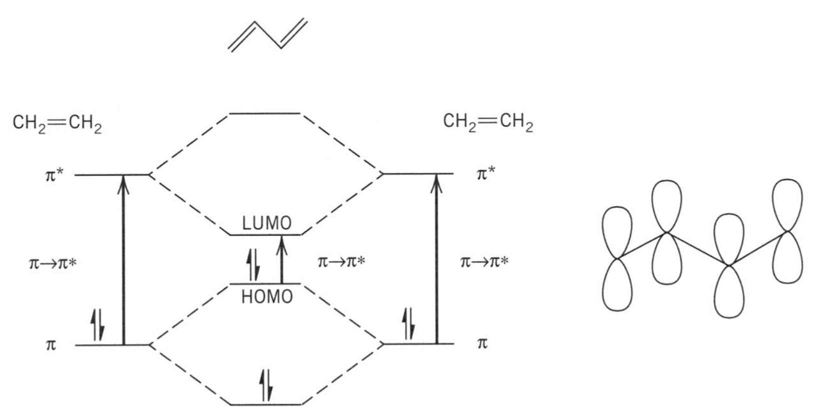

Figure 6.55 The energy diagram for 1,3-butadiene formed from the combination of two ethylene units.

A quantitative approach to structure determination using UV has been developed in steroid systems containing conjugated π-systems. Thus both dienes and enones show UV maxima predicted by the Woodward–Fieser rules. The position of the λ_{max} for these compounds can be predicted with a reasonable degree of accuracy using the information in Tables 6.12 (polyenes) and 6.13 (enones).

Table 6.12 Woodward–Fieser Rules for Diene Absorption

Base values	
Heteroannular diene	214
Homoannular diene	253
Increments for	
Double bond extending conjugation	+30
Alkyl substituent or ring residue	+5
Exocyclic double bond	+5
Polar groups	
acetoxy (OAc)	+0
alkoxy (OR)	+6
thioalkoxy (SR)	+30
halo (Cl, Br)	+5
amino (NR_2)	+60
Solvent correction	+0

endocyclic
exocyclic
exocyclic contribution = 2(+5)

$(\lambda_{max})_{calc}$ = sum
ϵ_{max} = 6,000–35,000

Table 6.13 Woodward–Fieser Rules for Enone Absorption (α,β-Unsaturated Carbonyl Compounds in Ethanol)

Base values

Ketones, $-\overset{\beta}{\underset{\mid}{C}}=\overset{\alpha}{\underset{\mid}{C}}-CO-$ acyclic or 6-membered ring	215 nm
5-membered ring	202 nm
Aldehydes, $-\overset{\beta}{C}=\overset{\alpha}{C}-CHO$	202 nm
Acids and esters, $-\overset{\beta}{C}=\overset{\alpha}{C}-CO_2H(R)$	197 nm

Additional conjugation

$-\overset{\delta}{C}=\overset{\gamma}{C}-\overset{\beta}{C}=\overset{\alpha}{C}-CO-$ etc.	+30 nm
If the second double bond is homoannular with the first,	+39 nm

Addition for each substituent	α	β	γ	δ
—R alkyl (including part of a carbocyclic ring)	+10 nm	12 nm	17 nm	17 nm
—OR	35 nm	30 nm	17 nm	31 nm
—OH	35 nm	30 nm	30 nm	50 nm
—SR	—	80 nm	—	—
—Cl	15 nm	12 nm	12 nm	12 nm
—Br	25 nm	30 nm	25 nm	25 nm
—OCOR (acyloxy)	6 nm	6 nm	6 nm	6 nm
$-NH_2$, —NHR, $-NR_2$	—	95 nm	—	—
If one double bond is exocyclic to one ring	+5 nm			
If exocyclic to two rings simultaneously	+10 nm			

ϵ_{max} 4,500–20,000

Solvent shifts

Water	+8 nm
Methanol	0 nm
Chloroform	−1 nm
Dioxane	−5 nm
Diethyl ether	−7 nm
Hexane	−11 nm
Cyclohexane	−11 nm

A pair of homoannular double bonds are both in the same ring; heteroannular double bonds are in different rings.

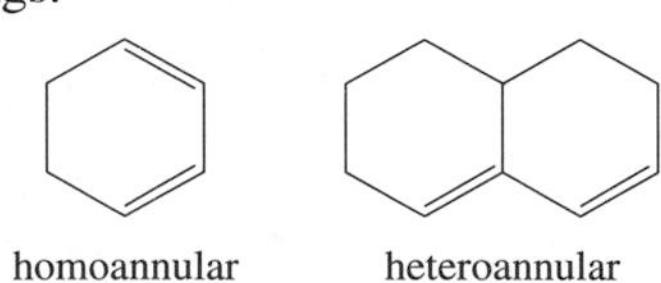

Let us calculate the λ_{max} expected for the following compound:

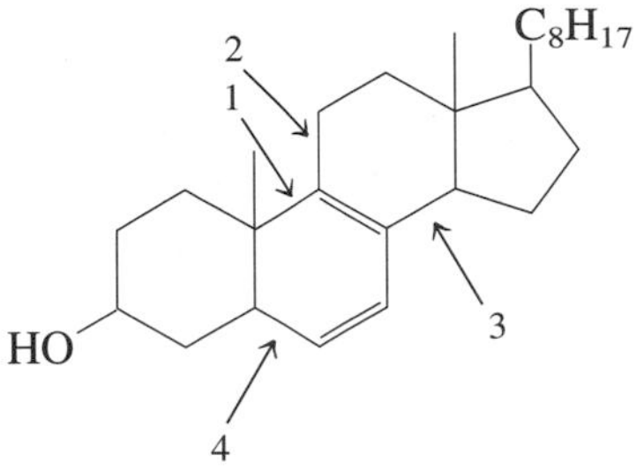

homoannular base	253
four ring residues 4(+5) =	20
$(\lambda_{max})_{calc}$ =	273 nm
$(\lambda_{max})_{obs}$ =	275 nm

Solvent corrections are done as follows: if the absorption was determined in hexane, +11 nm is added to that observed value in order to calculate the λ_{max} expected in ethanol. Let us calculate the λ_{max} expected for diosphenol:

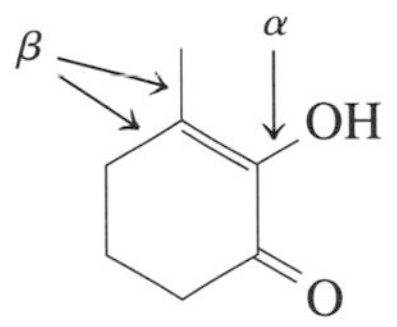

diosphenol

base		215
two β-alkyl/ring residue groups	2(12) =	24
one α-hydroxy group	1(35) =	35
	(λ_{max})calc =	274 nm (in alcohol)
	(λ_{max})obs =	270 nm (in alcohol)

Another correlation which is important for aromatic chemistry is that of Scott[12] who found that the effect of substituents on the spectrum of benzene could be correlated with structure (Table 6.14). Use of this table is illustrated by the following example:

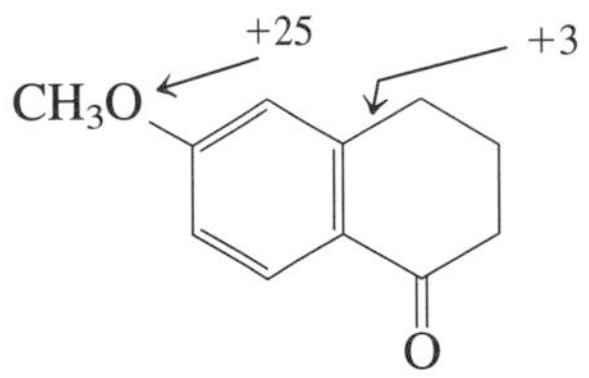

6-methoxytetralone

$(\lambda_{max})_{calc}$ = 246 + 3 + 25 (values from Table 6.14) = 274 nm
$(\lambda_{max})_{obs}$ = 276 nm

[12] A. I. Scott, *Interpretation of the Ultraviolet Spectra of Natural Products* (Pergamon Press, Elmsford, NY,

Table 6.14 Data for Calculation of the Principal Band of Substituted Benzene Derivatives, Ar—COG (in EtOH)

$ArCOR/ArCHO/ArCO_2H/ArCO_2R$		λ_{max} (nm)
Parent chromophore: $Ar{=}C_6H_5$—		
G = alkyl or ring residue (e.g., ArCOR)		246
G = H (ArCHO)		250
G = OH, OAlk ($ArCO_2H$, $ArCO_2R$)		230
Increment for each substituent on Ar		
—alkyl or ring residue	*o*-, *m*-	+3
	p-	+10
—OH, —OCH_3, —OR	*o*-, *m*-	+7
	p-	+25
—O^- (oxyanion)	*o*-	+11
	m-	+20
	p-	+78[a]
—Cl	*o*-, *m*-	+0
	p-	+10
—Br	*o*-, *m*-	+2
	p-	+15
—NH_2	*o*-, *m*-	+13
	p-	+58
—NHAc	*o*-, *m*-	+20
	p-	+45
—$NHCH_3$	*p*-	+73
—$N(CH_3)_2$	*o*-, *m*-	+20
	p-	+85

[a]This value may be decreased markedly by steric hindrance to coplanarity.

Other systematic studies on the variation of spectrum with structural change of such types of compounds as conjugated polyenes and acetylenes, semicarbazones, 2,4-dinitrophenylhydrazones, and oximes—conjugated and unconjugated—are discussed in the books by Gillam, Stern, and Timmons referred to on p. 199.

A definite and clear methodology as to how to use model compounds and the theory of additivity has developed for UV. First recall that as described above mesityl oxide and cholest-4-en-3-one (despite gross differences in their overall structures) have virtually superimposable UV spectra due to their nearly identical (circled) chromophores.

mesityl oxide

cholest-4-en-3-one

The theory and practice of UV lead to the following reasonable expectations. First the following diene and cyclohexene molecules are expected to yield virtually identical UV spectra, but the two isolated double bonds of the diene should give rise to an absorption that is twice as intense ($\epsilon_{diene} \approx 2\epsilon_{ene}$).

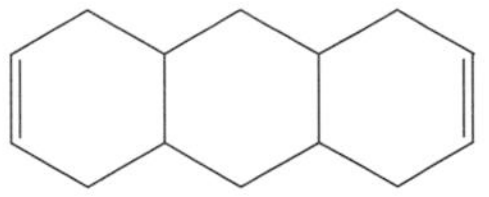

nonconjugated diene

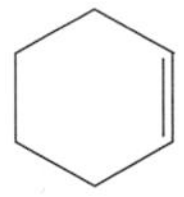

cyclohexene

Also the enone shown below should show virtually the same spectrum as that obtained by adding the UV spectrum of cyclohexanone to cyclohexene.

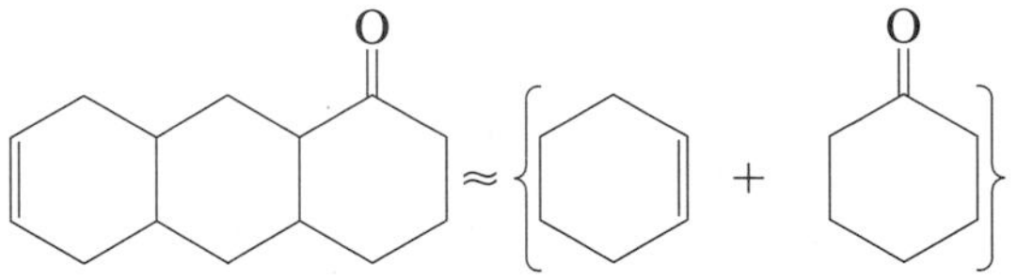
O
≈
+
O

nonconjugated enone   cyclohexene   cyclohexanone

Now let us apply these simple principles to research investigations. Terramycin is an antibiotic substance, and an important degradation product of terramycin was isolated:

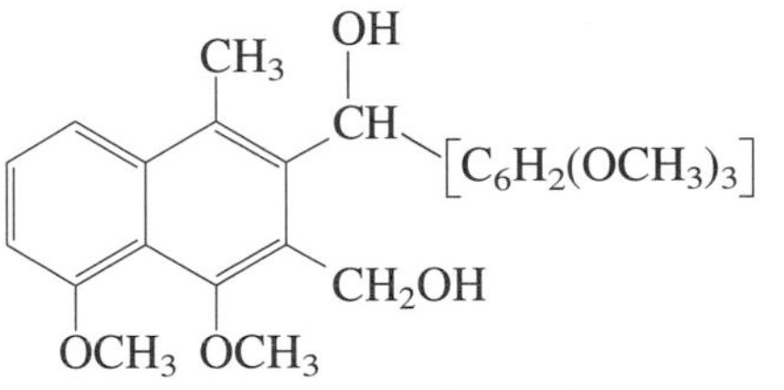
CH3
OH
CH
[C6H2(OCH3)3]
CH2OH
OCH3 OCH3

terramycin degradation product

The portion of the structure that is bracketed was known to be a trimethoxyphenyl group. We can assume that the CHOH group is sufficient to buffer the naphthalene ring system from the bracketed trimethoxybenzene system and thus cause these two units to act as isolated chromophores. The tacit assumption is made that the CH_3O groups will have virtually the same chromophoric effect here as do OH groups. Thus the following equivalence was assumed to allow use of the available model compound.

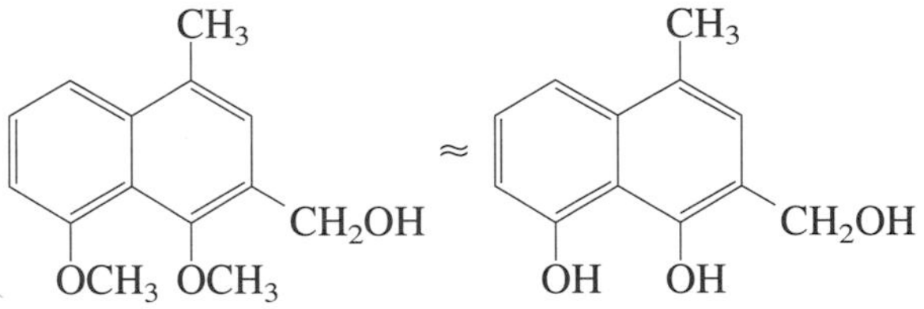
CH3
CH2OH
OCH3 OCH3
≈
CH3
CH2OH
OH OH

naphthalene portion of terramycin   naphthalene portion of model

As might be expected, 1,2,3-, 1,2,4-, and 1,3,5-trimethoxybenzenes were known compounds. The 1,2,4 compound provided by far the best fit when used as half of a 50:50 mixture with the naphthalene model shown:

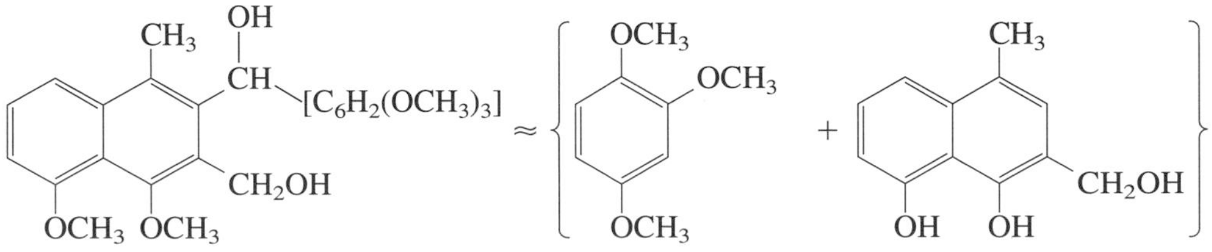
CH3
OH
CH
[C6H2(OCH3)3]
CH2OH
OCH3 OCH3
≈
OCH3
OCH3
OCH3
+
CH3
CH2OH
OH OH

degradation product   1,2,4-trimethoxybenzene   naphthalene portion of model

Thus an improved structural proposal for the terramycin degradation product could be made:

Terramycin degradation product is one of these three structures.

Another example of the use of UV to elucidate structure was the identification of the two compounds which arise from the treatment of quinoline with cyanogen bromide and hydrogen cyanide.

quinoline $\xrightarrow[HCN]{BrCN}$ dicyano product I + dicyano product II

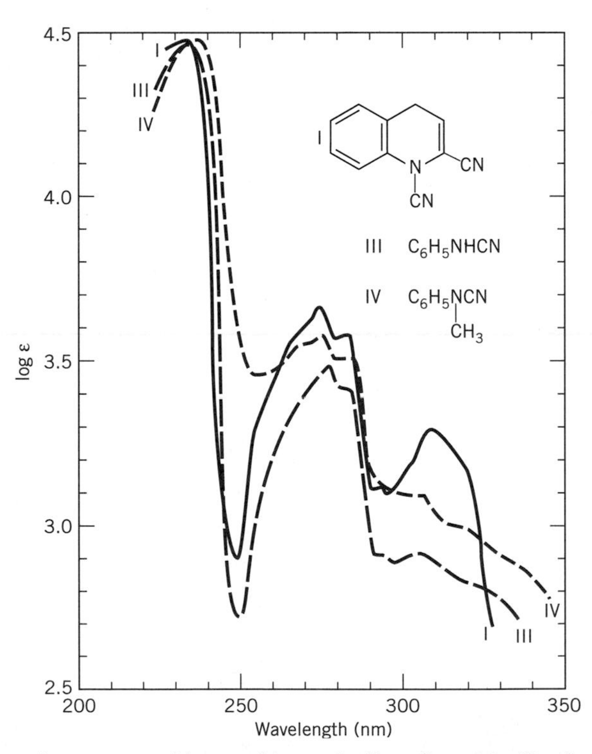

Figure 6.56 Absorption spectra of low-melting quinoline dicyanide (I), phenylcyanamide (III), and *N*-methylphenylcyanamide (IV). [Reproduced from the *Journal of the American Chemical Society* through the courtesy of the Society and Professor Carl R. Noller.]

Since this study preceded NMR, a simple analysis using the integrated area to identify the number of olefinic protons (and thus differentiate I and II) was not available. Moreover, chemical and synthetic confirmation methods were not available. It did seem possible that model compounds would allow a UV analysis of this system:

model compound III   model compound IV   model compound V

It seemed likely that both model compounds III and IV would serve as good models for dicyano product I since the vinyl cyanide portion (circled) of I showed negligible UV absorption compared to the substituted benzene system. Model compound V in turn should serve as a model for dicyano product II. The choices allowed structure identification as the UV spectra of I, III, and IV are nearly superimposable (Figure 6.56), as are the UV traces for II and V (Figure 6.57).

Steric inhibition of resonance can have a dramatic effect upon the appearance of UV spectra. For example, the UV spectrum of benzoquinuclidine resembles that of benzene rather than that of aniline ($C_6H_5NH_2$). This is because the geometric constraints of benzoquinuclidine prevent overlap of the orbital containing the nonbonded electrons of the amine with the orbitals on the benzene ring.

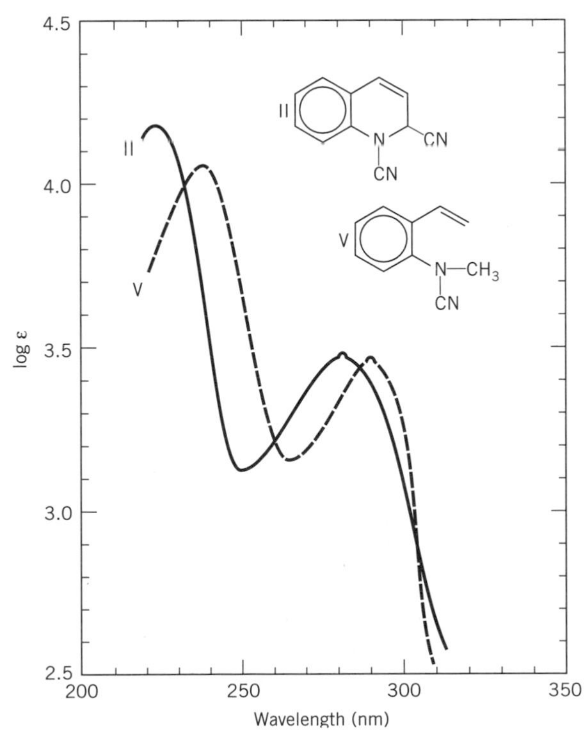

Figure 6.57 Absorption spectra of high-melting "quinoline dicyanide" (II) and *N*-methyl-*σ*-styrylcyanamide (V). [Reproduced from the *Journal of the American Chemical Society* through the courtesy of the Society and Professor Carl R. Noller.]

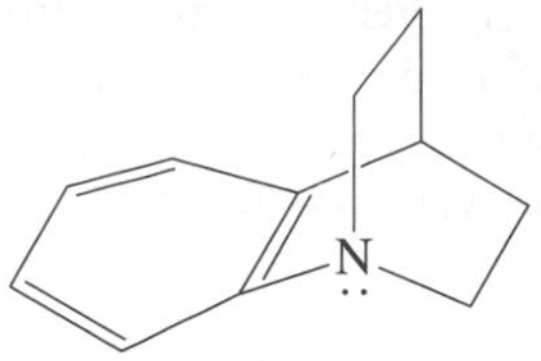

benzoquinuclidine

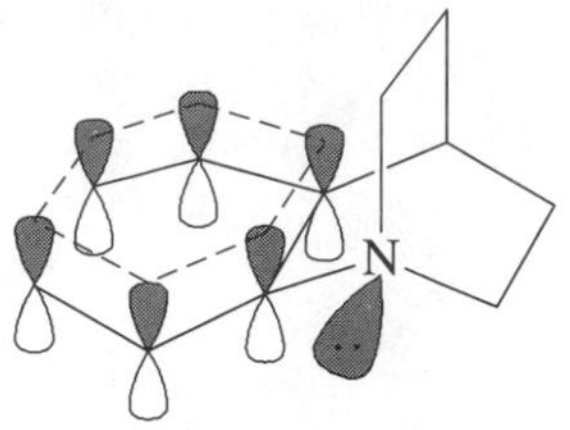

orbitals of benzoquinuclidine

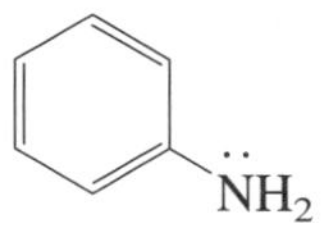

aniline

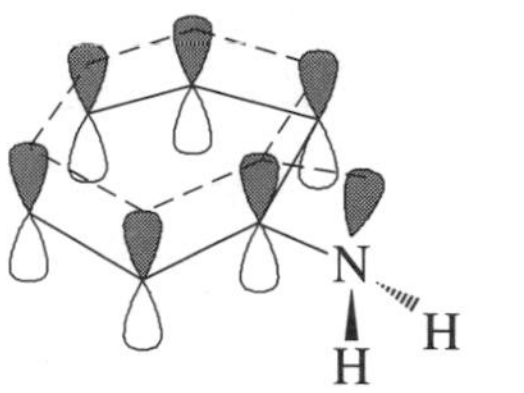

orbitals of aniline

Steric inhibition of resonance also plays a role with *t*-alkylnitrobenzenes. The UV spectrum of *p*-*t*-butylnitrobenzene reveals a peak at $\lambda_{max} = 265$ nm ($\epsilon_{max} > 10{,}000$), and the fact the *ortho* isomer has no such peak is explained by assuming that the steric bulk of an *ortho* *t*-butyl group disrupts the planarity necessary for resonance of the nitro group with the benzene ring.

This is depicted both in a molecular model rendition (Figure 6.58) showing a twisted nitro group, and by resonance forms:

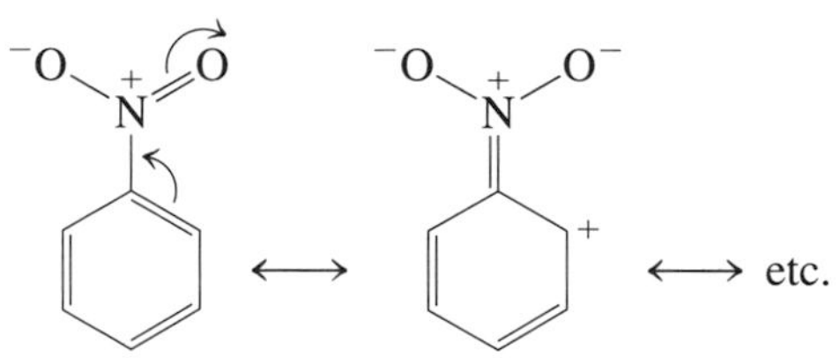

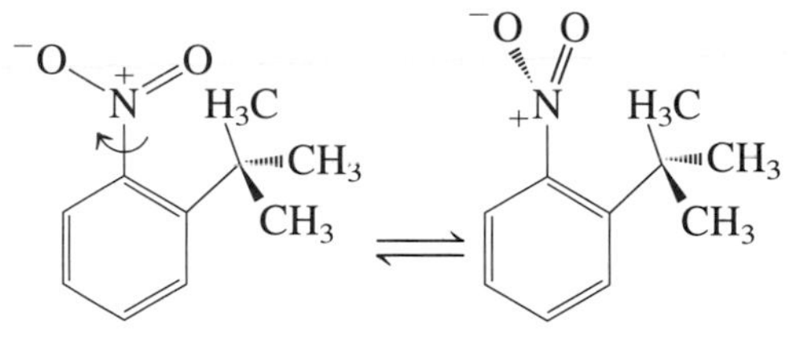

$\longleftrightarrow$ $\longleftrightarrow$ etc.

The biphenyl system also provides cases in which steric inhibition of resonance is a useful concept. Unsubstituted biphenyl has a UV spectrum that is clearly not simply the sum of the spectra due to two benzene rings. This is because the two rings are coplanar and the overlapping orbitals produce a conjugated π-system which is quite different from a simple benzene ring. If, however, the *ortho* positions of the biphenyl system are substituted by sufficiently bulky groups (e.g., the *ortho* methyl groups of dimesityl below), the coplanarity of the two rings is disrupted and, not surprisingly, the UV spectrum of dimesityl is different from that of biphenyl and is essentially identical to that of two isolated 1,3,5-trimethylbenzene (mesitylene) rings.

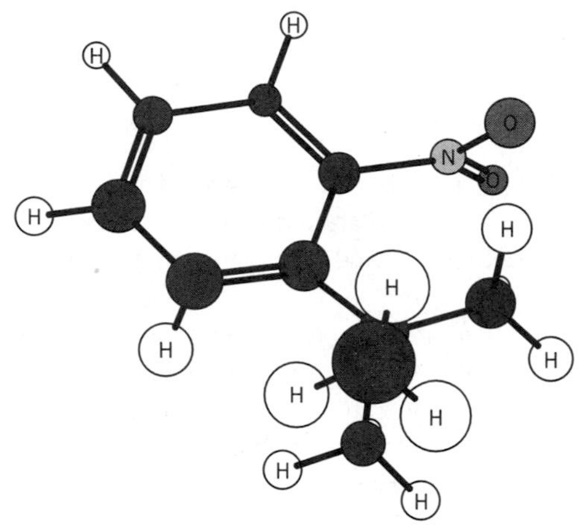

Figure 6.58 Model of *ortho-t*-butylnitrobenzene.

Table 6.15 Ultraviolet Solvents

Solvent	Lower wavelength limit (nm)
Water	205
Ethanol (95% or absolute)	210
Hexane	210
Cyclohexane	210
Methanol	210
Diethyl ether	210
Acetonitrile	210
Tetrahydrofuran	220
Dichloromethane	235
Chloroform	245
Carbon tetrachloride	265
Benzene	280

CH_3 H_3C
H_3C — — CH_3
CH_3 H_3C

biphenyl
(coplanar)

dimesityl
(rings not coplanar)

The diketo olefin shown below, although not formally conjugated, shows UV absorptions (λ_{max} = 223 nm, logϵ_{max} = 3.36; λ_{max} = 296 nm; λ_{max} – 307 nm, log ϵ_{max} – 2.41) consistent with an orbital interaction between the C=C and C=O portions of the molecule:

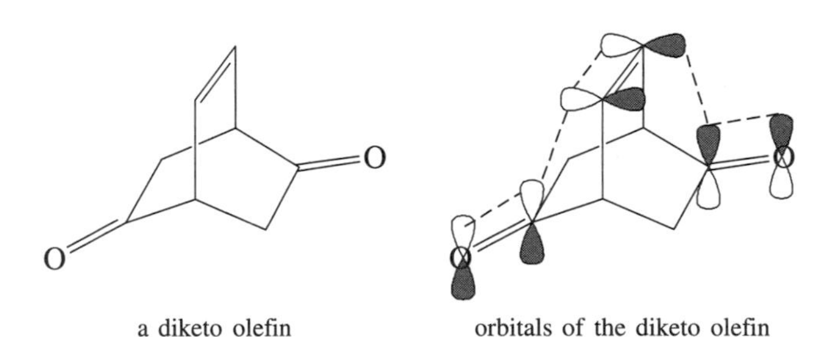

a diketo olefin

orbitals of the diketo olefin

The most common solvents for ultraviolet spectroscopy are cyclohexane, 95% ethanol, benzene-free absolute ethanol, and dioxane. A list of UV solvents, in decreasing order of useable range, is given in Table 6.15. The approximate concentration to use may be estimated if any information is available as to the expected value of the maximum value of ϵ. Thus the absorbance scale runs from 0.0 to 2.4 absorbance units, and, for a plot such that a maximum with $\epsilon = 10^4$ has an absorbance of 2.0, it is seen that the correct concentration is $2/10^4$ or 2×10^4 molar when a 1-cm cell is employed.

6.5 PROBLEMS

Sample Problem

A compound of molecular formula $C_5H_{12}O$ has a bp ≈124–128°C. The IR spectrum and both the ^1H and ^{13}C NMR spectra for this compound are listed below.[13] Deduce the structure of this compound and assign all NMR signals and the important IR signals.

Solution. The IR spectrum displays a broad, intense band at 3340 cm^{-1} which suggests that the compound is an alcohol. The ^1H NMR spectrum is also consistent with an alcohol: the signal at δ 3.65 (d, 2H) is consistent with protons on a carbon bearing the OH, and furthermore implies that the compound is a primary alcohol with another CH_2 at C-2, since the δ 3.65 signal is a triplet. The ^{13}C NMR spectrum suggests that the compound has

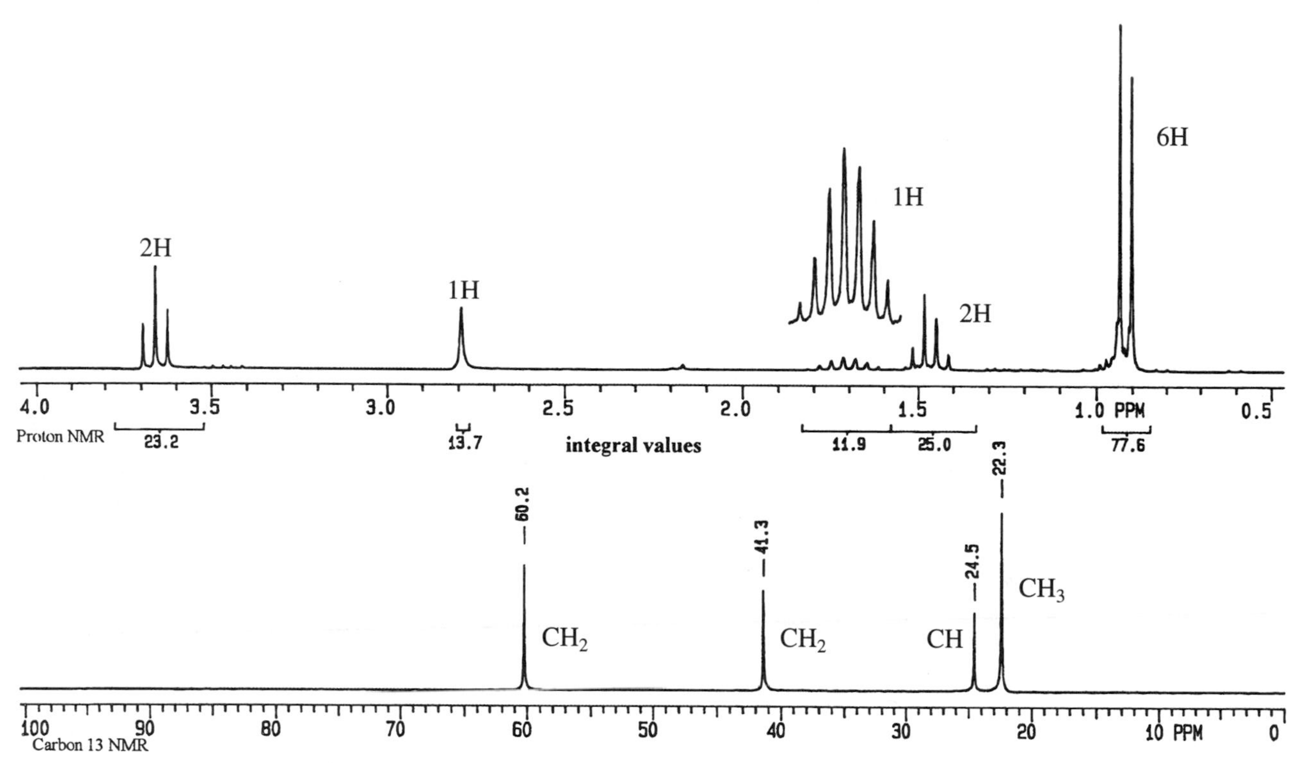

^1H NMR (top) and ^{13}C NMR spectra (bottom) for Sample Problem. Spectra courtesy of R. Tomasi, Sunbelt R&T, Inc., Tulsa, Oklahoma.

[13]Notations indicating CH_3, CH_2, and CH, and C (quaternary carbons) listed on the ^{13}C NMR spectrum are structural conclusions drawn from auxiliary information (off-resonance decoupled spectra, or 2D-NMR related techniques such as the Attached Proton Test, etc.).

symmetry, since we see four signals in that spectrum compared to the five carbons in the molecular formula. The 6H doublet at δ 0.91 and the septet at δ 1.7 reveal an isopropyl unit in this structure and thus we can piece together a preliminary structural proposal: 3-methyl-1-butanol (also called isopentyl or isoamyl alcohol). The C—O stretch is at 1059.05 cm^{-1} in the IR, and the NMR signals are assigned on the accompanying structures. A similar route to structure determination could be followed using the information about methyl, methylene, and methine carbons provided on the ^{13}C NMR spectrum. Confirmation of the OH signal could be made by looking for a concentration dependence for the δ 2.8 chemical shift.

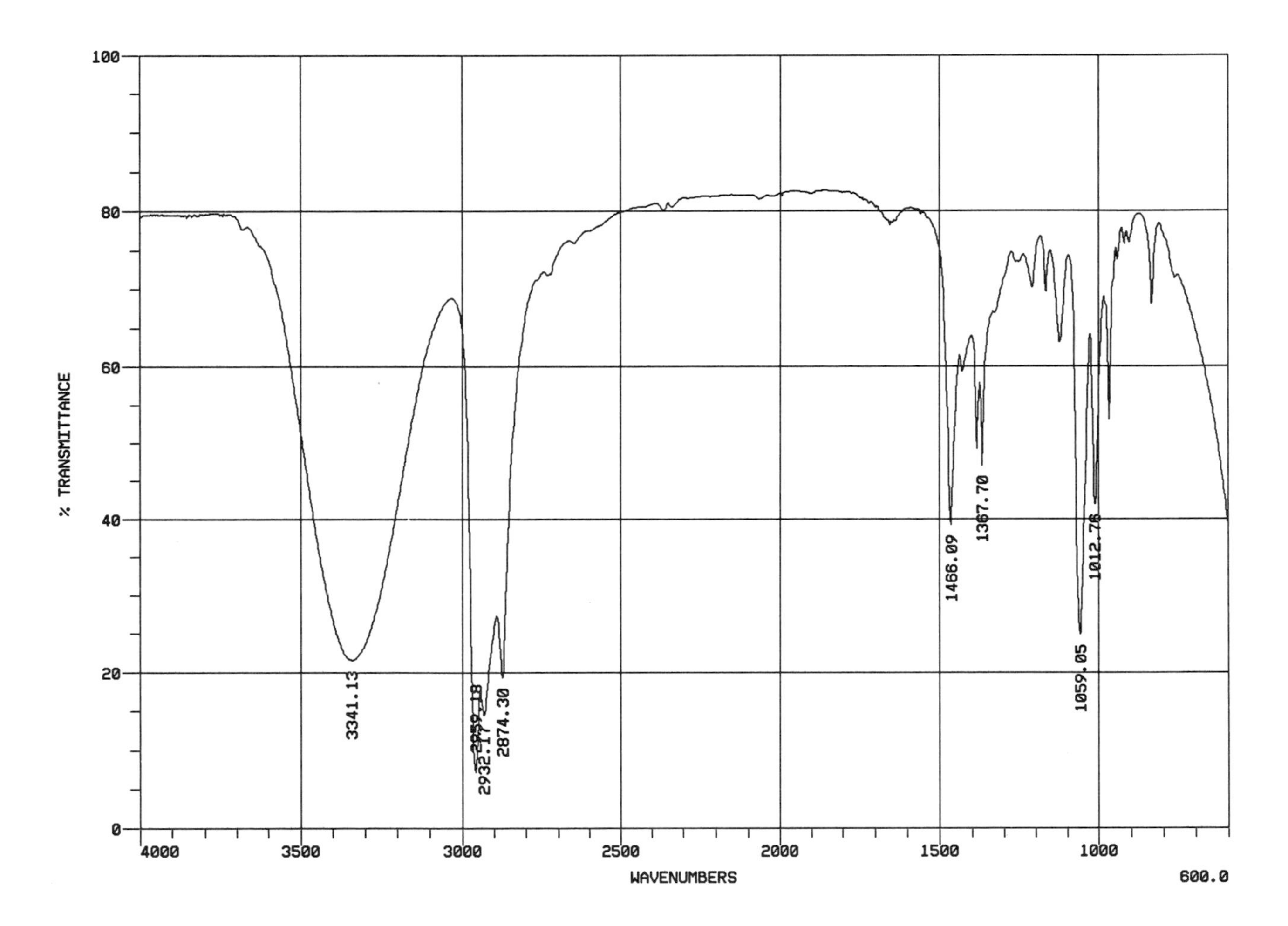

IR spectrum for Sample Problem. Spectrum courtesy of R. Tomasi, Sunbelt R&T, Inc., Tulsa, Oklahoma.

1. A compound of molecular formula $C_5H_{12}O$ has a bp ≈ 114–116°C. The IR spectrum and both the ^1H and ^{13}C NMR spectra for this compound are shown below. Deduce the structure of this compound and assign all NMR signals and the important IR signals.

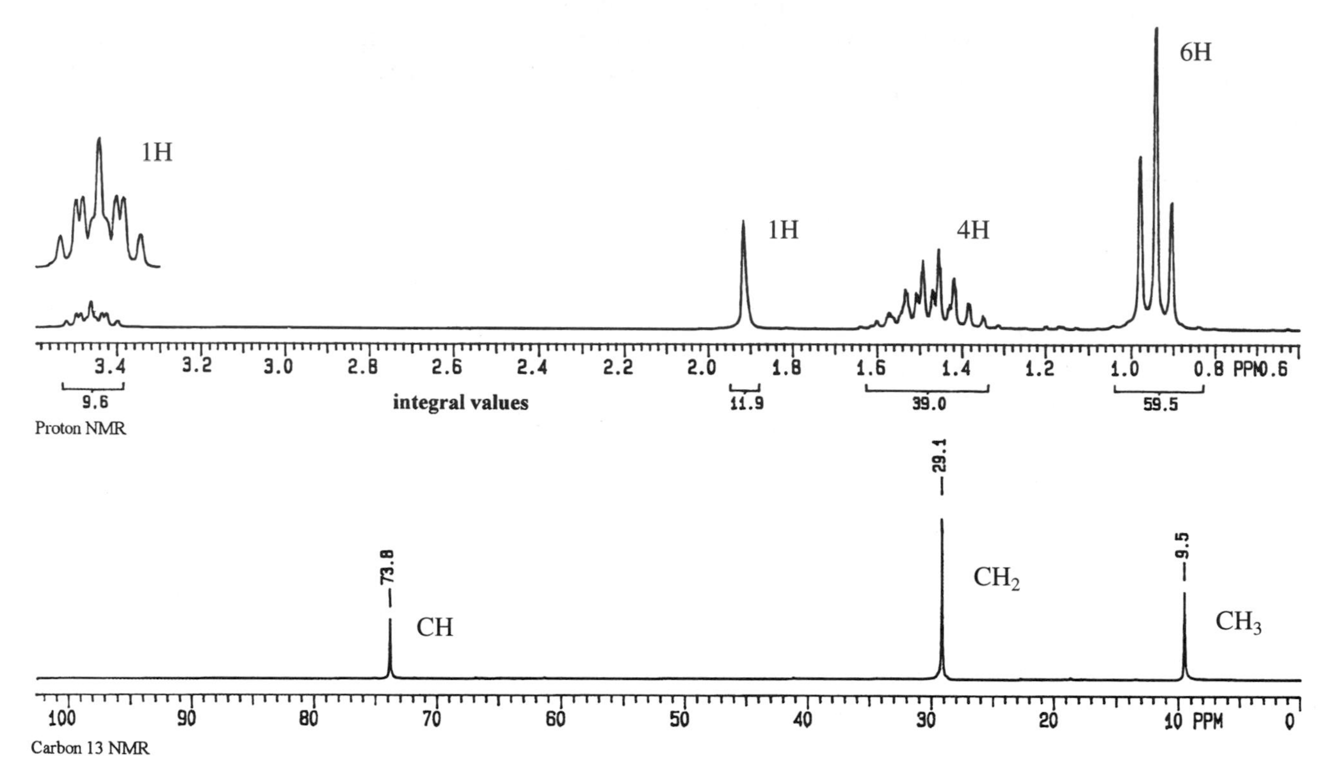

^1H NMR (top) and ^{13}C NMR (bottom) spectra for Problem 1. Spectra courtesy of R. Tomasi, Sunbelt R&T, Inc., Tulsa, Oklahoma.

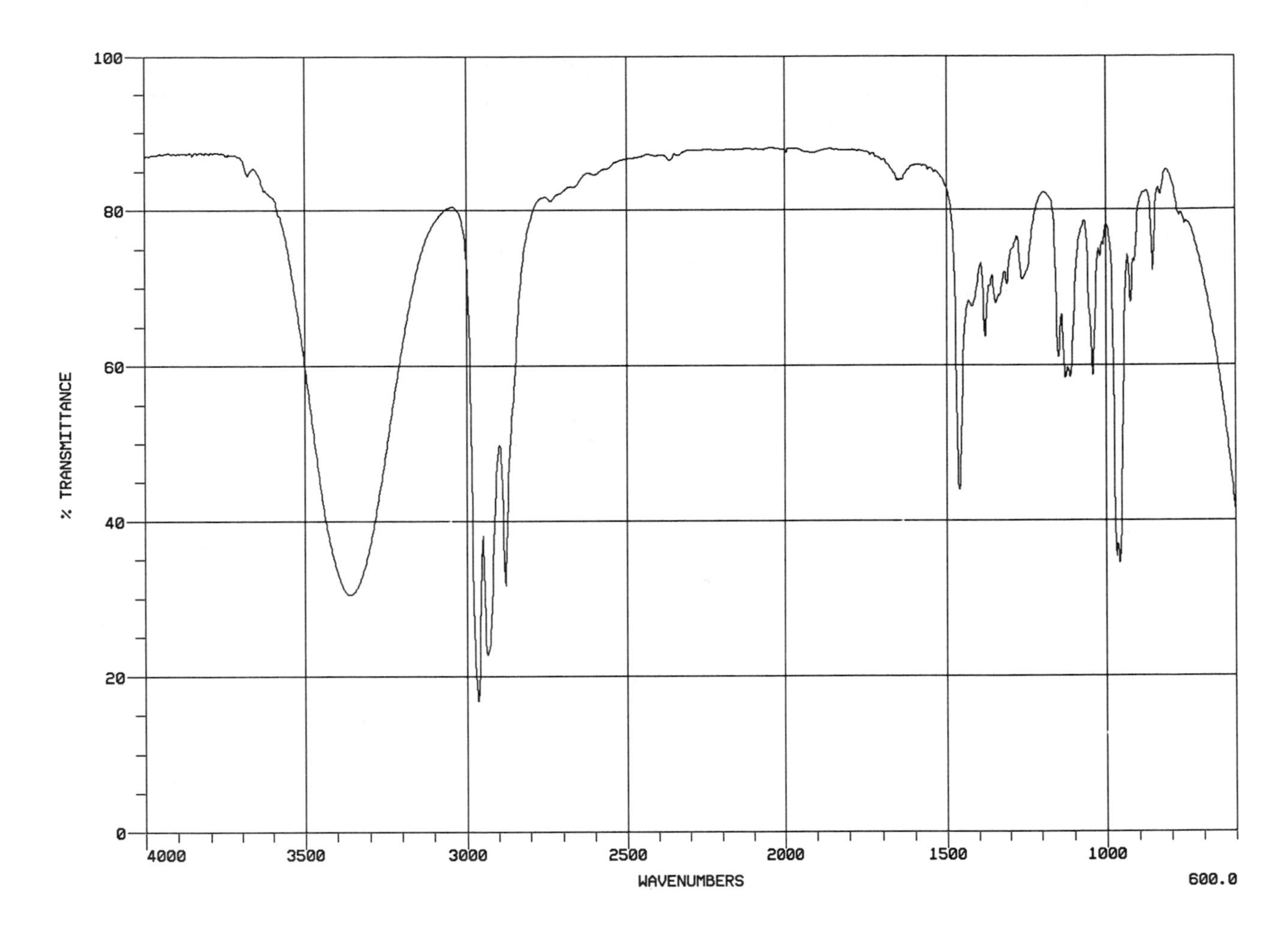

FT IR spectrum for Problem 1. Spectrum courtesy of R. Tomasi, Sunbelt R&T, Inc., Tulsa, Oklahoma.

2. A compound of molecular formula C_7H_8O has a bp ≈ 150–153°C. The IR spectrum and both the ^1H and ^{13}C NMR spectra for this compound are shown below. Deduce the structure of this compound and assign all NMR signals and the important IR signals.

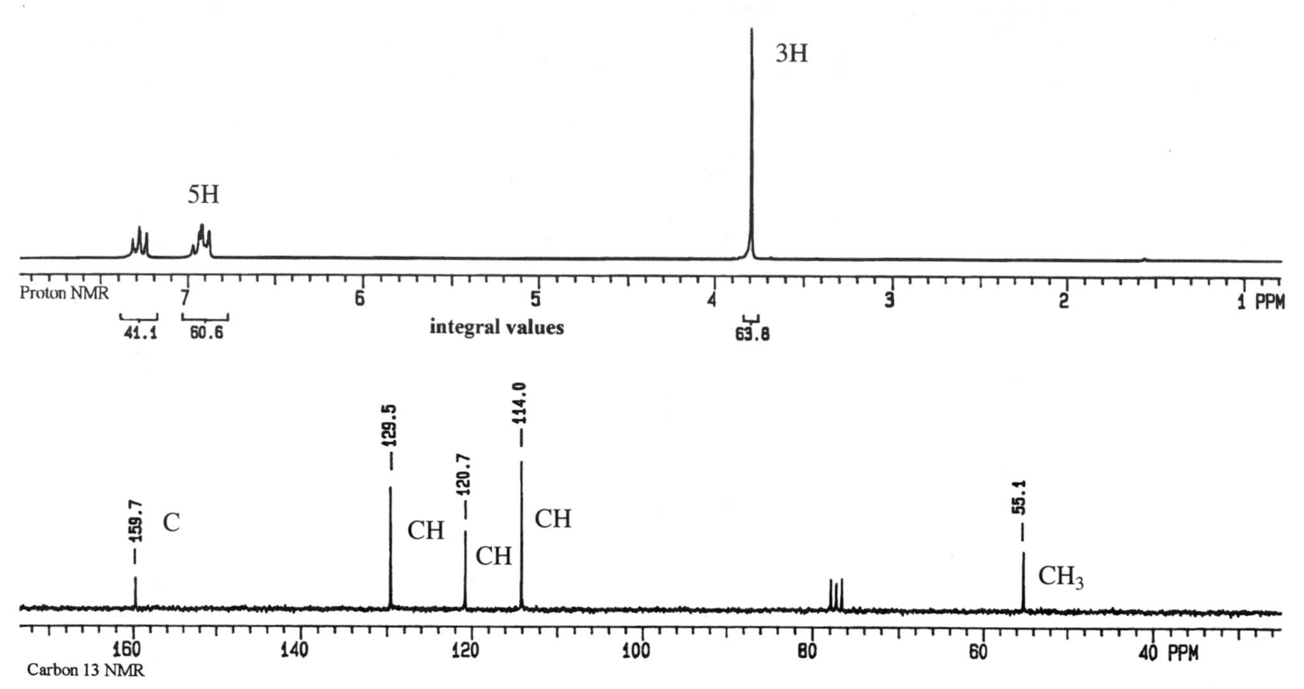

^1H NMR (top) and ^{13}C NMR spectra (bottom) for Problem 2. Spectra courtesy of R. Tomasi, Sunbelt R&T, Inc., Tulsa, Oklahoma.

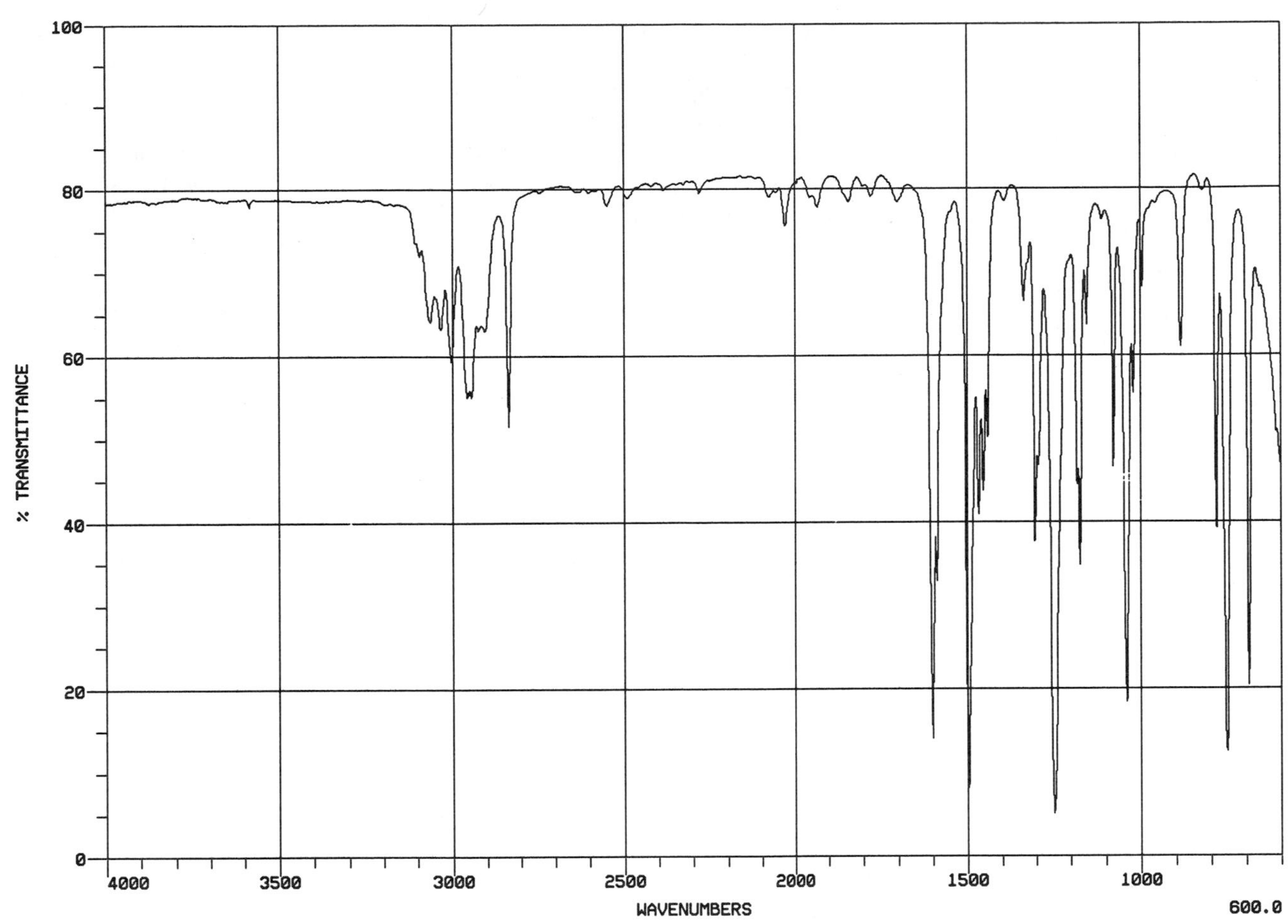

FT IR spectrum for Problem 2. Spectrum courtesy of R. Tomasi, Sunbelt R&T, Inc., Tulsa, Oklahoma.

3. A compound with a fishy odor burns with a sooty flame, gives a positive nitrogen test, and when treated with alkaline benzenesulfonyl chloride eventually gives a homogeneous solution. The compound when highly purified melted at 45°C and boiled near 200°C. Once melted, it was difficult to obtain the compound in a solid state. The IR spectrum and both the ^{1}H and ^{13}C NMR spectra for this compound are given below. Deduce the structure of this compound and assign all NMR signals and the important IR signals.

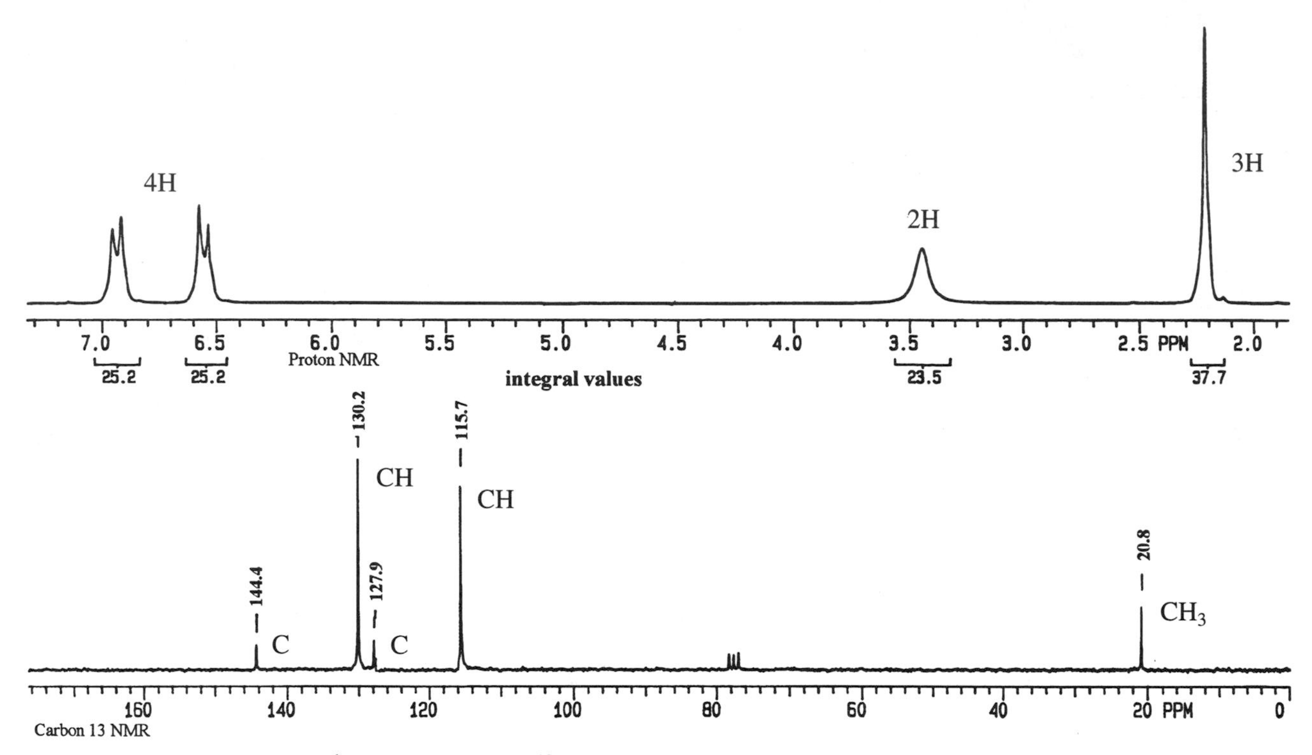

^{1}H NMR (top) and ^{13}C NMR (bottom) spectra for Problem 3. Spectra courtesy of R. Tomasi, Sunbelt R&T, Inc., Tulsa, Oklahoma.

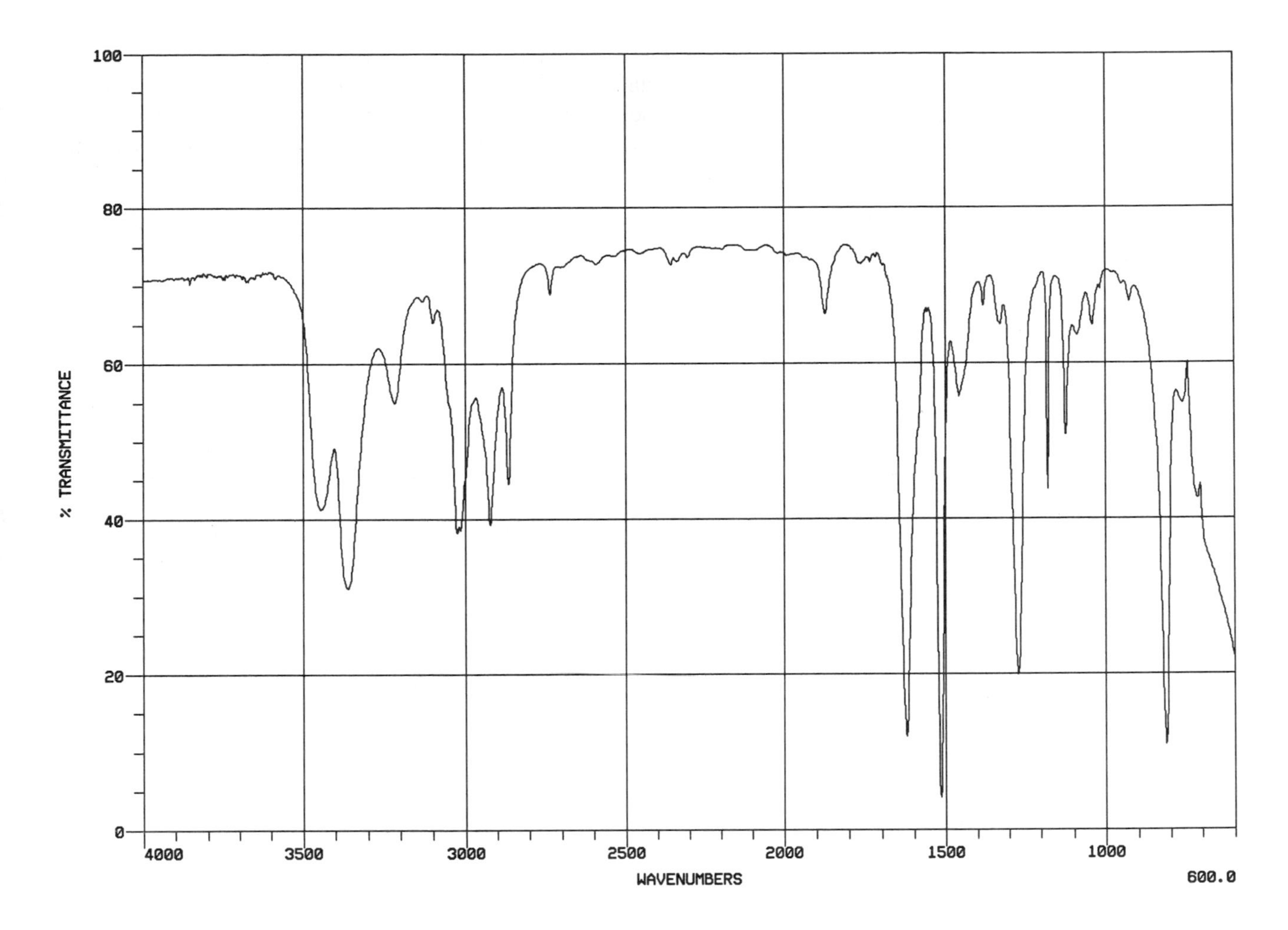

FT IR spectrum for Problem 3. Spectrum courtesy of R. Tomasi, Sunbelt R&T, Inc., Tulsa, Oklahoma.

4. A compound has a molecular formula of $C_8H_8O_2$. This compound is water insoluble, but gives a positive 2,4-dinitrophenylhydrazone test and a positive Tollens test. The IR spectrum and both the ^1H and ^{13}C NMR spectra for this compound are given below. Deduce the structure of this compound and assign all NMR signals and the important IR signals. Write reactions for the positive chemical tests.

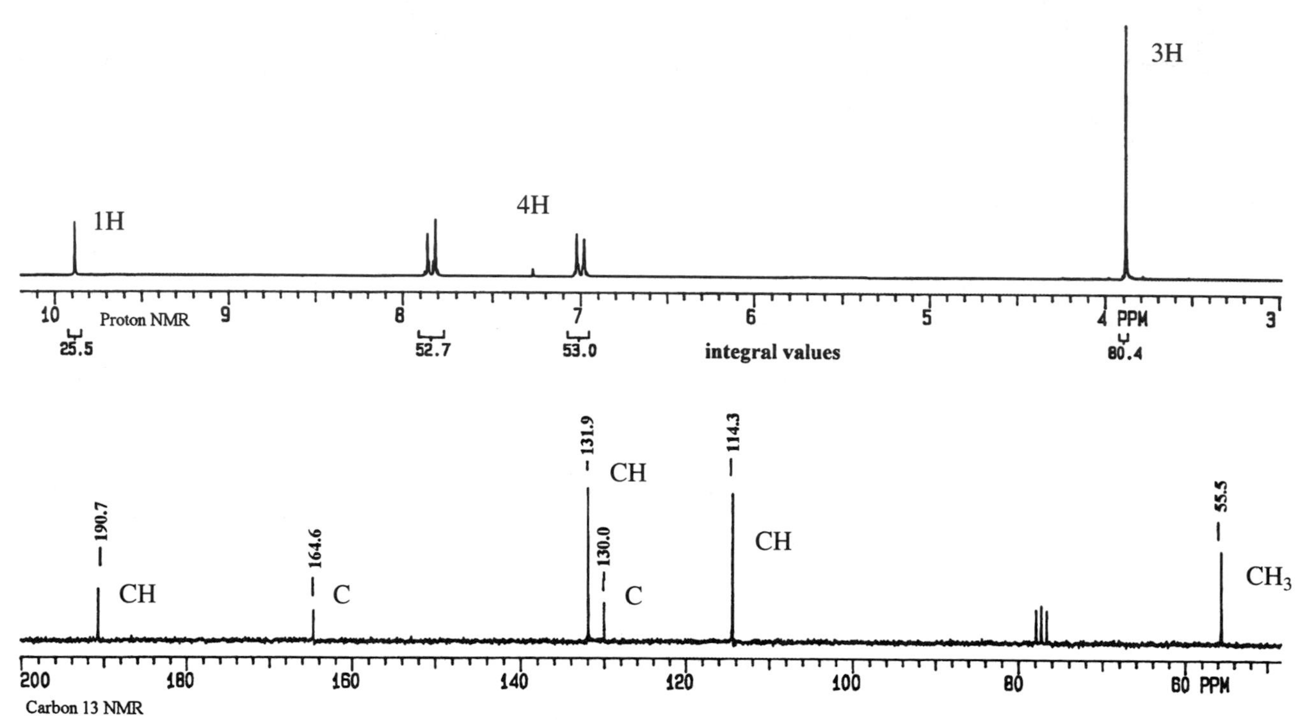

^1H NMR (top) and ^{13}C NMR (bottom) spectra for Problem 4. Spectra courtesy of R. Tomasi, Sunbelt R&T, Inc., Tulsa, Oklahoma.

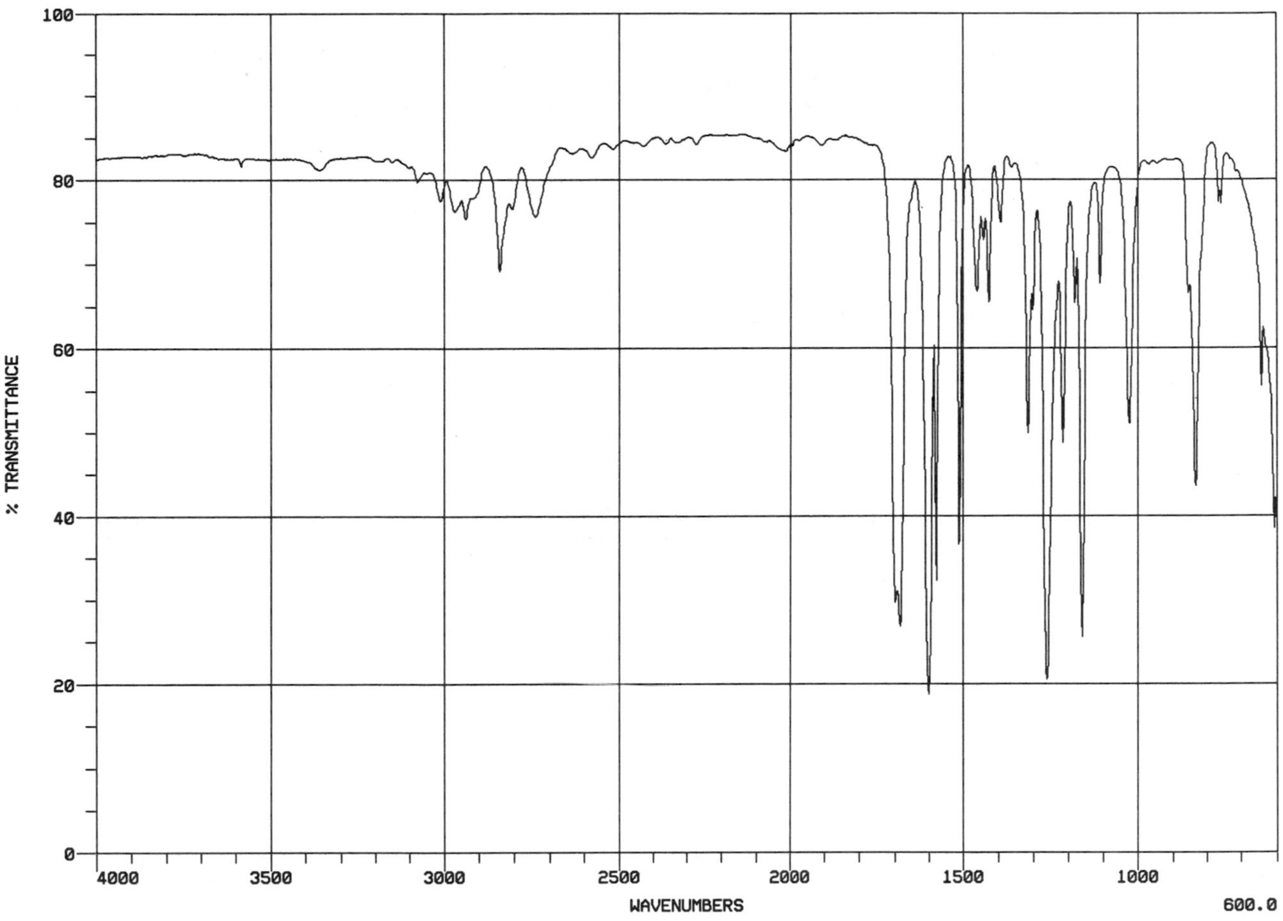

% TRANSMITTANCE

WAVENUMBERS

FT IR spectrum for Problem 4. Spectrum courtesy of R. Tomasi, Sunbelt R&T, Inc., Tulsa, Oklahoma.

5. A compound of mp 139° gave a positive chlorine test after sodium fusion, but gave a negative test with alcoholic silver nitrate. The compound was insoluble in water, but dissolved in aqueous NaOH or $NaHCO_3$. The IR spectrum and both the ^1H and ^{13}C NMR spectra for this compound are given below. Deduce the structure of this compound and assign all NMR signals and the important IR signals.

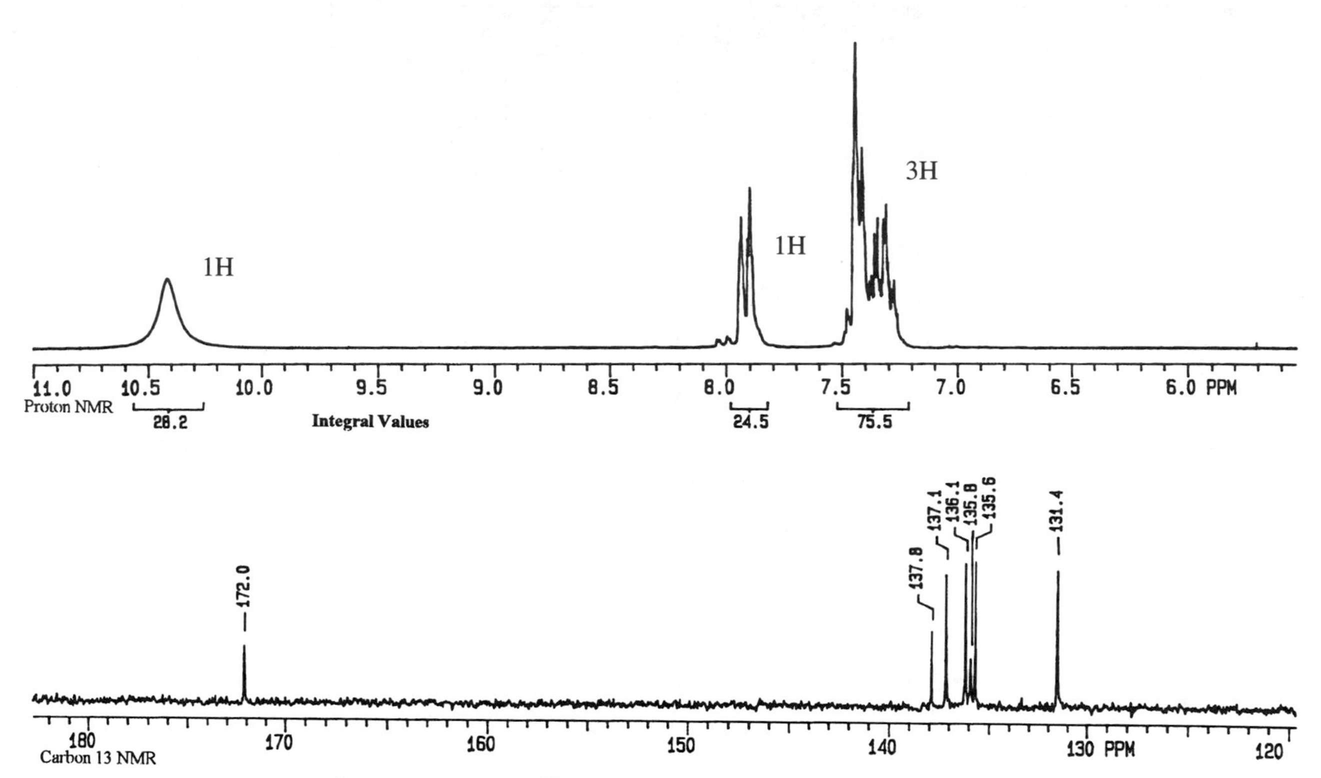

^1H NMR (top) and ^{13}C NMR (bottom) spectra for Problem 5. Spectra courtesy of R. Tomasi, Sunbelt R&T, Inc., Tulsa, Oklahoma.

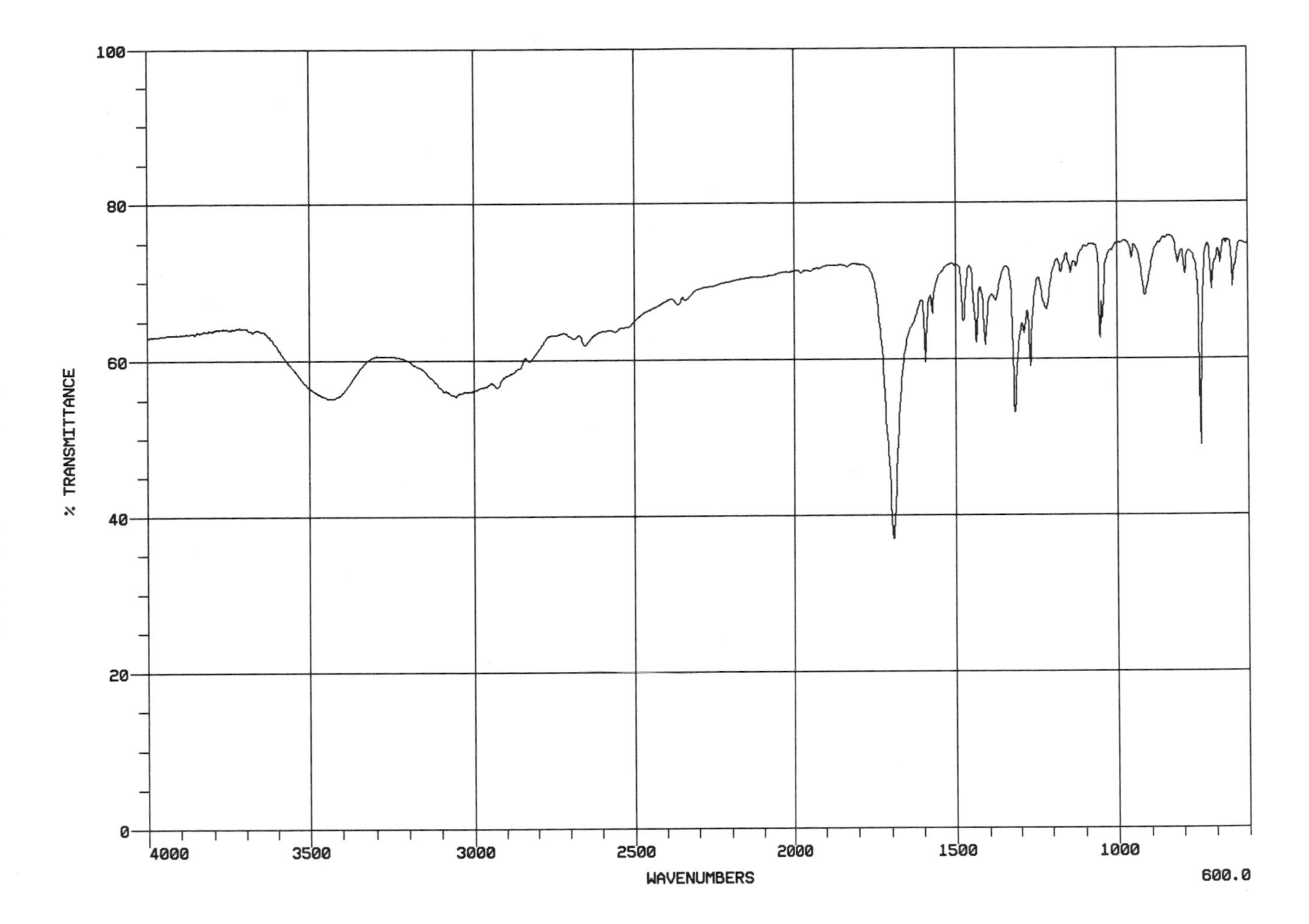

FT IR spectrum for Problem 5. Spectrum courtesy of R. Tomasi, Sunbelt R&T, Inc., Tulsa, Oklahoma.

UV References

E. A. Braude, *Determination of Organic Structures by Physical Methods,* edited by E. A. Braude and F. C. Nachod (Academic Press, New York, 1955), chap. 4.

A. E. Gillam and E. S. Stern, *An Introduction to Electronic Absorption Spectroscopy in Organic Chemistry,* 2nd ed. (Edward Arnold, London, 1957).

H. M. Hershenson, *Ultraviolet and Visible Absorption* (Academic Press, New York, 1955). This work provides a relatively simple route to spectra published between 1930 and 1957.

H. H. Jaffé and M. Orchin, *Theory and Applications of Ultraviolet Spectroscopy* (Wiley, New York, 1962).

S. F. Mason, *Quart. Rev.* **15**, 287 (1961).

E. S. Stern and T. C. J. Timmons, *Electronic Absorption Spectroscopy in Organic Chemistry* (St. Martin's Press, New York, 1971).

A. Streitwieser, Jr., *Molecular Orbital Theory for Organic Chemists* (Wiley, New York, 1961). Ultraviolet spectrometry is related to orbital theory.

D. H. Williams and I. Fleming, *Spectroscopic Methods in Organic Chemistry* (McGraw-Hill, NY, 1987).

CHAPTER 7

Chemical Tests for Functional Groups

As a result of earlier purity (Chapter 3), physical (Chapter 3), spectral (Chapter 6), and chemical tests (Chapters 3, 4, 5), the student probably has a reasonable idea regarding the identity of the unknown compound; it is, however, virtually certain that additional, thorough characterization is in order. A very large proportion of organic compounds lend themselves to final characterization by the chemical tests described in this chapter.

Despite the tremendous importance and ease of spectral analyses, chemical tests are indispensable to complete characterization. These "wet" tests are commonly used as a method of functional group and compound identification.

The classification tests are organized in two tables. Table 7.1 lists the classification tests alphabetically by the name of the test. Table 7.2 lists the tests alphabetically by the name of the functional group.

Each experiment contains a ***Cleaning Up***[1] procedure. If the same procedure applies to all parts of the experiment, then it is only listed once at the end of the experiment and before the discussion.

If a more microscale approach is desired, users of this book may wish to scale down the amount of unknown compounds and reagents by multiplying the amount used by 1/2, 1/5, or 1/10.

[1]The ***Cleaning Up*** procedures are based upon information found in K. L. Williamson, *Macroscale and Microscale Organic Experiments,* 2nd ed. (Heath, Lexington, MA 1994) and *Prudent Practices for Disposal of Chemicals from Laboratories* (National Academy Press, Washington, DC, 1983).

Table 7.1 Experiments

Classification test	Experiment number	Class tested for	Page
Acetyl chloride	6	Alcohols	217
		Amines	
		Carbohydrates	
		Phenols	
Anilide formation	4	Acid anhydrides	211
		Acyl halides	
Azoxybenzene and aluminum chloride	38	Hydrocarbons—Aromatic	287
Benedict's solution	26	Aldehydes	260
		Carbohydrates	
Benzenesulfonyl chloride (Hinsberg's method)	17	Amines	242
Borax test	24	Carbohydrates	240
Bromine solution	35	Hydrocarbons—Alkenes	276
		Hydrocarbons—Alkynes	
Bromine water	44	Phenols	298
Ceric ammonium nitrate	7	Alcohols	218
		Phenols	
Chloroform and aluminum chloride	39	Hydrocarbons—Aromatic	288
Chromic anhydride (chromium trioxide, Jones oxidation)	8	Alcohols	221
		Aldehydes	
Copper complex formation	23	Amino acids	254
2,4-Dinitrophenylhydrazine	11	Aldehydes	229
		Ketones	
Ester formation (Schotten–Baumann reaction)	3a	Acid anhydrides	209
		Acyl halides	
	3b	Carboxylic acids	210
Fehling's solution	27	Aldehydes	262
		Carbohydrates	
Ferric chloride–pyridine reagent	43	Phenols	296
Ferrous hydroxide reduction	40	Nitro compounds	292
Ferrox test	30	Ethers	267
Fuchsin–aldehyde reagent (Schiff's reagent)	15	Aldehydes	235
Fuming sulfuric acid	37	Hydrocarbons—Aromatic	285
Hinsberg's method (benzenesulfonyl chloride)	17	Amines	242
Hydrochloric acid–zinc chloride (Lucas test)	9	Alcohols	222
Hydroiodic acid (Zeisel's alkoxy method)	32	Esters	268
		Ethers	
Hydrolysis	1	Acid anhydrides	205
		Acyl halides	
Hydroxamic acid test	2b	Acid anhydrides	206
		Acyl halides	
	2c, 2d	Amides	207, 208
	2b	Esters	206
	2c	Nitriles	207
	2e	Sulfonic acids	208
		Sulfonyl chlorides	

Table 7.1 Experiments—(continued)

Classification test	Experiment number	Class tested for	Page
Hydroxylamine hydrochloride	12	Aldehydes	231
		Ketones	
Iodine test	31	Ethers	267
		Alkenes	
Iodoform test	10	Alcohols	224
		Ketones	
Jones oxidation (chromic anhydride, chromium trioxide)	8	Alcohols	221
		Aldehydes	
Liebermann's nitroso reaction	18d	Phenols	248
Lucas test (hydrochloric acid–zinc chloride)	9	Alcohols	222
Nickel chloride, carbon disulfide, and ammonium hydroxide	19	Amines	251
Nickel chloride and 5-nitrosalicylaldehyde	20	Amines	251
Ninhydrin test	22	Amino acids	254
Nitrous acid	18	Amines	245
	18e	Amino acids	248
Osazones	28	Carbohydrates	263
Periodic acid oxidation	25	Carbohydrates	258
Potassium permanganate solution	36	Hydrocarbons—Alkenes	279
		Hydrocarbons—Alkynes	
		Phenols	
Schiff's reagent (fuchsin–aldehyde reagent)	15	Aldehydes	235
Schotten–Baumann reaction (ester formation)	3a	Acid anhydrides	209
		Acyl halides	
Silver nitrate (ethanolic)	33	Acyl halides	270
		Carboxylic acids	
		Halides	
		Sulfonyl chlorides	
Sodium detection of active hydrogen	5	Alcohols	215
		Amines	
		Hydrocarbons—Alkynes	
Sodium bicarbonate test	29	Carboxylic acids	265
Sodium bisulfite addition complex	13	Aldehydes	232
		Ketones	
Sodium hydroxide fusion	45	Sulfonamides	300
Sodium hydroxide hydrolysis	16	Amides	237
		Nitriles	
Sodium hydroxide treatment	21	Amines	252
	42	Nitro compounds	293
Sodium iodide test	34	Halides	273
		Sulfonyl chlorides	
Tollens test	14	Aldehydes	234
		Carbohydrates	
Zeisel's alkoxy method (hydroiodic acid)	32	Esters	268
		Ethers	
Zinc and ammonium chloride reduction	41	Nitro compounds	293

Table 7.2 Classification Tests Listed by Functional Group

Functional group	Name of classification test	Experiment number	Page
Acid anhydride	Anilide formation	4	211
	Ester formation (Schotten–Baumann reaction)	3a	209
	Hydrolysis	1	205
	Hydroxamic acid test	2b	206
Acyl halide	Anilide formation	4	211
	Ester formation (Schotten–Baumann reaction)	3a	209
	Hydrolysis	1	205
	Hydroxamic acid test	2b	206
	Silver nitrate (ethanolic)	33	270
Alcohol	Acetyl chloride	6	217
	Ceric ammonium nitrate	7	218
	Chromic anhydride (chromium trioxide, Jones oxidation)	8	221
	Hydrochloric acid/zinc chloride (Lucas test)	9	222
	Iodoform test	10	224
	Sodium detection of active hydrogen	5	215
Aldehyde	Benedict's solution	26	260
	Chromic anhydride (chromium trioxide, Jones oxidation)	8	221
	2,4-Dinitrophenylhydrazine	11	229
	Fehling's solution	27	262
	Fuchsin–aldehyde reagent (Schiff's reagent)	15	235
	Hydroxylamine hydrochloride	12	231
	Sodium bisulfite addition complex	13	232
	Tollens test	14	234
Amide	Hydroxamic acid test	2c, 2d	207, 208
	Sodium hydroxide hydrolysis	16	237
Amine	Acetyl chloride	6	217
	Benzenesulfonyl chloride (Hinsberg's method)	17	242
	Nickel chloride, carbon disulfide, and ammonium hydroxide	19	251
	Nickel chloride and 5-nitrosalicylaldehyde	20	251
	Nitrous acid	18	245
	Sodium detection of active hydrogen	5	215
	Sodium hydroxide treatment	21	252
Amino acid	Copper complex formation	23	254
	Ninhydrin test	22	254
	Nitrous acid	18e	248
Carbohydrate	Acetyl chloride	6	217
	Benedict's solution	26	260
	Borax test	24	258
	Fehling's solution	27	262
	Osazones	28	263
	Periodic acid oxidation	25	258
	Tollens test	14	234
Carboxylic acid	Ester formation	3b	210
	Silver nitrate (ethanolic)	33	270
	Sodium bicarbonate test	29	265
Ester	Hydroiodic acid (Zeisel's alkoxy method)	32	268
	Hydroxamic acid test	2b	206
Ether	Ferrox test	30	267
	Hydroiodic acid (Zeisel's alkoxy method)	32	268
	Iodine test	31	267

Table 7.2—continued

Functional group	Name of classification test	Experiment number	Page
Halide	Silver nitrate test (ethanolic)	33	270
	Sodium iodide test	34	273
Hydrocarbon–Alkene	Bromine test	35	276
	Iodine	31	267
	Potassium permanganate solution	36	279
Hydrocarbon–Alkyne	Bromine test	35	276
	Potassium permanganate solution	36	279
	Sodium detection of active hydrogen	5	215
Hydrocarbon–Aromatic	Azoxybenzene and aluminum chloride	38	287
	Chloroform and aluminum chloride	39	288
	Fuming sulfuric acid	37	285
Ketone	2,4-Dinitrophenylhydrazine	11	229
	Hydroxylamine hydrochloride	12	231
	Iodoform test	10	224
	Sodium bisulfite addition complex	13	232
Nitrile	Hydroxamic acid test	2c	207
	Sodium hydroxide hydrolysis	16	237
Nitro compound	Ferrous hydroxide reduction	40	292
	Sodium hydroxide treatment	42	293
	Zinc and ammonium chloride reduction	41	293
Phenol	Acetyl chloride	6	217
	Bromine water	44	298
	Ceric ammonium nitrate	7	218
	Ferric chloride–pyridine reagent	43	296
	Liebermann's test	18d	248
	Potassium permanganate solution	36	279
Sulfonamide	Sodium hydroxide fusion test	45	300
Sulfonic acid	Hydroxamic acid test	2e	208
Sulfonyl chloride	Hydroxamic acid test	2e	208
	Silver nitrate test (ethanolic)	33	270
	Sodium iodide test	34	273

7.1 ACID ANHYDRIDES

The presence of acid anhydrides in an unknown sample can be demonstrated by several tests. Many acid anhydrides are rapidly converted by water to the corresponding carboxylic acids (Experiment 1, p. 205). Higher aliphatic anhydrides and aromatic anhydrides are not readily hydrolyzed with water. The presence of the carboxylic acid is detected by the addition of sodium bicarbonate, which results in the evolution of carbon dioxide gas, the sodium salt of the acid, and water.

$$\underset{\text{anhydride}}{RC(=O)OC(=O)R} + H_2O \longrightarrow 2\ \underset{\text{carboxylic acid}}{RC(=O)OH} + \text{heat}$$

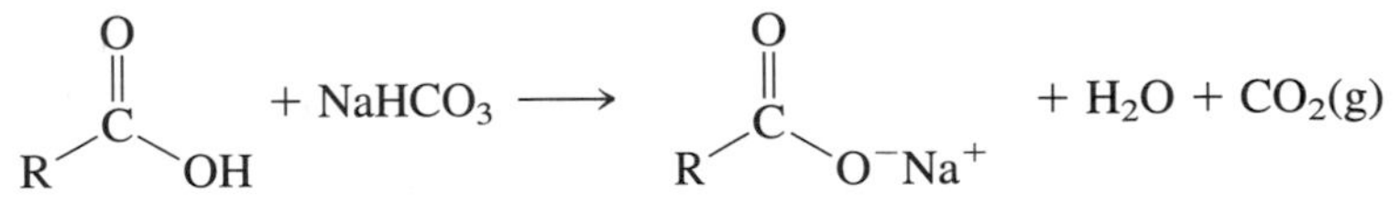

$$RCOOH + NaHCO_3 \longrightarrow RCOO^-Na^+ + H_2O + CO_2(g)$$

carboxylic acid — sodium salt of the acid

The hydroxamic acid test may be used to analyze for an acid anhydride (Experiment 2b, p. 206). The hydroxamic acid is treated with ferric chloride to form a ferric hydroxamate complex, which has a burgundy or magenta color.

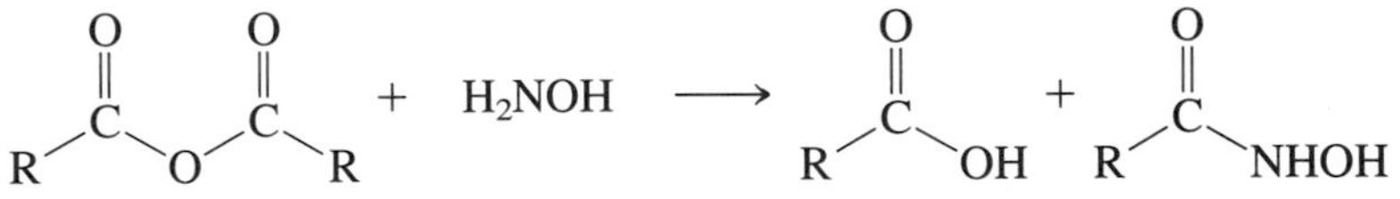

$$(RCO)_2O + H_2NOH \longrightarrow RCOOH + RCONHOH$$

anhydride — hydroxylamine — carboxylic acid — hydroxamic acid

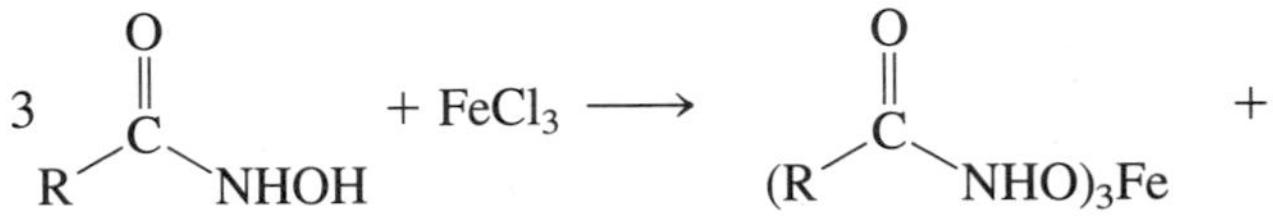

$$3\,RCONHOH + FeCl_3 \longrightarrow (RCONHO)_3Fe + 3HCl$$

hydroxamic acid — ferric hydroxamate complex (burgundy or magenta)

Acid anhydrides react quickly with alcohols to form esters and carboxylic acids (Experiment 3a, p. 209). The esters form an upper layer in the presence of the basic aqueous layer.

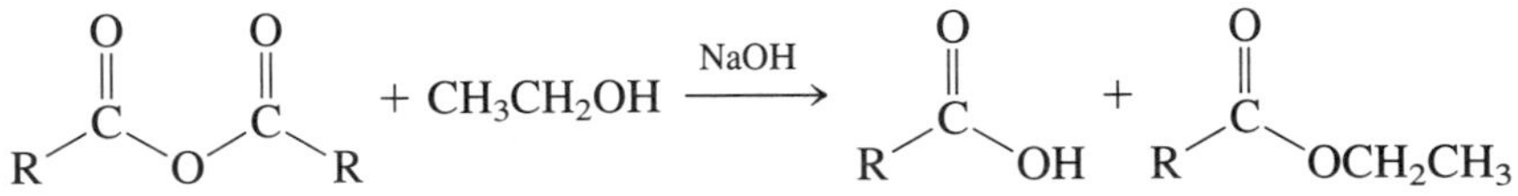

$$(RCO)_2O + CH_3CH_2OH \xrightarrow{NaOH} RCOOH + RCOOCH_2CH_3$$

anhydride — carboxylic acid — ester

Anilides are formed from acid anhydrides and aniline (Experiment 4, p. 211). The anilides precipitate from the solution.

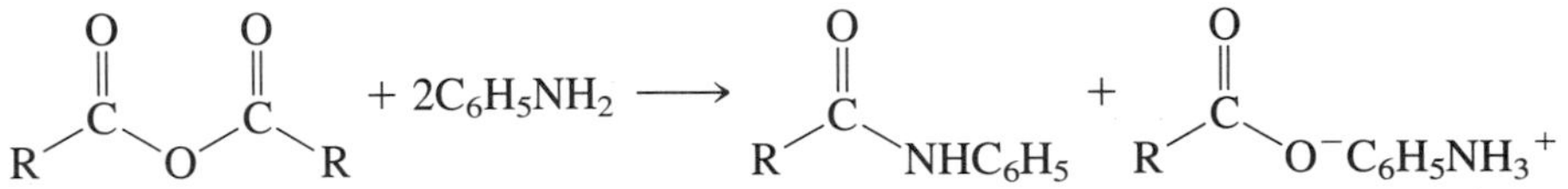

$$(RCO)_2O + 2C_6H_5NH_2 \longrightarrow RCONHC_6H_5 + RCOO^-C_6H_5NH_3^+$$

anhydride — aniline — anilide

Experiment 1. Hydrolysis of Acid Anhydrides and Acyl Halides

The anhydride and acid chloride tests are closely related and Section 7.2 (p. 211) discusses the latter.

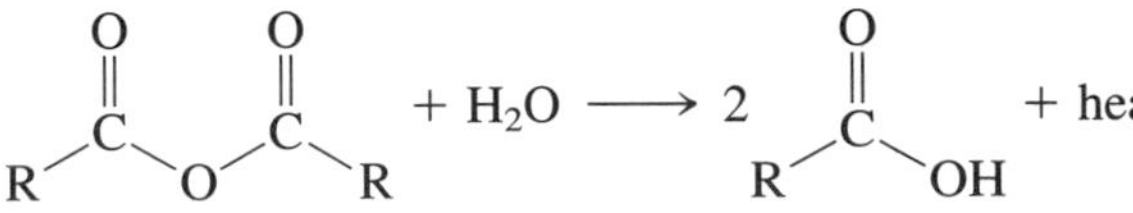

$$(RCO)_2O + H_2O \longrightarrow 2\,RCOOH + \text{heat}$$

anhydride — carboxylic acid

$$\underset{\text{acyl halide}}{RC(=O)X} + H_2O \longrightarrow \underset{\text{carboxylic acid}}{RC(=O)OH} + HX + \text{heat}$$

$$\underset{\text{carboxylic acid}}{RC(=O)OH} + NaHCO_3 \longrightarrow \underset{\text{sodium salt of the acid}}{RC(=O)O^-Na^+} + H_2O + CO_2(g)$$

Cautiously add a few drops or a few crystals of the unknown compound to 1 mL of water and touch the test tube to see if heat is evolved. If the test tube is warm, then the test is positive for an acid anhydride or an acyl halide, since water will react with these compounds to form the carboxylic acid with the evolution of heat. Add 1 mL of methanol to dissolve the sample. The solution is slowly poured into 1 mL of a saturated solution of sodium bicarbonate. Evolution of carbon dioxide gas is a positive test for the presence of the product carboxylic acid.

Cleaning Up. Place the reaction mixture in the aqueous solution container.

Discussion. Higher aliphatic anhydrides and aromatic anhydrides are not readily hydrolyzed with water and thus may not give a positive test.

Experiment 2. Hydroxamic Acid Test

(a) Preliminary Test. Dissolve a drop or a few crystals of the compound to be tested in 1 mL of 95% ethanol and add 1 mL of 1 M hydrochloric acid. Note the color produced when 1 drop of 5% ferric chloride solution is added to the solution. If a definite orange, red, blue, or violet color is produced, the following test for the acyl group is not applicable and should be omitted. Too much hydrochloric acid prevents the development of colored complexes of many phenols and all enols.

(b) Hydroxamic Acid Formation from Anhydrides, Acyl Halides, and Esters

$$\underset{\text{anhydride}}{RC(=O)OC(=O)R} + \underset{\text{hydroxylamine}}{H_2NOH} \longrightarrow \underset{\text{carboxylic acid}}{RC(=O)OH} + \underset{\text{hydroxamic acid}}{RC(=O)NHOH}$$

$$\underset{\text{acyl halide}}{RC(=O)X} + \underset{\text{hydroxylamine}}{H_2NOH} \longrightarrow \underset{\text{hydroxamic acid}}{RC(=O)NHOH} + HX$$

$$\underset{\text{ester}}{RC(=O)OR'} + \underset{\text{hydroxylamine}}{H_2NOH} \longrightarrow \underset{\text{hydroxamic acid}}{RC(=O)NHOH} + R'OH$$

$$3\ RC(=O)NHOH + FeCl_3 \longrightarrow (RC(=O)NHO)_3Fe + 3HCl$$

hydroxamic acid — ferric hydroxamate complex (burgundy or magenta)

Heat to boiling a mixture of 1 drop or about 40 mg of the compound, 1 mL of 0.5 M hydroxylamine hydrochloride in 95% ethanol, and 0.2 mL of 6 M sodium hydroxide. After the solution has cooled slightly, cautiously add 2 mL of 1 M hydrochloric acid. Anhydrides, acyl halides, and esters would have undergone the reaction with the hydroxylamine to form the hydroxamic acid, as indicated in the above equations. If the solution is cloudy, add 2 mL of 95% ethanol. Observe the color produced when one drop of 5% ferric chloride solution is added. If the color caused by the drop of ferric chloride solution does not persist, continue to add the ferric chloride solution dropwise until the observed color permeates the entire test solution. Compare the color with that produced in (a). A positive test will be a distinct burgundy or magenta color of the ferric hydroxamate complex, which is formed upon the reaction of the hydroxamic acid with the ferric chloride. Compare the color of this solution with the yellow observed when the original compound is tested with ferric chloride in the presence of acid.

(c) Hydroxamic Acid Formation from Nitriles and Amides

$$RC{\equiv}N + H_2NOH \longrightarrow RC(=NH)NHOH$$

nitrile — hydroxylamine

$$RC(=NH)NHOH + H_2O \xrightarrow{H^+} RC(=O)NHOH$$

hydroxamic acid

$$RC(=O)NH_2 + H_2NOH \cdot HCl \longrightarrow RC(=O)NHOH + NH_4Cl$$

amide — hydroxylamine hydrochloride — hydroxamic acid

$$3\ RC(=O)NHOH + FeCl_3 \longrightarrow (RC(=O)NHO)_3Fe + 3HCl$$

hydroxamic acid — ferric hydroxamate complex (burgundy or magenta color)

To 2 mL of a 1 M hydroxylamine hydrochloride solution in propylene glycol is added 1 drop or 30 mg of the compound dissolved in a minimum amount of propylene glycol. Add 1 mL of 1 M potassium hydroxide and the mixture is boiled gently for 2 min. Allow the mixture to cool to room temperature and add 0.5 to 1 mL of a solution of 5% alcoholic ferric chloride. A red to violet color constitutes a positive test. Yellow colors indicate negative tests, and brown colors or precipitates are indeterminate.

(d) Hydroxamic Acid Formation from Aromatic Primary Amides[2]

$$\underset{\text{aromatic amide}}{Ar{-}C({=}O){-}NH_2} + H_2O_2 \longrightarrow \underset{\text{hydroxamic acid}}{Ar{-}C({=}O){-}NHOH} + H_2O$$

$$3\ \underset{\text{hydroxamic acid}}{Ar{-}C({=}O){-}NHOH} + FeCl_3 \longrightarrow \underset{\substack{\text{ferric hydroxamate complex}\\ \text{(burgundy or magenta)}}}{(Ar{-}C({=}O){-}NHO)_3Fe} + 3HCl$$

This procedure tests for the presence of an aromatic primary amide, which would give a negative result when analyzed with procedure (c). Place 50 mg of the unknown in 5 mL of water. Add 0.5 mL of 3% hydrogen peroxide and 2 drops of 5% ferric chloride solution. Heat the solution to boiling. The hydrogen peroxide reacts with the aromatic amide to form the hydroxamic acid, which then reacts with the ferric chloride to form ferric hydroxamate complex. The characteristic magenta color should develop if the compound is indeed an aromatic primary amide.

(e) Hydroxamic Acid Formation from Sulfonic Acids and Sulfonyl Chlorides[3]

$$\underset{\substack{\text{sulfonic}\\ \text{acid}}}{ArSO_3H} + \underset{\substack{\text{thionyl}\\ \text{chloride}}}{SOCl_2} \longrightarrow \underset{\substack{\text{sulfonyl}\\ \text{chloride}}}{ArSO_2Cl} + HCl + SO_2$$

$$\underset{\substack{\text{sulfonyl}\\ \text{chloride}}}{ArSO_2Cl} + \underset{\substack{\text{hydroxylamine}\\ \text{hydrochloride}}}{H_2NOH \cdot HCl} \longrightarrow ArSO_2NHOH + 2HCl$$

$$ArSO_2NHOH + \underset{\text{acetaldehyde}}{H_3C{-}C({=}O){-}H} \longrightarrow \underset{\text{hydroxamic acid}}{H_3C{-}C({=}O){-}NHOH} + ArSO_2H$$

$$3\ \underset{\text{hydroxamic acid}}{H_3C{-}C({=}O){-}NHOH} + FeCl_3 + 3KOH \longrightarrow \underset{\substack{\text{ferric hydroxamate complex}\\ \text{(burgundy or magenta)}}}{(H_3C{-}C({=}O){-}NHO)_3Fe} + H_2O + 3KCl$$

To prepare the sulfonyl chloride from the sulfonic acid, combine 5 drops of thionyl chloride and 100 mg of the sulfonic acid in a test tube and heat in boiling water for 1 min. Allow the test tube to cool. To the test tube, 0.5 mL of a saturated solution of hydroxy-

[2]N. D. Cheronis, and J. B. Entrikin, *Semimicro Qualitative Organic Analysis,* 3rd ed. (Wiley, New York, 1965), p. 374.

[3]N. D. Cheronis, and J. B. Entrikin, *Semimicro Qualitative Organic Analysis* (Thomas Y. Cromwell Company, New York, 1947), p. 143.

lamine hydrochloride in methanol is added. A drop of acetaldehyde is added. The sulfonyl chloride undergoes the reaction with the hydroxylamine to form an intermediate, which when treated with the acetaldehyde, forms the hydroxamic acid. Add dropwise a solution of 2 M potassium hydroxide in methanol until the solution is slightly basic, when checked with pH paper. Heat the solution to boiling and then allow it to cool. Acidify the mixture by adding, dropwise, 0.5 M hydrochloric acid until blue litmus paper turns red. Add a drop of 5% ferric chloride solution. Ferric chloride converts the hydroxamic acid to the ferric hydroxamate complex. The magenta color of the ferric hydroxamate complex is a positive result.

Sulfonyl chlorides can be treated directly with the hydroxylamine hydrochloride. The salts of sulfonic acids are first neutralized with hydrochloric acid, then evaporated to dryness. The residue is then treated with thionyl chloride as described above.

Cleaning Up. For all parts of Experiment 2, neutralize the reaction mixture with sodium carbonate until the foaming ceases and place in the aqueous solution container.

Discussion. All esters of carboxylic acids, including polyesters and lactones, give definite magenta colors of varying degrees of intensity. Acid chlorides, acid anhydrides, and trihalo compounds such as trichloromethylbenzene and chloroform give positive magenta test results.

Formic acid produces a red color; with all other carboxylic acids the test is negative. Commercial lactic acid gives a positive test. Phthalic acid usually contains phthalic anhydride and thus gives a positive test.

Primary or secondary nitro compounds give a positive test.

Most imides give positive tests; aliphatic amides and salicylamide give light magenta colors, and most nitriles give a negative test with procedure (b) above. For procedure (c), all common nitriles and amides give positive tests. Benzanilide, diacetylbenzidine, and certain sterically hindered amides fail to give a positive test. Aromatic primary amides will only give a positive result with hydrogen peroxide in the presence of ferric chloride as described in (d).

Aldehydes with no α-hydrogen atoms may give weakly positive tests; with other aldehydes and ketones the test is negative.

Experiment 3. Ester Formation

(a) The Schotten–Baumann Reaction

$$\underset{\text{anhydride}}{RC(=O)OC(=O)R} + CH_3CH_2OH \xrightarrow{NaOH} \underset{\text{carboxylic acid}}{RC(=O)OH} + \underset{\text{ester}}{RC(=O)OCH_2CH_3}$$

$$\underset{\text{acyl halide}}{RC(=O)X} + CH_3CH_2OH \longrightarrow \underset{\text{ester}}{RC(=O)OCH_2CH_3} + HX$$

In a small flask, place 0.5 mL of ethanol, 1 mL of water, and 0.2 mL or 0.20 g of the unknown compound. To this solution add in portions, with vigorous shaking, 2 mL of 20% sodium hydroxide solution. Stopper the flask and shake the mixture for several minutes and then test the solution with litmus paper to make sure that it is still alkaline. The anhydrides and acyl halides will undergo a reaction with alcohols under basic conditions to form esters. Esters are both insoluble in water and less dense than water and thus will form a layer on top of the water.

(b) Esterification of a Carboxylic Acid

$$\underset{\text{carboxylic acid}}{RC(=O)OH} + \underset{\text{ethanol}}{CH_3CH_2OH} \longrightarrow \underset{\text{ester}}{RC(=O)OCH_2CH_3} + H_2O$$

A mixture of 0.20 g of the compound, 0.40 mL of absolute ethanol, and 0.20 mL of concentrated sulfuric acid is warmed over a steam bath for 2 min. The mixture is poured slowly into an evaporating dish containing 2 mL of saturated sodium bicarbonate solution. A second layer should be formed. Carefully smell the mixture. The presence of a sweet, fruity smell in the product, where no such smell existed in the original unknown, indicates that the original compound was a carboxylic acid and the acid was esterified. Large molecular weight carboxylic acids produce esters that are odorless.

Cleaning Up. For both Procedures (a) and (b), place the ester layer in the organic solvent container. Neutralize the aqueous layer with 10% hydrochloric acid, and place in the aqueous solution container.

Discussion. The Schotten–Baumann reaction in (a) is of particular interest because it might be expected that the water and hydroxyl ion present could compete with the alcohol to be acylated and reduce seriously the yield of the product desired. The success of the reaction is probably due to a combination of circumstances. It generally occurs in a heterogeneous medium with the organic reagent and the unknown anhydride or acyl halide in the same phase. One could speculate that the organic reagents combine in the organic phase and that only acid–base neutralization takes place in the inorganic layer.

$$R\overset{\delta+}{C}(=\overset{\delta-}{O})X + R'OH \longrightarrow R-C(O^-)(Z)-\overset{+}{O}(R')H \xrightarrow{B^-} RC(=O)OR' + Z^- + BH^+$$

Experiment 3a, $Z = O_2CR$ or halogen

The mechanism of the esterification carried out in (b) involves protonation as a method of enhancing the electrophilicity of the acid substrate toward nucleophilic attack by ethanol.

$$RC(=O)OH \underset{-H^+}{\overset{H^+}{\rightleftharpoons}} R\overset{+}{C}(OH)OH \underset{-CH_3CH_2OH}{\overset{CH_3CH_2OH}{\rightleftharpoons}} R-C(OH)_2-\overset{+}{O}(H)CH_2CH_3 \rightleftharpoons$$

$$R-C(\overset{+}{O}H_2)(OH)-OCH_2CH_3 \underset{H_2O}{\overset{-H_2O}{\rightleftharpoons}} R\overset{+}{C}(OH)OCH_2CH_3 \xrightarrow{-H^+} RC(=O)OCH_2CH_3$$

Experiment 4. Anilide Formation from Acid Anhydrides and Acyl Halides

$$\underset{\text{anhydride}}{RC(=O)OC(=O)R} + \underset{\text{aniline}}{2C_6H_5NH_2} \longrightarrow \underset{\text{anilide}}{RC(=O)NHC_6H_5} + RC(=O)O^- \quad C_6H_5NH_3^+$$

$$\underset{\text{acyl halide}}{RC(=O)X} + \underset{\text{aniline}}{2C_6H_5NH_2} \longrightarrow \underset{\text{anilide}}{RC(=O)NHC_6H_{18}} + C_6H_5NH_3^+X^-$$

(a) A few drops or a few crystals of the unknown sample are added to 0.5 mL of aniline. Pour the mixture into 5 mL of water. A positive test is the precipitation of the solid anilide.

(b) In a small flask, place 0.2 mL of aniline, 1 mL of water, and 0.2 mL or 0.2 g of the unknown. To this solution add in portions, with vigorous shaking, 10 mL of 20% sodium hydroxide solution. Shake the mixture in a stoppered flask for several minutes, and then test the solution with litmus paper to make sure that it is still alkaline. A positive test is the formation of a precipitate.

Cleaning Up. For both Procedures (a) and (b), the solid is filtered off and placed in the organic nonhazardous solid waste container. The filtrate is placed in the aqueous solution container.

Discussion. Anhydrides and acyl halides react with aniline to form anilides, which precipitate out of the solution. The mechanism involves the direct attack of the free amine (aniline) on the anhydride or acyl halide.

$$RC(=O)X + C_6H_5NH_2 \longrightarrow R-\overset{O^-}{\underset{^+NH_2C_6H_5}{C}}-X \xrightarrow{-X^-}$$

$$RC(=O)\overset{+}{N}H_2C_6H_5 \xrightarrow{-H^+} RC(=O)NHC_6H_5$$

Question

1. What is the role of the sodium hydroxide in Experiment 4b? (*Hint:* Note that two equivalents of aniline have been used in the equations at the start of the experiment.)

7.2 ACYL HALIDES

The same experiments used to detect the presence of acid anhydrides can be used for acyl halides. Theory predicts that the acyl halides are more reactive. Acid halides can be hydrolyzed to carboxylic acids (Experiment 1, p. 205). Addition of sodium bicarbonate to the solution produces the observable evolution of carbon dioxide gas, the sodium salt of the acid, and water.

$$\underset{\text{acyl halide}}{RC(=O)X} + H_2O \longrightarrow \underset{\text{carboxylic acid}}{RC(=O)OH} + HX + \text{heat}$$

$$\underset{\text{carboxylic acid}}{RC(=O)OH} + NaHCO_3 \longrightarrow \underset{\text{sodium salt of the acid}}{RC(=O)O^- Na^+} + H_2O + CO_2(g)$$

The hydroxamic acid test also will give a positive result for acid halides (Experiment 2b, p. 206). The acyl halide undergoes reaction with the hydroxylamine to form a hydroxamic acid, which is then treated with ferric chloride to form the magenta-colored ferric hydroxamate complex.

$$\underset{\text{acyl halide}}{RC(=O)X} + \underset{\text{hydroxylamine}}{H_2NOH} \longrightarrow \underset{\text{hydroxamic acid}}{RC(=O)NHOH} + HX$$

$$3\ \underset{\text{hydroxamic acid}}{RC(=O)NHOH} + FeCl_3 \longrightarrow \underset{\substack{\text{ferric hydroxamate complex}\\ \text{(burgundy or magenta)}}}{(RC(=O)NHO)_3Fe} + 3HCl$$

Acid halides undergo reaction with alcohols to form esters (Experiment 3a, p. 209). The esters form an upper layer with the basic aqueous layer.

$$\underset{\text{acyl halide}}{RC(=O)X} + CH_3CH_2OH \longrightarrow \underset{\text{ester}}{RC(=O)OCH_2CH_3} + HX$$

Aniline undergoes reaction with acid halides to form anilides, which precipitate from the solution (Experiment 4, p. 211).

$$\underset{\text{acyl halide}}{RC(=O)X} + \underset{\text{aniline}}{2C_6H_5NH_2} \longrightarrow \underset{\text{anilide}}{RC(=O)NHC_6H_5} + C_6H_5NH_3^+X^-$$

To distinguish between the acid anhydride and the acyl halide, the silver nitrate test for halides (Experiment 33, p. 270) is used. The acyl halide is hydrolyzed to form the carboxylic acid and hydrogen halide. The hydrogen halide undergoes reaction with silver nitrate to give an immediate precipitation of silver halide.

$$\underset{\text{acyl halide}}{\text{R}-\overset{\overset{\text{O}}{\|}}{\text{C}}-\text{X}} + H_2O \longrightarrow \underset{\text{carboxylic acid}}{\text{R}-\overset{\overset{\text{O}}{\|}}{\text{C}}-\text{OH}} + \text{HX} + \text{heat}$$

$$HX + AgNO_3 \longrightarrow AgX(s) + HNO_3$$

The more volatile acid halides can be detected merely by their obnoxious, lachrymatory odor. We do not recommend smelling compounds directly; however, some compounds such as volatile acid halides have distinctive odors that permeate the surroundings when the sample vial is opened.

Question

2. Why are the acyl halides generally more reactive than the anhydrides? (*Hint:* Write the mechanisms for the reaction of hydroxylamine with acetic anhydride and with acetyl chloride and discuss the leaving group ability of chloride vs. carboxylate.)

7.3 ALCOHOLS

Several methods are available for the analysis of the hydroxyl group, the functional group present in alcohols. Sodium metal undergoes reaction with hydroxyl groups of many alcohols to liberate hydrogen gas and form the salt of the alcohol (Experiment 5, p. 215). The rate is highly variable and depends upon the alcohol structure.

$$\underset{\text{alcohol}}{2ROH} + 2Na \longrightarrow 2RO^-Na^+ + H_2(g)$$

Another method of detecting such an active hydrogen is by adding acetyl chloride to the alcohol to form the ester (Experiment 6, p. 217), which is less dense than the aqueous layer.

$$\underset{\text{alcohol}}{\text{ROH}} + \underset{\text{acetyl chloride}}{\text{H}_3\text{C}-\overset{\overset{\text{O}}{\|}}{\text{C}}-\text{Cl}} \longrightarrow \underset{\text{ester}}{\text{H}_3\text{C}-\overset{\overset{\text{O}}{\|}}{\text{C}}-\text{OR}} + \text{HCl(g)}$$

Another method of testing for the presence of the alcoholic hydrogen involves ceric ammonium nitrate (Experiment 7, p. 218). The yellow ceric ammonium nitrate forms a red organometallic compound with alcohols. Positive results are obtained from alcohols of ten carbons or less.

$$\underset{\substack{\text{ceric ammonium nitrate}\\ \text{(yellow)}}}{(NH_4)_2Ce(NO_3)_6} + \underset{\text{alcohol}}{ROH} \longrightarrow \underset{\text{(red)}}{(NH_4)_2\overset{\overset{\text{OR}}{|}}{Ce}(NO_3)_5} + HNO_3$$

The Jones oxidation, in conjunction with the sodium metal test and the Lucas test, may be used to differentiate among primary (1°), secondary (2°), and tertiary (3°) alcohols. The Jones oxidation (Experiment 8, p. 221) only detects the presence of a hydroxyl substituent on a carbon bearing at least one hydrogen. Thus, only primary and secondary alcohols are oxidized to the corresponding carboxylic acids and ketones. Tertiary alcohols are not oxidized under these conditions. As the alcohol is oxidized, the solution changes from an orange-red color from the Cr^{6+} ion to a blue to green color from the Cr^{3+} ion.

$$\underset{\text{1° alcohol}}{3RCH_2OH} + \underset{\text{(orange-red)}}{4CrO_3} + 6H_2SO_4 \longrightarrow 3\ \underset{\text{carboxylic acid}}{RC(=O)OH} + 9H_2O + \underset{\text{(intense blue to green)}}{2Cr_2(SO_4)_3}$$

$$\underset{\text{2° alcohol}}{3R_2CHOH} + \underset{\text{(orange-red)}}{2CrO_3} + 3H_2SO_4 \longrightarrow \underset{\text{ketone}}{RC(=O)R} + 6H_2O + \underset{\text{(intense blue to green)}}{Cr_2(SO_4)_3}$$

Substrates which easily give rise to cationic character at the carbon bearing the hydroxyl group undergo the Lucas test (Experiment 9, p. 222) readily. Therefore, only secondary (2°) and tertiary (3°) alcohols form the alkyl halide, which appears as a second liquid layer, tertiary alcohols being the most reactive. Primary (1°) alcohols undergo reaction with the zinc chloride and hydrochloric acid either extremely slowly or not at all.

$$\underset{\text{2° alcohol}}{R_2CHOH} + HCl \xrightarrow{ZnCl_2} \underset{\text{alkyl halide}}{R_2CHCl} + H_2O$$

$$\underset{\text{3° alcohol}}{R_3COH} + HCl \xrightarrow{ZnCl_2} \underset{\text{alkyl halide}}{R_3CCl} + H_2O$$

The iodoform test (Experiment 10, p. 224) gives positive results for secondary alcohols in which a methyl group is attached to the carbon bearing the hydroxyl group. This type of alcohol is oxidized to a methyl ketone and under these basic conditions it forms a triiodo intermediate, which then is oxidized to the sodium salt of the acid and iodoform. Iodoform is a foul-smelling yellow precipitate.

$$\underset{\text{2° alcohol}}{RCH(OH)CH_3} + I_2 + 2NaOH \longrightarrow \underset{\text{methyl ketone}}{RC(=O)CH_3} + 2NaI + 2H_2O$$

$$\underset{\text{methyl ketone}}{RC(=O)CH_3} + 3I_2 + 3NaOH \longrightarrow RC(=O)CI_3 + 3NaI + 3H_2O$$

$$RC(=O)CI_3 + NaOH \longrightarrow \underset{\text{sodium salt of the carboxylic acid}}{RC(=O)O^-Na^+} + \underset{\text{iodoform (yellow solid)}}{CHI_3(s)}$$

Experiment 5. Sodium Detection of Active Hydrogen

$$2ROH + 2Na \longrightarrow 2RO^-Na^+ + H_2(g)$$
alcohol

$$2RNH_2 + 2Na \longrightarrow 2RNH^-Na^+ + H_2(g)$$
1° amine

$$2R_2NH + 2Na \longrightarrow 2R_2N^-Na^+ + H_2(g)$$
2° amine

$$H{-}C{\equiv}C{-}H + 2Na \longrightarrow Na^{+-}C{-}C{\equiv}C^-Na^+ + H_2(g)$$
terminal alkyne

$$2R{-}C{\equiv}C{-}H + 2Na \longrightarrow 2R{-}C{\equiv}C^-Na^+ + H_2(g)$$
terminal alkyne

To 0.25 mL or 0.25 g of the sample, add small thin slices of freshly cut sodium until no more will dissolve. Evolution of hydrogen gas indicates the presence of an acidic hydrogen, such as a hydroxyl group in an alcohol, a hydrogen attached to the nitrogen in a primary or secondary amine, or a hydrogen in a terminal alkyne. Cool the solution, and observe. Add an equal volume of ether. Another positive test is the formation of the solid salt. Liquid samples should be dried with calcium sulfate, prior to testing. Any residual water will undergo reaction with water. This test may be applied to solid compounds or very viscous liquids by dissolving them in an inert solvent such as anhydrous ligroin or toluene.

The order of reactivity of alcohols with sodium is known to decrease with increasing size of the alkyl portion of the molecule. This test is subject to many limitations, and the results should be interpreted with caution.

Cleaning Up. Enough ethanol is added dropwise to the solution until all of the sodium has reacted. Allow to stand for 1 hr. The reaction mixture is then diluted with 10 mL of water, neutralized with 10% hydrochloric acid, and placed in the aqueous solution container.

Discussion. This reagent is used in testing *neutral* compounds for the presence of groups which contain easily replaceable hydrogen atoms. Functional groups containing a hydrogen atom attached to oxygen, nitrogen, or sulfur may react with sodium to liberate hydrogen.

$$2ROH + 2Na \longrightarrow 2RO^-Na^+ + H_2(g)$$
alcohol

$$2RNH_2 + 2Na \longrightarrow 2RNH^-Na^+ + H_2(g)$$
1° amine

$$2R_2NH + 2Na \longrightarrow 2R_2N^-Na^+ + H_2(g)$$
2° amine

$$2RSH + 2Na \longrightarrow 2RS^-Na^+ + H_2(g)$$
sulfide

This test is most useful with alcohols of intermediate molecular weight, such as those containing from three to eight carbon atoms. Lower alcohols are difficult to obtain in anhydrous condition. The presence of traces of moisture causes the test to be positive. Alcohols of high molecular weight undergo reaction slowly with sodium, and the evolution of hydrogen gas is often so slow as to make the test of little value. Metallic sodium when cut in moist air adsorbs water on its surface so that, when placed in a perfectly dry solvent such as benzene, it gives off hydrogen gas produced by the interaction of the metal with the adsorbed moisture.

Hydrogen atoms attached to the carbon are not displaced by metals unless there are adjacent functional groups which activate the hydrogen atoms. Compounds with active methine groups, such as acetylene or monosubstituted acetylenes, undergo reaction with sodium.

$$H{-}C{\equiv}C{-}H + 2Na \longrightarrow Na^{+}\,{}^{-}C{-}C{\equiv}C^{-}Na^{+} + H_2(g)$$

terminal alkyne

$$2R{-}C{\equiv}C{-}H + 2Na \longrightarrow 2R{-}C{\equiv}C^{-}Na^{+} + H_2(g)$$

terminal alkyne

Frequently the hydrogen produced is not observed, as this hydrogen undergoes reaction with unsaturated functional groups as rapidly as it is produced.

A methylene group, adjacent to an activating group or between two activating groups, possesses hydrogen atoms which may be displaced by sodium. This hydrogen may also be difficult to observe due to its subsequent reaction with unsaturation in the original organic compound.

$$2\ H_3C{-}C({=}O){-}CH_2{-}C({=}O){-}OCH_2CH_3 + 2Na \longrightarrow 2\ H_3C{-}C({=}O){-}\overset{-}{C}H{-}C({=}O){-}OCH_2CH_3\ \ 2Na^{+} + H_2(g)$$

Reactive methyl groups are present in certain compounds, especially methyl ketones such as acetone and acetophenone. These react with sodium to give the sodium derivative of the ketone and a mixture of products formed by reducing and condensation. For example, acetone yields sodium acetonide, sodium isopropoxide, sodium pinacolate, mesityl oxide, and phorone.

Metallic sodium is thus a useful reagent for detecting the types of reactive hydrogen compounds that are not sufficiently active to produce hydrogen ions in an ionizing solvent. It is obviously unnecessary and dangerous to try the action of sodium on compounds known to be acids.

Structural effects upon acidity are complex. It is well known that liquid samples of alcohols follow this order of reactivity.

$$CH_3OH > CH_3CH_2OH > (CH_3)_2CHOH > (CH_3)_3COH$$

In the gas phase, the reverse order applies. It has been suggested that larger alkyl groups stabilize the alkoxide ion by polarization (thus the gas phase order), whereas bulky groups about the oxygen destabilize solvation (thus the solution order).

$$(CH_3)_2(CH_3)C{-}O^- \cdots HO{-}C(CH_3)_3$$

Questions

3. Predict the action of sodium on phenol, benzoic acid, oximes, nitromethane, and benzenesulfonamide. Why is this test never used with these compounds? What effect would the presence of moisture have on this test?
4. What is the principal problem of using metallic sodium as a classification reagent?

The "active" hydrogen of the hydroxyl group can often be detected by another procedure, involving the use of acid halides, described in (Experiment 6, below).

Experiment 6. Detection of Active Hydrogen with Acetyl Chloride

$$\underset{\text{alcohol}}{ROH} + \underset{\text{acetyl chloride}}{H_3C{-}C(=O){-}Cl} \longrightarrow \underset{\text{ester}}{H_3C{-}C(=O){-}OR} + HCl(g)$$

$$\underset{\text{phenol}}{ArOH} + \underset{\text{acetyl chloride}}{H_3C{-}C(=O){-}Cl} \longrightarrow \underset{\text{ester}}{H_3C{-}C(=O){-}OAr} + HCl(g)$$

$$\underset{1^\circ\text{ amine}}{RNH_2} + \underset{\text{acetyl chloride}}{H_3C{-}C(=O){-}Cl} \longrightarrow \underset{\text{amide}}{H_3C{-}C(=O){-}NHR} + HCl(g)$$

$$\underset{2^\circ\text{ amine}}{R_2NH} + \underset{\text{acetyl chloride}}{H_3C{-}C(=O){-}Cl} \longrightarrow \underset{\text{amide}}{H_3C{-}C(=O){-}NR_2} + HCl(g)$$

This test must be done in the hood. Add drop by drop 0.2 mL of acetyl chloride to 0.2 mL or 0.2 g of the unknown. Evolution of heat and hydrogen chloride gas is a positive test. Allow the mixture to stand for a minute or two and then pour it cautiously into 1 mL of water. Alcohols and phenols react with acetyl chloride to form esters, which is indicated by the formation of a top layer in the flask. Primary and secondary amines react with acetyl chloride to form amides, which precipitate from the solution.

Cleaning Up. The organic layer is separated and placed in the organic solvent container. Neutralize the aqueous layer with sodium carbonate and place in the aqueous solution container.

Discussion. Primary and secondary alcohols form esters, while tertiary alcohols form primarily the alkyl chloride, due to the reaction of the liberated hydrogen chloride on another molecule of the alcohol.

Only primary and secondary amines will react with acetyl chloride to form amides. However, some low molecular weight amides are water soluble. Many substituted anilines, especially those with nitro groups in the *ortho* or *para* position relative to the amino group, do not react with acetyl chloride. Tertiary amines do not have an active hydrogen and, therefore, give a negative result with acetyl chloride.

Question

5. Explain the difference in reactivity of the primary, secondary, and tertiary alcohols with acetyl chloride in the formation of the ester. Give the competing reaction for tertiary alcohols.

Experiment 7. Ceric Ammonium Nitrate

$$\underset{\substack{\text{ceric ammonium}\\\text{nitrate}\\\text{(yellow)}}}{(NH_4)_2Ce(NO_3)_6} + \underset{\text{alcohol}}{ROH} \longrightarrow \underset{\text{(red)}}{(NH_4)_2\overset{\overset{\displaystyle OR}{|}}{Ce}(NO_3)_5} + HNO_3$$

(a) For Water Soluble Compounds. To 1 mL of the ceric ammonium nitrate reagent add 4 to 5 drops of a liquid unknown or 0.1–0.2 g of a solid. Mix thoroughly and note if the yellow color of the reagent changes to red. Alcohols react with the reagent to form a red alkoxy cerium(IV) compound. If a red color develops, watch the solution carefully and note the time for the mixture to become colorless. If no change is noted in 15 min, the test tube may be stoppered and allowed to stand several hours or overnight. Also note if bubbles of carbon dioxide are liberated.

Try this test on (1) methanol; (2) isopropyl alcohol; (3) glycerol; (4) glucose; (5) lactic acid (85%).

(b) For Water Insoluble Compounds. Add 4 mL of dioxane[4] to 2 mL of the ceric ammonium nitrate reagent. If a red color develops or if the solution becomes colorless, the dioxane must be purified. If the mixture remains yellow or is only a light orange-yellow, it may be used to test water insoluble compounds. Divide the 6 mL of the solution in half, reserving 3 mL for observation as a control. To the other 3 mL of the dioxane containing reagent, add 4 or 5 drops of a liquid unknown or 0.1–0.2 g of a solid. Mix thoroughly and make the same observations as in procedure (a).

Try this test on (1) 1-heptanol; (2) benzyl alcohol; (3) (±)-mandelic acid.

Ceric Ammonium Nitrate Reagent. Add 1.3 mL of concentrated nitric acid to 40 mL of distilled water and then dissolve 10.96 g of yellow ceric ammonium nitrate in the dilute nitric acid solution. After the solid has dissolved, dilute to 50 mL. The test is

[4]The dioxane should be checked with ceric nitrate solution to be sure that it does *not* give a positive test. The dioxane sold as "histological grade" is usually pure enough so that it may be used. Pure dioxane does not give a red complex. Commercial dioxane sometimes contains glycols or antioxidants as preservatives and must be purified.

carried out at room temperature (20–25°C). Hot solutions (50–100°C) of Ce(IV) oxidize many types of organic compounds. This reagent is usable for about a month.

Cleaning Up. In Procedures (a) and (b), the reaction mixture is placed in the aqueous solution container.

(c) Alternate Procedure for Water Insoluble Compounds. The reagent consists of 0.43 g of ammonium hexanitrocerate dissolved in 2 mL of acetonitrile. Add about 0.1 g of the unknown compound to the reagent in a test tube. Stir the mixture with a glass rod and heat just to boiling. In 1–6 min the color will change from yellow to red. Even cholesterol, $C_{27}H_{45}OH$, gives a red to orange color. The red color disappears as oxidation of the alcohol group takes place.

Cleaning Up. The reaction mixture from Procedure (c) is treated with 1 mL of 10% sodium hydroxide. The solution is then neutralized with 10% hydrochloric acid and placed in the aqueous solution container.

Discussion. The ceric ammonium nitrate reagent forms red complexes with primary, secondary, and tertiary alcohols of up to ten carbons. Also, all types of glycols, polyols, carbohydrates, hydroxy acids, hydroxy aldehydes, and hydroxy ketones give red solutions. Phenols give a brown color or precipitate.

$$\underset{\substack{\text{ceric ammonium}\\\text{nitrate}\\\text{(yellow)}}}{(NH_4)_2Ce(NO_3)_6} + \underset{\text{phenol}}{ArOH} \longrightarrow \underset{\text{(brown or black)}}{(NH_4)_2\overset{\overset{OAr}{|}}{Ce}(NO_3)_5} + HNO_3$$

The red cerium(IV) compound has been shown to be the intermediate for the oxidation of alcohols by Ce(IV) solutions. Hence, *a second phase* of this test involves disappearance of the red color due to oxidation of the coordinated alcohol and reduction of the colored Ce(IV) complex to the colorless Ce(III) complex. Thus a positive test includes successively the formation, then the disappearance of the red color, assuming the oxidation step occurs within a reasonable time (see Table 7.3).

The overall sequence of reactions for a primary alcohol is as follows:

$$\text{(a)}\quad \underset{\text{(yellow)}}{(NH_4)_2Ce(NO_3)_6} + RCH_2OH \longrightarrow \underset{\text{(red)}}{(NH_4)_2\underset{\underset{OCH_2R}{|}}{Ce}(NO_3)_5} + HNO_3$$

$$\text{(b)}\quad \underset{\text{(red)}}{(NH_4)_2\underset{\underset{OCH_2R}{|}}{Ce}(NO_3)_5} \longrightarrow RCH_2O\cdot + \underset{\text{(colorless)}}{(NH_4)_2Ce(NO_3)_5} + HNO_3$$

$$\text{(c)}\quad RCH_2O\cdot + \underset{\text{(yellow)}}{(NH_4)_2Ce(NO_3)_6} \longrightarrow R{-}\overset{\overset{O}{\|}}{C}{-}H + \underset{\text{(colorless)}}{(NH_4)_2Ce(NO_3)_5} + HNO_3$$

The rates of the oxidation steps (b and c) depend upon the structure of the hydroxy compound.

Table 7.3 Approximate Times for Reduction of Red Ce(IV) Complexes at 20°C to Colorless Ce(III) Nitrato Anion with Oxidation of Alcohols to Aldehydes or Ketones

Compound	Time[a]	Compound	Time[a]
Primary Alcohols		**Diols, Triols, . . . , Polyols**	
Allyl alcohol	6 min	Pinacol	5 sec
Methyl cellosolve	1.2 hr	Mannitol	38 sec
1-Propanol	3.6 hr	2,3-Butanediol	1 min
Benzyl alcohol	4.0 hr	Glycerol	10 min
1-Butanol	4.1 hr	Propylene glycol	15 min
2-Methyl-1-propanol	4.1 hr	Diethylene glycol	3 hr
1-Heptanol	5.0 hr	Ethylene glycol	5 hr
Ethanol	5.5 hr		
Methanol	7.0 hr	1,4-Butanediol	1 hr
2-Methyl-1-butanol	7.0 hr	1,4-Butynediol	36 min
1-Decanol	12.0 hr	1,4-Butenediol (mostly cis)	3 min
Secondary Alcohols		**Carbohydrates**	
Cyclohexanol	3.7 hr	Glucose	1 min
2-Propanol	6.0 hr	Fructose	30 sec
2-Butanol	9.0 hr	Galactose	1 min
2-Pentanol	17.0 hr	Lactose	5 min
2-Octanol	16.0 hr	Maltose	8 min
Diphenylcarbinol	12.0 hr	Sucrose	12 min
		Cellulose—insoluble—no red	
		Starch—insoluble—no red	
Tertiary Alcohols		**Hydroxy Acids**	
tert-Butyl alcohol	>48 hr	Lactic acid	15 sec + CO_2
tert-Pentyl alcohol	>48 hr	Malic acid	30 sec + CO_2
3-Methyl-3-hydroxy-1-butyne	36 hr	Tartaric acid	1 min + CO_2
		Mandelic acid	1 min + CO_2
		Citric acid	1 min + CO_2
		Hydroxy Ketones	
		3-Hydroxy-2-butanone	15 sec
		3-Methyl-3-hydroxy-2-butanone	10 sec

[a]Variations in time of consideration can be expected due to variable size of reagent drops and to the age of the reagent.

The products from other hydroxylic compounds are listed below.

$$R-CH(OH)-R \longrightarrow R-C(=O)-R + H_2O$$

$$R-CH(OH)-CH(OH)-R \longrightarrow 2\ R-C(=O)-H + H_2O$$

$$R-\underset{OH}{\overset{R}{C}}-\underset{OH}{\overset{R}{C}}-R \longrightarrow 2\ R-\overset{O}{\overset{\|}{C}}-R + H_2O$$

$$R-\underset{OH}{\overset{H}{C}}-\underset{O}{\underset{\|}{C}}-OH \longrightarrow R-\overset{O}{\overset{\|}{C}}-H + H_2O + CO_2$$

Among simple hydroxy compounds, methanol gives the deepest red color. As the molecular weight of the alcohols increases, the color becomes less intense and somewhat brownish red.

A red color is produced by aqueous 40% formaldehyde (formalin). This is due to methanol present in the solution. Acetaldehyde frequently gives a red color due to the presence of 3-hydroxybutanal, acetaldol. Alternatively, these aldehydes may hydrate in aqueous solution to form gem diols, $RCH(OH)_2$, which may be the species that are oxidized.

Negative tests are indicated by the absence of the red complex with retention of the yellow color of the reagent. All pure aldehydes, ketones, saturated and unsaturated acids, ethers, esters, and dibasic and tribasic acids produce a negative test. The dibasic acids, oxalic and malonic, do *not* give a red color, but do reduce the yellow Ce(IV) to colorless Ce(III) solutions.

Basic aliphatic amines cause precipitation of white ceric hydroxide. If the amines are dissolved in dilute nitric acid, thus forming the amine nitrate, and this solution is treated with the ceric reagent, no red color develops provided that there are no alcoholic hydroxyl groups present in addition to the amino groups. If alcoholic groups are present, then dilute nitric acid solutions of such compounds do give red colors. For example, dilute nitric acid solutions of these compounds

$$HOCH_2CH_2NH_2 \qquad (HOOCH_2CH_2)_2NH \qquad (HOCH_2CH_2)_3N$$

all give positive tests.

Alcohols containing halogens give positive tests. For example, $ClCH_2CH_2OH$, $BrCH_2CH_2OH$, $ClCH_2CH_2CH_2OH$, and $CH_3CHOHCH_2Cl$ form red complexes.

Very insoluble alcohols of high molecular weight such as 1-hexadecanol, triphenylmethanol, or benzpinacol fail to react even in the dioxane solutions and do not give a red color.

Long-chain alcohols, C_{12} through C_{18}, will give a positive test when added to an acetonitrile solution of ammonium hexanitratocerate at the boiling point, 82°C. Procedure (c), listed above for water insoluble compounds, is especially useful for such long-chain alcohols.

Experiment 8. Chromic Anhydride (Chromium Trioxide, Jones Oxidation)

$$\underset{1°\ \text{alcohol}}{3RCH_2OH} + \underset{\text{(orange-red)}}{4CrO_3} + 6H_2SO_4 \longrightarrow \underset{\text{carboxylic acid}}{3\ R-\overset{O}{\overset{\|}{C}}-OH} + 9H_2O + \underset{\text{(intense blue to green)}}{2Cr_2(SO_4)_3}$$

$$3R_2CHOH + 2CrO_3 + 3H_2SO_4 \longrightarrow R_2C{=}O + 6H_2O + Cr_2(SO_4)_3$$

2° alcohol (orange-red) ketone (intense blue to green)

$$3\,RCHO + 2CrO_3 + 3H_2SO_4 \longrightarrow 3\,RCOOH + 3H_2O + Cr_2(SO_4)_3$$

aldehyde (orange-red) carboxylic acid (blue to green)

To 1 mL of acetone in a small test tube, add 1 drop of the liquid or about 10 mg of a solid compound. Then add 1 drop of the Jones reagent and note the result *within 2 sec.* Run a control test on the acetone and compare the result. A positive test for primary or secondary alcohols consists in the production of an opaque suspension with a green to blue color. Tertiary alcohols give no visible reaction within 2 sec, the solution remaining orange in color. *Disregard* any changes after 2 sec.

Jones Reagent. A suspension of 25 g of chromic anhydride (CrO_3) in 25 mL of concentrated sulfuric acid is poured slowly with stirring into 75 mL of water. The deep orange-red solution is cooled to room temperature before use.

A good grade of acetone should be used. Some samples of acetone may become cloudy in appearance in 20 sec, but this does not interfere, providing the test solution becomes yellow. If the acetone gives a positive test, it should be purified by adding a small amount of potassium permanganate and distilling.

Try this test on (1) 1-butanol; (2) 2-propanol; (3) 2-methyl-2-propanol; (4) benzaldehyde; (5) 2-propanone.

Cleaning Up. The reaction mixture is placed in the hazardous waste container.

Discussion. This test is a rapid method for distinguishing primary and secondary alcohols from tertiary alcohols. Positive tests are given by primary and secondary alcohols without restriction as to molecular weight. Even cholesterol ($C_{27}H_{46}O$) gives a positive test. Aldehydes give a positive test but would be detected by other classification experiments. Aldehydes produce the green color in 5–15 sec, with aliphatic aldehydes reacting more quickly than aromatic aldehydes. Ketones do not react. Olefins, acetylenes, amines, ethers, and ketones give negative tests within 2 sec provided that they are not contaminated with small amounts of alcohols. Enols may give a positive test, and phenols produce a dark-colored solution entirely unlike the characteristic green-blue color of a positive test.

Experiment 9. Hydrochloric Acid–Zinc Chloride (Lucas Test)

$$R_2CHOH + HCl \xrightarrow{ZnCl_2} R_2CHCl + H_2O$$

2° alcohol alkyl halide

$$R_3COH + HCl \xrightarrow{ZnCl_2} R_3CCl + H_2O$$

3° alcohol alkyl halide

(a) To 0.2 mL or 0.2 g of the sample in a test tube add 2 mL of the Lucas reagent at 26–27°C. Stopper the tube and shake; then allow the mixture to stand. Note the time required for the formation of the alkyl chloride, which appears as an insoluble layer or emulsion. Carry out the test on each of the following alcohols, and note by means of a watch the time required for the reaction to take place: (1) 1-butanol; (2) 2-methyl-2-propanol; (3) 2-pentanol; (4) 1-propanol; (5) 1-pentanol; (6) allyl alcohol; (7) benzyl alcohol.

Lucas Reagent. Dissolve 13.6 g (0.1 mole) of anhydrous zinc chloride in 10.5 g (0.1 mole) of concentrated hydrochloric acid, with cooling.

(b) To 0.2 mL or 0.2 g of the alcohol in a test tube add 1.2 mL of concentrated hydrochloric acid. Shake the mixture, and allow it to stand. Observe carefully during the first 2 min. Test the following alcohols, and record your results; (1) 1-propanol; (2) 2-pentanol; (3) benzyl alcohol; (4) 2-methyl-2-propanol.

Cleaning Up. Add sodium carbonate to the test solution until foaming no longer occurs and place the mixture in the aqueous solution container.

Discussion. The mechanism of the Lucas test is an S_N1-type process as follows:

$$(CH_3)_3C{-}OH \underset{}{\overset{H^+}{\rightleftharpoons}} (CH_3)_3C{-}OH_2^+ \overset{-H_2O}{\rightleftharpoons} (CH_3)_3C^+ \overset{Cl^-}{\rightleftharpoons} (CH_3)_3C{-}Cl$$

The role of the $ZnCl_2$ is to enhance the reactivity of the HCl by polar coordination:

$$\overset{\delta^-}{Cl_2Zn}\text{--}\overset{\delta^+}{H}{-}Cl$$

Since the Lucas test depends on the appearance of the alkyl chloride as a second liquid phase, it is normally applicable only to alcohols that are soluble in the reagent. This limits the test in general to monofunctional alcohols lower than hexyl and certain polyfunctional molecules.

The reaction of alcohols with halogen acids is a displacement reaction in which the reactive species is the conjugate acid of the alcohol, $R{-}OH_2^+$, and is analogous to the replacement reactions of organic halides and related compounds with silver nitrate and iodide ion (Experiments 33 and 34). The effects of structure on reactivity in these reactions are closely related. Thus primary alcohols do not react perceptibly with hydrochloric acid even in the presence of zinc chloride at ordinary temperatures. Chloride ion is too poor a nucleophilic agent to effect a concerted displacement reaction and, additionally, the primary carbonium ion is too unstable to serve as an intermediate in the carbonium ion mechanism. Hydrogen bromide and hydrogen iodide, which have greater nucleophilic reactivity, are also more reactive toward primary alcohols.

Tertiary alcohols react with concentrated hydrochloric acid so rapidly that the alkyl halide is visible within a few minutes at room temperature, first as a milky suspension and then as an oily layer. The acidity of the medium is increased by the addition of anhydrous zinc chloride, which is a strong Lewis acid, and, as a result, the reaction rate is increased. The high reactivity of tertiary alcohols is a consequence of the relatively great stability of the intermediate carbocation. Allyl alcohol, although a primary alcohol, yields a carbocation which is relatively stable because its charge is distributed equally on the two terminal carbon atoms.

$$\mathrm{H_2C{=}CH{-}CH_2{-}OH} \xrightarrow[-\mathrm{H_2O}]{\mathrm{H^+}} \left[\mathrm{H_2C{=}CH{-}CH_2^{+}} \longleftrightarrow \mathrm{{}^{+}CH_2{-}CH{=}CH_2}\right]$$

As a result, allyl alcohol reacts rapidly with the Lucas reagent, and the reaction is accompanied by the evolution of heat. Addition of ice water to the reaction causes the allyl chloride to form as a separate layer.

Secondary alcohols are intermediate in reactivity between primary and tertiary alcohols. Although they are not appreciably affected by concentrated hydrochloric acid alone, they react with it fairly rapidly in the presence of anhydrous zinc chloride. A cloudy appearance of the mixture is observed within 5 min, and in 10 min a distinct layer is usually visible.

For a more extended discussion of the effect of structure on reactivity in replacement reactions of this type see the discussion of the silver nitrate test (Experiment 33, p. 270).

Questions

6. Write the names and structures of all alcohols with the formula of $C_5H_{12}O$. How would they react with this reagent?
7. How would you account for the difference in the behavior of allyl alcohol and 1-propanol? Benzyl alcohol and 1-pentanol?

Experiment 10. Iodoform Test

The iodoform test is a test for methyl ketones and for secondary alcohols with a methyl group adjacent to the carbon bearing the hydroxyl group.

$$\left.\begin{array}{c}\mathrm{R{-}C({=}O){-}CH_3} \\ \text{or} \\ \mathrm{R{-}CH(OH){-}CH_3}\end{array}\right] \longrightarrow \mathrm{R{-}C({=}O){-}O^{-}Na^{+}} + \underset{\text{iodoform (yellow solid)}}{\mathrm{CHI_3(s)}}$$

The alcohols are oxidized to the methyl ketones by the "iodine bleach."

$$\underset{2°\ \text{alcohol}}{\mathrm{R{-}CH(OH){-}CH_3}} + \mathrm{I_2} + 2\mathrm{NaOH} \longrightarrow \underset{\text{methyl ketone}}{\mathrm{R{-}C({=}O){-}CH_3}} + 2\mathrm{NaI} + 2\mathrm{H_2O}$$

Iodination occurs preferentially and completely on the methyl group.

$$\underset{\text{methyl ketone}}{R-\overset{\overset{\displaystyle O}{\|}}{C}-CH_3} + 3I_2 + 3NaOH \longrightarrow R-\overset{\overset{\displaystyle O}{\|}}{C}-CI_3 + 3NaI + 3H_2O$$

Cleavage produces the carboxylate salt and iodoform.

$$R-\overset{\overset{\displaystyle O}{\|}}{C}-CI_3 + NaOH \longrightarrow \underset{\substack{\text{sodium salt of}\\ \text{the carboxylic acid}}}{R-\overset{\overset{\displaystyle O}{\|}}{C}-O^-Na^+} + \underset{\substack{\text{iodoform}\\ \text{(yellow solid)}}}{CHI_3(s)}$$

Place 4 drops of the liquid or 0.1 g of the solid to be tested in a test tube. Add 5 mL of dioxane,[5] and shake until all the sample has gone into solution. Add 1 mL of 10% sodium hydroxide solution, and then slowly add the iodine–potassium iodide solution with shaking, until a slight excess yields a definite dark color of iodine. If less than 2 mL of the iodine solution is decolorized, place the test tube in a water bath maintained at a temperature of 60°C. If the slight excess of iodine already present is decolorized, continue the addition of the iodine solution (keeping the iodine solution at 60°C), with shaking, until a slight excess of iodine solution again yields a definite dark color. The addition of iodine is continued until the dark color is not discharged by 2 min of heating at 60°C. The excess of iodine is removed by the addition of a few drops of 10% sodium hydroxide solution, with shaking. Now fill the test tube with water and allow to stand for 15 min. A positive test is indicated by the formation of a foul-smelling yellow precipitate (iodoform). The precipitate should be collected by filtration and dried and its melting point checked; iodoform (CHI_3) melts at 119–121°C (d) and has a distinctive foul odor. If the iodoform is reddish, dissolve in 3–4 mL of dioxane, add 1 mL of 10% sodium hydroxide solution, and shake until only a light lemon color remains. Dilute with water and filter.

Apply the test to (1) 2-propanol; (2) acetone; (3) ethyl acetate; (4) acetophenone; (5) methanol. (Note that some lower commercial grades of methanol give misleading positive results because of impurities.)

Iodine–Potassium Iodide Solution. Add 20.0 g of potassium iodide and 10.0 g of iodine to 80.0 mL of water and stir until the reaction is complete.

$$I_2 + KI \xrightarrow{H_2O} KI_3$$

The solution is deep brown due to the triiodide anion (I_3^-).

Cleaning Up. Add a few drops of acetone to the reaction mixture to destroy any unreacted iodine in potassium iodide solution. Remove the iodoform by suction filtration and place in the halogenated organic waste container. Place the filtrate in the aqueous solution container.

[5]Dioxane is appropriate for water *insoluble* compounds; water soluble compounds may be treated by substituting 2 mL of water for the dioxane solvent.

Discussion. This test is positive for compounds which contain the grouping $CH_3C(=O)—$, $ICH_2C(=O)—$, or $I_2CHC(=O)—$ joined to a hydrogen atom or to a carbon atom which does not have highly active hydrogens or groups which provide an excessive amount of steric hindrance. The test will, of course, be positive also for any compound which reacts with the reagent to give a derivative containing one of the requisite groupings. Conversely, compounds containing one of the requisite groupings will not give iodoform if that grouping is destroyed by the hydrolytic action of the reagent before iodination is complete.

Following are the principal types of compounds that give a positive test:

$$H_3C-C(=O)-H \qquad CH_3CH_2OH \qquad H_3C-C(=O)-R \qquad H_3C-CH(OH)-R$$

$$R-C(=O)-CH_2-C(=O)-R \qquad H_3C-CH(OH)-CH_2-CH(OH)-R$$

(R = any alkyl or aryl radical except an *ortho*-disubstituted aryl radical.) R, however, if large, will sterically inhibit this reaction. The test is negative for compounds of the following types:

$$H_3C-C(=O)-CH_2-C(=O)-OR \qquad H_3C-C(=O)-CH_2CN \qquad H_3C-C(=O)-CH_2NO_2 \qquad \text{etc.}$$

In such compounds the reagent removes the acetyl group and converts it to acetic acid, which resists iodination.[6]

A modified reagent[7] has been suggested for distinguishing methyl ketones from methyl carbinols. It consists of a solution of 1 g of potassium cyanide, 4 g of iodine, and 6 mL of concentrated ammonium hydroxide in 50 mL of water. *Potassium cyanide is extremely toxic; use only with the instructor's permission. Do not mix with acid.*

$$R-C(=O)-CH_3 \xrightarrow[KCN/NH_4OH]{I_2} CHI_3$$

$$R-CH(OH)-CH_3 \xrightarrow[KCN/NH_4OH]{I_2} \text{N.R.}$$

This reagent produces iodoform from methyl ketones but not from methyl carbinols.

The cleavage of trihalo ketones with base, exemplified by the second step of the iodoform test, is related to the reversal of the Claisen condensation. In each case the reaction can proceed because of the stability of the final anionic fragment.

[6]For a general discussion of this test, see R. C. Fuson and B. A. Bull, *Chem. Rev., 15,* 275 (1934).

[7]E. Rothlin, *Arch. Escuela Farm. Fac. Ci. Med. Cordoba [R. A.] Secc. ci.,* No. 10, p. 1 (1939). *C.A., 35,* 5091 (1941).

$$\mathrm{R{-}C({=}O){-}CI_3} + \mathrm{OH^-} \longrightarrow \left[\mathrm{R{-}C(O^-)(OH){-}CI_3}\right] \longrightarrow \mathrm{R{-}C({=}O){-}OH} + \mathrm{CI_3^-}$$

$$\mathrm{R{-}C({=}O){-}CH_2{-}C({=}O){-}R} + \mathrm{OH^-} \longrightarrow \left[\mathrm{R{-}C(O^-)(OH){-}CH_2{-}C({=}O){-}R}\right] \longrightarrow \mathrm{R{-}C({=}O){-}OH} + \mathrm{^-H_2C{-}C({=}O){-}R}$$

Secondary alcohols, and ketones of the structures

$$\mathrm{RCH_2{-}CH(OH){-}CH_2R} \quad \text{and} \quad \mathrm{RCH_2{-}C({=}O){-}CH_2R}$$

do not produce iodoform although they may undergo some halogenation on the methylene group adjacent to the carbonyl group. Occasionally commercial samples of diethyl ketone give a weak iodoform test. This is due to the presence of such impurities such as 2-pentanone.

Bifunctional alcohols and ketones of the following types give positive iodoform tests.

$$\mathrm{H_3C{-}CH(OH){-}CH(OH){-}CH_3} \qquad \mathrm{H_3C{-}C({=}O){-}CH(OH){-}CH_3}$$

$$\mathrm{H_3C{-}C({=}O){-}C({=}O){-}CH_3} \qquad \mathrm{H_3C{-}C({=}O){-}CH_2{-}CH_2{-}C({=}O){-}CH_3}$$

β-Keto esters do not produce iodoform by the test method, but their alkaline solutions do react with sodium hypoiodite.

Acetoacetic acid is unstable; acidic aqueous solutions decompose to give CO_2 and acetone.

$$\mathrm{H_3C{-}C({=}O){-}CH_2{-}C({=}O){-}OH} \xrightarrow{\mathrm{H^+}} \mathrm{H_3C{-}C({=}O){-}CH_3} + \mathrm{CO_2}$$

The acetone will give a positive iodoform test. This behavior is generally useful if a β-keto ester is one of the possibilities being considered, since these esters are hydrolyzed by boiling with 5% sulfuric acid (acid-induced retro condensation).

$$\mathrm{R{-}C({=}O){-}CH_2{-}C({=}O){-}OR'} + \mathrm{H_2O} \xrightarrow{\mathrm{H^+}} \mathrm{R{-}C({=}O){-}CH_3} + \mathrm{R'OH} + \mathrm{CO_2}$$

7.4 ALDEHYDES

The reaction of aldehydes and ketones with 2,4-dinitrophenylhydrazine (Experiment 11, p. 229) to form the 2,4-dinitrophenylhydrazone probably represents the most studied and most successful of all qualitative tests and derivatizing procedures. In addition, the general details of the reaction serve as a model for a number of other chemical reactions (osazone, semicarbazone, oxime, and other arylhydrazone preparations). The 2,4-dinitrophenylhydrazone precipitates from the solution.

$$>C{=}O + O_2N\text{-}C_6H_3(NO_2)\text{-}NHNH_2 \xrightarrow[\text{alcohol}]{H_2SO_4} O_2N\text{-}C_6H_3(NO_2)\text{-}NHN{=}C<$$

aldehyde or ketone | 2,4-dinitrophenylhydrazine | 2,4-dinitrophenylhydrazone

In the reaction of hydroxylamine hydrochloride with aldehydes (Experiment 12, p. 231), the formation of the oximes results in the liberation of hydrogen chloride, which can be detected by the change in color from orange to red of a pH indicator.

$$>C{=}O + H_2N\text{—}OH \cdot HCl \longrightarrow >C{=}N\text{—}OH + HCl + H_2O$$

aldehyde or ketone | hydroxylamine hydrochloride | oxime

The precipitation of a bisulfite addition complex (Experiment 13, p. 232) is indicative of a variety of carbonyl compounds reacting with sodium bisulfite. This reaction is greatly inhibited by the steric constraints about the carbonyl group.

$$>C{=}O + NaHSO_3 \longrightarrow \text{—}C(H)(OH)(SO_3^-Na^+)\text{—}$$

aldehyde or ketone | sodium bisulfite | sodium bisulfite addition complex

A simple chemical test for aldehydes involves the use of CrO_3 in Jones oxidation (Experiment 8, p. 221). As the aldehyde is oxidized to the carboxylic acid, the chromium is oxidized from a +3 oxidation state, which is an orange-red color, to a +6 oxidation state, which is a deep blue-green color.

$$3\ RCHO + 2CrO_3 + 3H_2SO_4 \longrightarrow 3\ RCOOH + 3H_2O + Cr_2(SO_4)_3$$

aldehyde | (orange-red) | carboxylic acid | (blue to green)

Aldehydes produce a silver mirror when mixed with Tollens reagent (Experiment 14, p. 234). As the aldehyde is oxidized to an acid, the silver is reduced from a +1 oxidation state to elemental silver and is deposited as a silver mirror or colloidal silver inside the reaction flask.

$$\underset{\text{aldehyde}}{R{-}\overset{\overset{O}{\|}}{C}{-}H} + \underset{\text{Tollens reagent}}{2Ag(NH_3)_2OH} \longrightarrow 2Ag(s) + \underset{\text{salt of the carboxylic acid}}{R{-}\overset{\overset{O}{\|}}{C}{-}O^-NH_4^+} + H_2O + 3NH_3$$

Schiff's reagent (Experiment 15, p. 235) undergoes reaction with aldehydes to form a violet-purple solution.

$$\underset{\text{aldehyde}}{R{-}\overset{\overset{O}{\|}}{C}{-}H} + \underset{\text{Schiff's reagent (colorless)}}{H_3\overset{+}{N}{-}C_6H_4{-}\underset{\underset{HO_3S}{|}}{C}\left[{-}C_6H_4{-}NHSO_2H\right]_2 Cl^-} \longrightarrow$$

$$\underset{\text{(violet-purple solution)}}{H_2\overset{+}{N}{=}C_6H_4{=}C\left[{-}C_6H_4{-}NH{-}\underset{\underset{O}{\downarrow}}{\overset{\overset{O}{\uparrow}}{S}}{-}\underset{\underset{R}{|}}{\overset{\overset{OH}{|}}{C}}{-}H\right]_2 Cl^-} + H_2SO_3$$

Benedict's solution (Experiment 26, p. 260) and Fehling's solution (Experiment 27, p. 262) will undergo reactions with aliphatic aldehydes, but not with aromatic aldehydes. These reagents oxidize the aliphatic aldehyde to a carboxylic acid, and the copper in the reagent is reduced from +2 to +1. The copper(I) oxide precipitates as a red, yellow, or yellowish green solid.

$$\underset{\text{aldehyde}}{R{-}\overset{\overset{O}{\|}}{C}{-}H} + \underset{\text{(blue)}}{2Cu^{2+}} \longrightarrow \underset{\text{carboxylic acid}}{R{-}\overset{\overset{O}{\|}}{C}{-}OH} + \underset{\text{(red, yellow, or yellowish green)}}{Cu_2O(s)}$$

Experiment 11. 2,4-Dinitrophenylhydrazine

$$\underset{\text{aldehyde or ketone}}{{>}C{=}O} + \underset{\text{2,4-dinitrophenylhydrazine}}{O_2N{-}C_6H_3(NO_2){-}NHNH_2} \xrightarrow[\text{alcohol}]{H_2SO_4} \underset{\text{2,4-dinitrophenylhydrazone}}{O_2N{-}C_6H_3(NO_2){-}NHN{=}C{<}}$$

Add a solution of 1 or 2 drops or about 50 mg of the compound to be tested in 2 mL of 95% ethanol to 3 mL of 2,4-dinitrophenylhydrazine reagent. Shake vigorously, and, if no precipitate forms immediately, allow the solution to stand for 15 min. If needed, the precipitate can be recrystallized from ethanol.

Try this test on (1) acetone; (2) acetophenone.

2,4-Dinitrophenylhydrazine Reagent.[8] Dissolve 3 g of 2,4-dinitrophenylhydrazine in 15 mL of concentrated sulfuric acid. This solution is then added, with stirring, to 20 mL of water and 70 mL of 95% ethanol. This solution is mixed thoroughly and filtered.

Cleaning Up. The test solution is placed in the hazardous waste container.

Discussion. Most aldehydes and ketones yield dinitrophenylhydrazones that are insoluble solids. The precipitate may be oily at first and become crystalline on standing. A number of ketones, however, give dinitrophenylhydrazones that are oils. For example, 2-decanone, 6-undecanone, and similar substances fail to form solid dinitrophenylhydrazones.

A further difficulty with the test is that certain allyl alcohol derivatives may be oxidized by the reagent to aldehydes or ketones, which then give a positive test. For example, the 2,4-dinitrophenylhydrazones of the corresponding carbonyl compounds have been obtained in yields of 10–25% from cinnamyl alcohol, 4-phenyl-3-buten-2-ol, and vitamin A_1. Benzhydryl alcohol also was found to be converted to benzophenone dinitrophenylhydrazone in low yield. Needless to say, there is always the further danger that an alcohol sample may be contaminated by enough of its aldehyde or ketone, formed by air oxidation, to give a positive test. If the dinitrophenylhydrazone appears to be formed in a very small amount, it may be desirable to carry out the reaction on the scale employed for the preparation of the derivative (Procedure 13, p. 321). The melting point of the solid should be checked to be sure it is different from that of 2,4-dinitrophenylhydrazine, mp 198°C.

If necessary, this hydrazone derivative can be recrystallized from a solvent such as ethanol. Recrystallization solvents containing reactive carbonyl groups, such as acetone, should not be used as they may result in the formation of another hydrazone.

The color of a 2,4-dinitrophenylhydrazone may give an indication as to the structure of the aldehyde or ketone from which it is derived. The dinitrophenylhydrazones of aldehydes and ketones in which the carbonyl group is not conjugated with another functional group are yellow. Conjugation with a carbon–carbon double bond or with a benzene ring shifts the absorption maximum toward the visible and is easily detected by an examination of the ultraviolet spectrum.[9] However, this shift is also responsible for a change in color from yellow to orange-red. In general, a yellow dinitrophenylhydrazone may be assumed to be unconjugated. However, an orange or red color should be interpreted with caution, since it may be due to contamination by an impurity.

In difficult cases, it may be desirable to try the preparation of a dinitrophenylhydrazone in diethylene glycol dimethyl ether (diglyme), ethylene glycol monomethyl ether (glyme), DMF, or DMSO. Difficulty in the workup, due to removal of a nonvolatile solvent, can be encountered in these cases. Methanol can be used as an alternative to ethyl alcohol; the more volatile alcohol may, however, result in mixtures that are difficult to purify.

[8]This reagent is sometimes called "Brady's reagent."

[9]Z. Rappoport and T. Sheradsky, *J. Chem. Soc. B,* 1968, 277; L. A. Jones, J. C. Holmes, and R. B. Seligman, *Anal. Chem., 28,* 191 (1956).

Experiment 12. Hydroxylamine Hydrochloride

$$\underset{\text{aldehyde or ketone}}{\overset{O}{\overset{\|}{\diagup C \diagdown}}} + \underset{\text{hydroxylamine hydrochloride}}{H_2N{-}OH \cdot HCl} \longrightarrow \underset{\text{oxime}}{\diagdown C{=}N{-}OH} + HCl + H_2O$$

(a) For Neutral Aldehydes. To 1 mL of the Bogen or Grammercy universal indicator–hydroxylamine hydrochloride reagent add a drop or a few crystals of the compound, and note the color change. If no pronounced change occurs at room temperature, heat the mixture to boiling. A change in color from orange to red constitutes a positive test. Try the test on (1) butanal; (2) acetone; (3) benzophenone; (4) glucose.

(b) For Acidic or Basic Aldehydes. To 1 mL of the indicator solution add about 0.2 g of the compound, and adjust the color of the mixture so that it matches 1 mL of the Bogen or Grammercy universal indicator–hydroxylamine hydrochloride reagent in a separate test tube of the same size. This is done by adding a few drops of 1% sodium hydroxide or 1% hydrochloric acid solution. Then add the resulting solution to 1 mL of the Bogen or Grammercy universal indicator–hydroxylamine hydrochloride reagent. and note if a red color is produced. Try this test on (1) *p*-(*N*,*N*-dimethylamino) benzaldehyde; (2) salicylaldehyde.

Try both (a) and (b) on tartaric acid.

Bogen or Grammercy Universal Indicator–Hydroxylamine Hydrochloride Reagent. To a solution of 50 mg of hydroxylamine hydrochloride in 100 mL of 95% ethanol is added 0.03 mL of Bogen or Grammercy universal indicator. The color of the solution is adjusted to a bright orange shade (pH 3.7–3.9) by adding 5% ethanolic sodium hydroxide dropwise. The reagent is stable for several months.

Indicator Solution. A solution of the indicator is made by adding 0.3 mL of either of the above solutions to 100 mL of 95% ethanol.

Cleaning Up. The test solutions are placed in the aqueous solution container.

Discussion. The change in color of the indicator is due to the hydrochloric acid liberated in the reaction of the carbonyl compound with hydroxylamine hydrochloride, with the oxime not being sufficiently basic to form a hydrochloride. All aldehydes and most ketones give an immediate change in color. Some higher molecular weight ketones such as benzophenone, benzil, benzoin, and camphor require heating. Sugars, quinones, and hindered ketones, such as 2-benzoylbenzoic acid, give a negative test.

Many aldehydes undergo autoxidation in the air and contain appreciable amounts of acids; hence the action of an aqueous solution or suspension on litmus must always be determined. If the solution is acidic, procedure (b) must be used; this is also true of compounds whose solubility behavior shows them to be acids or bases.

$$R{-}\overset{O}{\overset{\|}{C}}{-}H \xrightarrow{\text{air}} R{-}\overset{O}{\overset{\|}{C}}{-}OH$$

The mechanism of the reaction of carbonyl compounds with hydroxylamine is apparently very closely related to the mechanisms of the reaction with phenylhydrazine, dinitrophenylhydrazine, and semicarbazide. Semicarbazone formation is the reaction of this group that has been most thoroughly studied. The mechanism in aqueous solution, as catalyzed by the acid HA, is shown below.

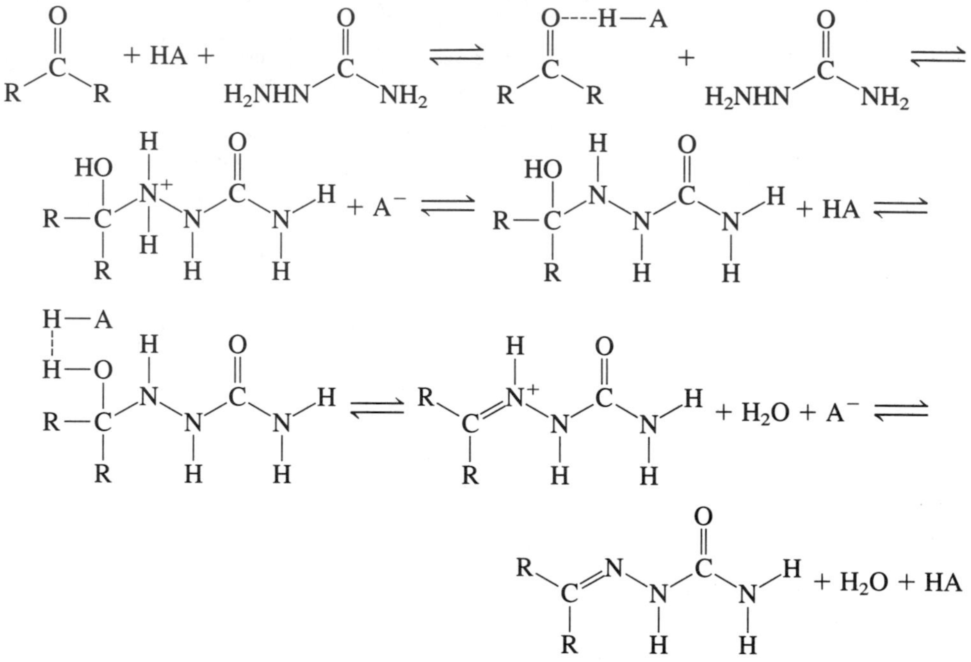

$$R_2C{=}O + HA + H_2NHN{-}C({=}O){-}NH_2 \rightleftharpoons R_2C{=}O\cdots H{-}A + H_2NHN{-}C({=}O){-}NH_2 \rightleftharpoons$$

$$R_2C(OH){-}N^{+}H_2{-}NH{-}C({=}O){-}NH{-}H + A^- \rightleftharpoons R_2C(OH){-}NH{-}NH{-}C({=}O){-}NH{-}H + HA \rightleftharpoons$$

$$R_2C(O(H)\cdots H{-}A){-}NH{-}NH{-}C({=}O){-}NH{-}H \rightleftharpoons R_2C{=}N^{+}H{-}NH{-}C({=}O){-}NH{-}H + H_2O + A^- \rightleftharpoons$$

$$R_2C{=}N{-}NH{-}C({=}O){-}NH{-}H + H_2O + HA$$

The mechanism shows that the reaction is strongly retarded by solutions that are either too acidic or too basic. Since the reaction requires both free semicarbazide and a proton source (HA), a too strongly acidic solution depresses the reaction rate by converting virtually all of the semicarbazide to its inactive conjugate acid, whereas a strongly basic solution depresses the rate by converting the acid HA to its inactive conjugate base. The most favorable conditions are those in which the product of the concentrations of HA and $NH_2NHCONH_2$ is a maximum.

The reaction is reversible. Although in relatively weakly acidic solutions the equilibrium may be made to lie far towards the right, strong acid removes semicarbazide by converting it to the conjugate acid and shifts the equilibrium toward the free ketone.

Experiment 13. Sodium Bisulfite Addition Complex

$$>C{=}O + NaHSO_3 \longrightarrow {-}C(H)(SO_3^{-}Na^{+}){-}OH$$

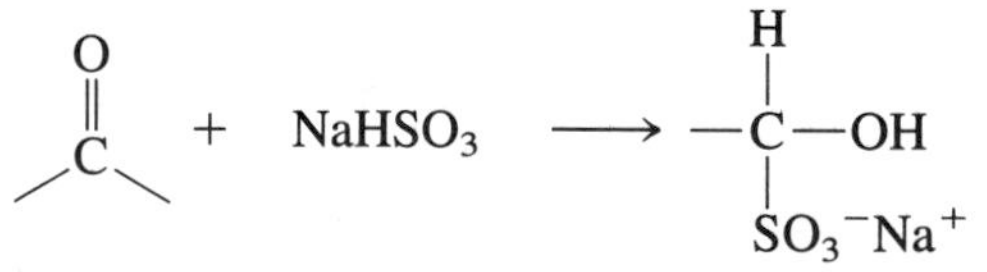

aldehyde or ketone — sodium bisulfite — sodium bisulfite addition complex

Prepare an alcoholic solution of sodium bisulfite by adding 1 mL of ethanol to 4 mL of a 40% aqueous solution of the salt. A small amount of salt will be precipitated by the alcohol and must be separated by decantation or filtration before the reagent is ready for use.

Place 1 mL of the reagent in a test tube and add 0.3 mL or 300 mg of the sample. Stopper the test tube and shake vigorously. Aldehydes and ketones react with sodium bisulfite to form a solid. Thus, the formation of a solid is a positive test. Try this test with (1) acetone; (2) benzaldehyde; (3) heptanal; (4) acetophenone.

Cleaning Up. Place the mixture in the aqueous solution container.

Discussion. The formation of bisulfite addition compounds, also known as α-hydroxyalkanesulfonates, is a general reaction of aldehydes. Most methyl ketones, low molecular weight cyclic ketones up to cyclooctanone, and certain other compounds having very active carbonyl groups behave similarly. Some methyl ketones, however, form the addition compounds slowly or not at all. Examples are aryl methyl ketones, pinacolone, and mesityl oxide. Cinnamaldehyde forms an addition compound containing two molecules of bisulfite.

The bisulfite addition compounds are in equilibrium with the carbonyl compound. These compounds are easily decomposed by either acids or alkalies to regenerate the original compounds and thus are stable only in neutral solutions. Compounds derived from low molecular weight carbonyl compounds are soluble in water. Another advantage of the bisulfite addition compounds is how easily they are purified.

The nitrogen analogs of aldehydes, imines (or Schiff bases), also undergo reaction with sodium bisulfite. The product is identical with that formed by the action of a primary amine on the aldehyde bisulfite compound.

$$R{-}CH{=}N{-}R' + NaHSO_3 \longrightarrow R{-}CH(NHR'){-}SO_3^-\,{}^+Na$$

$$R{-}CH(OH){-}SO_3^-Na^+ + R'NH_2 \longrightarrow R{-}CH(NHR'){-}SO_3^-\,{}^+Na$$

The carbon–sulfur bond in these compounds is reactive, with the sulfonate group being displaced by reactions with anions such as CN^-.

Questions

8. Suggest an explanation of the fact that cyclohexanone reacts with sodium bisulfite readily, whereas 3-pentanone does not.

9. What is the explanation of the failure of pinacolone to react? Compare the case with that of acetophenone.

10. Explain the behavior of cinnamaldehyde.

11. Why is an alcoholic solution of sodium bisulfite used? Try the test on acetone, using an aqueous solution.

Experiment 14. Tollens Test

$$\underset{\text{aldehyde}}{R{-}C({=}O){-}H} + \underset{\text{Tollens reagent}}{2Ag(NH_3)_2OH} \longrightarrow 2Ag(s) + \underset{\substack{\text{salt of the}\\\text{carboxylic acid}}}{R{-}C({=}O){-}O^-NH_4^+} + H_2O + 3NH_3$$

Add one drop or a few crystals of the sample to the freshly prepared Tollens reagent. A positive test is the formation of silver metal or colloidal silver. If no reaction takes place in the cold, the solution should be warmed slightly on a steam bath. However, excessive heating will cause the appearance of a false positive test by decomposition of the reagent.

Try this test with (1) formalin; (2) acetone; (3) benzaldehyde; (4) glucose; (5) hydroquinone; (6) 4-aminophenol.

Tollens Reagent. Into a test tube, which has been cleaned with 10% sodium hydroxide, place 2 mL of a 5% silver nitrate solution, and add a drop of 10% sodium hydroxide. Add 2% ammonia solution, drop by drop, with constant shaking, until the precipitate of silver oxide just dissolves. In order to obtain a sensitive reagent, it is necessary to avoid a large excess of ammonia.

This reagent should be prepared just before use and should not be stored because the solution decomposes on standing and deposits a highly explosive precipitate.

Cleaning Up. Pour the solution into a beaker. Add a few drops of 5% nitric acid to dissolve the silver mirror or colloidal silver. Combine all solutions. Make the solution acidic with 5% nitric acid, then neutralize with sodium carbonate. Five mL of saturated sodium chloride solution is added to precipitate the silver as silver chloride. The silver chloride is isolated by filtration and placed in the nonhazardous solid waste container. The filtrate is placed in the aqueous solution container.

Question

12. Would the presence of a reactive halogen atom interfere with this test?

Discussion. It should be noted that diphenylamine, aromatic amines, as well as 1-naphthols and certain other phenols, give a positive Tollens test. α-Alkoxy and α-dialkylamino ketones have been found to reduce ammoniacal silver nitrate. In addition, the stable hydrate of trifluoroacetaldehyde gives a positive test.

This test often results in a smooth deposit of silver metal on the inner surface of the test tube, hence the name the "silver mirror" test. In some cases, however, the metal forms merely as a granular gray or black precipitate, especially if the glass is not scrupulously clean.

The reaction is autocatalyzed by the silver metal and often involves an induction period of a few minutes.

False negative tests are common with water insoluble aldehydes.

Experiment 15. Fuchsin–Aldehyde Reagent (Schiff's Reagent)

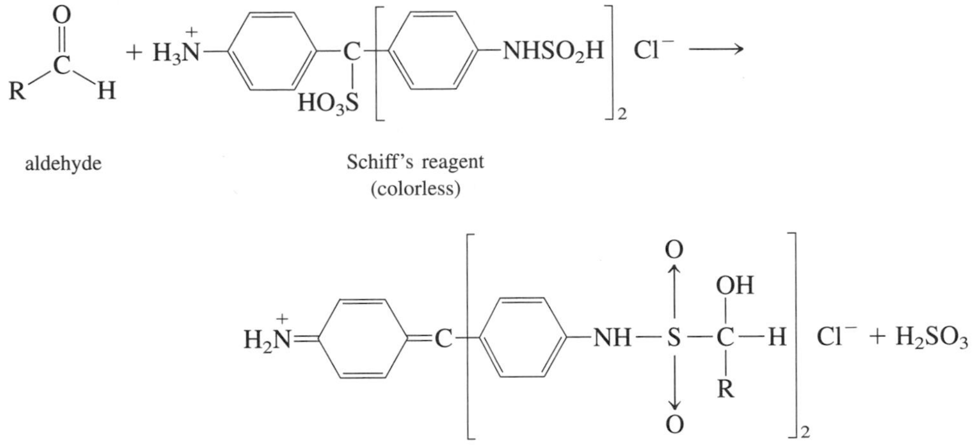

$$RCHO + H_3\overset{+}{N}-C_6H_4-C(SO_3H)\left[C_6H_4-NHSO_2H\right]_2\ Cl^- \longrightarrow$$

aldehyde — Schiff's reagent (colorless)

$$H_2\overset{+}{N}=C_6H_4=C\left[C_6H_4-NH-S(\rightarrow O)_2-C(OH)(R)-H\right]_2\ Cl^- + H_2SO_3$$

(violet-purple solution)

Place 2 mL of Schiff's reagent in a test tube and add 2 drops or a few crystals of the unknown. Shake the tube gently, and observe the color that is developed in 3–4 min. Aldehydes react with Schiff's reagent to form a complex which has a wine-purple color.

Try this test with (1) butanal; (2) benzaldehyde; (3) acetophenone; (4) acetone.

In this test the reagent should not be heated, and the solution tested should not be alkaline. When the test is used on an unknown, a simultaneous test on a known aldehyde and a known ketone should be performed for comparison.

Schiff's Reagent. Dissolve 0.05 g of pure fuchsin (4-rosaline hydrochloride) in 50 mL of distilled water. Add 2 mL of saturated sodium bisulfite solution. After allowing the solution to sit for 1 hr, add 1 mL of concentrated hydrochloric acid. Allow to stand overnight. This reagent is practically colorless and very sensitive.

Cleaning Up. Neutralize the test solution with sodium carbonate and place in the aqueous solution container.

Discussion. Fuchsin is a pink triphenylmethane dye which is converted to the colorless leucosulfonic acid by sulfurous acid. Apparently the reaction involves 1,6-addition of sulfurous acid to the quinoid nucleus of the dye.

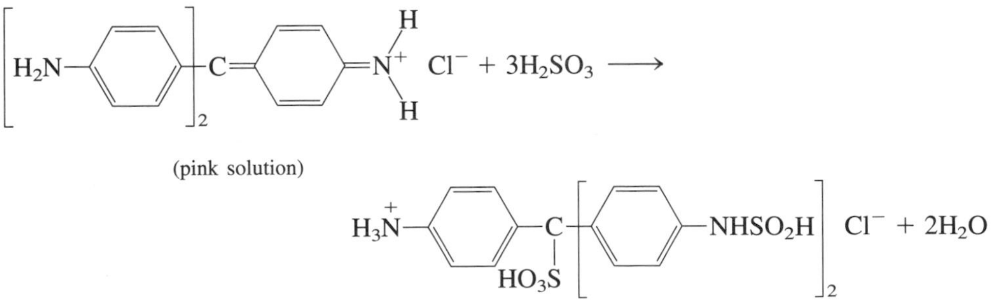

$$\left[H_2N-C_6H_4\right]_2C=C_6H_4=\overset{+}{N}H_2\ Cl^- + 3H_2SO_3 \longrightarrow$$

(pink solution)

$$H_3\overset{+}{N}-C_6H_4-C(SO_3H)\left[C_6H_4-NHSO_2H\right]_2\ Cl^- + 2H_2O$$

Schiff's reagent (colorless)

The leucosulfonic acid is unstable and loses sulfurous acid when treated with an aldehyde, resulting in a violet-purple quinoid dye.

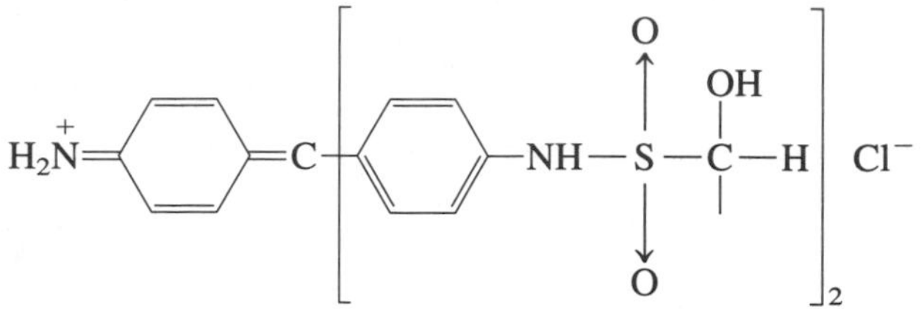

$$\left[H_2\overset{+}{N}{=}C_6H_4{=}C\left(C_6H_4{-}NH{-}\overset{O}{\underset{O}{S}}{-}\overset{OH}{C}{-}H \right)_2 \right] Cl^-$$

(violet-purple solution)

This violet-purple color is different from the color of the original fuchsin. It is not a light pink but has a blue cast bordering on a violet or purple. Some ketones and unsaturated compounds react with sulfurous acid to regenerate the pink color of the fuchsin. Therefore, the development of a light pink color in the reagent is not a positive test for aldehydes.

The fact that certain compounds cause the regeneration of the pink color of the original fuchsin has been made the basis of a test. When a specially prepared reagent is used and the reaction time is 1 hr, aldoses produce a pink color, whereas ketoses and disaccharides, except maltose, do not. This modification of the Schiff test must be employed with caution because many organic compounds produce a pink color with the reagent when shaken in the air. Other compounds such as α,β-unsaturated ketones combine with sulfurous acid and thus reverse the first reaction given above.

7.5 AMIDES

Aliphatic amides react with hydroxylamine hydrochloride to form hydroxamic acid (Experiment 2c, p. 207). Similarly, aromatic primary amides react with hydrogen peroxide to produce hydroxamic acid (Experiment 2d, p. 208). The hydroxamic acid then reacts with ferric chloride to form the ferric hydroxamate, which has a characteristic magenta color.

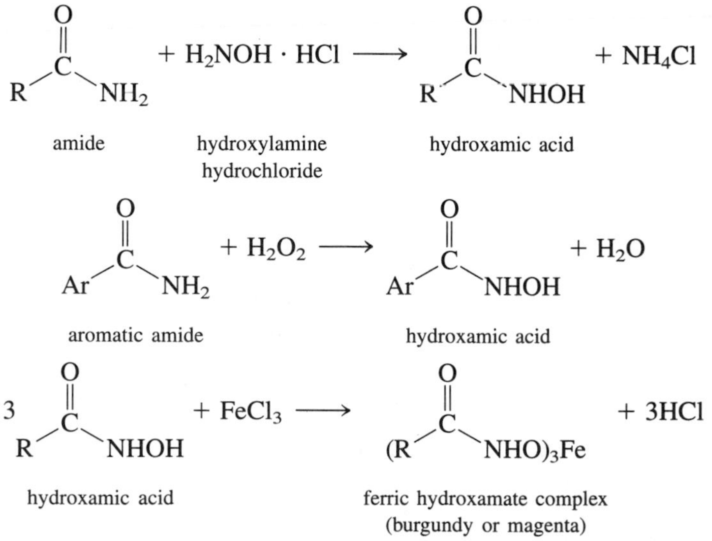

$$\underset{\text{amide}}{RC(=O)NH_2} + \underset{\text{hydroxylamine hydrochloride}}{H_2NOH \cdot HCl} \longrightarrow \underset{\text{hydroxamic acid}}{RC(=O)NHOH} + NH_4Cl$$

$$\underset{\text{aromatic amide}}{ArC(=O)NH_2} + H_2O_2 \longrightarrow \underset{\text{hydroxamic acid}}{ArC(=O)NHOH} + H_2O$$

$$3\ \underset{\text{hydroxamic acid}}{RC(=O)NHOH} + FeCl_3 \longrightarrow \underset{\text{ferric hydroxamate complex (burgundy or magenta)}}{(RC(=O)NHO)_3Fe} + 3HCl$$

Amides can be hydrolyzed to yield the salt of the carboxylic acid and ammonia or amine (Experiment 16, below). The presence of ammonia or a low molecular weight amine is detected with litmus paper.

$$\underset{\text{1° amide}}{RC(=O)NH_2} + NaOH \longrightarrow \underset{\text{sodium salt of the carboxylic acid}}{RC(=O)O^-Na^+} + \underset{\text{ammonia}}{NH_3}$$

$$\underset{\text{2° amide}}{RC(=O)NHR'} + NaOH \longrightarrow \underset{\text{sodium salt of the carboxylic acid}}{RC(=O)O^-Na^+} + \underset{\text{1° amine}}{R'NH_2}$$

$$\underset{\text{3° amide}}{RC(=O)NR'_2} + NaOH \longrightarrow \underset{\text{sodium salt of the carboxylic acid}}{RC(=O)O^-Na^+} + \underset{\text{2° amine}}{R'_2NH}$$

Experiment 16. Sodium Hydroxide Hydrolysis of Amides and Nitriles

$$\underset{\text{1° amide}}{RC(=O)NH_2} + NaOH \longrightarrow \underset{\text{sodium salt of the carboxylic acid}}{RC(=O)O^-Na^+} + \underset{\text{ammonia}}{NH_3}$$

$$\underset{\text{2° amide}}{RC(=O)NHR'} + NaOH \longrightarrow \underset{\text{sodium salt of the carboxylic acid}}{RC(=O)O^-Na^+} + \underset{\text{1° amine}}{R'NH_2}$$

$$\underset{\text{3° amide}}{RC(=O)NR'_2} + NaOH \longrightarrow \underset{\text{sodium salt of the carboxylic acid}}{RC(=O)O^-Na^+} + \underset{\text{2° amine}}{R'_2NH}$$

$$\underset{\text{nitrile}}{R{-}C{\equiv}N} + NaOH + H_2O \longrightarrow \underset{\text{salt of the carboxylic acid}}{RC(=O)O^-Na^+} + \underset{\text{ammonia}}{NH_3}$$

Treat 0.2 g of the unknown in a test tube with 5 mL of 10% sodium hydroxide solution. Shake the mixture and note whether or not ammonia is evolved. Heat the solution to boiling and note the odor. Test the action of the vapor on either pink moist litmus paper or filter paper moistened with a copper sulfate solution. If ammonia or amine is being evolved, the litmus paper turns blue. Ammonia, which is evolved only from primary amines, will turn the copper sulfate solution on the filter paper blue. Nitriles and ammonium salts will also give a positive test with the copper sulfate.

Try this test on (1) benzamide; (2) benzonitrile.

Cleaning Up. Place the mixture in the aqueous solution container.

Discussion. The ammonia or amine which is the product of this alkaline hydrolysis may be characterized by the Hinsberg test (Experiment 17).

Many substituted amides are hydrolyzed more easily by heating under reflux with 20% sulfuric acid.

$$2\ \mathrm{RC(=O)NHR'} + \mathrm{H_2SO_4} \xrightarrow{2\mathrm{H_2O}} 2\ \mathrm{RC(=O)OH} + (\mathrm{R'\overset{+}{N}H_3})_2\mathrm{SO_4}^{2-}$$

2° amide → carboxylic acid

$$2\ \mathrm{RC(=O)NR'_2} + \mathrm{H_2SO_4} \xrightarrow{2\mathrm{H_2O}} 2\ \mathrm{RC(=O)OH} + (\mathrm{R'_2\overset{+}{N}H_2})_2\mathrm{SO_4}^{2-}$$

3° amide → carboxylic acid

Nitriles, particularly cyanohydrins, are frequently hydrolyzed by acids. Treatment with concentrated hydrochloric acid, with heating, converts the nitriles to amides. The amides may be hydrolyzed further by diluting the mixture with water and heating for 0.5–2 hr.

$$\mathrm{R{-}C{\equiv}N} \xrightarrow[\mathrm{H_2SO_4}]{\mathrm{H_2O}} \mathrm{RC(=O)NH_2} \xrightarrow[\mathrm{H_2SO_4}]{\mathrm{H_2O}} \mathrm{RC(=O)OH} + \mathrm{NH_4^+HSO_4^-}$$

nitrile → 1° amide → carboxylic acid

7.6 AMINES AND AMINE SALTS

The active hydrogen on primary and secondary amines undergoes reaction with sodium to form the salt and to liberate hydrogen gas (Experiment 5, p. 215). A tertiary amine does not undergo reaction, since it does not have an active hydrogen.

$$2\mathrm{RNH_2} + 2\mathrm{Na} \longrightarrow 2\mathrm{RNH^-Na^+} + \mathrm{H_2(g)}$$

1° amine

$$2\mathrm{R_2NH} + 2\mathrm{Na} \longrightarrow 2\mathrm{R_2N^-Na^+} + \mathrm{H_2(g)}$$

2° amine

Primary and secondary amines react with acetyl chloride to produce amides, which often precipitate from the solution (Experiment 6, p. 217). This reaction is usually accompanied by the evolution of heat. Tertiary amines do not react with acetyl chloride because they lack a hydrogen on the nitrogen.

$$\underset{\text{1° amine}}{RNH_2} + \underset{\text{acetyl chloride}}{H_3C{-}C({=}O){-}Cl} \longrightarrow \underset{\text{amide}}{H_3C{-}C({=}O){-}NHR} + HCl(g)$$

$$\underset{\text{2° amine}}{R_2NH} + \underset{\text{acetyl chloride}}{H_3C{-}C({=}O){-}Cl} \longrightarrow \underset{\text{amide}}{H_3C{-}C({=}O){-}NR_2} + HCl(g)$$

The Hinsberg test (Experiment 17, p. 242) can be used to distinguish between primary, secondary, and tertiary amines. Benzenesulfonyl chloride undergoes reaction with primary amines in basic solution to form the sodium salts of the sulfonamide, which are soluble in the reaction mixture. Acidification of the reaction mixture results in the precipitation of the sulfonamides.

$$\underset{\text{1° amine}}{RNH_2} + \underset{\text{benzenesulfonyl chloride}}{C_6H_5SO_2Cl} + 2NaOH \longrightarrow \underset{\text{(soluble)}}{C_6H_5SO_2NR^-Na^+} + NaCl + 2H_2O$$

$$\xrightarrow{H^+} \underset{\text{sulfonamide (insoluble)}}{C_6H_5SO_2NHR}$$

Secondary amines, when treated with benzenesulfonyl chloride, yield the sulfonamides, which precipitate from the solution. Acidification of the solution does not dissolve the sulfonamide.

$$\underset{\text{2° amine}}{R_2NH} + \underset{\text{benzenesulfonyl chloride}}{C_6H_5SO_2Cl} + NaOH \longrightarrow \underset{\text{sulfonamide (insoluble)}}{C_6H_5SO_2NR_2} + NaCl + H_2O$$

$$\xrightarrow{H^+} \text{No Reaction}$$

Tertiary amines undergo reaction with benzenesulfonyl chloride to produce quaternary ammonium sulfonate salts, which yield sodium sulfonates and insoluble tertiary amines in basic solution. Acidification of the reaction mixture results in the formation of sulfonic acids and soluble amine salts.

$$\underset{\text{3° amine}}{R_3N} + \underset{\text{benzenesulfonyl chloride}}{C_6H_5SO_2Cl} \longrightarrow C_6H_5SO_2NR_3{}^+Cl^- \xrightarrow{2NaOH}$$

$$\underset{\text{(soluble)}}{C_6H_5SO_3{}^-Na^+} + \underset{\text{(insoluble)}}{NR_3} + NaCl + H_2O \xrightarrow{2HCl} C_6H_5SO_3H + \underset{\text{(soluble)}}{R_3NH^+Cl^-} + NaCl$$

The reaction of amines with nitrous acid (Experiment 18, p. 245) classifies the amine not only as primary, secondary, or tertiary, but also as aliphatic or aromatic. Primary aromatic and aliphatic amines react with nitrous acid to form an intermediate diazonium salt. The aliphatic diazonium salts decompose spontaneously by rapid loss of nitrogen, particularly when the original amino group is attached to a secondary or tertiary carbon. Most aromatic diazonium salts are stable at 0°C, but lose nitrogen slowly on warming to room temperature.

$$\underset{\text{1° aliphatic amine}}{RNH_2} + \underset{\text{nitrous acid}}{HONO} + 2HCl \longrightarrow \underset{\text{diazonium salt (unstable at 0°)}}{[RN_2^+Cl^-]} \xrightarrow[\text{spontaneous}]{H_2O} N_2(g) + ROH + RCl + ROR + \text{alkene}$$

$$\underset{\text{1° aromatic amine}}{ArNH_2} + \underset{\text{nitrous acid}}{HONO} + HCl \longrightarrow \underset{\text{diazonium salt (stable at 0°C)}}{ArN_2^+Cl^-} \xrightarrow{H_2O} N_2(g) + ArOH + HCl$$

The diazonium salt of the primary aromatic amine reacts with sodium 2-naphthol to produce a red-orange azo dye.

$$\underset{\text{diazonium salt of the 1° aromatic amine}}{ArN_2^-Cl^+} + \underset{\text{sodium 2-naphthol}}{C_{10}H_7O^-Na^+} + NaOH \longrightarrow \underset{\text{azo dye (red-orange)}}{Ar{-}N{=}N{-}C_{10}H_6O^-Na^+} + NaCl + H_2O$$

Secondary amines undergo a reaction with nitrous acid to form *N*-nitrosoamines, which are usually yellow solids.

$$\underset{\text{2° amine}}{R_2NH} + \underset{\text{nitrous acid}}{HONO} \longrightarrow \underset{\text{N-nitrosoamine (yellow oil or solid)}}{R_2N{-}N{=}O} + H_2O$$

Tertiary aliphatic amines do not react with nitrous acid, but form a soluble salt. The reaction mixture gives an immediate positive test on the starch-iodide paper for nitrous acid.

$$\underset{\text{3° aliphatic amine}}{R_3N} + H^+ \longrightarrow \underset{\text{(soluble)}}{R_3NH^+}$$

Tertiary aromatic amines react with nitrous acid to form the orange-colored hydrochloride salt of the C-nitrosoamine. Treating the solution with base liberates the blue or green C-nitrosoamine.

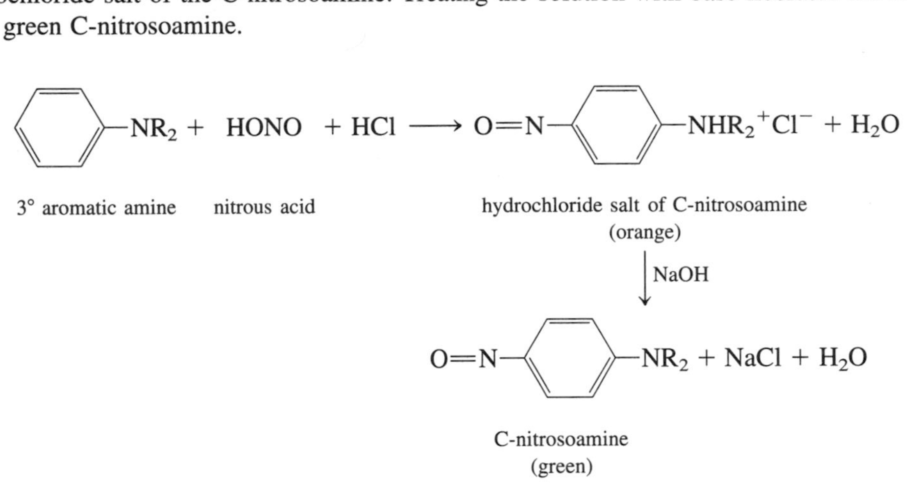

$$C_6H_5{-}NR_2 + HONO + HCl \longrightarrow O{=}N{-}C_6H_4{-}NHR_2^{+}Cl^{-} + H_2O$$

3° aromatic amine nitrous acid hydrochloride salt of C-nitrosoamine (orange)

$$\downarrow NaOH$$

$$O{=}N{-}C_6H_4{-}NR_2 + NaCl + H_2O$$

C-nitrosoamine (green)

Secondary amines combine with carbon disulfide and ammonium hydroxide, followed by nickel chloride to produce a solid product (Experiment 19, p. 251).

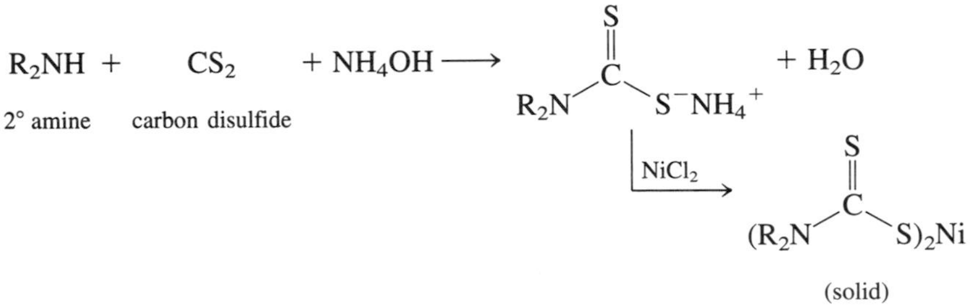

$$R_2NH + CS_2 + NH_4OH \longrightarrow R_2N{-}C({=}S){-}S^{-}NH_4^{+} + H_2O$$

2° amine carbon disulfide

$$\xrightarrow{NiCl_2} (R_2N{-}C({=}S){-}S)_2Ni$$

(solid)

Primary aliphatic amines react quickly with 5-nitrosalicylaldehyde, followed by nickel chloride to form a precipitate within several minutes (Experiment 20, p. 251).

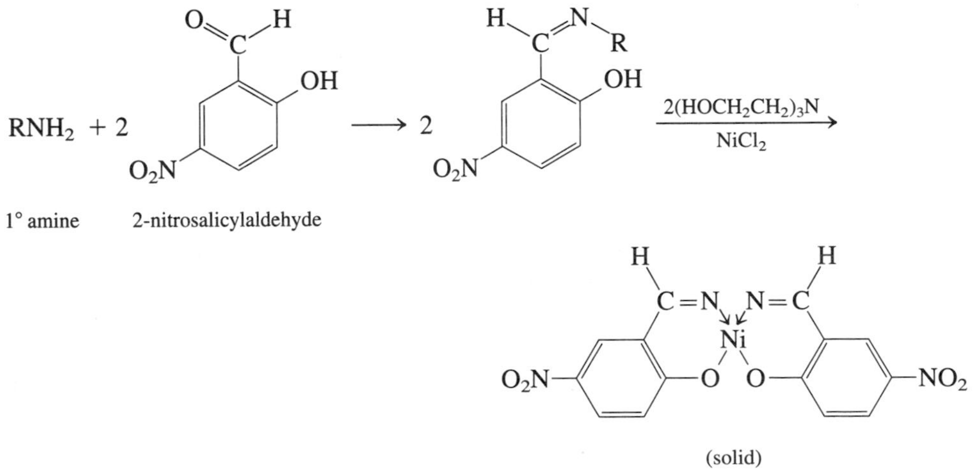

$$RNH_2 + 2\ O_2N{-}C_6H_3(OH){-}CHO \longrightarrow 2\ O_2N{-}C_6H_3(OH){-}CH{=}N{-}R \xrightarrow[NiCl_2]{2(HOCH_2CH_2)_3N}$$

1° amine 2-nitrosalicylaldehyde

$$(O_2N{-}C_6H_3(O^{-}){-}CH{=}N)_2Ni$$

(solid)

Amine salts can be detected by treating the salt with sodium hydroxide to liberate the ammonia or amine (Experiment 21, p. 252).

$$\underset{\text{ammonium salt}}{RCOO^-NH_4^+} + NaOH \longrightarrow \underset{\text{salt of a carboxylic acid}}{RCOO^-Na^+} + \underset{\text{ammonia}}{NH_3} + H_2O$$

$$\underset{\text{amine salt}}{RNH_3^+X^-} + NaOH \longrightarrow \underset{1^\circ\ \text{amine}}{RNH_2} + NaCl + H_2O$$

Experiment 17. Benzenesulfonyl Chloride (Hinsberg's Method for Characterizing Primary, Secondary, and Tertiary Amines)

$$\underset{1^\circ\ \text{amine}}{RNH_2} + \underset{\text{benzenesulfonyl chloride}}{C_6H_5SO_2Cl} + 2NaOH \longrightarrow \underset{\text{(soluble)}}{C_6H_5SO_2NR^-Na^+} + NaCl + 2H_2O$$

$$\xrightarrow{H^+} \underset{\text{sulfonamide (insoluble)}}{C_6H_5SO_2NHR}$$

$$\underset{2^\circ\ \text{amine}}{R_2NH} + \underset{\text{benzenesulfonyl chloride}}{C_6H_5SO_2Cl} + NaOH \longrightarrow \underset{\text{sulfonamide (insoluble)}}{C_6H_5SO_2NR_2} + NaCl + H_2O$$

$$\xrightarrow{H^+} \text{No Reaction}$$

$$\underset{3^\circ\ \text{amine}}{R_3N} + \underset{\text{benzenesulfonyl chloride}}{C_6H_5SO_2Cl} \longrightarrow C_6H_5SO_2NR_3^+Cl^- \xrightarrow{2NaOH}$$

$$\underset{\text{(soluble)}}{C_6H_5SO_3^-Na^+} + \underset{\text{(insoluble)}}{NR_3} + NaCl + H_2O$$

$$\downarrow 2HCl$$

$$C_6H_5SO_3H + \underset{\text{(soluble)}}{R_3NH^+Cl^-} + NaCl$$

To 0.3 mL or 300 mg of the unknown sample in a test tube, add 5 mL of 10% sodium hydroxide solution and 0.4 mL of benzenesulfonyl chloride. Stopper the test tube, and shake the mixture very vigorously. Test the solution to make sure that it is alkaline. After all the benzenesulfonyl chloride has reacted, cool the solution and separate the residue, if present, from the solution. Test the residue for solubility in 10% hydrochloric acid. If no residue remains, then treat the solution with 10% hydrochloric acid and observe whether a precipitate forms.

If all of the original compound dissolves in the base and no residue remains, and acidification produces a precipitate, the original unknown is a primary amine. Primary amines react with benzenesulfonyl chloride in basic solution to form the sodium salt of the sulfonamide, which is normally soluble in basic solution, but the sulfonamide precipitates upon acidification. If a residue is formed and it is insoluble in acid, the original unknown is a secondary amine. Secondary amines undergo reaction with benzenesulfonyl chloride to precipitate the sulfonamide, and acidification does not result in any change. If a residue is present which is soluble in acid, this indicates that the residue is the unreacted tertiary amine. Tertiary amines undergo reaction with benzenesulfonyl chloride to produce quaternary ammonium sulfonate salts, which yield sodium sulfonates and water insoluble tertiary amines in basic solution. Acidification of the reaction mixture results in the formation of sulfonic acids and water soluble amine salts. Any solid which is formed should be isolated and purified, and its melting point should be compared against that of the original amine.

If the amount of the solid is in sufficient quantity, it may be saved and used as a derivative for that unknown.

This test should be first tried with (1) aniline; (2) *N*-methylaniline; (3) *N,N*-dimethylaniline. Make note of all precipitate formations and dissolutions and use melting points to check identity of the product as an amine or a sulfonamide.

Cleaning Up. Any solids are isolated by filtration and placed in the organic nonhazardous solid waste container. If the original compound is identified as a primary or secondary amine, then dilute the solution with water and place in the aqueous solution container. If the unknown is a tertiary amine, then make the solution basic with 10% sodium hydroxide, and extract the amine with petroleum ether. Place the organic layer in the organic solvents container, and the aqueous layer in the aqueous solution container.

Discussion. The sodium salts of certain sulfonamides of cyclohexyl- through cyclodecylamine and certain high molecular weight amines are insoluble in 10% sodium hydroxide solution.[10] Usually they are soluble in water. Certain primary amines may yield alkali insoluble disulfonyl derivatives. These may be hydrolyzed by boiling for 30 min with 5% sodium ethoxide in absolute ethanol.

If the solution heats up considerably, it should be cooled. Certain *N,N*-dialkylanilines produce a purple dye if the mixture becomes too hot. This may be prevented by carrying out the reaction at 15–20°C.

When the Hinsberg method is used to separate a mixture of amines, it is necessary to recover the pure individual amines. The benzenesulfonamides may be hydrolyzed as follows.

$$C_6H_5{-}SO_2NHR + H_2O + HCl \longrightarrow C_6H_5{-}SO_3H + RNH_3^+Cl^-$$

2° benzenesulfonamide — benzenesulfonic acid — 1° amine salt

$$C_6H_5{-}SO_2NR_2 + H_2O + HCl \longrightarrow C_6H_5{-}SO_3H + R_2NH_2^+Cl^-$$

3° benzenesulfonamide — benzenesulfonic acid — 2° amine salt

Hydrolysis of the sulfonamide is accomplished by heating 1.0 g of the sulfonamide with 10 mL of 25% hydrochloric acid under reflux. Sulfonamides of primary amines require 24–36 hr refluxing, whereas sulfonamides of secondary amines may be hydrolyzed in 10–12 hr. After the reaction is complete, the mixture is cooled, made alkaline with 20% sodium hydroxide solution, and extracted with three 5-mL portions of ether. The ether so-

[10] P. E. Fanta and C. S Wang, *J. Chem. Educ, 41,* 280 (1964). Certain primary amines may yield alkali insoluble disulfonyl derivatives. These may be hydrolyzed by boiling for 30 min with 5% sodium ethoxide in absolute ethanol.

lution is dried, and the ether is removed by distillation. The resulting amine is distilled. With certain low- or high-boiling amines, the amine is recovered as a hydrochloride salt by passing dry hydrogen chloride gas through the dry ether solution.

Many sulfonamides are hydrolyzed only with great difficulty; a more satisfactory procedure involves 48% hydrobromic acid and phenol.[11] Instead of a simple hydrolysis, this reaction is a reductive cleavage in which the hydrogen bromide is oxidized to bromine and the sulfonamide reduced to the disulfide. The primary purpose of the phenol is to remove the bromine by the formation of 4-bromophenol.

$$2\,C_6H_5{-}SO_2NR_2 + 5HBr + 5\,C_6H_5{-}OH \longrightarrow$$

3° sulfonamide

$$C_6H_5{-}S{-}S{-}C_6H_5 + 2R_2NH + 4H_2O + 5Br{-}C_6H_4{-}OH$$

4-bromophenol

Arenesulfonyl chlorides can be useful in characterizing primary and secondary amines. The Hinsberg method for separating amines is based on the fact that the sulfonamides of primary amines are soluble in alkali whereas those of secondary amines are not. Since tertiary amines do not give amides, the method provides a means of classifying and separating the three types of amines. However, the results of the Hinsberg test must not be used alone in classifying amines. The solubility of the original compound must also be considered. If the original compound is amphoteric, which means that it is soluble in both acids and alkalies, the Hinsberg method fails to distinguish among the types of amines. For example, 4-(*N*-methylamino)benzoic acid undergoes reaction with benzenesulfonyl chloride and alkali to give a *solution* of the sodium salt of the *N*-benzenesulfonyl derivative.

$$HOOC{-}C_6H_4{-}N(H)CH_3 + C_6H_5{-}SO_2Cl + 2NaOH \longrightarrow$$

4-(*N*-methylamino)benzoic acid; benzenesulfonyl chloride

$$Na^+\,{}^-OOC{-}C_6H_4{-}N(CH_3){-}SO_2{-}C_6H_5 + NaCl + 2H_2O$$

benzenesulfonamide derivative

[11] H. R. Synder and R. E. Heckert, *J. Amer. Chem. Soc., 74,* 2006 (1952); H. R. Synder and H. C. Geller, *J. Amer. Chem. Soc., 74,* 4864 (1952).

Acidification of the solution precipitates the free acid. This observation, taken by itself, would incorrectly identify the original compound as a primary amine rather than a secondary amine.

Experiment 18. Nitrous Acid

(a) Diazotization

$$\underset{\text{sodium nitrite}}{NaNO_2} + HCl \longrightarrow \underset{\text{nitrous acid}}{HONO} + NaCl$$

$$\underset{\substack{1^\circ\text{ aliphatic}\\\text{amine}}}{RNH_2} + \underset{\substack{\text{nitrous}\\\text{acid}}}{HONO} + 2HCl \longrightarrow \underset{\substack{\text{diazonium salt}\\\text{(unstable at 0°C)}}}{[RN_2{}^+Cl^-]} \xrightarrow[\text{spontaneous}]{H_2O}$$

$$N_2(g) + ROH + RCl + ROR + \text{alkene}$$

$$\underset{\substack{1^\circ\text{ aromatic}\\\text{amine}}}{ArNH_2} + \underset{\substack{\text{nitrous}\\\text{acid}}}{HONO} + HCl \longrightarrow \underset{\substack{\text{diazonium salt}\\\text{(stable at 0°C)}}}{ArN_2{}^+Cl^-} \xrightarrow{H_2O} N_2(g) + ArOH + HCl$$

Dissolve 0.5 mL or 0.5 g of the sample in 1.5 mL of concentrated hydrochloric acid diluted with 2.5 mL of water, and cool the solution to 0°C in a beaker of ice. Dissolve 0.5 g of sodium nitrite in 2.5 mL of water, and add this solution dropwise, with shaking, to the cold solution of the amine hydrochloride. Continue the addition until the mixture gives a positive test for nitrous acid. The test is carried out by placing a drop of the solution on starch-iodide paper; a blue color indicates the presence of nitrous acid. If the test is positive, move 2 mL of the solution to another test tube, warm gently, and examine for evolution of gas.

The observation of rapid bubbling, foaming, or frothing, as the aqueous sodium nitrite is added at 0°C indicates the presence of a primary aliphatic amine. The evolution of gas upon warming indicates that the amine is a primary aromatic amine, and the solution should be subjected to the coupling reaction (b).

If a pale yellow oil or low-melting solid, which is the *N*-nitrosoamine, is formed with no evolution of gas, the original amine is a secondary amine. The oil or solid is isolated and treated under conditions of the Liebermann nitroso reaction (c) to provide confirmation of the presence of the *N*-nitrosoamine.

$$\underset{2^\circ\text{ amine}}{R_2NH} + \underset{\text{nitrous acid}}{HONO} \longrightarrow \underset{\substack{N\text{-nitrosoamine}\\\text{(yellow oil or solid)}}}{R_2N{-}N{=}O} + H_2O$$

An immediate positive test for nitrous acid with no evolution of gas indicates the presence of a tertiary aliphatic amine. The tertiary aliphatic amine is simply protonated to form a soluble salt under these conditions and does not react with the nitrous acid.

$$\underset{\substack{3^\circ\text{ aliphatic}\\\text{amine}}}{R_3N} + H^+ \longrightarrow \underset{\text{(soluble)}}{R_3NH^+}$$

The reaction of a tertiary aromatic amine (an aniline) with nitrous acid produces a dark-orange solution or an orange crystalline solid, which is the hydrochloride salt of the C-nitrosoamine. Treating 2 mL of the solution with 10% sodium hydroxide or sodium carbonate solution will produce the bright green or blue nitrosoamine base. The nitrosoamine base can be isolated, purified, and characterized.

$$C_6H_5\text{—}NR_2 + HONO + HCl \longrightarrow O{=}N\text{—}C_6H_4\text{—}NHR_2^{+}Cl^{-} + H_2O$$

3° aromatic amine — nitrous acid — hydrochloride salt of C-nitrosoamine (orange)

$$O{=}N\text{—}C_6H_4\text{—}NHR_2^{+}Cl^{-} \xrightarrow{NaOH} O{=}N\text{—}C_6H_4\text{—}NR_2 + NaCl + H_2O$$

C-nitrosoamine (green)

Cleaning Up. If the compound is identified as a primary aromatic amine, then the solution is diluted with 10 mL of water and then passed through 5 g of charcoal. The charcoal is then placed in the nonhazardous waste container and the aqueous solution is placed in the aqueous solution container. If the unknown is identified as a primary aliphatic amine or tertiary aliphatic amine, the solution can be diluted with water and placed in the aqueous solution container. For secondary amines and tertiary aromatic amines, the nitrosoamines are isolated by filtration and placed in the hazardous solid waste container. The filtrates are neutralized with 10% sodium hydroxide and placed in the aqueous solution container.

(b) Coupling

$$ArN_2^{-}Cl^{+} + C_{10}H_7O^{-}Na^{+} + NaOH \longrightarrow Ar\text{—}N{=}N\text{—}C_{10}H_6O^{-}Na^{+} + NaCl + H_2O$$

diazonium salt of the 1° aromatic amine — sodium 2-naphthol — azo dye (red-orange)

Add 2 mL of the cold diazonium solution to a solution of 0.1 g of 2-naphthol in 2 mL of 10% sodium hydroxide solution and 5 mL of water. The formation of the orange-red dye, with the evolution of gas only upon warming as noted in (a), indicates the original compound is a primary aromatic amine.

Cleaning Up. Separate any solids by filtration and place in the organic nonhazardous solid waste container. The remaining liquid is placed in the aqueous solution container.

(c) Liebermann's Nitroso Reaction

$$R_2N{-}N{=}O \xrightarrow{H_3O^+} R_2NH + HONO$$

N-nitrosoamine

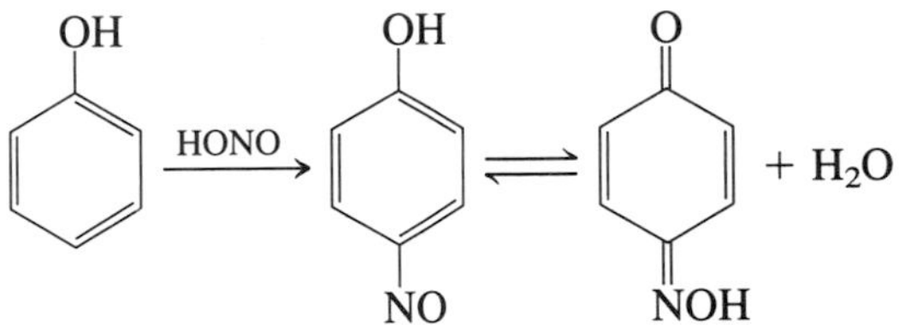

$$C_6H_5OH \xrightarrow{HONO} HO{-}C_6H_4{-}NO \rightleftharpoons O{=}C_6H_4{=}NOH + H_2O$$

(yellow)

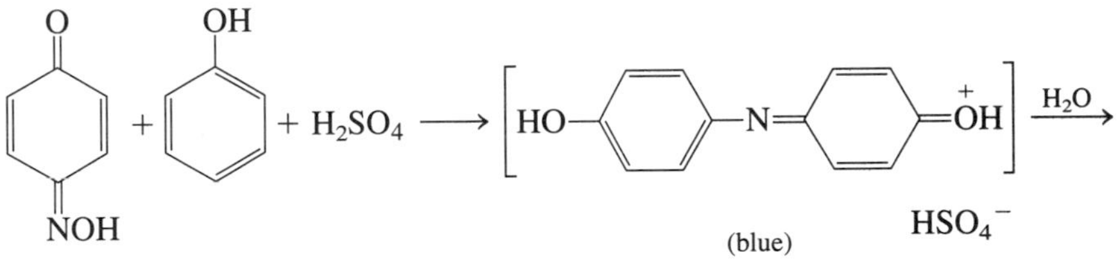

$$O{=}C_6H_4{=}NOH + C_6H_5OH + H_2SO_4 \longrightarrow \left[HO{-}C_6H_4{-}N{=}C_6H_4{=}\overset{+}{O}H\right] HSO_4^- \xrightarrow{H_2O}$$

(yellow) (blue)

$$O{=}C_6H_4{=}N{-}C_6H_4{-}OH \xrightarrow{NaOH} O{=}C_6H_4{=}N{-}C_6H_4{-}O^-Na^+$$

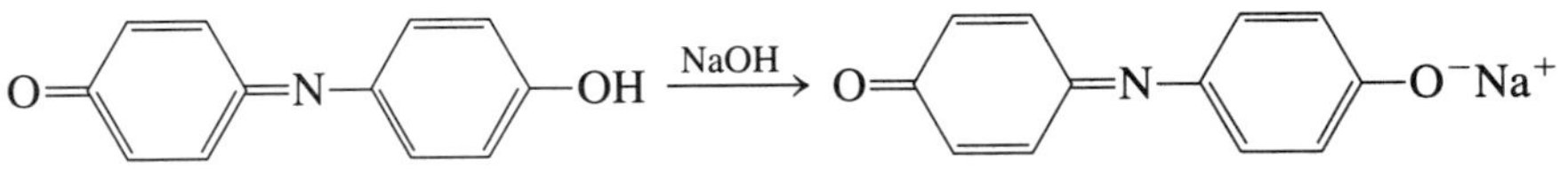

(red) (blue)

Add 0.05 g of the *N*-nitrosoamine, 0.05 g of phenol, and 2 mL of concentrated sulfuric acid to a test tube and warm gently for 20 sec. Cool the solution slightly. A blue color should develop, which changes to red when the solution is poured into 20 mL of ice water. Add 10% sodium hydroxide until the mixture is alkaline, and the blue color is produced again.

The *N*-nitrosoamine liberates nitrous acid in the presence of sulfuric acid. The nitrous acid then undergoes reaction with phenol to yield the yellow 4-nitrosophenol (quinone monoxime). The blue color observed in this reaction is due to indophenol formed from the reaction of the initially produced 4-nitrosophenol (quinone monoxime) with excess phenol. This reaction is characteristic of phenols in which an *ortho* or *para* position is unsubstituted.

To run a comparative test in order to check the colors, the following procedure may be used. Add a crystal of sodium nitrite to 2 mL of concentrated sulfuric acid, and shake until dissolved. Add 0.1 g of phenol, and a blue color will appear. The solution is poured into 20 mL of ice water, and the color of the solution changes to red. Addition of 10% sodium hydroxide, until the mixture is alkaline, results in the return of the blue color.

Try the above reactions on (1) aniline; (2) *N*-methylaniline; (3) *N,N*-dimethylaniline.

Cleaning Up. If necessary, make the solution basic with 10% sodium hydroxide. Place the mixture in the aqueous solution container.

(d) Test for Phenols

To test for the presence of a phenol, a modification of the Liebermann's test is used. Add a crystal of sodium nitrite to 2 mL of concentrated sulfuric acid, and shake until dissolved. Add 0.1 g of the unknown. A blue color indicates the presence of a phenol. Another indication of the presence of the phenol is the change of color of the solution to red, when it is poured into 20 mL of ice water, and the return of the blue color when the mixture becomes alkaline after the addition of 10% sodium hydroxide solution.

Cleaning Up. Place the basic aqueous layer in the aqueous solution container.

(e) Test for α-Amino Acids

$$R-CH(NH_3^+)-C(=O)-O^- + HONO + 2\ CH_3-C(=O)-OH \longrightarrow$$

α-amino acid, nitrous acid

$$\left[H_3C-C(=O)-O^-\ ^+N_2-CH(R)-C(=O)-OH \right] \xrightarrow{H_2O} N_2(g) + HO-CH(R)-C(=O)-OH$$

α-hydroxy acid

The procedure in (a) should be followed, except acetic acid is substituted for hydrochloric acid. A positive test is the evolution of nitrogen gas. The α-amino acid reacts with the nitrous acid to form an intermediate, which decomposes to nitrogen gas and an α-hydroxy acid.

Cleaning Up. Make the solution alkaline with 10% sodium hydroxide and place in the aqueous solution container.

Discussion. ***Reaction of primary amines.*** Both aliphatic and aromatic primary amines react with nitrous acid to give initially the corresponding diazonium ions. The aliphatic diazonium compounds are so unstable that their existence has not been directly detected. Nitrogen gas, alcohol, olefin, and products of other displacement and carbocation reactions are formed.

$$RNH_2 + HONO + 2HCl \longrightarrow [RN_2^+Cl^-] \xrightarrow{H_2O}$$

1° aliphatic amine; nitrous acid; diazonium salt (unstable at 0°C); spontaneous

$$N_2(g) + ROH + RCl + ROR + \text{alkene}$$

Aromatic diazonium salts, on the other hand, are generally stable in solution at 0°C. When heated in aqueous solution, they quickly lose nitrogen to give the aryl cation, Ar^+. This ion reacts rapidly with water to give phenol.

If allowed to dry, these compounds can be an explosion hazard, so the aryldiazonium ions should be used immediately upon preparation.

The coupling reaction between certain diazonium salts and phenols has been shown to involve reaction between diazonium ion and phenoxide ion.[12] If the solution is too acidic the phenoxide ion is converted to phenol, and thus reaction is retarded; if the solution is too basic the diazonium ion reacts with hydroxide ion to give diazotate, ArN_2O^-, which does not couple. The solution must, therefore, be properly buffered for a satisfactory coupling reaction.

The diazotization of anthranilic acid produces a zwitterion or a cyclic acyl diazotate. This compound is unstable and loses carbon dioxide and nitrogen to form a highly reactive benzyne intermediate. The reactive benzyne combines with ethanol to form phenetole, with diethyl maleate to form a substituted benzocyclobutene, and with anthracene to form triptycene. The best yields of products are obtained by carrying out the diazotization with amyl nitrite and the displacement reaction in an aprotic solvent such as methylene chloride, tetrahydrofuran, or acetonitrile. Again the diazonium ion intermediate should be handled with care as it is an explosion hazard.

$$\underset{\text{anthranilic acid}}{C_6H_4(COO^-)(NH_3^+)} \xrightarrow[\text{HCl}]{\text{NaNO}_2} \underset{\text{zwitterion}}{C_6H_4(COO^-)(N_2^+)} \rightleftharpoons \text{cyclic acyl diazotate}$$

$$\longrightarrow CO_2(g) + \underset{\text{benzyne}}{C_6H_4} + N_2(g)$$

Reaction of secondary amines. Both aliphatic and aromatic secondary amines react with nitrous acid to form *N*-nitroso compounds, commonly called nitrosamines. ***Caution:*** *Many of these compounds are carcinogenic and should be handled carefully.* They are pale yellow oils or solids.

$$\underset{2^\circ\text{ amine}}{R_2NH} + \underset{\text{nitrous acid}}{HONO} \longrightarrow \underset{\substack{N\text{-nitrosoamine} \\ \text{(yellow oil or solid)}}}{R_2N{-}N{=}O} + H_2O$$

Reaction of tertiary amines. The chemistry of the reaction of tertiary amines is quite complex.[13] Under certain conditions, it may appear that tertiary amines undergo no reaction; this is actually true only at low pH, low temperature, and dilute conditions. The amine is simply protonated to form salts under these mild conditions.

[12]C. K. Ingold, *Structure and Mechanism in Organic Chemistry,* 2nd ed. (Cornell University Press, Ithaca, NY, 1967), p. 387. See also N. A. Figero, *J. Chem. Educ., 43,* 142 (1966).

[13]G. E. Hein, *J. Chem. Educ., 40,* 181 (1963).

$$R_3N: \underset{\text{base}}{\overset{H^+}{\rightleftharpoons}} R_3NH^+$$

3° amine

These salts can be recognized by their reaction with base to regenerate the original amine.

Under higher temperatures, less acidic conditions, and other conditions, a variety of reactions occur when tertiary amines are treated with nitrous acid. For aliphatic amines, *N*-nitrosoamines are formed.

$$R_3N: + HONO \longrightarrow R_2N{-}N{=}O + ROH$$

3° aliphatic amine — nitrous acid — *N*-nitrosoamine (yellow oil or solid)

For aromatic amines, 4-nitroso-*N*,*N*-diarylanilines are produced.

$$(C_6H_5)_2N{-}C_6H_5 + HONO \longrightarrow (C_6H_5)_2N{-}C_6H_4{-}N{=}O$$

3° aromatic amine — nitrous acid — 4-nitroso-*N*,*N*-diarylaniline

N-Alkyl-*N*-nitrosoanilines and *N*,*N*-dialkyl-4-nitrosoanilines are produced from alkylarylamines.

$$C_6H_5{-}NR_2 + HONO \longrightarrow C_6H_5{-}N(R){-}N{=}O + O{=}N{-}C_6H_4{-}NR_2$$

3° aromatic amine — nitrous acid — *N*-alkyl-*N*-nitrosoaniline — *N*,*N*-dialkyl-4-nitrosoaniline

Although nitrous acid is useful for characterizing amines, other functional groups also react. A methylene group adjacent to a keto group is converted to an oximino group,[14] and alkyl mercaptans yield red *S*-alkyl thionitrites.

$$H_3C{-}C({=}O){-}CH_2CH_3 + HONO \longrightarrow H_3C{-}C({=}O){-}C({=}N{-}OH){-}CH_3 + H_2O$$

oxime

$$RCH_2SH + HONO \longrightarrow RCH_2{-}S{-}N{=}O + H_2O$$

S-alkyl thionitrite (red)

[14] W. L. Semon and V. R. Damrell, *Org. Syntheses Coll. Vol. II,* 204 (1943).

Experiment 19. Nickel Chloride, Carbon Disulfide, and Ammonium Hydroxide—Test for Secondary Aliphatic Amines

$$R_2NH + CS_2 + NH_4OH \longrightarrow R_2N{-}C({=}S){-}S^-NH_4^+ + H_2O \xrightarrow{NiCl_2} (R_2N{-}C({=}S){-}S)_2Ni\ (s)$$

2° amine carbon disulfide

Prepare an aqueous solution of the unknown by adding 1 or 2 drops or 50 mg of the unknown to 5 mL of water. If necessary, 1 or 2 drops of concentrated hydrochloric acid may be added to dissolve the amine. To 1 mL of the nickel chloride in carbon disulfide reagent in a test tube, add 0.5–1 mL of concentrated ammonium hydroxide, followed by 0.5–1 mL of the amine solution. A definite precipitation indicates that the unknown is a secondary amine. A slight turbidity is an indication of a trace of a secondary amine as an impurity. Try this test on (1) aniline; (2) *N*-methylaniline; (3) triethylamine.

Nickel Chloride in Carbon Disulfide Reagent. To 0.5 g of nickel chloride hexahydrate in 100 mL of water is added an amount of carbon disulfide such that, after the mixture has been shaken, a globule of carbon disulfide is left on the bottom of the bottle. If stored in a tightly stoppered bottle, the reagent is stable for long periods of time. When the undissolved carbon disulfide evaporates, more must be added.

Cleaning Up. Isolate the solid by suction filtration and place in the hazardous solid waste container. The liquid is placed in the hazardous waste container.

Discussion. This test is given by all secondary amines but not by primary amines. *It is very sensitive, and many commercial samples of tertiary amines produce turbidity because of the presence of small amounts of secondary amines.* This is true of substituted pyridines, quinolines, and isoquinolines separated from coal-tar distillates.

Experiment 20. Nickel Chloride and 5-Nitrosalicylaldehyde—Test for Primary Amines

$$RNH_2 + 2\ (\text{5-}O_2N\text{-}C_6H_3(OH)CHO) \longrightarrow 2\ (\text{5-}O_2N\text{-}C_6H_3(OH)CH{=}NR) \xrightarrow[NiCl_2]{2\ (HOCH_2CH_2)_3N} (O_2N\text{-}C_6H_3(O^-)CH{=}N)_2Ni\ (s)$$

1° amine 2-nitrosalicylaldehyde

To 5 mL of water, add 1 or 2 drops or 50 mg of the compound to be tested. If necessary, 1 or 2 drops of concentrated hydrochloric acid may be added to dissolve the amine. Add 0.5 mL of the amine solution to 3 mL of the nickel chloride and 5-nitrosalicylaldehyde reagent. An immediate, copious precipitate is produced by primary aliphatic amines, whereas primary aromatic amines usually require 2–3 min to give a definite precipitate. The appearance of a slight turbidity is not a positive test; it indicates that traces of primary amines may be present as impurities. Try the test on (1) butylamine; (2) diethylamine; (3) aniline; (4) *N,N*-dimethylaniline.

Nickel Chloride and 5-Nitrosalicylaldehyde Reagent. To 15 mL of triethanolamine is added a solution of 0.5 g of 5-nitrosalicylaldehyde dissolved in 25 mL of water. Then 0.5 g of nickel chloride hexahydrate dissolved in 10 mL of water is added, and the total volume of the solution is brought to 100 mL. If the triethanolamine contains ethanolamine, it may be necessary to add another 0.5 g of the aldehyde and remove the resulting precipitate by filtration.

Cleaning Up. Isolate the solid by suction filtration and place in the hazardous solid waste container. The liquid is placed in the hazardous waste container.

Discussion. *This test is so sensitive that care must be taken in interpreting it.* Only a definite precipitate produced in considerable quantity indicates a primary amine; a slight turbidity is merely indicative of impurities. Care must be taken to use the amounts specified above since the addition of large amounts of solutions of secondary amines will also give a precipitate. Many commercial samples of secondary and tertiary amines contain traces of primary amines and produce a turbidity.

The test is given by all primary amines capable of forming the Schiff base with 5-nitrosalicylaldehyde. Hydroxylamine and hydrazines substituted on only one nitrogen atom give positive tests. Amides do not give a precipitate. The test is not applicable to amino acids.

Experiment 21. Sodium Hydroxide Treatment of Ammonium Salts and Amine Salts

$$\underset{\text{ammonium salt}}{RC(=O)O^-NH_4^+} + NaOH \longrightarrow \underset{\text{salt of a carboxylic acid}}{RC(=O)O^-Na^+} + \underset{\text{ammonia}}{NH_3} + H_2O$$

$$\underset{\text{amine salt}}{RNH_3^+X^-} + NaOH \longrightarrow \underset{1^\circ\text{ amine}}{RNH_2} + NaCl + H_2O$$

Place 5 mL of 10% sodium hydroxide solution in a test tube, add 0.2–0.4 g of the compound, and shake the mixture vigorously. Note the odor of ammonia or the formation of an oily layer of the amine. Moistened pink litmus paper placed in the vapor above the solution will turn blue if ammonia or a volatile amine is present.

This test should be tried on (1) ammonium benzoate.

Cleaning Up. Any amine layer is separated and placed into the organic solvent container. Place the aqueous layer in the aqueous solution container.

7.7 AMINO ACIDS

The ninhydrin test (Experiment 22, p. 254) is the chemical basis for the amino acid analyzer, and it can be used to distinguish between different types of amino acids. α-Amino acids and β-amino acids react with ninhydrin to give a positive test, which is a blue to blue-violet color.

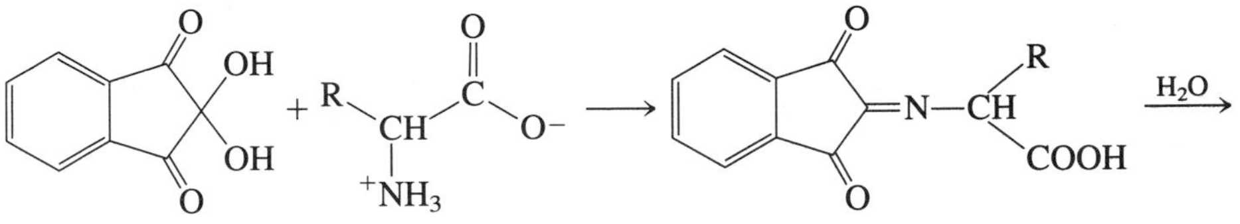

$$\text{ninhydrin (O, O, OH, OH)} + R\text{-}CH({}^{+}NH_3)\text{-}C(=O)\text{-}O^- \longrightarrow \text{(O, O)}{=}N\text{-}CH(R)\text{-}COOH \xrightarrow{H_2O}$$

ninhydrin α-amino acid

$$\text{(O, O)}\text{-}NH_2 + R\text{-}C(=O)\text{-}H + CO_2$$

$$\text{(O, O)}\text{-}NH_2 + \text{(O, O, OH, OH)} \longrightarrow \text{(O, O)}\text{-}N{=}\text{(O, O)}$$

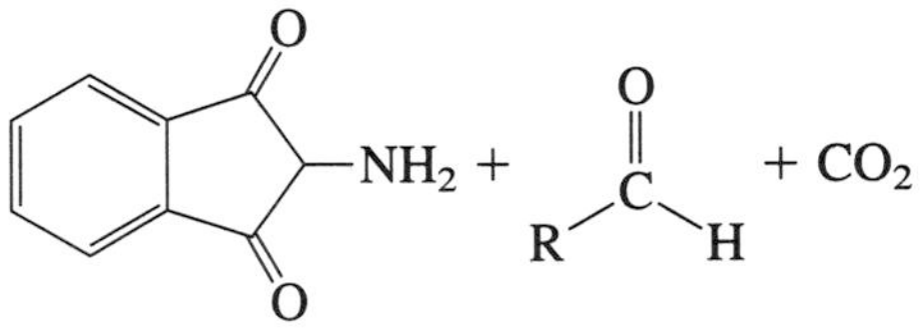

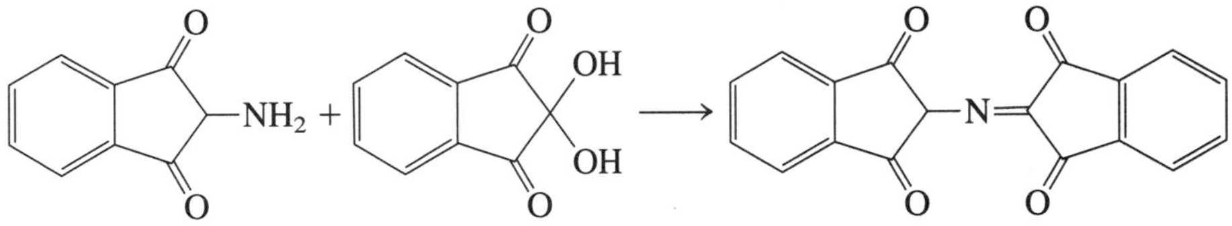

(blue to blue-violet)

α-Amino acids combine with nitrous acid to produce an intermediate which decomposes to form nitrogen gas and an α-hydroxy acid (Experiment 18e, p. 248).

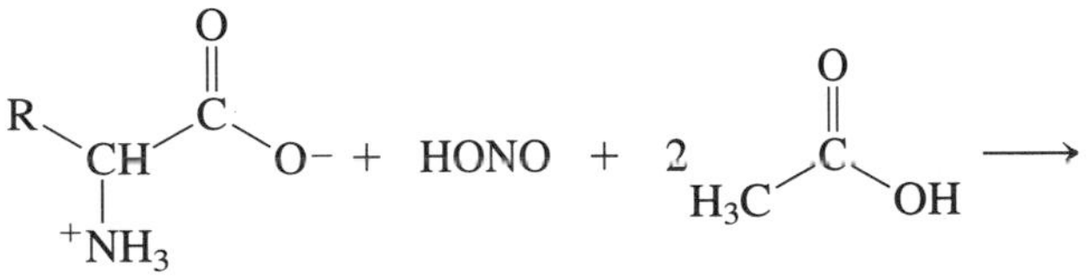

$$R\text{-}CH({}^{+}NH_3)\text{-}C(=O)\text{-}O^- + HONO + 2\ H_3C\text{-}C(=O)\text{-}OH \longrightarrow$$

α-amino acid nitrous acid

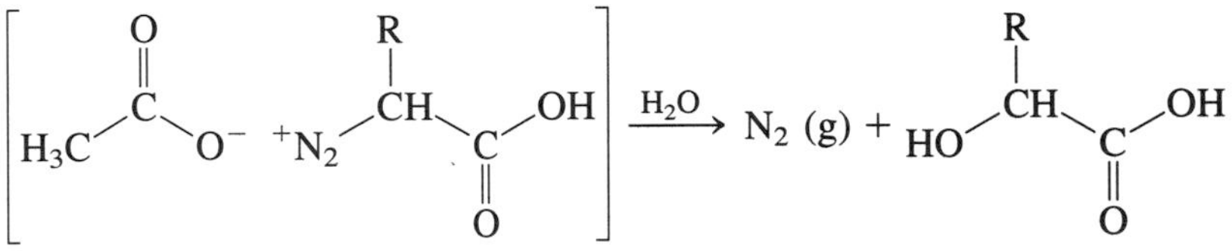

$$\left[H_3C\text{-}C(=O)\text{-}O^- \quad {}^{+}N_2\text{-}CH(R)\text{-}C(=O)\text{-}OH \right] \xrightarrow{H_2O} N_2\,(g) + HO\text{-}CH(R)\text{-}C(=O)\text{-}OH$$

Copper complexes are formed from the reaction of α-amino acids with copper sulfate (Experiment 23, p. 254). A deep blue color is produced, indicating the presence of the copper complex.

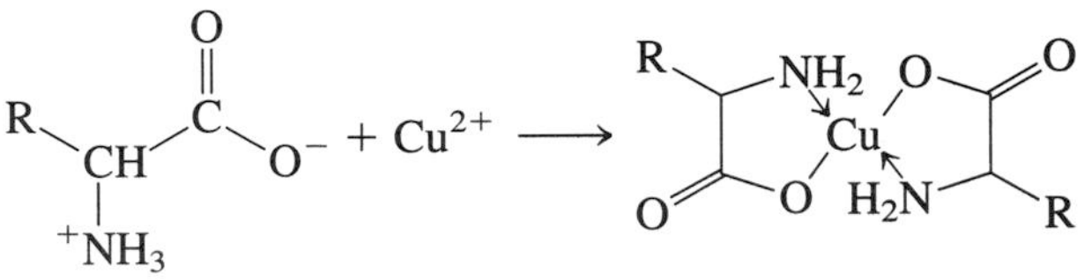

$$R\text{-}CH({}^{+}NH_3)\text{-}C(=O)\text{-}O^- + Cu^{2+} \longrightarrow \text{(R, }NH_2\text{, O, O; Cu; O, O, }H_2N\text{, R)}$$

α-amino acid

copper complex (blue)

Experiment 22. Ninhydrin Test

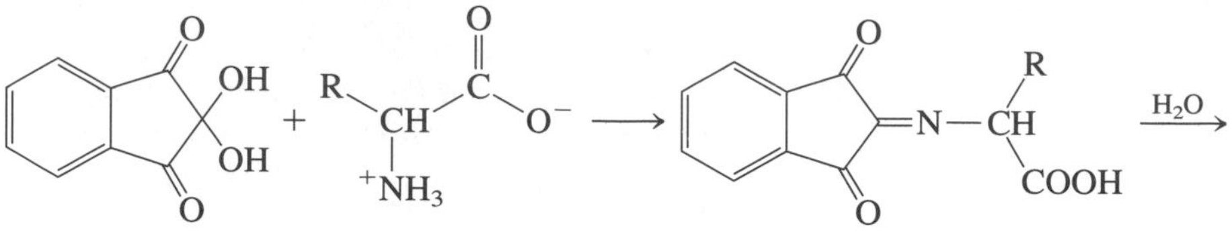

ninhydrin + α-amino acid → (imine) $\xrightarrow{H_2O}$ (amine) $+ RCHO + CO_2$

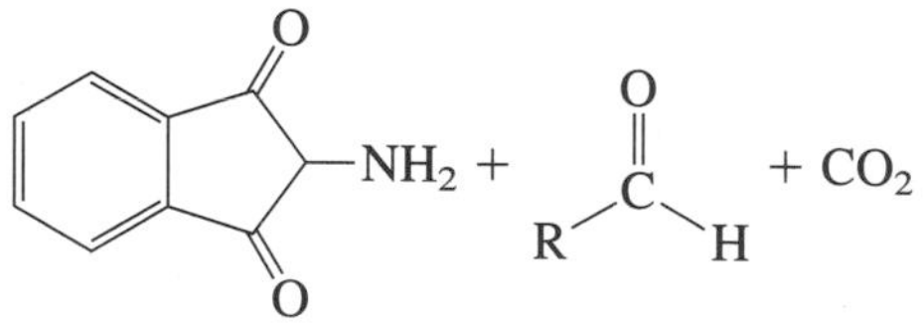

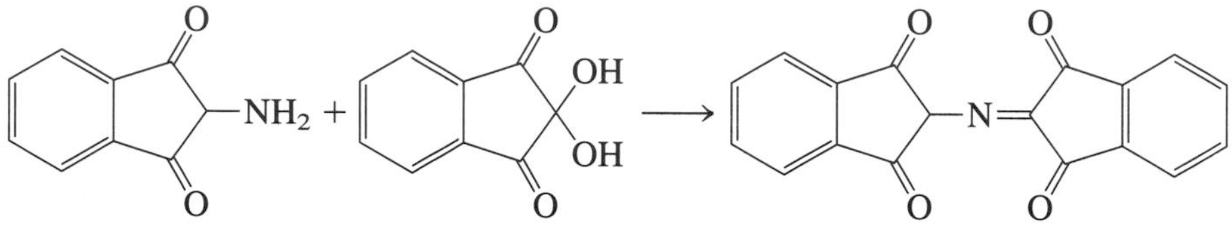

(blue to blue violet)

Add about 2 mg of the sample to 1 mL of a solution of 0.2 g of ninhydrin (1,2,3-indanetrione monohydrate) in 50 mL of water. The test mixture is heated to boiling for 15–20 sec; a blue to blue-violet color is given by α-amino acids and constitutes a positive test. Other colors (yellow, orange, red) are negative.

Cleaning Up. Place the mixture in the aqueous solution container.

Discussion. This reaction is important not only because it is a qualitative test, but also because it is the source of the absorbing material that can be measured quantitatively by an automatic amino acid analyzer. This color reaction is also used to detect the presence and position of amino acids after paper chromatographic separation.

Proline, hydroxyproline, and 2-, 3-, and 4-aminobenzoic acids fail to give a blue color but produce a yellow color instead. Ammonium salts give a positive test. Some amines, such as aniline, yield orange to red colors, which is a negative test.

Question

13. Give another name for ninhydrin.

Experiment 23. Copper Complex[15] Formation

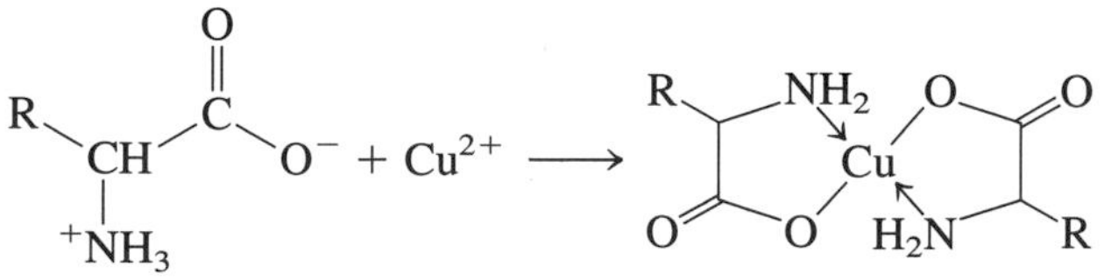

$RCH(NH_3^+)COO^- + Cu^{2+} \longrightarrow$ copper complex

α-amino acid

copper complex (blue color)

[15]B. S. Furniss, A. J. Hannaford, P. W. G. Smith, and A. R. Tatchell, *Vogel's Textbook of Practical Organic Chemistry,* 5th ed. (Wiley, New York, 1991), p. 1230.

A small amount of the compound is dissolved in 1 mL of water. Two drops of 1 M copper(II) sulfate are added. If a blue color is not formed immediately, then heat the test tube in a hot water bath for 5 min. A moderate to deep blue color of a liquid or a dark blue solid is a positive test.

Try this test on (1) alanine; (2) tyrosine; (3) lysine; (4) sucrose; (5) aspartic acid; (6) fructose; (7) aniline; (8) triethylamine.

Cleaning Up. Place the test solution in the aqueous solution container.

Discussion. Some α-amino acids are not very soluble in cold water. However, these amino acids are soluble in hot water and will give a positive test when the solution is heated. Aliphatic amines yield a blue precipitate. Anilines give a brown or green color, but other aromatic amines produce a blue-purple color.

A variety of tests are available that are specific for some amino acids (Table 7.4).

Table 7.4 Chemical Tests[a] for Amino Acids

Amino acid detected	Name of reaction	Reagents	Color
Arginine	Sakaguchi reaction	α-Naphthol and sodium hypochlorite	Red
Cysteine	Nitroprusside reaction	Sodium nitroprusside in dil. NH_3	Red
Cysteine	Sullivan reaction	Sodium 1,2-naphthoquinone-4-sulfonate and sodium hydrosulfite	Red
Histidine, tyrosine	Pauly reaction	Diazotized sulfanilic acid in alkaline solution	Red
Tryptophan	Ehrlich reaction	*p*-Dimethylaminobenzaldehyde in conc. HCl	Blue
Tryptophan	Glyoxylic acid reaction (Hopkins–Cole reaction)	Glyoxylic acid in conc. H_2SO_4	Purple
Tyrosine	Folin–Ciocalteu reaction	Phosphomolybdotungstic acid	Blue
Tyrosine	Millon reaction	$HgNO_3$ in nitric acid with a trace of nitrous acid	Red
Tyrosine, Tryptophan, Phenylalanine	Xanthoproteic reaction	Boiling conc. nitric acid	Yellow

[a]A number of the experimental procedures for these tests are described in J. P. Greenstein and M. Winitz, *Chemistry of the Amino Acids* (Wiley, New York, 1961).

7.8 CARBOHYDRATES

Carbohydrates or polysaccharides are polyhydroxy aldehydes and ketones; within this broad category are simpler compounds called sugars (or saccharides). The name carbohydrates is derived historically from the idea "hydrates of carbon," that is, the general formula $C_n(H_2O)_n$; for example, glucose has the formula $C_6H_{12}O_6$. Sucrose (common table sugar) is consistent with another general formula $C_n(H_2O)_m$, because it has the formula $C_{12}(H_2O)_{11}$. Carbohydrates do not have uniformly recognizable properties. These compounds are usually water soluble solids which melt with decomposition; these characteristics correctly point toward the presence of a large number of highly polar functional groups in these molecules.

The presence of the hydroxyl group on a carbohydrate can be detected with acetyl chloride (Experiment 6, p. 217). The reaction of the hydroxyl group with the acetyl chloride yields an ester, which appears as another liquid layer.

$$\underset{\text{alcohol}}{ROH} + \underset{\text{acetyl chloride}}{H_3C\text{-}C(=O)\text{-}Cl} \longrightarrow \underset{\text{ester}}{H_3C\text{-}C(=O)\text{-}OR} + HCl(g)$$

Tollens test (Experiment 14, p. 234) can be used to test for the presence of the aldehyde group. The aldehyde functional group is oxidized to the salt of the carboxylic acid, with the silver ion being reduced to elemental silver, which is deposited as a coating inside the reaction flask.

$$\underset{\text{aldehyde}}{R\text{-}C(=O)\text{-}H} + \underset{\text{Tollens reagent}}{2Ag(NH_3)_2OH} \longrightarrow 2Ag(s) + \underset{\text{salt of the carboxylic acid}}{R\text{-}C(=O)\text{-}O^-NH_4^+} + H_2O + 3NH_3$$

Vicinal diols can be detected with borax, sodium tetraborate decahydrate (Experiment 24, p. 258). Borax exists as $Na_2[B_4O_5(OH)_4]\cdot 8H_2O$. The resulting solution, with the addition of phenolphthalein indicator, is colorless at room temperature but yields a pink solution when warmed.

$$\underset{\text{vicinal diol}}{-C(OH)-C(OH)-} + \underset{\text{borax}}{[B_4O_5(OH)_4]^{2-}\ 2Na^+} \rightleftharpoons [(C_2O_2)B(O_2C_2)]^- \ H^+$$

Vicinal diols, 1,2-diketones, α-hydroxy ketones, and α-hydroxy aldehydes are oxidized with periodic acid (Experiment 25, p. 258).

$$\underset{\text{vicinal diol}}{R\text{-}CH(OH)\text{-}CH(OH)\text{-}R} + \underset{\text{periodic acid}}{HIO_4} \longrightarrow 2\ \underset{\text{aldehyde}}{R\text{-}C(=O)\text{-}H} + H_2O + \underset{\text{iodic acid}}{HIO_3}$$

$$\underset{\text{1,2-diketone}}{R{-}CO{-}CO{-}R} + \underset{\text{periodic acid}}{HIO_4} + H_2O \longrightarrow 2\ \underset{\text{carboxylic acid}}{R{-}COOH} + \underset{\text{iodic acid}}{HIO_3}$$

$$\underset{\alpha\text{-hydroxy acid}}{R{-}CH(OH){-}CO{-}R'} + \underset{\text{periodic acid}}{HIO_4} \longrightarrow \underset{\text{carboxylic acid}}{R'{-}COOH} + \underset{\text{aldehyde}}{R{-}CHO} + \underset{\text{iodic acid}}{HIO_3}$$

$$\underset{\alpha\text{-hydroxy aldehyde}}{R{-}CH(OH){-}CHO} + \underset{\text{periodic acid}}{HIO_4} \longrightarrow \underset{\text{carboxylic acid}}{H{-}COOH} + \underset{\text{aldehyde}}{R{-}CHO} + \underset{\text{iodic acid}}{HIO_3}$$

The iodic acid produced above is detected with 5% silver nitrate solution. An immediate precipitation of silver iodate occurs.

$$\underset{\text{iodic acid}}{HIO_3} + \underset{\text{silver nitrate}}{AgNO_3} \longrightarrow HNO_3 + \underset{\substack{\text{silver iodate}\\ \text{(white)}}}{AgIO_3(s)}$$

Benedict's solution (Experiment 26, p. 260) and Fehling's solution (Experiment 27, p. 262) will undergo reaction with reducing sugars such as α-hydroxy ketones and α-hydroxy aldehydes. The solution is initially a blue color from Cu^{2+} complex, but as the reaction proceeds, copper(I) oxide precipitates as a red, yellow, or yellowish green solid.

$$\underset{\alpha\text{-hydroxy ketone}}{R{-}CH(OH){-}CO{-}R'} + \underset{\text{(blue)}}{2Cu^{2+}} \longrightarrow \underset{\text{1,2-diketone}}{R{-}CO{-}CO{-}R'} + \underset{\substack{\text{(red, yellow, or}\\ \text{yellowish green)}}}{Cu_2O(s)}$$

$$\underset{\alpha\text{-hydroxy aldehyde}}{R{-}CH(OH){-}CHO} + \underset{\text{(blue)}}{2Cu^{2+}} \longrightarrow \underset{\alpha\text{-keto acid}}{R{-}CO{-}COOH} + \underset{\substack{\text{(red, yellow, or}\\ \text{yellowish green)}}}{Cu_2O(s)}$$

The preparation of osazones (Experiment 28, p. 263) from sugars and phenylhydrazine along with the time required for the solid osazone to form can be used in distinguishing among the various sugars.

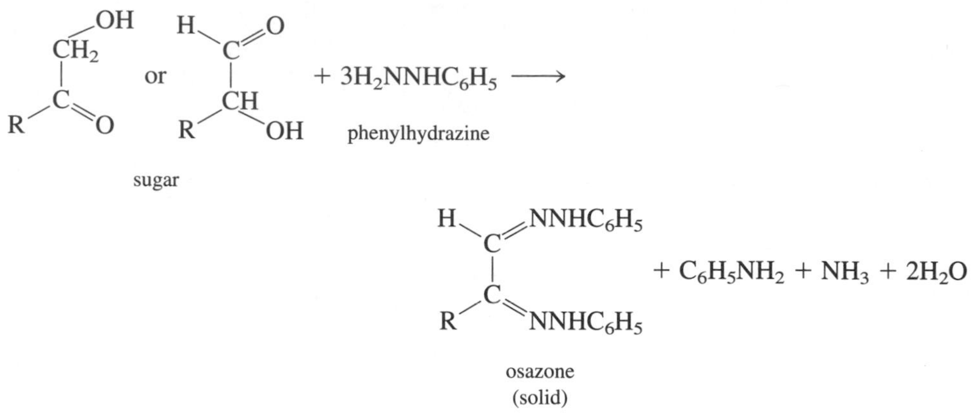

$$\underset{\text{sugar}}{RC(=O)CH_2OH \text{ or } RCH(OH)CHO} + \underset{\text{phenylhydrazine}}{3H_2NNHC_6H_5} \longrightarrow \underset{\substack{\text{osazone}\\ \text{(solid)}}}{HC(=NNHC_6H_5)C(R)=NNHC_6H_5} + C_6H_5NH_2 + NH_3 + 2H_2O$$

Experiment 24. Borax Test

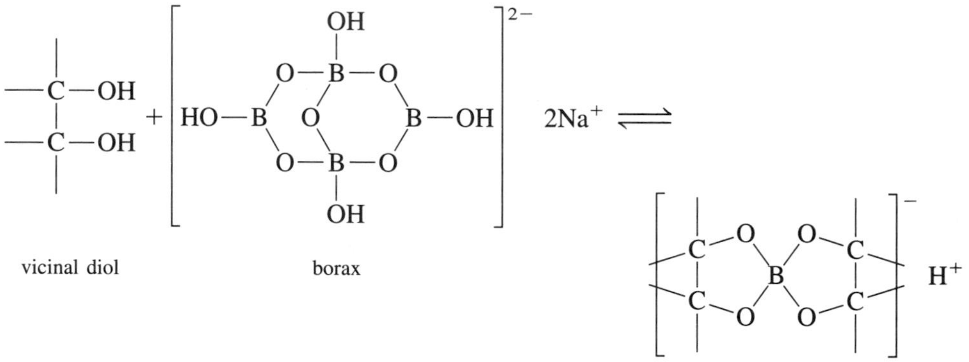

$$\underset{\text{vicinal diol}}{-\overset{|}{\underset{|}{C}}-OH,\ -\overset{|}{\underset{|}{C}}-OH} + \underset{\text{borax}}{\left[(HO)_4B_4O_5\right]^{2-}}\ 2Na^+ \rightleftharpoons \left[\text{(diol)}_2B\right]^- H^+$$

In a test tube, a few drops of phenolphthalein are added to 0.5 mL of a 1% solution of borax. A pink solution is formed. A couple of drops or a few crystals of the unknown are added. If the pink color begins to fade after the unknown and the reagent have been mixed together, then continue to add small amounts of the unknown until the pink color fades completely. Place the test tube in a hot water bath. If the pink color reappears on warming, and dissipates again on cooling, then the unknown is a polyhydric alcohol.

Try this test on (1) fructose; (2) lactose; (3) sucrose; (4) ethylene glycol.

Cleaning Up. Place the test solution with water in the aqueous solution container.

Discussion. Carbohydrates and 1,2-diols give a positive test.

Experiment 25. Periodic Acid Oxidation of Vicinal Diols and Related Compounds

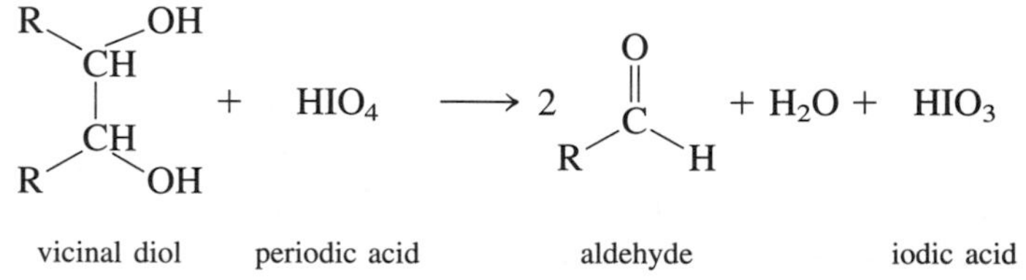

$$\underset{\text{vicinal diol}}{RCH(OH)CH(OH)R} + \underset{\text{periodic acid}}{HIO_4} \longrightarrow 2\ \underset{\text{aldehyde}}{RC(=O)H} + H_2O + \underset{\text{iodic acid}}{HIO_3}$$

$$\underset{\text{1,2-diketone}}{R-CO-CO-R} + \underset{\text{periodic acid}}{HIO_4} + H_2O \longrightarrow 2\ \underset{\text{carboxylic acid}}{R-COOH} + \underset{\text{iodic acid}}{HIO_3}$$

$$\underset{\alpha\text{-hydroxy acid}}{R-CH(OH)-CO-R'} + \underset{\text{periodic acid}}{HIO_4} \longrightarrow \underset{\text{carboxylic acid}}{R'-COOH} + \underset{\text{aldehyde}}{R-CHO} + \underset{\text{iodic acid}}{HIO_3}$$

$$\underset{\alpha\text{-hydroxy aldehyde}}{R-CH(OH)-CHO} + \underset{\text{periodic acid}}{HIO_4} \longrightarrow \underset{\text{carboxylic acid}}{H-COOH} + \underset{\text{aldehyde}}{R-CHO} + \underset{\text{iodic acid}}{HIO_3}$$

$$\underset{\beta\text{-hydroxylamine}}{R-CH(OH)-CH_2-NHR'} + \underset{\text{periodic acid}}{HIO_4} \longrightarrow \underset{\text{aldehydes}}{R'-CHO + R-CHO} + \underset{\text{iodic acid}}{HIO_3} + NH_3$$

Place 2 mL of the periodic acid reagent in a small test tube, add 1 drop (no more) of concentrated nitric acid, and shake thoroughly. Then add 1 drop or a small crystal of the compound to be tested. Shake the mixture for 10–15 sec, and add 1 or 2 drops of 5% aqueous silver nitrate solution. The instantaneous formation of a *white* precipitate (silver iodate) indicates that the organic compound has been oxidized by the periodate, which is thereby reduced to iodate. This constitutes a positive test. Failure to form a precipitate, or the appearance of a brown precipitate that redissolves on shaking, constitutes a negative test.

Dioxane may be added to facilitate the reaction of water insoluble polyols.

$$\underset{\text{iodic acid}}{HIO_3} + \underset{\text{silver nitrate}}{AgNO_3} \longrightarrow HNO_3 + \underset{\text{silver iodate (white)}}{AgIO_3(s)}$$

Apply the test to (1) 2-propanol; (2) acetone; (3) ethylene glycol; (4) glycerol; (5) glucose; (6) tartaric acid; (7) lactic acid.

Periodic Acid Reagent. Dissolve 0.5 g of paraperiodic acid (H_5IO_6) in 100 mL of distilled water.

Cleaning Up. Place the test solution in the aqueous solution container.

Discussion. Periodic acid has a very selective oxidizing action on 1,2-glycols, α-hydroxy aldehydes, α-hydroxy ketones, 1,2-diketones, α-hydroxy acids, and β-hydroxylamines. The rate of the reaction decreases in the order mentioned. Under the conditions specified above, α-hydroxy acids sometimes give a negative test. β-Dicarbonyl compounds and other active methylene compounds also react.

It is important that the exact amounts of reagent and nitric acid be used. The test depends on the fact that silver iodate is only slightly soluble in dilute nitric acid whereas silver periodate is very soluble. If too much nitric acid is present, however, the silver iodate will fail to precipitate.

Olefins, secondary alcohols, 1,3-glycols, ketones, and aldehydes are not affected by periodic acid under the above conditions. The periodic acid test is best suited for water soluble compounds.

The following mechanism has been proposed to account for the oxidation of vicinal diols:

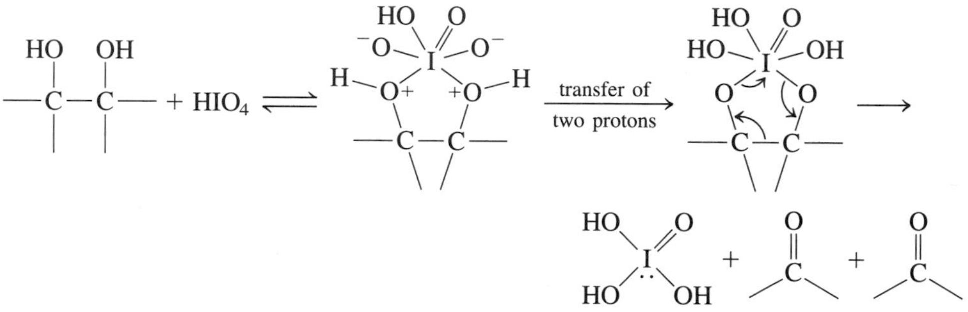

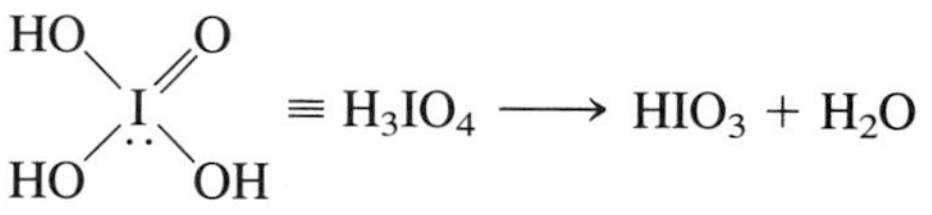

HO OH —C—C— + HIO_4 ⇌ [cyclic periodate ester: HO, O, $^-$O, I, O$^-$, H—O$^+$, $^+$O—H, —C—C—] $\xrightarrow{\text{transfer of two protons}}$ [HO, O, HO—I—OH, O, O, —C—C—] ⟶ (HO)(HO)(HO)I=O + C=O + C=O

$$(HO)_3I{=}O \equiv H_3IO_4 \longrightarrow HIO_3 + H_2O$$

Experiment 26. Benedict's Solution

Compounds Containing No Sulfur

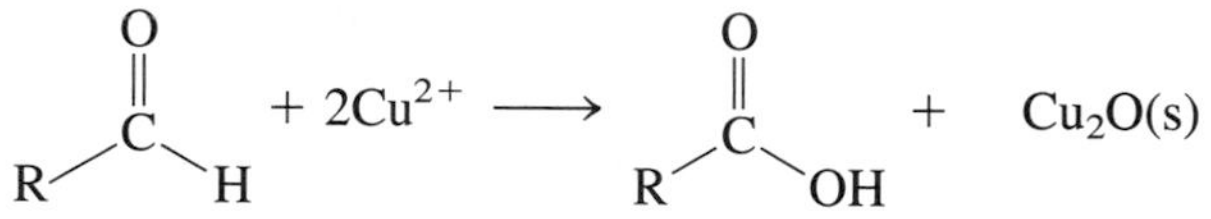

$$\text{RCHO} + 2Cu^{2+} \longrightarrow \text{RCOOH} + Cu_2O(s)$$

aldehyde (blue) carboxylic acid (red, yellow, or yellowish green)

$$\text{RCH(OH)C(=O)R}' + 2Cu^{2+} \longrightarrow \text{RC(=O)C(=O)R}' + Cu_2O(s)$$

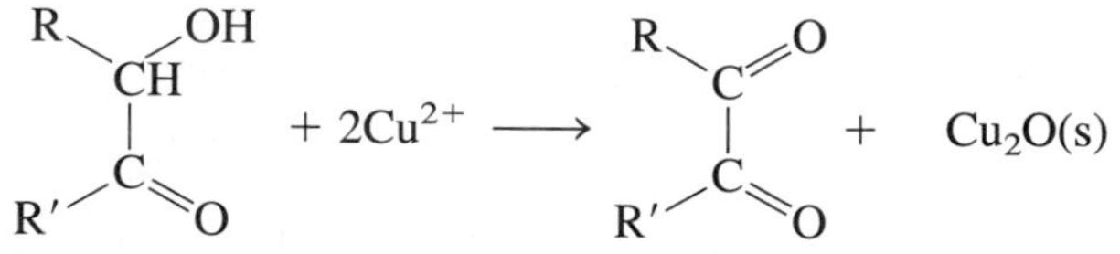

α-hydroxyketone (blue) 1,2-diketone (red, yellow, or yellowish green)

$$\text{R-CH(OH)-CHO} + 2Cu^{2+} \longrightarrow \text{R-CO-COOH} + Cu_2O(s)$$

α-hydroxyaldehyde (blue) α-keto acid (red, yellow, or yellowish green)

$$\text{ArNH-NHAr} + 2Cu^{2+} \longrightarrow \text{ArN=NAr} + Cu_2O(s)$$

hydrazobenzene (blue) azo compound (red, yellow, or yellowish green)

$$ArNHNH_2 + 2Cu^{2+} \longrightarrow ArH + ArOH + N_2(g) + Cu_2O(s)$$

arylhydrazine (blue) phenol (red, yellow, or yellowish green)

To a solution or suspension of 0.2 g of the compound in 5 mL of water, add 5 mL of Benedict's solution. Benedict's solution oxidizes a variety of compounds, with the corresponding reduction of Cu^{2+} to Cu^{1+}. The precipitation of the copper(I) oxide as a red, yellow, or yellowish green solid is a positive test. If no precipitate is formed, heat the mixture to boiling and cool. Note if any solid is formed. Try this test on (1) butanal; (2) acetoin; (3) benzoin; (4) glucose; (5) sucrose; (6) glycerol; (7) 2-butanol; (8) acetone; (9) phenylhydrazine.

To a solution of 0.2 g of sucrose in 5 mL of water, add 2 drops of concentrated hydrochloric acid and boil the solution for a minute. Cool the solution, neutralize the acid with dilute sodium hydroxide solution, and try the action of Benedict's solution. Explain the result.

Benedict's Solution. A solution of 17.3 g of sodium citrate and 10.0 g of anhydrous sodium carbonate in 80.0 mL of water is heated until the salts are dissolved. Additional water is added to bring the volume up to 85.0 mL. A solution of 1.73 g of hydrated copper sulfate in 10.0 mL of water is poured slowly with stirring into the solution of the citrate and carbonate. The final solution is made up to 100 mL by the addition of water.

Cleaning Up. Remove the colored copper complex by suction filtration and place in the nonhazardous solid waste container. Place the filtrate in the aqueous solution container.

Discussion. Benedict's solution, which contains the copper bound in the complex anion, functions as a selective oxidizing agent. It was introduced as a reagent for reducing sugars to replace Fehling's solution, which is very strongly alkaline. Benedict's reagent will detect 0.01% of glucose in water. The color of the precipitate may be red, yellow, or yellowish green, depending on the nature and amount of the reducing agent present.

Benedict's reagent is reduced by α-hydroxy aldehydes, α-hydroxy ketones, and α-keto aldehydes. It does not oxidize simple aromatic aldehydes. Molecules containing only the alcohol functional group or only the keto grouping are not oxidized by Benedict's solution.

Hydrazine derivatives, as exemplified by phenylhydrazine and hydrazobenzene, are oxidized by this reagent. Other easily oxidizable systems, such as phenylhydroxylamine, aminophenol, and related photographic developers, also reduce Benedict's solution.

Experiment 27. Fehling's Solution

$$R{-}C({=}O){-}H + 2Cu^{2+} \longrightarrow R{-}C({=}O){-}OH + Cu_2O(s)$$

aldehyde (blue) carboxylic acid (red, yellow, or yellowish green)

$$R{-}CH(OH){-}C({=}O){-}R' + 2Cu^{2+} \longrightarrow R{-}C({=}O){-}C({=}O){-}R' + Cu_2O(s)$$

α-hydroxyketone (blue) 1,2-diketone (red, yellow, or yellowish green)

$$R{-}CH(OH){-}C({=}O){-}H + 2Cu^{2+} \longrightarrow R{-}C({=}O){-}C({=}O){-}OH + Cu_2O(s)$$

α-hydroxyaldehyde (blue) α-keto acid (red, yellow, or yellowish green)

$$Ar{-}NH{-}NH{-}Ar + 2Cu^{2+} \longrightarrow Ar{-}N{=}N{-}Ar + Cu_2O(s)$$

hydrazobenzene (blue) azo compound (red, yellow, or yellowish green)

$$ArNHNH_2 + 2Cu^{2+} \longrightarrow ArH + ArOH + N_2(g) + Cu_2O(s)$$

arylhydrazine (blue) phenol (red, yellow, or yellowish green)

To a solution of 0.2 g of the compound in 5 mL of water, add 5 mL of Fehling's solution, and heat the mixture to boiling. Cool the solution. Fehling's solution oxidizes many compounds and copper in the reagent is reduced from Cu^{2+} to Cu^{1+}. The precipitation of the copper(I) oxide as a red, yellow, or yellowish green solid is a positive test. Try this test on (1) glucose; (2) sucrose; (3) glycerol; (4) maltose; (5) lactose; (6) phenylhydrazine.

Fehling's Solution. Mix 2.5 mL of the following two solutions immediately before use. *Fehling's Solution #1.* Dissolve 17.32 g of hydrated copper sulfate crystals in 200 mL of water and dilute the solution to 250 mL. *Fehling's Solution #2.* Dissolve 86.5 g of sodium potassium tartrate and 35 g of sodium hydroxide in 100 mL of water and dilute the solution to 250 mL.

Cleaning Up. Remove the colored copper complex by suction filtration and place in the nonhazardous solid waste container. Place the filtrate in the aqueous solution container.

Discussion. Benedict's and Fehling's solutions serve as a test for reducing sugars. Nonreducing sugars are hydrolyzed by heating with a small amount of 10% hydrochloric acid, then neutralized with 10% sodium hydroxide. The resulting solution will then give a positive test with Benedict's and Fehling's solutions.

Experiment 28. Osazones

$$\underset{\text{sugar}}{R{-}C({=}O){-}CH_2OH \quad \text{or} \quad R{-}CH(OH){-}CH{=}O} + \underset{\text{phenylhydrazine}}{3H_2NNHC_6H_5} \longrightarrow$$

$$\underset{\substack{\text{osazone}\\ \text{(solid)}}}{R{-}C({=}NNHC_6H_5){-}CH{=}NNHC_6H_5} + C_6H_5NH_2 + NH_3 + 2H_2O$$

Place 0.2 g of the unknown sample in a test tube and add 0.4 g of phenylhydrazine hydrochloride, 0.6 g of crystallized sodium acetate, and 4 mL of distilled water. Place the test tube in a beaker of boiling water. Note the time that the test tube was immersed and the time of the precipitation.

After 20 min, remove the test tube from the hot water bath and set it aside to cool. A small amount of the liquid and solid is poured on a watch glass. Tip the watch glass from side to side to spread out the crystals, and absorb some of the mother liquid with a piece of filter paper, taking care not to crush or break up the clumps of crystals. Examine the crystals under a low-power microscope (about 80–100×), and compare with photomicrographs.[16]

This experiment should be tried first with glucose, maltose, sucrose, and galactose in four separate test tubes.

The formation of tarry products due to oxidation of the phenylhydrazine may be prevented by the addition of 0.5 mL of saturated sodium bisulfite solution. This should be done before heating if it is desired to isolate the osazone and determine its melting point.

Cleaning Up. The solid products are placed in the organic nonhazardous solid waste. Add 8 mL of 5.25% sodium hypochlorite (household bleach) to the filtrate. Heat the so-

[16] W. Z. Hassid and R. M. McCready, *Ind. Eng. Chem., Anal. Ed., 14,* 683 (1942).

lution at 45–50° for 2 hr to oxidize the amine. Dilute the solution with 10 mL of water and place in the aqueous solution container.

Discussion. The times required for the formation of the osazones can be a valuable aid in distinguishing among various sugars. The following figures are the times required for the osazone to precipitate from the hot solution: fructose, 2 min; glucose, 4–5 min; xylose, 7 min; arabinose, 10 min; galactose, 15–19 min; raffinose, 60 min; lactose, osazone soluble in hot water; maltose, osazone soluble in hot water; mannose, 0.5 min (hydrazone); sucrose, 30 min (owing to hydrolysis and formation of glucosazone).

Osazone formation involves hydrazone formation at C-1 of an aldose (or C-2 of a ketose) and oxidation of C-2 (or C-1) of an alcohol group to a ketone (or an aldehyde). The new carbonyl group is also converted to a hydrazone. It has been suggested that the reaction stops here (rather than further oxidation at C-3, etc.) because of hydrogen-bonding stabilization of the osazone:

$$\begin{array}{c} \text{H}-\text{C}=\text{N}-\text{N}-\text{C}_6\text{H}_5 \\ | \qquad\qquad | \\ \text{R}-\text{C}=\text{N}^{\delta-}\cdots\text{H}^{\delta+} \\ | \\ \text{NHC}_6\text{H}_5 \end{array}$$

7.9 CARBOXYLIC ACIDS

Carboxylic acids are primarily identified by spectroscopy and solubility tests. However, a few classification tests can be used to confirm the presence of the carboxylic group.

Silver nitrate reacts with carboxylic acids to form silver salts of the carboxylic acid (Experiment 33, p. 270). These silver salts are soluble in dilute nitric acid, whereas silver halides are insoluble in nitric acid.

$$\underset{\text{carboxylic acid}}{\text{RC(=O)OH}} + \underset{\text{silver nitrate}}{AgNO_3} \longrightarrow \underset{\text{silver salt}}{\text{RC(=O)O}^-\text{Ag}^+} + HNO_3$$

Carboxylic acids react with a sodium bicarbonate solution to form the carboxylate anion and carbon dioxide gas (Experiment 29, p. 265).

$$\underset{\text{carboxylic acid}}{\text{RC(=O)OH}} + NaHCO_3 \longrightarrow \underset{\text{sodium salt of the acid}}{\text{RC(=O)O}^-\text{Na}^+} + H_2O + CO_2(g)$$

Another test for carboxylic acids involves the esterification of the acid (Experiment 3b, p. 210). The ester forms another layer and has a sweet, fruity smell.

$$\underset{\text{carboxylic acid}}{\text{RC(=O)OH}} + \underset{\text{ethanol}}{CH_3CH_2OH} \longrightarrow \underset{\text{ester}}{\text{RC(=O)OCH}_2\text{CH}_3} + H_2O$$

Question

14. Volatility contributes greatly to odor. Briefly explain why ethyl esters are normally more volatile than the corresponding carboxylic acids.

Experiment 29. Sodium Bicarbonate Test

$$RC(=O)OH + NaHCO_3 \longrightarrow RC(=O)O^-Na^+ + H_2O + CO_2(g)$$

carboxylic acid — sodium salt of the acid

A few drops or a few crystals of the unknown sample are dissolved in 1 mL of methanol and slowly added to 1 mL of a saturated solution of sodium bicarbonate. Evolution of carbon dioxide gas is a positive test for the presence of the carboxylic acid.

Cleaning Up. Place the test solution in the aqueous solution container.

7.10 ESTERS

Esters characteristically have a sweet fruity smell. Esters combine with hydroxylamine to yield an alcohol and hydroxamic acid (Experiment 2, p. 206). The solution is then treated with ferric chloride to produce the ferric hydroxamate complex, which has a characteristic burgundy or magenta color.

$$RC(=O)OR' + H_2NOH \longrightarrow RC(=O)NHOH + R'OH$$

ester — hydroxylamine — hydroxamic acid

$$3\,RC(=O)NHOH + FeCl_3 \longrightarrow (RC(=O)NHO)_3Fe + 3HCl$$

hydroxamic acid — ferric hydroxamate complex (burgundy or magenta)

Esters are cleaved by hydroiodic acid (Experiment 32, p. 268) to form an alkyl iodide and a carboxylic acid. The alkyl iodide is treated with mercuric nitrate to yield mercuric iodide, which is an orange color.

$$RC(=O)OR' + HI \longrightarrow R'I + RC(=O)OH$$

ester — alkyl iodide — carboxylic acid

$$Hg(NO_3)_2 + 2R'I \xrightarrow{H_2O} HgI_2 + 2R'OH + 2HNO_3$$

mercuric acetate — alkyl iodide — mercuric iodide (orange)

7.11 ETHERS

Ethers are only a little more polar and slightly more reactive than either saturated hydrocarbons or alkyl halides. The ether oxygen can be protonated by concentrated sulfuric acid.

$$\underset{\text{ether}}{R_2\ddot{O}:} + H_2SO_4 \underset{M}{\overset{Z}{\rightleftharpoons}} R_2\ddot{O}H^+HSO_4^-$$

Caution: *Ethers form extremely explosive peroxides upon standing, especially when exposed to air and/or light. Liquid ether that shows solid precipitates should not be handled at all.*

Peroxides can be detected by treating the ether with starch-iodide paper moistened with dilute hydrochloric acid; peroxides will cause the paper to turn blue. Methods of removing peroxides have been described.[17] Many laboratories have substituted *t*-butyl methyl ether for diethyl ether because of the much greater peroxide danger from the latter.

Pure ethers are more likely to be initially diagnosed by their failure to undergo reactions rather than by their ability to undergo chemical reactions.

The ferrox test is used to distinguish ethers from hydrocarbons (Experiment 30, p. 267). Ferric ammonium sulfate reacts with potassium thiocyanate to form iron hexathiocyanatoferrate, which reacts with oxygen-containing compounds to form a reddish purple solution.

$$\underset{\text{ferric ammonium sulfate}}{2Fe(NH_4)_2(SO_4)_2} + \underset{\text{potassium thiocyanate}}{6KSCN} \longrightarrow \underset{\text{iron hexathiocyanatoferrate}}{Fe[Fe(SCN)_6]} + 3K_2SO_4 + (NH_4)_2SO_4$$

Ethers can be detected by the iodine test (Experiment 31, p. 267); as the ether undergoes reaction with iodine, the color of the solution changes from purple to tan.

$$\underset{\text{ether}}{R_2\ddot{O}:} + I_2 \longrightarrow R_2\ddot{O}:\!\longrightarrow I_2$$

Treatment of an ether with hydroiodic acid results in the cleavage of the ether (Experiment 32, p. 268). The alkyl iodide product is heated and its vapors undergo reaction with mercuric nitrate to form mercuric iodide, which is orange.

$$\underset{\text{ether}}{R'OR} + \underset{\text{hydroiodic acid}}{2HI} \longrightarrow \underset{\text{alkyl iodides}}{R'I + RI} + H_2O$$

$$\underset{\text{ether}}{ArOR} + \underset{\text{hydroiodic acid}}{HI} \longrightarrow \underset{\text{alkyl iodide}}{RI} + \underset{\text{phenol}}{ArOH}$$

[17] B. S. Furniss, A. J. Hannaford, P. W. G. Smith, and A. R. Tatchell, *Vogel's Textbook of Practical Organic Chemistry,* 5th ed. (Wiley, New York, 1991), p. 404.

$$\underset{\text{mercuric acetate}}{Hg(NO_3)_2} + \underset{\text{alkyl iodide}}{2RI} \xrightarrow{H_2O} \underset{\text{mercuric iodide (orange)}}{HgI_2} + 2ROH + 2HNO_3$$

Experiment 30. Ferrox Test

$$\underset{\text{ferric ammonium sulfate}}{2Fe(NH_4)_2(SO_4)_2} + \underset{\text{potassium thiocyanate}}{6KSCN} \longrightarrow \underset{\text{iron hexathiocyanatoferrate}}{Fe[Fe(SCN)_6]} + 3K_2SO_4 + (NH_4)_2SO_4$$

In a dry test tube, a crystal of ferric ammonium sulfate and a crystal of potassium thiocyanate are carefully ground together with a glass stirring rod. The iron hexathiocyanatoferrate that is formed sticks to the stirring rod.

In another test tube, dissolve 30 mg or 3 drops of the unknown in a minimal amount of toluene. The stirring rod bearing the iron hexathiocyanatoferrate solid is used to stir the unknown. If the solid dissolves and a reddish purple color develops, the compound contains oxygen.

Try this test on (1) diethyl ether; (2) anisole; (3) salicylic acid; (4) acetone.

Cleaning Up. The mixture is placed in the hazardous waste container.

Discussion. Hydrocarbons, alkyl halides, diaryl ethers, and other high molecular weight ethers give a negative test. Most compounds in which the oxygen is in a carbonyl group in conjugation with an aromatic ring or a double bond produce a negative test. Such compounds include cinnamic acid, aromatic acids, aromatic esters, aromatic anhydrides, or aromatic ketones. Phenols also give a negative test. However, aromatic aldehydes give a positive test. Other compounds that produce a positive test are aliphatic acids, aliphatic aldehydes, aliphatic ketones, aliphatic esters, aliphatic anhydrides, and aliphatic ethers.

This test should be used in conjunction with other tests. If the unknown produces a positive test, but yields a negative test for an aldehyde (Experiments 8, p. 221; 11–15, pp. 229–235; 26, p. 260; 27, p. 262), a ketone (Experiments 10–13, pp. 224–232), an ester (Experiment 2b, p. 206; 32, p. 268), a carboxylic acid (Experiments 3b, p. 210; 29, p. 265; 33, p. 270), or an anhydride (Experiments 1, p. 205; 2b, p. 206; 3a, p. 209; 4, p. 211), then the compound is an aliphatic or a low molecular weight ether.

Experiment 31. Iodine Test for Ethers and Unsaturated Hydrocarbons[18]

$$\underset{\text{alkene}}{>C{=}C<} + I_2 \longrightarrow >C{=}C< \longrightarrow I_2$$

[18]D. J. Pasto, C. R. Johnson, and M. J. Miller, *Experiments and Techniques in Organic Chemistry* (Prentice Hall, Upper Saddle River, NJ, 1992) pp. 277–278.

$$\underset{\text{ether}}{R_2\ddot{O}:} + I_2 \longrightarrow R_2\ddot{O}:\!\longrightarrow I_2$$

Add 0.25 mL or 0.25 g of the unknown sample to 0.5 mL of the iodine in methylene chloride solution. If an ether is present, the purple solution becomes tan in color. Aromatic hydrocarbons, saturated hydrocarbons, fluorinated hydrocarbons, and chlorinated hydrocarbons do not react. Unsaturated hydrocarbons produce a light tan solid, while retaining the purple color of the iodine solution.

Try this test on (1) ethyl ether; (2) cyclohexane; (3) cyclohexene; (4) toluene.

Iodine in Methylene Chloride Solution. Add a couple of crystals of iodine to 100 mL of methylene chloride. Stopper the flask tightly.

Cleaning Up. Place test solution in the halogenated organic waste container.

Discussion. Some compounds with nonbonded electron pairs or π-bonded electrons form charge-transfer complexes with iodine yielding a brown solution. The brown color is from the iodine forming a charge-transfer-complex with the nonbonded or π electrons. These compounds are formed quickly and, sometimes, reversibly. They are often unstable so conclusions should be drawn on the results observed in a short period of time.

Some alcohols and ketones may also give a positive test. Thus, this test should be used in addition to other tests for compounds containing oxygen.

Experiment 32. Hydroiodic Acid (Zeisel's Alkoxyl Method)

$$\underset{\text{ether}}{R'OR} + \underset{\text{hydroiodic acid}}{2HI} \longrightarrow \underset{\text{alkyl iodides}}{R'I + RI} + H_2O$$

$$\underset{\text{ether}}{ArOR} + \underset{\text{hydroiodic acid}}{HI} \longrightarrow \underset{\text{alkyl iodide}}{RI} + \underset{\text{phenol}}{ArOH}$$

$$\underset{\text{alcohol}}{ROH} + \underset{\text{hydroiodic acid}}{HI} \longrightarrow \underset{\text{alkyl iodide}}{RI} + H_2O$$

$$\underset{\text{ester}}{RC(=O)OR'} + HI \longrightarrow \underset{\text{alkyl iodide}}{R'I} + \underset{\text{carboxylic acid}}{RC(=O)OH}$$

$$RCH(OR')_2 + 2HI \longrightarrow 2R'I + RC(=O)H + H_2O$$

$$\underset{\substack{\text{mercurie}\\\text{acetate}}}{Hg(NO_3)_2} + \underset{\substack{\text{alkyl}\\\text{iodide}}}{2R'I} \xrightarrow{H_2O} \underset{\substack{\text{meruric}\\\text{iodide}\\\text{(orange)}}}{HgI_2} + 2R'OH + 2HNO_3$$

Place about 0.1 g or 0.1 mL of the compound in a 16 × 150 mm test tube. Carefully add, by means of a pipet, 1 mL of glacial acetic acid and 1 mL of 57% hydroiodic acid (sp gr 1.7). Add a boiling chip, and insert into the mouth of the test tube a gauze plug as described below. The gauze plug is twisted so as to make a good fit and pushed down so that it is 4 cm from the mouth of the test tube. A small piece of nonabsorbent cotton is gently pushed on top of the plug by means of a glass rod so as to make a disk of cotton 2–3 mm thick. A piece of filter paper 2 × 10 cm is folded longitudinally, moistened with a solution of mercuric nitrate, and placed on the cotton disk. The test tube is immersed to a depth of 4–5 cm in an oil bath kept at 120–130°C. *Do this reaction in the hood.*

When the reaction mixture boils, vapors rise through the porous plug, which usually turns gray. The volatile alkyl halide, rising through the plugs, reacts with the mercuric nitrate to produce a light orange or vermilion color due to the formation of the mercuric iodide. A positive test consists in the formation of an orange or vermilion color on the test paper within a 10-min heating period. A yellow color constitutes a negative or doubtful test.

Try this test on (1) anisole; (2) methyl benzoate.

Gauze Plugs. A solution of 0.10 g of lead acetate in 1.0 mL of water is added to 6.0 mL of 1 M sodium hydroxide solution and stirred until the precipitate dissolves. To this solution is added a solution of 0.5 g of hydrated sodium thiosulfate in 1.0 mL of water. About 0.5 mL of glycerol is added, and the solution is diluted to 10.0 mL. About 5 mL of this solution is pipetted on strips of double cheesecloth 2 × 45 cm. The strips of cloth are dried and rolled to fit the test tube.

Mercuric Nitrate Solution. A saturated solution of mercuric nitrate is prepared in 24.5 mL of distilled water to which has been added 0.5 mL of concentrated nitric acid. Mercury compounds are very toxic and require handling with special care.

Cleaning Up. Place the gauze plugs and filter paper in the hazardous solid waste container. Place the test solution in the halogenated organic waste container.

Discussion. This test is based on the classic Zeisel method for estimating quantitatively the percentage of methoxyl or ethoxyl groups. Functional groups containing methyl, ethyl, 1-propyl, or 2-propyl radicals attached to oxygen are cleaved by the hydroiodic acid with the formation of a volatile alkyl halide. Alkoxy derivatives in which the group is butyl or larger are difficult to cleave, and the iodide is too high boiling to be volatilized. Some butoxy compounds give a positive test, but the procedure is not reliable (the boiling point of butyl iodide is 131°C).

This class reaction is most useful for ethers, esters, and acetals in which the groups are methyl or ethyl. Methanol, ethanol, 1-propanol, 2-propanol, and even higher alcohols such as 1-butanol and 3-methyl-1-butanol will also give a positive test. The test has been applied to numerous alkaloids and methylated sugars. The chief interference is caused by the presence of a sulfur-containing functional group which liberates hydrogen sulfide when heated with hydroiodic acid.

Some ethers may require a more vigorous reagent for cleavage. It is sometimes advantageous to use 2 mL of the hydroiodic acid, 0.1 g of the phenol, and 1 mL of propanoic anhydride for 0.1 g of the sample.

Hydroiodic acid is a strong acid and it protonates the ether; iodide ion nucleophilically displaces the protonated alkoxy group, giving alkyl iodide. This process is expected to predominate when the R groups are primary.

$$\underset{\text{ether}}{ROR} + \underset{\text{hydroiodic acid}}{HI} \longrightarrow R_2\overset{+}{O}H \xrightarrow{I^-} RI + H_2O \xrightarrow{HI} R\overset{+}{O}H_2 \xrightarrow{I^-} RI + H_2O$$

Question

15. Modify the mechanism just above to explain how ethers with tertiary alkyl groups are hydrolyzed.

7.12 HALIDES

Since aliphatic halides are often detected initially by qualitative halogen analysis, it should not be surprising that further characterization takes advantage of the wide range of processes by which halogens can be displaced. The two tests for displaceable halogen which are discussed below are complementary and are thus often very useful in classifying the structures of alkyl halides. The silver nitrate reaction proceeds by a carbocation (S_N1) process and the sodium iodide reaction by a direct displacement (S_N2). Thus the following reactivity orders:

$$AgNO_3\text{–ethanol test:} \quad R_3CX > R_2CHX > RCH_2X$$

$$NaI\text{–acetone test:} \quad R_3CX < R_2CHX < RCH_2X$$

In both the silver nitrate test (Experiment 33, below) and the sodium iodide test (Experiment 34, p. 273), the halide is displaced from the alkyl halide to form an insoluble salt. The reaction of the alkyl halide with silver nitrate yields a silver halide precipitate.

$$\underset{\text{alkyl halide}}{RX} + \underset{\text{silver nitrate}}{AgNO_3} \xrightarrow{CH_3CH_2OH} \underset{\text{silver halide}}{AgX(s)} + RONO_2$$

The sodium iodide test can be used to test for the presence of bromine or chlorine. Sodium halide precipitates from the solution.

$$\underset{\text{alkyl chloride}}{RCl} + \underset{\text{sodium iodide}}{NaI} \xrightarrow{\text{acetone}} \underset{\text{alkyl iodide}}{RI} + \underset{\text{sodium chloride}}{NaCl(s)}$$

$$\underset{\text{alkyl bromide}}{RBr} + \underset{\text{sodium iodide}}{NaI} \xrightarrow{\text{acetone}} \underset{\text{alkyl iodide}}{RI} + \underset{\text{sodium bromide}}{NaBr(s)}$$

Experiment 33. Silver Nitrate Solution (Ethanolic)

$$\underset{\text{alkyl halide}}{RX} + \underset{\text{silver nitrate}}{AgNO_3} \xrightarrow{CH_3CH_2OH} \underset{\text{silver halide}}{AgX(s)} + RONO_2$$

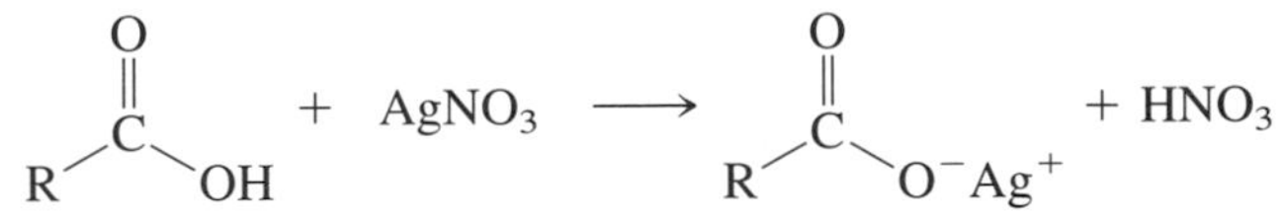

$$\mathrm{RC(=O)OH} + \mathrm{AgNO_3} \longrightarrow \mathrm{RC(=O)O^-Ag^+} + \mathrm{HNO_3}$$

carboxylic acid silver nitrate silver salt

This reagent is useful for classifying compounds known to contain halogen. Add 1 drop or a couple of crystals of the unknown to 2 mL of the 2% ethanolic silver nitrate solution. If no reaction is observed after 5 min standing at room temperature, heat the solution to boiling and note if a precipitate is formed. If there is a precipitate, note its color. Add 2 drops of 5% nitric acid, and note if the precipitate dissolves. Silver halides are insoluble in dilute nitric acid; silver salts of organic acids are soluble.

Apply this test to (1) benzoyl chloride; (2) benzyl chloride; (3) ethyl bromide; (4) bromobenzene; (5) chloroform; (6) chloroacetic acid.

Cleaning Up. Add a saturated solution of sodium chloride. The silver halide is isolated by filtration and placed in the nonhazardous solid waste container. The filtrate is placed in the aqueous solution container.

Discussion. Since alkyl halides often contain small amounts of isomeric impurities, it may be advisable in certain borderline cases to collect and dry the silver halide obtained from a weighed sample of unknown. Generally an approximate value for the molecular weight can be arrived at from a consideration of its physical constants and from an inspection of the list of possibilities; the estimated theoretical yield of the silver halide can thus be compared with the amount obtained. If the observed yield amounts to only a few percent, the test is negative. An alkyl halide that gives only a small amount of silver halide because it reacts slowly may be distinguished from a mixture of an inert halide with a small amount of reactive impurity by collecting the halide initially precipitated and then testing the filtrate with more silver nitrate.

Many halogen-containing substances react with silver nitrate to give an insoluble silver halide, and the rate of this reaction is an index of the degree of reactivity of the halogen atom in question. This information is valuable because it permits certain deductions to be drawn concerning the structure of the molecule. Moreover, the identity of the halogen can sometimes be determined from the color of the silver halide (silver chloride is white, silver bromide is pale yellow, and silver iodide is yellow), although impurities easily cause ambiguous results.

The most reactive halides are those that are ionic. Among organic compounds, the amine salts of the halogen acids constitute the most common examples.

$$[\mathrm{RNH_3}]^+\mathrm{X}^-$$

Less frequently encountered are oxonium salts and "carbonium" salts which contain ionic halogen.

$$\left[\mathrm{R:\ddot{O}:R} \atop \mathrm{R'}\right]^+\mathrm{X}^- \qquad \left[\mathrm{(H_3C)_2N{-}C_6H_4}\right]_3\mathrm{C^+Cl^-}$$

R′ = alkyl, H 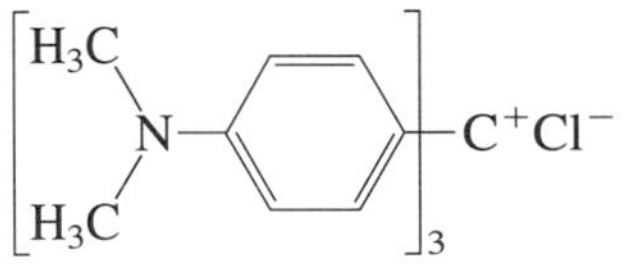 violet

Aqueous solutions of these salts give an immediate precipitate of the silver halide with aqueous silver nitrate solution.

A summary of the results to be expected in the alcoholic silver nitrate test is given below.

I. The following water soluble compounds give an immediate precipitate with aqueous silver nitrate.

1. Amine salts of halogen acids.

$$RNH_3^+Cl^- + Ag^+NO_3^- \longrightarrow AgX(s) + RNH_3^+NO_3^-$$

2. Oxonium salts.
3. Carbonium halides.
4. Low molecular weight acid chlorides. Many of these are easily hydrolyzed by water and produce the halide ion.

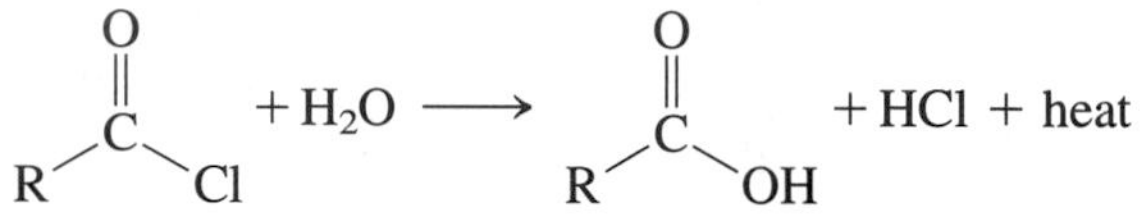

$$\underset{\text{acyl halide}}{RC(=O)Cl} + H_2O \longrightarrow \underset{\text{carboxylic acid}}{RC(=O)OH} + HCl + \text{heat}$$

II. Water insoluble compounds fall roughly into three groups with respect to their behavior toward alcoholic silver nitrate solutions.

1. Compounds in the first group give an immediate precipitation at room temperature.

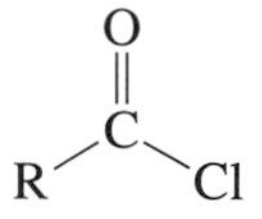

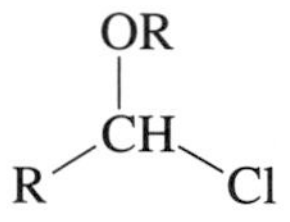

$$RC(=O)Cl \qquad RCH(OR)Cl \qquad R_3CCl \qquad RI$$

$$RCH{=}CHCH_2X \text{ (R, H trans)} \qquad RCH{=}CHCH_2X \text{ (R, CH}_2\text{X cis)} \qquad RCHBrCH_2Br$$

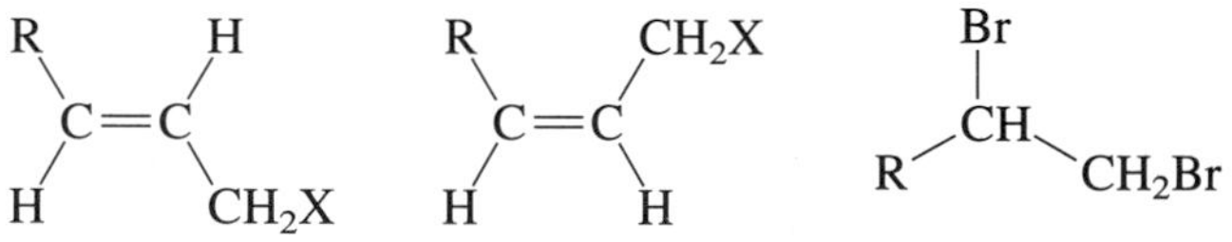

2. The second group includes compounds which react slowly or not at all at room temperature but give a precipitate readily at higher temperatures.

$$RCH_2Cl \qquad R_2CHCl \qquad RCHBr_2 \qquad O_2N\text{-}C_6H_3(NO_2)\text{-}Cl$$

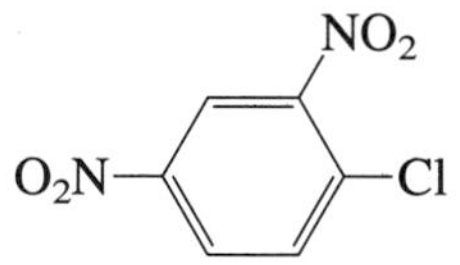

3. A final group is made up of compounds which are usually inert toward hot alcoholic silver nitrate solutions.

$$ArX \qquad RCH{=}CHX \text{ (R, X trans)} \qquad RCH{=}CHX \text{ (R, X cis)} \qquad HCCl_3$$

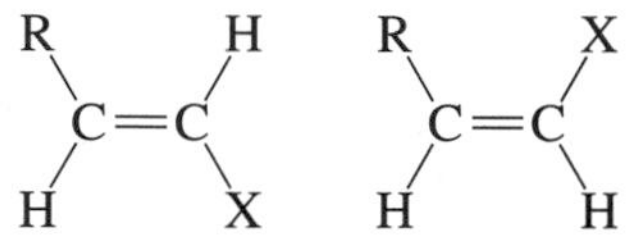

In the reaction with silver nitrate, cyclohexyl halides exhibit a decreased reactivity when compared with the corresponding open-chain secondary halides. Cyclohexyl chloride is inactive, and cyclohexyl bromide is less reactive than 2-bromohexane, although it will give a precipitate with alcoholic silver nitrate. Similarly, 1-methylcyclohexyl chloride is considerably less reactive than acyclic tertiary chlorides. However, both 1-methylcyclopentyl and 1-methylcycloheptyl chloride are more reactive than the open-chain analogs.

Since, as emphasized above, reactivity toward alcoholic silver nitrate is often very different from reactivity toward sodium iodide in acetone (Experiment 34, below), *both* tests should be used with any halogen compound.

A major factor in the reactivity of an alkyl halide (RX) toward silver nitrate is the stability of the carbocation (R^+) which is formed. Specifically, the more stable the carbocation, the faster the reaction with ethanolic $AgNO_3$.

$$R{-}X \xrightarrow{Ag^+} \overset{\delta^+}{R}{-}{-}{-}\overset{\delta^-}{X}{-}{-}{-}\overset{\delta^+}{Ag} \longrightarrow AgX + R^+$$

$$R^+ \xrightarrow{^-NO_3} R{-}ONO_2$$

Questions

16. Suggest a reason why $ClCH_2OR$ compounds should be reasonably reactive toward $AgNO_3$. (*Hint:* Resonance theory supports a stable, although primary, carbocation.)

17. Suggest a reason why benzyl chloride reacts faster than cyclohexylmethyl chloride.

18. Suggest a reason why vinyl and aryl halides are quite inert toward ethanolic $AgNO_3$.

19. The compound shown below is expected to be inert toward ethanolic $AgNO_3$, despite the fact that it is tertiary. Explain your reasoning.

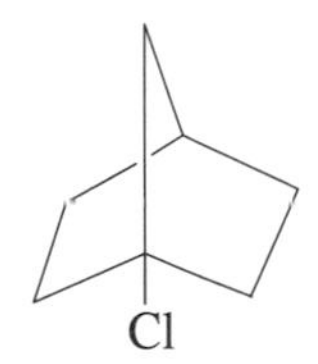

1-chloronorbornane
(1-chlorobicyclo[2.2.1]heptane)

Experiment 34. Sodium Iodide in Acetone Test

$$\underset{\text{alkyl chloride}}{RCl} + \underset{\text{sodium iodide}}{NaI} \xrightarrow{\text{acetone}} \underset{\text{alkyl iodide}}{RI} + \underset{\text{sodium chloride}}{NaCl(s)}$$

$$\underset{\text{alkyl bromide}}{RBr} + \underset{\text{sodium iodide}}{NaI} \xrightarrow{\text{acetone}} \underset{\text{alkyl iodide}}{RI} + \underset{\text{sodium bromide}}{NaBr(s)}$$

To 1 mL of the sodium iodide in acetone reagent in a test tube add 2 drops of the compound whose elemental analysis showed the presence of chlorine or bromine. If the compound is a solid, dissolve about 0.1 g in the smallest possible volume of acetone, and add the solution to the reagent. Shake the test tube, and allow the solution to stand at room temperature for 3 min. Note if a precipitate is formed and also if the solution turns reddish brown because of the liberation of free iodine. If no change occurs at room temperature, place the test tube in a beaker of water at 50°C. Excessive heating causes loss of acetone and precipitation of sodium iodide, which can lead to false positive results. At the end of 6 min, cool to room temperature and note if a reaction has occurred.

Try this test on (1) 1-bromobutane; (2) 2-bromobutane; (3) 2-chloro-2-methylbutane; (4) 1,2-dibromoethane; (5) benzyl chloride; (6) benzoyl chloride; (7) benzenesulfonyl chloride.

Sodium Iodide in Acetone Reagent. Fifteen grams of sodium iodide is dissolved in 100 mL of pure acetone. The solution, colorless at first, becomes a pale lemon yellow. It should be kept in a dark bottle and discarded as soon as a definite red-brown color develops.

Cleaning Up. Place the test solution in the aqueous solution container.

Discussion. This test is used to classify aliphatic chlorides and bromides as primary, secondary, or tertiary. It depends on the fact that sodium chloride and sodium bromide are only very slightly soluble in acetone. As might be anticipated for a direct displacement (S_N2) process, the order of reactivity of simple halides is primary > secondary > tertiary. With sodium iodide, primary bromides give a precipitate of sodium bromide within 3 min at 25°C, whereas the chlorides give no precipitate and must be heated to 50°C in order to effect a reaction. Secondary and tertiary bromides react at 50°C, but the tertiary chlorides fail to react within the time specified. Tertiary chlorides will react if the test solutions are allowed to stand for a day or two. These results are consistent with the following S_N2 process.

$$\text{R—X} \xrightarrow{\text{I}^-} \left[\overset{\delta^-}{\text{I}} \text{---} \text{C} \text{---} \overset{\delta^-}{\text{X}} \right] \longrightarrow \text{I—R}$$

In the reaction with sodium iodide in acetone, cyclopentyl chloride undergoes reaction at a rate comparable with that of acyclic secondary chlorides, whereas the reaction with cyclohexyl chloride is considerably slower. Thus, cyclopentyl chloride and bromide and similar compounds fail to react appreciably with sodium iodide at 50°C within 6 min.

Benzyl ($ArCH_2X$) and allyl $\left(\text{>C=C—C—X} \right)$ halides are extremely reactive toward sodium iodide in acetone and give a precipitate of sodium halide within 3 min at 25°C; the reason for this reactivity is discussed below.

Although triphenylmethyl chloride would be expected to be too hindered to undergo a displacement reaction with iodide ion, it has been found to react much faster than benzyl chloride. However, the reaction is not a simple replacement to form trityl iodide, since the color of *iodine* is observed. This example emphasizes the care that needs to be taken in the interpretation of qualitative tests of this type. It must be remembered that members of a class of compounds under consideration may react with the same reagent in several ways and that for each kind of reaction the effect of structure on reactivity may be different.

Polybromo compounds such as bromoform and 1,1,2,2-tetrabromoethane undergo reaction with sodium iodide at 50°C to give a precipitate and liberate iodine. Carbon tetrabromide undergoes reaction at 25°C.

Sulfonyl bromides and chlorides give an immediate precipitate and also liberate free iodine. Presumably the iodine is formed by the action of sodium iodide on the sulfonyl iodide.[19]

$$ArSO_2X + NaI \longrightarrow ArSO_2I + NaX(s)$$

$$ArSO_2I \xrightarrow{NaI} ArSO_2Na + I_2$$

Benzenesulfonyl chlorides gave a 60% yield of sodium benzenesulfinate; other products obtained were diphenyl disulfone (27%) and phenyl thiosulfinate (10%).

Alkyl sulfonates also undergo reaction, producing the corresponding sodium sulfonates as precipitates. Groups (R) that give stable carbocations undergo S_N1 reactions; otherwise S_N2 displacement can compete.

$$ArSO_2OR + NaI \longrightarrow ArSO_2ONa + RI$$

This reaction must be kept in mind in the event that one of the groups in the sulfonic ester contains halogen.

1,2-Dichloro and 1,2-dibromo compounds not only give a precipitate of sodium chloride or bromide but also liberate free iodine.

$$RHC(Br){-}CHR(Br) + 2NaI \longrightarrow RHC{=}CHR + 2NaBr + I_2$$

Stepwise:

$$I^- \curvearrowright RHC(Br){-}CHR(Br) + 2NaI \longrightarrow RHC{=}CHR + IBr + Br^-$$

$$IBr + I^- \longrightarrow I_2 + Br^-$$

A comparison of ethylene halides gave the following results:

	***ppt* at 25°C**
$BrCH_2CH_2Br$	1.5 min
$BrCH_2CH_2Cl$	3 min
$ClCH_2CH_2Cl$	none (ppt at 50°C in 2.5 min)

These results support the standard leaving group effects upon RX reactivity in S_N2 displacements: I > Br > Cl.

7.13 HYDROCARBONS—ALKANES

Alkanes are not usually characterized chemically because they are quite inert to most reactions discussed in this book. Since chemists usually rely heavily on physical and spec-

[19]B. S. Furniss, A. J. Hannaford, P. W. G. Smith, and A. R. Tatchell, *Vogel's Textbook of Practical Organic Chemistry,* 5th ed. (Wiley, New York, 1991), p. 1234.

tral characterization, we can use the lack of reaction to conclude that our compound is not in a more reactive class. This is consistent with the fact that alkanes fall into solubility class I.

7.14 HYDROCARBONS—ALKENES

The carbon–carbon double bond of alkenes (olefins) can be detected very easily by chemical tests. Alkenes can be detected by the use of iodine (Experiment 31, p. 267). If an alkene is present, a tan solid is formed while the purple color of the iodine solution is retained.

$$\text{>C=C<} + I_2 \longrightarrow \text{>C=C<} \longrightarrow I_2$$

alkene

Addition across the double bond occurs in both the bromine test (Experiment 35, below) and the potassium permanganate test (Experiment 36, p. 279). Since the ionic characters of the bromine and potassium permanganate reactions are very different, there is some complementary character between the two tests. For example, some alkenes bearing electron-withdrawing groups undergo rapid reaction with potassium permanganate, but often slow or negligible reaction with bromine. Bromine adds across the carbon–carbon double bond, with dissipation of the brown-red bromine color.

$$\text{>C=C<} + Br_2 \longrightarrow -\overset{Br}{\underset{|}{\overset{|}{C}}}-\underset{Br}{\underset{|}{\overset{|}{C}}}-$$

alkene (red-brown) 1,2-dibromoalkane (colorless)

Alkenes are oxidized to 1,2-diols with potassium permanganate, with reduction of the manganese from +7, which is purple, to +4, which is brown.

$$3\ \text{>C=C<} + 2KMnO_4 + 4H_2O \longrightarrow 3\ -\underset{OH}{\underset{|}{\overset{|}{C}}}-\underset{OH}{\underset{|}{\overset{|}{C}}}- + 2KOH + 2MnO_2(s)$$

alkene (purple) 1,2-diol (brown)

Experiment 35. Bromine in Carbon Tetrachloride Solution

$$\text{>C=C<} + Br_2 \longrightarrow -\overset{Br}{\underset{|}{\overset{|}{C}}}-\underset{Br}{\underset{|}{\overset{|}{C}}}-$$

alkene (red-brown) 1,2-dibromoalkane (colorless)

$$\text{—C}{\equiv}\text{C—} \xrightarrow[\text{(brown)}]{Br_2} \text{(Br)C=C(Br)} \xrightarrow[\text{(brown)}]{Br_2} \text{—CBr}_2\text{—CBr}_2\text{—}$$

alkyne — 1,2-dibromoalkene (colorless) — 1,1,2,2-tetrabromoalkane (colorless)

In the hood, 0.1 g or 0.2 mL of the unknown compound is added to 2 mL of carbon tetrachloride, and a 5% solution of bromine in carbon tetrachloride is added drop by drop, with shaking, until the bromine color persists. Methylene chloride (dichloromethane) can be used as a solvent as an alternative to the much more toxic carbon tetrachloride. It should be kept in mind, however, that the methylene chloride solutions of bromine are less stable. Apply this test to (1) 2-pentene; (2) hexane; (3) benzene; (4) phenol; (5) formic acid; (6) benzaldehyde; (7) ethanol; (8) allyl alcohol; (9) acetophenone; (10) aniline. For a control, add the same number of drops to the solvent alone, and note the color.

A positive test for a carbon–carbon double bond or triple bond involves the disappearance of the bromine color without the evolution of hydrogen bromide. If the bromine color is discharged and hydrogen bromide is evolved, then substitution has occurred on the unknown. Hydrogen bromide gas can be detected by placing moistened blue litmus paper across the mouth of the test tube and noting if it turns red, indicating the presence of an acidic gas.

Cleaning Up. Place the test solution in the hazardous waste container.

Discussion. This reagent is widely used to test for the presence of an olefinic or acetylenic linkage. It should be employed in conjunction with the potassium permanganate test (Experiment 36, p. 279). Many laboratories forbid the use of carbon tetrachloride, so consult your instructor on the choice of solvent prior to performing this test.

Carbon tetrachloride is a good solvent for bromine and for many organic compounds, but it does not dissolve hydrogen bromide. The evolution of hydrogen bromide is, accordingly, accepted as evidence that the reaction is substitution rather than addition. When employed in detecting unsaturation, this reagent may lead to erroneous conclusions for two reasons. The first is that not all olefinic compounds take up bromine. The presence of electron-withdrawing groups on the carbon atoms of the ethylenic bond causes the addition to be slow and in extreme cases prevents the reaction. The following illustrates this point:

$$C_6H_5CH{=}CH_2 + Br_2 \xrightarrow{\text{rapid}} C_6H_5CH(Br)CH_2Br$$

$$C_6H_5CH{=}CHCOOH + Br_2 \xrightarrow{\text{slow}} C_6H_5CH(Br)CH(Br)COOH$$

In some cases, electron-withdrawing groups may require the use of potassium permanganate to reveal the alkene double bond.

A positive test for unsaturation is one in which the bromine color is discharged *without the evolution of hydrogen bromide.*

When bromination is employed as a qualitative test, carbon tetrachloride has been the preferred solvent; it is, however, an unfortunate choice for a study of the mechanism of the reaction, because in nonpolar solvents the reaction is complicated. The reaction is powerfully catalyzed by traces of water or acids and may be inhibited by oxygen. It could be anticipated that such effects would make it difficult to obtain reproducible results and so render uncertain the interpretation of any negative qualitative test.

Methylene chloride has also been used as a solvent for bromine. This suffers from the disadvantage that the bromine–methylene chloride solution has a limited shelf life. Fresh solutions should be prepared each year.

In a nonpolar solvent, the addition reaction has been shown to be a two-stage process. The first is the reaction of the alkene with bromine to give a bromonium ion. The second step is the attack of the bromine ion on either carbon of the three-membered-ring on the side away from the bromine. The final products contain the two bromine atoms on opposite sides, in an anti orientation.

$$\mathrm{R_2C{=}CR_2 + Br{-}Br \longrightarrow R_2\overset{\overset{+}{Br}}{C{-}C}R_2 + Br^-}$$

$$\mathrm{R_2\overset{\overset{+}{Br}}{C{-}C}R_2 + Br^- \longrightarrow R{-}\underset{R}{\overset{Br}{C}}{-}\underset{Br}{\overset{R}{C}}{-}R}$$

$$\mathrm{R_2\overset{\overset{+}{Br}}{C{-}C}R_2 + Br^- \longrightarrow R{-}\underset{Br}{\overset{R}{C}}{-}\underset{R}{\overset{Br}{C}}{-}R}$$

In a polar solvent such as water, the bromohydrin is the major product accompanied by the evolution of hydrogen bromide.

Substituents that help to stabilize the positively charged bromonium ion increase the rate; those that destabilize the positive ion retard the reaction. For example, substitution of an alkyl group on either of the doubly bonded carbon atoms in alkenes of the type $RCH{=}CH_2$ increases the rate of addition in acetic acid solution by a factor of 30–40. Attached carboxyl groups (COOH) slow the reaction.

Discharge of the bromine color *accompanied by the evolution of hydrogen bromide,* indicating that substitution has occurred, is characteristic of many compounds. In this category are enols, many phenols, and enolizable compounds. Ketones, like other carbonyl compounds, may exhibit an induction period because the hydrobromic acid liberated acts as a catalyst for the enolization step of the bromination. Simple esters do not give this test. Ethyl acetoacetate decolorizes the solution immediately, whereas ethyl malonate may require as much as a minute. A number of active methylene compounds which do not discharge the color at room temperature will readily give a positive test at 70°C. Among these substances are propanal and cyclopentanone. Benzyl cyanide, even at 70°C, may require several minutes.

Aromatic amines are exceptional because the first mole of hydrogen bromide is not evolved, but reacts with the amine to produce a salt. For this reason this reaction could be mistaken for simple addition.

$$C_6H_5NH_2 \xrightarrow{Br_2} p\text{-}BrC_6H_4NH_3^+Br^-$$

Benzylamine represents an unusual type that reacts readily with bromine. Substitution of the hydrogen atoms on the nitrogen atom appears to take place, followed by decomposition to benzonitrile.

$$3C_6H_5CH_2NH_2 + 2Br_2 \longrightarrow [C_6H_5CH_2NBr_2] + 2C_6H_5CH_2NH_3{}^+Br^-$$

$$[C_6H_5CH_2NBr_2] \longrightarrow C_6H_5CN + 2HBr$$

Certain tertiary amines such as pyridine form perbromides upon treatment with bromine.

$$C_5H_5N \xrightarrow{Br_2} C_5H_5N^+\text{–}Br\ Br^-$$

The bromine color is also discharged by aliphatic amines of all types.

Experiment 36. Potassium Permanganate Solution

$$3\ \gt C{=}C\lt + 2KMnO_4 + 4H_2O \longrightarrow 3\ {-}\underset{OH}{\overset{|}{C}}{-}\underset{OH}{\overset{|}{C}}{-} + 2KOH + 2MnO_2(s)$$

alkene (purple) 1,2-diol manganese dioxide (brown)

$$R{-}C{\equiv}C{-}R' + 2KMnO_4 + 2H_2O \longrightarrow$$

alkyne potassium permanganate (purple)

$$RC(=O)O^-K^+ + R'C(=O)O^-K^+ + 2H_2O + 2MnO_2$$

salts of carboxylic acids manganese dioxide (brown)

Method A. Baeyer Test—Aqueous Solutions. To 2 mL of water or ethanol add 0.1 g or 0.2 mL of the compound to be examined. Then add a 2% aqueous potassium permanganate solution drop by drop with shaking until the purple color of the permanganate persists. Apply this test to (1) 2-pentene; (2) toluene; (3) phenol; (4) benzaldehyde; (5) aniline; (6) formic acid; (7) cinnamic acid; (8) glucose. If the permanganate color is not changed in 0.5–1 min, allow the tubes to stand for 5 min with occasional vigorous shaking. Do not be deceived by a slight reaction, which may be due to the presence of impurities.

The disappearance of the purple color and the formation of a brown suspension, which is manganese(II) oxide, at the bottom of the test tube is a positive indication for a carbon–carbon double or triple bond.

Cleaning Up. Place the test solution in the aqueous solution container.

Method B. Phase Transfer Method Using Quaternary Ammonium Salts. Place 5 mL of the "Purple Benzene" reagent in a test tube, add 1 drop of distilled water and add about 0.01 g of the unknown. Stopper the tube, shake, and then rock the tube back and forth holding it over a sheet of white paper to follow the change in color from purple to brown. The time required ranges from 30 sec to 5–15 min. Carry out this test on (1) cyclohexene; (2) stilbene; (3) 1,1-diphenylethene; (4) triphenylethene; (5) glyceryl trioleate (corn or cottonseed oil); (6) diphenylacetylene.

Compare the times required for these reactions against those for Method A for the same compounds.

"Purple Benzene" Reagent. This reagent is prepared prior to use. Dissolve 2 g of sodium chloride and 10 mg of potassium permanganate in 10 mL of distilled water. Place the solution in a 125-mL separatory funnel, and add 10 mL of benzene. ***Caution:*** *Benzene is a known carcinogen. Use benzene in the hood, do not breathe the vapors, and avoid contact with the skin. Benzene is toxic; check with your instructor to see if a different solvent could be used.* Agitate the mixture gently, and allow the layers to separate. Note that the benzene layer is colorless; only the lower aqueous layer is purple. Next add 20 mg of tetrabutyl ammonium bromide to the mixture, agitate and allow the layers to separate. The benzene layer should become deep purple owing to the phase transfer of the permanganate anion along with the tetrabutyl ammonium cation from the aqueous layer into the benzene layer. The transfer is not complete; however, an equilibrium is established. Separate the layers and note the appearance of the purple benzene over 5 min. If the purple color changes to brown, the presence of impurities in the quaternary ammonium salt is indicated; the reagent is prepared again after the quaternary salt is purified. Benzene is not oxidized by dilute permanganate at 20°C, since it is not olefinic but is aromatic.

In addition to tetrabutyl ammonium bromide, tetrapentyl or tetrahexyl ammonium chloride or bromide may be used. Other long-chain quaternary ammonium halides have also been used.

Cleaning Up. Place the benzene layer in the hazardous waste container for benzene. Place the aqueous layer in the aqueous solution container.

Discussion. ***Procedure A.*** A solution of potassium permanganate is decolorized by compounds having ethylenic or acetylenic linkages. This is known as Baeyer's test for unsaturation. In cold dilute solutions the chief product of the action of potassium permanganate on an olefin is the glycol. If the reaction mixture is heated, further oxidation takes place, leading ultimately to cleavage of the carbon chain:

$$RHC{=}CHR' \longrightarrow R{-}CH(OH){-}CH(OH){-}R' \longrightarrow R{-}C({=}O){-}OH + HO{-}C({=}O){-}R'$$

The mechanism for this dihydroxylation is well known and is supported by the fact that a *syn* addition takes place; that is, both hydroxyl groups are added from the same side.

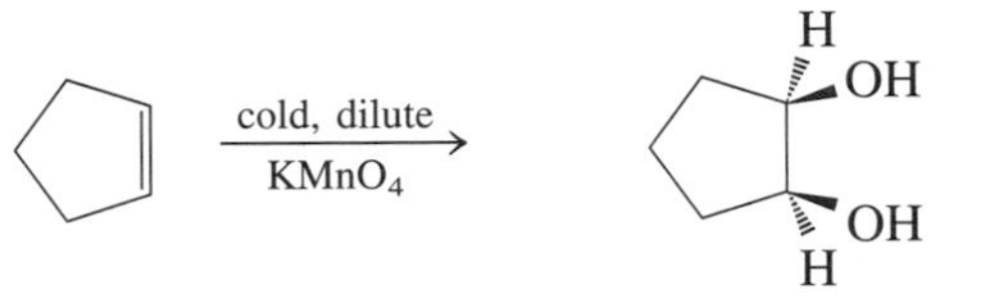

cold, dilute $KMnO_4$

cyclopentene

cis-1,2-dihydroxycyclopentanediol

This mechanism involves a concerted addition of two oxygen atoms of the permanganate ion to the face of the alkene.

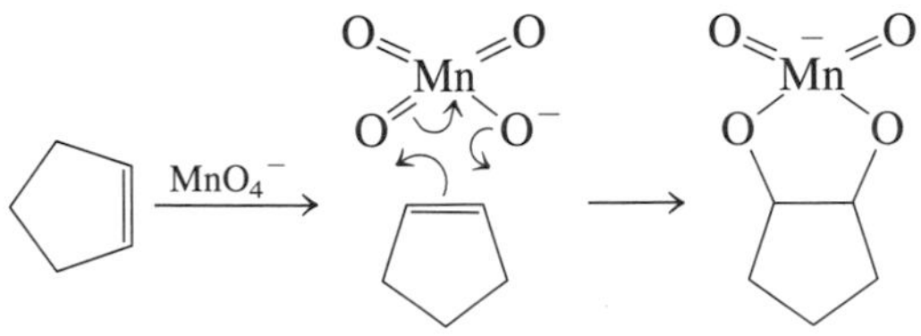

MnO_4^-

Hydrolysis of this bridged intermediate leaving the C—O bonds intact gives the *cis* diol.

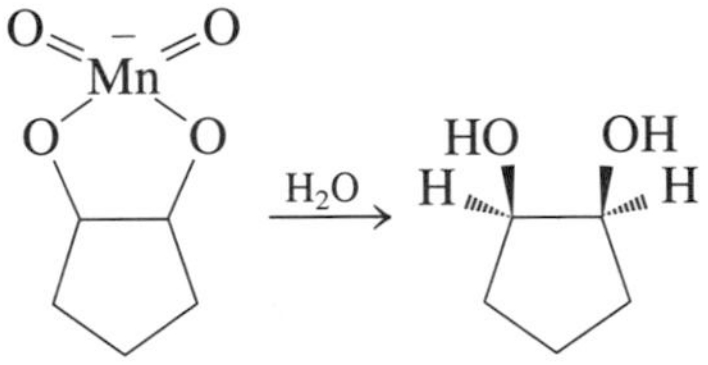

H_2O

Although acetone rather than ethanol has sometimes been employed as a solvent for water insoluble compounds, it has been found that certain carefully prepared olefins show a negative test in acetone but a positive one in ethanol. Ethanol does not react with neutral, dilute potassium permanganate at room temperature within 5 min.

Potassium permanganate in aqueous acetic acid has been used to distinguish among simple primary, secondary, and tertiary alcohols. Under the conditions of the test, primary and secondary alcohols undergo reaction but tertiary alcohols do not. In fact, *tert*-butyl alcohol can be a co-solvent for use in the Baeyer test. It is frequently used in place of ethanol or acetone.

Acetylenic linkages are usually cleaved by oxidation and yield acids.

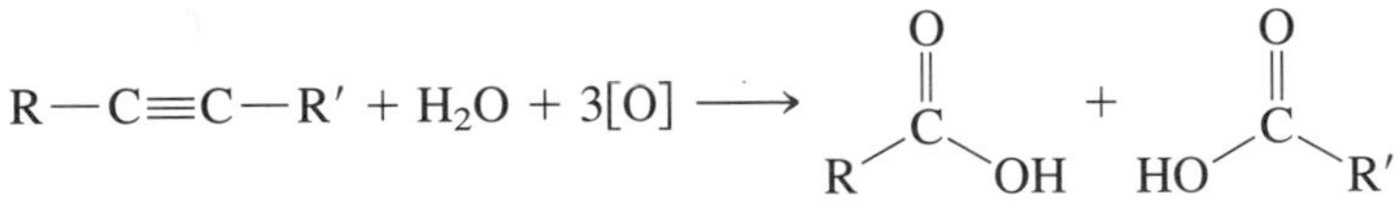

$$R-C{\equiv}C-R' + H_2O + 3[O] \longrightarrow R-C(=O)-OH + HO-C(=O)-R'$$

The speed with which unsaturated compounds decolorize potassium permanganate depends on the solubility of the organic compound. If the compound is very insoluble, it is necessary to powder the compound and shake the mixture vigorously for several minutes or to dissolve the substance in a solvent unaffected by permanganate. A few tetrasubstituted olefins such as $C_6H_5CBr{=}CBrC_6H_5$ and $(C_6H_5)_2C{=}C(C_6H_5)_2$ fail to give positive tests with bromine in CCl_4 or the permanganate solutions above.

Inspection of the following equation shows that, as the reaction proceeds, the solution becomes alkaline:

$$3\ \mathrm{>C{=}C<} + 2KMnO_4 + 4H_2O \longrightarrow 3\ \mathrm{-\underset{OH}{\overset{|}{C}}-\underset{OH}{\overset{|}{C}}-} + 2KOH + 2MnO_2(s)$$

alkene (purple) 1,2-diol (brown)

It is necessary, however, to avoid using a solution that is strongly alkaline, as this changes the nature of the test. In sodium carbonate solution, for example, even acetone gives a positive test. Frequently no actual precipitate of manganese dioxide is observed; the purple color gradually changes to a reddish brown.

However, the use of permanganate in neutral media is feasible. Thus, with zinc permanganate, the zinc hydroxide produced is only slightly soluble and the solution remains practically neutral. Also, it is possible to use potassium permanganate in the presence of magnesium sulfate to accomplish this objective. In this case, the hydroxyl ion is precipitated in the form of insoluble magnesium hydroxide.

The Baeyer test, though superior to the bromine test for unsaturated compounds, offers certain complications in its turn. All easily oxidizable substances give this test. Carbonyl compounds that decolorize bromine solutions generally give a negative Baeyer test. Acetone is a good example; although it decolorizes bromine solutions rapidly, it can be used as a solvent in the Baeyer test. Aldehydes give a positive Baeyer test; however, many of them, such as benzaldehyde and formaldehyde, do not decolorize bromine solutions. Formic acid and its esters, which have the O=CH— group, also reduce permanganate.

Alcohols form another important class of compounds that decolorize permanganate solutions but not bromine solutions. Pure alcohols do not give the test readily; however, they often contain impurities that are easily oxidized. Other types of compounds also are likely to contain slight amounts of impurities that may decolorize permanganate solutions. For this reason, the decolorization of only a drop or two of the permanganate solution cannot always be accepted as a positive test.

Phenols and arylamines also reduce permanganate solution and undergo oxidation to quinones; these may be further oxidized with an excess of the reagent to yield a series of oxidation products, among which are maleic acid, oxalic acid, and carbon dioxide.

Procedure B. The phase transfer technique, using either long-chain quaternary ammonium salts or crown ethers, has become a valuable method for carrying out many types of reactions. For this qualitative Baeyer test, the use of the quaternary ammonium salts has given good results, whereas tests using samples of available crown ethers were erratic.

When a polar solvent such as water is placed in a flask with a nonpolar solvent such as benzene or another hydrocarbon, two layers will form; the less dense hydrocarbon layer will be the top layer. Normally, inorganic compounds such as $KMnO_4$ are highly ionic and polar, thus favoring the polar solvent. Since we desire to carry out a reaction of a nonpolar organic compound, the alkene, we must deal with the fact that this organic compound will much more readily dissolve in the nonpolar organic solvent. It is the role of the phase transfer catalyst to convert the inorganic reagent to a modified form which is soluble in the organic solvent. Quaternary ammonium salts are very effective at this conversion and less expensive than crown ethers. Here the "quat" salt, Q^+X^-, is tetrabutyl ammonium bromide, $(CH_3CH_2CH_2CH_2)_4NBr$. The ionic character of this reagent causes it to be sufficiently soluble in the water layer where it can undergo an ion exchange to form a new quat salt, $Q^+MnO_4^-$, which contains the permanganate oxidizing agent:

$$[Q^+ + Br^-] + [K^+ + MnO_4^-] \rightleftharpoons [Q^+ + MnO_4^-] + [K^+ + Br^-]$$

$$Q^+(aq) + MnO_4^-(aq) \rightleftharpoons Q^+MnO_4^- \text{ (nonpolar solvent)}$$

In the organic layer, the permanganate oxidizes the alkene to release the hydroxide. The hydroxide is soluble in the aqueous layer and reacts with more permanganate to repeat the cycle:

$$Q^+MnO_4^- + \text{>C=C<} \longrightarrow \text{—C(OH)—C(OH)—} + MnO_2 + Q^+OH^-$$

$$Q^+OH^- \xrightarrow[\text{transfer}]{\text{phase}} Q^+ + OH^-$$

$$[Q^+ + OH^-] + [K^+ + MnO_4^-] \rightleftharpoons [Q^+ + MnO_4^-] + [K^+ + OH^-]$$

In a similar way, crown ethers mask the K^+ ion to form complexes that are soluble in nonpolar solvents.

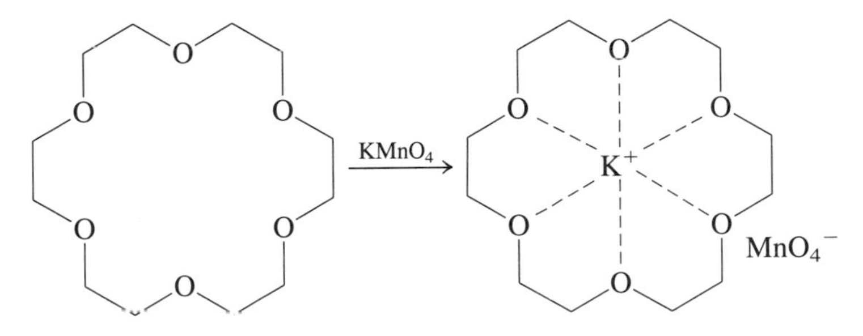

(soluble in nonpolar solvents)

Questions

20. What functional groups respond to both the bromine and the permanganate tests?

21. Which of these tests is better for detecting the presence of multiple bonds? Explain.

22. In what instances is it helpful to use both reagents?

7.15 HYDROCARBONS—ALKYNES

Alkynes (acetylenes) give a positive test with bromine (Experiment 35, p. 276) and potassium permanganate (Experiment 36, p. 279). Bromine adds across the carbon-carbon triple bond, with the bromine color of the solution dissipating. Potassium permanganate oxidizes alkynes to carboxylic acids, concurrently with the reduction of the manganese from an oxidation state of +7, a purple color, to +4, a brown color.

$$-C\equiv C- \xrightarrow[\text{(brown)}]{Br_2} \underset{\text{1,2-dibromoalkene (colorless)}}{(Br)C{=}C(Br)} \xrightarrow[\text{(brown)}]{Br_2} \underset{\text{1,1,2,2-tetrabromoalkane (colorless)}}{-CBr_2-CBr_2-}$$

alkyne

$$\underset{\text{alkyne}}{R-C\equiv C-R'} + \underset{\text{potassium permanganate (purple)}}{2\ KMnO_4} + 2\ H_2O \longrightarrow$$

$$\underset{\text{salts of carboxylic acids}}{RC(=O)O^-K^+ + R'C(=O)O^-K^+} + 2\ H_2O + \underset{\text{manganese dioxide (brown)}}{2\ MnO_2}$$

Terminal alkynes react with sodium to liberate hydrogen gas and form a salt (Experiment 5, p. 215).

$$\underset{\text{terminal alkyne}}{H-C\equiv C-H} + 2Na \longrightarrow Na^{+\,-}C-C\equiv C^-Na^+ + H_2(g)$$

$$\underset{\text{terminal alkyne}}{2R-C\equiv C-H} + 2Na \longrightarrow 2R-C\equiv C^-Na^+ + H_2(g)$$

7.16 HYDROCARBONS—AROMATIC

If the results of preliminary chemical tests suggest aromatic character for an unknown, then a variety of tests can be used to characterize chemically this class of organic compound. Specifically, we can consider introducing new substituents onto the aromatic ring, or modifying existing substituents, such that we may more readily characterize the new compound. If the molecule already contains reactive chemical substituents (acids, amines, ethers, carbonyl compounds, etc.), the chemist is referred to other sections for that particular group.

The most vigorous test will be described first, and the tests that follow will be described in decreasing order of the severity of conditions. A few of the most inert aromatic compounds may remain unchanged after even the most vigorous test; characterization of those compounds may rely more on the spectral and physical tests than usual.

Fuming sulfuric acid converts aromatic compounds to arylsulfonic acids (Experiment 37, p. 285). The aromatic compound dissolves completely with the evolution of heat.

$$\underset{\text{aromatic compound}}{ArH} \xrightarrow[SO_3]{H_2SO_4} \underset{\text{arylsulfonic acid}}{ArSO_3H} + \text{heat}$$

The reaction of aromatic compounds with azoxybenzene and aluminum chloride gives rise to 4-arylazobenzene, with the color of the solution or precipitate indicating particular functional groups (Experiment 38, p. 287).

$$\text{Ar—H} + \text{C}_6\text{H}_5\text{—}\overset{+}{\underset{\text{O}^-}{\text{N}}}\text{=N—C}_6\text{H}_5 \xrightarrow{AlCl_3} [\text{Ar—C}_6\text{H}_4\text{—N=N—C}_6\text{H}_5]\ AlCl_3$$

aromatic compound — azobenzene — 4-arylazobenzene (colored complex)

Aromatic compounds react with chloroform and aluminum chloride to produce triaryl carbocations in a variety of colors, depending upon the functional groups on the aryl ring (Experiment 39, p. 288).

$$3ArH + CHCl_3 \xrightarrow{AlCl_3} Ar_3CH + 3HCl$$

aromatic compound — triarylmethane

$$Ar_3CH + R^+ \longrightarrow Ar_3C^+ + RH$$

triarylmethane — triaryl carbocation (colored)

The triaryl carbocations are in solution as $Ar_3C^+\ AlCl_4^-$ salts and are responsible for the colors observed.

Experiment 37. Fuming Sulfuric Acid

$$ArH \xrightarrow[SO_3]{H_2SO_4} ArSO_3H + \text{heat}$$

aromatic compound — arylsulfonic acid

Caution: *Use this reagent with relatively inert compounds only, such as those compounds that do not dissolve in the solubility tests with concentrated sulfuric acid. Compounds for which preliminary tests indicate strongly activating groups (OH, NH_2, etc.) may be decomposed violently by fuming sulfuric acid.*

This test must be done in a hood. Place 0.5 mL of 20% fuming sulfuric acid *(hazardous)* in a clean, dry test tube, and add 0.25 mL or 0.25 g of the unknown. Shake the mixture vigorously, and allow it to stand for a few minutes. A positive test for the presence of an aromatic ring is complete dissolution of the unknown, evolution of heat, and minimal charring. Repeat the experiment using (1) benzene; (2) bromobenzene; (3) 1,2-dibromoethane; (4) cyclohexane.

Cleaning Up. In the hood, the test solution is poured into a large beaker containing 10 mL of water. Sodium carbonate is added, a few crystals at a time, until the foaming ceases. The solution is then placed in the aqueous solution container.

Discussion. Before considering fuming sulfuric acid, let us review some of the properties of sulfuric acid. Concentrated (100%) sulfuric acid is a remarkable solvent in two respects. Its dielectric constant appears to be very much greater than that of many other compounds for which this property has been measured.[20] Thus, forces of attraction between dissolved ions are so small in dilute solution that activity coefficients may be taken as unity.

$$2H_2SO_4 \rightleftharpoons H_3SO_4^+ + HSO_4^-$$

The second unusual property is that, in addition to the autoprotolysis like that found in hydroxylic solvents (such as water), there is a self-dissociation resulting initially in the formation of sulfur trioxide and water. However, at the concentrations concerned, each of these reacts essentially completely with sulfuric acid so that the overall equilibrium is as follows:

$$\begin{aligned} H_2SO_4 &\rightleftharpoons H_2O + SO_3 \\ SO_3 + H_2SO_4 &\rightleftharpoons HSO_3^+ + HSO_4^- \\ H_2O + H_2SO_4 &\rightleftharpoons H_3O^+ + HSO_4^- \\ \hline 3H_2SO_4 &\rightleftharpoons HSO_3^+ + H_3O^+ + 2HSO_4^- \end{aligned}$$

Also:

$$H_2SO_4 + HSO_3^+ + HSO_4^- \rightleftharpoons H_3SO_4^+ + H_2SO_7^-$$

Concentrated sulfuric acid converts ethylene to ethyl hydrogen sulfate, but sulfuric acid containing added sulfur trioxide (fuming sulfuric acid) yields ethionic acid as shown below. The reason for the difference between the two sets of conditions is understood if it is realized that 100% sulfuric acid contains sulfonating species such as SO_3 or HSO_3^+ in small concentration and therefore the sulfonating reagent fails to compete with proton addition, the first step in alkyl sulfate formation:

$$H_2C{=}CH_2 + H_2SO_4 \longrightarrow CH_3CH_2^+ + HSO_4^-$$

However, in fuming sulfuric acid, the addition of HSO_3^+, SO_3, or some other sulfonating agent to the double bond becomes important, so that the principal reaction is the following:

$$H_2C{=}CH_2 + HSO_3^+ \longrightarrow {}^+CH_2CH_2SO_3H$$

In each case the second step is the same; the reaction of the carbocation with the bisulfate ion.

$$CH_3CH_2^+ + HSO_4^- \longrightarrow \underset{\text{ethyl hydrogen sulfate}}{CH_3CH_2OSO_3H}$$

$${}^+CH_2CH_2SO_3H + HSO_4^- \longrightarrow \underset{\text{ethionic acid}}{HO_3SOCH_2CH_2SO_3H}$$

The first step in the formation of the ethionic acid is, at least formally, like the first step in aromatic substitution. The exact nature of the electrophile is probably complex.

The action of fuming sulfuric acid on 1,2-dihalo compounds is complex. The mixture turns dark, and some free halogen is liberated. It seems probable that loss of hydro-

[20]R. J. Gillespie, E. D. Hughes, and C. K. Ingold, *J. Chem. Soc.*, 1950, 2473.

gen halide occurs followed by polymerization of the vinyl halide. For example, 1,2-dibromoethane probably undergoes the following changes.

$$BrH_2C—CH_2Br \longrightarrow H_2C{=}CHBr + HBr$$

$$nH_2C{=}CHBr \longrightarrow —(CH_2—\underset{\displaystyle Br}{\underset{|}{CH}})_n—$$

$$2HBr + H_2SO_4 \longrightarrow Br_2 + SO_2 + 2H_2O$$

Questions

23. Note that this test is useful only for compounds insoluble in sulfuric acid. Why?

24. Write equations for the reactions involved with 1-hexene. Compare the products of this reaction with those formed in your solubility test with concentrated sulfuric acid (pp. 111–112).

Experiment 38. Azoxybenzene and Aluminum Chloride

$$\underset{\text{aromatic compound}}{Ar—H} + \underset{\text{azobenzene}}{C_6H_5—\overset{O^-}{\overset{|}{\underset{+}{N}}}{=}N—C_6H_5} \xrightarrow{AlCl_3} \underset{\text{4-arylazobenzene (colored complex)}}{[Ar—C_6H_4—N{=}N—C_6H_5]\,AlCl_3}$$

Place 0.5 mL or 0.4 g of the dry compound in a clean, dry test tube; add one crystal of azoxybenzene and about 25 mg of anhydrous aluminum chloride. Note the color. If no color is produced immediately, warm the mixture for a few minutes. Wait for up to 30 min to observe any color change. Try the test on (1) petroleum ether; (2) chlorobenzene; (3) ethyl bromide; (4) naphthalene. If the hydrocarbon is a solid, a solution of 0.5 g of it in 2 mL of dry carbon disulfide may be used.

Cleaning Up. The reaction mixture is placed in the organic solvent container.

Discussion. The color produced is due to an addition compound formed from the 4-arylazobenzene and aluminum chloride. Many oxygen functions interfere with this test and may give confusing color changes.

The 4-arylazobenzene is formed from the aromatic substitution of the aromatic compound by aluminum chloride followed by the elimination of water and aluminum chloride.

$$C_6H_5—N{=}\underset{+}{\overset{O^-}{\overset{|}{N}}}—C_6H_5 \xrightarrow{AlCl_3} (C_6H_5)^{+}{=}{=}N{=}{=}\overset{O—AlCl_3^-}{\overset{|}{N}}—C_6H_5$$

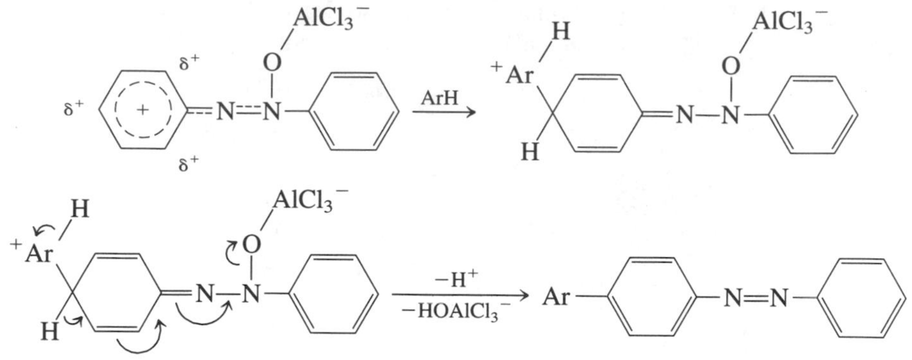

4-arylazobenzene

Aromatic hydrocarbons and their halogen derivatives produce a deep orange to dark red color in solution or give a precipitate. Fused aromatic rings such as naphthalene, anthracene, and phenanthrene produce brown colors. Aliphatic hydrocarbons give no color or, at most, a pale yellow.

Experiment 39. Chloroform and Aluminum Chloride

$$\underset{\text{aromatic compound}}{3\text{ArH}} + \text{CHCl}_3 \xrightarrow{\text{AlCl}_3} \underset{\text{triarylmethane}}{\text{Ar}_3\text{CH}} + 3\text{HCl}$$

$$\underset{\text{triarylmethane}}{\text{Ar}_3\text{CH}} + \text{R}^+ \longrightarrow \underset{\text{triaryl carbocation (colored)}}{\text{Ar}_3\text{C}^+} + \text{RH}$$

To 2 mL of dry chloroform in a test tube add 0.1 mL or 0.1 g of the unknown compound. Mix thoroughly, and incline the test tube so as to moisten the wall. Then add 0.5–1.0 g of anhydrous aluminum chloride so that some of the powder strikes the side of the test tube. Note the color of the powder on the side, as well as the solution. Try the test on (1) petroleum ether; (2) biphenyl; (3) chlorobenzene; (4) toluene; (5) naphthalene.

Cleaning Up. The test solution is placed in the halogenated organic waste container.

Discussion. The colors produced by the reaction of aromatic compounds with chloroform and aluminum chloride are quite characteristic. Aliphatic compounds, which are insoluble in sulfuric acid, give no color or only a very light yellow. Typical colors produced are the following.

Compound	Color
Benzene and its homologs	Orange to red
Aryl halides	Orange to red
Naphthalene	Blue
Biphenyl	Purple
Phenanthrene	Purple
Anthracene	Green

With time the colors change to various shades of brown. Similar colors are obtained when chloroform is replaced by carbon tetrachloride.

Aromatic esters, ketones, amines, and other oxygen- or nitrogen-containing compounds may also give blue or green colors. This test should be used in conjunction with other tests to confirm the presence or absence of an aromatic structure.

The reaction begins as three successive (1–3 below) Friedel–Crafts reactions between the aromatic and chlorinated hydrocarbons. These alkylation reactions are promoted by aluminum chloride, a Lewis acid, and facilitated by positive delocalization by the chlorine and the aryl groups.

$$CHCl_3 + AlCl_3 \longrightarrow \overset{\delta+}{C}HCl_2\text{------}Cl\text{------}\overset{\delta-}{Al}Cl_3 \xrightarrow{-AlCl_4^-} \left[H-\overset{\delta+}{C}\begin{matrix} Cl^{\delta-} \\ Cl^{\delta-} \end{matrix} \right] \xrightarrow[(1)]{ArH}$$

$$\overset{\delta+}{Ar}\begin{matrix} H \\ CHCl_2 \end{matrix} \xrightarrow{-H^+} ArCHCl_2 \xrightarrow{AlCl_3} Ar-\overset{+}{C}\begin{matrix} H \\ Cl \end{matrix} \xrightarrow[(2)]{ArH} Ar-\underset{Cl}{CH}-\overset{H}{Ar^+} \xrightarrow{-H^+}$$

$$Ar-\underset{Cl}{CH}-Ar \xrightarrow{AlCl_3} Ar-\overset{+}{C}H-Ar \xrightarrow[(3)]{ArH} Ar-\underset{Ar}{CH}-\overset{H}{Ar^+} \xrightarrow{-H^+} Ar_3CH$$

Partially substituted chlorides (e.g., Ar_2CHCl or $ArCHCl_2$) may undergo reaction with aluminum chloride to give mono or diaryl carbocations:

$$Ar_2CHCl + AlCl_3 \longrightarrow Ar_2CH^+ + AlCl_4^-$$

These carbocations abstract hydride from the triarylmethane to give rise to stable triaryl carbocations:

$$Ar_3CH + Ar_2CH^+ \longrightarrow \underset{\text{(colored)}}{Ar_3C^+} + Ar_2CH_2$$

Question

25. Propose a reason why tetraarylmethanes are not formed.

7.17 KETONES

Many of the same reactions described in Section 7.4, pp. 228–236, for the classification of aldehydes can also be used to classify ketones. The addition of 2,4-dinitrophenylhydrazine to ketones to precipitate the 2,4-dinitrophenylhydrazones (Experiment 11, p. 229) is probably the most useful of these reactions.

$$\underset{\text{aldehyde or ketone}}{\overset{O}{\overset{\|}{C}}} + \underset{\text{2,4-dinitrophenylhydrazine}}{O_2N\text{-}C_6H_3(NO_2)\text{-}NHNH_2} \xrightarrow[\text{alcohol}]{H_2SO_4} \underset{\text{2,4-dinitrophenylhydrazone}}{O_2N\text{-}C_6H_3(NO_2)\text{-}NHN{=}C{<}}$$

The reaction of hydroxylamine hydrochloride with ketones (Experiment 12, p. 231) produces oximes and results in the liberation of hydrogen chloride, which can be detected by an indicator.

$$\underset{\text{aldehyde or ketone}}{\overset{O}{\overset{\|}{C}}} + \underset{\text{hydroxylamine hydrochloride}}{H_2N\text{—}OH \cdot HCl} \longrightarrow \underset{\text{oxime}}{{>}C{=}N\text{—}OH} + HCl + H_2O$$

The precipitation of a bisulfite addition complex (Experiment 13, p. 232) is indicative of a variety of carbonyl compounds. This reaction is greatly influenced by the steric environment of the carbonyl group.

$$\underset{\text{aldehyde or ketone}}{\overset{O}{\overset{\|}{C}}} + \underset{\text{sodium bisulfite}}{NaHSO_3} \longrightarrow \underset{\text{sodium bisulfite addition complex}}{\text{—}\overset{H}{\overset{|}{\underset{SO_3^-Na^+}{\underset{|}{C}}}}\text{—}OH}$$

The iodoform test (Experiment 10, p. 224) will give positive results with methyl ketones. A positive test is indicated by the precipitation of iodoform, a foul-smelling yellow solid.

$$\underset{\text{methyl ketone}}{R\overset{O}{\overset{\|}{C}}CH_3} + 3I_2 + 3NaOH \longrightarrow R\overset{O}{\overset{\|}{C}}CI_3 + 3NaI + 3H_2O$$

$$R\overset{O}{\overset{\|}{C}}CI_3 + NaOH \longrightarrow \underset{\text{sodium salt of the carboxylic acid}}{R\overset{O}{\overset{\|}{C}}O^-Na^+} + \underset{\text{iodoform (yellow solid)}}{CHI_3(s)}$$

7.18 NITRILES

Nitriles can be hydrolyzed under basic conditions to yield the salt of the carboxylic acid and ammonia (Experiment 16, p. 237). The ammonia vapor is detected by litmus paper.

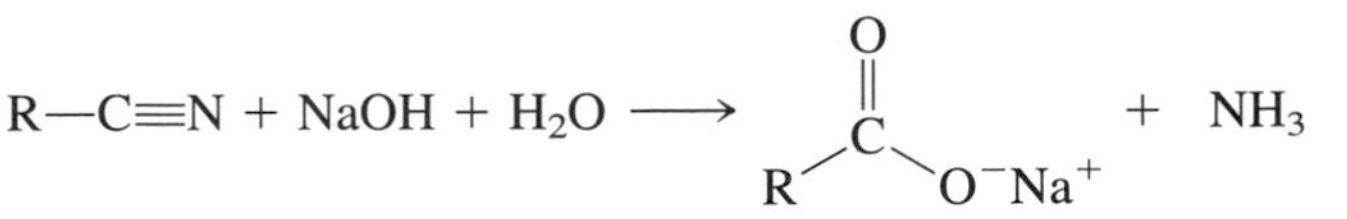

$$R\text{—}C{\equiv}N + NaOH + H_2O \longrightarrow RC(=O)O^-Na^+ + NH_3$$

nitrile — salt of the carboxylic acid — ammonia

Nitriles, along with many other compounds, give a positive hydroxamic acid test (Experiment 2c, p. 207) The hydroxamic acid is detected with ferric chloride to form the ferric hydroxamate complex, which has a burgundy or magenta color.

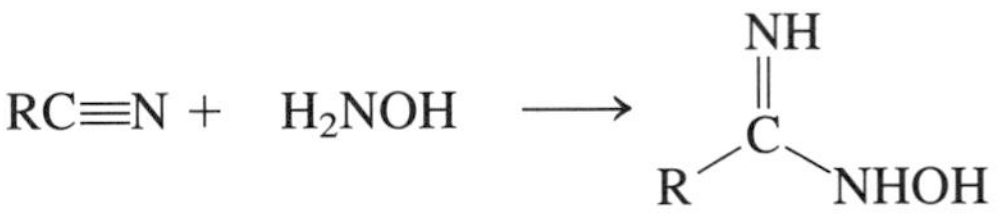

$$RC{\equiv}N + H_2NOH \longrightarrow RC(=NH)NHOH$$

nitrile — hydroxylamine

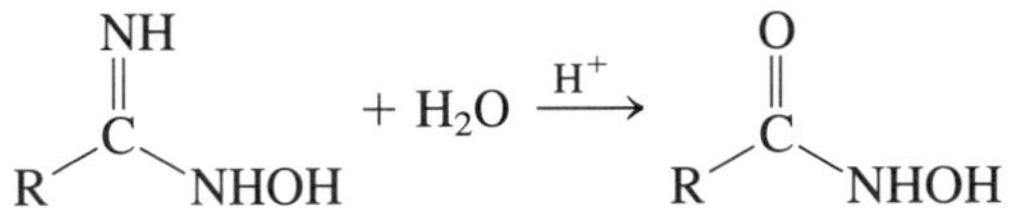

$$RC(=NH)NHOH + H_2O \xrightarrow{H^+} RC(=O)NHOH$$

hydroxamic acid

$$3\ RC(=O)NHOH + FeCl_3 \longrightarrow (RC(=O)NHO)_3Fe + 3HCl$$

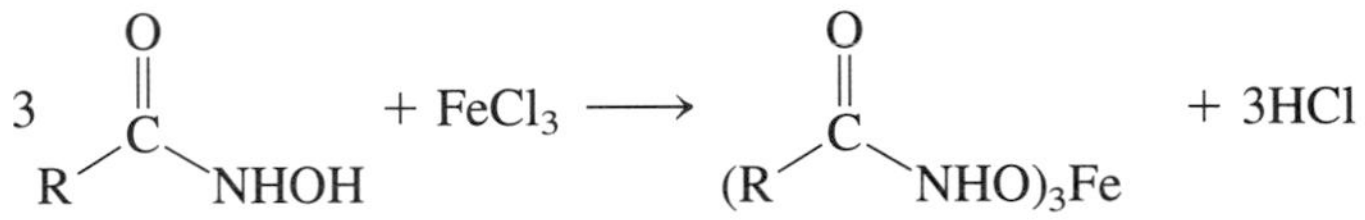

hydroxamic acid — ferric hydroxamate complex (burgundy or magenta)

7.19 NITRO COMPOUNDS

The presence of a nitro group is detected in several different ways. In the ferrous hydroxide reduction (Experiment 40, p. 292), a positive test is noted by the change in color from green to red-brown or brown due to the oxidation of iron from +2 to +3.

$$RNO_2 + 6Fe(OH)_2 + 4H_2O \longrightarrow RNH_2 + 6Fe(OH)_3$$

nitroalkane — ferrous hydroxide (green) — ferric hydroxide (red-brown to brown)

Of the nitro compounds, only tertiary aliphatic nitro compounds and aromatic nitro compounds are reduced by zinc and ammonium chloride (Experiment 41, p. 293) to the hydroxylamine. The hydroxylamine is then detected by the formation of metallic silver in the Tollens test (Experiment 14, p. 234).

$$RNO_2 + 4[H] \xrightarrow[NH_4Cl]{Zn} RNHOH + H_2O$$

nitro compound — hydroxylamine

$$RNHOH + 2Ag(NH_3)_2OH \longrightarrow RNO + 2H_2O + 2Ag(s) + 4NH_3$$

Tollens reagent

The number of nitro groups on an aromatic ring can be determined by the reaction of the unknown with sodium hydroxide (Experiment 42, p. 293). In the reaction with sodium hydroxide, the mononitro aromatic compounds yield no color change; dinitro aromatic compounds produce a bluish purple color; trinitro aromatic compounds give a red color. The color of the solution is due to a Meisenheimer complex.

O_2N–$C_6H_3(NO_2)$–G $\xrightarrow{OH^-}$ O_2N–[ring]$^-$(OH)(G)–NO_2

Meisenheimer complex

Experiment 40. Ferrous Hydroxide Reduction

$$\underset{\text{nitroalkane}}{RNO_2} + \underset{\substack{\text{ferrous hydroxide}\\ \text{(green)}}}{6Fe(OH)_2} + 4H_2O \longrightarrow RNH_2 + \underset{\substack{\text{ferric hydroxide}\\ \text{(red-brown to brown)}}}{6Fe(OH)_3}$$

Add about 10 mg of the compound to 1 mL of the ferrous sulfate reagent in a test tube, and then add 0.7 mL of the alcoholic potassium hydroxide reagent. Insert a glass tube so that it reaches the bottom of the test tube, and pass a stream of inert gas through the tube for about 30 sec in order to remove air. Stopper the tube quickly, and shake. Note the color of the precipitate after 1 min. A positive test is the formation of the red-brown to brown precipitate of iron(III) hydroxide. Try this test on (1) nitrobenzene; (2) 3-nitroaniline; (3) ethanol; (4) 2-propanol; (5) nitromethane.

Ferrous Sulfate Reagent. To 100 mL of recently boiled, distilled water add 5.0 g of ferrous ammonium sulfate crystals and 0.4 mL of concentrated sulfuric acid. An iron nail is introduced to retard air oxidation.

Alcoholic Potassium Hydroxide Reagent. Three grams of potassium hydroxide is dissolved in 3 mL of distilled water, and this solution is added to 100 mL of 95% ethanol.

Cleaning Up. The solid is isolated by filtration and placed in the nonhazardous solid waste container; the filtrate is neutralized with 10% hydrochloric acid and placed in the aqueous solution container.

Discussion. The red-brown to brown precipitate[21] of iron(III) hydroxide (ferric hydroxide) is formed by the oxidation of iron(II) hydroxide (ferrous hydroxide) by the nitro compound, which in turn is reduced to the primary amine. A negative test is indicated by a greenish precipitate. In some cases partial oxidation may cause a darkening of the ferrous hydroxide.

Practically all nitro compound give a positive test in 30 sec. The speed with which the nitro compound is reduced depends on its solubility. 4-Nitrobenzoic acid, which is soluble in the alkaline reagent, gives a test almost immediately, whereas 2-nitronaphthalene must be shaken for 30 sec.

[21] W. M. Hearson and R. G. Gustavson, *Ind. Eng. Chem., Anal. Ed., 9,* 352 (1937).

A positive test is also given by other compounds that oxidize ferrous hydroxide. Nitroso compounds, quinones, hydroxylamines, alkyl nitrates, and alkyl nitrites are in this group. Highly colored compounds cannot be tested.

Experiment 41. Zinc and Ammonium Chloride Reduction

$$\underset{\text{nitro compound}}{RNO_2} + 4[H] \xrightarrow[NH_4Cl]{Zn} \underset{\text{hydroxylamine}}{RNHOH} + H_2O$$

$$RNHOH + \underset{\text{Tollens reagent}}{2Ag(NH_3)_2OH} \longrightarrow RNO + 2H_2O + 2Ag(s) + 4NH_3$$

Dissolve 0.2 mL or 0.2 g of the unknown in 4 mL of 50% ethanol, and add 0.2 g of ammonium chloride and 0.2 g of zinc dust. Shake, and heat to boiling. Allow to stand for 5 min, filter, and test the action of the filtrate on Tollens reagent (Experiment 14, p. 234). A positive test is the formation of a black or grey precipitate or a silver mirror. Try this test on (1) nitrobenzene; (2) 3-nitroaniline; (3) ethanol.

Cleaning Up. Pour the solution into a beaker. Add a few drops of 5% nitric acid to dissolve the silver mirror or colloidal silver. Combine all solutions. Make the solution acidic with 5% nitric acid, then neutralize with sodium carbonate. Add 2 mL of saturated sodium chloride solution to precipitate the silver as silver chloride. The silver chloride is isolated by filtration and placed in the nonhazardous solid waste container. The filtrate is placed in the aqueous solution container.

Discussion. This test depends on the reduction of the unknown to a hydrazine, a hydroxylamine, or an aminophenol; all these compounds are oxidized by Tollens reagent.

This test cannot be applied if the original compound reduces Tollens reagent.

Tertiary aliphatic compounds and aromatic nitro compounds give a positive test. Nitroso, azoxy, and azo compounds are reduced with zinc and ammonium chloride, with the products being oxidized in the Tollens test.

Experiment 42. Treatment of Aromatic Nitro Compounds with Sodium Hydroxide

To 5 mL of 20% sodium hydroxide solution add 2 mL of ethanol and a drop or a crystal of the unknown, and shake vigorously. Note the color of the solution.

Alternatively, 0.1 g of the unknown is dissolved in 10 mL of acetone, and 2–3 mL of 10% sodium hydroxide solution is added with shaking. Note the color of the solution.

Try this test on (1) nitrobenzene; (2) 1,3-dinitrobenzene.

Cleaning Up. The test solution is placed in the aqueous solution container.

Discussion. Mononitro benzene compounds give no color or a very light yellow with these reagents. If two nitro groups on the same ring are present, a bluish purple color develops; the presence of three nitro groups produces a blood red color. The presence of an amino, substituted amino, or hydroxyl group in the molecules inhibits the formation of the characteristic red and purple colors.

Polynitro compounds can form Meisenheimer complexes, which may lead to a colored solution.

$$O_2N\text{-}C_6H_3(NO_2)\text{-}G \xrightarrow{OH^-} O_2N\text{-}[C_6H_3(NO_2)(OH)(G)]^-$$

Meisenheimer complex

Nitrophenols can form highly conjugated and stable phenoxide anions, which may be a source of color.

$$p\text{-}HOC_6H_4NO_2 \xrightarrow{NaOH} \left[{}^-O\text{-}C_6H_4\text{-}N^+(=O)O^- \longleftrightarrow O=C_6H_4=N^+(O^-)O^- \right] Na^+$$

7.20 PHENOLS

As with alcohols, the acidic hydrogen in a phenol can be detected with acetyl chloride (Experiment 6, p. 217). An ester layer is formed.

$$\underset{\text{phenol}}{ArOH} + \underset{\text{acetyl chloride}}{H_3C\text{-}C(=O)\text{-}Cl} \longrightarrow \underset{\text{ester}}{H_3C\text{-}C(=O)\text{-}OAr} + HCl(g)$$

Phenols undergo reaction with the yellow ceric ammonium nitrate (Experiment 7, p. 218) to produce brown or black products.

$$\underset{\substack{\text{ceric ammonium nitrate}\\ \text{(yellow)}}}{(NH_4)_2Ce(NO_3)_6} + \underset{\text{phenol}}{ArOH} \longrightarrow \underset{\text{(brown or black)}}{(NH_4)_2Ce(OAr)(NO_3)_5} + HNO_3$$

A modification of Liebermann's test (Experiment 18d, p. 248) can be used to test for the presence of a phenol. A blue intermediate is formed, which changes to red when diluted, and blue when the solution is made basic.

$$C_6H_5OH \xrightarrow{HONO} p\text{-}HOC_6H_4NO \rightleftharpoons \underset{\text{(yellow)}}{O=C_6H_4=NOH} + H_2O$$

$$\text{O=C}_6\text{H}_4\text{=NOH (yellow)} + \text{C}_6\text{H}_5\text{OH} + H_2SO_4 \longrightarrow \left[\text{HO}-\text{C}_6\text{H}_4-\text{N}=\text{C}_6\text{H}_4=\overset{+}{\text{O}}\text{H}\right] \text{HSO}_4^- \text{ (blue)} \xrightarrow{H_2O}$$

$$\text{HO}=\text{C}_6\text{H}_4=\text{N}-\text{C}_6\text{H}_4-\text{OH (red)} \xrightarrow{\text{NaOH}} \text{HO}=\text{C}_6\text{H}_4=\text{N}-\text{C}_6\text{H}_4-\text{O}^-\text{Na}^+ \text{ (blue)}$$

Phenols reduce potassium permanganate solutions and undergo oxidation to quinones; an excess of the reagent yields a series of oxidation products, including maleic acid, oxalic acid, and carbon dioxide (Experiment 36, p. 279). The manganese is reduced from +7, which gives a purple solution, to +4, which is brown.

$$3\,\underset{\text{phenol}}{\text{C}_6\text{H}_5\text{OH}} + \underset{\text{(purple)}}{4KMnO_4} \longrightarrow 3\,\underset{\text{quinone}}{\text{O}=\text{C}_6\text{H}_4=\text{O}} + \underset{\text{(brown)}}{4MnO_2} + 4KOH + H_2O$$

Phenols can be detected by treatment with ferric chloride (Experiment 43, p. 296). The procedure using pyridine solvent has resulted in accurate results in 90% of the phenolic substrates tested; previous procedures using water or alcohol–water solvents had only a 50% success rate. The color of the solution changes immediately to blue, violet, purple, green, or red-brown.

$$\underset{\text{phenol}}{3ArOH} + 3\,\text{C}_5\text{H}_5\text{N} + FeCl_3 \longrightarrow \underset{\text{(colored complex)}}{Fe(OAr)_3} + 3\,\text{C}_5\text{H}_5\text{N}^+\text{H}\ Cl^-$$

Since the aromatic nucleus of a phenol is substantially more reactive toward electrophilic aromatic substitution than benzene, bromination of phenols should be carried out under mild conditions (Experiment 44, p. 298). The discharge of the bromine color is a positive test.

$$\underset{\text{phenol}}{\text{C}_6\text{H}_5\text{OH}} + 3Br_2 \longrightarrow \underset{\text{2,4,6-tribromophenol}}{\text{C}_6\text{H}_2\text{Br}_3\text{OH}} + 3HBr$$

Experiment 43. Ferric Chloride–Pyridine Reagent

$$3ArOH + 3\,C_5H_5N + FeCl_3 \longrightarrow Fe(OAr)_3 + 3\,C_5H_5NH^+\,Cl^-$$

phenol (colored complex)

This test must be done in the hood. Add 30–50 mg of the solid unknown or 4 or 5 drops of the liquid unknown to 2 mL or pure chloroform in a clean, dry test tube. Stir the solution. If the unknown does not seem to dissolve, even partially, add 2–3 mL more chloroform and warm gently. Cool to 25°C and add 2 drops of 1% solution of anhydrous ferric chloride in chloroform followed by 3 drops of pyridine. Shake the tube and note the color produced *immediately.* A positive test is shown by production of a blue, violet, purple, green, or red-brown solution. Frequently the colors change in a few minutes. Try the test on (1) phenol; (2) 4-cresol; (3) salicylaldehyde; (4) 2-nitrophenol; (5) 3-bromophenol.

1% Ferric Chloride in Chloroform. Pure chloroform, free from ethanol, should be used as a solvent and for preparing the ferric chloride solution. Add 1 g of the black crystals of *anhydrous* ferric chloride to 100 mL of pure chloroform. The mixture is shaken occasionally for about an hour, and allowed to stand to permit the insoluble material to settle. Decant the pale yellow solution into a screw-cap bottle fitted with a medicine dropper.

Cleaning Up. The test solution is placed in the halogenated organic waste container.

Discussion. This reagent is useful for detecting compounds containing a hydroxyl group directly attached to an aromatic nucleus. Treatment of chloroform solutions of phenols, naphthols, and their ring-substituted derivatives with a chloroform solution of anhydrous ferric chloride and pyridine produces a characteristic blue, violet, purple, green, or red-brown complexes.

Alcohols, ethers, aldehydes, acids, ketones, hydrocarbons, and their halogen derivatives give negative results of colorless, pale yellow, or tan solutions.

This method is especially valuable for substituted phenols and naphthols that are very insoluble in water. Even 2,4,6-trichlorophenol, 2,4,6-tribromophenol, nonylphenol, phenolphthalein, and thymolphthalein give positive tests provided sufficient chloroform is used (about 5 mL) to get them into solution.

Phenolic compounds which have failed to give positive tests are picric acid, 2,6-di-*tert*-butylphenol, phenolsulfonic acid, naphtholsulfonic acid, hydroquinone, *dl*-tyrosine, 4-hydroxylphenylglycine, and 4-hydroxybenzoic acid. The 4-hydroxybenzoic acid gives a distinct yellow color, a negative result, whereas salicylic acid gives a violet color, a positive test. The esters of 4-hydroxybenzoic acid produce purple colors and 4-hydroxybenzaldehyde a violet-purple color.

It is of interest that 5,5-dimethyl-1,3-cyclohexanedione (dimedon, methone) gives a beautiful purple color. Resorcinol gives a blue-violet color. Note that several of the tautomeric forms of these compounds are similar in structure to tautomeric forms of phenols.

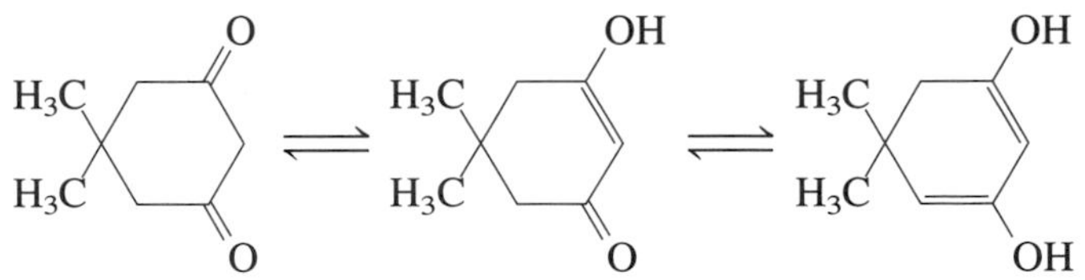

dimedon

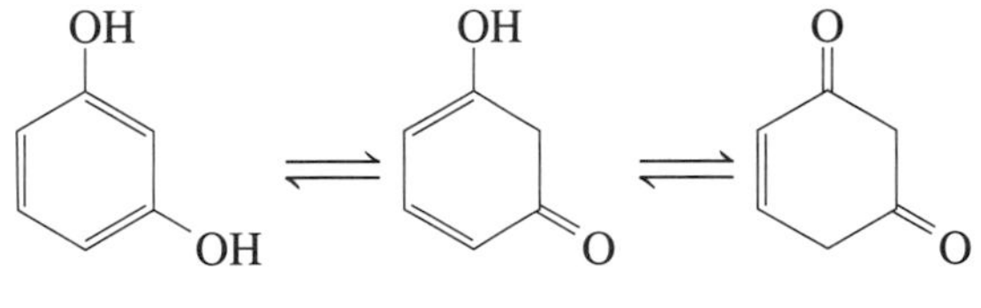

resorcinol

Salicylaldehyde forms a highly colored complex with ferric chloride:

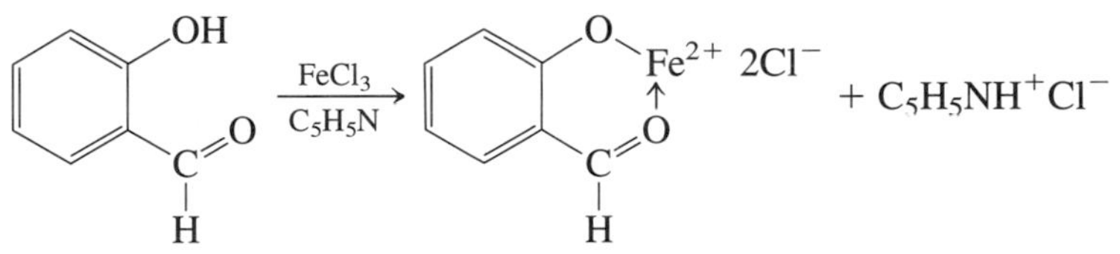

$$\text{salicylaldehyde} \xrightarrow[C_5H_5N]{FeCl_3} \text{(colored complex)}\ Fe^{2+}\ 2Cl^- + C_5H_5NH^+Cl^-$$

salicylaldehyde (colored complex)

In aqueous or aqueous alcoholic solutions, some enols, oximes, and hydroxamic acids produce red-, brown-, or magenta-colored complexes with *aqueous* ferric chloride. In aqueous solutions, aldehydes and ketones with α-hydrogens may tautomerize to the enol form, which will then give a violet-, red-, or tan-colored complex with the ferric chloride.

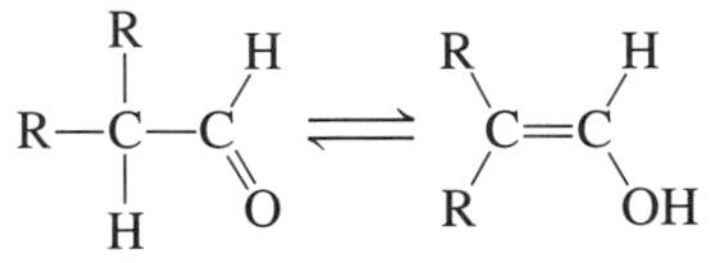

$$R_2CH{-}CH{=}O \rightleftharpoons R_2C{=}CH{-}OH$$

aldo form enol form

$$R_2CH{-}C(R){=}O \rightleftharpoons R_2C{=}C(R){-}OH$$

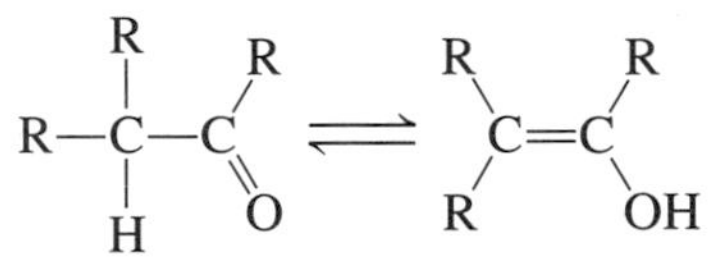

keto form enol form

However, in this anhydrous chloroform test, these compounds give yellow or pale tan solutions quite different from the phenols.

Experiment 44. Bromine Water

phenol + $3Br_2$ ⟶ 2,4,6,-tribromophenol + 3HBr

Dissolve 0.1 g of the unknown in 10 mL of water. Add bromine water drop by drop until the bromine color is no longer discharged. The discharge of the bromine color is a positive test. In some cases, a white precipitate (the brominated phenol) may also be formed.

Try this test on (1) phenol; (2) aniline; (3) salicylic acid; (4) 4-nitrophenol.

Cleaning Up. Place the test solution in the halogenated organic waste container.

Discussion. It has been shown that, in the bromination of benzene and 2-nitroanisole with bromine water, the brominating agent operates by complex mechanisms.

Mercaptans are oxidized readily by bromine water to disulfides.

$$2RSH + Br_2 \longrightarrow RSSR + 2HBr$$

The advantage of bromine in water over bromine in carbon tetrachloride is that the more polar solvent greatly increases the rate of bromination by the ionic mechanism. Of course, it is impossible with this solvent to observe the evolution of hydrogen bromide. An excess of bromine water converts tribromophenol to a yellow tetrabromo derivative, 2,4,4,6-tetrabromocyclohexadienone. The tetrabromo compound is readily converted to the tribromophenol by washing with 2% hydroiodic acid.

2,4,6,-tribromophenol $\underset{2\%\ HI}{\overset{Br_2}{\rightleftharpoons}}$ 2,4,4,6,-tetrabromocyclohexadienone

Questions

26. Will 2,4,6-tribromoaniline give a positive test? Explain your answer.

27. Could the decolorization of the bromine water result from the presence of an inorganic compound?

28. Is bromine hydrolyzed in water?

29. Explain the following order of reactivity toward electrophilic bromine: $C_6H_5O^- > C_6H_5OH >> C_6H_6$.

For more discussion of the reaction of bromine with organic compounds, consult (Experiment 35, p. 276).

7.21 SULFONAMIDES, SULFONIC ACIDS, SULFONYL CHLORIDES

The presence of sulfonamides can be detected by fusing with sodium hydroxide (Experiment 45, p. 300) and testing for the evolution of amine or ammonia and of sulfur dioxide.

$$\underset{\text{sulfonamide}}{ArSO_2NH_2} + 2NaOH \longrightarrow ArONa + NaHSO_3 + \underset{\text{ammonia}}{NH_3(g)}$$

$$NaHSO_3 + HCl \longrightarrow \underset{\text{sulfur dioxide}}{SO_2(g)} + NaCl + H_2O$$

Sulfonic acids are structurally different from sulfuric acid only in that an organic group has been substituted for one hydroxyl group, and thus the high acid strength of sulfonic acids is not surprising. Sulfonic acids and their metal salts are usually soluble in water.

Sulfonyl chlorides and sulfonic acids can be detected through the hydroxamic acid test (Experiment 2e, p. 208). The sulfonyl chloride is produced from the sulfonic acid and thionyl chloride. The sulfonyl chloride is treated with hydroxylamine, which undergoes reaction with acetaldehyde to form the hydroxamic acid. The hydroxamic acid undergoes reaction with ferric chloride to form the ferric hydroxamate complex, which is a burgundy or magenta color.

$$\underset{\text{sulfonic acid}}{ArSO_3H} + \underset{\text{thionyl chloride}}{SOCl_2} \longrightarrow \underset{\text{sulfonyl chloride}}{ArSO_2Cl} + HCl + SO_2$$

$$\underset{\text{sulfonyl chloride}}{ArSO_2Cl} + \underset{\text{hydroxylamine hydrochloride}}{H_2NOH{\bullet}HCl} \longrightarrow ArSO_2NHOH + 2HCl$$

$$ArSO_2NHOH + \underset{\text{acetaldehyde}}{H_3C{-}C({=}O){-}H} \longrightarrow \underset{\text{hydroxamic acid}}{H_3C{-}C({=}O){-}NHOH} + ArSO_2H$$

$$3\ \underset{\text{hydroxamic acid}}{H_3C{-}C({=}O){-}NHOH} + FeCl_3 + 3\ KOH \longrightarrow \underset{\text{ferric hydroxamate complex (burgundy or magenta color)}}{(H_3C{-}C({=}O){-}NHO)_3Fe} + H_2O + 3KCl$$

The presence of the halogen in the sulfonyl halides can be detected by ethanolic silver nitrate solution (Experiment 33, p. 270). The sulfonyl halide is converted to the sulfonic ester and hydrogen halide. The halide ion reacts with the silver cation to form the insoluble silver halide.

$$\underset{\text{sulfonyl halide}}{RSO_2X} + CH_3CH_2OH \longrightarrow \underset{\text{sulfonyl ester}}{RSO_3CH_2CH_3} + HX$$

$$HX + AgNO_3 \longrightarrow AgX(s) + HONO_2$$

The halogen in sulfonyl bromides or chlorides can also be detected by sodium iodide in acetone (Experiment 34, p. 273), with the formation of the solid sodium chloride and the liberation of iodine.

$$ArSO_2X + NaI \longrightarrow ArSO_2I + NaX(s)$$

$$ArSO_2I \xrightarrow{NaI} ArSO_2Na + I_2$$

Experiment 45. Sodium Hydroxide Fusion of Sulfonamides

$$\underset{\text{sulfonamide}}{ArSO_2NH_2} + 2NaOH \longrightarrow ArONa + NaHSO_3 + \underset{\text{ammonia}}{NH_3(g)}$$

$$NaHSO_3 + HCl \longrightarrow \underset{\text{sulfur dioxide}}{SO_2(g)} + NaCl + H_2O$$

In a test tube, fuse 0.25 g of the unknown with 1.5 g of powdered sodium hydroxide by heating with a bunsen burner. The escaping gas should be tested for the presence of ammonia or amines. Place moist pink litmus paper in the test tube, being careful to avoid touching the sides of the test tube with the paper. If ammonia or amine is being evolved, the litmus paper turns blue.

After the tube has cooled, add just enough distilled water to dissolve the sample. The solution is then acidified with 2 M hydrochloric acid. A filter paper that has been covered with a paste of nickel(II) hydroxide is suspended over the test tube. The tube should be gently warmed to speed up the production of sulfur dioxide. If sulfur dioxide is present, the green nickel(II) hydroxide is oxidized to grey-black nickel(IV) oxyhydrate.

A positive test for the presence of a sulfonamide is the evolution both of ammonia or amine and of sulfur dioxide.

Nickel(II) Hydroxide Reagent. The nickel(II) hydroxide reagent is prepared immediately before use by slowly adding 1 M sodium hydroxide to 0.20 g of nickel(II) chloride until no more solid precipitates. The precipitate is washed with 10-mL portions of cold water until the washings are no longer basic. The nickel(II) hydroxide is moistened with water and applied as a paste to a strip of filter paper.

Cleaning Up. Place the solid nickel(II) hydroxide and the filter paper impregnated with the nickel(II) hydroxide solution in the hazardous solid waste container. Place the reaction mixture in the aqueous solution container.

Discussion. The sulfur dioxide undergoes reaction with the green nickel(II) hydroxide to yield black nickel(IV) oxyhydrate.[22]

$$\underset{\substack{\text{nickel(II) hydroxide}\\ \text{(green)}}}{2Ni(OH)_2} + SO_2 \longrightarrow (NiOH)_2SO_3 + H_2O$$

$$(NiOH)_2SO_3 + O_2 \longrightarrow \underset{\substack{\text{nickel(IV) oxyhydrate}\\ \text{(black)}}}{NiO(OH)_2} + NiSO_4$$

[22]F. Feigl, *Spot Tests in Organic Analysis* (Elsevier Scientific Publishing Company, New York, 1966), p. 87.

CHAPTER 8

The Preparation of Derivatives

Derivatization procedures have somewhat diminished in significance with the advent of organic spectrometry. However, these procedures still provide both physical data and an insight to the chemistry of the unknown, especially when the possibilities for the identity of the unknown have similar boiling or melting points and similar spectra. Chemists should also remember that certain "derivatizations" are really "conversions" of one common organic compound into another. Conversion (e.g., oxidizing a secondary alcohol to a ketone) may yield a compound that should also be thoroughly characterized. Indeed, many derivatizations are really syntheses or "preps".

In this edition of the book, care was taken that, for a particular type of compound, both the procedure for preparing derivatives and the melting points of those derivatives are given.

As in Chapter 7, each procedure in this chapter contains a ***Cleaning Up*** procedure. Users of this book may wish to scale down some procedures by multiplying the amount of reagent used by 1/2, 1/5, or 1/10, bearing in mind that these experiments have not been tested in the smaller amounts.

In the derivative tables in Appendix III, compounds were listed only if they had two derivatives or more.

Table 8.1 Index to Derivatization Procedures by Functional Group Class

Compound	Derivative	Procedure Number	Page
Acid anhydride	Acid	2	308
	Amide	3b	309
	Anilide	4c	311
	4-Toluidide	4d	311
Acyl Halide	Acid	2	308
	Amide	3b	309
	Anilide	4d	311
	4-Toluidide	4d	311
Alcohol	Phenylurethane	8	315
	1-Naphthylurethane	8	315
	4-Nitrobenzoate	9	316
	3,5-Dinitrobenzoate	10	317
	Hydrogen 3-nitrophthalate	11	318
Aldehyde	Semicarbazone	12	321
	2,4-Dinitrophenylhydrazone	13	321
	4-Nitrophenylhydrazone	14	322
	Phenylhydrazone	14	322
	Oxime	15	323
	Dimedon derivative	16	323
	Oxidation to an acid	17	324
Amide	9-Acylamidoxanthene	18	328
	Hydrolysis to acid and amine	19	329
	4-Nitrobenzyl ester	19, then 5	329, 312
	4-Bromophenacyl ester	19, then 5	329, 312
	Acetamide	19, then 20a	329, 333
	Benzamide	19, then 20b or 20c	329, 333 or 334
Amine—1′ and 2′	Acetamide	20a	333
	Benzamide	20b or 20c	333 or 334
	Benzenesulfonamide	21	335
	4-Toluenesulfonamide	21	335
	Phenylthiourea	22	336
	Amine hydrochloride	26	338
Amine—3′	Chloroplatinate	24	337
	Methyl 4-toluenesulfonate	23b	337
	Methyl iodide	23a	337
	Picrate acid	25	338
	Amine hydrochloride	26	338
Amino acid	4-Toluenesulfonamide	27	343
	Phenylureido acid	28	343
	Acetamide	20d	335
	Benzamide	20d	335
	3,5-Dinitrobenzamide	29	344
	2,4-Dinitrophenyl derivative	30	344
Carbohydrate	Phenylosazone	31	346
	4-Nitrophenylhydrazone	32	347
	4-Bromophenylhydrazone	32	347
	Acetate	33	347

Compound	Derivative	Procedure Number	Page
Carboxylic acid	Neutralization equivalent	1	307
	4-Toluidide	4a or 4b	310
	Anilide	4a or 4b	310
	4-Nitrobenzyl ester	5	312
	4-Bromophenacyl ester	5	312
	Amide	3a	309
	S-Benzylthiuronium salt	6	313
	Phenylhydrazide	7	313
Ester	Saponification and hydrolysis	34	351
	Saponification equivalent	35	354
	Amide	34, then 3a	351, 309
	4-Toluidide	36	356
		34, then 4a or 4b	351, 310
	3,5-Dinitrobenzoate	37	358
	N-Benzylamide	38	358
	Acid Hydrazide	39	359
Ether—Alkyl	3,5-Dinitrobenzoate	40	359
Ether—Aromatic	Picrate	41	361
	Sulfonamide	42, then 43	362, 363
	Nitro derivative	51	373
	Bromo derivative	44	364
Halide—Alkyl	Anilide	46	366
	N-Naphthylamide	46	366
	Alkylmercuric halide	45	366
	Alkyl 2-naphthyl ether	47	367
	Alkyl 2-naphthyl ether picrate	48	368
	S-Alkylthiuronium picrate	49	368
Halide—Aromatic	Nitration	51	373
	Sulfonamide	42, then 43b	362, 363
	Oxidation	50	370
Hydrocarbon—Aromatic	Nitration	51	373
	Aroylbenzoic acid	52	374
	Picrate	41	361
Ketone	Semicarbazone	12	321
	2,4-Dinitrophenylhydrazone	13	321
	4-Nitrophenylhydrazone	14	322
	Phenylhydrazone	14	322
	Oxime	15	323
Nitrile	Hydrolysis of nitrile	53	377
	Amide	53, then 3; 54	377, 309 378
	Anilide	53, then 4a or 4b	377, 310
	Reduction of nitrile	55	378
	Benzamide	55, then 20b or 20c	378, 333 or 334
	Benzenesulfonamide	55, then 21	378, 335
	Phenylthiourea	55, then 22	378, 336
	a-(Imidioylthio)acetic acid hydrochloride	56	379

Compound	Derivative	Procedure Number	Page
Nitro compound	Reduction to amine	57	381
	Acetamide	57, then 20a	381, 333
	Benzamide	57, then 20b or 20c	381, 333 or 334
	Benzenesulfonamide	57, then 21	381, 335
Phenol	Phenylurethane	8; 58	315 383
	1-Naphthylurethane	8; 58	315 383
	4-Nitrobenzoate	9	316
	3,5-Dinitrobenzoate	10	317
	Acetate	59	383
	Benzoate	9	316
	Aryloxyacetic acid	60	384
	Bromo derivative	61	384
Sulfonamide	Sulfonic acid	64	388
	Sulfonyl chloride	64, then 62	388, 387
	Sulfanilide	64, then 62, then 63	388, 387, 387
	N-Xanthylsulfonamide	65	389
Sulfonic acid	Sulfonyl chloride	62	387
	Sulfonamide	62, then 43	387, 363
	Sulfanilide	62, then 63	387, 387
	Benzylthiuronium sulfonate	66	389
	4-Toluidine salt	67	390
Sulfonyl chloride	Sulfonic acid	64	388
	Sulfonamide	43	363
	Sulfanilide	63	387

8.1 CARBOXYLIC ACIDS, ACID ANHYDRIDES, ACID HALIDES

One of the most useful pieces of information about a carboxylic acid is its *neutralization equivalent* (NE), which is obtained by quantitative titration with a standardized base (Procedure 1, p. 307). The molecular weight of the acid, within experimental error, is an integral (usually 1, 2, 3, etc.) multiple of the neutralization equivalent.

Acid anhydrides, acyl halides, and carboxylic acids can be derivatized by using some of the same procedures. Acid anhydrides and acyl halides can be hydrolyzed to the carboxylic acids or the sodium salts of the acid (Procedure 2, p. 308). Symmetrical anhydrides ($\mathrm{R\underset{\underset{O}{\|}}{C}O\underset{\underset{O}{\|}}{C}R}$) lead to just one kind of carboxylic acid. Unsymmetrical anhydrides ($\mathrm{R\underset{\underset{O}{\|}}{C}O\underset{\underset{O}{\|}}{C}R'}$) are far more difficult to characterize. If the acid is a solid, it frequently will serve as a derivative of anhydrides and acyl halides. Otherwise, the mixture of the sodium salt of the acid and sodium chloride obtained from the basic hydrolysis of the acyl halide may be used for preparing other solid derivatives such as amides (Procedure 4b, p. 310).

$$\text{RC(=O)OC(=O)R} + H_2O \xrightarrow{2NaOH} 2\ \text{RC(=O)O}^-Na^+$$

acid anhydride — sodium salt of the acid

$$\text{RC(=O)X} + H_2O \xrightarrow{NaOH} \text{RC(=O)O}^-Na^+ \xrightarrow{H^+} \text{RC(=O)OH}$$

acyl halide — sodium salt of the acid — carboxylic acid

Acids may be converted to amides by reaction with thionyl chloride, followed by treatment of the intermediate acyl chloride with concentrated aqueous ammonia. Acid anhydrides undergo reaction with aqueous ammonia to yield amides and carboxylic acids (Procedure 3, p. 309).

$$\text{RC(=O)OH} + SOCl_2 \longrightarrow \text{RC(=O)Cl} + SO_2 + HCl$$

carboxylic acid — thionyl chloride — acyl chloride

$$\text{RC(=O)Cl} + 2NH_3 \longrightarrow \text{RC(=O)NH}_2 + NH_4Cl$$

acyl chloride — amide

$$\text{RC(=O)OC(=O)R} + NH_3 \longrightarrow \text{RC(=O)NH}_2 + \text{RC(=O)OH}$$

acid anhydride — amide — carboxylic acid

This method is particularly suitable if the amide is insoluble in water.

Anilides and 4-toluidides are excellent derivatives because of the ease with which they may be made and purified. They may be prepared from the free acid, the salt of the acid, the acyl halide, or the anhydride (Procedure 4, p. 310).

$$\text{RC(=O)OH} \xrightarrow{SOCl_2} \text{RC(=O)Cl} \xrightarrow{ArNH_2} \text{RC(=O)NHAr}$$

carboxylic acid — acyl chloride — anilide

$$\text{RC(=O)OH} \xrightarrow{NaOH} \text{RC(=O)O}^-Na^+ \xrightarrow[ArNH_2]{HCl} \text{RC(=O)NHAr}$$

carboxylic acid — sodium salt of the acid — anilide

$$RC(=O)X + ArNH_2 \longrightarrow RC(=O)NHAr + ArNH_3X$$

acyl halide — aniline or 4-toluidine — anilide

$$RC(=O)OC(=O)R + 2ArNH_2 \longrightarrow RC(=O)NHAr + RC(=O)O^-\,ArNH_3^+$$

acid anhydride — aniline or 4-toluidine — anilide

Solid esters furnish a useful means for characterizing the acids. Some methyl esters are solid, but in most cases the 4-nitrobenzyl or the 4-bromophenacyl esters are preferred. These are prepared by treating the salts of the acids with either 4-nitrobenzyl chloride or 4-bromophenacyl bromide (Procedure 5, p. 312).

$$RC(=O)O^-Na^+ + ClCH_2-C_6H_4-NO_2 \longrightarrow RC(=O)O-CH_2-C_6H_4-NO_2 + NaCl$$

salt of the carboxylic acid — 4-nitrobenzyl chloride — 4-nitrobenzyl ester

$$RC(=O)O^-Na^+ + BrCH_2-C(=O)-C_6H_4-NO_2 \longrightarrow RC(=O)O-CH_2-C(=O)-C_6H_4-NO_2 + NaBr$$

salt of the carboxylic acid — 4-bromophenacyl bromide — 4-bromophenacyl ester

This method is particularly advantageous because it does not require an anhydrous sample of the acid.

S-Benzylthiuronium halide undergoes reaction with the salt of the carboxylic acid to yield the corresponding *S*-benzylthiuronium salt (Procedure 6, p. 313).

$$\mathrm{RC(=O)O^-Na^+} + \mathrm{C_6H_5CH_2SC(NH_2)_2{}^+X^-} \longrightarrow$$

salt of the acid   *S*-benzylthiuronium salt

$$\mathrm{C_6H_5CH_2SC(NH_2)_2{}^+\ {}^-OC(=O)R} + \mathrm{NaX}$$

S-benzylthiuronium salt

The reaction products of phenylhydrazine and acids are also good derivatives. At the boiling point (243°C) of phenylhydrazine, simple unsubstituted aliphatic mono- and dibasic acids form phenylhydrazides (Procedure 7, p. 313).

$$\mathrm{RC(=O)OH} + \mathrm{C_6H_5NHNH_2} \longrightarrow \mathrm{RC(=O)NHNHC_6H_5} + \mathrm{H_2O}$$

carboxylic acid   phenylhydrazine   phenylhydrazide

Procedure 1. Neutralization Equivalents (NE) of Carboxylic Acids

$$\mathrm{RC(=O)OH} + \mathrm{NaOH} \longrightarrow \mathrm{RC(=O)O^-Na^+} + \mathrm{H_2O}$$

carboxylic acid   salt of the acid

A sample of the acid (about 0.2 g) is weighed to at least three figures on an analytical balance and dissolved in 50–100 mL of water or ethanol. The mixture may be heated if necessary to dissolve all the compound. This solution is titrated with a previously standardized sodium hydroxide solution having a molarity of about 0.1, determined to three significant figures, phenolphthalein or bromothymol blue being used as an indicator. The neutralization equivalent of the acid is calculated according to the formula

$$\text{neutralization equivalent} = \frac{\text{weight of the sample} \times 1000}{\text{volume of alkali (mL)} \times \text{M}}$$

Cleaning Up. Place the mixture in the aqueous solution container.

Discussion. The molecular weight (MW) of an acid may be determined from the neutralization equivalent (NE) by multiplying that value by the number of acidic groups (x) in the molecule:

$$\text{MW} = x(\text{NE})$$

The change in medium, even from pure water to pure ethanol, affects the pK of both the organic acid and the indicator. For this reason best results are obtained in water or aqueous ethanol with only enough ethanol to dissolve the organic acid. In absolute or 95% ethanol it is often impossible to obtain a sharp end point with phenolphthalein. In such cases bromothymol blue should be employed as the indicator. Acids may also be titrated in a solvent composed of ethanol and benzene or toluene.

A blank must always be run on the solvent; the same amount of phenolphthalein that was used in the titration must be employed in the blank. In ordinary work the neutralization equivalents should agree with the calculated values within $\pm 1\%$. By using carefully purified and dried samples and good technique, the error may be reduced to $\pm 0.3\%$.

In order to give an accurate neutralization equivalent, the substance titrated must be pure and anhydrous. If the value obtained for the neutralization equivalent does not agree with the theoretical value, the compound should be recrystallized from a suitable solvent and carefully dried.

Amine salts of strong acids may be titrated by the same procedure.

Questions

1. Calculate the neutralization equivalent of benzoic acid and of phthalic acid.
2. If an acid is not completely dried, will the neutralization equivalent be high or low?
3. Would the presence of an aromatic amino group interfere in the determination of the neutralization equivalent? What would be the effect of an aliphatic amino group?
4. What types of phenols may be titrated quantitatively?[1]
5. From a theoretical point of view, what should the ionization constant of an acid be in order that the acid may be titrated with phenolphthalein?[1]

Procedure 2. Hydrolysis of Acid Anhydrides and Acyl Halides to Carboxylic Acids

$$\underset{\text{acid anhydride}}{RC(=O)OC(=O)R} + H_2O \xrightarrow{2NaOH} 2\,\underset{\text{sodium salt of the acid}}{RC(=O)O^-Na^+} \xrightarrow{H^+} 2\,\underset{\text{carboxylic acid}}{RC(=O)OH}$$

$$\underset{\text{acyl halide}}{RC(=O)X} + H_2O \xrightarrow{NaOH} \underset{\text{sodium salt of the acid}}{RC(=O)O^-Na^+} \xrightarrow{H^+} \underset{\text{carboxylic acid}}{RC(=O)OH}$$

One gram of the acid anhydride or acyl halide is added to 5 mL of water in a small flask and hydrolysis is induced by slowly adding 10% sodium hydroxide solution until the solution is alkaline to litmus. The flask is gently warmed for a few minutes. The resulting solution is then acidified with 10% hydrochloric acid until it is acidic. Insoluble carboxylic acids can then be removed by filtration. If no solid is obtained, the solution is

[1] An analytical book such as those cited in Chapter 11 should be consulted.

then neutralized and evaporated to dryness. The resulting mixture of the sodium salt of the acid and sodium chloride may be used for preparing anilides, 4-toluidides (Procedure 4, p. 310), or esters (Procedure 5, p. 312).

Cleaning Up. Neutralize the filtrate with 10% sodium hydroxide and place in the aqueous solution container.

Procedure 3. Preparation of Amides

(a) From the Carboxylic Acid

$$\underset{\text{carboxylic acid}}{RC(=O)OH} + \underset{\text{thionyl chloride}}{SOCl_2} \longrightarrow \underset{\text{acyl chloride}}{RC(=O)Cl} + SO_2 + HCl$$

$$\underset{\text{acyl chloride}}{RC(=O)Cl} + 2NH_3 \longrightarrow \underset{\text{amide}}{RC(=O)NH_2} + NH_4Cl$$

$$\underset{\text{acyl chloride}}{RC(=O)Cl} + \underset{\text{1° amine}}{2R'NH_2} \longrightarrow \underset{\text{2° amide}}{RC(=O)NHR'} + R'NH_3Cl$$

One gram of the acid is heated under reflux with 5 mL of thionyl chloride for 15–30 min. The mixture is poured cautiously into 15 mL of ice-cold, concentrated aqueous ammonia. The solution is extracted three 10-mL portions of methylene chloride. The methylene chloride layer is isolated and evaporated to dryness. The crude amide is purified by recrystallization from ethanol.

(b) From the Acyl Halide or Acid Anhydride

$$\underset{\text{acyl chloride}}{RC(=O)Cl} + 2NH_3 \longrightarrow \underset{\text{amide}}{RC(=O)NH_2} + NH_4Cl$$

$$\underset{\text{acid anhydride}}{RC(=O)OC(=O)R} + NH_3 \longrightarrow \underset{\text{amide}}{RC(=O)NH_2} + \underset{\text{carboxylic acid}}{RC(=O)OH}$$

One gram of the acyl halide or acid anhydride is poured into 15 mL of concentrated aqueous ammonia. The solution is extracted with three 10-mL portions of methylene chloride. The methylene chloride layer is isolated and evaporated to dryness. The crude amide is purified by recrystallization from ethanol.

Cleaning Up. Neutralize the filtrate with 10% hydrochloric acid and place in the aqueous solution container.

Procedure 4. Anilides and 4-Toluidides

(a) From the Carboxylic Acid

$$\underset{\text{carboxylic acid}}{\mathrm{R{-}C({=}O){-}OH}} \xrightarrow{SOCl_2} \underset{\text{acyl chloride}}{\mathrm{R{-}C({=}O){-}Cl}} \xrightarrow{ArNH_2} \underset{\text{anilide}}{\mathrm{R{-}C({=}O){-}NHAr}}$$

One gram of the acid is mixed with 2 mL of thionyl chloride, and the mixture is heated at reflux for 30 min. The mixture is cooled, and a solution of 1–2 g of aniline or 4-toluidine in 30 mL of benzene[2] is added, and the mixture is heated on a steam bath for 2 min. **Caution:** *Benzene is a known carcinogen. Use benzene in the hood, do not breathe the vapors, and avoid contact with the skin.* The benzene solution is poured into a separatory funnel and washed successively with 2 mL of water, 5 mL of 5% hydrochloric acid, 5 mL of 5% sodium hydroxide solution, and 2 mL of water. In the hood, benzene is removed by distillation using a steam bath and the resulting amide is recrystallized from water or ethanol.

(b) From the Sodium Salt of the Carboxylic Acid

$$\underset{\text{sodium salt of the acid}}{\mathrm{R{-}C({=}O){-}O^{-}Na^{+}}} \xrightarrow[ArNH_2]{HCl} \underset{\text{anilide}}{\mathrm{R{-}C({=}O){-}NHAr}}$$

A mixture of 0.4 g of the dry powdered sodium salt of the acid, 0.5 mL of aniline or 4-toluidine, and 0.15 mL of concentrated hydrochloric acid is placed in a test tube. The test tube is placed in an oil bath which is heated, the temperature being kept between 150° and 160°C for 45–60 min. The test tube is then removed and the product purified by one of the following methods.

1. If the acid under consideration has fewer than six carbon atoms, 5 mL of 95% ethanol is added. The solution is heated to boiling and decanted into 15 mL of hot water. The

[2]Benzene is highly toxic and should only be used with the instructor's permission. Other solvents such as toluene or tetrachloroethane may be substituted for benzene.

resulting solution is evaporated to a volume of 10–12 mL and cooled in an ice bath. The crystals are removed by filtration and recrystallized from a small amount of water or dilute ethanol.

2. If the acid contains six or more carbon atoms, the crude reaction product is powdered and washed with 15 mL of 5% hydrochloric acid and then with 15 mL of cold water. Thirty milliliters of 95% ethanol is added, the solution is heated to boiling, and then is filtered. The filtrate is chilled in an ice bath, and the crystals of the amide are removed by filtration. The product is then recrystallized from aqueous ethanol.

(c) From the Anhydride

$$\underset{\text{acid anhydride}}{RC(=O)OC(=O)R} + \underset{\text{aniline}}{C_6H_5NH_2} \longrightarrow \underset{\text{anilide}}{RC(=O)NHC_6H_5} + RC(=O)OH$$

One gram of the anhydride is mixed with 1 g of aniline, and the mixture is heated in a boiling water bath for 5 min. Next, 5 mL of water is added, and the solution is heated to boiling, then cooled. The anilide is recrystallized from ethanol or aqueous ethanol.

(d) From the Acyl Anhydride or the Acid Halide

$$\underset{\text{acyl halide}}{RC(=O)X} + \underset{\text{aniline or 4-toluidine}}{ArNH_2} \longrightarrow \underset{\text{anilide}}{RC(=O)NHAr} + ArNH_3X$$

$$\underset{\text{acid anhydride}}{RC(=O)OC(=O)R} + \underset{\text{aniline or 4-toluidine}}{2ArNH_2} \longrightarrow \underset{\text{anilide}}{RC(=O)NHAr} + RC(=O)O^-\,ArNH_3^+$$

One gram of the acid anhydride or acyl halide may be mixed with the aniline or 4-toluidine and the remainder of procedure (a) followed.

Cleaning Up. Place benzene in the hazardous waste container for benzene and any unreacted amine in the aromatic organic solvent container. Neutralize the aqueous filtrate, if needed, with sodium carbonate, and place in the aqueous solution container.

Procedure 5. 4-Nitrobenzyl and 4-Bromophenacyl esters from Carboxylic Acids

$$R{-}C({=}O){-}O^-Na^+ \; + \; ClCH_2{-}C_6H_4{-}NO_2 \longrightarrow R{-}C({=}O){-}O{-}CH_2{-}C_6H_4{-}NO_2 + NaCl$$

salt of the carboxylic acid; 4-nitrobenzyl chloride; 4-nitrobenzyl ester

$$R{-}C({=}O){-}O^-Na^+ \; + \; BrCH_2{-}C({=}O){-}C_6H_4{-}NO_2 \longrightarrow R{-}C({=}O){-}O{-}CH_2{-}C({=}O){-}C_6H_4{-}NO_2 + NaBr$$

salt of the carboxylic acid; 4-bromophenacyl bromide; 4-bromophenacyl ester

One gram of the acid is added to 5 mL of water in a small flask and is neutralized carefully with 10% sodium hydroxide solution. A little of the acid is then added until the solution is just acid to litmus. If the original acid is obtained as a sodium salt, 1 g of the salt is dissolved in 5–10 mL of water. If this solution is alkaline, a drop or two of 10% hydrochloric acid is added. Ten milliliters of ethanol and 1 g of 4-nitrobenzyl chloride or 4-bromophenacyl bromide are added. ***Caution:*** *Phenacyl halides are lachrymators.* The mixture is then heated under reflux for 1 hr if the acid is monobasic, 2 hr if dibasic, and 3 hr if tribasic. Occasionally the addition of a few more milliliters of ethanol may be necessary if a solid separates during reflux. The solution is allowed to cool, and the precipitated ester is purified by recrystallization from ethanol.

Cleaning Up. Place any unreacted halide in the halogenated organic waste container. The aqueous filtrate is placed in the aqueous solution container.

Discussion. In preparing derivatives with these reagents, care must be taken that the original reaction mixture is not alkaline. Alkalies cause hydrolysis of the phenacyl halides to phenacyl alcohols. In addition, 4-bromophenacyl bromide should not be used if considerable amounts of sodium chloride are present in the sodium salt of the acid.

Crown ethers can be used to improve the ease with which esters are formed. For example, potassium acetate will convert 1-bromoheptane to heptyl acetate in near quantitative yield.[3]

Procedure 6. S-Benzylthiuronium Salts of Carboxylic Acids

$$RC(=O)O^-Na^+ + C_6H_5CH_2SC(NH_2)_2{}^+X^- \longrightarrow C_6H_5CH_2SC(NH_2)_2{}^+ \; {}^-OC(=O)R + NaX$$

salt of the acid — *S*-benzylthiuronium salt — *S*-benzylthiuronium salt

About 0.3 g of the acid (or 0.5 g of the salt) is added to 3 or 4 mL of water, a drop of phenolphthalein indicator solution is added, and the solution is neutralized by the dropwise addition of 5% sodium hydroxide solution. An excess of alkali must be avoided. If too much is used, dilute hydrochloric acid is added until the solution is just a pale pink. To this aqueous solution of the sodium salt is added a hot solution of 1 g of the benzylthiuronium chloride or bromide (prepared according to Procedure 49, p. 368, using benzyl chloride or benzyl bromide) in 10 mL of 95% ethanol. The mixture is cooled, and the salt is collected on a filter. A few salts (e.g., from formic acid) fail to precipitate, and part of the ethanol must be evaporated to obtain the salt.

The thiuronium salts of organic acids separate in a state of high purity and usually do not require crystallization. If necessary they may be recrystallized from a small amount of dioxane.

Cleaning Up. The aqueous filtrate is placed in the aqueous solution container.

Procedure 7. Phenylhydrazine and Phenylhydrazonium Salts from Carboxylic Acids

$$RC(=O)OH + C_6H_5NHNH_2 \longrightarrow RC(=O)NHNHC_6H_5 + H_2O$$

carboxylic acid — phenylhydrazine — phenylhydrazide

(a) With No Solvent. One gram of the acid is dissolved in 2 mL of phenylhydrazine, and the solution is refluxed gently for 30 min. The crystalline product, which separates when the solution cools, is isolated by filtration and washed with small quantities of toluene or ether until the crystals are white. When a large excess of phenylhydrazine is

[3]G. Gokel and H. Durst, *Synthesis,* 178 (1976).

used, it is sometimes necessary to dilute the mixture with toluene in order to bring about precipitation of the product. The derivatives of the lower monobasic acids are recrystallized from hot toluene, whereas those of the higher acids and dibasic acids are best recrystallized from ethanol or ethanol–water mixtures. The derivatives obtained from dibasic acids by this method are bis-β-phenylhydrazides.

(b) Using Toluene as a Solvent. One gram of the acid is mixed with 2 mL of phenylhydrazine dissolved in 5 mL of toluene. Sometimes a white solid precipitates immediately; it is recrystallized from ethanol. If no solid separates, the mixture is heated under reflux for 30 min, and the product that precipitates upon cooling is collected on a filter, washed with ether, and recrystallized from toluene or ethanol. Sulfonic acids, halogen-substituted aliphatic acids, and aliphatic dibasic acids yield salts by this procedure, whereas simple unsubstituted aliphatic acids give phenylhydrazides.

Cleaning Up. The initial filtrates, any unreacted phenylhydrazine, and toluene filtrates are placed in the aromatic organic solvent container; the ether filtrate is placed in the organic solvent container. The ethanol filtrate is diluted with 10 mL of water and placed in the aqueous solution container.

8.2 ALCOHOLS

The most general derivatives of primary and secondary alcohols are the phenylurethanes and 1-naphthylurethanes. Urethane derivatives are prepared when the alcohol is treated with either phenyl isocyanate or naphthyl isocyanate (Procedure 8, p. 315).

$$\underset{\text{alcohol}}{ROH} + \underset{\text{isocyanate}}{ArN{=}C{=}O} \longrightarrow \underset{\text{urethane}}{ArNH{-}C({=}O){-}OR}$$

The presence of water as an impurity in the alcohol causes difficulty in obtaining the urethane. Water hydrolyzes the isocyanates to give arylamines, which combine with the excess reagent to produce disubstituted ureas.

$$\underset{\text{isocyanate}}{ArN{=}C{=}O} + H_2O \longrightarrow ArNH{-}C({=}O){-}OH \longrightarrow \underset{\text{arylamine}}{ArNH_2} + CO_2$$

$$\underset{\text{arylamine}}{ArNH_2} + \underset{\text{isocyanate}}{ArN{=}C{=}O} \longrightarrow \underset{\text{disubstituted urea}}{ArNH{-}C({=}O){-}NHAr}$$

The ureas are higher melting and less soluble than the corresponding urethanes; and ureas, even in small amounts, make the isolation and purification of the urethanes a matter of considerable difficulty. For this reason, this procedure is most useful for alcohols which are insoluble in water and, therefore, easily obtained in anhydrous conditions.

Urethanes can be obtained from tertiary alcohols only with great difficulty. The isocyanates cause dehydration to occur with the formation of the alkene and diarylurea.

4-Nitrobenzoates (Procedure 9, p. 316) or 3,5-dinitrobenzoates (Procedure 10, p. 317) are easily prepared from the reaction of the alcohol with the corresponding acyl halide.

$$\text{ROH} + \text{ArC(=O)Cl} \longrightarrow \text{ArC(=O)OR}$$

alcohol acyl halide benzoate

For water soluble alcohols, which are likely to contain traces of moisture, the 3,5-dinitrobenzoates are generally more satisfactory as derivatives than the urethanes.

The reaction of alcohols with 3-nitrophthalic anhydride produces hydrogen 3-nitrophthalate derivatives (Procedure 11, p. 318).

ROH + [3-nitrophthalic anhydride, NO_2] ⟶ [ring with C(=O)OH, C(=O)OR, NO_2]

alcohol 3-nitrophthalic anhydride hydrogen 3-nitronaphthalate

Procedure 8. Phenylurethanes and 1-Naphthylurethanes of Alcohols and Phenols

$$\text{ROH} + \text{C}_6\text{H}_5\text{N=C=O} \longrightarrow \text{C}_6\text{H}_5\text{NHC(=O)OR}$$

alcohol or phenol phenyl isocyanate phenylurethane

$$\text{ROH} + \text{C}_{10}\text{H}_7\text{N=C=O} \longrightarrow \text{C}_{10}\text{H}_7\text{NHC(=O)OR}$$

alcohol or phenol 1-naphthyl isocyanate 1-naphthylurethane

One gram of the anhydrous alcohol or phenol is placed in a test tube, and 0.5 mL of phenyl isocyanate or 1-naphthyl isocyanate is added. ***Caution:*** *The isocyanates are lachrymators.* If the reactant is a phenol, the reaction should be catalyzed by the addition of 2 or 3 drops of anhydrous pyridine or triethylamine. If a spontaneous reaction does not take place, the solution is stoppered loosely and warmed on a steam bath for 5 min. It is then cooled in a beaker of ice, and the sides of the tube are scratched with a glass rod to induce crystallization. The urethane is purified by dissolving it in 5 mL of petroleum ether or carbon tetrachloride (*toxic*), filtering the hot solution to remove the unwanted urea by-product, and cooling the filtrate in an ice bath. The crystals which are formed in the filtrate are collected by filtration.

Cleaning Up. Treat any unreacted phenyl isocyanate with an excess of 5.25% sodium hypochlorite (household bleach), dilute with 10 mL of water, and place in the aqueous solution container. The petroleum ether filtrate is placed in the organic solvent waste container. The carbon tetrachloride filtrate is placed in the halogenated organic waste container. The urea by-product is placed in the hazardous waste container.

Procedure 9. 4-Nitrobenzoates of Alcohols and Phenols; Benzoates of Phenols

$$ROH + 4\text{-}O_2NC_6H_4COCl \longrightarrow 4\text{-}O_2NC_6H_4COOR + HCl$$

alcohol or phenol — 4-nitrobenzoyl chloride — 4-nitrobenzoate

$$ArOH + C_6H_5COCl \longrightarrow C_6H_5COOAr + HCl$$

phenol — benzoyl chloride — benzoate

(a) With Pyridine. One milliliter or 1 g of the alcohol or phenol is dissolved in 3 mL of anhydrous pyridine, and 0.5 g of 4-nitrobenzoyl or benzoyl chloride is added. After the initial reaction has subsided, the mixture is warmed over low heat for a minute and poured, with vigorous stirring, into 10 mL of water. The precipitate is allowed to settle, and the supernatant liquid is decanted. The residue is stirred with 5 mL of 5% sodium carbonate solution, removed by filtration, and purified by recrystallization from ethanol.

(b) Without Pyridine. One milliliter or 1 g of the alcohol or phenol, mixed with 0.5 g of 4-nitrobenzoyl or benzoyl chloride, is heated to boiling over low heat. The mixture is poured into water and purified as in (a).

Cleaning Up. Place the organic layer in the organic solvent container. Place the aqueous layer in the aqueous solution container.

Procedure 10. 3,5-Dinitrobenzoates of Alcohols and Phenols

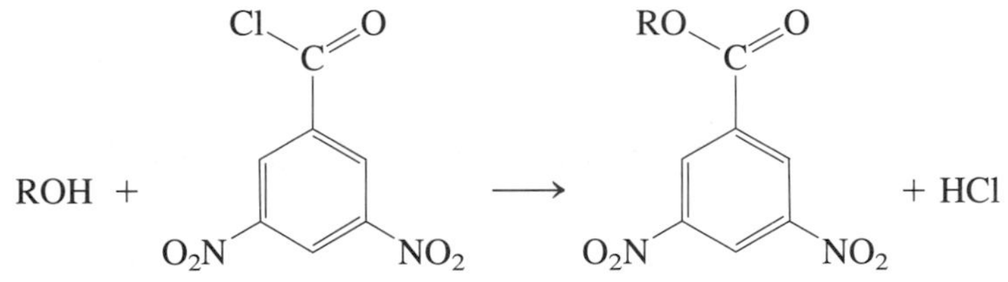

$$ROH + 3,5\text{-}(O_2N)_2C_6H_3COCl \longrightarrow 3,5\text{-}(O_2N)_2C_6H_3COOR + HCl$$

alcohol or phenol — 3,5-dinitrobenzoyl chloride — 3,5-dinitrobenzoate

(a) With Pyridine. About 0.5 g of 3,5-dinitrobenzoyl chloride is mixed with 1 mL or 0.8 g of the alcohol or phenol in a test tube and the mixture boiled gently for 5 min. Then 10 mL of distilled water is added and the solution cooled in an ice bath until the product solidifies. The precipitate is collected on a filter, washed with 10 mL of 2% sodium carbonate solution, and recrystallized from 5–10 mL of a mixture of ethyl alcohol and water of such composition that the ester will dissolve in the hot solution but will separate when the solution is cooled. The crystals are isolated by filtration and dried.

If 3,5-dinitrobenzoyl chloride is not available, it may be made by mixing 0.5 g of 3,5-dinitrobenzoic acid with 1 g of phosphorus pentachloride in a test tube. In the hood, the mixture is warmed gently to start the reaction. After the initial rapid reaction has subsided, the mixture is heated for about 4 min at such a rate as to cause vigorous bubbling. While still liquid the mixture is poured onto a watch glass and allowed to solidify. The solid material is isolated and used immediately for the preparation of the derivative as described above.

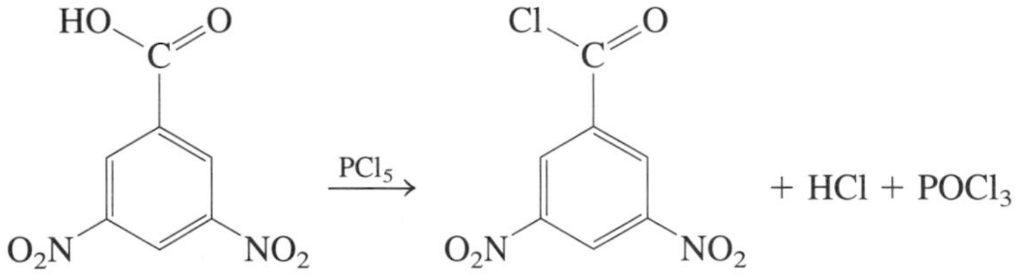

$$3,5\text{-}(O_2N)_2C_6H_3COOH \xrightarrow{PCl_5} 3,5\text{-}(O_2N)_2C_6H_3COCl + HCl + POCl_3$$

3,5-dinitrobenzoic acid — 3,5-dinitrobenzoyl chloride

(b) With Pyridine and Tosyl Chloride.[4] In a small flask, 95 mg of 4-toluenesulfonyl chloride (tosyl chloride) is added to a mixture of 106 mg of 2,4-dinitrobenzoic acid dissolved in 0.5 mL of dry pyridine. The mixture is stirred vigorously and placed in an ice bath. Once the mixture is cold, 100 mg or 1 mL of the alcohol or phenol is added and the solution is stirred vigorously. The mixture is allowed to cool in the ice bath for 10 min.

The alcohol derivatives are precipitated by the addition of 2 mL of water. The product is isolated by suction filtration, washed with 1 mL of cold water, and recrystallized from ethanol.

[4]R. F. Smith and G. M. Cristalli, *J. Chem. Educ.*, 72, A164 (1995).

The phenol derivatives are precipitated by the addition of 2 mL of 1 M sodium hydroxide solution. The derivative is isolated by vacuum filtration and washed with 1 mL of cold water. The derivative is recrystallized by placing the product in 1 mL of boiling water, adding enough *N,N*-dimethylformamide until the solid is dissolved. After dissolution occurs, the solution is chilled to reprecipitate the crystals, and the crystals are isolated by suction filtration.

Cleaning Up. Place the filtrate in the aqueous solution container.

Procedure 11. Hydrogen 3-Nitrophthalates of Alcohols

ROH + [3-nitrophthalic anhydride, NO_2] ⟶ [hydrogen 3-nitrophthalate: COOH, COOR, NO_2]

alcohol 3-nitrophthalic anhydride hydrogen 3-nitronaphthalate

An extremely useful application of this derivative would be the determination of the neutralization equivalent (Procedure 1, p. 307) of such an acid ester; this could very possibly lead to an estimate of the molecular weight of the alcohol.

(a) From Alcohols with Boiling Points Below 150°C. A mixture of 0.4 g of 3-nitrophthalic anhydride and 0.5 mL or 0.4 g of the alcohol is heated gently in a test tube fitted with a glass tube for a condenser. The heating is continued for 5–10 min after the mixture liquefies. The mixture is cooled, diluted with 5 mL of water, and heated to boiling. If solution is not complete, another 5–10 mL of hot water is added. The solution is cooled and the ester allowed to crystallize. Sometimes the derivative separates as an oil and must be allowed to stand overnight to crystallize. The product is recrystallized once or twice from hot water.

(b) From Alcohols with Boiling Points Above 150°C. A mixture of 0.4 g of 3-nitrophthalic anhydride, 0.5 g of the alcohol, and 5 mL of dry toluene is heated until all the anhydride has dissolved, and then heated for 15 min more. The toluene is then decanted off, and the residue is extracted twice with 5 mL of hot water. The residual oil is dissolved in 10 mL of 95% ethanol, and the solution is heated to boiling. If the hot solution is not clear, it should be filtered. Water is added to the hot solution until a turbidity is produced which is cleared up with a drop or two of ethanol. The solution is allowed to cool slowly; sometimes several days are needed for the solution to solidify. The crystals are isolated by filtration.

Cleaning Up. Place the toluene in the aromatic organic solvent container. Neutralize the aqueous filtrate with sodium carbonate and place in the aqueous solution container.

Discussion. Many of the higher alkyl 3-nitrophthalates derived from the monoalkyl ethers of ethylene glycol and diethylene glycol separate as oils and must be allowed to stand several days to solidify. Occasionally toluene may be substituted for the ethanol–water mixture for recrystallization. It is sometimes useful to determine the neutralization equivalent of the alkyl acid phthalate as well as the melting point.

8.3 ALDEHYDES AND KETONES

For low molecular weight, water soluble aldehydes and ketones, it is often advantageous to prepare the semicarbazone from the reaction of the aldehyde or ketone with semicarbazide hydrochloride (Procedure 12, p. 321). All semicarbazones are solids and generally can be obtained nearly pure without recrystallization. Sometimes these derivatives form slowly, and care must be taken to allow sufficient time for the reaction to go to completion.

$$\mathrm{R_2C{=}O} + {}^{-}\mathrm{Cl}^{+}\mathrm{H_3N{-}NH{-}C({=}O){-}NH_2} \xrightarrow{\mathrm{H_3C{-}C({=}O){-}O^{-}Na^{+}}} \mathrm{R_2C{=}N{-}NH{-}C({=}O){-}NH_2} + \mathrm{H_3C{-}C({=}O){-}OH} + H_2O + NaCl$$

aldehyde or ketone; semicarbazide hydrochloride; semicarbazone

The most useful derivatives of aldehydes are the 2,4-dinitrophenylhydrazones (Procedure 13, p. 321), phenylhydrazones (Procedure 14, p. 322), and 4-nitrophenylhydrazones (Procedure 14, p. 322). Of these, the 2,4-dinitrophenylhydrazones are recommended because they are most likely to be solids. Low molecular weight ketones may also be derivatized by 2,4-dinitrophenylhydrazones (Procedure 13, p. 321) or 4-nitrophenylhydrazones (Procedure 14, p. 322). For high molecular weight ketones, phenylhydrazones are suitable (Procedure 14, p. 322). The 2,4-dinitrophenylhydrazone, phenylhydrazone, and 4-nitrophenylhydrazone are prepared from the reaction of the aldehyde or ketone with 2,4-dinitrophenylhydrazine, phenylhydrazine, or 4-nitrophenylhydrazine, respectively.

$$\mathrm{R_2C{=}O} + \mathrm{2,4\text{-}(O_2N)_2C_6H_3{-}NHNH_2} \xrightarrow[\text{alcohol}]{H_2SO_4} \mathrm{2,4\text{-}(O_2N)_2C_6H_3{-}NHN{=}CR_2}$$

aldehyde or ketone; 2,4-dinitrophenylhydrazine; 2,4-dinitrophenylhydrazone

$$\mathrm{R_2C{=}O} + \mathrm{C_6H_5{-}NHNH_2} \xrightarrow[\text{alcohol}]{CH_3COOH} \mathrm{C_6H_5{-}NHN{=}CR_2}$$

aldehyde or ketone; phenylhydrazine; phenylhydrazone

$$\mathrm{R_2C{=}O} + \mathrm{4\text{-}O_2N{-}C_6H_4{-}NHNH_2} \xrightarrow[\text{alcohol}]{CH_3COOH} \mathrm{4\text{-}O_2N{-}C_6H_4{-}NHN{=}CR_2}$$

aldehyde or ketone; 4-nitrophenylhydrazine; 4-nitrophenylhydrazone

Oximes (Procedure 15, p. 323) are prepared from the reaction of the aldehyde or ketone with hydroxylamine hydrochloride. However, these derivatives are likely to melt lower than the corresponding 2,4-dinitrophenylhydrazones and semicarbazones. The reaction between carbonyl compound and hydroxylamine is reversible, and care must be taken to avoid unnecessary contact with strongly acid solutions; under these conditions the oxime may be hydrolyzed to the original compound.

$$\mathrm{>C{=}O} + \mathrm{H_2N{-}OH \cdot HCl} \xrightarrow{\mathrm{NaOH}} \mathrm{>C{=}N{-}OH} + \mathrm{NaCl} + \mathrm{H_2O}$$

aldehyde or ketone — hydroxylamine hydrochloride — oxime

Aldehydes react with methone to yield dimedon derivatives (Procedure 16, p. 323).

$$\mathrm{RCHO} + 2\ \text{methone} \longrightarrow \text{bis-methone}$$

aldehyde — methone or "dimedon" — "methone derivative" or "bis-methone"

The CHO group in the aldehyde can be oxidized to a carboxylic acid (Procedure 17, p. 324), then the acid can be derivatized as described in Procedures 3 (p. 309), 4 (p. 310), or 5 (p. 312). Oxidizing agents such as potassium permanganate, hydrogen peroxide, and silver nitrate can be used.

$$\mathrm{RCHO} \xrightarrow[\mathrm{OH^-}]{[\mathrm{O}]} \mathrm{RCOO^-Na^+} \xrightarrow{\mathrm{H^+}} \mathrm{RCOOH}$$

aldehyde — carboxylic acid

Methyl ketones may be oxidized to acids selectively by means of sodium hypochlorite.[5] This reagent is particularly useful for unsaturated methyl ketones because many other oxidizing agents attack the double bond.

$$\mathrm{RCH{=}CH{-}CO{-}CH_3} + 3\mathrm{NaOCl} \longrightarrow$$

α,β-unsaturated ketone — sodium hypochlorite

$$\mathrm{RCH{=}CH{-}CO{-}CO{-}O^-Na^+} + \mathrm{CHCl_3} + 2\mathrm{NaOH}$$

α,β-unsaturated acid

[5]A. M. Van Arendonk and M. E. Cuperty, *J. Amer. Chem. Soc.*, *53*, 3184 (1931); C. D. Hurd and C. L. Thomas, *J. Amer. Chem. Soc.*, *55*, 1646 (1933).

Procedure 12. Semicarbazones of Aldehydes and Ketones

$$R_2C{=}O + {}^{-}Cl\,{}^{+}H_3N{-}NH{-}C({=}O){-}NH_2 \xrightarrow{H_3C{-}C({=}O){-}O^{-}Na^{+}} R_2C{=}N{-}NH{-}C({=}O){-}NH_2 + H_3C{-}C({=}O){-}OH + H_2O + NaCl$$

aldehyde or ketone — semicarbazide hydrochloride — semicarbazone

(a) For Water Soluble Compounds. One milliliter or 1 g of the aldehyde or ketone, 1 g of semicarbazide hydrochloride, and 1.5 g of sodium acetate are dissolved in 10 mL of water in a test tube. The mixture is shaken vigorously, and the test tube is placed in a beaker of boiling water for 5 min. The test tube is removed from the beaker, allowed to cool, and then placed in a beaker of ice. The sides of the tube are scratched with a glass rod. The crystals of the semicarbazone are removed by filtration and recrystallized from water or 25–50% ethanol.

(b) For Water Insoluble Compounds. One milliliter or 1 g of the aldehyde or ketone is dissolved in 10 mL of ethanol. Water is added until the solution is faintly turbid; the turbidity is removed with a few drops of ethanol. Then 1 g of semicarbazide hydrochloride and 1.5 g of sodium acetate are added, and from this point (a) is followed.

Cleaning Up. The filtrates are combined and made slightly acidic with 10% hydrochloric acid and placed in the aqueous solution container.

Procedure 13. 2,4-Dinitrophenylhydrazones of Aldehydes and Ketones

$$R_2C{=}O + O_2N{-}C_6H_3(NO_2){-}NHNH_2 \xrightarrow[\text{alcohol}]{H_2SO_4} O_2N{-}C_6H_3(NO_2){-}NHN{=}CR_2$$

aldehyde or ketone — 2,4-dinitrophenylhydrazine — 2,4-dinitrophenylhydrazone

(a) In Ethanol. In a small flask, 0.4 g of 2,4-dinitrophenylhydrazine is added to 2 mL of concentrated sulfuric acid. Three milliliters of water is added dropwise, with swirling or stirring, until the 2,4-dinitrophenylhydrazine is dissolved. To this warm solution is added 10 mL of 95% ethanol.

The freshly prepared 2,4-dinitrophenylhydrazine solution is added to a mixture of 0.5 g of the aldehyde or ketone in 20 mL of 95% ethanol, and the resulting mixture is allowed to stand at room temperature. Crystallization of the 2,4-dinitrophenylhydrazone usually occurs within 5–10 min. If no precipitate is formed, the mixture is allowed to stand overnight.

The 2,4-dinitrophenylhydrazone is recrystallized from 30 mL of 95% ethanol, with heating on a steam bath. If the derivative dissolves immediately, water is added slowly until the cloud point is reached or up to a maximum of 5 mL of water. If the derivative does not dissolve, then ethyl acetate is added to the hot mixture until solution is attained. The hot solution is filtered through a fluted filter and allowed to stand at room temperature until crystallization is complete (about 12 hr). The crystals are then isolated by suction filtration.

(b) In Diethylene Glycol Dimethyl Ether (Diglyme). A solution of 0.17 g of 2,4-dinitrophenylhydrazine dissolved in 5 mL of diethylene glycol dimethyl ether (diglyme) is warmed and then allowed to stand at room temperature for several days. To this solution at room temperature is added 0.1 g of the carbonyl compound in 1 mL of 95% ethanol or in 1 mL of diethylene glycol dimethyl ether. Three drops of concentrated hydrochloric acid are then added. If there is not an immediate precipitation, the solution should be diluted with water and allowed to stand. The crystals of 2,4-dinitrophenylhydrazine are isolated by suction filtration and are recrystallized as described in (a).

Cleaning Up. Any unreacted 2,4-dinitrophenylhydrazine is placed in the hazardous waste container. The filtrates from (a) and the ethanol from recrystallization are combined and diluted with 10 mL of water and placed in the aqueous solution container. The filtrate from (b) is placed in the organic solvent container.

Procedure 14. Phenylhydrazones and 4-Nitrophenylhydrazones of Aldehydes and Ketones

$$\underset{\text{aldehyde or ketone}}{\text{C=O}} + \underset{\text{phenylhydrazine}}{C_6H_5\text{—NHNH}_2} \xrightarrow[\text{alcohol}]{CH_3COOH} \underset{\text{phenylhydrazone}}{C_6H_5\text{—NHN=C}}$$

$$\underset{\text{aldehyde or ketone}}{\text{C=O}} + \underset{\text{4-nitrophenylhydrazine}}{O_2N\text{—}C_6H_4\text{—NHNH}_2} \xrightarrow[\text{alcohol}]{CH_3COOH} \underset{\text{4-nitrophenylhydrazone}}{O_2N\text{—}C_6H_4\text{—NHN=C}}$$

A mixture of 0.5 mL of phenylhydrazine or 0.5 g of 4-nitrophenylhydrazine and 0.5 g of the aldehyde or ketone in 10–15 mL of ethanol is heated to boiling, and a drop of glacial acetic acid is added. The mixture is kept hot for a few minutes, and more ethanol is added if necessary to obtain a clear solution. The solution is cooled, and the hydrazone is collected by filtration. It may be recrystallized from a small amount of ethanol.

If the derivative does not separate from the solution on cooling, the mixture is heated to the boiling point, water is added until the solution is cloudy, and then a drop or two of ethanol is added to clarify it. The phenylhydrazone or 4-nitrophenylhydrazone that separates on cooling is recrystallized from a water–ethanol mixture.

Cleaning Up. Any unreacted 2,4-dinitrophenylhydrazine is placed in the hazardous waste container. The filtrate is placed in the aqueous solution container.

Procedure 15. Oximes of Aldehydes and Ketones

$$>C{=}O + H_2N{-}OH \cdot HCl \xrightarrow{NaOH} >C{=}N{-}OH + NaCl + H_2O$$

aldehyde or ketone — hydroxylamine hydrochloride — oxime

(a) In Pyridine. This procedure is used for water insoluble aldehydes and ketones. A mixture of 1 g of the aldehyde or ketone, 1 g of hydroxylamine hydrochloride, 5 mL of pyridine, and 5 mL of absolute ethanol is heated under reflux for 2 hr on a steam bath. The solvents are removed by distillation using a steam bath. The residue is mixed thoroughly with 5 mL of cold water, and the mixture is filtered. The oxime is recrystallized from methanol, ethanol, or an ethanol–water mixture.

(b) In Water. About 0.5 g of the hydroxylamine hydrochloride is dissolved in 3 mL of water. Two milliliters of 10% sodium hydroxide solution and 0.2 g of the aldehyde or ketone are then added. If the carbonyl compound is water insoluble, just sufficient ethanol is added to the mixture to give a clear solution. The mixture is warmed on the steam bath for 10 min and cooled in an ice bath. In order to hasten crystallization, the sides of the flask are scratched with a glass rod. Occasionally the addition of a few milliliters of water will assist separation of the oxime. The product may be recrystallized from water or dilute ethanol.

(c) For Larger or Cyclic Ketones. Certain cyclic ketones, such as camphor, require an excess of alkali and a longer time of heating. If a ketone fails to yield an oxime by (a) or (b), 0.5 g of it should be treated with 0.5 g of hydroxylamine hydrochloride, 2 g of potassium hydroxide, and 10-mL of 95% ethanol. The mixture is heated under reflux for 2 hr and poured into 75 mL of water. The suspension is stirred and allowed to stand to permit the unchanged ketone to separate. The solution is filtered, acidified with 10% hydrochloric acid, and allowed to stand to permit the oxime to crystallize. The product is recrystallized from ethanol or an ethanol–water mixture.

Cleaning Up. The distilled solvents from (a) are placed in the organic solvent container. The filtrates are placed in the aqueous solution container.

Procedure 16. Dimedon Derivatives of Aldehydes

$$RCHO + 2\ (H_3C)_2C_6H_6O_2 \longrightarrow R{-}CH(C_6H_5O_2(CH_3)_2)_2$$

aldehyde — methone or "dimedon" — "methone derivative" or "bis-methone"

To a solution of 0.1 g of the aldehyde in 4 mL of 50% ethanol is added 0.4 g of methone (dimedon), if the aldehyde is aliphatic. If an aromatic aldehyde is used, only 0.3 g of the methone is used. One drop of piperidine is added, and the mixture is boiled gently for 5 min. If the hot solution is clear at the end of this time, water is added dropwise until the solution just begins to turn cloudy. The mixture is then chilled, and the aldehyde bis-methone condensation product, after being separated by filtration, is washed with 2 mL of cold 50% ethanol. The derivative is recrystallized from mixtures of methanol and water.

Cleaning Up. Place the filtrate in the aqueous solution container.

Procedure 17. Oxidation of an Aldehyde to an Acid

(a) Permanganate Method

$$\underset{\text{aldehyde}}{R-\overset{O}{\overset{\|}{C}}-H} \xrightarrow[\text{NaOH/H}_2\text{O}]{\text{KMnO}_4} R-\overset{O}{\overset{\|}{C}}-O^-Na^+ \xrightarrow{H^+} \underset{\text{carboxylic acid}}{R-\overset{O}{\overset{\|}{C}}-OH}$$

A few drops of 10% sodium hydroxide solution are added to a solution or suspension of 1 g of the aldehyde in 10–20 mL of water. A saturated solution of potassium permanganate in water is added dropwise until a definite purple color remains after shaking the solution. The mixture is acidified with 10% sulfuric acid, and sodium bisulfite is added until the permanganate and manganese dioxide have been converted to manganese sulfate, as evidenced by the loss of the purple color in the solution. The carboxylic acid is removed by filtration and recrystallized from water or a water–acetone mixture. If the acid does not separate, it may be recovered by extraction with three 15-mL portions of chloroform, methylene chloride, or diethyl ether. The organic layer is dried with magnesium sulfate, filtered, and the organic solvent removed by distillation using a steam bath to leave the crude carboxylic acid.

(b) Hydrogen Peroxide Method

$$\underset{\text{aldehyde}}{R-\overset{O}{\overset{\|}{C}}-H} \xrightarrow[\text{NaOH/H}_2\text{O}]{\text{H}_2\text{O}_2} R-\overset{O}{\overset{\|}{C}}-O^-Na^+ \xrightarrow{H^+} \underset{\text{carboxylic acid}}{R-\overset{O}{\overset{\|}{C}}-OH}$$

In a 250-mL flask are placed 10 mL of 5% sodium hydroxide solution and 15 mL of 3% hydrogen peroxide solution. The solution is warmed to 65–70°C, and 0.5 g of the aldehyde is added. The mixture is shaken and kept at 65°C for 15 min. If the aldehyde has not dissolved, a few milliliters of ethanol may be added. Another 5 mL of hydrogen peroxide is added, and the mixture is warmed for 10 min more. The solution is made acidic with 5% hydrochloric acid, and the acid which separates is removed by filtration.

If the acid is either a liquid or water soluble, it is best to make the solution neutral to phenolphthalein and evaporate to dryness.

(c) Silver(I) Oxide Method[6]

$$\underset{\text{aldehyde}}{\text{RCHO}} \xrightarrow[\text{NaOH/H}_2\text{O}]{\text{AgNO}_3} \text{RCO}_2^-\text{Na}^+ \xrightarrow{\text{H}^+} \underset{\text{carboxylic acid}}{\text{RCOOH}}$$

Dissolve 3.4 g of silver nitrate in 10 mL of distilled water in a 50-mL beaker. Add 10% sodium hydroxide solution dropwise with vigorous stirring until no further precipitation of silver oxide occurs (approximately 4 mL). Then add 1 g of the aldehyde with vigorous stirring and an additional 5 mL of 10% sodium hydroxide solution. The reaction mixture usually warms up as the oxidation proceeds and the silver oxide is converted to metallic silver. After 10–15 min of stirring, the silver and unchanged silver oxide are removed by filtration. The filtrate is acidified with 20% nitric acid and the solution cooled in an ice bath. The precipitated solid carboxylic acid is collected on a filter, dried, and its melting point taken. If necessary, it may be recrystallized from a small amount of hot water or a 1 : 1 water–2-propanol mixture. If the acid is so soluble in water that it does not precipitate at the acidification step, it may be extracted with three 15-mL portions of chloroform, methylene chloride, or diethyl ether. The organic layer is dried with magnesium sulfate, filtered, and the organic solvent removed by distillation using a steam bath to leave the crude carboxylic acid.

Cleaning Up. Place the filtrate from (a) in the hazardous waste container. Recrystallization solvents from (a), (b), and (c), and the filtrate from (b) are placed in the aqueous solution container. Chloroform and diethyl ether from (a) and (c) are placed in the organic solvent container. The silver and silver oxide from (c) are placed in a beaker, made acidic with 5% nitric acid, and then neutralized with sodium carbonate. Saturated sodium chloride solution is then added to precipitate the silver chloride. The silver chloride is isolated by suction filtration and placed in the nonhazardous solid waste container. The filtrate is placed in the aqueous solution container.

Discussion. The carboxylic acids formed from these procedures are characterized by Procedures 3–7 (pp. 309–313).

8.4 AMIDES

Xanthydrol reacts readily with unsubstituted amides and imides to form 9-acylamidoxanthenes, which are good derivatives (Procedure 18, p. 328). These are the only derivatives that are prepared directly from the amide.

$$\underset{\text{amide}}{\text{RCONH}_2} + \underset{\text{xanthydrol}}{\text{xanthydrol (H, OH)}} \xrightarrow{\text{CH}_3\text{COOH}} \underset{\text{9-acylamidoxanthene}}{\text{9-(RCONH)xanthene (H, NH–C(=O)R)}}$$

[6]S. C. Thomason and D. G. Kugler, *J. Chem. Educ., 45,* 546 (1968). These authors describe the use of silver(I) oxide and silver(II) oxide.

The most general method for chemically characterizing primary amides consists in hydrolyzing them with alkali to the salt of the carboxylic acid and ammonia (Procedure 19a, p. 329). Acidification of the salt produces a carboxylic acid. Either the salt or the acid can be characterized.

$$\underset{\text{1° amide}}{RC(=O)NH_2} + NaOH \longrightarrow RC(=O)O^-Na^+ + \underset{\text{ammonia}}{NH_3}$$

$$RC(=O)O^-Na^+ \xrightarrow{H^+} \underset{\text{carboxylic acid}}{RC(=O)OH}$$

Hydrolysis of substituted amides yields carboxylic acids and primary or secondary amines instead of ammonia. The hydrolysis occurs faster in acidic conditions than in basic conditions. In basic hydrolysis, the amide is hydrolyzed to the amine and the salt of the carboxylic acid. The carboxylic acid is liberated by acidification (Procedure 19b, p. 329).

$$\underset{\text{2° amide}}{RC(=O)NHR'} + NaOH \longrightarrow RC(=O)O^-Na^+ + \underset{\text{1° amine}}{R'NH_2}$$

$$RC(=O)O^-Na^+ \xrightarrow{H^+} \underset{\text{carboxylic acid}}{RC(=O)OH}$$

$$\underset{\text{3° amide}}{RC(=O)NR'_2} + NaOH \longrightarrow RC(=O)O^-Na^+ + \underset{\text{2° amine}}{R'_2NH}$$

$$RC(=O)O^-Na^+ \xrightarrow{H^+} \underset{\text{carboxylic acid}}{RC(=O)OH}$$

In acidic hydrolysis, the amide yields an ammonium salt and the carboxylic acid. The solution of the ammonium salts are made basic and the amine is liberated (Procedure 19c, p. 330).

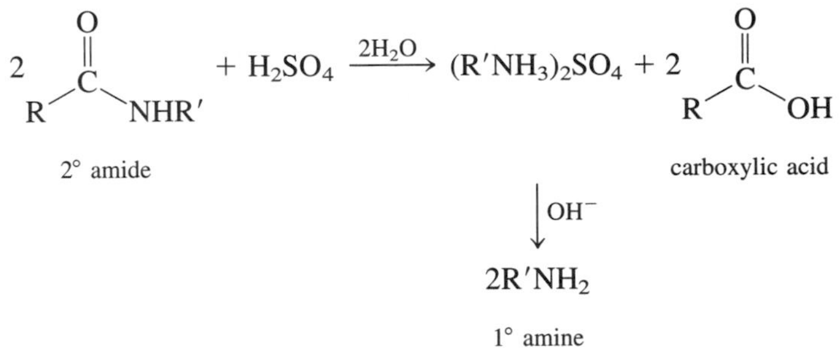

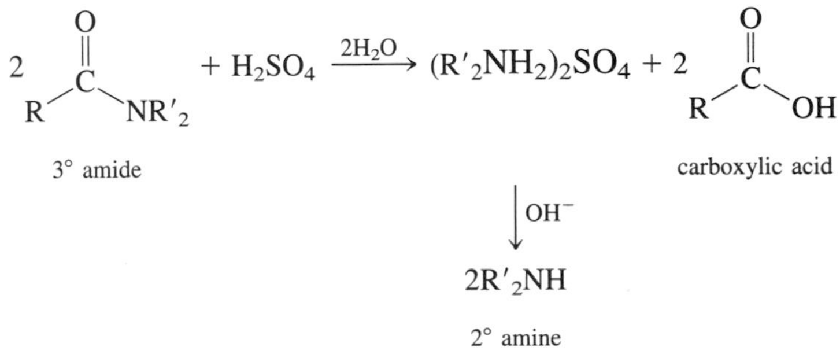

$$2\,RC(=O)NHR' + H_2SO_4 \xrightarrow{2H_2O} (R'NH_3)_2SO_4 + 2\,RC(=O)OH$$

2° amide; carboxylic acid

$$(R'NH_3)_2SO_4 \xrightarrow{OH^-} 2R'NH_2$$

1° amine

$$2\,RC(=O)NR'_2 + H_2SO_4 \xrightarrow{2H_2O} (R'_2NH_2)_2SO_4 + 2\,RC(=O)OH$$

3° amide; carboxylic acid

$$(R'_2NH_2)_2SO_4 \xrightarrow{OH^-} 2R'_2NH$$

2° amine

Once the carboxylic acid and the amine are isolated from the amide, derivatives of these compounds can be prepared. The 4-nitrobenzyl ester and the 4-bromophenacyl ester (Procedure 5, p. 312) are recommended as derivatives for the carboxylic acid.

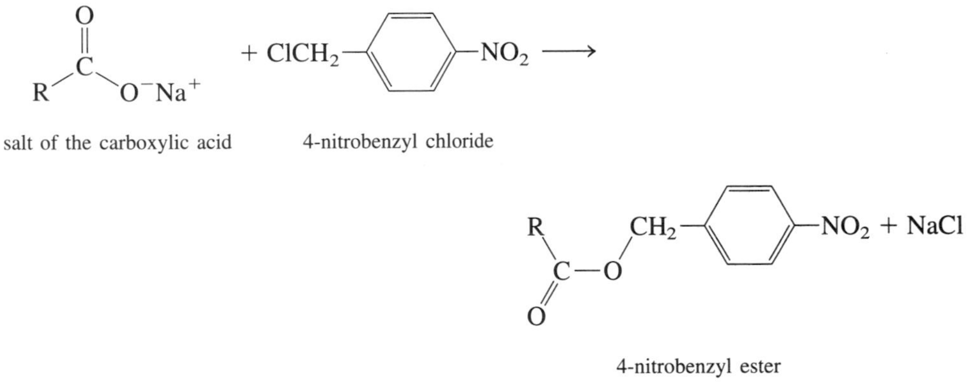

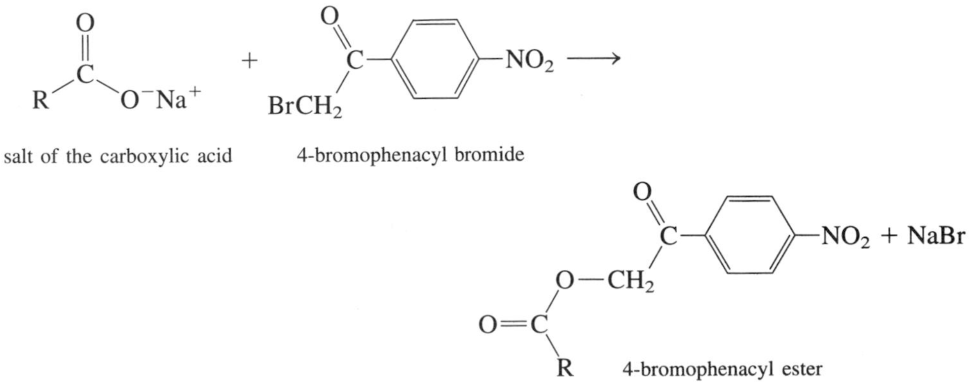

$$RC(=O)O^-Na^+ + ClCH_2-C_6H_4-NO_2 \longrightarrow RC(=O)O-CH_2-C_6H_4-NO_2 + NaCl$$

salt of the carboxylic acid; 4-nitrobenzyl chloride; 4-nitrobenzyl ester

$$RC(=O)O^-Na^+ + BrCH_2-C(=O)-C_6H_4-NO_2 \longrightarrow RC(=O)O-CH_2-C(=O)-C_6H_4-NO_2 + NaBr$$

salt of the carboxylic acid; 4-bromophenacyl bromide; 4-bromophenacyl ester

The acetamide (Procedure 20a, p. 333) and benzamide derivatives (Procedure 20b or 20c, p. 333 or 334) of the amine are also excellent derivatives. The acetamide is prepared by treating the amine with acetic anhydride. The benzamide is synthesized from the reaction of the amine with benzoyl chloride.

$$R{-}NH{-} + H_3C{-}C(=O){-}O{-}C(=O){-}CH_3 \longrightarrow {-}N(R){-}C(=O){-}CH_3 + H_3C{-}C(=O){-}OH$$

amine | acetic anhydride | acetamide | acetic acid

$$2\ R{-}NH{-} + C_6H_5{-}C(=O){-}Cl \longrightarrow {-}N(R){-}C(=O){-}C_6H_5 + R{-}N^{+}H{-}\quad Cl^{-}$$

amine | benzoyl chloride | benzamide

Procedure 18. 9-Acylamidoxanthenes from Primary Amides and Imides

$$R{-}C(=O){-}NH_2 + \text{xanthydrol} \xrightarrow{CH_3COOH} \text{9-acylamidoxanthene}$$

amide | xanthydrol | 9-acylamidoxanthene

About 0.5 g of xanhydrol is dissolved in 7 mL of glacial acetic acid. If the solution is not clear, it is allowed to stand a few minutes or is centrifuged and the clear solution decanted into a clean test tube. To this solution is added 0.5 g of the amide, and the mixture is warmed at 85°C in a beaker of water for 20–30 min. Upon cooling, the acylamidoxanthene is collected on a filter and recrystallized from a mixture of 2 parts of dioxane and 1 part water.

Some amides fail to dissolve in the acetic acid and may be converted to the derivative by using a mixture of 5 mL of ethanol, 2 mL of acetic acid, and 3 mL of water as the solvent for the reaction.

Cleaning Up. Place the filtrate in the aqueous solution container.

Procedure 19. Hydrolysis of Amides

(a) Basic Conditions—Primary Amides

$$\underset{\text{1° amide}}{RCONH_2} + NaOH \longrightarrow RCOO^-Na^+ + \underset{\text{ammonia}}{NH_3}$$

$$RCOO^-Na^+ \xrightarrow{H^+} \underset{\text{carboxylic acid}}{RCOOH}$$

A mixture of 25 mL of 10% sodium hydroxide solution is allowed to reflux with 1.0 g of the amide for 30 min. The odor of ammonia will be very apparent. The aqueous solution is cooled in an ice bath, and concentrated hydrochloric acid is added until the solution is acidic to litmus. A solid carboxylic acid is removed by filtration. If the carboxylic acid does not precipitate, the solution is extracted with three 15-mL portions of chloroform, methylene chloride, or diethyl ether. The organic layer is dried with magnesium sulfate, filtered, and the organic solvent removed by distillation using a steam bath to leave the crude carboxylic acid.

(b) Basic Conditions—Substituted Amides

$$\underset{\text{2° amide}}{RCONHR'} + NaOH \longrightarrow RCOO^-Na^+ + \underset{\text{1° amine}}{R'NH_2}$$

$$RCOO^-Na^+ \xrightarrow{H^+} \underset{\text{carboxylic acid}}{RCOOH}$$

$$\underset{\text{3° amide}}{RCONR'_2} + NaOH \longrightarrow RCOO^-Na^+ + \underset{\text{2° amine}}{R'_2NH}$$

$$RCOO^-Na^+ \xrightarrow{H^+} \underset{\text{carboxylic acid}}{RCOOH}$$

Twenty-five mL of 10% sodium hydroxide solution is refluxed with 1.0 g of the amide for 1 hr and the solution is cooled. The aqueous layer is extracted twice with 15 mL of ethyl ether. The combined ether layers are dried with magnesium sulfate, and then filtered. The ether is removed by distillation, using a steam bath, to yield the primary or secondary amine. The aqueous solution is cooled in an ice bath, and concentrated hydrochloric acid is added until the solution is acidic to litmus. Solid carboxylic acids are removed by filtration. Volatile carboxylic acids are extracted with three 15-mL portions of chloroform, methylene chloride, or diethyl ether. The organic layer is dried with magnesium sulfate, filtered, and the organic solvent removed by distillation using a steam bath to leave the crude carboxylic acid.

(c) Acidic Conditions—Substituted Amides

$$2\ \underset{\text{2° amide}}{R\text{-}C(=O)\text{-}NHR'} + H_2SO_4 \xrightarrow{2H_2O} (R'NH_3)_2SO_4 + 2\ \underset{\text{carboxylic acid}}{R\text{-}C(=O)\text{-}OH}$$

$$(R'NH_3)_2SO_4 \xrightarrow{OH^-} \underset{\text{1° amine}}{2R'NH_2}$$

$$2\ \underset{\text{3° amide}}{R\text{-}C(=O)\text{-}NR'_2} + H_2SO_4 \xrightarrow{2H_2O} (R'_2NH_2)_2SO_4 + 2\ \underset{\text{carboxylic acid}}{R\text{-}C(=O)\text{-}OH}$$

$$(R'_2NH_2)_2SO_4 \xrightarrow{OH^-} \underset{\text{2° amine}}{2R'_2NH}$$

One gram of the amide is heated with 20 mL of 20% sulfuric acid for 1–2 hr and the solution is then cooled. The carboxylic acid may be removed by filtration if insoluble in water. Water soluble or liquid carboxylic acids are extracted with three 15-mL portions of chloroform, methylene chloride, or diethyl ether. The organic layer is dried with magnesium sulfate, filtered, and the organic solvent removed by distillation using a steam bath to leave the crude carboxylic acid. Cool the solution, and add 20% sodium hydroxide solution until the solution is alkaline. Extract with two 15-mL portions of ethyl ether, dry the combined ether layers with magnesium sulfate, and filter. The ether is removed by distillation, using the steam bath, to yield the amine.

Cleaning Up. Place the diethyl ether in the organic solvent container. Neutralize the acidic filtrates from (a) and (b) with sodium carbonate and the basic filtrate from (c) with 10% hydrochloric acid. These aqueous solutions are then placed in the aqueous solution container.

Discussion. A solid amine or carboxylic acid is purified by recrystallization and used as a derivative. Derivatives of any amine or carboxylic acid can also be prepared.

The amine can undergo reaction with acetic anhydride to yield the acetamide (Procedure 20a, p. 333) or with benzoyl chloride to produce the benzamide (Procedure 20b or 20c, p. 333 or 334).

The carboxylic acid or the salt of the carboxylic acid can undergo reaction with 4-nitrobenzyl chloride to give the 4-nitrobenzyl ester or with 4-bromophenacyl bromide to yield the 4-bromophenacyl ester (Procedure 5, p. 312).

8.5 AMINES

The most useful derivatives of primary and secondary amines take advantage of their reactive N—H bond. The amides of acetic and benzoic acids (Procedure 20, p. 333) are conveniently prepared by treatment of the amine, respectively, with acetic anhydride or benzoyl chloride. Acetyl and benzoyl derivatives are known for most primary and secondary amines, and for this reason these derivatives are useful.

$$\underset{\text{amine}}{R(-)N{-}H} + \underset{\text{acetic anhydride}}{H_3C{-}C(=O){-}O{-}C(=O){-}CH_3} \longrightarrow \underset{\text{acetamide}}{-N(R){-}C(=O){-}CH_3} + \underset{\text{acetic acid}}{H_3C{-}C(=O){-}OH}$$

$$2\,\underset{\text{amine}}{R(-)N{-}H} + \underset{\text{benzoyl chloride}}{C_6H_5{-}C(=O){-}Cl} \longrightarrow \underset{\text{benzamide}}{-N(R){-}C(=O){-}C_6H_5} + R{-}\overset{+}{N}(-)(H){-}\ Cl^-$$

Arylsulfonamides are frequently used as derivatives. The benzenesulfonamides are frequently used and their preparation is related to the Hinsberg method for classifying amines (compare Experiment 17, p. 242, and Procedure 21, p. 335). The reaction of a primary amine with alkaline benzenesulfonyl chloride or 4-toluenesulfonyl chloride produces the soluble salt of the arylsulfonamide. Acidification results in precipitation of the arylsulfonamide.

$$\underset{1^\circ\ \text{amine}}{RNH_2} + \underset{\text{arylsulfonyl chloride}}{ArSO_2Cl} + 2NaOH \longrightarrow \underset{\text{(soluble)}}{ArSO_2NR^-Na^+} + NaCl + 2H_2O$$

$$\xrightarrow{HCl} \underset{\text{arylsulfonamide}}{ArSO_2NHR} + NaCl$$

A secondary amine reacts with benzenesulfonyl chloride or 4-toluenesulfonyl chloride to give the arylsulfonamide, which is usually insoluble.

$$\underset{2^\circ\ \text{amine}}{R_2NH} + \underset{\text{arylsulfonyl chloride}}{ArSO_2Cl} + NaOH \longrightarrow \underset{\text{arylsulfonamide}}{ArSO_2NR_2} + NaCl + H_2O$$

The phenylthioureas are especially valuable for characterizing low molecular weight, water soluble amines. They are formed by treatment of amines with phenyl isothiocyanate (Procedure 22, p. 336).

$$RNH_2 + C_6H_5N{=}C{=}S \longrightarrow C_6H_5NH{-}C({=}S){-}NHR$$

1° amine phenyl isocyanate phenylthiourea

$$R_2NH + C_6H_5N{=}C{=}S \longrightarrow C_6H_5NH{-}C({=}S){-}NR_2$$

2° amine phenyl isocyanate phenylthiourea

The reagent phenyl isothiocyanate is not sensitive to water; in fact, this reaction may be carried out with dilute aqueous solutions of the amines.

Tertiary amines vary so greatly in nature that no type of derivative has been found to be generally applicable. Perhaps the most useful derivatives are the quaternary ammonium salts formed by the reaction of the amine with methyl iodide or methyl 4-toluenesulfonate (Procedure 23, p. 337).

$$R_3N + CH_3I \longrightarrow (R_3NCH_3)^+I^-$$

3° amine methyl iodide quaternary salt

$$R_3N + H_3C{-}C_6H_4{-}SO_3CH_3 \longrightarrow H_3C{-}C_6H_4{-}SO_3^-\ R_3\overset{+}{N}CH_3$$

3° amine methyl 4-toluenesulfonate quaternary salt

Ammonium salts of chloroplatinic acid (Procedure 24, p. 337) and picric acid (Procedure 25, p. 338) are also employed frequently as derivatives of tertiary amines. Chloroplatinic salts are prepared from the tertiary amine and chloroplatinic acid.

$$R_3N + H_2PtCl_6{\cdot}6H_2O \xrightarrow[H_2O]{HCl} R_3NH^+HPtCl_6^-$$

3° amine chloroplatinic acid chloroplatinic salt

Picric acid undergoes reaction with tertiary amines to yield the picrates.

$$R_3N + \text{2,4,6-}(O_2N)_3C_6H_2OH \longrightarrow R_3NH^+\ \text{2,4,6-}(O_2N)_3C_6H_2O^-$$

3° amine picric acid picrate

Primary, secondary, and tertiary amines are readily converted to hydrochloride salts (Procedure 26, p. 338). These hydrochlorides are almost always easily purified solids with sharp melting points, which make them useful for purposes of characterization. The amine hydrochlorides are prepared by passing hydrogen chloride gas through an ether solution of the amine.

$$\underset{1^\circ \text{ amine}}{RNH_2} + HCl(g) \longrightarrow \underset{\text{amine hydrochloride}}{RNH_3^+Cl^-}$$

$$\underset{2^\circ \text{ amine}}{R_2NH} + HCl(g) \longrightarrow \underset{\text{amine hydrochloride}}{R_2NH_2^+Cl^-}$$

$$\underset{3^\circ \text{ amine}}{R_3N} + HCl(g) \longrightarrow \underset{\text{amine hydrochloride}}{R_3NH^+Cl^-}$$

Procedure 20. Substituted Amides from Amines

(a) Substituted Acetamides from Water Insoluble Amines

$$\underset{1^\circ \text{ amine}}{RNH_2} + \underset{\text{acetic anhydride}}{H_3C{-}C(=O){-}O{-}C(=O){-}CH_3} \longrightarrow \underset{\text{acetamide}}{H{-}N(R){-}C(=O){-}CH_3} + \underset{\text{acetic acid}}{H_3C{-}C(=O){-}OH}$$

$$\underset{2^\circ \text{ amine}}{RR'NH} + \underset{\text{acetic anhydride}}{H_3C{-}C(=O){-}O{-}C(=O){-}CH_3} \longrightarrow \underset{\text{acetamide}}{R'{-}N(R){-}C(=O){-}CH_3} + \underset{\text{acetic acid}}{H_3C{-}C(=O){-}OH}$$

A solution of the amine is prepared by dissolving 0.5 g of the compound in 15 mL of 5% hydrochloric acid. A few chips of ice are added, followed by 3 mL of acetic anhydride. The mixture is stirred or swirled vigorously, and a previously prepared solution of 2.5 g of sodium acetate trihydrate in 2.5 mL of water is added in one portion. If the product does not crystallize, the mixture is chilled overnight. The crystals are isolated by filtration.

Recrystallization may be effected from cyclohexane or from a mixture of cyclohexane and benzene. The acetamide must be thoroughly dry before recrystallization is attempted from these solvents. An ethanol–water mixture may also be used for recrystallization.

(b) Benzamides: Pyridine Method

$$2\ \underset{1^\circ \text{ amine}}{RNH_2} + \underset{\text{benzoyl chloride}}{C_6H_5{-}C(=O){-}Cl} \xrightarrow{C_5H_5N} \underset{\text{benzamide}}{H{-}N(R){-}C(=O){-}C_6H_5} + R{-}NH_3^+\ Cl^-$$

$$2\,RR'NH + C_6H_5COCl \xrightarrow{C_5H_5N} C_6H_5CONRR' + RR'NH_2^+\ Cl^-$$

2° amine benzoyl chloride benzamide

To a solution of 0.5 g of the compound in 5 mL of dry pyridine and 10 mL of dry benzene is added, dropwise, 0.50 mL of benzoyl chloride. ***Caution:*** *Benzene is a known carcinogen. Use benzene in the hood, do not breathe the vapors, and avoid contact with the skin.* The resulting mixture is heated in a water bath at 60–70°C for 30 min and is then poured into 100 mL of water. The benzene layer is separated, and the aqueous layer is washed successively with 10 mL of 5% hydrochloric acid, 10 mL of water, and 10 mL of 5% sodium carbonate solution. The benzene layers are dried with a little anhydrous magnesium sulfate. The drying agent is removed by filtration through a fluted filter, and the benzene is removed by distillation to a volume of 3–4 mL. Twenty milliliters of hexane is stirred into the mixture, and the crystalline benzamide is removed by suction filtration and washed with hexane. Recrystallization may usually be effected from a mixture of cyclohexane and hexane or from a mixture of cyclohexane and ethyl acetate. Ethanol or aqueous ethanol may also be used with many compounds.

(c) Benzamides: Sodium Hydroxide Method

$$2\,RNH_2 + C_6H_5COCl \xrightarrow{NaOH} C_6H_5CONHR + NaCl + H_2O$$

1° amine benzoyl chloride benzamide

$$2\,RR'NH + C_6H_5COCl \xrightarrow{NaOH} C_6H_5CONRR' + NaCl + H_2O$$

2° amine benzoyl chloride benzamide

The regular procedure for the Schotten–Baumann reaction described under Experiment 3a may be used. Two modified procedures are described here.

1. A mixture of 20 mL of 5% sodium hydroxide solution, 5 mL of chloroform, 0.5 g of the compound, and 0.5 g of benzoyl chloride is shaken or stirred for about 20 min and then allowed to stand for 12 hr. The chloroform layer is separated, and the aqueous layer is washed with 10 mL of chloroform. The combined chloroform solutions are washed with water, dried with anhydrous magnesium sulfate, and evaporated to a volume of 2–3 mL. Twenty milliliters of hexane is stirred into the solution, and the derivative is removed by filtration and washed with hexane.
2. About 1 mL of the amine is added to a solution of 1 g of benzoyl chloride in 20 mL of dry benzene. ***Caution:*** *Benzene is a known carcinogen. Use benzene in the hood, do not breathe the vapors, and avoid contact with the skin.* The resulting solution is refluxed for 15 min and is then allowed to cool. The solution is filtered, and the pre-

cipitate is washed with 10 mL of warm benzene, the washings being added to the original filtrate. The benzene solution is next washed with 10 mL of 2% sodium carbonate solution, then with 10 mL of 2% hydrochloric acid (***Caution:*** *foaming*), and finally with 10 mL of distilled water. The benzene is removed by distillation, and the residue is recrystallized from 60% ethanol.

(d) Benzamides and Acetamides from Amino Acids

$$2\,R{-}CH(NH_3^+){-}C(=O){-}O^- + C_6H_5{-}C(=O){-}Cl \longrightarrow C_6H_5{-}C(=O){-}NH{-}CH(R){-}C(=O){-}OH$$

α-amino acid — benzoyl chloride — benzamide

$$R{-}CH(NH_3^+){-}C(=O){-}O^- + H_3C{-}C(=O){-}O{-}C(=O){-}CH_3 \longrightarrow H_3C{-}C(=O){-}NH{-}CH(R){-}C(=O){-}OH + H_3C{-}C(=O){-}OH$$

α-amino acid — acetic anhydride — acetamide — acetic acid

Benzamide. In a test tube, dissolve 0.5 g of the amino acid in 10 mL of 10% sodium bicarbonate solution. Add 1 g of benzoyl chloride. Stopper and shake the mixture vigorously, venting it periodically to allow the carbon dioxide gas to escape. When the odor of benzoyl chloride has disappeared, acidify the solution with 10% hydrochloric acid to Congo Red at a pH of 4. Isolate the crystals by filtration and rinse the crystals with a small amount of diethyl ether to remove any benzoic acid. Recrystallize the product from 60% ethanol.

Acetamide. A solution of the amino acid is prepared by dissolving 0.5 g of the compound in 15 mL of 5% hydrochloric acid. A few chips of ice are added, followed by 3 mL of acetic anhydride. The mixture is refluxed for 1 hr and is then allowed to cool. The solution is filtered, and the precipitate is recrystallized from 60% ethanol.

Cleaning Up. Place the reaction filtrates from (a) and (d) in the aqueous solution container. The chloroform from (c) and (d) are placed in the halogenated organic waste container. The distilled benzene from (b) and (c) are placed in the hazardous waste container for benzene. The nonhalogenated filtrates and other solvents are placed in the organic solvent container. The ethanol–water recrystallization solvents and aqueous layers are placed in the aqueous solution container.

Procedure 21. Benzenesulfonamides and 4-Toluenesulfonamides from Primary and Secondary Amines (Hinsberg's Method)

$$\underset{1^\circ\ \text{amine}}{RNH_2} + \underset{\text{arylsulfonyl chloride}}{ArSO_2Cl} + 2NaOH \longrightarrow \underset{\text{(soluble)}}{ArSO_2NR^-} + NaCl + 2H_2O$$

$$ArSO_2NR^- \xrightarrow{HCl} \underset{\text{arylsulfonamide}}{ArSO_2NHR} + NaCl$$

$$\underset{\text{2° amine}}{R_2NH} + \underset{\text{arylsulfonyl chloride}}{ArSO_2Cl} + NaOH \longrightarrow \underset{\text{arylsulfonamide}}{ArSO_2NR_2} + NaCl + H_2O$$

To 1 mL or 1 g of the amine in a test tube, add 17 mL of 10% sodium hydroxide solution and 2.0 g of benzenesulfonyl chloride or 4-toluenesulfonyl chloride. Stopper the test tube, and shake the mixture very vigorously. Test the solution to make sure that it is alkaline. Cool the solution and acidify with 10% hydrochloric acid. The benzenesulfonamide is isolated by filtration and recrystallized from 95% ethanol.

Cleaning Up. Place the filtrate in the aqueous solution container.

Procedure 22. Phenylthioureas from Amines

$$\underset{\text{1° amine}}{RNH_2} + \underset{\text{phenyl isocyanate}}{C_6H_5N{=}C{=}S} \longrightarrow \underset{\text{phenylthiourea}}{C_6H_5NH{-}C({=}S){-}NHR}$$

$$\underset{\text{2° amine}}{R_2NH} + \underset{\text{phenyl isocyanate}}{C_6H_5N{=}C{=}S} \longrightarrow \underset{\text{phenylthiourea}}{C_6H_5NH{-}C({=}S){-}NR_2}$$

Equal amounts of the amine and phenyl isothiocyanate are mixed in a test tube and shaken for 2 min. If no reaction occurs spontaneously, the mixture is heated for 3 min over a low flame. ***Caution:*** *The isothiocyanates are lachrymators.* The aliphatic amines usually react immediately, whereas the aromatic amines require heating. The mixture is kept in a beaker of ice until the mass solidifies. The solid is powdered and washed with petroleum ether and 50% ethanol in order to remove any excess of either reactant. The residue is then recrystallized from 95% ethanol.

Cleaning Up. Treat any unreacted phenyl isothiocyanate with an excess of 5.25% sodium hypochlorite (household bleach). Place the filtrates in the aqueous solution container.

Discussion. Occasionally the thiourea derivative reversibly undergoes a complicating disproportionation reaction with the original amine.

$$C_6H_5NH{-}C({=}S){-}NHR + RNH_2 \rightleftharpoons RNH{-}C({=}S){-}NHR + C_6H_5NH_2$$

$$C_6H_5NH{-}C({=}S){-}NHR + C_6H_5NH_2 \rightleftharpoons C_6H_5NH{-}C({=}S){-}NHC_6H_5 + RNH_2$$

This complication is averted if long heating times are avoided.

Procedure 23. Quaternary Ammonium Salts of Tertiary Amines

(a) Addition Product with Methyl Iodide

$$\underset{\text{3° amine}}{R_3N} + \underset{\text{methyl iodide}}{CH_3I} \longrightarrow \underset{\text{quaternary salt}}{(R_3NCH_3)^+I^-}$$

A mixture of 0.5 g of the amine and 0.5 mL of methyl iodide is warmed in a test tube over a low flame for a few minutes and is then cooled in an ice bath. ***Caution:*** *Methyl iodide is a suspected cancer agent.* The tube is scratched with a glass rod to hasten crystallization. The product is purified by recrystallization from absolute ethanol, methanol, or ethyl acetate.

Cleaning Up. Place the filtrate in the hazardous waste container.

(b) Addition Product with Methyl 4-Toluenesulfonate

$$\underset{\text{3° amine}}{R_3N} + \underset{\text{methyl 4-toluenesulfonate}}{H_3C\text{—}C_6H_4\text{—}SO_3CH_3} \longrightarrow \underset{\text{quaternary salt}}{H_3C\text{—}C_6H_4\text{—}SO_3^-\ R_3\overset{+}{N}CH_3}$$

One gram of the amine is added to a solution of 2–3 g of methyl 4-toluenesulfonate in 10 mL of dry benzene. ***Caution:*** *Benzene is a known carcinogen. Use benzene in the hood, do not breathe the vapors, and avoid contact with the skin.* The solution is refluxed for 10–20 min and cooled. The products are recrystallized by dissolving them in the least possible amount of boiling ethanol; ethyl acetate is added until precipitation starts, and the mixture is cooled. The product is removed by filtration and quickly dried; the melting point is determined immediately.

Cleaning Up. Place the filtrate in the hazardous waste container for benzene.

Procedure 24. Chloroplatinate Salts from Tertiary Amines

$$\underset{\text{3° amine}}{R_3N} + \underset{\text{chloroplatinic acid}}{H_2PtCl_6 \cdot 6H_2O} \xrightarrow[H_2O]{HCl} \underset{\text{chloroplatinic salt}}{R_3NH^+HPtCl_6^-}$$

Ten milliliters of a 25% aqueous solution of chloroplatinic acid ($H_2PtCl_6 \cdot 6H_2O$) is added slowly with shaking to a solution of 0.5 g of the amine in 10 mL of 10% hydrochloric acid. The crystalline chloroplatinate which separates is collected on a filter and washed with 10% hydrochloric acid. It may be recrystallized from ethanol containing a drop of concentrated hydrochloric acid to prevent hydrolysis.

Cleaning Up. Place the filtrate in the hazardous waste container.

Procedure 25. Picrates from Tertiary Amines

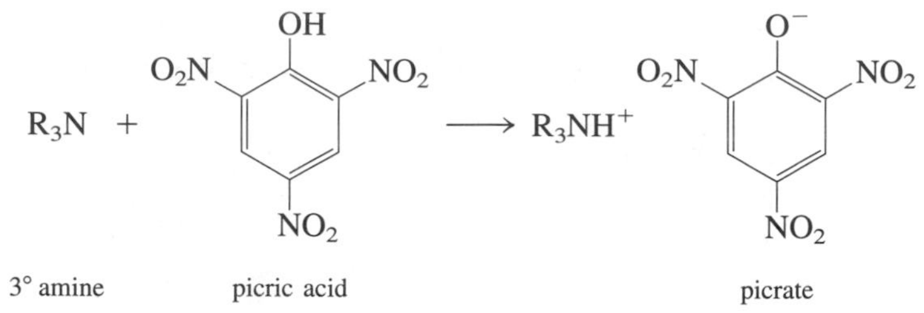

$$R_3N + \text{picric acid (2,4,6-}(O_2N)_3C_6H_2OH) \longrightarrow R_3NH^+ \; \text{picrate (2,4,6-}(O_2N)_3C_6H_2O^-)$$

3° amine picric acid picrate

Caution: *Picric acid can explode if it is allowed to dry. Do not allow the reagent to dry out.* A sample of the compound (0.3–0.5 g) is added to 10 mL of 95% ethanol. If the sample does not dissolve, the mixture is shaken until a saturated solution results and is then filtered. The filtrate is added to 10 mL of a saturated solution of picric acid (2,4,6-trinitrophenol) in 95% ethanol, and the solution is heated to boiling. The solution is allowed to cool slowly, and the yellow crystals of the picrate are removed by filtration. Most picrates are pure enough that recrystallization is usually not needed. ***Caution:*** *Some picrates explode when heated.* Certain picrates, especially those of hydrocarbons, dissociate when heated and consequently cannot be recrystallized. In such cases the original precipitate should be washed with a very small amount of ether and dried in preparation for the melting point determination. If recrystallization is needed, then ethanol is used as a solvent.

Cleaning Up. Place the filtrate in the aqueous solution container.

Procedure 26. Amine Hydrochloride Salts[7]

$$RNH_2 + HCl(g) \longrightarrow RNH_3^+Cl^-$$

1° amine amine hydrochloride

$$R_2NH + HCl(g) \longrightarrow R_2NH_2^+Cl^-$$

2° amine amine hydrochloride

$$R_3N + HCl(g) \longrightarrow R_3NH^+Cl^-$$

3° amine amine hydrochloride

A hydrogen chloride generator (Figure 8.1) is assembled in a fume hood using a three-neck 250-mL round-bottom flask equipped with a stoppered 100 mL pressure-equalizing addition funnel and a glass tube leading to a 250-mL Erlenmeyer flask. The addition funnel contains a long tube which is below the solution in the round-bottom flask. The glass tube connecting the two flasks is above the solution in the round-bottom flask, but below the solution in the Erlenmeyer flask, so that the hydrogen chloride gas is bubbled through the reaction mixture. The round-bottom flask is charged with 50 mL of concentrated sulfuric acid and 30 mL of concentrated hydrochloric acid is placed in the dropping funnel. One gram of the amine is dissolved in 75 mL of anhydrous diethyl ether

[7]Personal communication with Rogers Lambert, Professor of Chemistry, Radford University, Radford, Virginia.

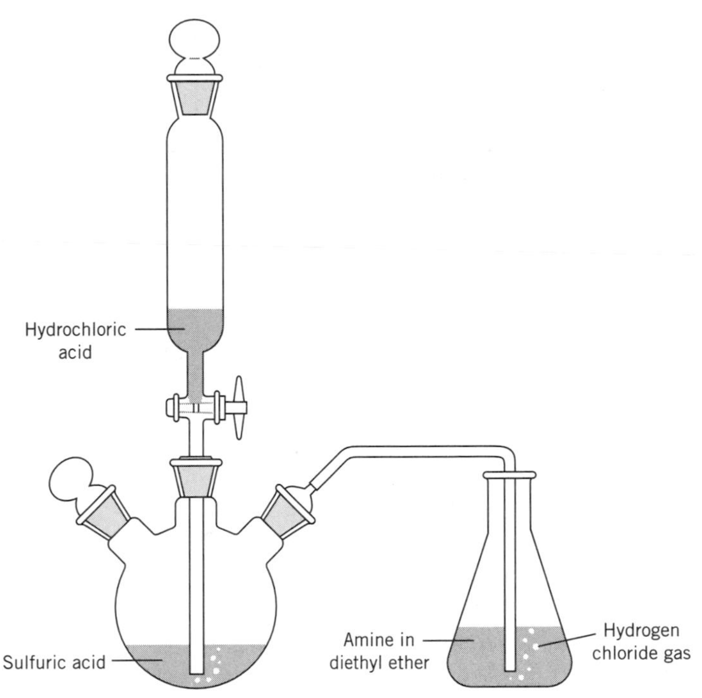

Figure 8.1 Apparatus for Procedure 26.

and placed in the Erlenmeyer flask. The hydrochloric acid is slowly added to the sulfuric acid and the hydrogen chloride gas that is generated is bubbled into the ether solution. Care must be taken in the slow addition of the hydrochloric acid to the sulfuric acid so that no sulfuric acid is carried over into the amine solution. After the precipitate stops forming in the ether solution, the amine hydrochloride salt is filtered from the ether, washed with anhydrous ether, and recrystallized from a mixture of hexane and 2-propanol.

Cleaning Up. Place the sulfuric acid–hydrochloric acid mixture in a 600-mL beaker and slowly add solid sodium carbonate until the foaming ceases. Place the solution in the aqueous solution container. The organic solvents and filtrates are placed in the organic solvent container.

8.6 AMINO ACIDS

The melting points or decomposition points of amino acids are not exact. The values depend upon the rate of heating. Hence, in using these constants to prepare a list of possibilities, extra allowance must be made for their inaccuracy.

The α-amino acids occurring naturally in plants and animals or obtained from the acid or enzymatic hydrolysis of proteins and peptides are optically active (except glycine) and belong to the *configurational* L-series. The specific rotations are valuable constants for identification (Table 8.2).

Table 8.2 Specific Rotations of α-Amino Acids[a]

Amino acid configuration	Solvent	C (g/100 mL)	Temp. (°C)	$[\alpha]_o$ (deg.)	R/S
L-Alanine	1.0 N HCl	5.8	15	+14.7	2(*S*)
L-Arginine	6.0 N HCl	1.6	23	+26.9	2(*S*)
L-Aspartic acid	6.0 N HCl	2.0	24	+24.6	2(*S*)
L-Cystine	1.0 N HCl	1.0	24	−214.4	2(*R*), 2′(*R*)
L-Glutamic acid	6.0 N HCl	1.0	22	+31.2	2(*S*)
L-Histidine	6.0 N HCl	1.5	22	+13.0	2(*S*)
Hydroxy-L-proline	1.0 N HCl	1.3	20	−47.3	2(*S*), 4(*R*)
Allohydroxy-L-proline	Water	2.6	18	−58.1	2(*S*), 4(*S*)
L-Isoleucine	6.0 N HCl	5.1	20	+40.6	2(*S*), 3(*S*)
L-Alloisoleucine	6.0 N HCl	3.9	20	+38.1	2(*S*), 3(*R*)
L-Leucine	6.0 N HCl	2.0	26	+15.1	2(*S*)
L-Lysine	6.0 N HCl	2.0	23	+25.9	2(*S*)
L-Methionine	0.2 N HCl	0.8	25	+21.2	2(*S*)
L-Phenylalanine	5.4 N HCl	3.8	20	−7.1	2(*S*)
L-Proline	0.5 N HCl	0.6	20	−52.6	2(*S*)
L-Serine	1.0 N HCl	9.3	25	+14.4	2(*S*)
L-Threonine	Water	1.3	26	+28.4	2(*S*), 3(*R*)
L-Allothreonine	Water	1.6	26	+9.6	2(*S*), 3(*S*)
L-Tryptophan	Water	1.0	22	−31.5	2(*S*)
L-Tyrosine	6.3 N HCl	4.4	20	−8.6	2(*S*)
L-Valine	6.0 N HCl	3.4	20	+28.8	2(*S*)

[a] A leading reference for amino acids is *The Merck Index,* 9th ed. (Merck & Co., Rahway, NJ. 12th ed., 1996).

Amino acids can be divided into four general categories; this division is based on their acid–base and charge properties.

1. *Hydrophobic:* Amino acids which are substantially less water soluble than class 2; for example, phenylalanine,

$$C_6H_5{-}CH_2{-}CH(NH_3^+){-}C(=O){-}O^-$$

which is in solubility class $A_2(B)$, is virtually insoluble in water.

2. *Hydrophilic (polar, no net charge):* These amino acids have polar functional groups (OH) which, despite not having a positive or negative charge, are reasonably soluble in water, for example, threonine.

$$HO{-}CH(CH_3){-}CH(NH_3^+){-}C(=O){-}O^-$$

These compounds are differentiated from those in succeeding classes in that the polar groups are not appreciably acidic or basic (in the proton-transfer sense).

3. *Positively charged (basic):* Such amino acids have basic, usually nitrogenous, functions which are protonated at intracellular[8] pH. Lysine,

$$\underset{}{NH_2}-\overset{\epsilon}{C}H_2-\overset{\delta}{C}H_2-\overset{\gamma}{C}H_2-\overset{\beta}{C}H_2-\overset{\alpha}{C}H(^{+}NH_3)-C(=O)-O^{-}$$

would also have its ϵ-amino group protonated under these conditions.

4. *Negatively charged (acidic):* These amino acids have an acidic, thus ionized, carboxyl group at intracellular pH. For example, the β-carboxyl group of aspartic acid,

$$HO-C(=O)-\overset{\beta}{C}H_2-\overset{\alpha}{C}H(^{+}NH_3)-C(=O)-O^{-}$$

would also be ionized to a carboxylate group under these conditions.

Solid derivatives of the amino acids are usually obtained from reaction of the amino groups rather than the carbonyl groups. The Hinsberg reaction (Procedure 27, p. 343) furnishes good derivatives for a considerable number of the amino acids. In the Hinsberg reaction, 4-toluenesulfonyl chloride reacts with the amino acid to yield a 4-toluenesulfonamide.

$$\underset{\alpha\text{-amino acid}}{R-CH(^{+}NH_3)-C(=O)-O^{-}} + NaOH \longrightarrow R-CH(NH_2)-C(=O)-O^{-} \xrightarrow[\text{4-toluenesulfonyl chloride (}p\text{-toluenesulfonyl chloride)}]{H_3C-C_6H_4-SO_2-Cl} \xrightarrow{H_3O^+}$$

$$H_3C-C_6H_4-S(=O)_2-NH-CH(R)-C(=O)OH$$

4-toluenesulfonamide
(*p*-toluenesulfonamide)

Phenyl isocyanate reacts with the amino acids to produce the corresponding substituted phenylureas, also known as phenylureido acids (Procedure 28, p. 343).

[8]Intracellular pH: 6.0–7.0.

$$\underset{\alpha\text{-amino acid}}{RCH(\overset{+}{N}H_3)COO^-} + \underset{\text{phenyl isocyanate}}{C_6H_5N{=}C{=}O} \longrightarrow \underset{\text{phenylureido acid}}{C_6H_5NHC(=O)NHCH(R)COOH}$$

Procedures similar to ones used for the preparation of amides are used to synthesize benzamides and acetamides from amino acids (Procedure 20d, p. 335).

$$2\,\underset{\alpha\text{-amino acid}}{RCH(\overset{+}{N}H_3)COO^-} + \underset{\text{benzoyl chloride}}{C_6H_5C(=O)Cl} \longrightarrow \underset{\text{benzamide}}{C_6H_5C(=O)NHCH(R)COOH}$$

$$\underset{\alpha\text{-amino acid}}{RCH(\overset{+}{N}H_3)COO^-} + \underset{\text{acetic anhydride}}{H_3CC(=O)OC(=O)CH_3} \longrightarrow \underset{\text{acetamide}}{H_3CC(=O)NHCH(R)COOH} + \underset{\text{acetic acid}}{H_3CCOOH}$$

3,5-Dinitrobenzamides are prepared by treating the amino acid with 3,5-dinitrobenzoyl chloride in the presence of base (Procedure 29, p. 344).

$$\underset{\alpha\text{-amino acid}}{RCH(\overset{+}{N}H_3)COO^-} + \underset{\text{3,5-dinitrobenzoyl chloride}}{3,5\text{-}(O_2N)_2C_6H_3C(=O)Cl} \longrightarrow \underset{\text{3,5-dinitrobenzamide}}{3,5\text{-}(O_2N)_2C_6H_3C(=O)NHCH(R)C(=O)OH}$$

The N-substituted 2,4-dinitroanilines obtained by the action of 2,4-dinitrofluorobenzene on amino acids (Procedure 30, p. 344) have found extensive use as derivatives of amino acids, peptides, and proteins.

$$\underset{\alpha\text{-amino acid}}{RCH(\overset{+}{N}H_3)COO^-} + \underset{\substack{\text{2,4-dinitrofluorobenzene}\\ \text{(Sanger's reagent)}}}{2,4\text{-}(O_2N)_2C_6H_3F} \longrightarrow \underset{\text{N-substituted 2,4-dinitroaniline}}{2,4\text{-}(O_2N)_2C_6H_3NHCH(R)C(=O)OH} + NaF$$

Procedure 27. 4-Toluenesulfonamides from Amino Acids

$$R{-}CH(^{+}NH_3){-}C({=}O){-}O^{-} + NaOH \longrightarrow R{-}CH(NH_2){-}C({=}O){-}O^{-} \xrightarrow[\text{4-toluenesulfonyl chloride (}p\text{-toluenesulfonyl chloride)}]{H_3C{-}C_6H_4{-}SO_2{-}Cl} \xrightarrow{H_3O^{+}} H_3C{-}C_6H_4{-}SO_2{-}NH{-}CH(R){-}C({=}O){-}OH$$

α-amino acid

4-toluenesulfonamide
(p-toluenesulfonamide)

About 1 g of the amino acid is dissolved in 20 mL of 1 M sodium hydroxide solution, a solution of 2 g of 4-toluenesulfonyl chloride in 25 mL of diethyl ether is added, and the mixture is shaken mechanically or vigorously stirred for 3–4 hr. The ether layer is separated, and the aqueous layer is acidified to Congo Red at a pH of about 4 using 10% hydrochloric acid. The derivative usually separates as a solid, which is removed by filtration and recrystallized from 4–5 mL of 60% ethanol. If an oil is obtained upon acidification, the mixture is placed in a refrigerator overnight to induce crystallization.

The sodium salts of the derivatives of phenylalanine and tyrosine are sparingly soluble in water and separate during the initial reaction. The resulting suspension is acidified. The salts go into solution and the mixture separates into two layers. The 4-toluenesulfonyl derivatives then crystallize from the ether layer and are removed by filtration.

The derivatives of glutamic and aspartic acids, arginine, lysine, tryptophan, and proline crystallize with difficulty; other derivatives should be tried in the event that oils are produced.

Cleaning Up. Place the ether layer in the organic solvent container. The aqueous filtrate is neutralized with sodium carbonate and placed in the aqueous solution container.

Procedure 28. Phenylureido Acids from Amino Acids[9]

$$R{-}CH(^{+}NH_3){-}C({=}O){-}O^{-} + C_6H_5N{=}C{=}O \longrightarrow C_6H_5{-}NH{-}C({=}O){-}NH{-}CH(R){-}C({=}O){-}OH$$

α-amino acid — phenyl isocyanate — phenylureido acid

One gram of the phenyl isocyanate and 0.5 g of the amino acid are combined with 25 mL of 2% sodium hydroxide solution in a small Erlenmeyer flask. ***Caution:*** *Phenyl iso-*

[9]B. S. Furniss, A. J. Hannaford, P. W. G. Smith, and A. R. Tatchell, *Vogel's Textbook of Practical Organic Chemistry,* 5th ed. (Wiley, New York, 1989), p. 1281.

cyanate is a lachrymator. The resulting mixture is vigorously shaken for 5 min, then left undisturbed for 30 min. Any solid diarylurea is filtered off, and the filtrate is acidified with 5% hydrochloric acid to Congo Red at a pH of about 4. The crystals are washed with cold water, then recrystallized from water or ethanol.

Cleaning Up. Treat any unreacted phenyl isocyanate with an excess of 5.25% sodium hypochlorite (household bleach), dilute with 10 mL of water, and place in the aqueous solution container. Neutralize the aqueous filtrate with sodium carbonate and place in the aqueous solution container.

Procedure 29. 3,5-Dinitrobenzamides from Amino Acids[10]

α-amino acid + 3,5-dinitrobenzoyl chloride ⟶ 3,5-dinitrobenzamide

In a small flask, 1.0 g of the amino acid is dissolved in 10 mL of 1 M sodium hydroxide. Next, 1.6 g of powdered 3,5-dinitrobenzoyl chloride is added and the mixture is shaken vigorously until the acid chloride dissolves. The solution is then shaken for another 2 min, and any undissolved solids are filtered off. The mixture is acidified to Congo Red at a pH of 4 with 5% hydrochloric acid. The 3,5-dinitrobenzamide is filtered off and recrystallized from water or 50% ethanol.

Cleaning Up. The aqueous layer is made slightly basic with sodium carbonate and placed in the aqueous solution container.

Procedure 30. 2,4-Dinitrophenyl Derivatives of Amino Acids

α-amino acid + 2,4-dinitrofluorobenzene (Sanger's reagent) ⟶ N-substituted 2,4-dinitroaniline + NaF

To a solution or suspension of 0.5 g of the amino acid in 10 mL of water and 1.0 g of sodium bicarbonate is added a solution of 0.8 g of 2,4-dinitrofluorobenzene (*highly toxic*) in 5 mL of ethanol. The mixture is shaken very vigorously and allowed to stand at room

[10] B. S. Furniss, A. J. Hannaford, P. W. G. Smith, and A. R. Tatchell, *Vogel's Textbook of Practical Organic Chemistry,* 5th ed. (Wiley, New York, 1989), p. 1279.

temperature for 1 hr with occasional vigorous shaking. Then 5-mL of saturated sodium chloride solution is added, and the mixture extracted twice with 10 mL of ether to remove unchanged reactants. The aqueous layer is then poured into 25 mL of cold 5% hydrochloric acid with vigorous stirring. This mixture should be distinctly acidic to Congo Red or below a pH of 4. The product sometimes separates as an oil; a solid is induced to form by stirring or by scratching. The derivative is collected on a filter and recrystallized from 50% ethanol.

This procedure may be used for amines and for proteins or peptides with free primary amino groups on the chain or at the end.

Cleaning Up. Place the ether layer in the organic solvent container. Neutralize the aqueous layer with sodium carbonate and place in the aqueous solution container.

8.7 CARBOHYDRATES

The osazones give useful information concerning the common sugars. The relative rates of formation of the osazone are significant. Since the melting points of the individual osazones often lie too close together to serve as a means of identification, the shapes of the crystals are usually checked against known photomicrographs.

Treating the carbohydrate with at least three equivalents of phenylhydrazine produces the phenylosazone (Procedure 31, p. 346).

$$R{-}CH(OH){-}CH{=}O \text{ or } R{-}C({=}O){-}CH_2OH + 3H_2NNHC_6H_5 \longrightarrow$$

sugar (aldose or ketose) — phenylhydrazine

$$R{-}C({=}NNHC_6H_5){-}CH{=}NNHC_6H_5 + C_6H_5NH_2 + NH_3 + 2\ H_2O$$

osazone (solid)

4-Nitrophenylhydrazones and 4-bromophenylhydrazones are prepared by reacting the carbohydrate with one equivalent of the substituted phenylhydrazine (Procedure 32, p. 347).

$$R{-}CH(OH){-}CH{=}O + ArNHNH_2 \longrightarrow R{-}CH(OH){-}CH{=}NNHAr$$

aldose — substituted phenylhydrazine — substituted phenylhydrazone

$$Ar = O_2N{-}C_6H_4{-} \text{ or } Br{-}C_6H_4{-}$$

$$R\text{-}C(=O)\text{-}CH_2OH + ArNHNH_2 \longrightarrow R\text{-}C(=NNHAr)\text{-}CH_2OH$$

ketose — substituted phenylhydrazine — substituted phenylhydrazone

Ar = $O_2N\text{-}C_6H_4\text{-}$ or $Br\text{-}C_6H_4\text{-}$

Acetates also make good derivatives for carbohydrates (Procedure 33, p. 347). Excess acetic anhydride is used so that all free hydroxyl groups are acetylated.

Procedure 31. Preparation of Osazones from Carbohydrates

$$H\text{-}C(=O)\text{-}CH(OH)\text{-}R \text{ or } R\text{-}C(=O)\text{-}CH_2OH + 3\ H_2NNHC_6H_5 \longrightarrow$$

sugar (aldose or ketose) — phenylhydrazine

$$H\text{-}C(=NNHC_6H_5)\text{-}C(=NNHC_6H_5)\text{-}R\ (s) + C_6H_5NH_2 + NH_3 + 2\ H_2O$$

osazone

To a 0.20-g sample of the carbohydrate in a test tube is added 0.40 g of phenylhydrazine hydrochloride, 0.60 g of crystalline sodium acetate, and 4 mL of distilled water. Place the test tube in a beaker of boiling water and note the time of precipitation. The test tube must be shaken occasionally to avoid supersaturation.

After 20 min, remove the tube from the hot water bath and set it aside to cool. Place a small amount of the crystals and liquid on a watch glass, spreading out the crystals by tipping the watch glass from side to side. Absorb some of the mother liquor with a piece of filter paper, taking care not to crush or break up the crystals. Isolate the remainder of the crystals by filtration. Examine the crystals under a low-power microscope (about 80–100×), and compare with photomicrographs.[11]

The formation of tarry products owing to oxidation of the phenylhydrazine may be prevented by the addition of 0.5 mL of saturated sodium bisulfite solution prior to heating.

Cleaning Up. The solid products are placed in the organic nonhazardous solid waste. Add 8 mL of 5.25% sodium hypochlorite (household bleach) to the filtrate. Heat the solution at 45–50°C for 2 hr to oxidize the amine. Dilute the solution with 10 mL of water and place in the aqueous solution container.

[11]W. Z. Hassid and R. M. McCready, *Ind. Eng. Chem., Anal. Ed., 14,* 683 (1942).

Discussion. The times required for the formation of the osazone are discussed in Experiment 28 (p. 263) in Chapter 7.

Procedure 32. Substituted Phenylhydrazones from Carbohydrates[12]

$$\mathrm{H(C{=}O)\!-\!CH(OH)R} + \mathrm{ArNHNH_2} \longrightarrow \mathrm{H(C{=}NNHAr)\!-\!CH(OH)R}$$

aldose, substituted phenylhydrazine, substituted phenylhydrazone

$$\mathrm{HOCH_2\!-\!C(=O)R} + \mathrm{ArNHNH_2} \longrightarrow \mathrm{HOCH_2\!-\!C(=NNHAr)R}$$

ketose, substituted phenylhydrazine, substituted phenylhydrazone

Ar = $O_2N-C_6H_4-$ or $Br-C_6H_4-$

Place 0.25 g of the sugar, 3 mL of ethanol, and 0.25 g of 4-bromophenylhydrazine or 4-nitrophenylhydrazine in a test tube. Place the test tube in a hot water bath and heat until no more reaction occurs. Cool the solution in an ice bath. The crystals are isolated by filtration, washed with a small amount of cold ethanol, and recrystallized from ethanol.

Cleaning Up. Add 2 mL of 5.25% sodium hypochlorite (household bleach) to the filtrate. Heat the mixture for 2 hr to hydrolyze the amine, cool the mixture, and place in the aqueous solution container.

Procedure 33. Acetates of Polyhydroxy Compounds

This procedure is useful for enhancing the volatility of carbohydrates and related compounds.

(a) Without Pyridine. One gram of the anhydrous polyhydroxy compound is mixed with 0.5 g of powdered fused sodium acetate and 5 mL of acetic anhydride. The mixture is heated on a steam bath, with occasional shaking, for 2 hr. At the end of this time the warm solution is poured, with vigorous stirring, into 30 mL of ice water. The mixture is allowed to stand, with occasional stirring, until the excess of acetic anhydride has been hydrolyzed. The crystals are removed by filtration, washed with water, and purified by recrystallization from ethanol.

[12] B. S. Furniss, A. J. Hannaford, P. W. G. Smith, and A. R. Tatchell, *Vogel's Textbook of Practical Organic Chemistry,* 5th ed. (Wiley, New York, 1989), p. 1247.

(b) With Pyridine. One gram of the polyhydroxy compound is added to 10 mL of anhydrous pyridine. Four grams of acetic anhydride is added, with shaking, and after any initial reaction has subsided the solution is refluxed for 3–5 min. The mixture is cooled and poured into 25–35 mL of ice water. The acetyl derivative is removed by filtration, washed with cold 2% hydrochloric acid, and then washed with water. It is purified by recrystallization from ethanol.

Cleaning Up. Place the filtrates in the aqueous solution container.

8.8 ESTERS

The most fundamental reaction of esters is the saponification reaction. Saponification converts an ester to an alcohol and the salt of a carboxylic acid (Procedure 34, p. 351). The carboxylic acid is liberated by acidification of the salt solution. Acid hydrolysis of the ester yields the carboxylic acid and the alcohol. Although the carboxylic acids and the alcohols can each be characterized, experience has shown that direct derivatization of the ester is a more efficient approach.

$$\underset{\text{ester}}{\mathrm{R{-}C({=}O){-}OR'}} + \mathrm{NaOH} \xrightarrow{H_2O} \underset{\text{salt of the acid}}{\mathrm{R{-}C({=}O){-}O^-Na^+}} + \underset{\text{alcohol}}{\mathrm{R'OH}}$$

$$\underset{\text{salt of the acid}}{\mathrm{R{-}C({=}O){-}O^-Na^+}} \xrightarrow{H^+} \underset{\text{carboxylic acid}}{\mathrm{R{-}C({=}O){-}OH}}$$

$$\underset{\text{ester}}{\mathrm{R{-}C({=}O){-}OR'}} + \mathrm{H_2O} \xrightarrow{H_2SO_4} \underset{\text{carboxylic acid}}{\mathrm{R{-}C({=}O){-}OH}} + \underset{\text{alcohol}}{\mathrm{R'OH}}$$

Saponification equivalents of esters (Procedure 35, p. 354) are extremely useful, especially for samples where previous molecular weight determinations were unsuccessful. The saponification equivalent is simply the equivalent weight of the ester determined by a titrimetric procedure; the procedure is conceptually similar to that used for carboxylic acids (Procedure 1, p. 307). Either the saponification equivalent or a small integral ($x = 1, 2, 3, \ldots$) multiple of the saponification equivalent, within experimental error, will be the molecular weight.

Because of difficulties involved in the separation and purification of the hydrolysis products, it is best to prepare ester derivatives by reactions using the original ester. Esters containing other functional groups may often be identified by reference to solid derivatives obtained by reactions such as halogenation of aromatic rings, nitration of aromatic rings, or acylation of alcoholic or phenolic hydroxy groups.

Some simple esters undergo reaction with aqueous or alcoholic ammonia to produce amide derivatives.

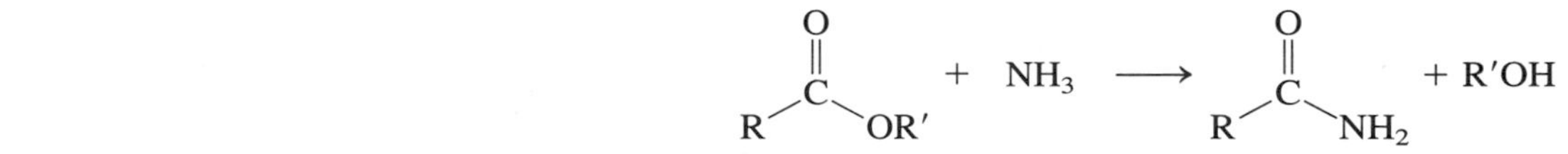

$$RC(=O)OR' + NH_3 \longrightarrow RC(=O)NH_2 + R'OH$$

ester ammonia amide alcohol

However, most esters must be heated under pressure in order to effect this reaction. A better way to prepare the amide would be to hydrolyze the ester (Procedure 34, p. 351), followed by treatment of the acid with thionyl chloride to form the acyl halide. The acyl halide is then treated with aqueous ammonia to form the amide (Procedure 31, p. 346).

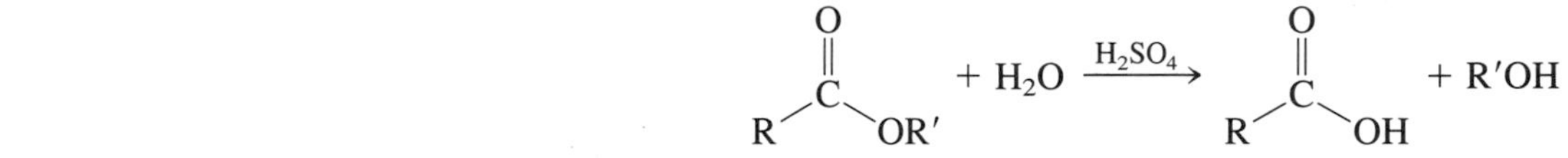

$$RC(=O)OR' + H_2O \xrightarrow{H_2SO_4} RC(=O)OH + R'OH$$

ester carboxylic acid alcohol

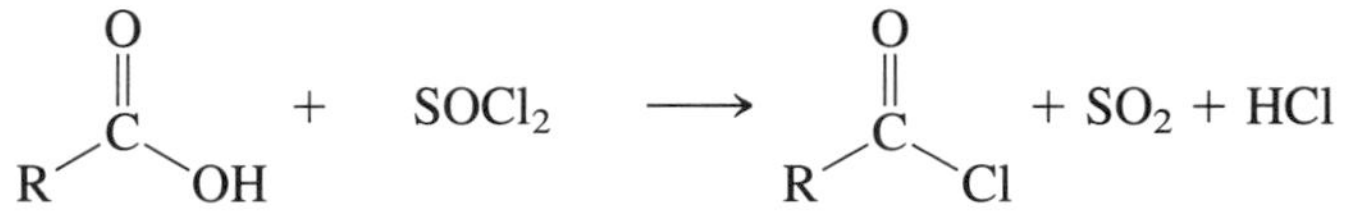

$$RC(=O)OH + SOCl_2 \longrightarrow RC(=O)Cl + SO_2 + HCl$$

carboxylic acid thionyl chloride acid chloride

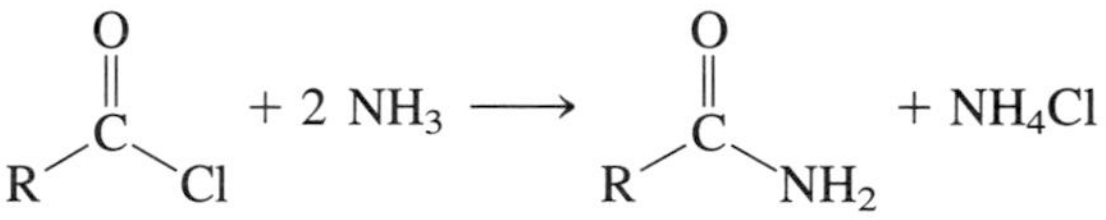

$$RC(=O)Cl + 2\ NH_3 \longrightarrow RC(=O)NH_2 + NH_4Cl$$

acyl chloride amide

The 4-toluidide of the acidic portion of the ester may be obtained through a Grignard reaction (Procedure 36, p. 356). Ethylmagnesium bromide is prepared from ethyl bromide and magnesium. The 4-toluidine (4-aminotoluene) is converted by the ethyl Grignard compound to an aminomagnesium compound, which then undergoes reaction with the ester of interest. Hydrolysis of the resulting oxymagnesium compound yields the 4-toluidide derivative.

$$CH_3CH_2Br + Mg \longrightarrow CH_3CH_2MgBr$$

ethylmagnesium bromide

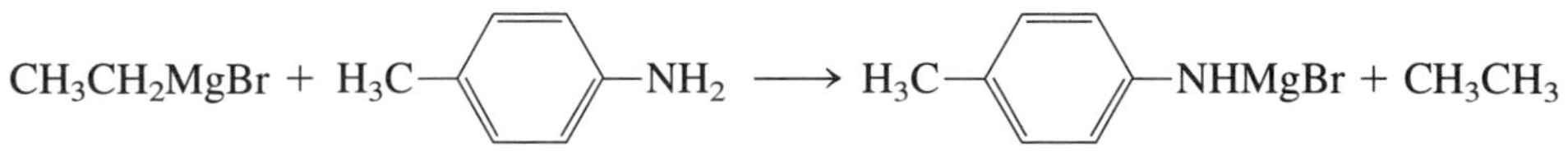

$$CH_3CH_2MgBr + H_3C{-}C_6H_4{-}NH_2 \longrightarrow H_3C{-}C_6H_4{-}NHMgBr + CH_3CH_3$$

4-toluidine (4-aminotoluene) aminomagnesium compound

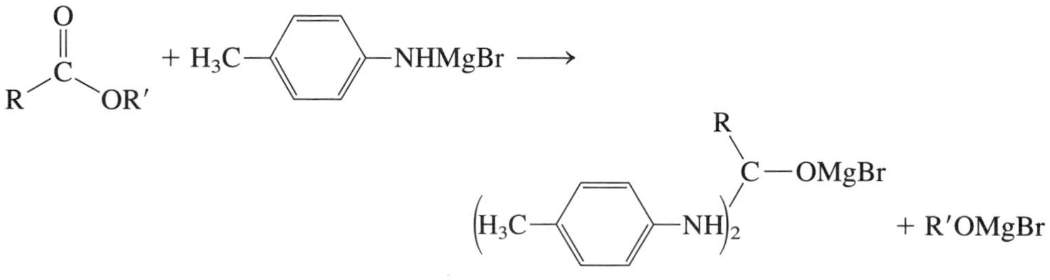

$$RC(=O)OR' + H_3C{-}C_6H_4{-}NHMgBr \longrightarrow (H_3C{-}C_6H_4{-}NH)_2C(R){-}OMgBr + R'OMgBr$$

oxymagnesium compound

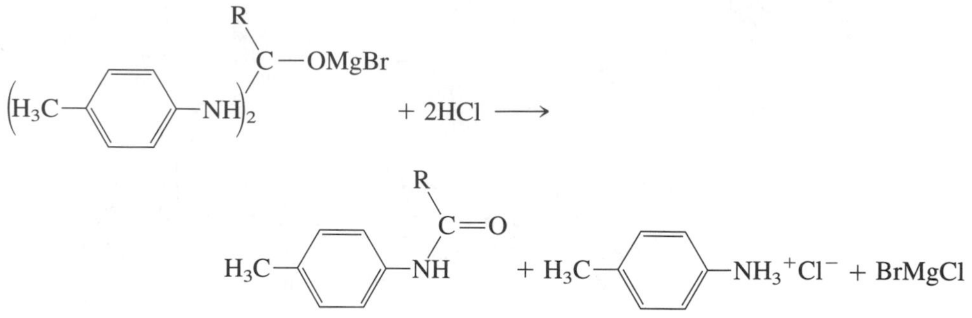

$$(H_3C-C_6H_4-NH)_2C(R)-OMgBr + 2HCl \longrightarrow H_3C-C_6H_4-NH-C(=O)R + H_3C-C_6H_4-NH_3^+Cl^- + BrMgCl$$

4-toluidide

The 3,5-dinitrobenzoates, formed from the alcohol portion of a simple ester, are produced by effecting an interchange reaction between 3,5-dinitrobenzoic acid and the ester in the presence of concentrated sulfuric acid (Procedure 37, p. 358).

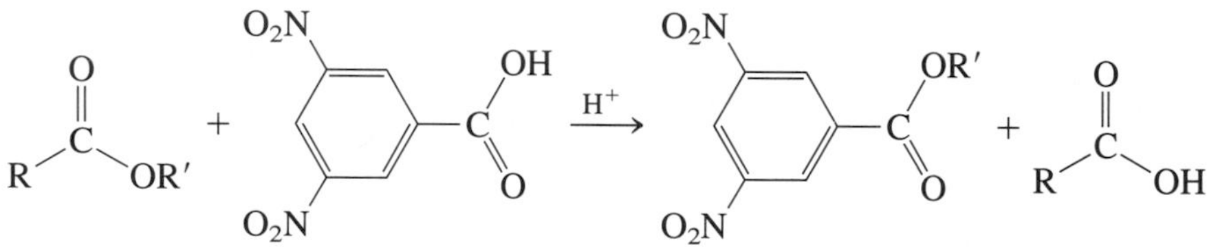

$$RCOOR' + (O_2N)_2C_6H_3COOH \xrightarrow{H^+} (O_2N)_2C_6H_3COOR' + RCOOH$$

ester — 3,5-dinitrobenzoic acid — 3,5-dinitrobenzamide

The reaction of esters with benzylamine in the presence of a little ammonium chloride affords *N*-benzylamides (Procedure 38, p. 358).

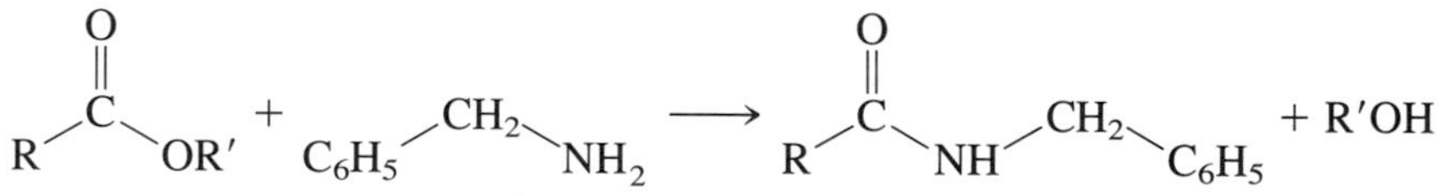

$$RCOOR' + C_6H_5CH_2NH_2 \longrightarrow RCONHCH_2C_6H_5 + R'OH$$

ester — benzylamine — *N*-benzylamide

The reaction proceeds well when R′ is methyl or ethyl. Esters of higher alcohols should be subjected to a preliminary methanolysis, followed by treatment with benzylamine.

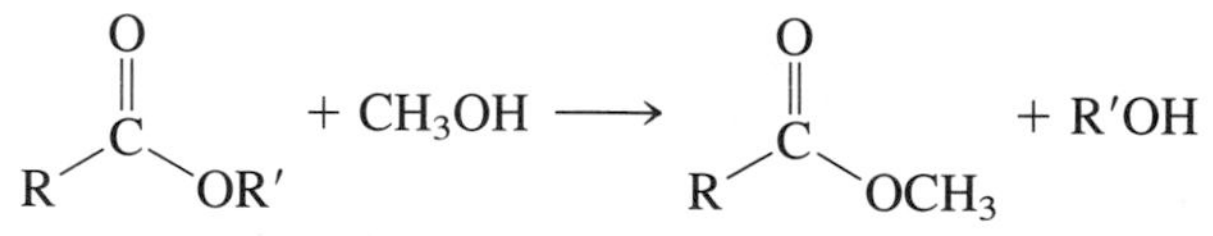

$$RCOOR' + CH_3OH \longrightarrow RCOOCH_3 + R'OH$$

ester — methyl ester

Hydrazine undergoes reaction readily with esters to produce acid hydrazides, which serve as satisfactory derivatives (Procedure 39, p. 359).

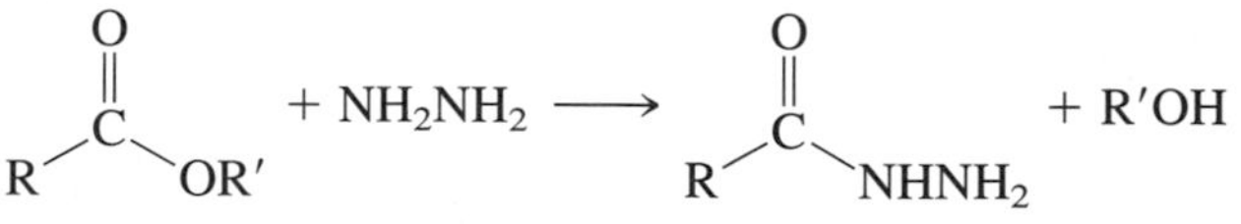

$$RCOOR' + NH_2NH_2 \longrightarrow RCONHNH_2 + R'OH$$

ester — hydrazine — acid hydrazide

Esters of higher alcohols should be converted to methyl esters, prior to reacting with hydrazine.

Procedure 34. Saponification and Hydrolysis of Esters

(a) Refluxing Water Solvent

$$\underset{\text{ester}}{R{-}C({=}O){-}OR'} + NaOH \xrightarrow{H_2O} \underset{\text{salt of the acid}}{R{-}C({=}O){-}O^-Na^+} + \underset{\text{alcohol}}{R'OH}$$

$$\underset{\text{salt of the acid}}{R{-}C({=}O){-}O^-Na^+} \xrightarrow{H^+} \underset{\text{carboxylic acid}}{R{-}C({=}O){-}OH}$$

In a round-bottom flask fitted with a reflux condenser place 12 mL of a 10% sodium hydroxide solution. Add 1 mL or 0.8 g of an ester and heat to boiling. Continue refluxing the solution until the ester layer or the characteristic odor disappears (about 1–2 hr). Reverse the condenser, distill about 2 mL, and saturate the distillate with potassium carbonate. Note the formation of two layers. The amount of sample required will obviously depend on the molecular weight of the alcohol to be isolated as well as on the molecular weight of the original ester. The alcohol layer is isolated from the aqueous layer.

Cool the residue in the flask, and acidify with 10% hydrochloric acid. Remove by suction filtration any solid acid that separates. Liquid carboxylic acids are isolated by extraction with three 15-mL portions of chloroform, methylene chloride, or diethyl ether. The organic layer is dried with magnesium sulfate, filtered, and the organic solvent removed by distillation using a steam bath to leave the crude carboxylic acid.

(b) Refluxing Diethylene Glycol

$$\underset{\text{ester}}{R{-}C({=}O){-}OR'} + NaOH \xrightarrow{(HOCH_2CH_2)_2O} \underset{\text{salt of the acid}}{R{-}C({=}O){-}O^-Na^+} + \underset{\text{alcohol}}{R'OH}$$

$$\underset{\text{salt of the acid}}{R{-}C({=}O){-}O^-Na^+} \xrightarrow{H^+} \underset{\text{carboxylic acid}}{R{-}C({=}O){-}OH}$$

In a small distilling flask place 3 mL of diethylene glycol, 0.5 g of potassium hydroxide pellets, and 0.5 mL of water. Heat the mixture over a low flame until the alkali has dissolved, and cool. Add 1 g of the ester, and mix thoroughly. Distill the mixture and place the receiving flask in an ice-water bath. The flask is heated over a small flame at first, and the contents are mixed by shaking. When only one liquid phase or one liquid and one solid phase are present, the mixture is heated more strongly so that the alcohol distills.

The residue in the flask is either a solution or a suspension of the potassium salt of the acid derived from the ester. Add 10 mL of water and 10 mL of ethanol to the residue, and shake thoroughly. Add 3 M sulfuric acid until the solution is slightly acidic to phenolphthalein. Allow the mixture to stand about 5 min and then filter. The filtrate is used directly for the preparation of a derivative. If the original ester was so high boiling that an accurate boiling point could not be obtained, it may be desirable to divide the filtrate in half and make two derivatives.

(c) Concentrated Sulfuric Acid

$$\underset{\text{ester}}{RC(=O)OR'} + H_2O \xrightarrow{H_2SO_4} \underset{\text{carboxylic acid}}{RC(=O)OH} + \underset{\text{alcohol}}{R'OH}$$

One gram of the sterically hindered ester is dissolved in the minimum amount of 100% sulfuric acid and the resulting solution is diluted with 20 mL of ice water. The alcohol is removed by distillation; the carboxylic acid is isolated by suction filtration and recrystallized.

Cleaning Up. The aqueous layer from (a) and the solvent from (c) are placed in the aqueous solution container. The sulfuric acid from (c) is neutralized with sodium carbonate and placed in the aqueous solution container.

Discussion. The carboxylic acids can be characterized further by the preparation of derivatives described in Procedures 3, p. 309, or 4, p. 310. The alcohols can also be isolated and characterized by derivatives.

Esters vary considerably in the ease with which they may be saponified. Most simple esters boiling below 110°C will be saponified completely by refluxing with 10% sodium hydroxide solution as described in (a). Esters boiling between 100 and 200°C require a longer time of 1–2 hr for complete saponification.

The hydrolysis of water insoluble esters may be accelerated by the addition of 0.1 g of sodium lauryl sulfate (Gardinol) to the alkali and the ester. The mixture is shaken vigorously to emulsify the ester and is then heated to refluxing. A large flask must be used because the emulsifying agent causes considerable foaming.

Very high boiling esters (above 200°C) which are insoluble in water hydrolyze slowly, and prolonged refluxing may result in the loss of a volatile alcohol. A solution of potassium hydroxide in diethylene glycol (bp 244°C) is used in (b). Diethylene glycol is not only an excellent solvent for esters but also permits the use of a higher reaction temperature, and all but high-boiling alcohols—(boiling points over 180°C)—can be distilled from the reaction mixture in a pure state.

Saponification represents the most useful procedure for characterizing esters. However, it must be remembered that hot concentrated alkali also affects other functional groups. Aldehydes that have α-hydrogen atoms undergo the aldol condensation:

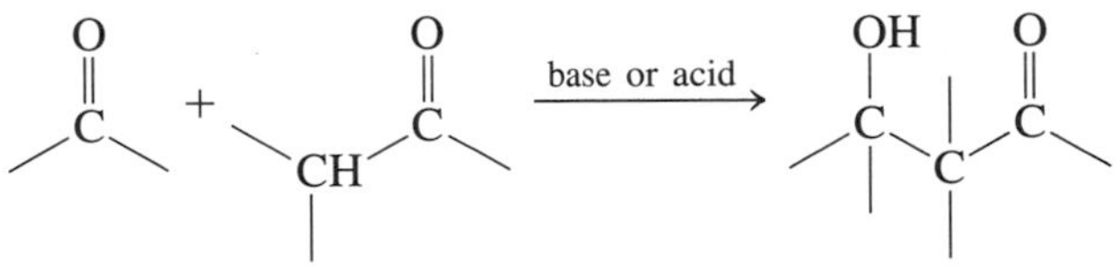

$$\text{R}_2\text{C}{=}\text{O} + \text{R}_2\text{CH--C(=O)R} \xrightarrow{\text{base or acid}} \text{R}_2\text{C(OH)--CR}_2\text{--C(=O)R}$$

Aldehydes with no α-hydrogen atoms undergo the Cannizzaro reaction and form an alcohol and the sodium salt of the acid:

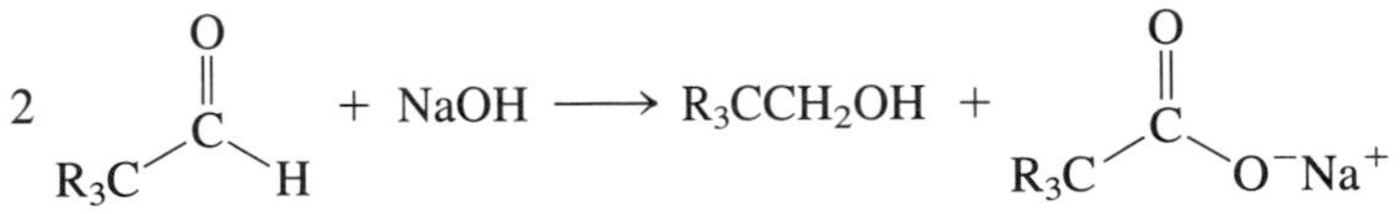

$$2\ R_3C\text{--}\overset{O}{\overset{\|}{C}}\text{--}H + NaOH \longrightarrow R_3CCH_2OH + R_3C\text{--}\overset{O}{\overset{\|}{C}}\text{--}O^-Na^+$$

Polyfunctional compounds such as β-diketones and β-keto esters undergo cleavage under the influence of hot alkalies. The possibility of such interfering reactions is detected by means of other classification reagents mentioned in Chapter 7 and emphasizes the fact that *a single classification reagent cannot be taken as proof of the presence of a certain functional group.* It is important to correlate all tests in attempting to draw conclusions concerning the structure of an unknown compound.

In procedure (c), esters of sterically hindered acids undergo the following reaction with 100% sulfuric acid:

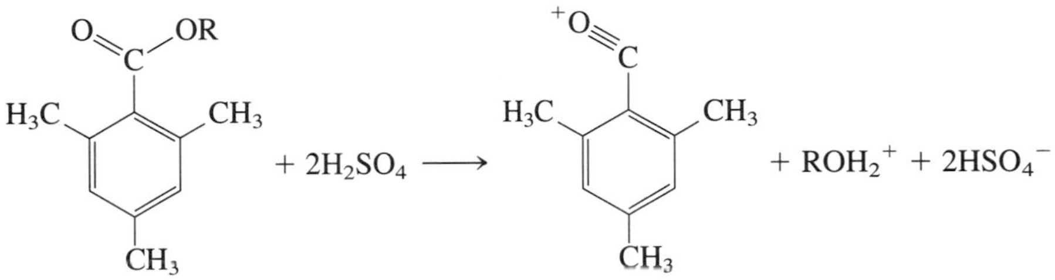

$$2,4,6\text{-}(CH_3)_3C_6H_2\text{--}C(=O)OR + 2H_2SO_4 \longrightarrow 2,4,6\text{-}(CH_3)_3C_6H_2\text{--}C{\equiv}O^+ + ROH_2^+ + 2HSO_4^-$$

When water is added, the intermediate ion forms the acid.

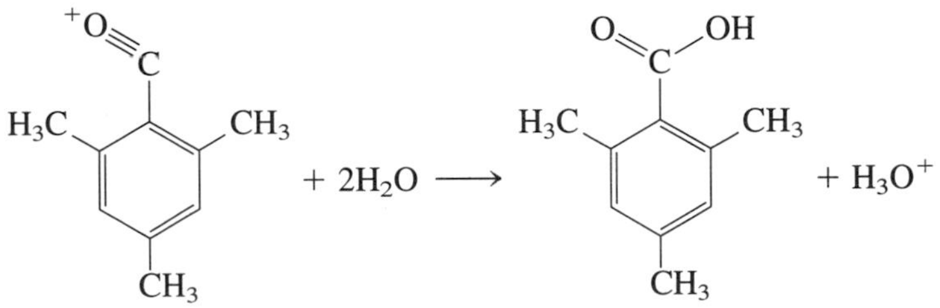

$$2,4,6\text{-}(CH_3)_3C_6H_2\text{--}C{\equiv}O^+ + 2H_2O \longrightarrow 2,4,6\text{-}(CH_3)_3C_6H_2\text{--}C(=O)OH + H_3O^+$$

Conversely, the sterically hindered acid may be converted readily to an ester by dissolving it in 100% sulfuric acid and treating the solution with the alcohol.

Unhindered esters do not undergo these reactions. They dissolve in 100% sulfuric acid and are recovered unchanged when the solution is poured into ice water.

Esters of sterically hindered acids and alcohols such as *tert*-butyl 2,4,6-trimethylbenzoate, though extremely resistant to hydrolysis by alkalies, may be hydrolyzed readily by boiling for 1 hr with 18% hydrochloric acid.[13]

[13] S. G. Cohen and A. Schneider *J. Amer. Chem. Soc. 63,* 3382 (1941).

Questions

6. Describe the structural characteristics and physical properties for esters in which procedure (a) would be the most likely method.
7. Show by equations the products formed by the alkaline hydrolysis of (a) 4-phenylphenacyl acetate; (b) ethylene glycol dibenzoate; (c) dibutyl oxalate; (d) glycerol triacetate and (e) diethyl phthalate. List a suitable derivative for detecting the products formed in these reactions.

Procedure 35. Saponification Equivalents (SE) of Esters

(a) Diethylene Glycol Method

$$\underset{\text{ester}}{RC(=O)OR'} + NaOH \xrightarrow{(HOCH_2CH_2)_2O} \underset{\text{salt of the acid}}{RC(=O)O^-Na^+} + \underset{\text{alcohol}}{R'OH}$$

$$RC(=O)O^-Na^+ \xrightarrow{H^+} \underset{\text{carboxylic acid}}{RC(=O)OH}$$

The weight of a small round-bottom flask is determined with three-figure accuracy. A sample of 0.4–0.6 g of the ester is added and the flask is weighed again. The difference in weights is the exact weight of the ester, with three significant figures. Exactly 10 mL of the potassium hydroxide in diethylene glycol reagent, measured from a buret, is pipetted into the flask.

The flask is stoppered with a greased glass stopper and the ester is mixed thoroughly with the reagent for 10 min by constant rotation of the flask. The flask is then equipped with a reflux condenser and heated in an oil bath at a temperature of 70–80°C for 2–3 min with constant agitation during heating. The flask is removed from the heating bath, stoppered, and shaken vigorously. The condenser is placed on the flask and the mixture is heated at 120–130°C for 3 min. The flask and its contents are then cooled to 80–90°C and 15 mL of distilled water is added through the condenser, rinsing down any condensation of alcohol or carboxylic acid vapors. The contents of the flask are mixed thoroughly and then the solution is allowed to cool to room temperature.

The solution is titrated with 0.25 M hydrochloric acid, which has been previously standardized to three-figure accuracy, using phenolphthalein as the indicator.

Potassium Hydroxide in Diethylene Glycol Reagent. The reagent is made by dissolving 3 g of potassium hydroxide pellets in 15 mL of diethylene glycol. It is necessary to warm the mixture gently to effect solution. A thermometer should be used for stirring, and the mixture should not be heated above 130°C; higher temperatures may cause the reagent to be colored. After all the solid has dissolved, the warm solution is poured into 35 mL of diethylene glycol. The solution is mixed thoroughly and allowed to cool.

It is approximately 1.0 M and is standardized to three significant figures by pipetting 10 mL in a flask, adding 10 mL of water, and titrating with 0.25 M hydrochloric acid solution, which has been previously standardized to a three-figure accuracy.

(b) Alcoholic Sodium Hydroxide Method

$$\underset{\text{ester}}{R-\overset{\overset{O}{\|}}{C}-OR'} + NaOH \xrightarrow{CH_3CH_2OH} \underset{\text{salt of the acid}}{R-\overset{\overset{O}{\|}}{C}-O^-Na^+} + \underset{\text{alcohol}}{R'OH}$$

$$\xrightarrow{H^+} \underset{\text{carboxylic acid}}{R-\overset{\overset{O}{\|}}{C}-OH}$$

The weight of a small round-bottom flask is determined with three-figure accuracy. A sample of 0.2–0.4 g of the ester is added and the flask is weighed again. The difference in weights is the exact weight of the ester with three significant figures. Fifteen milliliters of the alcoholic sodium hydroxide reagent, measured from a buret, is added to the flask. The flask is equipped with a reflux condenser and the mixture is gently heated under reflux for 1.25–1.5 hr. At the end of this time it is allowed to cool slightly. Ten milliliters of distilled water is added through the condenser into the solution, rinsing down any condensation along the sides of the tube. Two drops of phenolphthalein indicator are added to the flask and the solution is titrated with 0.25 M hydrochloric acid, which has been previously standardized with three-figure accuracy. The end point should be a faint pink.

Alcoholic Sodium Hydroxide Reagent. In the hood, 8 g of sodium is dissolved in 250 mL of absolute ethanol, and after the sodium has dissolved, 25 mL of water is added. *Care must be taken when working with sodium. Only dry equipment is used to weigh out the sodium. A dry knife is used to expose the shiny metal; only the shiny metal is used in any reaction. Any equipment used should be placed in a beaker; ethanol is slowly added. The beaker is allowed to sit in the hood for several hours before disposing of the liquid.* This solution is standardized to three-figure accuracy by titration against a weighed sample of pure potassium acid phthalate.

Cleaning Up. Place the neutralized solution in the aqueous solution container. The contents of the beaker used in the disposal of waste sodium are placed in the aqueous solution container.

Discussion. The saponification equivalent (SE) is calculated according to the following equation:

$$\text{saponification equivalent (SE)} = \frac{\text{weight of the sample (mg)}}{\text{volume of alkali (mL)} \times M_{KOH} - \text{volume of acid (mL)} \times M_{HCl}}$$

Procedure (a) gives complete saponification of esters which are insoluble in water. Esters such as benzyl acetate, butyl phthalate, ethyl sebacate, butyl oleate, and glycol and glycerol esters are completely saponified.

The ester must be *pure* and *anhydrous* in order to give an accurate saponification equivalent. The following precautions must be observed in order to obtain accurate results.

1. The alcoholic sodium hydroxide solution should be standardized immediately before use and its molarity (M) recorded.
2. The standard solutions should be measured accurately from a buret because a slight error in the amount of alkali will cause a large error in the saponification equivalent. This is especially noticeable with high molecular weight esters.
3. Heating for 1.5 hr will saponify most esters. For some, a longer time (2–24 hr) may be necessary.
4. The glassware should be lubricated lightly with stopcock grease prior to assembling. Hot alcoholic alkali will "freeze" two ground glass joints together within a few minutes if the grease is not present.
5. The end point should be a *faint* pink. This is the color assumed by phenolphthalein at pH 9, which represents the hydrogen ion concentration of solutions of the sodium salts of most organic acids.
6. The molecular weight of the ester is equal to x times the saponification equivalent, where x is the number of ester groups in the molecule.

Questions

8. What saponification equivalent (SE) value would be obtained for ethyl acetoacetate? Ethyl hydrogen phthalate? Diethyl propanedioate? Ethyl cyanoacetate? Dibutyl phthalate?
9. What would happen if Procedure 35 was applied to benzaldehyde?
10. If an ester had already been partially hydrolyzed, what effect would this have on the saponification equivalent?

Procedure 36. 4-Toluidides (4-Methylanilides) from Esters

$$CH_3CH_2Br + Mg \longrightarrow CH_3CH_2MgBr$$

ethylmagnesium bromide

$$CH_3CH_2MgBr + H_3C{-}C_6H_4{-}NH_2 \longrightarrow H_3C{-}C_6H_4{-}NHMgBr + CH_3CH_3$$

4-toluidine (4-aminotoluene) aminomagnesium compound

$$R{-}C({=}O){-}OR' + H_3C{-}C_6H_4{-}NHMgBr \longrightarrow (H_3C{-}C_6H_4{-}NH)_2C(R){-}OMgBr + R'OMgBr$$

oxymagnesium compound

$$(H_3C{-}C_6H_4{-}NH)_2C(R){-}OMgBr + 2HCl \longrightarrow H_3C{-}C_6H_4{-}NH{-}C(R){=}O + H_3C{-}C_6H_4{-}NH_3^+Cl^- + BrMgCl$$

4-toluidide

All glassware used to prepare the Grignard reagent must be dry. It is recommended that the glassware be baked in the oven for a few hours prior to usage. A 50-mL round-bottom flask is equipped with a Y-adapter, which holds a condenser and an addition funnel. Drying tubes are attached to the condenser and addition funnel.

One-half gram of magnesium is weighed out and a few of these magnesium turnings are ground up with a mortar and pestle. The magnesium is placed in the round-bottom flask. Next, 2.5 g (1.75 mL) of ethyl bromide and 15 mL of anhydrous diethyl ether are placed in the addition funnel. One-half milliliter of the ethyl bromide solution is slowly added. It may take several minutes before the Grignard reaction actually begins, which is indicated by the appearance of bubbles. If no reaction begins, then add 2 drops of 1,2-dibromoethane to initiate the reaction. Once the reaction has definitely begun, the remainder of the ethyl bromide is slowly added dropwise. Allow the mixture to stand until the reaction has ceased. Almost all of the magnesium metal should have been consumed to form the Grignard reagent, thus leaving very little, if any, of the magnesium turnings in the reaction flask.

A solution of 2.2 g of 4-toluidine (4-methylaniline) dissolved in 10 mL of anhydrous diethyl ether is placed in the addition funnel and slowly added to the ethylmagnesium bromide. After the vigorous evolution of ethane has ceased, 1 g of the ester in 5 mL of anhydrous diethyl ether is added dropwise from the addition funnel.

After the addition of the ester solution is complete, heat the solution on a steam bath for 10 min. Transfer the reaction mixture into an Erlenmeyer flask and place the flask in an ice bath. Add 10% hydrochloric acid until the solution is acidic to litmus. Acidification hydrolyzes the Grignard reagent and makes an ammonium salt from the excess 4-toluidine. Separate the ether layer and dry it with magnesium sulfate. Remove the ether by distillation, using a steam bath. The resulting solid 4-toluidide may be recrystallized from aqueous ethanol.

Cleaning Up. Place the recovered ether in the organic solvent container. Neutralize the aqueous layer with sodium carbonate and place in the aqueous solution container.

Discussion. Ethyl magnesium bromide is commercially available. Thus this sequence can be started with the second step.

Question

11. Calculate the number of moles of magnesium and ethyl bromide used in the procedure above. Which is the limiting reagent: magnesium or ethyl bromide?

Procedure 37. Alkyl 3,5-Dinitrobenzoates from Esters

$$\underset{\text{ester}}{RCOOR'} + \underset{\text{3,5-dinitrobenzoic acid}}{3,5\text{-}(O_2N)_2C_6H_3COOH} \xrightarrow{H^+} \underset{\text{3,5-dinitrobenzamide}}{3,5\text{-}(O_2N)_2C_6H_3COOR'} + RCOOH$$

About 1 mL or 0.9 g of the ester is mixed with 0.8 g of 3,5-dinitrobenzoic acid, and 1 drop of concentrated sulfuric acid is added. If the original ester boiled below 150°C, the mixture is heated gently under reflux. If the ester boiled above 150°C, the mixture is heated in an oil bath at 150°C, with frequent stirring. The time required varies from 30 min to 1 hr, the longer time being used in those cases in which the 3,5-dinitrobenzoic acid fails to dissolve in about 15 min. After the mixture has cooled, 15 mL of absolute ether is added, and the solution is extracted with two 10-mL portions of 5% sodium carbonate solution (***Caution:*** *foaming*) to neutralize the sulfuric acid and remove unreacted 3,5-dinitrobenzoic acid. The ether layer is washed with 10 mL of water, and the solvent is evaporated. The residue (usually an oil) is dissolved in 3 mL of boiling ethanol. After the solution has been filtered, water is added dropwise until the solution begins to get cloudy. The mixture is cooled and stirred to induce crystallization of the derivative.

Cleaning Up. Place the filtrates in the aqueous solution container. Any recovered ether is placed in the organic solvent container.

Procedure 38. *N*-Benzylamides from Esters

$$\underset{\text{ester}}{RCOOR'} + \underset{\text{benzylamine}}{C_6H_5CH_2NH_2} \longrightarrow \underset{\text{N-benzylamide}}{RCONHCH_2C_6H_5} + R'OH$$

A mixture of 1 g of the ester, 2 mL of benzylamine, and 0.1 g of powdered ammonium chloride is heated for 1 hr in a small flask fitted with a reflux condenser. After being cooled, the reaction mixture is washed with 10 mL of water to remove excess benzylamine and to induce crystallization. Often the addition of a few drops of 10% hydrochloric acid will promote crystallization; an excess of hydrochloric acid is to be avoided because it dissolves *N*-benzylamides. Occasionally the presence of unchanged ester may prevent crystallization. In that case it is best to boil the solution for a few minutes with 10 mL of water in an evaporating dish in the hood to volatilize the ester. The solid amide is collected on a filter, washed with a little ligroin, and recrystallized from a mixture of ethanol and water or of acetone and water.

Esters of alcohols higher than ethanol should be heated for 30 min with 5 mL of absolute methanol in which a small piece of sodium (0.1 g) has been dissolved. At the end of the reflux period, the methanol is evaporated and the residue treated by the above procedure.

Cleaning Up. Add 10% hydrochloric acid to the initial filtrate until the solution is slightly acidic. Combine all filtrates and place them in the aqueous solution container.

Procedure 39. Acid Hydrazides from Esters

$$\underset{\text{ester}}{\mathrm{R{-}C({=}O){-}OR'}} + \underset{\text{hydrazine}}{\mathrm{NH_2NH_2}} \longrightarrow \underset{\text{acid hydrazide}}{\mathrm{R{-}C({=}O){-}NHNH_2}} + \mathrm{R'OH}$$

Wear gloves when handling hydrazine, since hydrazine is a carcinogen. Do this experiment in the hood. One gram of the methyl or ethyl ester and 1 mL of 85% hydrazine hydrate are mixed, and the mixture is heated under reflux for 15 min. Just enough absolute ethanol is then added, through the top of the condenser, to obtain a clear solution. After the mixture has been heated under reflux for 2 hr, the alcohol is evaporated and the residue cooled. The crystals of the hydrazide are collected on a filter and recrystallized from water or a mixture of water and ethanol.

Higher esters must be subjected to methanolysis, as described in Procedure 38, p. 349, before treatment with hydrazine.

Cleaning Up. Combine all of the filtrates, dilute with 10 mL of water, and neutralize with sodium carbonate. Add 10 mL of 5.25% sodium hypochlorite (household bleach), and heat in a water bath at 50°C for 1 hr to oxidize any unreacted hydrazine. Dilute the mixture with 10 mL of water and place in the aqueous solution container.

8.9 ETHERS—ALIPHATIC

Derivatives of symmetrical aliphatic ethers can be prepared by treating the ether with 3,5-dinitrobenzoyl chloride in the presence of zinc chloride (Procedure 40, below). This method cleaves symmetrical ethers and forms the corresponding 3,5-dinitrobenzoate. The melting points of the 3,5-dinitrobenzoates can be found in the derivative tables under alcohols.

$$\underset{\text{ether}}{\mathrm{ROR}} + \underset{\text{3,5-dinitrobenzoyl chloride}}{\mathrm{3,5\text{-}(O_2N)_2C_6H_3COCl}} \longrightarrow \underset{\text{3,5-dinitrobenzoate}}{\mathrm{3,5\text{-}(O_2N)_2C_6H_3COOR}} + \mathrm{RCl}$$

Procedure 40. Alkyl 3,5-Dinitrobenzoates from Ethers

$$\underset{\text{ether}}{\mathrm{ROR}} + \underset{\text{3,5-dinitrobenzoyl chloride}}{\mathrm{3,5\text{-}(O_2N)_2C_6H_3COCl}} \longrightarrow \underset{\text{3,5-dinitrobenzoate}}{\mathrm{3,5\text{-}(O_2N)_2C_6H_3COOR}} + \mathrm{RCl}$$

A mixture of 1 mL or 0.9 g of the ether, 0.15 g of anhydrous zinc chloride, and 0.5 g of 3,5-dinitrobenzoyl chloride is placed in a small flask connected to a reflux condenser. The mixture is refluxed for 1 hr and then cooled. To this mixture is then added 10 mL of 5% sodium carbonate solution. The mixture is warmed to 90°C in a water bath, cooled, and filtered. The precipitate is washed with 5 mL of 5% sodium carbonate solution and 10 mL of distilled water. The residue is dissolved in 10 mL of hot chloroform, and the solution is filtered while hot; the filtrate is then cooled in an ice bath. Any precipitated ester is removed by suction filtration. If the ester does not separate, the chloroform is evaporated. The residue is allowed to dry on a watch glass, and the melting point is determined.

Cleaning Up. Place the aqueous filtrates in the aqueous solution container. Place the chloroform filtrate in the halogenated organic waste container.

Question

12. Describe the problem that would arise if an unsymmetrical ether such as ethyl methyl ether was subjected to 3,5-dinitrobenzoate derivatization.

8.10 ETHERS—AROMATIC

Picrates are used as derivatives of aromatic ethers (Procedure 41, p. 361). The picrates are prepared by the treatment of the aromatic ether with picric acid.

OR (aromatic ether) + O_2N, OH, NO_2, NO_2 (picric acid) ⟶ picrate

aromatic ether picric acid picrate

Aromatic ethers undergo reaction smoothly with chlorosulfonic acid at 0°C to produce arylsulfonyl chlorides (Procedure 42, p. 362).

OR (aromatic ether) + $2ClSO_3H$ ⟶ OR / O=S=O / Cl + HCl + H_2SO_4

aromatic ether chlorosulfonic acid arylsulfonyl chloride

Since the arylsulfonyl chlorides are usually oils or low-melting solids, they then undergo reaction with ammonium carbonate or ammonia to form the arylsulfonamides (Procedure 43, p. 363). Arylsulfonamides obtained in this way are useful derivatives.

$$\text{arylsulfonyl chloride } (ROC_6H_4SO_2Cl) \xrightarrow[\text{or } NH_4OH]{(NH_4)_2CO_3} \text{arylsulfonamide } (ROC_6H_4SO_2NH_2)$$

The derivatives of aromatic ethers which are employed most frequently are those obtained by nitration (Procedure 51, p. 373) and bromination (Procedure 44, p. 364).

$$\underset{\text{aromatic ether}}{RO{-}Ar{-}H} \xrightarrow[H_2SO_4]{HNO_3} \underset{\text{nitro compound}}{RO{-}Ar{-}NO_2}$$

$$\underset{\text{aromatic ether}}{RO{-}Ar{-}H} \xrightarrow{Br_2} \underset{\text{bromo compound}}{RO{-}Ar{-}Br}$$

The position of the nitro or bromo substituent on the aromatic ring depends upon the directing effect of the ether group and of the other groups attached to the aromatic ring. Frequently the bromo or the nitro group substitutes in more than one position on the aromatic ring.

The formation of picrate, nitro, and bromo compounds takes advantage of the ability of the aromatic ring to undergo attack by electron-deficient species.

Procedure 41. Picrates of Aromatic Ethers

$$\underset{\text{aromatic ether}}{C_6H_5OR} + \underset{\text{picric acid}}{2,4,6\text{-}(O_2N)_3C_6H_2OH} \longrightarrow \underset{\text{picrate}}{C_6H_5OR \rightarrow (O_2N)_3C_6H_2OH}$$

A solution of 0.5–1 g of the ether in 5 mL of chloroform is added to a boiling solution of 6 mL of picric acid in chloroform. ***Caution:*** *The picric acid may explode if it becomes dry; do not let the solvent evaporate.* The mixture is stirred thoroughly, set aside, and allowed to cool. The picrate crystallizes when the mixture is allowed to stand. The melting point should be determined as soon as possible because some picrates decompose.

Picric Acid in Chloroform. One gram of picric acid is dissolved in 15 mL of chloroform. This solution is prepared by the instructor.

Cleaning Up. Add 50 mL of water to the chloroform filtrate. Shake thoroughly and separate the layers. Pour the aqueous layer into the aqueous solution container. Place the chloroform layer in the halogenated organic waste container.

Procedure 42. Sulfonyl Chlorides from Aromatic Ethers

$$\text{ArOR (aromatic ether)} + 2ClSO_3H \text{ (chlorosulfonic acid)} \longrightarrow p\text{-}RO\text{-}C_6H_4\text{-}SO_2Cl \text{ (arylsulfonyl chloride)} + HCl + H_2SO_4$$

***Caution:** This reaction must be done in a hood.* A solution of 1 g of the compound in 5 mL of dry chloroform in a clean, dry test tube is cooled in a beaker of ice to 0°C. About 3 mL of chlorosulfonic acid is added dropwise, and after the initial evolution of the hydrogen chloride has subsided, the tube is removed from the ice bath and allowed to warm up to room temperature (about 20 min). The contents of the tube are poured into a 50 mL beaker full of cracked ice. The chloroform layer is removed and washed with water. The chloroform is evaporated, and the residual sulfonyl chloride is recrystallized from low-boiling petroleum ether or chloroform.

Cleaning Up. Place any recovered chloroform in the halogenated waste container. The aqueous solution is placed in the aqueous solution container. Recovered petroleum ether is placed in the organic solvent container.

Discussion. Since the arylsulfonyl chlorides are usually oils or low-melting solids, they are used to form sulfonamide derivatives (Procedure 43, p. 363).

This procedure may also be used for the preparation of sulfonyl chlorides from most of the simple aryl halides and alkylbenzenes. For the halotoluenes, it is desirable to warm the chloroform solution to 50°C for 10 min and then pour the mixture on cracked ice.

Polyhalogen derivatives require more drastic treatment. For such compounds the sulfonation is carried out without any solvent, and the sulfonation mixture is warmed to 100°C for 1 hr under a reflux condenser.

Two other reactions may take place during the sulfonation. Fluorobenzene, iodobenzene, and 1,2-dibromobenzene yield the corresponding sulfones when treated with chlorosulfonic acid at 50°C in the absence of a solvent.

$$2\,C_6H_5X \text{ (aryl halide)} \xrightarrow[50°C]{ClSO_3H} X\text{-}C_6H_4\text{-}SO_2\text{-}C_6H_4\text{-}X$$

chlorosulfonic acid

These sulfones are solids and will serve as derivatives. In other cases (e.g., during Procedure 43, p. 363), small amounts of sulfones may be produced and these are separable from the sulfonamide because they are insoluble in alkali. A second reaction is ring chlorination. This reaction takes place with 1,4-diiodobenzene and 1,2,4,5-tetrachlorobenzene. Unsatisfactory results are frequently obtained with aryl iodides.

Procedure 43. Sulfonamides from Aromatic Ethers

(a) With Ammonium Carbonate

$$\text{ROC}_6\text{H}_4\text{-}(O{=}S{=}O)\text{-Cl} \xrightarrow{(NH_4)_2CO_3} \text{ROC}_6\text{H}_4\text{-}(O{=}S{=}O)\text{-NH}_2 + NH_4Cl + H_2O + CO_2$$

arylsulfonyl chloride   arylsulfonamide

The sulfonyl chloride (0.5 g) is mixed with 2.0 g of dry powdered ammonium carbonate and heated at 100°C for 30 min. The mixture is cooled and washed with three 10-mL portions of cold water.

The crude sulfonamide is dissolved in 10 mL of 5% sodium hydroxide solution, gentle heating being used if necessary, and the solution is filtered to remove any sulfone or chlorinated products. The filtrate is acidified with 5% hydrochloric acid, and the sulfonamide is removed by filtration. It is purified by recrystallization from dilute ethanol and dried at 100°C.

(b) With Ammonium Hydroxide

$$\text{ROC}_6\text{H}_4\text{-}(O{=}S{=}O)\text{-Cl} \xrightarrow{NH_4OH} \text{ROC}_6\text{H}_4\text{-}(O{=}S{=}O)\text{-NH}_2$$

arylsulfonyl chloride   arylsulfonamide

About 0.5 g of the sulfonyl chloride is boiled with 5 mL of concentrated ammonium hydroxide for 10 min. The mixture is diluted with 10 mL of water, cooled, and filtered.

The crude sulfonamide is purified according to the directions given in (a).

Cleaning Up. Pour the aqueous rinsings and filtrates from (a) and (b) into the aqueous solution container.

Discussion. This procedure, or the Hinsberg procedure (Procedure 21, p. 335), can be used to prepare N-substituted sulfonamides by using primary and secondary amines instead of ammonia.

Procedure 44. Bromination of Ethers

$$\underset{\text{aromatic ether}}{\text{RO—Ar—H}} \xrightarrow{Br_2} \underset{\text{bromo compound}}{\text{RO—Ar—Br}}$$

One gram of the compound is dissolved in 15 mL of glacial acetic acid, and 1–1.5 mL of liquid bromine is added. *Bromine is very toxic; bromine is measured out in the hood and the experiment is done in the hood.* The mixture is allowed to stand for 15–30 min and is then poured onto 50–100 mL of water. The bromo compound that separates is removed by filtration and purified by recrystallization from dilute ethanol. In some cases carbon tetrachloride may be substituted for acetic acid as the solvent. The carbon tetrachloride is distilled, and the residue is recrystallized. *Carbon tetrachloride is a known carcinogen. Do not allow any carbon tetrachloride to contact the skin.*

Cleaning Up. Place the filtrates in the halogenated organic waste container.

8.11 HALIDES—ALKYL

The best methods for making derivatives of alkyl and cycloalkyl halides depend on their conversion into the corresponding Grignard reagents. The anilides and 1-naphthalides, prepared from the Grignard reagents (Procedure 46, p. 366) by treatment with phenyl and 1-naphthyl isocyanate, respectively, are the derivatives most frequently used.

$$\underset{\text{alkyl halide}}{\text{RX}} + \text{Mg} \longrightarrow \underset{\substack{\text{alkylmagnesium halide}\\ \text{(Grignard reagent)}}}{\text{RMgX}}$$

$$\text{RMgX} + \underset{\substack{\text{phenyl isocyanate or}\\ \text{naphthyl isocyanate}}}{\text{Ar—N=C=O}} \longrightarrow \text{Ar—N=C(R)(OMgX)} \xrightarrow{H_2O} \text{Ar—N=C(R)(OH)}$$

$$\longrightarrow \underset{\substack{\text{anilide or}\\ \text{1-naphthalide}}}{\text{Ar—NH—C(R)=O}}$$

The Grignard reagent may be converted to the corresponding alkylmercuric halide by treatment with a mercuric halide (Procedure 45, p. 366).

$$\underset{\substack{\text{alkylmagnesium}\\ \text{halide}}}{\text{RMgX}} + \underset{\substack{\text{mercuric}\\ \text{halide}}}{HgX_2} \longrightarrow \underset{\substack{\text{alkylmercuric}\\ \text{halide}}}{\text{RHgX}} + MgX_2$$

Alkyl 2-naphthyl ethers are prepared by the reaction of the alkyl halides with 2-naphthol (Procedure 47, p. 367).

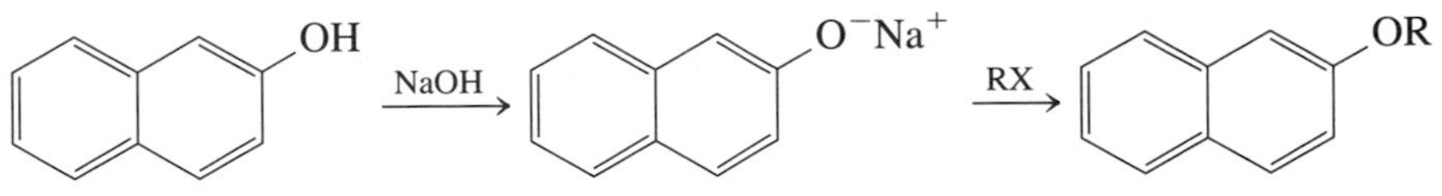

2-naphthol $\xrightarrow{NaOH}$ $\xrightarrow{RX}$ alkyl 2-naphthyl ether

The alkyl 2-naphthyl ethers can then be treated with picric acid to form another derivative, the alkyl 2-naphthyl ether picrates (Procedure 48, p. 368).

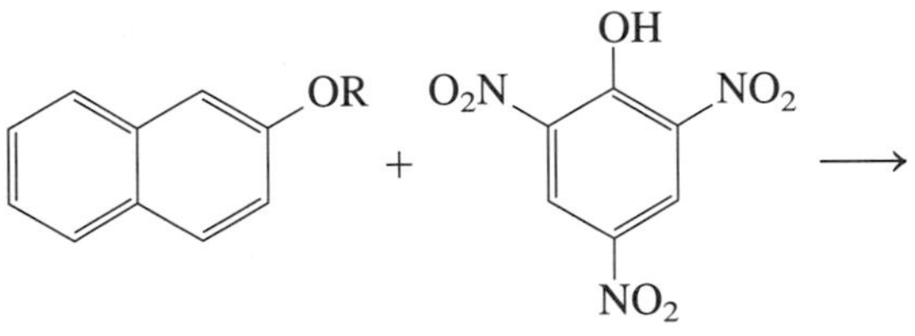

alkyl 2-naphthyl ether + picric acid $\longrightarrow$

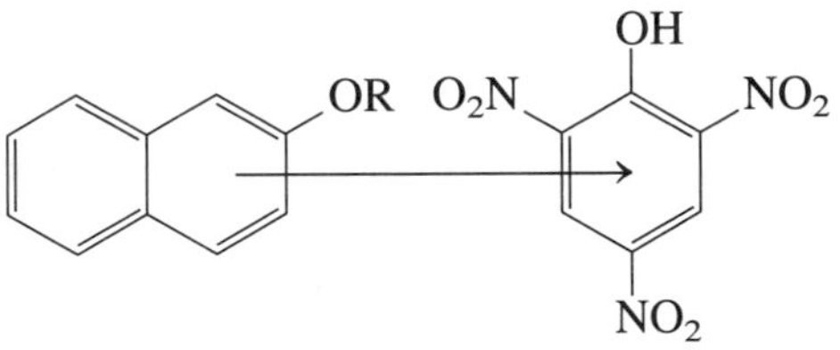

alkyl 2-naphthyl picrate

S-Alkylthiuronium picrates are also derivatives of alkyl halides (Procedure 49, p. 368). These derivatives are prepared by the treatment of the alkyl halide to form the intermediate, which is treated with picric acid to produce the *S*-alkylthiuronium picrate.

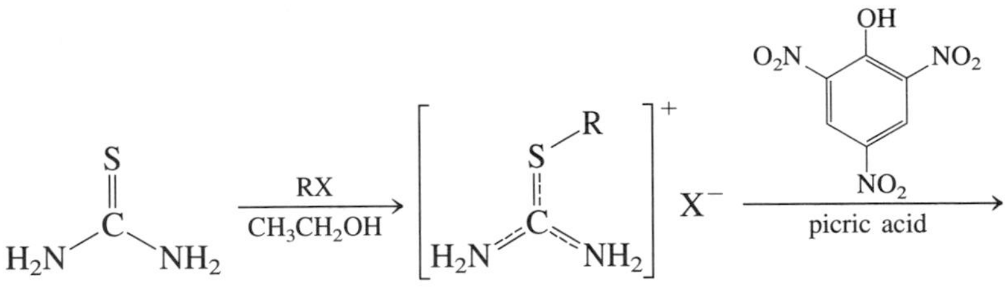

thiourea $\xrightarrow[CH_3CH_2OH]{RX}$ $[H_2N{=}C(SR){=}NH_2]^+ X^-$ $\xrightarrow{\text{picric acid}}$

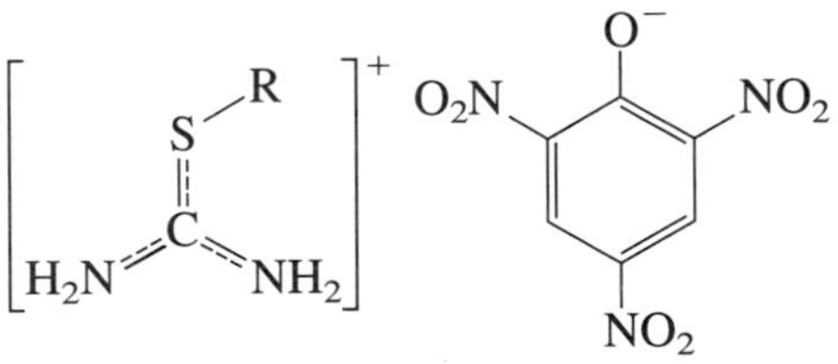

S-alkylthiuronium picrate

Derivatives of alkyl and cycloalkyl halides are particularly useful not only because these compounds are encountered frequently but also because they are readily made from alcohols, and so furnish an indirect way of identifying the alcohols. All the preceding methods are to be used with caution in view of the fact that rearrangements sometimes occur.

Procedure 45. Grignard Reagents and Alkylmercuric Halides

$$\underset{\text{alkyl halide}}{RX} + Mg \longrightarrow \underset{\substack{\text{alkylmagnesium halide}\\ \text{(Grignard reagent)}}}{RMgX}$$

$$\underset{\substack{\text{alkylmagnesium}\\ \text{halide}}}{RMgX} + \underset{\substack{\text{mercuric}\\ \text{halide}}}{HgX_2} \longrightarrow \underset{\substack{\text{alkylmercuric}\\ \text{halide}}}{RHgX} + MgX_2$$

In a 50-mL round-bottom flask, the Grignard reagent is prepared by treating 0.3 g of magnesium with 1 mL of the alkyl halide in 15 mL of dry diethyl ether, following the directions given in Procedure 36, p. 356.

When the reaction is complete, the solution is filtered through a little glass wool and the filtrate allowed to flow into a test tube containing 4–5 g of mercuric chloride, bromide, or iodide, depending on the halogen in the original alkyl halide. The reaction mixture is shaken vigorously, warmed on a steam cone for a few minutes, and the solvent removed by distillation using a steam bath. The residue is boiled with 20 mL of 95% ethanol, and the solution is filtered. The filtrate is diluted with 10 mL of water and cooled in an ice bath. The alkylmercuric halide that separates is collected on a filter and recrystallized from 60% ethanol.

Cleaning Up. Place the recovered ether in the organic solvent container. Distill off the ethanol from the filtrate, using a steam bath. The ethanol is placed in the aqueous solution container. The liquid that remains after distillation may contain some mercuric halides and is placed in the hazardous waste container.

Procedure 46. Anilides and *N*-Naphthylamides

$$\underset{\text{alkyl halide}}{RX} + Mg \longrightarrow \underset{\substack{\text{alkylmagnesium halide}\\ \text{(Grignard reagent)}}}{RMgX}$$

$$RMgX + \underset{\text{phenyl isocyanate}}{C_6H_5N{=}C{=}O} \longrightarrow C_6H_5N{=}C(OMgX)R \xrightarrow{H_2O} C_6H_5N{=}C(OH)R \longrightarrow \underset{\text{anilide}}{C_6H_5NH{-}C({=}O){-}R}$$

RMgX + naphthyl isocyanate → (OMgX adduct) $\xrightarrow{H_2O}$ (OH adduct) → 1-naphthalide

naphthyl isocyanate

1-naphthalide

The Grignard reagent, prepared with the amounts as listed in Procedure 45, p. 366 and following the directions in Procedure 36, p. 356, is treated with 0.5 mL of phenyl or 1-naphthyl isocyanate dissolved in 10 mL of absolute diethyl ether. The mixture is shaken and allowed to stand for 10 min. About 25 mL of 2% hydrochloric acid is added, with very vigorous shaking. The ether layer is separated and dried with magnesium sulfate, and the ether is distilled. The residue is recrystallized from methanol, diethyl ether, or petroleum ether.

Cleaning Up. Treat any unreacted phenyl isocyanate with an excess of 5.25% sodium hypochlorite (household bleach), dilute with 10 mL of water, and place in the aqueous solution container. Place the recovered ether and recrystallization solvents in the organic solvent container.

Procedure 47. Alkyl 2-Naphthyl Ethers from Alkyl Halides

2-naphthol $\xrightarrow{NaOH}$ sodium 2-naphthoxide (O^-Na^+) $\xrightarrow{RX}$ alkyl 2-naphthyl ether (OR)

2-naphthol

alkyl 2-naphthyl ether

To a solution of 0.3 g of sodium hydroxide in 12 mL of ethanol are added 1 g of 2-naphthol and 1 g of the alkyl halide. *2-Naphthol is a carcinogen and must only be used in the hood.* If the halide is a chloride, 0.25 g of potassium iodide is added also. The mixture is heated under reflux for 30 min and poured into 40 mL of cold water. If the mixture is not distinctly alkaline to phenolphthalein, then add dropwise 10% sodium hydroxide solution as needed, and stir vigorously. The alkyl 2-naphthyl ether is removed by filtration and recrystallized from ethanol or an ethanol–water mixture.

Cleaning Up. Place any unreacted alkyl halide in the halogenated organic waste container. Place the aqueous layer in the aqueous solution container.

Discussion. Since the reaction is S_N2, the alkyl halide is restricted to primary and some secondary alkyl halides.

Question

13. This procedure is useful occasionally for making derivatives of dihalides of the type $X(CH_2)_nX$. Potassium iodide must not be added if the compound is a 1,2-dihalide ($n = 2$). Why?

Procedure 48. Picrates of Alkyl 2-Naphthyl Ethers

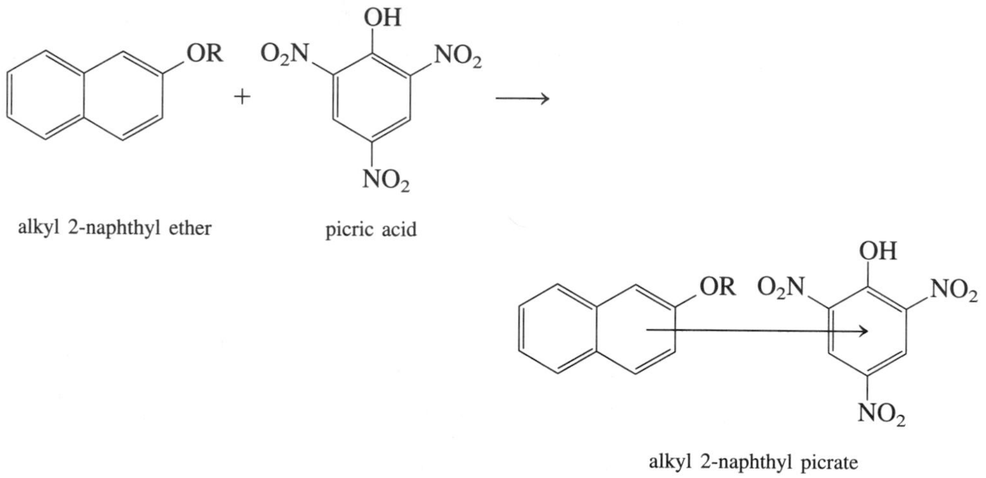

OR O₂N OH NO₂ + NO₂ →

alkyl 2-naphthyl ether picric acid

OR O₂N OH NO₂ NO₂

alkyl 2-naphthyl picrate

The alkyl 2-naphthyl ether from Procedure 47 is dissolved in a minimum volume of hot ethanol. A hot solution of 6 mL of 20% picric acid in ethanol is added. *The picric acid may explode if it becomes dry; do not let the solvent evaporate.* The picrate separates out on cooling, and can be recrystallized from ethanol.

20% Picric Acid in Ethanol. One gram of picric acid is dissolved in 5 mL of ethanol. The solution is prepared by the instructor.

Cleaning Up. Pour the filtrates into the aqueous solution container.

Procedure 49. *S*-Alkylthiuronium Picrates from Alkyl Halides

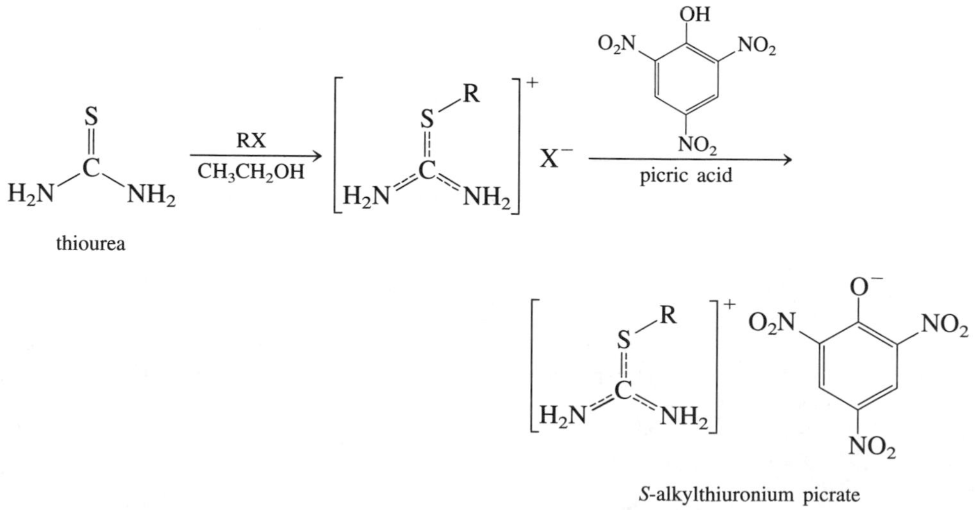

$H_2N-C(=S)-NH_2$ (thiourea) $\xrightarrow[CH_3CH_2OH]{RX}$ $[H_2N{=}C(SR){=}NH_2]^+ X^-$ $\xrightarrow{\text{picric acid}}$ $[H_2N{=}C(SR){=}NH_2]^+$ picrate anion (O⁻, O₂N, NO₂, NO₂)

thiourea

picric acid

S-alkylthiuronium picrate

A mixture of 0.5 powdered thiourea, 0.5 g of the alkyl halide, and 5 mL of 95% ethanol is placed in a reflux apparatus. The reflux time is dependent upon the type of alkyl halide: primary alkyl bromides and iodides, 10–20 min; secondary alkyl bromides and iodides, 2–3 hr; alkyl chlorides, with the addition of 1 g of potassium iodide, 50–60 min; and dihalides, 20–50 min.[14]

Six milliliters of 10% picric acid in ethanol is added to the above solution, and the mixture is heated at reflux until a clear solution is obtained. *The picric acid may explode if it becomes dry; do not let the solvent evaporate.* After cooling, the *S*-alkylthiuronium picrate is removed by filtration and recrystallized from ethanol.

10% Picric Acid in Ethanol. Dissolve 0.4 g of picric acid in 5 mL of ethanol. This solution is prepared by the instructor.

Cleaning Up. Place the aqueous filtrate in the aqueous solution container.

8.12 HALIDES—AROMATIC

Many excellent methods are available for making suitable derivatives of halogen derivatives of aromatic hydrocarbons. The most general method of these procedures is nitration (Procedure 51, p. 373). The aromatic ring is nitrated in the presence of nitric and sulfuric acids.

$$\underset{\text{aryl halide}}{X{-}Ar{-}H} \xrightarrow[H_2SO_4]{HNO_3} \underset{\text{nitro compound}}{X{-}Ar{-}NO_2}$$

Aryl halides undergo reaction readily with chlorosulfonic acid to produce arylsulfonyl chlorides (Procedure 42, p. 362), which yield sulfonamides when treated with ammonia solution (Procedure 43b, p. 362).

$$\underset{\text{aryl halide}}{X{-}C_6H_5} + \underset{\text{chlorosulfonic acid}}{2ClSO_3H} \longrightarrow \underset{\text{arylsulfonyl chloride}}{X{-}C_6H_4{-}SO_2Cl} + HCl + H_2SO_4$$

$$\underset{\text{arylsulfonyl chloride}}{X{-}C_6H_4{-}SO_2Cl} \xrightarrow{NH_4OH} \underset{\text{arylsulfonamide}}{X{-}C_6H_4{-}SO_2NH_2}$$

[14]B. S. Furniss, A. J. Hannaford, P. W. G. Smith, and A. R. Tatchell, *Vogel's Textbook of Practical Organic Chemistry*, 5th ed. (Wiley, New York, 1989), p. 1253.

Many aryl halides contain other groups attached to the aromatic ring. Aryl halides containing side chains are frequently oxidized to the corresponding substituted aromatic acid (Procedure 50, below). Oxidizing agents such as sodium dichromate or potassium permanganate can be used.

$$\underset{\text{aryl compound}}{\text{ArR}} \xrightarrow[H_2SO_4]{Na_2Cr_2O_7} \underset{\text{carboxylic acid}}{\text{Ar}-\overset{\overset{O}{\|}}{C}-\text{OH}}$$

$$\underset{\text{aryl compound}}{\text{ArR}} \xrightarrow[OH^-]{KMnO_4} \text{Ar}-\overset{\overset{O}{\|}}{C}-O^- \xrightarrow{H^+} \underset{\text{carboxylic acid}}{\text{Ar}-\overset{\overset{O}{\|}}{C}-\text{OH}}$$

Procedure 50. Oxidation of a Side Chain of an Aromatic Compound

(a) Dichromate Oxidation

$$\underset{\text{aryl compound}}{\text{ArR}} \xrightarrow[H_2SO_4]{Na_2Cr_2O_7} \underset{\text{carboxylic acid}}{\text{Ar}-\overset{\overset{O}{\|}}{C}-\text{OH}}$$

In a small flask are placed 8 mL of water, 3.5 g of sodium dichromate, and 1 g of the compound to be oxidized. Five milliliters of concentrated sulfuric acid is added, the flask is attached to a reflux condenser, and the apparatus thoroughly shaken. The flask is heated carefully until the reaction starts; then the flame should be removed and the flask cooled if necessary. After the mixture has ceased to boil from the heat of the reaction, it is heated under reflux for 2 hr. The contents of the flask are cooled into an ice bath and poured into 15 mL of water. The precipitate is isolated by suction filtration. The precipitate is mixed with 10 mL of 5% sulfuric acid, and the mixture is warmed on a steam bath with vigorous stirring. It is cooled, and the precipitate is separated and washed with 10 mL of cold water. The precipitate is dissolved in 10 mL of 5% sodium hydroxide solution, and the solution is filtered. The filtrate is poured, with vigorous stirring, into 15 mL of cold 10% sulfuric acid. The precipitate is collected on a filter, washed with water, and purified by recrystallization from ethanol or 60% ethanol.

Cleaning Up. Since the filtrate probably contains unreacted sodium dichromate, it must be treated. To the filtrate add 10% sulfuric acid until the pH is 1. Slowly add solid sodium thiosulfate until the solution becomes a cloudy blue color. Neutralize with 10% sodium carbonate. Collect the precipitate of chromium hydroxide through Celite in a Buchner funnel. Place the Celite and the chromium hydroxide solid in the hazardous waste container for heavy metals. Place the filtrate in the aqueous solution container.

(b) Permanganate Oxidation

$$\underset{\text{aryl compound}}{ArR} \xrightarrow[OH^-]{KMnO_4} Ar-\overset{\overset{O}{\parallel}}{C}-O^- \xrightarrow{H^+} \underset{\text{carboxylic acid}}{Ar-\overset{\overset{O}{\parallel}}{C}-OH}$$

One gram of the compound is added to 80 mL of water containing 4 g of potassium permanganate. One milliliter of 10% sodium hydroxide solution is added, and the mixture is heated under reflux for 3 hr. At the end of this time the mixture is allowed to cool and the purple color is discharged by adding a small amount of sodium bisulfite. The brown suspension of manganese dioxide is removed by suction filtration through a Buchner funnel containing a thin layer of Celite (filter aid). The filtrate is acidified carefully with 10% sulfuric acid and the resulting mixture is heated for 0.5 hr and cooled. The precipitated acid is collected on a filter and recrystallized from ethanol. If the acid is appreciably soluble in water, it may not separate from this dilute acid solution. In this event the acid may be extracted with chloroform or diethyl ether. A slight precipitate of silicic acid sometimes appears on acidification; hence it is important to recrystallize the acid before taking the melting point.

Cleaning Up. Place the manganese dioxide in the hazardous waste container for heavy metals. Neutralize the aqueous filtrate with sodium carbonate and place in the aqueous solution container.

Discussion. Aromatic hydrocarbons which have side chains may be oxidized to the corresponding acids. Aromatic acids having several carboxyl groups are sometimes difficult to oxidize, and for this reason the utility of the oxidation method is limited. If there are two side chains situated in adjacent positions on the ring, oxidation is recommended because the resulting acid (phthalic acid) is easy to identify. 1,2-Dialkylbenzenes, as well as compounds substituted on the 1,2-dialkyl chains, undergo complete oxidation with Cr(VI) (Procedure 50a). The melting points of the acids from the oxidation of the alkylbenzenes may be obtained by reference to the tables in Appendix III.

Aromatic rings which are substituted with electron-withdrawing groups (nitro, halo, etc.) easily survive even the more vigorous oxidation procedures, whereas rings substituted with electron-donating groups may be oxidized on the ring more readily than on the side chain. For example, the oxidation of a substituted phenol can give the aliphatic acid in sufficient yield to be characterized.

$$\text{HO-C}_6\text{H}_4\text{-R} \xrightarrow[H_2O,\ 25^\circ C]{KMnO_4} R-\overset{\overset{O}{\parallel}}{C}-OH$$

It is clear that hydrogens α to the ring (benzylic hydrogens) facilitate this oxidation. The lack of hydrogens α to the ring prevents the oxidation.

$$(CH_3)_3C\text{-}C_6H_5 \xrightarrow[OH^-,\ heat]{KMnO_4} \text{No Reaction}$$

Substitution of a second aromatic ring onto the alkyl group can decrease the degree of oxidation.

$$ArCH_2Ar' \xrightarrow{oxidation} Ar\text{-}C(=O)\text{-}Ar'$$

8.13 HYDROCARBON—AROMATIC

The most useful derivatives of aromatic hydrocarbons are usually those obtained by nitration (Procedure 51, p. 373). Nitration of the aromatic ring is effected by a combination of nitric and sulfuric acids or by fuming nitric acid. Highly alkylated nitrated benzenes can then be reduced to the amines (Procedure 57, p. 381), which are then acetylated or benzoylated to give mono- or diacetamino or benzamino derivatives (Procedure 20, p. 333).

$$\underset{\text{aryl compound}}{Ar\text{—}H} \xrightarrow[H_2SO_4]{HNO_3} \underset{\text{nitro compound}}{Ar\text{—}NO_2}$$

$$\underset{\text{aryl compound}}{Ar\text{—}H} \xrightarrow[HNO_3]{\text{fuming}} \underset{\text{nitro compound}}{Ar\text{—}NO_2}$$

$$\underset{\text{nitro compound}}{ArNO_2} \xrightarrow[HCl]{Sn} \underset{\text{amine}}{ArNH_2} \xrightarrow{R\text{-}C(=O)\text{-}Cl} \underset{\text{amide}}{ArNH\text{-}C(=O)\text{-}R}$$

Aromatic hydrocarbons and their halogen derivatives undergo the Friedel–Crafts reaction with phthalic anhydride, producing aroylbenzoic acids in good yield (Procedure 52, p. 374).

$$\underset{\text{aryl compound}}{ArH} + \underset{\text{phthalic anhydride}}{C_6H_4(CO)_2O} \xrightarrow[CS_2,\ heat]{AlCl_3,} \underset{\text{aroylbenzoic acid}}{C_6H_4(C(=O)Ar)(C(=O)OH)}$$

Picric acid combines with some aromatic hydrocarbons to yield picrates (Procedure 41, p. 361). The stability of the picrates varies considerably; many of them dissociate readily into the original reactants.

$$\text{ArH} + \text{(OH)}(O_2N)_2C_6H_2\text{-}NO_2 \longrightarrow \text{ArH} \rightarrow \text{(OH)}(O_2N)_2C_6H_2\text{-}NO_2$$

aryl compound — picric acid — picrate

Procedure 51. Aromatic Nitration

Caution: *The nitration of an aromatic compound, especially if it is an unknown substance, should be carried out with special care because many of these compounds react violently. A small-scale test behind a shield should be tried first.*

(a) With Nitric and Sulfuric Acids

$$\underset{\text{aryl compound}}{\text{Ar—H}} \xrightarrow[H_2SO_4]{HNO_3} \underset{\text{nitro compound}}{\text{Ar—}NO_2}$$

The nitro compound may be prepared by either method.

1. About 1 g of the compound is added to 4 mL of concentrated sulfuric acid. Four milliliters of concentrated nitric acid is added drop by drop, with shaking after each addition. The flask is connected to a reflux condenser and kept in a beaker of water at 45°C for 5 min. The reaction mixture is poured on 25 g of cracked ice and the precipitate collected on a filter. It may be recrystallized from 60% ethanol.
2. Four mL of concentrated nitric acid is placed in a 50-mL round-bottom flask, which is placed in an ice bath. Four mL of concentrated sulfuric acid is added dropwise. To this solution is added 1 g of the unknown a few drops at a time, with shaking after each addition. The solution is refluxed and worked up as described above.

(b) With Fuming Nitric Acid

$$\underset{\text{aryl compound}}{\text{Ar—H}} \xrightarrow[HNO_3]{\text{fuming}} \underset{\text{nitro compound}}{\text{Ar—}NO_2}$$

The procedure outlined above is followed except that 4 mL of fuming nitric acid[15] is used instead of concentrated nitric acid, and the mixture is warmed on a steam bath for 10 min. Occasionally, with compounds which are difficult to nitrate, fuming sulfuric acid may be substituted for the concentrated sulfuric acid.

[15]This corresponds to highly concentrated (ca. 90%) nitric acid, sometimes called "white" or "yellow" fuming nitric acid; this is a powerful nitrating agent. It is not to be confused with "red" nitric acid, which contains dissolved NO_2 and is a vigorous oxidizing agent.

Cleaning Up. Place the solution in a large beaker. Slowly dilute the filtrate with 20 mL of water, add sodium carbonate until the foaming ceases, and place in the aqueous solution container.

Discussion. Procedure (a) yields 1,3-dinitrobenzene from benzene or nitrobenzene and the 4-nitro derivative of chlorobenzene, bromobenzene, benzyl chloride, or toluene. Phenol, acetanilide, naphthalene, and biphenyl yield dinitro derivatives. It is best to employ Procedure (b) for halogenated benzenes because it produces dinitro derivatives which are easier to purify than the mononitro derivatives formed in (a). Mesitylene, the xylenes, and 1,2,4-trimethylbenzene yield trinitro derivatives.

Polynuclear aromatic hydrocarbons, which are easily oxidized, are not nitrated successfully by either procedure. Experiment 39, p. 288, should be used to determine the possibility of a polynuclear aromatic hydrocarbon. These compounds require mild conditions to nitrate without oxidative decomposition.

Procedure 52. Aroylbenzoic Acids from Aromatic Hydrocarbons

$$\text{ArH} + \text{phthalic anhydride} \xrightarrow[\text{CS}_2,\ \text{heat}]{\text{AlCl}_3,} o\text{-}C_6H_4(\text{C(=O)Ar})(\text{C(=O)OH})$$

aryl compound phthalic anhydride aroylbenzoic acid

To a solution of 1 g of the dry aromatic hydrocarbon and 1.2 g of phthalic anhydride in 10 mL of dry carbon disulfide is added 2.4 g of anhydrous aluminum chloride. *Aluminum chloride is very hygroscopic and releases hydrogen chloride gas when it undergoes reaction with water.* The mixture is heated under a reflux condenser in a boiling water bath for 30 min and cooled. The carbon disulfide layer is decanted, and 10 mL of concentrated hydrochloric acid and 10 mL of water are added to the residue. The acid should be added slowly at first, with cooling by ice if necessary, and the final mixture should be thoroughly shaken.

If the aroylbenzoic acid separates as a solid, it is immediately collected on a filter and washed with cold water. If an oil separates, the mixture is cooled in an ice bath for some time to induce crystallization. If the product remains oily, the supernatant liquid is decanted and the oil is washed with cold water. The crude product, whether a solid or an oil, is boiled for 1 min with 30 mL of 10% ammonium hydroxide solution to which has been added about 0.1 g of Norite. The hot solution is filtered and cooled; 25 g of crushed ice is then added, and the solution is acidified with concentrated hydrochloric acid. The aroylbenzoic acid is removed by filtration and is recrystallized from dilute ethanol. Sometimes it is necessary to allow the product to stand overnight in order to obtain crystals.

Cleaning Up. Place the carbon disulfide in the hazardous waste container. Neutralize the aqueous filtrates with sodium carbonate and place in the aqueous solution container.

Discussion. This procedure produces a derivative with a functional group that can be easily characterized. For example, the neutralization equivalent (Procedure 1, p. 307) of the acid may be determined; this will yield the molecular weight of the keto acid derivative. The molecular weight of the original aromatic compound may be calculated:

$$\text{mol. wt. of ArH} = \text{mol. wt. of acid} - 148$$

Question

14. Explain the origin of the value 148 in the preceding formula.

8.14 NITRILES

The nitriles may be hydrolyzed to the corresponding carboxylic acid by means of mineral acid or aqueous base (Procedure 53, p. 377).

$$\underset{\text{nitrile}}{R{-}C{\equiv}N} \xrightarrow[H_2O]{H_2SO_4} \underset{\text{carboxylic acid}}{R{-}C(=O){-}OH} + NH_4{}^+HSO_4{}^-$$

$$\underset{\text{nitrile}}{R{-}C{\equiv}N} \xrightarrow[H_2O]{NaOH} R{-}C(=O){-}O^-Na^+ + NH_3$$

$$R{-}C(=O){-}O^-Na^+ \xrightarrow{H^+} \underset{\text{carboxylic acid}}{R{-}C(=O){-}OH}$$

If the resulting acid is a solid, it serves as an excellent derivative. The carboxylic acid can be derivatized by preparing the acyl chloride, followed by treatment with ammonia or aniline to form the amide (Procedure 3, p. 309) or the anilide (Procedure 4a or 4b, p. 310), respectively.

$$\underset{\text{carboxylic acid}}{R{-}C(=O){-}OH} + \underset{\text{thionyl chloride}}{SOCl_2} \longrightarrow \underset{\text{acyl chloride}}{R{-}C(=O){-}Cl} + SO_2 + HCl$$

$$\underset{\text{acyl chloride}}{R{-}C(=O){-}Cl} + 2NH_3 \longrightarrow \underset{\text{amide}}{R{-}C(=O){-}NH_2} + NH_4Cl$$

$$\underset{\text{acyl halide}}{R{-}C(=O){-}X} + \underset{\text{aniline or 4-toluidine}}{ArNH_2} \longrightarrow \underset{\text{anilide}}{R{-}C(=O){-}NHAr} + ArNH_3X$$

Controlled hydrolysis of the nitrile with a boron trifluoride–acetic acid complex results in the corresponding amide (Procedure 54, p. 378). This procedure can be used instead of the multistep process listed above.

$$R{-}C{\equiv}N \xrightarrow[BF_3,\ CH_3COOH]{H_2O} \left[R{-}C(OH){=}NH \right] \longrightarrow R{-}C({=}O)NH_2$$

nitrile amide

Reduction of nitriles with sodium and an alcohol forms primary amines (Procedure 55, p. 378).

$$R{-}C{\equiv}N \xrightarrow[CH_3CH_2OH]{Na} \xrightarrow{HCl} RCH_2NH_3^+Cl^- \xrightarrow{NaOH} RCH_2NH_2$$

nitrile 1° amine

The primary amines can then be used to prepare the benzamide (Procedure 20b or 20c, p. 333 or 334), benzenesulfonamide (Procedure 21, p. 335), and phenylthiourea (Procedure 22, p. 336) derivatives. The benzamide is prepared from the reaction of the primary amine with benzoyl chloride.

$$2\ R{-}NH_2 + C_6H_5{-}C({=}O){-}Cl \longrightarrow H{-}N(R){-}C({=}O){-}C_6H_5 + R{-}NH_3^+\ Cl^-$$

1° amine benzoyl chloride benzamide

Treatment of the primary amine with benzenesulfonyl chloride yields the benzamide.

$$RNH_2 + C_6H_5SO_2Cl + 2NaOH \longrightarrow C_6H_5SO_2NR^-Na^+ + NaCl + 2H_2O$$

1° amine benzenesulfonyl chloride (soluble)

$$\xrightarrow{HCl} C_6H_5SO_2NHR + NaCl$$

benzenesulfonamide

The phenylthiourea is prepared by the reaction of the amine with phenyl isothiocyanate.

$$RNH_2 + C_6H_5N{=}C{=}S \longrightarrow C_6H_5{-}NH{-}C({=}S){-}NHR$$

1° amine phenyl isocyanate phenylthiourea

Mercaptoacetic acid (thioglycolic acid) condenses with nitriles in the presence of hydrogen chloride to produce α-(imidioylthio)acetic acid hydrochlorides (Procedure 56, p. 379).

$$R{-}C{\equiv}N + HS{-}CH_2{-}C({=}O){-}OH + HCl \longrightarrow R{-}C({=}NH \cdot HCl){-}S{-}CH_2{-}C({=}O){-}OH$$

nitrile thioglycolic acid α-(imidioylthio)acetic acid hydrochloride

Procedure 53. Hydrolysis of a Nitrile to a Carboxylic Acid

(a) Acidic Conditions for Aromatic Nitriles

$$\underset{\text{aromatic nitrile}}{Ar{-}C{\equiv}N} \xrightarrow[H_2O]{H_2SO_4} \underset{\text{carboxylic acid}}{Ar{-}C(=O){-}OH} + NH_4^+ HSO_4^-$$

About 5 mL of 75% sulfuric acid and 0.2 g of sodium chloride are placed in a small round-bottom flask fitted with a reflux condenser. The flask is heated to 150–160°C by means of an oil bath, and 1 g of the nitrile is added dropwise through the top of the condenser, with vigorous shaking after the addition of each portion. The mixture is heated, with stirring, at 160°C for 30 min and at 190°C for another 30 min. It is then cooled and poured on 20 g of cracked ice in a beaker, and the precipitate is collected on a filter. The precipitate is treated with a slight excess of 10% sodium hydroxide solution, and any insoluble amide is removed by filtration. Acidification of the filtrate with 10% sulfuric acid yields the carboxylic acid, which may be purified by recrystallization from an acetone–water mixture.

(b) Basic Conditions

$$\underset{\text{nitrile}}{R{-}C{\equiv}N} \xrightarrow[H_2O]{NaOH} R{-}C(=O){-}O^- Na^+ + NH_3$$

$$R{-}C(=O){-}O^- Na^+ \xrightarrow{H^+} \underset{\text{carboxylic acid}}{R{-}C(=O){-}OH}$$

One gram of the nitrile is refluxed with 5 mL of 40% sodium hydroxide for 1–3 hr, or until the evolution of ammonia ceases. The mixture is cooled in an ice bath, and 25% sulfuric acid is added slowly until the solution is acidic. Insoluble carboxylic acids can then be removed by filtration and recrystallized. If no solid is obtained, the solution is extracted three times with 15 mL of diethyl ether. The combined ether layers are dried, filtered, and concentrated to yield the carboxylic acid.

Cleaning Up. Neutralize the aqueous filtrates with sodium carbonate and place in the aqueous solution container. Place the recovered ether in the organic solvent container.

Discussion. The solid carboxylic acid may be used as a derivative. A solid or liquid carboxylic acid can be converted to an amide (Procedure 3, p. 309) or an anilide (Procedure 4a or 4b, p. 310).

Hydrochloric acid is more effective than sulfuric for some nitriles. However, for nitriles which are difficult to hydrolyze, it is advisable to choose sulfuric acid because of its higher boiling point. The addition of a small amount of hydrochloric acid (as sodium chloride) increases the rate of reaction.

Prior to using Procedure (a), a few drops or crystals of the sample are placed in a test tube containing 0.2 mL of 75% sulfuric acid. In the hood, the sample is heated to boiling. If charring occurs, Procedure (b) should be used.

Procedure 54. Controlled Hydration of a Nitrile to an Amide

$$\underset{\text{nitrile}}{R{-}C{\equiv}N} \xrightarrow[\substack{BF_3,\\ CH_3COOH}]{H_2O} \left[R{-}C\begin{matrix} /OH \\ \backslash\backslash NH \end{matrix} \right] \longrightarrow \underset{\text{amide}}{R{-}C\begin{matrix} //O \\ \backslash NH_2 \end{matrix}}$$

A mixture of 1 g of the nitrile, 1.33 g of water, and 6.7 g of boron trifluoride–acetic acid complex[16] is placed in a 100-mL flask fitted with a reflux condenser. The mixture is heated to 115–120°C in an oil bath, held at this temperature range for 10 min, and then cooled in an ice bath to 15–20°C. A solution of 6 M sodium hydroxide is added slowly, keeping the temperature below 20°C, until the mixture is just alkaline to litmus paper. About 30–35 mL of the sodium hydroxide solution will be needed. The cold solution is extracted with three 35-mL portions of 1:1 ethyl ether–ethyl acetate solvent. The combined extracts are dried with 1.5 g of anhydrous sodium sulfate and filtered into a distilling flask; the ethyl ether–ethyl acetate solvent is removed by distillation from a water bath. The residual amide is purified by crystallization from water or a mixture of water and methanol. The amide is dried in a vacuum desiccator before the melting point is taken.

Cleaning Up. Place the recovered ethyl ether–ethyl acetate mixture in the organic solvent container. Place the aqueous layer in the aqueous solution container.

Discussion. This controlled hydration of a nitrile to an amide may be accomplished in a short time in yields of 90–95% by use of boron trifluoride–acetic acid catalyst using a limited amount of water.[17]

Diamides (from dinitriles) may be extracted from the cold alkaline mixture more efficiently by use of three 100-mL portions of methylene chloride or chloroform.

Procedure 55. Reduction of Nitriles to Primary Amines

$$\underset{\text{nitrile}}{R{-}C{\equiv}N} \xrightarrow[CH_3CH_2OH]{Na} \xrightarrow{HCl} RCH_2NH_3{}^+Cl^- \xrightarrow{NaOH} \underset{1^\circ\text{ amine}}{RCH_2NH_2}$$

In a clean, dry 100-mL round-bottom flask, fitted with a reflux condenser, are placed 20 mL of absolute ethanol and 1 g of an aliphatic nitrile or 2 g of an aromatic nitrile.

[16]The liquid complex, $BF_3{\cdot}2CH_3COOH$ may be purchased from the Aldrich Chemical Company or prepared according to the procedure described in L. F. Fieser and M. Feiser, *Reagents for Organic Synthesis* (Wiley, New York, 1967), p. 69, and *Org. Syn., Coll. Vol. II,* 586 (1943).

[17]C. R. Hauser and D. S. Hoffenberg, *J. Org. Chem. 20,* 1448 (1955).

Through the top of the condenser 1.5 g of finely cut sodium slices is added as rapidly as possible without causing the reaction to become too vigorous. When the reduction is complete (10–15 min), the mixture is cooled to 20°C, and 10 mL of concentrated hydrochloric acid is added dropwise through the condenser and with vigorous swirling of the contents of the flask. The reaction mixture is tested to make certain that it is acid to litmus. The glassware is then rearranged to a simple distillation apparatus and about 20 mL of the ethanol–water mixture are removed by distillation. The flask and contents are cooled, and a small dropping funnel containing 15 mL of 40% sodium hydroxide is fitted to the top of the distillation apparatus. An adapter is attached to the end of the condenser and arranged so that it dips into 3 mL of water in a 50-mL Erlenmeyer flask. The alkali is added drop by drop, with shaking. The reaction is vigorous, and care must be exercised to avoid adding the alkali too fast. After all the alkali has been added, the mixture is heated until the distillation of the amine is complete. Distillation is stopped when the contents of the flask are very viscous.

Cleaning Up. Rinse the initial distillate of ethanol–water into an aqueous solution container. Neutralize the aqueous layer with 10% hydrochloric acid and place in the aqueous solution container.

Discussion. The primary amine isolated from this reaction is used to prepare derivatives such as a benzamide (Procedure 20b or 20c, p. 333 or 334), a benzenesulfonamide (Procedure 21, p. 335), or a phenylthiourea (Procedure 22, p. 336).

Procedure 56. α-(Imidioylthio)acetic Acids from Nitriles

$$R{-}C{\equiv}N + HS{-}CH_2{-}C(=O){-}OH + HCl \longrightarrow R{-}C(=NH){-}S{-}CH_2{-}C(=O){-}OH \cdot HCl$$

nitrile — thioglycolic acid — α-(imidioylthio)acetic acid hydrochloride

One gram of the nitrile and 2.0 g of mercaptoacetic acid (thioglycolic acid) are dissolved in 15 mL of absolute ethyl ether in a clean, dry test tube. The solution is cooled in an ice bath and thoroughly saturated with dry hydrogen chloride. The hydrogen chloride gas is prepared as previously described (Procedure 26, p. 338) and is dried by passing the gas through a trap containing concentrated sulfuric acid. The tube is tightly stoppered and kept in the ice bath or refrigerator until crystals of the derivative separate. Aliphatic nitriles form the addition compound in 15–30 min, whereas aromatic nitriles usually have to stand overnight in the refrigerator.

The crystals are removed by filtration, washed thoroughly with absolute ether, and placed in a vacuum desiccator containing sulfuric acid in the bottom and small beakers of potassium hydroxide pellets and paraffin wax in the top. The decomposition point is determined in the usual melting point apparatus and, if necessary, the neutralization equivalent by titration with standard alkali, thymol blue being used as the indicator.

Determination of the neutralization equivalent (Procedure 1, p. 307) can lead to the mass of the R group; one should keep in mind that there are two acid groups in such a derivative.

Cleaning Up. Add 100 mL of 5.25% sodium hypochlorite (household bleach) to the aqueous filtrate. The mixture is heated at 45–50°C for 2 hr in a hot water bath. At this point all of the thioglycolic acid should have been oxidized. Place the mixture in the aqueous solution container.

8.15 NITRO COMPOUNDS

The reduction of nitro compounds in acidic media leads to the formation of primary (1°) amines (Procedure 57, p. 381).

$$\underset{\text{nitro compound}}{RNO_2} \xrightarrow[HCl]{Sn} NH_3^+Cl^- \xrightarrow{NaOH} \underset{1^\circ \text{ amine}}{RNH_2}$$

These primary amines can then be converted into suitable derivatives such as acetamides (Procedure 20a, p. 333), benzamides (Procedure 20b or 20c, p. 333 or 334), or benzenesulfonamides (Procedure 21, p. 335). The acetamide is prepared by the reaction of the amine with acetic anhydride.

$$\underset{1^\circ \text{ amine}}{R{-}NH{-}H} + \underset{\text{acetic anhydride}}{H_3C{-}C(=O){-}O{-}C(=O){-}CH_3} \longrightarrow \underset{\text{acetamide}}{H{-}N(R){-}C(=O){-}CH_3} + \underset{\text{acetic acid}}{H_3C{-}C(=O){-}OH}$$

The reaction of the amine with benzoyl chloride produces the benzamide.

$$2\ \underset{1^\circ \text{ amine}}{R{-}NH{-}H} + \underset{\text{benzoyl chloride}}{C_6H_5{-}C(=O){-}Cl} \longrightarrow \underset{\text{benzamide}}{H{-}N(R){-}C(=O){-}C_6H_5} + R{-}NH_3^+\ Cl^-$$

Amines undergo reaction with benzenesulfonyl chloride to yield benzenesulfonamides.

$$\underset{1^\circ \text{ amine}}{RNH_2} + \underset{\text{benzenesulfonyl chloride}}{C_6H_5SO_2Cl} + 2NaOH \longrightarrow \underset{\text{(soluble)}}{C_6H_5SO_2NR^-Na^+} + NaCl + 2H_2O$$

$$\xrightarrow{HCl} \underset{\text{benzenesulfonamide}}{C_6H_5SO_2NHR} + NaCl$$

Procedure 57. Reduction of a Nitro Compound with Tin and Hydrochloric Acid

$$\underset{\text{nitro compound}}{RNO_2} \xrightarrow[HCl]{Sn} NH_3^+Cl^- \xrightarrow{NaOH} \underset{1^\circ \text{ amine}}{RNH_2}$$

One gram of the nitrogen-containing compound (nitro, nitroso, azo, azoxy, or hydrazo) is added to 2 g of granulated tin in a small flask. The flask is connected to a reflux condenser, and 20 mL of 10% hydrochloric acid is added, in small portions, with vigorous shaking after each addition. Then 5 mL of ethanol is added. Finally the mixture is warmed on the steam bath for 10 min. The solution is decanted while it is still hot into 10 mL of water, and sufficient 40% sodium hydroxide solution is added to dissolve the tin hydroxide. The solution is extracted several times with 10-mL portions of ether. The ether extract is dried with magnesium sulfate, filtered, and the ether removed by distillation. The remaining residue is the amine, which can then be made into other derivatives or recrystallized and a melting point taken if the amine is a solid.

Cleaning Up. Neutralize the aqueous filtrate with 10% hydrochloric acid. Separate the tin hydroxide by filtration, and place in the nonhazardous solid waste container. Place the aqueous filtrate in the aqueous solution container. Place the recovered ether in the organic solvent container.

Discussion. These primary amines can be used to prepare derivatives such as acetamides (Procedure 20a, p. 333), benzamides (Procedure 20b or 20c, p. 333 or 334), or benzenesulfonamides (Procedure 21, p. 335).

8.16 PHENOLS

Phenols, like alcohols, yield urethanes when treated with isocyanates (Procedures 8, p. 315, or 58, p. 383). Both the phenylurethanes and the 1-naphthylurethanes (α-naphthylurethanes) are generally useful derivatives in identifying phenols.

$$\underset{\text{phenol}}{ArOH} + \underset{\text{isocyanate}}{Ar'{-}N{=}C{=}O} \longrightarrow \underset{\text{phenylurethane}}{Ar'{-}NH{-}C(=O){-}OAr}$$

Benzoates (Procedure 9, p. 316), 4-nitrobenzoates (Procedure 9, p. 316), or 3,5-dinitrobenzoates (Procedure 10, p. 317) are easily prepared from the reaction of the phenol with the corresponding acyl halide.

$$\underset{\text{phenol}}{ArOH} + \underset{\text{acyl halide}}{Ar'{-}C(=O){-}Cl} \longrightarrow \underset{\text{benzoate}}{Ar'{-}C(=O){-}O{-}Ar}$$

The acetates of monohydroxy aromatic compounds are usually liquids. Aromatic compounds containing two or more hydroxy groups, as well as substituted phenols, are frequently solids. Preparation of the acetates from phenols is done with acetic anhydride in basic media (Procedure 59, p. 383).

$$\text{ArOH} + \text{H}_3\text{C}-\overset{\overset{\text{O}}{\|}}{\text{C}}-\text{O}-\overset{\overset{\text{O}}{\|}}{\text{C}}-\text{CH}_3 \xrightarrow{\text{NaOH}} \xrightarrow{\text{HCl}} \text{H}_3\text{C}-\overset{\overset{\text{O}}{\|}}{\text{C}}-\text{O}-\text{Ar} + \text{H}_3\text{C}-\overset{\overset{\text{O}}{\|}}{\text{C}}-\text{OH}$$

phenol acetic anhydride acetate

In presence of alkali, phenols react readily with chloroacetic acid to give aryloxyacetic acids. These derivatives crystallize well from water and have proved to be exceedingly useful in characterization work (Procedure 60, p. 384).

$$\text{ArOH} + \text{NaOH} \longrightarrow \text{ArO}^-\text{Na}^+$$

phenol

$$\text{ArO}^-\text{Na}^+ + \text{Cl}-\text{CH}_2-\overset{\overset{\text{O}}{\|}}{\text{C}}-\text{OH} \longrightarrow \text{Ar}-\text{O}-\text{CH}_2-\overset{\overset{\text{O}}{\|}}{\text{C}}-\text{OH} + \text{NaCl}$$

chloroacetic acid aryloxyacetic acid

The aryloxyacetic acids can be compared not only by melting point determinations, but also by their neutralization equivalents (Procedure 1, p. 307). Accurate determination of the neutralization equivalent can lead to an estimate of the total mass of substituents on the aryl ring.

Similar to the use of bromine as a qualitative test for phenols (Experiment 44, p. 298), bromine can be used to form derivatives of phenol (Procedure 61, p. 384). Since the phenolic ring is reactive toward such electrophilic reagents, every proton atom in an *ortho* or *para* position is displaced by bromine; in fact, bromine often substitutes for groups other than protons.

$$\text{ArOH} + \text{Br}_2 \longrightarrow \text{Ar(Br)}_x\text{OH}$$

phenol

For example, bromine reacts with both phenol and 4-hydroxybenzoic acid to form 2,4,6-tribromophenol.

$$\text{phenol} \xrightarrow{3\text{Br}_2} \text{2,4,6-tribromophenol} \xleftarrow{3\text{Br}_2} \text{4-hydroxybenzoic acid}$$

phenol

4-hydroxybenzoic acid

2,4,6-tribromophenol

Procedure 58. Phenylurethanes and 1-Naphthylurethanes from Phenols

$$ArOH + C_6H_5{-}N{=}C{=}O \longrightarrow C_6H_5{-}NH{-}C({=}O){-}OAr$$

phenol phenyl isocyanate phenylurethane

$$ArOH + C_{10}H_7{-}N{=}C{=}O \longrightarrow C_{10}H_7{-}NH{-}C({=}O){-}OAr$$

phenol 1-naphthyl isocyanate 1-naphthylurethane

To a mixture of 0.5 g of the dry phenol and 0.5 mL of phenyl isocyanate or 1-naphthyl isocyanate in a dry 25-mL flask is added 1 drop of dry pyridine. The flask is loosely stoppered with a plug of cotton and heated on a steam bath for 15 min. If the derivative does not solidify during this time, crystallization is induced by cooling the flask and scratching the walls. When crystals have formed, 10 mL of dry ethyl acetate is added. The mixture is heated on a steam bath and filtered through a fluted filter. Hexane is now added until a turbidity or crystals are obtained. Crystallization is allowed to proceed overnight; the product is then removed by filtration, washed with hexane, and air dried.

If any water is present in the original sample or reagents, the product will be contaminated with *N,N'*-diphenylurea (mp 238°C) or *N,N'*-dinaphthylurea (mp 287°C). Purification may be effected by warming the product with 10 mL of carbon tetrachloride and removing the insoluble diphenylurea by filtration. The urethanes may be obtained by cooling the filtrate. Occasionally the filtrate may have to be evaporated to 2 or 3 mL to cause the crystallization to occur.

Cleaning Up. Treat any unreacted phenyl isocyanate or 1-naphthyl isocyanate with an excess of 5.25% sodium hypochlorite (household bleach) and place in the aqueous solution container. The organic solvents are placed in the organic solvent container.

Procedure 59. Preparation of Acetates from Phenols

$$ArOH + H_3C{-}C({=}O){-}O{-}C({=}O){-}CH_3 \xrightarrow{NaOH} \xrightarrow{HCl} H_3C{-}C({=}O){-}O{-}Ar + H_3C{-}C({=}O){-}OH$$

phenol acetic anhydride acetate

One gram of the phenol is dissolved in 5 mL of 3 M sodium hydroxide solution. If the solution is not slightly basic, add 3 M sodium hydroxide as needed. Approximately 15 g of crushed ice is added quickly, followed immediately by 1.5 g (1.5 mL) of acetic anhydride. Shake the mixture vigorously for 1 min. If the acetate does not separate immediately, then add 10% sulfuric acid until the solution is acidic to litmus. The acetate is isolated by filtration and recrystallized from hot water or dilute ethanol.

Cleaning Up. Place the filtrate in the aqueous solution container.

Procedure 60. Aryloxyacetic Acids of Phenols

$$\underset{\text{phenol}}{\text{ArOH}} + \text{NaOH} \longrightarrow \text{ArO}^-\text{Na}^+$$

$$\text{ArO}^-\text{Na}^+ + \underset{\text{chloroacetic acid}}{\text{Cl}-\text{CH}_2-\text{C}(=\text{O})-\text{OH}} \longrightarrow \underset{\text{aryloxyacetic acid}}{\text{Ar}-\text{O}-\text{CH}_2-\text{C}(=\text{O})-\text{OH}} + \text{NaCl}$$

To a mixture of 1 g of the phenol with 5 mL of a 33% sodium hydroxide solution is added 1.5 g of chloroacetic acid. The mixture is shaken thoroughly, and 1–5 mL of water may be added if necessary in order to dissolve the sodium salt of the phenol. The test tube containing the mixture is then kept in a beaker of boiling water for 1 hr. The solution is cooled, diluted with 10–15 mL of water, acidified to Congo Red at a pH of 4 with 10% hydrochloric acid, and extracted with 50 mL of ether. The ether solution is washed with 10 mL of cold water and is then shaken with 25 mL of 5% sodium carbonate solution. The sodium carbonate solution is acidified with 10% hydrochloric acid (***Caution:*** *foaming*); the aryloxyacetic acid is collected by filtration and recrystallized from hot water.

Cleaning Up. Place the ether layer in the organic solvent container. Neutralize the aqueous filtrate with sodium carbonate and place in the aqueous solution container.

Procedure 61. Bromination of Phenols

$$\underset{\text{phenol}}{\text{ArOH}} + \text{Br}_2 \longrightarrow \text{Ar(Br)}_x\text{OH}$$

(a) With Bromine and Potassium Bromide. *This reaction must be done in the hood.* The potassium bromide and bromine solution is added slowly, with shaking, to a solution of 0.5 g of the phenol dissolved in water, ethanol, or dioxane until a yellow color persists. About 25 mL of water is then added, and the mixture is shaken vigorously to break up the lumps. The bromo derivative is removed by filtration and washed with a dilute solution of sodium bisulfite. It is recrystallized from ethanol or a water–ethanol mixture.

Potassium Bromide and Bromine Solution. A brominating solution is prepared by dissolving 15 g of potassium bromide in 100 mL of water and adding 10 g of bromine. This solution is prepared by the instructor.

(b) With Bromine in Acetic Acid. *This reaction must be done in the hood.* One gram of the compound is dissolved in 10–15 mL of acetic acid. A solution of 3–4 mL of bromine dissolved in 10–15 mL of acetic acid is added dropwise until the yellow to brown color of bromine persists. The mixture is allowed to stand for 20 min. Pour the mixture into 75 mL of water. Isolate the solid product by suction filtration and rinse the product with 10 mL of cold water. The product is recrystallized from 60% ethanol.

Cleaning Up. The filtrates and any excess bromine in acetic acid solution are extracted with three 5-mL portions of methylene chloride to remove the unreacted bromine. The methylene chloride extract is treated with sufficient cyclohexene to discharge the bromine color and is then placed in the halogenated organic waste container. The aqueous layer is placed in the aqueous solution container.

8.17 SULFONIC ACIDS, SULFONYL CHLORIDES, SULFONAMIDES

Sulfonic acids and their salts are converted into sulfonyl chlorides by heating with phosphorus pentachloride (Procedure 62, p. 387).

$$R-SO_2-OH + PCl_5 \longrightarrow R-SO_2-Cl + POCl_3 + HCl$$

sulfonic acid — phosphorus pentachloride — sulfonyl chloride

$$R-SO_2-O^-Na^+ + PCl_5 \longrightarrow R-SO_2-Cl + POCl_3 + NaCl$$

salt of the sulfonic acid — phosphorus pentachloride — sulfonyl chloride

The sulfonyl chloride is rarely used as a derivative itself. The sulfonyl chloride undergoes reaction with aqueous ammonium carbonate or aqueous ammonia to obtain the sulfonamide (Procedure 43, p. 363).

$$R-SO_2-Cl \xrightarrow[\text{or } NH_4OH]{(NH_4)_2CO_3} R-SO_2-NH_2$$

sulfonyl chloride — sulfonamide

The sulfanilide is prepared by reacting sulfonyl chloride with aniline (Procedure 63, p. 387).

$$R-SO_2-Cl + C_6H_5NH_2 \longrightarrow R-SO_2-NH-C_6H_5$$

sulfonyl chloride — aniline — sulfanilide

The sulfonamides and sulfonyl chlorides can be hydrolyzed to sulfonic acids under acidic conditions (Procedure 64, p. 388).

$$R{-}SO_2{-}N\!\!<\; + H_2O + HCl \longrightarrow R{-}SO_2{-}OH + H{-}\overset{H}{\underset{|}{N^+}}{-}\quad Cl^-$$

sulfonamide sulfonic acid

$$R{-}SO_2{-}Cl + H_2O + HCl \longrightarrow R{-}SO_2{-}OH + 2HCl$$

sulfonyl chloride sulfonic acid

The resulting sulfonic acids can then be treated with phosphorus pentachloride to produce the sulfonyl chloride (Procedure 62, p. 387), which is then used as an intermediate in the preparation of other derivatives (Procedures 43, p. 363, or 63, p. 387).

Primary sulfonamides react with xanthydrol to form *N*-xanthylsulfonamides, which are satisfactory derivatives (Procedure 65, p. 389).

$$RSO_2NH_2 + \text{xanthydrol (HO–CH}<\text{)} \xrightarrow{CH_3COOH} N\text{-xanthylsulfonamide (RSO}_2\text{NH–CH}<\text{)} + H_2O$$

sulfonamide xanthydrol *N*-xanthylsulfonamide

Sulfonic acids and their salts readily react with *S*-benzylthiuronium chloride to give *S*-benzylthiuronium sulfonates. This reaction represents the shortest and most direct method for obtaining derivatives of these compounds (Procedure 66, p. 389).

$$RSO_3H + NaOH \longrightarrow RSO_3^-Na^+ + H_2O$$

sulfonic acid; sodium salt of sulfonic acid

$$RSO_3^-Na^+ + [C_6H_5CH_2SC(NH_2)_2]^+Cl^- \longrightarrow [C_6H_5CH_2SC(NH_2)_2]^+RSO_3^- + NaCl$$

sodium salt of the sulfonic acid; *S*-benzylthiuronium chloride; *S*-benzylthiuronium sulfonate

The 4-toluidine salts are prepared by treating the sulfonic acid with 4-toluidine or the sodium salt of the sulfonic acid with hydrochloric acid and 4-toluidine (Procedure 67, p. 390). These are useful derivatives.

$$ArSO_3H + H_3C{-}C_6H_4{-}NH_2 \longrightarrow H_3C{-}C_6H_4{-}NH_3^+\ {}^-O_3SAr$$

sulfonic acid; 4-toluidine (4-methylaniline); 4-toluidine salt

$$ArSO_3^-Na^+ + H_3C\text{—}C_6H_4\text{—}NH_2 \xrightarrow{HCl} H_3C\text{—}C_6H_4\text{—}NH_3^+ \ ^-O_3SAr + NaCl$$

sodium salt of the sulfonic acid; 4-toluidine (4-methylaniline); 4-toluidine salt

Procedure 62. Sulfonyl Chlorides from Sulfonic Acids and Their Salts

$$R\text{—}SO_2\text{—}OH + PCl_5 \longrightarrow R\text{—}SO_2\text{—}Cl + POCl_3 + HCl$$

sulfonic acid; phosphorus pentachloride; sulfonyl chloride

$$R\text{—}SO_2\text{—}O^-Na^+ + PCl_5 \longrightarrow R\text{—}SO_2\text{—}Cl + POCl_3 + NaCl$$

salt of the sulfonic acid; phosphorus pentachloride; sulfonyl chloride

One gram of the dry sulfonic acid or the anhydrous salt is mixed with 2.5 g of phosphorus pentachloride in a clean, dry round-bottom flask. A reflux condenser is attached, and the flask is heated in an oil bath at 150°C for 30 min. The mixture is cooled, and 10 mL of dry benzene is added. ***Caution:*** *Benzene is a known carcinogen. Use benzene in the hood, do not breathe the vapors, and avoid contact with the skin.* The mixture is then warmed on a steam bath, and the solid mass is stirred thoroughly. The solution is filtered through a dry filter paper and the filtrate is washed rapidly with two 10-mL portions of water. The benzene layer is separated, dried with calcium chloride, and the solvent is removed by distillation using a steam bath. A solid sulfonyl chloride may be recrystallized from petroleum ether or chloroform. Liquid or recrystallized sulfonyl chlorides then can be used to prepare other derivatives, such as those described in Procedures 43, p. 363, or 63, below.

Cleaning Up. Place the recovered benzene in the hazardous waste container for benzene. Place the aqueous layer in the aqueous solution container.

Procedure 63. Sulfanilides

$$R\text{—}SO_2\text{—}Cl + C_6H_5NH_2 \longrightarrow R\text{—}SO_2\text{—}NH\text{—}C_6H_5$$

sulfonyl chloride; aniline; sulfanilide

A mixture of 1.0–1.5 g of the sulfonyl chloride in 10 mL of benzene and 2.5 g of aniline is heated under reflux for 1 hr. ***Caution:*** *Benzene is a known carcinogen. Use benzene in the hood, do not breathe the vapors, and avoid contact with the skin.* The solution is concentrated to half its volume and chilled. The solid is collected on a filter, washed thoroughly with warm water, and recrystallized from ethanol. If a solid dissolves in warm water, then concentrate the original benzene filtrate and treat it as above to isolate the sulfanilide.

Cleaning Up. Place the benzene filtrate in the hazardous waste container for benzene. The water and ethanol filtrates are placed in the aqueous solution container.

Procedure 64. Hydrolysis of Sulfonamides and Sulfonyl Chlorides

$$R{-}S(=O)_2{-}N\langle \; + H_2O + HCl \longrightarrow R{-}S(=O)_2{-}OH + H{-}\overset{+}{N}(H){-} \quad Cl^-$$

sulfonamide — sulfonic acid

$$R{-}S(=O)_2{-}Cl + H_2O + HCl \longrightarrow R{-}S(=O)_2{-}OH + 2HCl$$

sulfonyl chloride — sulfonic acid

$$R{-}S(=O)_2{-}OH + NaOH \longrightarrow R{-}S(=O)_2{-}O^- \, Na^+ + H_2O$$

sulfonic acid — sodium salt of the sulfonic acid

To hydrolyze sulfonamides or sulfonyl chlorides, 1 g of the compound is heated with 10 mL of 25% hydrochloric acid under reflux. For unsubstituted sulfonamides or sulfonyl chlorides the reaction is complete in 1–2 hr. Sulfonamides of primary amines require 24–36 hr refluxing, whereas sulfonamides of secondary amines may be hydrolyzed in 10–12 hr. After the reaction is complete, the mixture is cooled. The solution is made alkaline with 20% sodium hydroxide solution and extracted with three 25-mL portions of diethyl ether. The ether solution is dried with magnesium sulfate, filtered, and concentrated to yield the primary or secondary amine. With certain very low or very high boiling amines it is often more convenient to recover them as hydrochlorides by passing dry hydrogen chloride gas into the dry ether solution (Procedure 26, p. 338). The amines can then be characterized by a suitable derivative (Procedures 20–22, pp. 333–336).

The aqueous layer contains the sodium salt of the sulfonic acid. The solution is acidified with 25% hydrochloric acid, and the sulfonic acid is isolated by filtration. If no solid is formed, the solution is neutralized with sodium bicarbonate and an excess of sodium chloride added. The solid which is formed contains the sodium salt of the sulfonic acid and sodium chloride. Recrystallization from ethanol, including hot filtration to remove the insoluble sodium chloride, affords the sulfonate salt. Concentration of the ethanol yields more of the sulfonate salt.

Cleaning Up. The recovered ether is placed in the organic solvent container. The aqueous layer is placed in the aqueous solution container.

Procedure 65. *N*-Xanthylsulfonamides from Sulfonamides

$$RSO_2NH_2 + \text{xanthydrol (H, HO on C)} \xrightarrow{CH_3COOH} N\text{-xanthylsulfonamide (H, RSO}_2\text{NH on C)} + H_2O$$

sulfonamide xanthydrol *N*-xanthylsulfonamide

About 0.2 g of xanthydrol is dissolved in 10 mL of glacial acetic acid. If the mixture is not clear, it is filtered or centrifuged, and to the clear solution is added 0.2 g of the sulfonamide. The mixture is shaken and allowed to stand at room temperature until the derivative separates; this may require as long as 1.5 hr. The *N*-xanthylsulfonamide is removed by filtration and recrystallized from a dioxane–water mixture.

Cleaning Up. Place the filtrates in the aqueous solution container.

Procedure 66. Benzylthiuronium Sulfonates

$$\underset{\text{sulfonic acid}}{RSO_3H} + NaOH \longrightarrow \underset{\text{sodium salt of sulfonic acid}}{RSO_3^-Na^+} + H_2O$$

$$\underset{\text{sodium salt of the sulfonic acid}}{RSO_3^-Na^+} + \underset{S\text{-benzylthiuronium chloride}}{[C_6H_5CH_2SC(NH_2)_2]^+Cl^-} \longrightarrow \underset{S\text{-benzylthiuronium sulfonate}}{[C_6H_5CH_2SC(NH_2)_2]^+RSO_3^-} + NaCl$$

About 1 g of the sodium or potassium salt of the sulfonic acid is dissolved in the smallest amount of water, heat being used if necessary to effect solution. If the free sulfonic acid is the starting material, it is dissolved in 2 M sodium hydroxide solution, and any excess alkali is neutralized with hydrochloric acid, phenolphthalein being used as the indicator.

A solution of 1 g of the benzylthiuronium chloride is dissolved in the smallest possible amount of water. This solution and that of the sulfonate are chilled in an ice bath, mixed, and shaken thoroughly. Occasionally it is necessary to scratch the tube and cool in an ice bath to induce crystallization. The benzylthiuronium sulfonate crystals are collected on a filter, washed with a little cold water, and recrystallized from 50% ethanol.

Cleaning Up. Place the filtrate in the aqueous solution container.

Procedure 67. 4-Toluidine Salts of Sulfonic Acid

(a) From Free Sulfonic Acids

$$ArSO_3H + H_3C\text{—}C_6H_4\text{—}NH_2 \longrightarrow H_3C\text{—}C_6H_4\text{—}NH_3^{+}\ {}^{-}O_3SAr$$

sulfonic acid — 4-toluidine (4-methylaniline) — 4-toluidine salt

One gram of the sulfonic acid is dissolved in the minimum amount of boiling water and 1 g of 4-toluidine is added. More water or an additional portion of the sulfonic acid is added to obtain a clear solution. The solution is cooled and the flask scratched to induce crystallization of the salt. The salt is removed by filtration and recrystallized from the minimum amount of water.

(b) From Soluble Salts of Sulfonic Acids

$$ArSO_3^{-}Na^{+} + H_3C\text{—}C_6H_4\text{—}NH_2 \xrightarrow{HCl} H_3C\text{—}C_6H_4\text{—}NH_3^{+}\ {}^{-}O_3SAr + NaCl$$

sodium salt of the sulfonic acid — 4-toluidine (4-methylaniline) — 4-toluidine salt

About 1 g of the sodium, potassium, or ammonium salt is dissolved in the minimum amount of boiling water, and 0.5 g of 4-toluidine and 1–2 mL of concentrated hydrochloric acid are added. If a precipitate separates or if the 4-toluidine is not completely dissolved, more hot water and a few drops of concentrated hydrochloric acid are added until a clear solution is obtained at the boiling point. The solution is cooled, and the walls of the flask are scratched to induce crystallization of the salt. The product is removed by filtration and recrystallized from a small amount of water or dilute ethanol.

Cleaning Up. Make the filtrate basic with 10% sodium hydroxide solution. Extract with two 10-mL portions of diethyl ether. Place the ether layer in the organic solvent waste container. The aqueous layer is placed in the aqueous solution container.

CHAPTER 9

Structural Problems—Solution Methods and Exercises

The laboratory examination of an organic compound results in the accumulation of the physical properties, elements present or absent, solubility, spectral data, and behavior toward certain classification reagents and in various special tests. All these observed facts must be correlated and interpreted in order to arrive at possible structural formulas for the compound in question. It is necessary to show what functional groups are present, to determine the nature of the nucleus to which they are attached, and to find the positions of attachment.

It is the purpose of this discussion to point out, by means of several specific examples, the mode of attack and reasoning involved in deducing information concerning the structure of a molecule from experimental data.

SAMPLE PROBLEMS

9.1 COMPOUNDS WITH STRUCTURES PREVIOUSLY DESCRIBED IN THE LITERATURE

The identification of these compounds does not require quantitative analysis for the elements present, molecular weight determination, or calculation of molecular formulas. The identification is based on the *matching* of the physical and chemical properties of the substance being studied and the data on its derivatives. The laboratory work in this course as described in the preceding chapters is concerned with these previously described compounds.

Two very helpful physical constants described in Chapter 6 are the neutralization equivalents of acids and bases and the saponification equivalents of esters. These numerical data, in conjunction with the solubility class and behavior toward reagents, frequently give valuable clues concerning the molecular structure of the compound. Their use may best be explained by reference to examples.

Example 1 *An organic acid has a neutralization equivalent of* 45 ± 1.

As pointed out on page 307, the neutralization equivalent of an acid is dependent on the number of carboxyl groups in the molecule. If one carboxyl group is present, the neutralization equivalent is equal to the molecular weight. If the present compound is monobasic, its molecular weight[1] must be 44, 45, or 46. A carboxyl group weighs 45; hence, if the molecular weight were 45, nothing could be attached to the carboxyl group. A molecular weight of 44 is obviously impossible, but a molecular weight of 46 leaves a residue of 1 after the weight of the carboxyl radical is subtracted. Only one element has the atomic weight of 1; thus, formic acid (HCOOH) is one possibility.

However, the compound might be dibasic, in which event the molecular weight would be 90 ± 2. Two carboxyl groups equals $2 \times 45 = 90$. Hence, a possible residue of 0, 1, or 2 units remains. There are no bivalent atoms of this atomic weight, so the only possible dibasic acid is oxalic acid, in which the two carboxyl groups are united:

COOH
|
COOH

Thus, by assuming first a monobasic acid and then a dibasic acid, two possible structures have been deduced from the neutralization equivalent alone. In order to decide between the two, the physical state or the solubility class of the compound must be used. If this compound, with a neutralization equivalent of 45 ± 1, is a liquid, soluble in water and in pure ether (solubility group S_1), it must be formic acid. If it is a solid, soluble in water but insoluble in ether (group S_2), it is anhydrous oxalic acid.

Consideration of the molecular weights indicates in a similar fashion that the compound could not be tribasic (mol wt 135 ± 3) or tetrabasic (mol wt 180 ± 4).

Example 2 *An acid (A) possessed a neutralization equivalent of* 136 ± 1. *It gave negative tests for halogen, nitrogen, and sulfur. It did not decolorize cold potassium permanganate solution; but when an alkaline solution of the compound was heated with this reagent for an hour and acidified, a new compound (B) was precipitated. This compound had a neutralization equivalent of* 83 ± 1.

First consider the compound B. Assume it to be monobasic.

$$\begin{aligned} \text{molecular weight} &= 83 \pm 1 \\ \text{less one —COOH} &= \underline{45} \\ \text{residue} &= 38 \pm 1 \end{aligned}$$

[1]For purposes of illustration and calculation in this chapter, whole numbers have been used for the atomic weights of the elements carbon, hydrogen, oxygen, nitrogen, and bromine. The actual atomic weights (which must be used in all precise quantitative analyses) differ from these rounded-off values by an amount less than the experimental error involved in the determination of neutralization and saponification equivalents.

This residue to which the carboxyl group is attached must be made up of some combination of carbon, hydrogen, and perhaps oxygen that is stable to hot potassium permanganate solution. Examination shows that this is not possible.

$$\begin{aligned} \text{residue} &= 38 \pm 1 \\ \text{three carbon atoms} = 3 \times 12 &= \underline{36} \\ \text{remainder} &= 2 \pm 1 \end{aligned}$$

The residue might be C_3H, C_3H_2, or C_3H_3, none of which corresponds to a compound that would be stable to permanganate. The alkane would require C_3H_8 as the parent compound, and the alkyl group would have to be C_3H_7—; similarly, the cycloalkane would have to be C_3H_6 and the cyclopropyl group C_3H_5—.

The presence of oxygen in this residue is also excluded. If it is assumed to be present, the following figures are obtained:

$$\begin{aligned} \text{residue} &= 38 \pm 1 \\ \text{one oxygen atom} &= \underline{16} \\ & 22 \pm 1 \\ \text{one carbon atom} &= \underline{12} \\ \text{remainder} &= 10 \pm 1 \end{aligned} \qquad \textit{or} \qquad \begin{aligned} \text{residue} &= 38 \pm 1 \\ \text{two oxygen atoms} &= \underline{32} \\ \text{remainder} &= 6 \pm 1 \end{aligned}$$

Neither of these remainders corresponds to any atom or group of atoms that forms a reasonable organic compound. For example, the former suggest $CH_{10}OCO_2H$ for the compound and the latter, $H_6O_2CO_2H$, both of which are unreasonable. Thus, it is now safe to conclude that compound B *cannot be monobasic.*

$$\begin{aligned} \text{molecular weight} = 2 \times 83 \pm 1 &= 166 \pm 2 \\ \text{two carboxyl groups} &= \underline{90} \\ \text{residue} &= 76 \pm 2 \end{aligned}$$

If this residue is saturated and aliphatic, it must be made up of $—CH_2—$ units.

$$\begin{aligned} \text{five } —CH_2— &= 5 \times 14 = 70 \\ \text{six } —CH_2— &= 6 \times 14 = 84 \end{aligned}$$

Neither of these corresponds to the weight of the residual radical, 76 ± 2.

A grouping that is stable to hot permanganate is the benzene nucleus. This is an arrangement of six CH groups or $6 \times 13 = 78$ = molecular weight of benzene itself. If two carboxyl groups are present, two of the hydrogen atoms are replaced and the residue becomes $78 - 2 = 76$. This value checks that calculated above for the residue, and hence a possible structure for B is $C_6H_4(COOH)_2$; that is, it may be one of the three phthalic acids.

The question now arises whether compound B could be tribasic. If so we have the following values:

$$\begin{aligned} \text{molecular weight} = 3 \times 83 \pm 1 &= 249 \pm 3 \\ \text{three carboxyl groups} = 3 \times 45 &= \underline{135} \\ \text{residue} &= 114 \pm 3 \end{aligned}$$

Inspection shows that this residue cannot be aromatic because it does not correspond to one or more benzene rings. A benzene ring plus a side chain is excluded because the side chain would be oxidized by the permanganate. The value 114 ± 3 does correspond, however, to eight CH_2 groups ($8 \times 14 = 112$) within experimental error. Moreover, $C_8H_{15}(COOH)_3$, with a molecular weight of 246, falls within the limit of 249 ± 3. Although this tricarboxylic acid represents a possible structure for a compound with a molecular weight of 249 ± 3, it would be impossible to produce it from compound A, which has a neutralization equivalent of 136 ± 1.

Let us assume that A is monobasic.

$$\begin{aligned} \text{molecular weight A} &= 136 \pm 1 \\ \text{one —COOH} &= \underline{45} \\ \text{residue} &= 91 \pm 1 \end{aligned}$$

Since B has a C_6H_4 grouping stable to permanganate, this same group must also be present in A.

$$\begin{aligned} \text{residue} &= 91 \pm 1 \\ C_6H_4 &= \underline{76} \\ \text{remainder} &= 15 \pm 1 \end{aligned}$$

This remainder of 15 ± 1 corresponds to a methyl group that must be attached to the ring.

$$C_6H_4\begin{matrix} \diagup COOH \\ \diagdown CH_3 \end{matrix} \longrightarrow C_6H_4\begin{matrix} \diagup COOH \\ \diagdown COOH \end{matrix}$$

compound A NE = 136 compound B NE = 83

The original must be 2-, 3-, or 4-methylbenzoic acid (*o*-, *m*-, or *p*-toluic acid), each of which would give the reactions cited. Additional data such as a melting point, a derivative, or spectra are necessary to distinguish among them.

This example also illustrates the fact that oxidation almost invariably converts a compound with a given neutralization equivalent to a product that has a lower neutralization equivalent. This generalization follows naturally from the increase in the number of carboxyl groups or cleavage of the molecule into smaller fragments.

Example 3 *A colorless crystalline compound (A) gave a positive test for nitrogen but not for halogens or sulfur. It was insoluble in water, dilute acids, and alkalies. It produced a red-colored complex with ammonium hexanitratocerate reagent but did not react with phenylhydrazine. Compound A dissolved in hot sodium hydroxide solution with the liberation of ammonia and the formation of a clear solution. Acidification of this solution produced compound B, which contained no nitrogen and gave a neutralization equivalent of* 182 ± 1. *Oxidation of B by hot permanganate solution produced C, which had a neutralization equivalent of* 98 ± 1. *When either A or B was heated with hydrobromic acid for some time, a compound D separated. This compound contained bromine but no nitrogen. It gave a precipitate with bromine water and a violet color with ferric chloride, and it readily reduced dilute potassium permanganate. It was soluble in sodium bicarbonate solution.*

These reactions may be summarized as follows.

$$\begin{array}{ccc} \text{compound A} & \xrightarrow[H_2O]{NaOH} & NH_3 + \text{solution} \\ \downarrow \scriptstyle{HBr} & & \downarrow \scriptstyle{H^+} \\ \text{compound D} & \xleftarrow{HBr} & \text{compound B (NE} = 98 \pm 1) \\ & & \downarrow \scriptstyle{KMnO_4,\ \Delta} \\ & & \text{compound C} \end{array}$$

The elimination of nitrogen from compound A by alkaline hydrolysis suggests the presence of a nitrile or amide grouping, because these functional groups liberate ammonia when they undergo hydrolysis.

$$RCN + NaOH + H_2O \longrightarrow RCOONa + NH_3$$

$$RCONH_2 + NaOH \longrightarrow RCOONa + NH_3$$

The imide grouping, —CONHCO—, which also liberates ammonia, is excluded by the fact that compound A is not soluble in sodium hydroxide solution. Since compound B contained no nitrogen, certain substituted anilines, such as 2,4-dinitroaniline, are excluded. The absence of nitrogen in B and the fact that it was acidic and not basic likewise eliminates a substituted urea, which also liberates ammonia when hydrolyzed.

$$RNHCONH_2 + H_2O \longrightarrow RNH_2 + CO_2 + NH_3$$

The positive red color with ammonium hexanitratocerate reagent suggests A is an alcohol; an amino or phenolic group is excluded by the fact that compound A is neutral. Further evidence for the presence of a hydroxyl group is furnished by the fact that compound D, produced from A by the action of hydrobromic acid, contained bromine.

$$ROH + HBr \longrightarrow RBr + H_2O$$

The properties of compound D strongly suggest the presence of a phenolic hydroxyl group. Ease of bromination, sensitivity to permanganate, and color with ferric chloride are properties characteristic of substituted phenols. This phenolic hydroxyl group was produced by the action of hydrobromic acid on some functional group present in A and B, because neither of these originally contained the phenol grouping. One type of compound that produces a phenol when treated with hydrobromic acid is an aryl alkyl ether.

$$ArOR + HBr \longrightarrow ArOH + RBr$$

If the alkyl group is small, it would be lost as alkyl bromide during the treatment with hydrobromic acid. Thus compounds A, B, and C probably contain such a mixed ether group and also an aromatic nucleus, since the substituted phenol D contains one. The solubility of D in sodium bicarbonate solution is probably due to the presence of a carboxyl group, because both the nitrile and amide groups are hydrolyzed to carboxyl groups by acids as well as alkalies. Hence compound D is a hydroxybenzoic acid with the bromine attached to a side chain. This side chain must also be present in compound A with an alcoholic group in place of the bromine atom.

The neutralization equivalents of compounds B and C may now be considered. It will be noted that the neutralization equivalent of compound C, produced by permanganate oxidation, is lower than that of compound B, which acquired its acidic properties by a hydrolysis reaction only. This oxidation obviously affects the side chain, and compound C must have more carboxyl groups than B.

Assume compound C to be dibasic.

molecular weight = $2 \times 98 \pm 1$ =	196 ± 2
two carboxyl groups = 2×45 =	90
	106 ± 2
one oxygen atom in ether linkage =	16
	90 ± 2
benzene minus three hydrogen atoms (C_6H_3) =	75
residue =	15 ± 2

This residue of 15 corresponds to a CH_3— group, and hence the ether grouping must have been CH_3O—.

It is now necessary to find the length of the side chain to which the alcohol group in A and B is attached.

If C is dibasic, B is monobasic.

NE = mol wt =	182 ± 1
carboxyl group =	45
	137 ± 1
methoxyl group =	31
	106 ± 1
hydroxyl group =	17
	89 ± 1
benzene nucleus (C_6H_3) =	75
residue =	14 ± 1

This residue represents the weight of the aliphatic side chain and obviously corresponds to —CH_2—. Hence, possible structures for A, B, C, and D are

C_6H_3(—$CONH_2$ (or CN))(—CH_2OH)(—OCH_3) (compound A) $\xrightarrow{NaOH}$ C_6H_3(—COOH)(—CH_2OH)(—OCH_3) (compound B)

compound A $\xrightarrow{HBr}$ C_6H_3(—COOH)(—CH_2Br)(—OH) (compound D)

compound B $\xrightarrow{HBr}$ compound D

compound B $\xrightarrow{KMnO_4}$ C_6H_3(—COOH)(—COOH)(—OCH_3) (compound C)

This example illustrates the fact that a given reagent may affect more than one functional group. Thus, boiling hydrobromic acid affected three functional groups in compound A and two in compound B.

Example 4 *A solid compound, melting at* 60–61°C, *gave no tests for sulfur, nitrogen, or halogen. It was insoluble in water, dilute acid, and alkali, but reacted with concentrated sulfuric acid and was placed in solubility group* N. *The unknown gave an orange-red precipitate with* 2,4-*dinitrophenylhydrazine and a positive hydroiodic reagent test.*

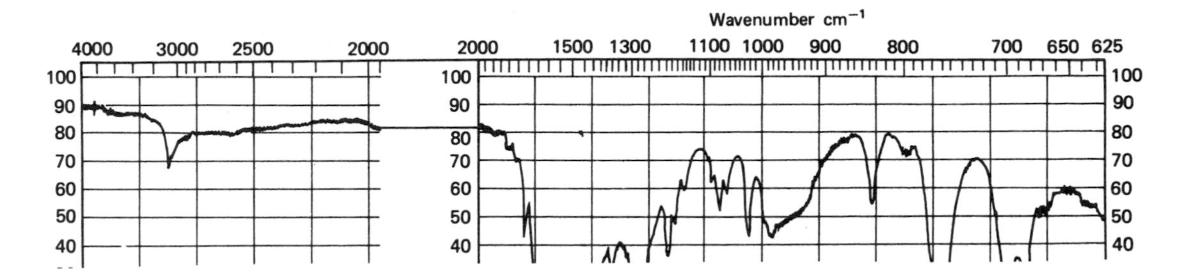

Example 4. Infrared spectrum. [Reprinted from "The Aldrich Collection of IR Spectra," edited by C. J. Pouchert, copyright © 1971, 1975; with permission.]

The oxime derivative had a melting point of 95°C. *A thin film (melt) was prepared and the following infrared spectrum was recorded.*

The following infrared bands have immediate implications as to functional group features for this compound:

1680 cm^{-1}	C=O stretch (conjugated with a π-system)
1590 cm^{-1}	aromatic C⩦C stretch
1500 cm^{-1}	aromatic C⩦C stretch

These features are suggested by tables in Chapter 6. The strong bands at 1270 cm^{-1}, 1140 cm^{-1}, and 1020 cm^{-1} imply that the molecule possesses a C—O—C unit; ether and/or ester functional groups would provide this unit.

Since a sample of the original compound gave an orange-red precipitate with the 2,4-dinitrophenylhydrazine reagent, it confirmed the presence of an aldehyde or ketone group, but did not exclude the possibility of an ester group *in addition* to the aldehyde or ketone group.

Ms. Mary Four, who was working on the identification of this compound, then consulted the tables of solid ketones, aldehydes, and esters with the melting point range 60–61 ± 2°C.[2] The following list of possible compounds was obtained:

Solid ketones	**mp (°C)**		**mp (°C)**
(A) $C_6H_5COCH_2C_6H_5$	60	(D) $C_6H_5CH{=}CHCOC_6H_5$	62
(B) $CH_3-C_6H_4-COC_6H_5$ (para-substituted benzene ring)	60	(E) $CH_3O-C_6H_4-COC_6H_5$ (para-substituted benzene ring)	62
(C) $C_6H_5COCH_2COCH_3$ ⇅ $C_6H_5-C(OH){=}CHCOCH_3$	61		

[2]This overall range of 58–63°C was used to take into consideration problems of possible inpurities and the differences in techniques and thermometer use by prior chemists. By using this wide range, the chances of missing a compound are reduced.

Solid aldehydes	mp (°C)		mp (°C)
(F) 3,4-(CH_3O)$_2$$C_6H_3$—CHO	58	(I) C_6H_5—C_6H_4—CHO	60
(G) phenanthrene—CHO	59	(J) naphthalene with CHO and OCH_3	60
(H) 2-naphthalene—CHO	60		

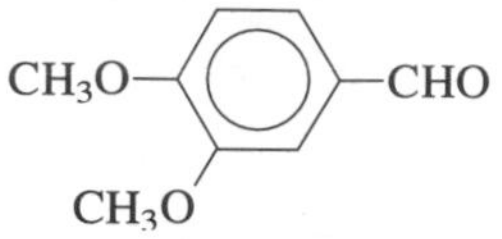

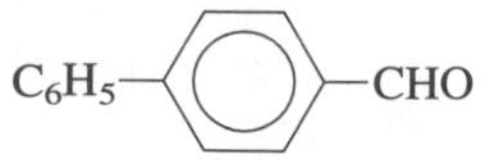

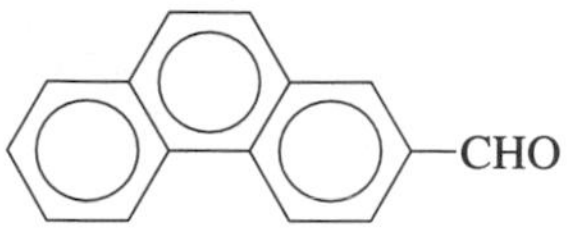

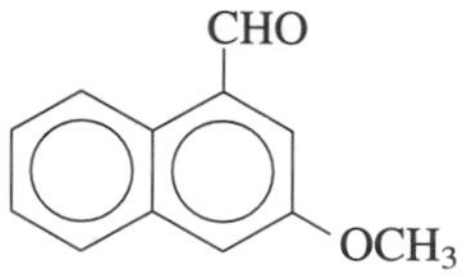

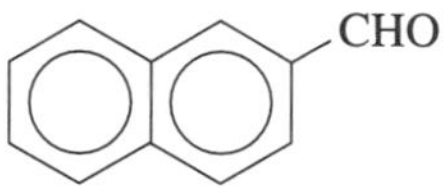

Note that the above list contains some compounds that would not give the C—O—C stretching that the IR seems to display; it is, however, clearly better to list too many compounds initially than presumptuously to limit the list and inadvertently omit the proper structure.

A search of tables of solid esters (mp 58–63°C) that *also* had an aldehyde or keto group resulted in no additional possibilities. Ms. Four now reexamined the IR spectrum in conjunction with the above structural formulas. The moderate absorption bands at 2940 cm^{-1} and 2860 cm^{-1} suggest an aromatic aldehyde (aldehydic C—H stretching bands); the IR bands for C—O—C stretch (see above) suggest aromatic methoxy groups.

Since the hydroiodic acid test gave a positive test, the presence of a methoxy group is confirmed. This eliminated compounds A, B, C, D, G, H, and I from consideration and left only compounds E, F, and J; those containing methoxyl groups in addition to a carbonyl group. Although the IR spectrum already favored an aromatic aldehyde, the ketone E was eliminated by showing that the original compound was oxidized by potassium permanganate solution and the acidic chromium trioxide reagent.

To distinguish between the remaining possibilities (compounds F and J), the oxime derivatives are compared. 3,4-Dimethoxybenzaldehyde is the unknown since it has an oxime with a melting point of 95°C.

In summary, the melting point, and the IR and NMR spectra lead to F. Fingerprint spectra[3] would solidify this choice.

Note that although they were not used here, NMR and other chemical tests could have been helpful. A pair of NMR singlets near δ 3.8 would have supported the aromatic methoxyl groups, and a positive Tollens test would have supported the aldehyde group.

Example 5 *One of Mr. John Five's unknowns melted at* 60–61°C *and gave no tests for sulfur, nitrogen, or halogen. It was in solubility group* N *and gave an orange-red precipitate of melting point* 180–181°C *with the* 2,4-*dinitrophenylhydrazine reagent. The unknown did not reduce potassium permanganate or chromium trioxide; but it gave a positive hydroiodic acid test. Its infrared spectrum was the following.*

[3]Fingerprinting means careful comparison of the unknown's spectra to literature spectra.

spectrum of his unknown gave the following:

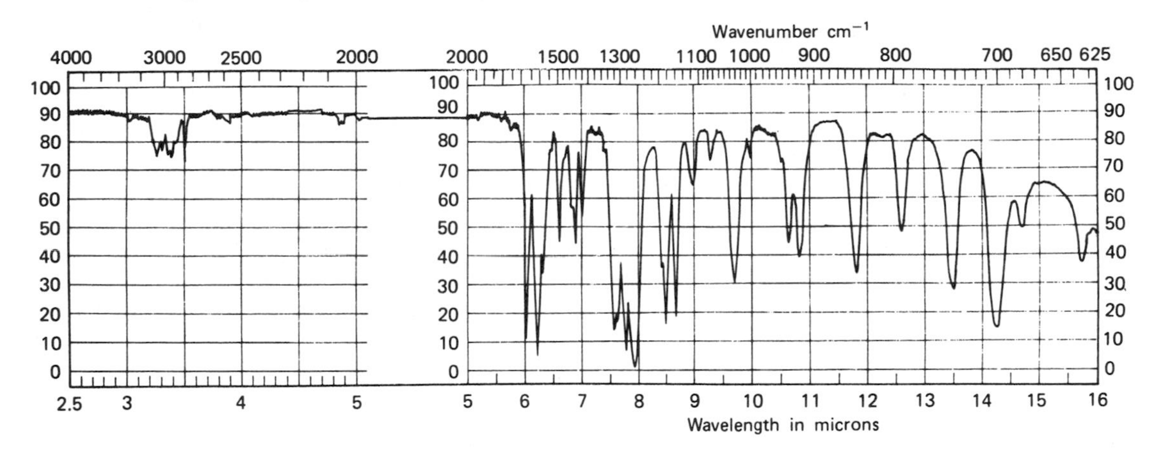

Example 5. Infrared spectrum. [Reprinted from "The Aldrich Collection of IR Spectra," edited by C. J. Pouchert, copyright © 1971, 1975; with permission.]

The NMR spectrum of this unknown ($CDCl_3$, 60 MHz, TMS $\delta = 0.00$) showed

δ 3.83, 3H, sharp singlet
δ 6.95, 2H, doublet
δ 7.3–8.0, 7H, multiplet

The first NMR signal strongly implies a methyl group; the chemical shift suggests that the methyl is on oxygen. All remaining signals imply aromaticity.

Observing the carbonyl band at 1667 cm^{-1}, Mr. Five prepared a list of possible solid aldehydes and ketones melting over the range 58–63°C, and his list was identical to that assembled by Ms. Mary Four shown on pp. 397–398. Since his unknown did not reduce potassium permanganate or the chromium trioxide reagent, all oxidizable compounds (C, D, F, G, H, I, and J)[4] are eliminated. The remaining compounds possible were A, B, and E. Inspecting these structures, he noted that only compound E contained a methoxyl group and that the IR spectrum (asymmetric C—O—C stretch of aryl alkyl ether, 1266 cm^{-1}, symmetric C—O—C stretch, 1030 cm^{-1}) indicated this group. Since the compound gave a positive hydriodic acid test, it contains a methoxy group. Final identification was made by the 2,4-dinitrophenylhydrazone, which confirmed the original as 4-methoxybenzophenone.

Example 6 *Mr. Henry Six received a compound for identification, that melted at* 60–61°C; *it contained no nitrogen, sulfur, or halogen and was in solubility group* N. *The unknown reduced potassium permanganate and chromium trioxide, but gave a negative Zeisel test. It gave a magenta color with aqueous ferric chloride. The* 2,4*-dinitrophenylhydrazone derivative melted at* 150–151°C. *The infrared spectrum is shown on the next page.*

The NMR spectrum of this compound (determined in deuteriochloroform, TMS δ = 0.00, 60 MHz) showed

δ 2.22, 3H sharp singlet
δ 6.22, 2H, singlet
δ 7.25–7.50, 3H, multiplet
δ 7.7–7.95, 2H, multiplet

[4]These compounds contain either a C=C or a CHO unit.

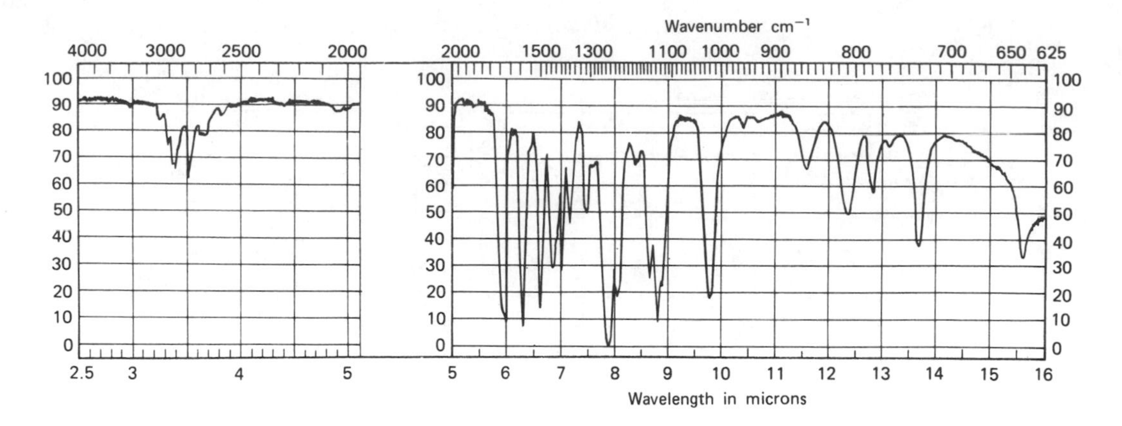

Example 6. Infrared spectrum. [Reprinted from "The Aldrich Collection of IR Spectra," edited by C. J. Pouchert, copyright © 1971, 1975; with permission.]

The rather broad IR band at 1667–1538 cm^{-1} was puzzling, but Mr. Six prepared a list of solid aldehydes and ketones melting at 58–63°C, and his list of possibilities was identical with those made (pp. 397–398) by Ms. Mary Four and Mr. John Five. Since the unknown reduced potassium permanganate solution and chromium trioxide reagent, the nonoxidizable compounds of A, B, and E were eliminated. The Zeisel test for methoxy groups was negative, thus eliminating compounds F and J; also, E was eliminated (again). The lack of NMR evidence for a methoxyl group supports these eliminations. The remaining possibilities were compounds C, D, G, H, and J. Mr. Six restudied his IR spectrum and compared the bands with the tables and discussion in Chapter 6. It appeared that the wide band at 1667–1538 cm^{-1} suggested the C=C stretch of an enolized ketone. The O—H stretch band of enols is broad and here runs from 3200 to 2500 cm^{-1} (broad because of hydrogen bonding). A magenta color with aqueous ferric chloride is indicative of an enol. The 2,4-dinitrophenylhydrazone derivative agreed with the literature value for that derivative of benzoylacetone, compound C.

In retrospect, the NMR signals can be discussed further. The singlet δ 6.22 is due to the methylene group and the most downfield aromatic signals are due to the protons *ortho* to the benzoyl groups.[5]

$$C_6H_5-\overset{\overset{O}{\|}}{C}CH_2\overset{\overset{O}{\|}}{C}-CH_3 \rightleftharpoons C_6H_5-\overset{\overset{O\cdots H\cdots O}{|\qquad\|}}{C{=}CH{-}C}-CH_3 \quad \text{or} \quad C_6H_5-\overset{\overset{O\cdots H\cdots O}{\vdots\qquad\vdots}}{C{\cdots}CH{\cdots}C}-CH_3$$

Example 7 *An ester, containing only carbon, hydrogen, and oxygen, possessed a saponification equivalent of* 74 ± 1.

The first step is to work out the possibilities on the assumption that the molecule contains only one ester group. In that event the molecular weight is equal to the saponification equivalent. The type of formula for an ester is R—COO—R′, and therefore the first step is to subtract the weight of —COO— from the molecular weight.

[5]It is of interest that the NMR sample is not observably enolized and yet the IR sample is highly enolized.

$$\begin{aligned} \text{molecular weight} &= 74 \pm 1 \\ \text{—COO—} &= \underline{44} \\ \text{residue} &= 30 \pm 1 \end{aligned}$$

This residue represents the combined weight of R and R′. In saturated esters[6] containing only carbon, hydrogen, and oxygen, this residue must always be equal to C_nH_{2n+2} and thus must always be an even number. Thus residual weights of 31 and 29 are impossible, and the value 30 represents the molecular weight of C_nH_{2n+2}. Mere inspection in this case shows that the hydrocarbon residue is C_2H_6, but the general approach is to solve for the value of n by multiplying its value by the atomic weights of the elements in the formula and setting this equal to the residual weight.

$$\begin{aligned} 12n + 1(2n + 2) &= 30 \\ 14n &= 28 \\ n &= 2 \end{aligned}$$

This residue of C_2H_6 represents the sum of R and R′, and it is now necessary to write the possibilities.

$$R + R' = C_2H_6 = H + C_2H_5 \textit{ or } CH_3 + CH_3$$

Assuming that the compound was a monoester, we have two possibilities.

$HCOOC_2H_5$	CH_3COOCH_3
ethyl formate	methyl acetate

If two ester groups are present in the molecule, then the molecular weight is twice the saponification equivalent. Two ester groups will contain two —COO— combinations; hence

$$\begin{aligned} \text{molecular weight} &= 2 \times 74 \pm 1 = 148 \pm 2 \\ \text{two —COO—} &= 2 \times 44 = \underline{\;88\;} \\ \text{residue} &= \quad 60 \pm 2 \end{aligned}$$

The value of 60 ± 2 represents the summation of those portions of the molecule other than the two —COO— groups. Again, $C_nH_{2n+2} = 60 \pm 2$ and

$$\begin{aligned} 12n + 1(2n + 2) &= 60 \pm 2 \\ 14n &= 58 \pm 2 \end{aligned}$$

Since n must be an integer, the only value of the right-hand side of the equation that will fulfill this requirement is 56, and hence $n = 4$. The residue must be C_4H_{10} and has to be divided among the various hydrocarbon radicals present in the type formulas for a compound with two ester groups. Some of the possible type formulas are the following.

COOR	COOR	RCOO	RCOO
|	|	|	|
$(CH_2)_x$	$(RCH)_x$	$(CH_2)_x$	$(CHR)_x$
|	|	|	|
COOR′	COOR	RCOO	RCOO

Using these type formulas for esters of a dicarboxylic acid, or a dihydroxy alcohol, possible structures may now be written in which the four carbon atoms and ten hydrogen atoms are distributed properly in each of the above formulas.

This example illustrates the use of saponification equivalents in deducing possible structures. It also shows that a saponification equivalent is not quite so useful as the neutraliza-

[6]In an olefinic ester the residue is C_nH_{2n}; in an acetylenic ester, C_nH_{2n-2}.

tion equivalent of an acid. It is always desirable to have some additional data concerning either the acid or alcohol or both produced by saponification of the ester in order to reduce the number of isomeric esters which possess the required saponification equivalent.

9.2 DETERMINATION OF THE STRUCTURE OF NEW COMPOUNDS NOT DESCRIBED IN THE CHEMICAL LITERATURE

Interpretation of Molecular Formulas

In research, quantitative analyses together with molecular weight determinations routinely yield the molecular formula of any unknown substance. Much information about possible functional groups can often be deduced from such information alone, and for this reason it is pertinent to consider the significance of the molecular formula.

A saturated hydrocarbon without any rings has the general formula C_nH_{2n+2}. Introduction of oxygen to give an alcohol, ether, acetal, or any other saturated acyclic compound does not change the carbon-to-hydrogen ratio, and the molecular formula is $C_nH_{2n+2}O_m$. The introduction of a double bond or a ring into a saturated molecule requires the removal of two hydrogen atoms, and the introduction of a triple bond involves the removal of four hydrogen atoms. By an examination of the carbon-to-hydrogen ratio, then, it is possible to draw conclusions concerning the possible number of multiple bonds or rings in a molecule.

For example, the substance C_3H_6O must have either one double bond (either olefinic or carbonyl) or one ring but cannot have a triple bond because it is only two hydrogen atoms short of saturation. The compound $C_8H_{12}O$ must have three double bonds or three rings or some combination of double bonds, triple bonds, and rings that accounts for the three pairs of missing hydrogen atoms. It cannot contain a benzene ring, however, because such a ring requires a shortage from saturation of four pairs of hydrogen atoms. (Thus the possibility that the oxygen function is a phenolic hydroxyl group is immediately excluded.)

Furthermore, the introduction of a halogen atom into a saturated molecule necessitates the removal of one hydrogen atom, and consequently the general formula of a saturated acyclic monohalide is $C_nH_{2n+1}X$. On the other hand, introduction of a nitrogen atom to give an acyclic saturated amine requires also the addition of an extra hydrogen atom so that the formula is $C_nH_{2n+3}N$. A consequence of these generalizations, which is of value in deriving molecular formula from the analyst's carbon and hydrogen determination, is that a molecule with no elements other than carbon, hydrogen, and oxygen must contain an even number of hydrogen atoms; an odd number of halogen or nitrogen atoms requires an odd number of hydrogen atoms; and an even number of halogen or nitrogen atoms requires an even number of hydrogen atoms. Thus a molecular formula such as $C_5H_{11}O_3$, calculated from analytical data, is obviously incorrect; the correct formula is likely either $C_5H_{10}O_3$ or $C_5H_{12}O_3$.

The preceding discussion can be summarized in equation form. For the generalized molecular formula $I_yII_nIII_zIV_x$ (e.g., $C_xH_yN_zO_n$), where

I can be H, F, Cl, Br, I, D, etc. (i.e., any monovalent atom),
II can be O, S, or any other bivalent atom,
III can be N, P, or any other trivalent atom,
IV can be C, Si, or any other tetravalent atom,

the index of hydrogen deficiency $= x - y/2 + z/2 + 1$.

This index represents "missing pairs" of monovalent atoms that correspond to double bonds, triple bonds, and/or cyclic features in the structure of interest. This formula should be used with care when dealing with other than simple oxidation states of elements in the second and lower rows of the periodic table. For example, the index for the formula C_3H_8OS is zero. Possible structures for this formula include the following:

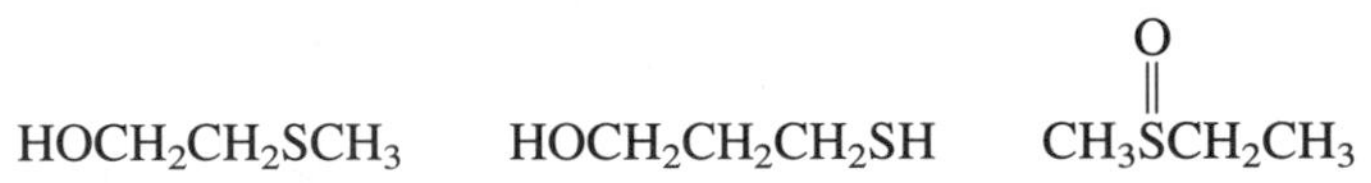

$$HOCH_2CH_2SCH_3 \qquad HOCH_2CH_2CH_2SH \qquad CH_3\overset{\overset{\displaystyle O}{\|}}{S}CH_2CH_3$$

The last formula, a sulfoxide, seems to present a contradiction. We can, however, include this if we visualize it in terms of its polar covalent resonance form:

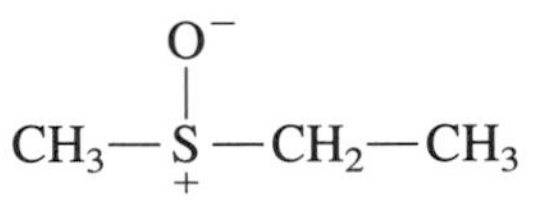

$$CH_3-\underset{+}{\overset{\overset{\displaystyle O^-}{|}}{S}}-CH_2-CH_3$$

Thus, all three structures would have an index of zero.

Problem

Write all the structures possible for the formula $C_3H_9O_4P$.

Example 8 *A substance (A) has the formula* C_6H_{10}. *On hydrogenation over platinum under mild conditions it is converted to* B (C_6H_{14}). *When compound A was ozonized and the ozonide reductively cleaved, two products, acetaldehyde and glyoxal, were formed. These known compounds were identified by means of their* 2,4-*dinitrophenylhydrazones.*

The molecular formula (it has an index of hydrogen deficiency of $6 - 5 + 1 = 2$) shows that the maximum number of double bonds and rings is two, or else there is one triple bond. Uptake of two moles of hydrogen indicates that there are, in fact, either two double bonds or one triple bond.[7] Ozonolysis to give two-carbon fragments suggests that the six-carbon skeleton originally present must have been cleaved at two points to give three two-carbon fragments rather than at one. This means that two double bonds rather than one triple bond must have been present. Finally, the probable arrangement of the three two-carbon pieces is as shown.

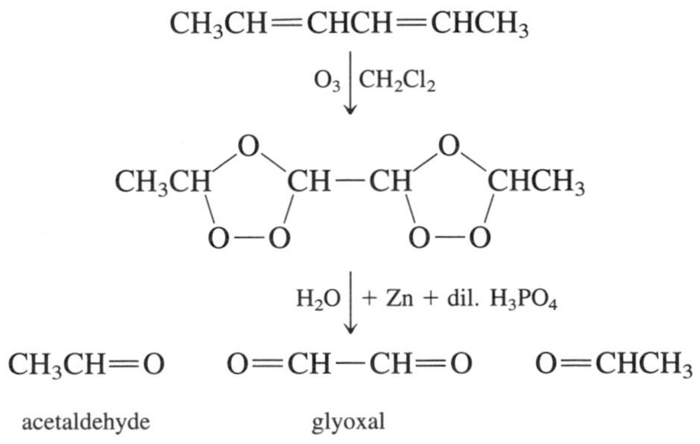

$$CH_3CH{=}CHCH{=}CHCH_3$$

$$\downarrow O_3 \mid CH_2Cl_2$$

(bis-ozonide: CH_3CH and CH joined through O and O—O in a five-membered ring, $CH-CH$, second ring joined through O and O—O to $CHCH_3$)

$$\downarrow H_2O \mid + Zn + \text{dil. } H_3PO_4$$

$$\underset{\text{acetaldehyde}}{CH_3CH{=}O} \qquad \underset{\text{glyoxal}}{O{=}CH-CH{=}O} \qquad O{=}CHCH_3$$

[7]Less frequently, hydrogen will add across a *strained* ring of a compound containing no multiple bonds; we have assumed that this is not so here in order to allow initial hypotheses to be made.

By considering the molecular formulas of several compounds and the reactions that produced then, one is frequently able to deduce possible formulas. The following example involves the deduction of a large amount of information from only a few clues.

Example 9 *A neutral compound* A ($C_{15}H_{14}O$) *gives a negative Baeyer permanganate test and is not attacked by hydrogen bromide; it is oxidized to an acid* B ($C_{14}H_{10}O_3$) *by hot chromic acid solution.*

First, note that the oxidation has caused a *loss* of one carbon atom and four hydrogen atoms and a *gain* of two oxygen atoms. It is necessary to find a functional group or several groups that will do this. In this connection it is useful to tabulate the behavior of the common functional groups on oxidation, making a note of the gain and loss in composition. From the table below it will be noted that the oxidation of an ethyl side chain corresponds exactly to the oxidation of A to B; in both cases there are gains of two oxygen atoms and a loss of four hydrogen atoms and one carbon atom.

Functional group	Oxidation product	Gain	Loss	
RCHO	RCOOH	1O		
RCH_2OH	RCOOH	1O	2H	
R_2CHOH	R_2CO		2H	
$R_2C(OH)CH_3$	R_2CO		4H	1C
$RCH{=}CH_2$	RCOOH	2O	2H	1C
$RC{\equiv}CH$	RCOOH	2O		1C
$ArCH_2CH_2OH$	ArCOOH	1O	4H	1C
$ArCOCH_3$	ArCOOH	1O	2H	1C
$ArCH_3$	ArCOOH	2O	2H	
$ArCH_2CH_3$	ArCOOH	2O	4H	1C
ArC_nH_{2n+1}	ArCOOH	2O	$2n$ H	$(n-1)$C
Ar_2CHCH_3	Ar_2CO	1O	4H	1C
Ar_2CH_2	Ar_2CO	1O	2H	
From Example 9:				
(A) $C_{15}H_{14}O$	(B) $C_{14}H_{10}O_3$	2O	4H	1C

Extracting an ethyl group from A ($C_{15}H_{14}O$) or a carboxyl group from B leaves a unit $C_{13}H_9O$. This unit is derived from some parent compound C ($C_{13}H_{10}O$) which is stable to oxidation.

The next question concerns the character of the functional group containing the oxygen atom. What functional groups containing only one oxygen atom are stable to oxidation? Consideration of various functional groups leads to the conclusion that the ether linkage is one possibility and that a properly substituted ketone is a second. Ketones with no hydrogen atoms on the α-carbon atoms are usually stable to oxidation. The most common examples are diaryl ketones.

The next step consists of considering the ratio of carbon to hydrogen in the compounds A and B and the hypothetical parent compound, C ($C_{13}H_{10}O$). These carbon and hydrogen atoms must be combined so that the resulting compound will be stable to oxidation. A completely saturated compound, C_nH_{2n+2}, would require a formula $C_{13}H_{28}$ for a 13-carbon-atom compound, C. Even allowing two hydrogen atoms as equivalent to the oxygen atom, it is obvious that the compound has no such ratio of carbon and hydrogen atoms. An alicyclic compound would require C_nH_{2n} or $C_{13}H_{26}$. Ordinary olefinic compounds and acetylenic compounds with enough double or triple bonds to lower the ratio of hydrogen to carbon are excluded by the stability to oxidation.

The only large class of compounds with such a low ratio of hydrogen to carbon atoms is aromatic in nature. Since benzene has six carbon atoms whereas the parent compound (C) has 13, the possibility of two benzene rings is suggested.[8] This leaves one carbon atom to be accounted for.

Subtracting two phenyl groups, $(C_6H_5—)_2$, from compound C ($C_{13}H_{10}O$) leaves a residue of CO. It will be remembered that diaryl ketones are stable to oxidation. The parent compound (C) is evidently benzophenone, $C_6H_5COC_6H_5$, and the compounds A and B are probably

$$C_6H_5COC_6H_4CH_2CH_3 \longrightarrow C_6H_5COC_6H_4CO_2H$$

compound A (one of the isomers of ethylbenzophenone) — compound B (one of the isomers of benzoylbenzoic acid)

9.3 PROBLEMS

The following problems are designed to give the student added experience in the types of reasoning illustrated by the examples above. It is of the utmost importance to seek the answers by systematic procedures, and students are urged to avoid a random attack on the problems.

After determining structures, equations should be written for all reactions and all major spectral bands should be assigned to the appropriate structural features.

For spectra, the following abbreviations are used:

NMR

Integration: 1H = one proton, 2H = 2 protons, etc
Multiplicity: s = singlet, d = doublet, t = triplet, q = quartet, b = broad, m = multiplet (undetermined).

IR

Band intensities: s = strong, m = medium, w = weak, b = broad.

Problem Set A

Most of the ^{1}H NMR spectra (200 MHz) are supplemented with signal integrations in terms of the number of protons (1H, 2H, etc. . .) causing a given signal. For most of the carbon signals (^{13}C NMR, 50 MHz) the carbons (as determined by APT studies, off-resonance decoupling, etc.) have been identified as to whether they correspond to methyl (CH_3), methylene (CH_2), methine (CH), or quaternary (C) carbons.

1. Three compounds (A–C) yield the solubilities and chemical test results described below. Assign a name from the list of three below to each of A–C.

Compound A: insoluble in water; insoluble in sulfuric acid; sodium fusion, followed by silver nitrate treatment, gives a white precipitate.

[8] Another approach focuses on the fact that A ($C_{15}H_{14}O$) has $15 - 7 + 1 = 9°$ of unsaturation (index of hydrogen deficiency). Each benzene ring provides 4° of unsaturation (3π-bonds and one ring) and the 9th turns out to be due to a carbonyl group.

Compound B: insoluble in water; insoluble in sulfuric acid; sodium fusion, followed by silver nitrate treatment, gives no precipitate.

Compound C: insoluble in water; soluble in sulfuric acid; sodium fusion, followed by silver nitrate treatment, gives no precipitate.

Choices: chlorocyclohexane, diethyl ether, cyclohexane.

2. Five compounds (A–E) yield the solubilities and chemical test results described below. Assign a name from the list of five below to each of A–E.

Compound A: insoluble in water; does not decolorize bromine; does not undergo a reaction with acetyl chloride; does not form a precipitate with 2,4-dinitrophenylhydrazine; does not form a precipitate when treated with excess iodine in aqueous sodium hydroxide (iodoform test).

Compound B: insoluble in water; quickly decolorizes bromine, does not undergo a reaction with acetyl chloride; does not form a precipitate with 2,4-dinitrophenylhydrazine; does not form a precipitate when treated with excess iodine in aqueous sodium hydroxide.

Compound C: soluble in water; undergoes a rapid reaction with acetyl chloride; does not form a precipitate with 2,4-dinitrophenylhydrazine; forms a yellow precipitate with a noxious odor when treated with excess iodine in aqueous sodium hydroxide.

Compound D: soluble in water; does not undergo a reaction with acetyl chloride if carefully dried; forms a yellow precipitate with 2,4-dinitrophenylhydrazine; forms a yellow precipitate with a noxious odor when treated with excess iodine in aqueous sodium hydroxide.

Compound E: soluble in water; undergoes a rapid reaction with acetyl chloride; does not form a precipitate with 2,4-dinitrophenylhydrazine; does not form a precipitate when treated with excess iodine in aqueous sodium hydroxide.

Choices: acetone (propanone), *t*-butyl alcohol (2-methyl-2-propanol), cyclohexane, cyclohexene, ethanol.

3. A compound of boiling point 97–98°C is known to contain none of the following elements: sulfur, halogen, and nitrogen. Upon treatment of this compound with 3,5-dinitrobenzoyl chloride, a solid of melting point 74–74.5°C is obtained. Treatment of the original compound with CrO_3 in aqueous sulfuric acid gives an immediate blue-green color. Moreover, treatment of the original compound with iodine in aqueous sodium hydroxide (iodoform test) gives a yellow precipitate. The IR, ^{1}H NMR, and ^{13}C NMR spectra are provided here for the original compound. Deduce the structure of the compound and assign the major IR bands as well as all of the NMR absorptions, and write reactions for all positive chemical tests.

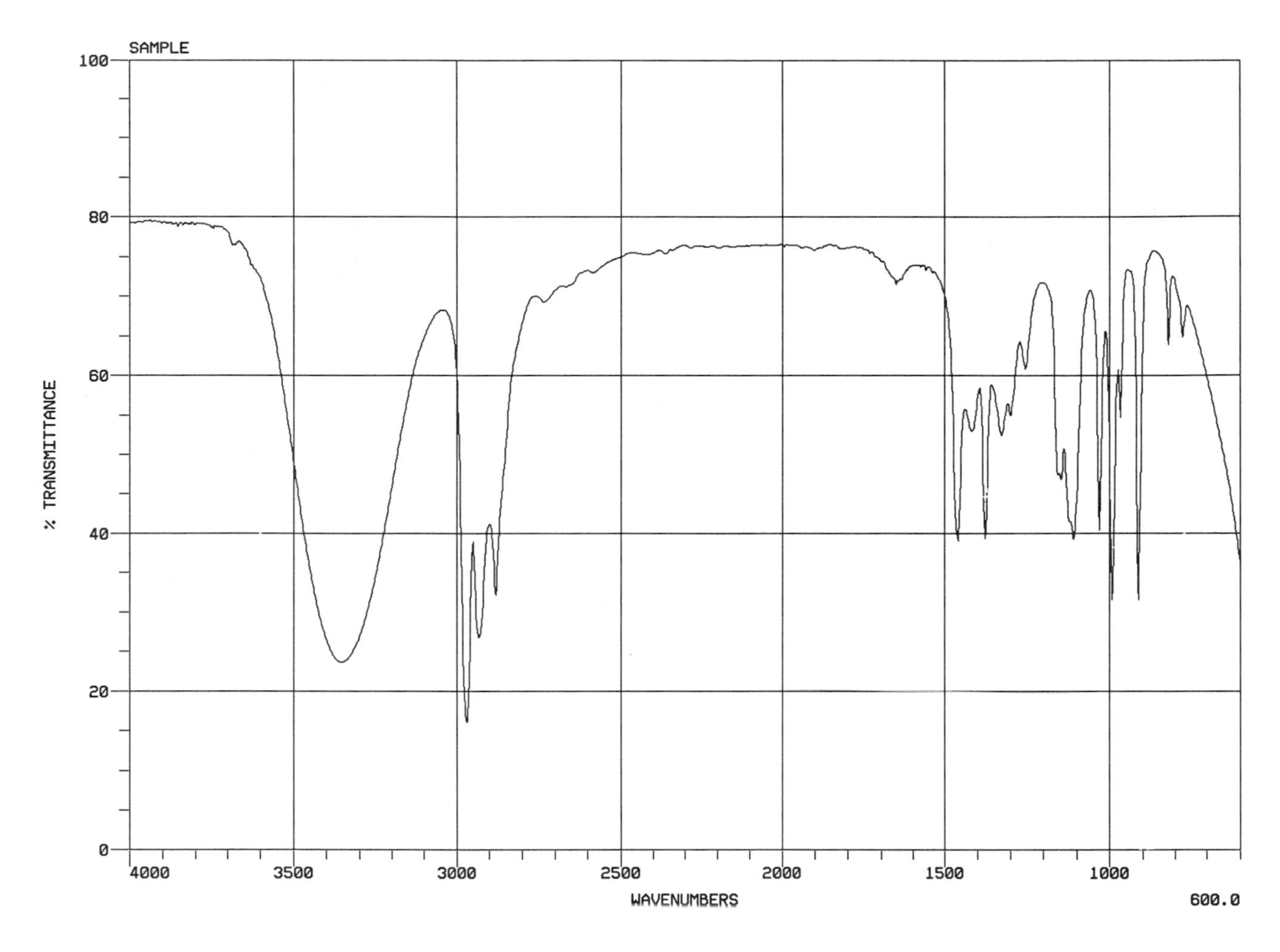

Problem 3. Infrared spectrum. (Reprinted from "A Spectrum of Spectra" by Richard A. Tomasi, Sunbelt R&T, Inc. Tulsa, Oklahoma; with permission.)

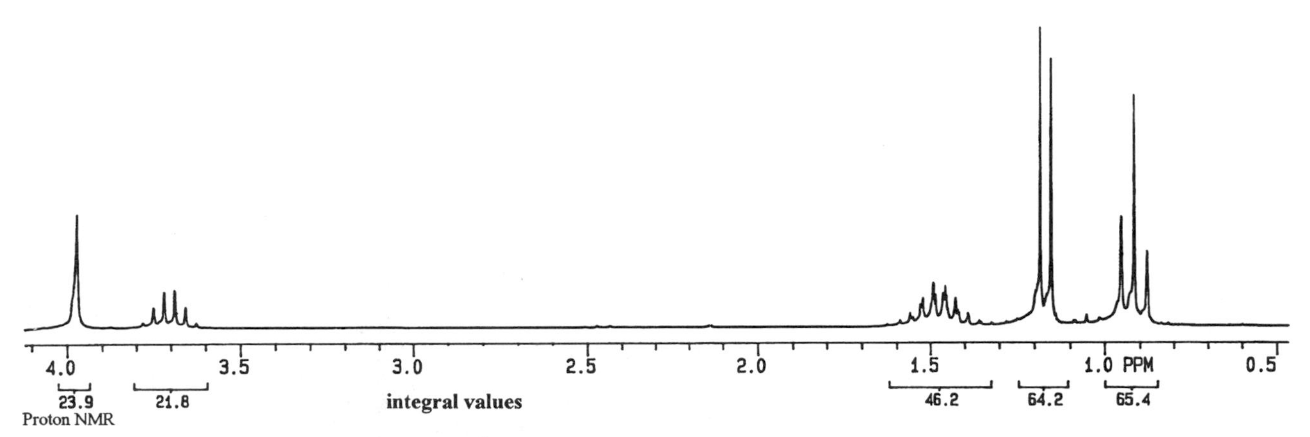

Problem 3. ^{1}H NMR spectrum (200 MHz). (Reprinted from "A Spectrum of Spectra" by Richard A. Tomasi, Sunbelt R&T, Inc. Tulsa, Oklahoma; with permission.) Integration: δ 0.93 (3H), δ 1.18 (3H), δ 1.49 (2H), δ 3.70 (1H), δ 3.93 (1H).

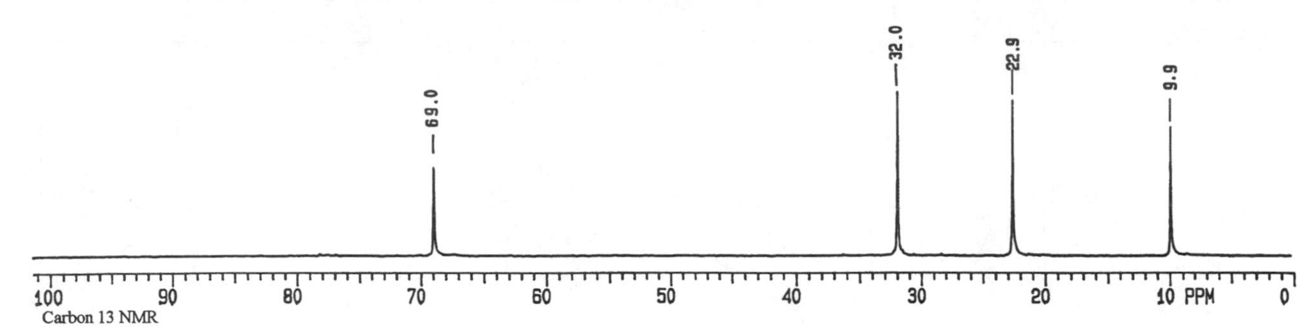

Problem 3. ^{13}C NMR spectrum (50 MHz). (Reprinted from "A Spectrum of Spectra" by Richard A. Tomasi, Sunbelt R&T, Inc. Tulsa, Oklahoma; with permission.) Integration: δ 9.9 (CH_3), δ 22.9 (CH_3), δ 32.1 (CH_2), δ 69.0 (CH).

4. A compound of boiling point 153–157°C is known to contain none of the following elements: sulfur, halogen, and nitrogen. It burns with a clean flame. Upon treatment with acetyl chloride this compound shows evidence of a chemical reaction and a pleasant-smelling compound is formed. Treatment of the original compound with acidic CrO_3 in sulfuric acid gives an immediate blue-green color. In addition, treatment of the original compound with iodine in aqueous sodium hydroxide (iodoform test) does not give a precipitate. Upon treatment of the original compound with 3,5-dinitrobenzoyl chloride, a solid of melting point 60°C is obtained. The IR, ^1H NMR, and ^{13}C NMR spectra are provided here for the original compound. Deduce the structure of the compound and assign the major IR bands as well as all of the NMR absorptions, and write reactions for all positive chemical tests.

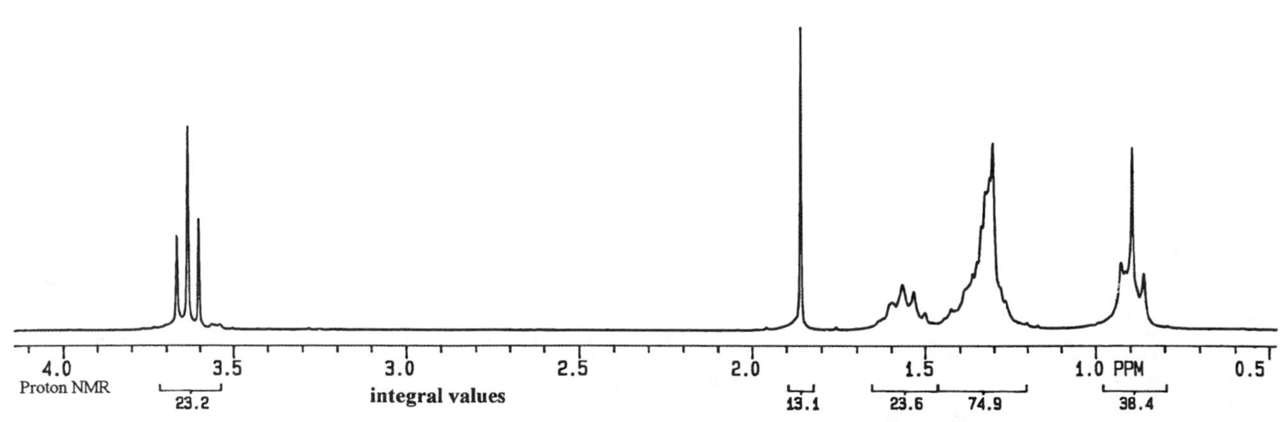

Problem 4. ^1H NMR spectrum (200 MHz). (Reprinted from "A Spectrum of Spectra" by Richard A. Tomasi, Sunbelt R&T, Inc. Tulsa, Oklahoma; with permission.) Integration: δ 0.91 (3H), δ 1.31 (6H), δ 1.57 (2H), δ 1.86 (1H), δ 3.63 (2H).

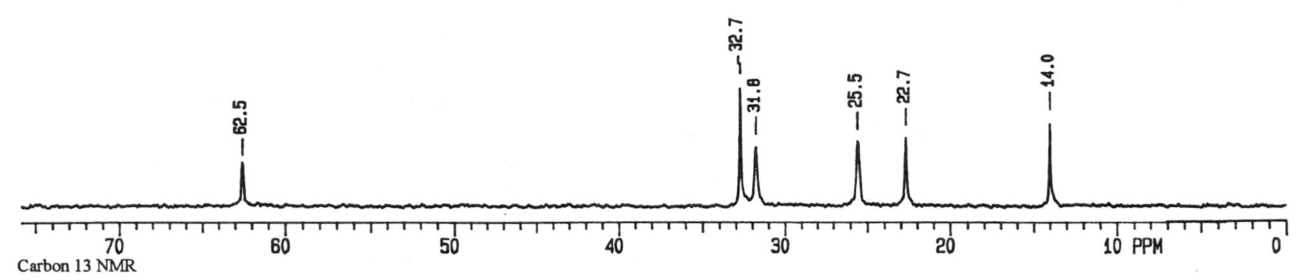

Problem 4. ^{13}C NMR spectrum (50 MHz). (Reprinted from "A Spectrum of Spectra" by Richard A. Tomasi, Sunbelt R&T, Inc. Tulsa, Oklahoma; with permission.) Integration: δ 14.0 (CH_3), δ 22.7 (CH_2), δ 25.5 (CH_2), δ 31.8 (CH_2), δ 32.7 (CH_2), δ 62.5 (CH_2).

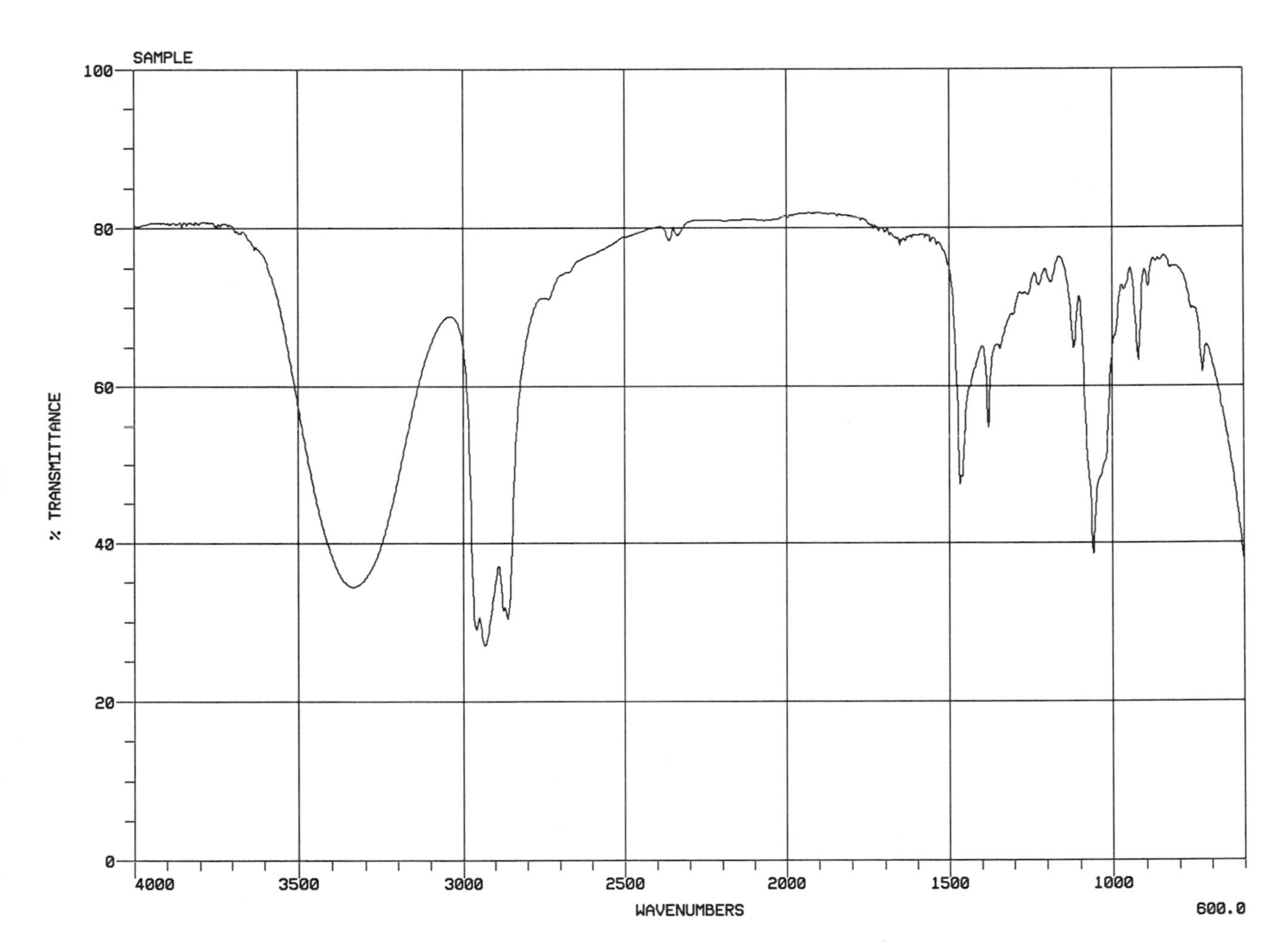

Problem 4. Infrared spectrum. (Reprinted from "A Spectrum of Spectra" by Richard A. Tomasi, Sunbelt R&T, Inc. Tulsa, Oklahoma; with permission.)

5. A compound of melting point 128°C upon sodium fusion gives evidence only for nitrogen. Upon treatment with acetyl chloride this compound shows no evidence of a chemical reaction. Acid hydrolysis of this compound produced another compound of neutralization equivalent 121 ± 1. Mass spectrometric (MS) analysis of the original compound reveals the most intense peaks to be at m/z 121, 105, and 77. The IR, ^{1}H NMR, and ^{13}C NMR spectra are provided here for the original compound. Deduce the structure of the compound and assign the major IR bands as well as all of the NMR absorptions, and write reactions for all positive chemical tests.

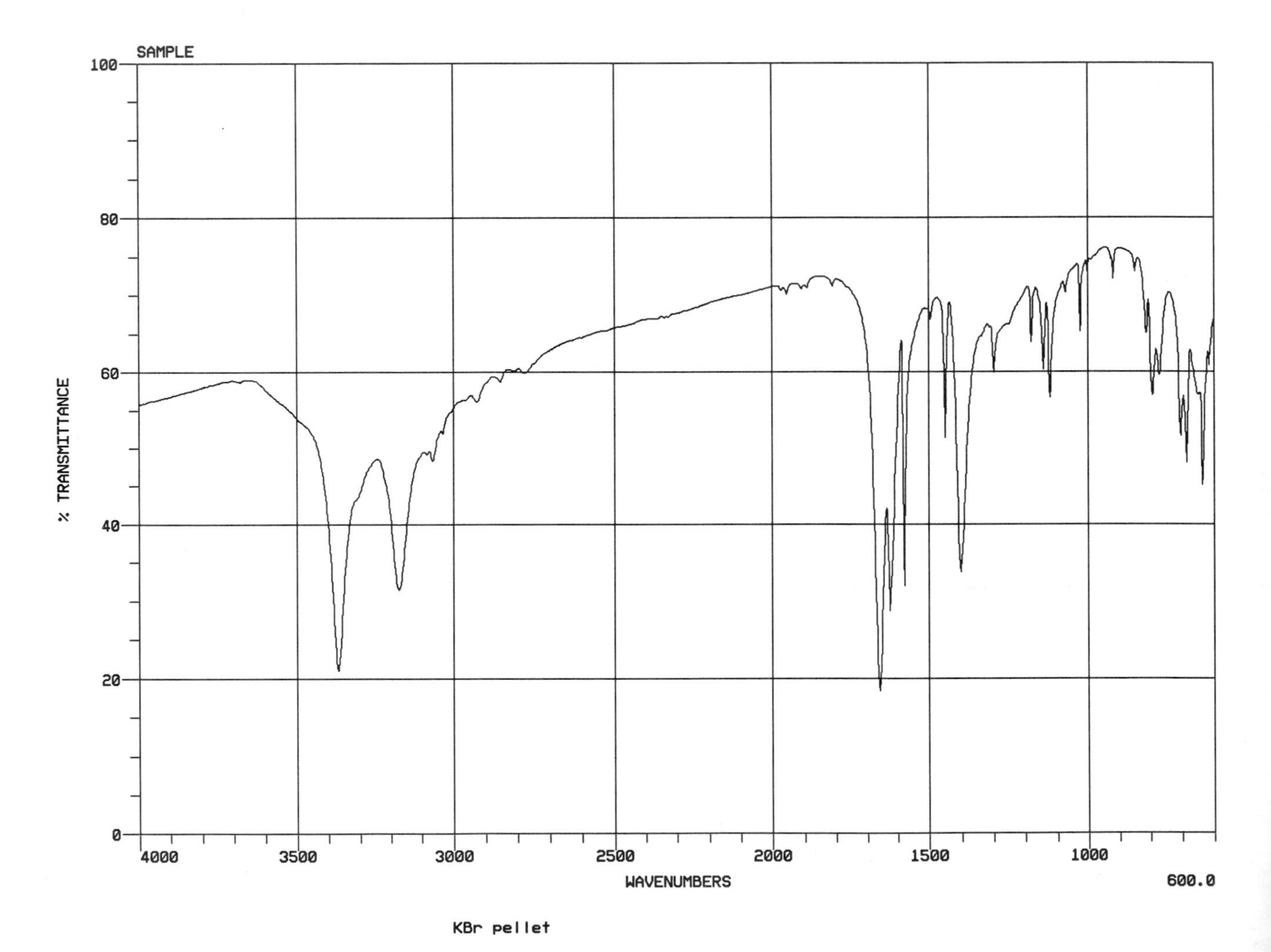

Problem 5. Infrared spectrum. (Reprinted from "A Spectrum of Spectra" by Richard A. Tomasi, Sunbelt R&T, Inc. Tulsa, Oklahoma; with permission.)

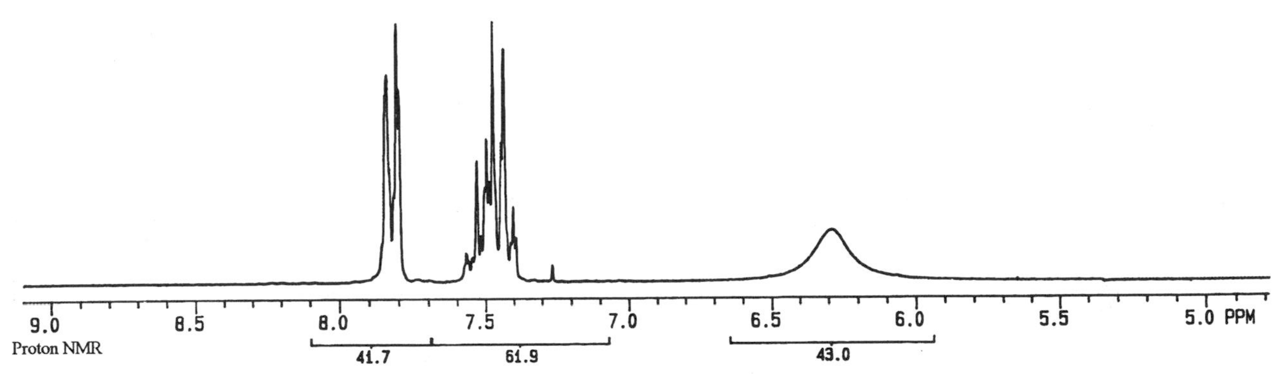

Problem 5. ^{1}H NMR spectrum (200 MHz). (Reprinted from "A Spectrum of Spectra" by Richard A. Tomasi, Sunbelt R&T, Inc. Tulsa, Oklahoma; with permission.) Integration: δ 6.30 (2H), δ 7.49 (3H), δ 7.81 (2H).

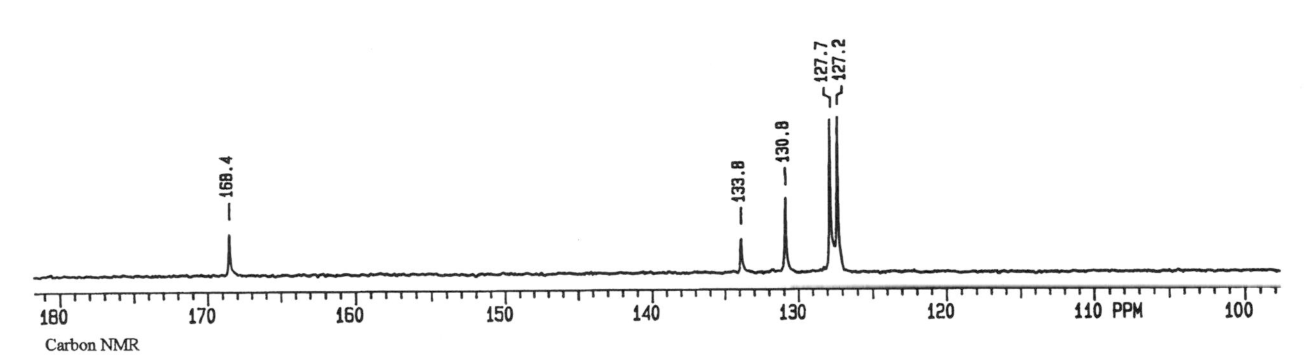

Problem 5. ^{13}C NMR spectrum (50 MHz). (Reprinted from "A Spectrum of Spectra" by Richard A. Tomasi, Sunbelt R&T, Inc. Tulsa, Oklahoma; with permission.) Integration: δ 127.2 (CH), δ 127.7 (CH), δ 130.8 (CH), δ 133.8 (C), δ 168.4 (C).

6. A compound of boiling point 153–156°C upon sodium fusion gives evidence only for nitrogen. Treatment with acetyl chloride provides no evidence of a chemical reaction. Treatment with hydrochloric acid, however, results in immediate dissolution and the evolution of heat. Treatment of the original compound with an alkaline solution of benzenesulfonyl chloride (Hinsberg test) gives no sign of reaction. The IR, ^{1}H NMR, and ^{13}C NMR spectra are provided here for the original compound. Deduce the structure of the compound and assign the major IR bands as well as all of the NMR absorptions, and write reactions for all positive chemical tests.

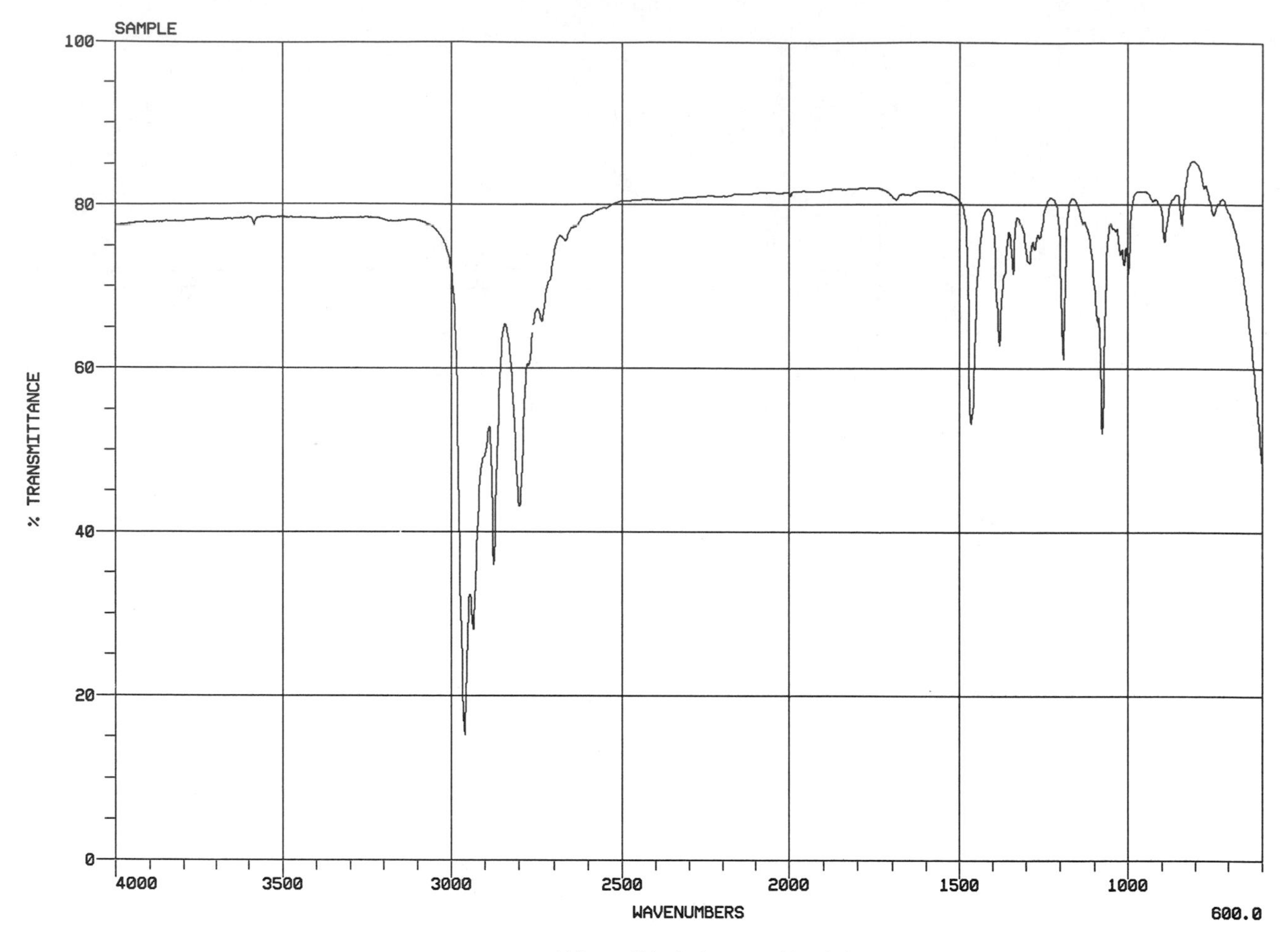

Problem 6. Infrared spectrum. (Reprinted from "A Spectrum of Spectra" by Richard A. Tomasi, Sunbelt R&T, Inc. Tulsa, Oklahoma; with permission.)

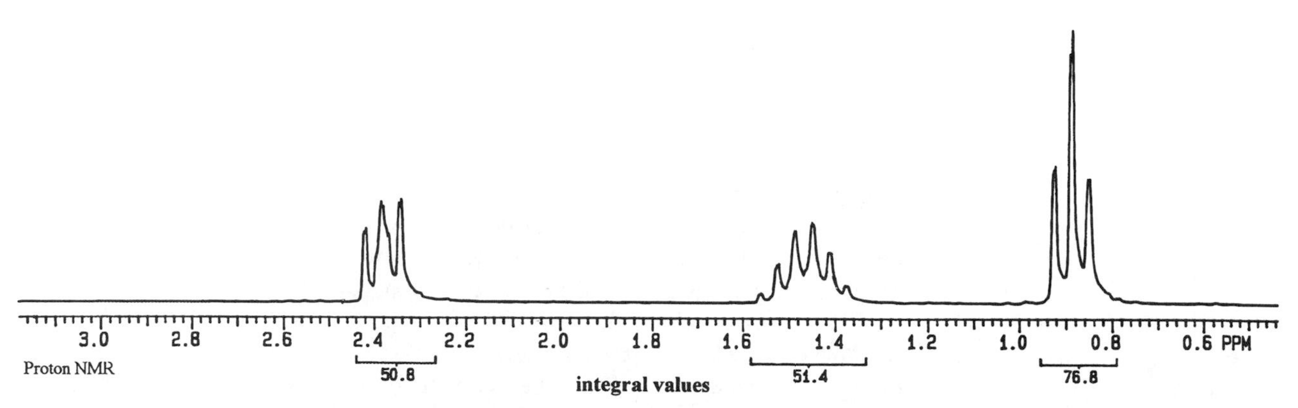

Problem 6. ^{1}H NMR spectrum (200 MHz). (Reprinted from "A Spectrum of Spectra" by Richard A. Tomasi, Sunbelt R&T, Inc. Tulsa, Oklahoma; with permission.) Integration: δ 0.89 (9H), δ 1.47 (6H), δ 2.39 (6H).

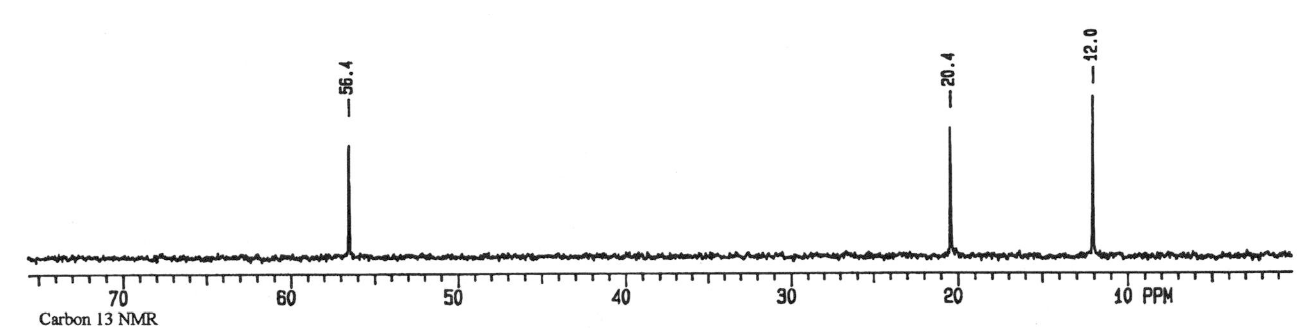

Problem 6. ^{13}C NMR spectrum (50 MHz). (Reprinted from "A Spectrum of Spectra" by Richard A. Tomasi, Sunbelt R&T, Inc. Tulsa, Oklahoma; with permission.) Integration: δ 12.0 (CH_3), δ 20.4 (CH_2), δ 56.4 (CH_2).

7. A compound of boiling point 250–252°C is known to contain none of the following: nitrogen, sulfur, and halogen. Upon treatment with acetyl chloride there is no evidence of a chemical reaction. The original compound does give a silver mirror when treated with Tollens reagent. Treatment with $KMnO_4$ or with bromine in an organic solvent results in an immediate reaction in each case. Treatment of the original compound under the conditions of the 2,4-dinitrophenylhydrazine test results in rapid formation of a red precipitate of mp 251–254°C. The IR, ^1H NMR, and ^{13}C NMR spectra are provided here for the original compound. Deduce the structure of the compound and assign the major IR bands as well as all of the NMR absorptions, and write reactions for all positive chemical tests. Also explain the apparent discrepancy between the ^{13}C NMR spectrum and the numerically tabulated chemical shifts.

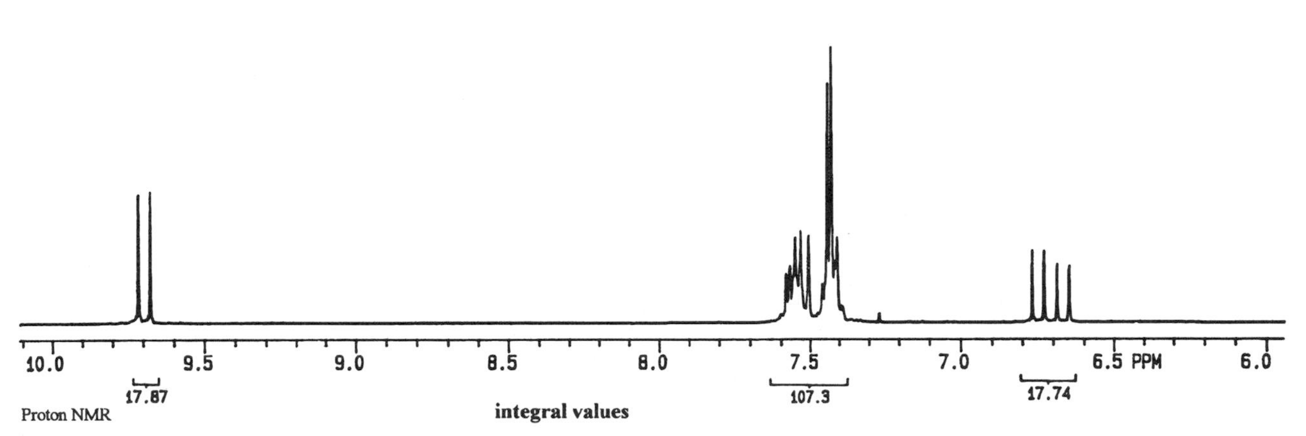

Problem 7. ^1H NMR spectrum (200 MHz). (Reprinted from "A Spectrum of Spectra" by Richard A. Tomasi, Sunbelt R&T, Inc. Tulsa, Oklahoma; with permission.) Integration: δ 6.70 (1H), δ 7.40–7.60 (1H, 5H), δ 9.70 (1H).

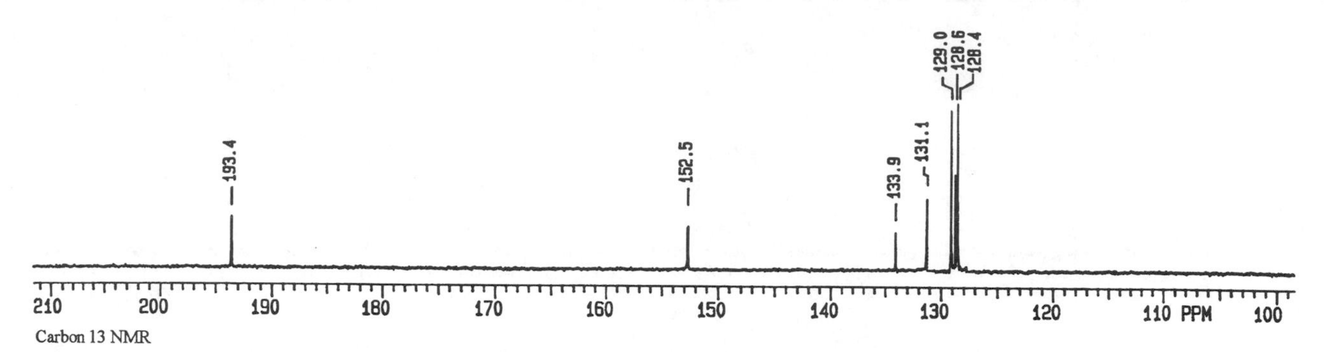

Problem 7. ^{13}C NMR spectrum (50 MHz). (Reprinted from "A Spectrum of Spectra" by Richard A. Tomasi, Sunbelt R&T, Inc. Tulsa, Oklahoma; with permission. Integration: δ 128.4 (CH), δ 128.6 (CH), δ 129.0 (CH), δ 131.1 (CH), δ 133.9 (C), δ 152.5 (CH), δ 193.4 (CH).

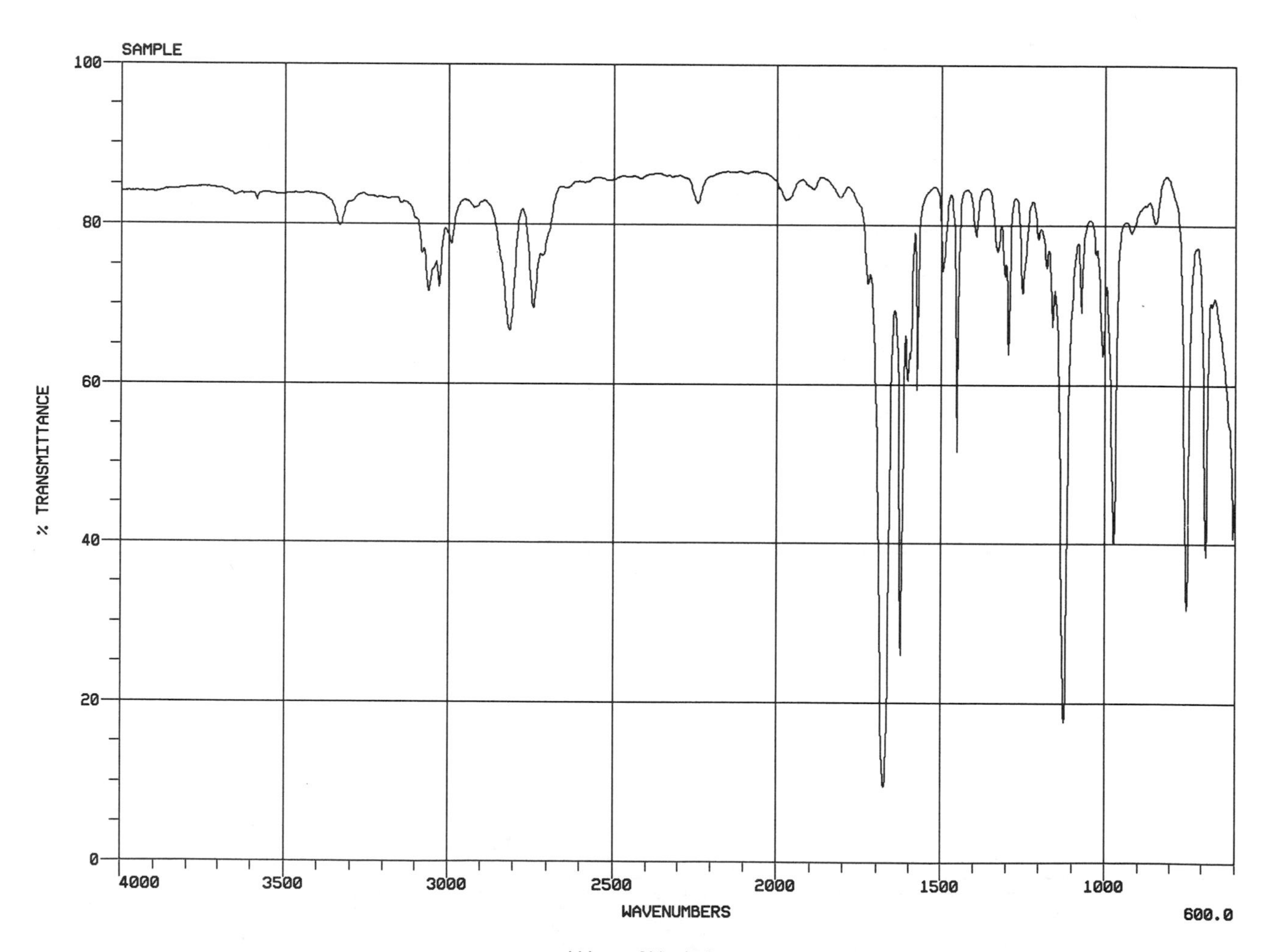

Problem 7. Infrared spectrum. (Reprinted from "A Spectrum of Spectra" by Richard A. Tomasi, Sunbelt R&T, Inc. Tulsa, Oklahoma; with permission.)

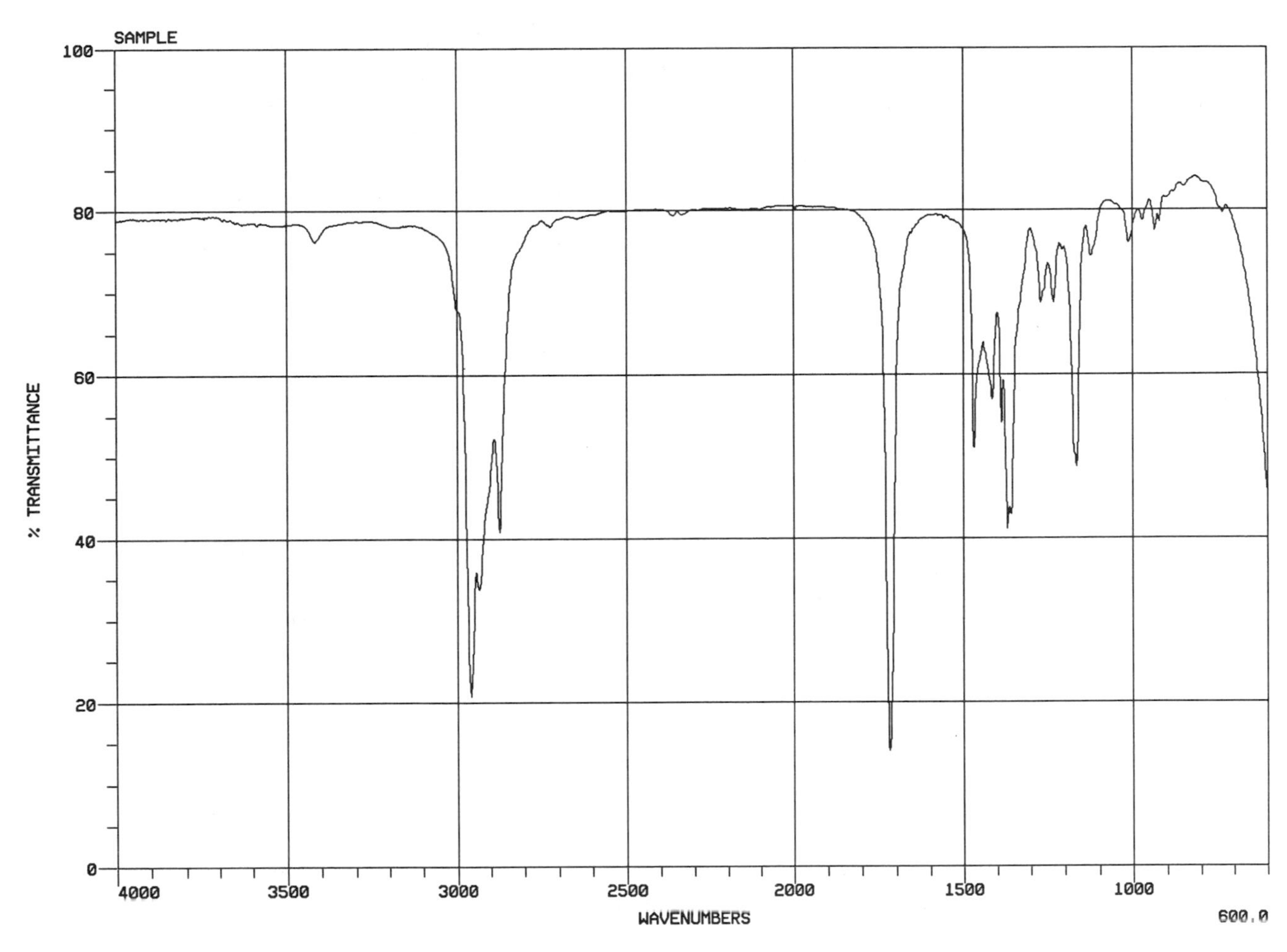

Problem 8. Infrared spectrum. (Reprinted from "A Spectrum of Spectra" by Richard A. Tomasi, Sunbelt R&T, Inc. Tulsa, Oklahoma; with permission.)

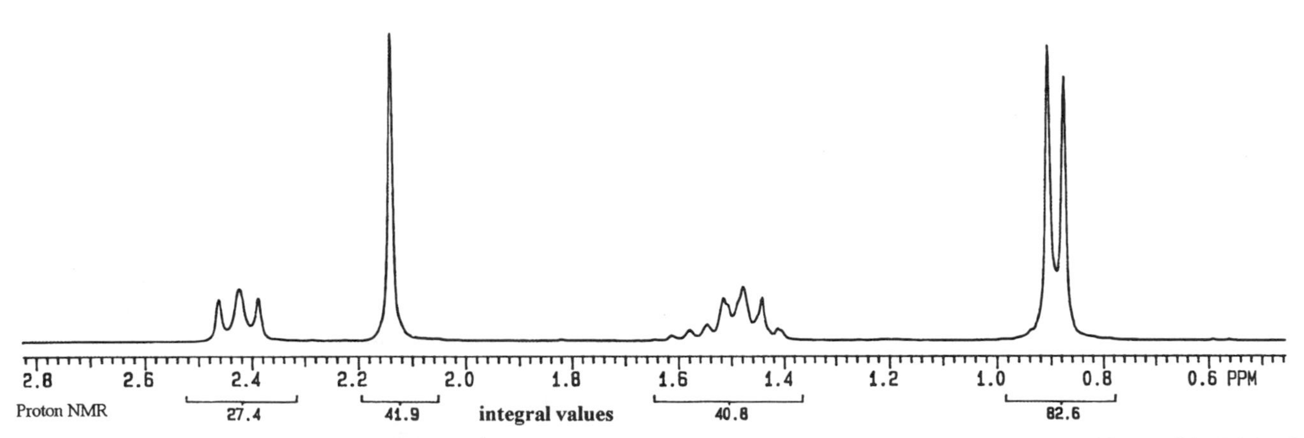

Problem 8. ^{1}H NMR spectrum (200 MHz). (Reprinted from "A Spectrum of Spectra" by Richard A. Tomasi, Sunbelt R&T, Inc. Tulsa, Oklahoma; with permission.) Integration: δ 0.90 (6H), δ 1.49 (3H), δ 2.14 (3H), δ 2.43 (2H).

8. A compound of boiling point 140–145°C is known to contain none of the following: sulfur, nitrogen, and halogen. Treatment with acetyl chloride shows no evidence of a chemical reaction. This compound gives a 2,4-dinitrophenylhydrazone, but the Tollens test is negative. The IR, ^{1}H NMR, and ^{13}C NMR spectra are provided here for the original compound. Deduce the structure of the compound and assign the major IR bands as well as all of the NMR absorptions, and write reactions for all positive chemical tests.

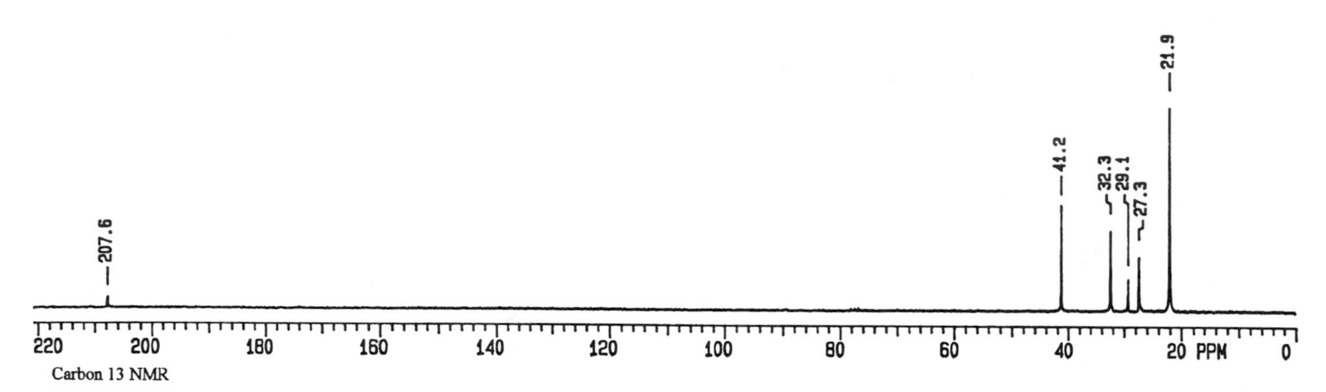

Problem 8. ^{13}C NMR spectrum (50 MHz). (Reprinted from "A Spectrum of Spectra" by Richard A. Tomasi, Sunbelt R&T, Inc. Tulsa, Oklahoma; with permission.) Integration: δ 21.9 (CH_3), δ 27.3 (CH), δ 29.1 (CH_3), δ 32.3 (CH_2), δ 41.2 (CH_2), δ 207.6 (C).

9. A compound is known to contain none of the following: sulfur, nitrogen, and halogen. Moreover, quantitative elemental analysis revealed that the compound was 85.64% carbon and 6.78% hydrogen. This compound gives both a phenylhydrazone and a semicarbazone, but the Tollens test is negative. The IR, ^{1}H NMR, and ^{13}C NMR spectra are provided here for the original compound. Deduce the structure of the compound and assign the major IR bands as well as all of the NMR absorptions, and write reactions for all positive chemical tests.

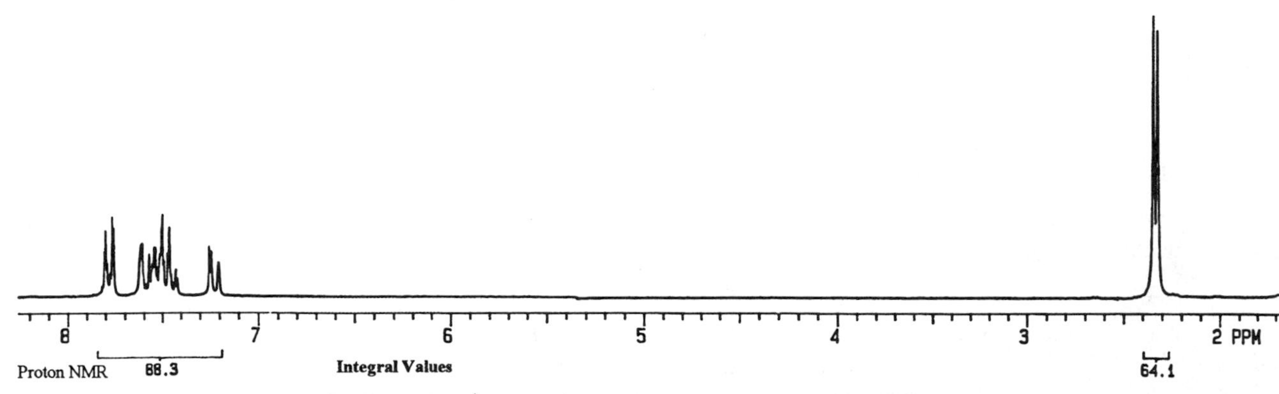

Problem 9. ^{1}H NMR spectrum (200 MHz). (Reprinted from "A Spectrum of Spectra" by Richard A. Tomasi, Sunbelt R&T, Inc. Tulsa, Oklahoma; with permission.) Integration: δ 2.35 (6H), δ 7.20–7.80 (8H).

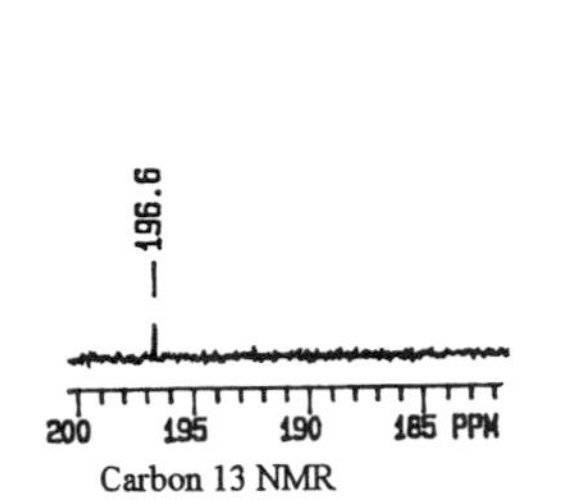

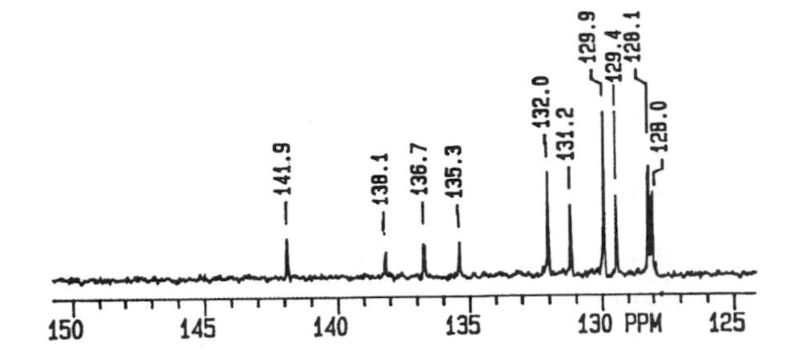

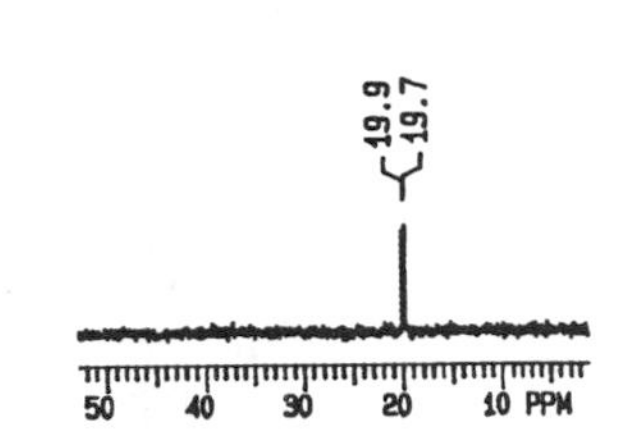

Problem 9. ^{13}C NMR spectrum (50 MHz). (Reprinted from "A Spectrum of Spectra" by Richard A. Tomasi, Sunbelt R&T, Inc. Tulsa, Oklahoma; with permission.) Integration: δ 19.7 (CH_3), δ 19.9 (CH_3), δ 196.6 (C).

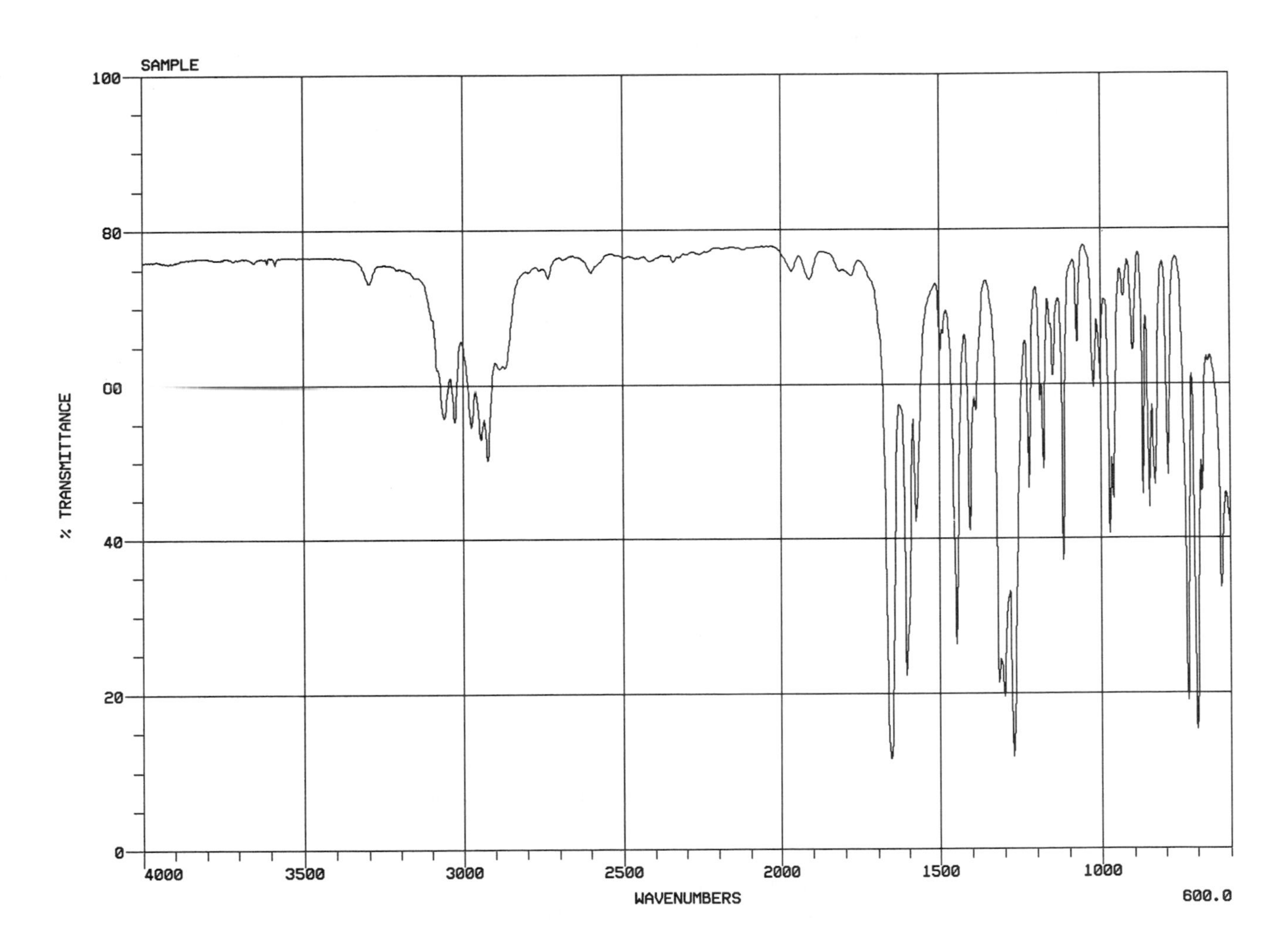

Problem 9. Infrared spectrum. (Reprinted from "A Spectrum of Spectra" by Richard A. Tomasi, Sunbelt R&T, Inc. Tulsa, Oklahoma; with permission.)

Problem Set 1

In the investigation of unknown compounds, the following types of behavior are observed frequently. Indicate in each instance the deductions which may be made as to the nature of the compound.

1. A yellow, neutral compound containing only carbon, hydrogen, and oxygen was changed to an acid by the action of hydrogen peroxide. The original yellow compound displayed a singlet in the ^{1}H NMR spectrum at $\approx \delta$ 9.0 ppm, as well as other signals.
2. A neutral compound reacted with phenylhydrazine to yield a product that differed from the expected phenylhydrazone by the elements of ethanol; that is, the condensation involved the elimination not only of the elements of water but also of those of ethanol.
3. A compound containing only carbon, hydrogen, and oxygen reacted with acetyl chloride but not with phenylhydrazine. Treatment with periodic acid converted it into a compound that reacted with phenylhydrazine but not with acetyl chloride.
4. A yellow, neutral compound formed a derivative with *o*-phenylenediamine. The original compound showed an IR band at 1710 cm^{-1}.
5. An alcohol gave a positive iodoform test and a negative Lucas test.
6. A neutral compound containing only carbon, hydrogen, and oxygen reacted with acetyl chloride but not with phenylhydrazine. Heating with mineral acids converted it to a compound that failed to react with acetyl chloride but that gave positive tests with phenylhydrazine and bromine in carbon tetrachloride. The ^{1}H NMR spectrum of the original compound showed only two singlets.
7. A nitrogen-containing compound gave a positive nitrous acid test for secondary amines, but its derivative with benzenesulfonyl chloride was soluble in alkalies.
8. A compound, when treated with excess ethanol and a trace of acid, was found to take up the elements of ethyl ether. One of the two possible classes of compounds that serve as an answer showed a ^{1}H NMR signal at ca. δ 9.0–10.0 before treatment; treatment moved this signal to ca. δ 5.0.
9. A neutral compound containing carbon, hydrogen, and oxygen underwent dimerization in an ethanolic solution of sodium cyanide. The IR spectrum of the reactant showed bands at ca. 2800 and 2700 cm^{-1}.
10. A basic compound failed to react with benzenesulfonyl chloride but yielded a derivative with nitrous acid. The original basic compound showed no appreciable IR bands near 3333 cm^{-1}. The product of nitrous acid treatment showed an IR band at ca. 1550 cm^{-1}.
11. An ester had a saponification equivalent of 59 $\pm$ 1. The ^{1}H NMR spectrum showed two singlets with an integration ratio of 3:1.
12. A solid water soluble acid had a neutralization equivalent of 54 $\pm$ 1. At temperatures above its melting point, the compound lost carbon dioxide and formed a new solid acid with a neutralization equivalent of 59 $\pm$ 1. The nonexchangeable protons of the original compound resulted in a multiplet ^{1}H NMR signal; the product formed upon heating showed the nonexchangeable protons as a singlet.

13. A water soluble acidic compound containing nitrogen and sulfur gave a neutralization equivalent of 142 ± 1. Addition of barium chloride to an aqueous solution produced a precipitate insoluble in acids. Alkali caused the separation of a basic compound. The hydrochloride of this basic compound had a neutralization equivalent of 130 ± 1. Both the first and last compounds (the acids) showed broad IR bands at ca. 3333–2500 cm^{-1}.

14. A water soluble compound containing nitrogen gave a neutralization equivalent of 73 ± 1 when titrated with standard hydrochloric acid and methyl red as the indicator. It formed a precipitate when treated with benzenesulfonyl chloride and sodium hydroxide solution. The original compound showed a quartet (4H), a triplet (6H), as well as a broadened singlet (1H) in the ^{1}H NMR spectrum.

15. A 12-g sample of a compound, C_8H_8O, was treated with a carbon tetrachloride solution of bromine containing 60 g of bromine. Hydrobromic acid was evolved, and a quantitative determination showed that 16.2 g were liberated. After the reaction was complete the solution was still red, and it was found that 12 g of bromine remained unused. Calculate the number of atoms of bromine per molecule that were introduced by substitution and, if addition also took place, the number of atoms of bromine which were added. Use Br = 80, H = 1, C = 12, O = 16. Also describe important NMR and IR features you would anticipate for both the reactant and the product; emphasize the important spectral changes expected as the result of the bromine treatment.

Problem Set 2

In solving the problems of this set and those in sets 3 and 4, follow the general procedure described in the sample problems above. Also the sample report forms at the end of Chapter 2 are largely consistent with the information provided below. Make the customary allowances for experimental error in boiling points and melting points. Assign spectral data to structural features. Give equations for derivatives.

I. (1) White crystals, mp 117–118°C.

(2) Elemental analysis for X, N, S—negative. (X = halogen)

(3) Solubility—water (+).

(4) Classification tests:

$C_6H_5NHNH_2$—negative CH_3COCl—positive

$KMnO_4$—positive HIO_4—positive

Br_2 in CCl_4—negative

Neutralization equivalent = 151±1

(5) Derivative: *p*-Nitrobenzyl ester, mp 123°C.

(6) Spectra:

IR (mineral oil mull): 3333–2400 cm^{-1} (s, b); 1706 cm^{-1} (s); 1449, 1379, 1300, 1200, 1190, 943, 865, 730 cm^{-1} (all m).

^{1}H NMR (60 HMz, acetone): δ 5.22, 1H, s; δ 6.93–7.88, 7H, m.

II. (1) Colorless liquid, bp 259–261°C

(2) Elemental analysis for Br—positive; for Cl, I, N, S—negative.

(3) Solubility—water (−), NaOH (−), HCl (−), H_2SO_4 (+).

(4) Classification tests:

$H_2NOH{\cdot}HCl$—negative $KMnO_4$—negative
CH_3COCl—negative Br_2 in CCl_4—negative
$AgNO_3$—negative NaI—negative

Hot sodium hydroxide—clear solution which on acidification gave white crystals, mp 250°C, containing bromine.

Saponification equivalent of the original compound = 229 ± 2.

(5) Derivatives:

(a) Treatment with hydrazine gave colorless crystals, mp 164°C.

(b) Treatment with 3,5-dinitrobenzoic acid and sulfuric acid gave pale yellow crystals, mp 92°C.

III. (1) Brown liquid, bp 198–200°C.

(2) Elemental analysis for X, N, S—negative. (X = halogen)

(3) Solubility—water (−), NaOH (+), $NaHCO_3$ (−), HCl (−).

(4) Classification tests:

CH_3COCl—positive Br_2 in H_2O—precipitate
$C_6H_5NHNH_2$—negative $FeCl_3$—violet
Ce(IV)—negative

(5) Derivative: Treatment with chloroacetic acid gave white crystals, mp 102–103°C.

(6) Spectra:

IR (neat): 3390 cm^{-1} (s, b); 1163, 935, 780, 690 cm^{-1} (all s).

^1H NMR (60 MHz, $CDCl_3$): δ 2.25, 3H, s; δ 5.67, 1H, s; δ 6.5–7.3, 4H, m.

IV. (1) Colorless liquid, bp 194–195°C.

(2) Elemental analysis for X, N, S—negative. (X = halogen)

(3) Solubility—all (−).

(4) Classification tests:

$H_2SO_4 \cdot SO_3$—negative $Br_2 + CCl_4$—negative
$AlCl_3 + CHCl_3$—light yellow

(5) Derivatives: none.

sp gr $^{20}_{4}$ = 0.8963 n^{20}_{D} = 1.4811

(6) Spectra:

Mass spectrometry: molecular formula of $C_{10}H_{18}$.

^{13}C NMR: δ 24.6 (CH_2), 29.6 (CH_2), and 36.9 (CH).

V. (1) White crystals, mp 187–188°C.

(1) Elemental analysis for X, N, S—negative. (X = halogen)

(3) Solubility—water (+). Aqueous solution is acid to litmus.

(4) Classification tests:

$KMnO_4$—negative $C_6H_5NHNH_2$—negative
Br_2 in CCl_4—negative CH_3COCl—negative

Neutralization equivalent = 59

(5) Derivative: Heating with 4-bromophenacyl bromide gave white crystals, mp 209–210°C.

(6) Spectra:

IR (Nujol mull): 3333–3222 cm^{-1} (s, b); 1695, 1418, 1307, 1198, 913 cm^{-1} (all s).

^1H NMR (60 MHz, DMSO-d_6): δ 2.43, s, δ 11.80, bs; areas 2:1 (upfield singlet to downfield singlet). δ 11.80 shift is concentration dependent.

VI. (1) Reddish brown solid, mp 72°C.

(2) Elemental analysis for N—positive; for X, S—negative.

(3) Solubility—water (−), NaOH (−), HCl (+). A solution of the compound in dilute hydrochloric acid was decolorized with Norit and alkali added. The compound purified in this manner melted at 73°C.

(4) Classification tests:

$KMnO_4$—positive	$C_6H_5SO_2Cl$ + NaOH—residue soluble in hydrochloric acid
Br_2 in CCl_4—precipitate	$Fe(OH)_2$—negative
Bromine water—precipitate	Tollens reagent—positive
$C_6H_5NHNH_2$—positive	

Hot sodium hydroxide decomposed the compound. After removal of a dark brown solid by filtration, the filtrate was neutralized and a light tan precipitate obtained. After recrystallization from a water–alcohol mixture, it melted with decomposition at 236–240°C. It was soluble in HCl and NaOH, insoluble in sodium bicarbonate. The original compound reacted with acetone and sodium hydroxide to give a yellow precipitate, mp 134–135°C.

(5) Derivatives:

Phenylhydrazone, mp 148°C.

Semicarbazone, mp 224°C.

(6) Spectra:

IR (Nujol mull): no absorption near 3333 cm^{-1}; absorptions at 1653 cm^{-1} (m) and 1587 cm^{-1} (s).

^1H NMR (60 MHz, $CDCl_3$) δ 3.05, 6H, s; δ 6.69, 2H, d; δ 7.71, 2H, d; δ 9.70, 1H, s.

Problem Set 3

1. A brown liquid (I) boiled at 193–195°C. It contained nitrogen but gave negative tests for sulfur and the halogens. It was insoluble in water but soluble in dilute acid. It did not react with acetyl chloride or benzenesulfonyl chloride. Treatment of a hydrochloric acid solution of the unknown compound with sodium nitrite, followed by neutralization, gave a compound (II) that melted at 83–84°C. Compound II was insoluble in alkalies but dissolved in boiling concentrated sodium hydroxide solution, with the liberation of a gas (III). This gas (III) was absorbed in water, and the aqueous solution was treated with phenyl isothiocyanate to form a compound (IV) with a melting point of 134–135°C. Careful acidification of the alkaline solution, followed by extraction, gave a compound (V) that melted at 125–126°C.

The ^1H NMR spectrum of compound V (60 MHz, acetone) showed δ 6.63, 2H, d; δ 7.67, 2H, d; δ 8.7, 1H, bs. The δ 8.7 signal was so broad as to be detectable only by electronic integration.

2. A colorless liquid was found to be soluble in water and in ether. It boiled at 94–96°C, and gave negative tests for the halogens, nitrogen, and sulfur. It reduced a dilute potassium permanganate solution, decolorized bromine in carbon tetrachloride, reacted with acetyl chloride, and liberated hydrogen upon treatment with sodium. It did not give iodoform when treated with sodium hypoiodite and did not react with phenylhydrazine. Treatment with 3,5-dinitrobenzoyl chloride transformed it into a compound melting at 47–48°C.

The ^1H NMR spectrum of the original compound (60 MHz, $CDCl_3$) showed δ 3.58, 1H, s; δ 4.13, 2H, m; δ 5.13, 1H, m; δ 5.25, 1H, m; δ 6.0, 1H, 10 lines. The δ 3.58 signal showed a chemical shift that was concentration-dependent.

3. A yellow solid (I), melting at 113–114°C, contained nitrogen but no halogens, sulfur, or metals. It was insoluble in water and alkalies but soluble in dilute acids. The acid solution of I was treated with sodium nitrite in the cold and then boiled. The product (II) of this reaction separated when the solution was cooled. It contained nitrogen and melted at 95–96°C; it was insoluble in acids and sodium bicarbonate solution but soluble in sodium hydroxide solution. The products obtained by treating compounds I and II with zinc and a boiling solution of ammonium chloride readily reduced Tollens reagent. The original compound (I) was treated with benzenesulfonyl chloride and alkali. Acidification of the resulting solution gave a compound (III) that melted at 135–136°C.

The ^1H NMR spectrum of compound II (60 MHz, DMSO-d_6) showed δ 7.0–7.75, 4H, m; δ 9.8, 1H, s.

4. A colorless crystalline compound (I) melted at 186–187°C; it contained nitrogen but no halogens or sulfur. It was insoluble in water and dilute acids but soluble in dilute sodium bicarbonate solution. It gave a neutralization equivalent of 180 $\pm$ 2 but did not react with bromine in carbon tetrachloride, dilute potassium permanganate solution, acetyl chloride, or phenylhydrazine. It was treated for some time with boiling hydrochloric acid. When this reaction mixture was cooled, a compound (II) separated that melted at 120–121°C and gave a neutralization equivalent of 121 $\pm$ 1. The filtrate remaining after the removal of II was evaporated to dryness, and the residue (III) was purified by recrystallization. It contained nitrogen and chlorine, was rather hygroscopic, and decomposed when an attempt was made to determine is melting point. It was insoluble in ether, and its aqueous solution gave a precipitate with silver nitrate. A solution of III was treated with nitrous acid in the cold. A vigorous evolution of a gas was observed. Compound III was treated with benzenesulfonyl chloride and sodium hydroxide solution. Acidification of the resulting solution gave a new product (IV) that melted at 164–165°C.

^1H NMR spectra (60 MHz): I (NaOD, D_2O) showed δ 4.06, 2H, s; δ 4.77, HDO impurity, s; δ 7.5–8.0, 5H, m. II (CCl_4) showed δ 7.3–7.7, 3H, m; δ 8.0–8.25, 2H, m; δ 12.8, 1H, s. III (D_2O) showed δ 3.58, s; δ 4.9, HDO.

5. A colorless crystalline compound (I) melted at 168°C, with decomposition. It gave negative tests for nitrogen, halogens, sulfur, and metals. It was soluble in water but insoluble in ether. It reacted with acetyl chloride, decolorized permanganate solution and reduced Fehling's solution and Tollens reagent. It reacted with phenylhydrazine to give a product (II) which melted, with decomposition, at 199–201°C. When I was warmed with concentrated nitric acid, a vigorous reaction took place, and a compound (III) separated when the reaction mixture was cooled. This compound (III) was insoluble in water but readily soluble in alkalies; it gave a neutralization equivalent of 104 $\pm$ 1. Compound III reacted with acetyl chloride but not with phenylhydrazine, and melted at about 212–213°C, with decomposition. If kept above its melting point for some time, it was converted into a new compound (IV) which melted at 132–133°C, after recrystallization. Compound IV was insoluble in water but soluble in sodium bicarbonate solution, and it gave a neutralization equivalent of 111 ± 1. Treatment of the sodium salt of IV with *p*-bromophenacyl bromide gave a compound (V) melting at 137–138°C.

The original compound (I) was optically active, having a specific rotation of +81°, but the degradation products III, IV, and V were optically inactive.

A solution of the potassium salt of IV (in D_2O) gave rise to these ^1H NMR signals: δ 6.59, 1H, m; δ 7.05, 1H, m; δ 7.64, 1H, m. A Nujol mull of IV gave these IR bands: 3100–2400 cm^{-1} (s); 1675 cm^{-1} (s); several bands in the 1667–1000 cm^{-1} region; 932 cm^{-1} (m); 885 cm^{-1} (m); 758 cm^{-1} (m).

Problem Set 4

1. A compound insoluble in water, sodium hydroxide, and hydrochloric acid but soluble in concentrated sulfuric acid and containing nitrogen melted at 68°C. When treated with tin and hydrochloric acid, it yielded a substance that reacted with benzenesulfonyl chloride to give an alkali soluble derivative. When the original compound was treated with zinc and hot sodium hydroxide solution, it was converted to a new substance melting at 130°C.

   The product of the reaction with tin and hydrochloric acid was neutralized with base and then distilled. This distillate (in CCl_4) gave rise to the following ^1H NMR (60 MHz) spectrum: δ 3.32, 2H, s; δ 6.44, 2H, m; δ 6.6, 1H, m; δ 7.0, 2H, m.

2. A compound boiled at 166–169°C and contained sulfur but no nitrogen or halogen. It was insoluble in water and dilute acids but dissolved in sodium hydroxide solutions. Its sodium derivative reacted with 2,4-dinitrochlorobenzene to give a compound melting at 118–119°C. When allowed to stand in air, the original compound was slowly oxidized to a derivative melting at 60–61°C.

   The ^1H NMR spectrum of the air oxidation product (60 MHz, $CDCl_3$) showed only δ 7.2–7.7, m. The ^1H NMR spectrum of the original compound showed δ 3.39, 1H, s; δ 7.12, 5H, s.

3. A compound melted at 141–142°C and contained nitrogen but no halogen or sulfur. It was insoluble in water, dilute acids, and dilute alkalies. It was unaffected by treatment with tin and hydrochloric acid. When treated for a long time with hot sodium hydroxide solution, it reacted, forming an insoluble oil (I). The oil was soluble in dilute hydrochloric acid and reacted with acetyl chloride to give a solid melting at 111–112°C. Acidification of the alkaline solution from which I was removed gave a solid melting at 120–121°C, whose neutralization equivalent was 122 ± 1.

   A ^1H NMR spectrum of the mp 120–121°C solid (60 MHz, $CDCl_3$) showed δ 7.4–7.7, 3H, m; δ 8.0–8.3, 2H, m; δ 12.8, 1H, s. The ^1H NMR spectrum of compound I (60 MHz, $CDCl_3$) showed: δ 2.15, 3H, s; δ 3.48, 2H, bs; δ 6.45–6.8, 2H, m; δ 6.8–7.15, 2H, m.

   Identify I and II and the original (mp 141–142°C) compound.

4. A compound boiled at 159–161°C and contained chlorine but no nitrogen or sulfur. It was insoluble in water, in dilute acids and alkalies, and in cold concentrated sulfuric acid. It dissolved in fuming sulfuric acid. It gave no precipitate with hot alcoholic silver nitrate solution. Treatment with a hot solution of potassium permanganate caused the compound to dissolve slowly. The resulting solution, when acidified with

sulfuric acid, gave a precipitate which melted at 138–139°C and had a neutralization equivalent of 157 ± 1.

A ^{1}H NMR spectrum of the original compound (60 MHz, $CDCl_3$) showed δ 2.37, 3H, s; δ 7.0–7.35, 4H, m. An IR spectrum of the material melting at 138–139°C (in a Nujol mull) showed bands at 3333–2381 cm^{-1} (s); 1678 cm^{-1} (s); several bands in the region 1587–1389 cm^{-1} (m); 1316 cm^{-1} (s); 1053, 1042 cm^{-1} (m); 913 cm^{-1} (bm); 742 cm^{-1} (s).

5. A colorless liquid boiled at 188–192°C. It contained only carbon, hydrogen, and oxygen. It was insoluble in water, dilute acids, and alkalies, but dissolved readily in cold concentrated sulfuric acid. It did not react with phenylhydrazine or acetyl chloride and did not decolorize a carbon tetrachloride solution of bromine. Boiling alkalies dissolved it slowly. The resulting mixture was subjected to steam distillation. The distillate contained a compound which when pure, boiled at 129–130°C and reacted with α-naphthyl isocyanate to give a derivative melting at 65–66°C. IR and ^{1}H NMR spectra were obtained from the bp 129–130°C material. The alkaline residue, left after the steam distillation, was acidified with phosphoric acid and steam-distilled. The distillate contained an acid which yielded an anilide melting at 108–109°C.

Treatment of the original compound with lithium aluminum hydride (followed by the usual careful work-up) resulted in only one compound, which showed NMR and IR spectra identical to those obtained from material of boiling point 129–130°C described above. The 60 MHz ^{1}H NMR spectrum showed: δ 0.92, 6H, d; δ 1.2–1.7, 3H, m; δ 2.13, 1H, bs; δ 3.63, 2H, t.

6. An unknown compound was a pink solid that melted at 109–112°C. Treatment with decolorizing carbon and recrystallization removed the color and brought the melting point to 112–114°C. The compound burned with a smoky flame and left no residue. Elemental analysis showed nitrogen to be present and sulfur and halogens to be absent. The compound was insoluble in water and dilute alkalies but dissolved in ether and dilute acids. It reacted with benzenesulfonyl chloride to give a derivative that was soluble in alkali and melted at 101–102°C. The acetyl derivative melted at 132°C.

Infrared bands for the original compound (5000–1250 cm^{-1} in $CHCl_3$, 1250–650 cm^{-1} in CS_2) were found at 3400, 3350, 3200, 3050, 1640, 1610, 1520, 1480, 1400, 1290, 1280, 1230, 1190, 1130, 970, 890, 870, 850, 820, 750, and 720 cm^{-1}.

Problem Set 5

For each of the problems in this and the following sets, give the structure of an organic compound which will fulfill the conditions stated and show by equations the changes it undergoes. Associate all major spectral bands with appropriate components of your answers.

1. An acid (A) containing only carbon, hydrogen, and oxygen had a neutralization equivalent of 103 ± 1. It gave a negative test with phenylhydrazine. Treatment with sulfuric acid converted it to a new acid (B) that decolorized permanganate and bromine solutions and had a neutralization equivalent of 87 ± 1. The original acid (A) was transformed by hypoiodite to iodoform and a new acid (C), the neutralization equivalent of which was 52 ± 1.

The ^{1}H NMR spectrum of compound B (60 MHz, $CDCl_3$) showed δ 1.90, 3H, d of d (J = 8, 2 Hz); δ 5.83, 1H, d of q (J = 15, 2Hz); δ 7.10, 1H, m; δ 12.18, 1H, s.

2. An acid had a neutralization equivalent of 97. It could not be made to undergo substitution of bromine for hydrogen readily, even in the presence of phosphorus tribromide. Vigorous oxidation transformed it into a new acid whose neutralization equivalent was 83.

The ^1H NMR spectrum of the second acid (60 MHz, $CDCl_3$) showed δ 8.08, s; δ 11.0, s; relative areas = 2:1 for the δ 8.08 to 11.0 signals.

3. An optically active hydrocarbon dissolved in cold, concentrated sulfuric acid, decolorized permanganate solutions, and readily absorbed bromine. Oxidation converted it to an acid containing the same number of carbon atoms as the parent substance and having a neutralization equivalent of 66. Mass spectrometry indicated the molecular ion of the hydrocarbon to be m/z 68.

4. A compound had a neutralization equivalent of 66. The substance was not affected by bromine in carbon tetrachloride, but heat transformed it into an acid whose neutralization equivalent was 88. The ^1H NMR spectrum of the latter acid (60 MHz, $CDCl_3$) showed δ 1.20, 6H, d; δ 2.57, 1H, septet; δ 12.4, 1H, bs.

5. A base had a neutralization equivalent of 121 $\pm$ 1. Vigorous oxidation converted it to an acid having a neutralization equivalent of 121 $\pm$ 1. The ^1H NMR spectrum of the acid (60 MHz, CCl_4) showed δ 7.52, 3H, m; δ 8.14, 2H, m; δ 12.82, 1H, s. The ^1H NMR spectrum of the base (60 MHz, $CDCl_3$) showed δ 0.91, 2H, s; δ 2.78, 4H, 8 lines; δ 7.23, 5H, s.

6. An acid whose neutralization equivalent was 166 was unaffected by bromine in carbon tetrachloride but gave a positive iodoform test.

7. A compound (A) gave negative tests for nitrogen, sulfur, and halogens. It was insoluble in water but dissolved in dilute sodium hydroxide solution. Compound A gave no color with ferric chloride and did not decolorize a solution of potassium permanganate. Treatment with concentrated hydrobromic acid converted A into two compounds. One was a compound (insoluble in all tests) containing bromine, which gave a precipitate with sodium iodide in acetone. The other was a new acid (B), which decolorized bromine solutions and gave a color with ferric chloride. Compound B contained no halogen. The neutralization equivalents of compounds A and B were, respectively, 180 $\pm$ 2 and 137 $\pm$ 1.

The ^1H NMR spectrum of acid B (60 MHz, $CDCl_3$ containing DMSO-d_6) showed δ 6.84, 2H, d (J = 9 Hz); δ 7.86, 2H, d (J = 9 Hz); δ 8.2, 2H, bs.

8. An optically active acid had a molecular weight of 98.

9. A compound (I) gave a precipitate with 2,4-dinitrophenylhydrazine. Compound I was heated with 25% aqueous sodium hydroxide, and the mixture was partially distilled. The distillate contained a compound that reacted with sodium and gave a positive iodoform test. It gave a negative Lucas test. The residue from the distillation was acidified with phosphoric acid and the mixture stream-distilled. A volatile acid was isolated. Its *p*-bromophenacyl ester had a saponification number of 257 $\pm$ 1.

Mass spectral analysis of the original compound (I) gave rise to the following data (only peaks of intensity greater than or equal to 10% of the base peak are reported):

m/z	Percent of base	*m/z*	Percent of base
15	13.7	45	13.2
27	22.6	85	18.0
29	42.7	88	12.3
31	28.9	130	10.4
42	13.5	131	0.8
43	(100, base)		

The IR spectrum of a neat sample of compound I showed bands at 3000 cm^{-1} (m); 2933 cm^{-1} (w); 1715 cm^{-1} (s); 1634 cm^{-1} (w); 1408 cm^{-1} (w); 1364 cm^{-1} (m); 1316 cm^{-1} (s); 1250 cm^{-1} (s); 1149 cm^{-1} (s); 1042 cm^{-1} (s).

Problem Set 6

1. An acid (I) contained nitrogen and had a neutralization equivalent of 197 ± 2. Treatment with thionyl chloride followed by ammonium hydroxide converted I to a neutral compound (II) which reacted with an alkaline hypobromite solution to yield a basic compound (III). Hydrolysis of compound III produced a new compound (IV), soluble in both acids and bases and having a neutralization equivalent of 186 ± 2. Treatment of this compound with nitrous acid in the presence of sulfuric acid gave a clear solution. When copper cyanide was added to this solution, compound I was regenerated. Hydrolysis of I gave an acid (V) having a neutralization equivalent of 107 ± 1. Heat converted V into a neutral substance (VI), which could be reconverted to V by hydrolysis. Oxidation of compound IV gave a new acid (VII) having a neutralization equivalent of 71 ± 1.

   The ^1H NMR spectrum of an isomer of V (60 MHz, $CDCl_3$ plus DMSO-d_6) showed the following signals (of area 1:1:1:1): δ 7.6, m; δ 7.9, m; δ 8.28, s; δ 11.8, bs. The δ 11.8 signal showed a concentration-dependent chemical shift. The two upfield multiplets, although complex, showed symmetry; the two signals were nearly mirror images of each other.

2. An unknown (I) was insoluble in water but soluble in both dilute acid and dilute alkali. It contained nitrogen and bromine. No satisfactory neutralization equivalent could be obtained. Treatment with acetic anhydride converted it into an acid (II) which gave a neutralization equivalent of 270 ± 3. When I was treated with cold nitrous acid, a gas was liberated and a compound (III) was produced which gave a neutralization equivalent of 230 ± 2. Compounds I, II, and III when vigorously oxidized gave the same product—a bromine-containing acid whose neutralization equivalent was found to be 199 ± 2.

   The ^1H NMR spectrum of the product of vigorous oxidation (60 MHz, $CDCl_3$ containing DMSO-d_6) showed δ 7.60, 2H, d; δ 7.95, 2H, d; δ 12.18, 1H, bs. The δ 12.8 signal showed a chemical shift which was concentration dependent.[9]

3. A solid ester (I) was saponified (saponification equivalent 173 ± 2), and the alkaline aqueous solution was evaporated nearly to dryness. The distillate was pure water. The residue was acidified and distilled; a colorless oil (II) was isolated. Substance II reacted with acetyl chloride and gave a violet color with ferric chloride solution. The residue from the distillation yielded a solid (III) which was alkali soluble. When III was heated above its melting point, it changed to IV; IV was dissolved by shaking with warm aqueous alkali, and the solution was acidified; the solid that separated was identical with III.

   The ^1H NMR spectrum of compound II (60 MHz, $CDCl_3$) showed δ 2.25, 3H, s; δ 5.67, 1H, s; δ 6.5–7.3, 4H, m. The δ 5.67 signal had a concentration-dependent chemical shift.

   The ^1H NMR spectrum of III (60 MHz, $CDCl_3$) showed δ 7.4–7.9, m, and δ 12.08, s; the singlet had an area one-half that of the multiplet. In addition, the chemical shift of the singlet was concentration dependent.

[9]The two doublets are separated from one another so much that the signal could be mistaken for a quartet centered at δ ca. 7.75.

4. A solid (I) giving a positive test for nitrogen was soluble in water and insoluble in ether. Addition of cold alkali liberated a water soluble, ether soluble compound (II) which possessed an ammonia odor and reacted with benzenesulfonyl chloride to give a derivative insoluble in alkali. Addition of hydrochloric acid to a dilute aqueous solution of I gave a solid acid (III) which possessed a neutralization equivalent of 167 $\pm$ 2. After it had been boiled with zinc dust and ammonium chloride solution, gave a compound IV that reduced Tollens reagent.

   The ^{1}H NMR spectrum of II (60 MHz, CCl_4) showed δ 0.53, 1H, s; δ 0.90, 6H, t, distorted; δ 1.2–1.7, 8H, m; δ 2.52, 4H, t.

   The ^{1}H NMR spectrum of III (60 MHz, $CDCl_3/CF_3CO_2H$) showed δ 7.77, 1H, t (J = 7 Hz); ca. δ 8.54, 2H, d of m (J = 7 Hz, ca. 2 Hz and smaller splittings); δ 8.96, 1H, t (J = ca. 2 Hz) δ 12.0, 1H, bs.

5. An optically active compound ($C_5H_{10}O$) was found to be slightly soluble in water. Its solubility was not increased appreciably by sodium hydroxide or hydrochloric acid. It gave negative tests with phenylhydrazine, Lucas reagent, and hypoiodite but decolorized solutions of bromine and permanganate. It reacted with acetyl chloride. Oxidation with permanganate converted it to an acid having a neutralization equivalent of 59 $\pm$ 1. When heated, the acid lost carbon dioxide and was converted to a new acid having a neutralization equivalent of 73 $\pm$ 1.

6. A compound (I) containing carbon, hydrogen, oxygen, and nitrogen was soluble in dilute sodium hydroxide solution and in dilute hydrochloric acid but insoluble in sodium bicarbonate solution. It reacted with an excess of acetic anhydride to give a product (II) which was insoluble in water, in dilute acids, and in dilute alkalies. Compound I decolorized bromine water; when I was dissolved in an excess of dilute hydrochloric acid and the cold solution was treated with sodium nitrite, a new product (III) separated without the evolution of nitrogen.

   The ^{1}H NMR spectrum of I (CF_3COOH) showed δ 3.31, 1H, bs; δ 3.65, 3H, s; δ 5.51, 1H, bs; δ 7.12, 2H, d (J = 10 Hz); δ 7.49, 2H, d (J = 10 Hz);

Problem Set 7

1. When a solid, $C_9H_6O_2$ (I), was heated with a dilute solution of potassium carbonate, it was converted to the potassium salt of an acid, $C_9H_8O_3$ (II). The salt reverted to I when treated with acids. Compound I reacted with bromine in carbon tetrachloride to yield $C_9H_6O_2Br_2$ (III). When heated with a solution of sodium hydroxide, compound III gradually went into solution. Acidification of the solution precipitated a compound, $C_9H_6O_3$ (IV).

   The ^{1}H NMR spectrum of acid II (60 MHz, $CDCl_3$ containing DMSO-d_6) showed δ 6.4–8.2, m; δ 10.3, very broad s. The δ 10.3 signal showed a chemical shift that was dependent on sample concentration; the area of this signal was one-third that of the δ 6.4–8.2 signal.

2. A compound, C_8H_8ONBr, when treated with boiling potassium hydroxide solution, gave the potassium salt of an acid, $C_8H_8O_3$, which was resolvable into (+) and (−) forms.

   A ^{1}H NMR spectrum of the acid of formula $C_8H_8O_3$ (60 MHz, acetone) showed, δ 5.22, 1H, s; δ 7.2–7.7, 7H, m. The chemical shift of the δ 5.22 signal was not concentration dependent.

3. A compound, $C_{11}H_{12}O_4$ (I), reacted with hot sodium hydroxide solution to yield a salt, $C_{11}H_{13}O_5Na$ (II). Treatment of the salt with hot dilute sulfuric acid converted it

to an acid, $C_9H_8O_4$ (III). Heat converted compound III to a high-melting compound, $C_{18}H_{12}O_6$ (IV).

4. A compound, $C_5H_8O_2$ (I), was changed to $C_5H_9O_2Cl$ (II) by treatment with boiling concentrated hydrochloric acid. Compound II reacted slowly with a solution of potassium hydroxide to yield $C_5H_9O_3K$ (III). This compound was converted into $C_5H_8O_3$ (IV) by treatment with an alkaline permanganate solution. Compound IV, when treated with a solution of sodium hypochlorite, was transformed into $C_4H_6O_4$ (V), an acid whose neutralization equivalent was 58 ± 1. This acid, when heated, gave $C_4H_4O_3$ (VI). When the aqueous solution of II was made exactly neutral with sodium hydroxide solution, the original compound (I) was slowly regenerated.

   The ^1H NMR spectrum of V (60 MHz, DMSO-d_6) showed only two singlets, one at δ 2.43 and the other at δ 11.8. The two singlets were of area 2 : 1, respectively, and the δ 11.8 signal was very broad.

5. A liquid neutral compound had the formula $C_7H_7NO_3$. The ^1H NMR spectrum of this compound (60 MHz, CCl_4) showed δ 3.89, 3H, s; δ 6.91, 2H, d ($J = 10$ Hz); δ 8.12, 2H, d ($J = 10$ Hz).

6. A compound, $C_{16}H_{13}N$, formed salts with strong mineral acids. The salts were hydrolyzed by water. The ^1H NMR spectrum of this compound (in CCl_4) showed δ 5.61, 1H, bs; δ 6.60–7.55, 10H, m; δ 7.80, 2H, m.

7. A compound, $C_{14}H_{10}O$, gave $C_{13}H_{10}O_3$ when oxidized by alkaline permanganate. The original compound reacted with sodium to give $C_{14}H_9ONa$.

8. An acid of neutralization equivalent 57 was unaffected by bromine in carbon tetrachloride.

Problem Set 8

1. A liquid, $C_5H_4O_2$ (I), decolorized a permanganate solution and also reduced Tollens solution. When heated with an alkali cyanide, compound I dimerized. Treatment of compound I with excess ethanol in the presence of an acid catalyst converted it to a new substance, $C_9H_{14}O_3$.

   The ^1H NMR spectrum of compound I (60 MHz, $CDCl_3$) showed δ 6.63, 1H, d of d ($J = 5$, 2 Hz); δ 7.28, 1H, d ($J = 5$ Hz); δ 7.72, 1H, m; δ 9.67, 1H, s. The chemical shift position of the δ 9.67 signal was not concentration dependent.

2. A compound, $C_4H_4O_4$ (I), was soluble in dilute sodium hydroxide solution and when treated with bromine was converted into $C_4H_4Br_2O_4$ (II). Compound I was regenerated from II by treatment of the latter with zinc dust. Compound I was converted into compound III by hydrogenation in the presence of nickel. Compound III possessed a neutralization equivalent of 59 and lost a molecule of water when heated. The anhydride (IV) thus produced reacted with benzene in the presence of aluminum chloride to give $C_{10}H_{10}O_3$ (V), which was soluble in alkali, reacted with phenylhydrazine, and was converted into benzoic acid by vigorous oxidation.

   The ^1H NMR spectrum of the compound of formula $C_{10}H_{10}O_3$ (60 MHz, $CDCl_3$) showed δ 2.8, 2H, t ($J = 7$ Hz); δ 3.3, 2H, t ($J = 7$ Hz); δ 7.2–7.6, 3H, m; δ 8.0, 2H, m; δ 11.7, 1H, s. The chemical shift of the δ 11.7 signal was concentration dependent.

3. A compound, $C_{11}H_{10}N_2$, was converted by vigorous oxidation to $C_{11}H_8N_2O$.

4. A compound, $C_5H_{10}O$, decolorizes an alkaline solution of potassium permanganate but is not affected by bromine in carbon tetrachloride. The mass spectrum of this

compound (bp 75°C) results in the following peaks of intensity greater than or equal to 5% of the base peak:

m/z	Percent of base	m/z	Percent of base
86	34	41	83
57	(100, base)	39	17
55	7	29	40
43	18	27	11

5. A compound, $C_{14}H_{12}O$, is converted by chromic acid oxidation into an acid whose neutralization equivalent is 226. The ^{1}H NMR spectrum of the oxidation product (60 MHz, $CDCl_3$ containing DMSO-d_6) showed δ 7.45–7.95, 7H, m; δ 8.19, 2H, d (J = 7 Hz); δ 10.8, 1H, bs.

6. A neutral compound, $C_{10}H_6O_4$ (I), when heated with a sodium hydroxide solution, was converted to the salt of an acid (II) having a neutralization equivalent of 209 ± 2. Compound I reacted with alkaline hydrogen peroxide to yield a new acid (III) having a neutralization equivalent of 111 ± 1.

The ^{1}H NMR spectrum of compound I (60 MHz, $CDCl_3$) showed δ 6.64, 2H, d of d (J = 5 Hz, 2 Hz); δ 7.63, 2H, d (J = 5 Hz); δ 7.78, 2H, d (J = 2 Hz).

7. A compound, $C_{10}H_7NO_2$ (I), when treated with iron and hydrochloric acid, was converted into $C_{10}H_9N$ (II), which was soluble in dilute hydrochloric acid. By vigorous oxidation of I and II, $C_8H_5O_6N$ and $C_8H_6O_4$, respectively, were produced. Both oxidation products were soluble in alkali.

The ^{1}H NMR spectrum of I (60 MHz, $CDCl_3$) showed a complex multiplet extending from δ 7.15 to δ 8.5. The product of oxidation of II, $C_8H_6O_4$, when heated, lost the elements of water.

Problem Set 9

1. A naturally occurring compound, $C_{10}H_{10}O_2$ (I), was found to be unreactive toward acetyl chloride and phenylhydrazine. Heating with potassium hydroxide, however, brought about isomerization. The isomer (II) also failed to react with acetyl chloride and phenylhydrazine. Both isomers decolorized solutions of bromine and permanganate. Ozone converted isomer II to a compound (III) which formed a phenylhydrazone. Oxidation of compound II or III with alkaline potassium permanganate yielded an acid, $C_8H_6O_4$ (IV).

The ^{1}H NMR spectrum of I (60 MHz, $CDCl_3$) showed δ 3.30, 2H, d of m (J = 7 Hz, additional small splittings); δ 4.90, 1H, m; δ 5.15, 1H, m; δ 5.88, 2H, s; ca. δ 5.6–6.2, 1H, m (a wide band, highly split); δ 6.67, 3H, s (with small splittings).

2. A compound, $C_8H_5ClO_2$ (I), when heated with absolute alcohol, gave $C_{14}H_{20}O_4$ (II), which by oxidation with alkaline permanganate was converted after acidification into $C_8H_6O_4$. Treatment with aniline converted I into $C_{20}H_{16}N_2O$. The product of permanganate oxidation was converted by treatment with excess thionyl chloride to compound III $C_8H_4Cl_2O_2$. Compound III yielded a ^{1}H NMR spectrum (60 MHz, $CDCl_3$) showing δ 7.72, 1H, t (J = 7 Hz); δ 8.33, 2H, t of d (J = 7, 2 Hz); δ 8.82, 1H, t (J = 2 Hz).

3. A compound, $C_8H_{14}O_3$ (I), was soluble in dilute sodium hydroxide solution. Phenylhydrazine converted I into $C_{12}H_{14}N_2O$(II), and heating with 20% hydrochloric acid transformed I into $C_5H_{10}O$ (III). Compound III gave a crystalline precipitate when treated with semicarbazide but, when treated with a solution of sodium hypoiodite, gave no iodoform.

The compound arising from semicarbazide treatment of III, when dissolved in $CDCl_3$, yielded a ^1H NMR spectrum (60 MHz) that showed δ 1.09, 6H, t; δ 2.26, 4H, q; δ 5.89, 2H, bs; δ 8.59, 1H, bs.

4. A compound, C_9H_7N (I), was converted by catalytic reduction into $C_9H_{11}N$ (II). When compound II was treated with an excess of methyl iodide followed by wet silver oxide and the reaction product heated, a compound, $C_{11}H_{15}N$ (III), was produced. This compound was converted (a) by vigorous oxidation into $C_8H_6O_4$ (IV); (b) by treatment with ozone and hydrolysis of the reaction product into $C_{10}H_{13}NO$ (V). The ozonization product yielded $C_{20}H_{26}N_2O_2$ (VI) when heated when a dilute solution of potassium cyanide. Compound VI was converted into IV by vigorous oxidation.

   Compound I (60 MHz, $CDCl_3$) yielded a ^1H NMR spectrum showing δ 7.25–8.1, 5H, m; δ 8.52, 1H, d ($J = 7$ Hz); δ 9.26, 1H, s.

5. A compound, $C_{10}H_{10}O$ (I), reacted with sodium to give a derivative, $C_{10}H_9ONa$, which gave I when treated with water. Treatment of compound I with cold 80% sulfuric acid in the presence of mercuric sulfate and then with water gave $C_{10}H_{12}O_2$ (II). This compound dissolved slowly in a solution of sodium hypochlorite, and by acidification of the solution $C_9H_{10}O_3$ (III) was obtained. Boiling 48% hydrobromic acid converted compound III into $C_8H_7BrO_2$ (IV), a compound transformed by vigorous oxidation into $C_8H_6O_4$.

   The ^1H NMR spectrum of the compound of formula $C_8H_6O_4$ (60 MHz, $CDCl_3$ containing DMSO-d_6) showed only two singlets of area 2 : 1 at, respectively, δ 8.08 and δ 11.0.

6. A compound having the molecular formula $C_{14}H_{12}$ is converted by permanganate oxidation into a derivative of molecular formula $C_{13}H_{10}O$ which is not affected by further treatment with permanganate.

   The ^1H NMR spectrum of the original compound (100 MHz, CCl_4) showed δ 5.35, 2H, s; δ 7.21, 10H, s (slightly distorted near the base).

7. An optically active compound, $C_9H_{13}N$, was converted by vigorous oxidation into an acid whose neutralization equivalent is 83. A closely related isomer of the original compound yielded a ^1H NMR spectrum (60 MHz, $CDCl_3$) which showed δ 0.95, 2H, s; δ 2.30, 3H, s; δ 2.5–3.0, 4H, m (8 lines, symmetrical); δ 7.05, 4H, s. Oxidation of this isomer (which was not optically active) resulted in the same acid as was obtained upon oxidation of the original compound.

8. A compound does not decolorize bromine water but reacts with sodium to give $C_8H_6O_3Na_2$. It has a neutralization equivalent of 152. This compound is not optically active, and vigorous oxidation converts it to an acid, $C_8H_6O_4$, of neutralization equivalent 82 $\pm$ 2. A ^1H NMR spectrum of this acid (60 MHz, $CDCl_3$ containing DMSO-d_6) shows δ 8.08, 4H, s; δ 11.0, 2H, bs.

Problem Set 10

1. A colorless, crystalline solid (I) contained nitrogen but no halogen or sulfur. It was soluble in water but insoluble in ether. Its aqueous solution was acidic and gave a neutralization equivalent of 123 $\pm$ 1. It was treated with sodium nitrite and hydrochloric acid. A gas evolved, one-third of which was soluble in potassium hydroxide solution. The solution from the nitrous acid treatment was evaporated to dryness; only sodium

chloride, sodium nitrate, and sodium nitrite were left. Compound I was heated with dilute sodium hydroxide solution; ammonia was evolved, and the solution on acidification gave off a gas. Evaporation of this solution left only inorganic salts.

The original compound was carefully neutralized with base under mild conditions to give a solid of melting point 135°C. The UV spectrum of this solid, in the presence of sodium hydroxide, showed no features above 220 nm. The solid resulted in the following spectra.

Mass spectrum (only peaks of intensity greater than or equal to 10% of base are reported):

m/z	Percent of base
60	(100, base)
44	60
43	18
28	17

The IR spectrum showed peaks at 3450, 3350, 1690, 1640, 1600, 1470, 1160, 1050, 1000, 790, and 710 cm^{-1}.

2. An unknown compound containing carbon, hydrogen, oxygen, and nitrogen was insoluble in water, in dilute acids, and in dilute alkalies. It did not react with acetyl chloride or phenylhydrazine and was not easily reduced. When treated with hot aqueous alkali, the substance slowly dissolved. Distillation of this alkaline solution gave a distillate containing a compound which was salted out by means of potassium carbonate. This compound gave the iodoform test but no reaction with a hydrochloric acid solution of zinc chloride. The original alkaline solution (residue from the distillation) was acidified with sulfuric acid, and the solution was again distilled; a volatile acid was obtained in the distillate. This distillate reduced permanganate. The residual liquor from this second distillation was exactly neutralized, and a solid was obtained. This solid contained nitrogen and possessed a neutralization equivalent of 137 ± 1.

A 100-mg sample of the volatile acid (in 0.5 mL of CCl_4) yielded a ^1H NMR spectrum which showed δ 5.8–6.75, 3H, m; δ 12.4, 1H, s. The ^1H NMR spectrum of a 60-mg sample of the solid of neutralization equivalent 137 (dissolved in 0.5 mL of acetone) showed δ 6.55, 3H, s (shift position was concentration dependent); δ 6.75, 2H, d (J = 8 Hz); δ 7.83, 2H, d (J = 8 Hz). The compound that salted out with potassium carbonate showed, among other bands, one ^1H NMR signal whose chemical shift was concentration dependent and which appeared as a singlet when the analysis was done in a $CDCl_3$ solution and as a triplet (J = ca. 7 Hz) when done in a DMSO-d_6 solution.

3. An acid, $C_6H_8O_4$ (I), was converted into $C_6H_6O_2Cl_2$ (II) by phosphorus pentachloride. The chlorine derivative, when treated with benzene in the presence of anhydrous aluminum chloride, gave $C_{18}H_{16}O_2$ (III). This compound did not decolorize permanganate but decolorized a bromine solution and readily formed a dioxime. The dioxime rearranged under the influence of phosphorus pentachloride to yield $C_{18}H_{18}O_2N_2$ (IV), a compound which gave the original acid (I) when hydrolyzed.

The ^1H NMR spectrum of compound II (60 MHz, CCl_4) showed δ 2.2–2.6, 4H, m; δ 3.90, 2H, t (distorted, J = ca. 7 Hz); δ 11.9, 2H, bs.

4. A compound (I) contained carbon, hydrogen, oxygen, nitrogen, and chlorine. It was soluble in water but insoluble in ether. The aqueous solution immediately gave a precipitate with silver nitrate. When the aqueous solution of I was exactly neutralized, a new compound (II) free from chlorine separated. It reacted with acetic anhydride to give an alkali soluble, acid insoluble compound (III) possessing a neutralization equivalent of 207 ± 2. Compound II reacted with benzenesulfonyl chloride to give an alkali soluble product and with nitrous acid without evolution of any gas even when heated. The product from the latter treatment still contained nitrogen and was soluble in alkalies and insoluble in acids. Vigorous oxidation of either I, II, or III gave a nitrogen-free acid, insoluble in water, and possessing a neutralization equivalent of 82 ± 2.

The ^1H NMR spectrum of this nitrogen-free acid (60 MHz, $CDCl_3$ containing DMSO-d_6) showed δ 7.62, 1H, t ($J = 7$ Hz); δ 8.25, 2H, distorted doublet ($J =$ ca. 2 Hz); δ 8.70, 1H, distorted s; δ 12.28, 2H, s.

5. A compound, $C_{10}H_6O_3$ (I), decolorized alkaline permanganate and reacted with hydroxylamine. It decomposed when distilled at ordinary pressure to give $C_9H_6O_2$ (II), a compound which yielded a monosodium derivative and was readily oxidized to $C_8H_6O_4$ (III). Compound III was an acid which, when heated with soda lime, was converted into $C_7H_6O_2$ (IV). Compound IV was decomposed by heating with dilute hydrochloric acid under pressure and yielded a weakly acidic compound having the formula $C_6H_6O_2$.

The ^1H NMR spectrum of compound IV (60 MHz, $CDCl_3$) showed δ 5.90, 2H, s; δ 6.83, 4H, s (slight distortions at the bottom of this signal). The spectrum of compound III ($CDCl_3$) showed δ 6.00, 2H, s; δ 6.8, 1H, d ($J = 7$ Hz); δ 7.38, 1H, d ($J = 2$ Hz); δ 7.55, 1H, d of d ($J = 7, 2$ Hz); δ 7.6, 1H, bs. The chemical shift of the δ 7.6 signal was concentration dependent.

Problem Set 11

1. A compound (A) contained only carbon, hydrogen, and oxygen. It was insoluble in water, dilute acids, and dilute alkalies but dissolved in cold concentrated sulfuric acid. It gave negative tests with phenylhydrazine and acetyl chloride. When heated with alkali and neutralized, it yielded an oil (B) and a volatile acid having a neutralization equivalent of 59 ± 1. Compound B reacted with acetyl chloride but not with phenylhydrazine. Treatment with a concentrated solution of hydrogen bromide transformed it into a bromine-containing compound (C) which gave positive tests with silver nitrate, ferric chloride, and bromine water.

Mild oxidation converted B to an acid (D) having a neutralization equivalent of 164 ± 2.

Treatment of acid D with ethyl alcohol and a trace of sulfuric acid resulted in compound E. The ^1H NMR spectrum of compound E (60 MHz, CCl_4) showed δ 1.35, t ($J = 7$ Hz); δ 1.40, t ($J = 7$ Hz); δ 3.8–4.5, 4H, 6 lines (relative intensities 1:3:4:4:3:1, spaced by ca. 7 Hz each); δ 6.79, 2H, d ($J = 9$ Hz); δ 7.89, 2H, d ($J = 9$ Hz). The signals at δ 1.35 and δ 1.40 had a combined area of 6H.

2. A compound (I), giving positive tests for nitrogen and chlorine, was insoluble in water and hydrochloric acid but soluble in sodium bicarbonate solution. A neutralization equivalent of 210 ± 2 was obtained. Compound I reacted with acetyl chloride but not with hot alcoholic silver nitrate. The acetyl derivative (II) had a neutralization equivalent of 253 ± 2. Boiling alkali liberated ammonia from I, and acidification of the resulting solution precipitated a new acid (III), which had a neutralization equivalent of 115 ± 1. Compound III contained chlorine but no nitrogen. When compound III was boiled with potassium permanganate solution, a new compound having the same solubility characteristics as I was produced which still contained chlorine and gave a neutralization equivalent of 81 ± 1.

Compound IV (neutralization equivalent 71) could be chlorinated to give the compound of neutralization equivalent 81; compound IV yielded a ^{1}H NMR spectrum (60 MHz, $CDCl_3/DMSO$-d_6) displaying δ 7.57, 1H, t ($J = 7$ Hz); δ 8.18, 2H, d ($J = 7$ Hz); δ 11.2, 3H, bs.

The IR spectrum of I (in Nujol) showed, among other bands, strong absorptions at 1779 cm^{-1} and 1712 cm^{-1}.

3. A neutral compound (A) was a colorless solid that gave negative tests for nitrogen, sulfur, and the halogens. It gave positive tests with phenylhydrazine and acetyl chloride. Mild oxidation converted it to a new compound, a yellow solid (B). The new compound (B) gave a positive test with phenylhydrazine, but a negative test with acetyl chloride. Vigorous oxidation of B converted it into an acid (C) with a neutralization equivalent of 121 ± 1.

The ^{1}H NMR spectrum of compound A (60 MHz, $CDCl_3$) showed δ 4.5, 1H, bs; δ 5.9, 1H, bs; δ 7.2–7.5, 8H, m (the major part of which was a sharp singlet); δ 7.85, 2H, d of d.

4. A liquid (I) containing chlorine was insoluble in water, dilute hydrochloric acid, and dilute sodium hydroxide. It dissolved in cold concentrated sulfuric acid. It gave no precipitate with warm alcoholic silver nitrate and did not react with acetyl chloride. It gave a precipitate with phenylhydrazine but gave a negative Tollens test. When boiled with concentrated sodium hydroxide solution, compound (I) dissolved. The distillate from this alkaline solution gave a positive iodoform test. Acidification of the alkaline solution precipitated a compound (II) which contained chlorine and had a neutralization equivalent of 156 ± 1. Compound II was not affected by permanganate solution. After removal of compound II, a portion of the acidic filtrate was distilled, and the distillate was found to be acid to litmus. The original compound (I) had a saponification equivalent of 113 ± 1. The acidic filtrate yielded a compound with a neutralization equivalent of 61 ± 1.

The ^{1}H NMR spectrum of compound II (60 MHz, $CDCl_3/DMSO$-d_6) showed δ 7.1–7.5, 3H, m (primarily a large singlet); δ 7.82, 1H, m; δ 9.0, 1H, bs.

5. A liquid had a saponification equivalent of 163 ± 1. Saponification yielded an oil, which gave a positive ferric chloride test, and an acid of neutralization equivalent 60 ± 1.

The ^{1}H NMR spectrum of 49 mg of the oil (60 MHz) in 0.5 mL of CCl_4 showed δ 2.18, 6H, s; δ 5.73, 1H, s; δ 6.33, 2H, s; δ 6.45, 1H, s. Upon addition of more CCl_4, the δ 5.73 signal moved upfield.

Problem Set 12

1. An ether insoluble compound (I), containing nitrogen, dissolved in water, giving an alkaline solution. Titration of this solution with standard acid gave a neutralization equivalent of 37 ± 1. Treatment of a cold solution of I with sodium nitrite and hydrochloric acid liberated a gas. The solution of NE = 37 was made distinctly alkaline, and benzoyl chloride was added. A neutral nitrogen-containing compound (II) separated. Compound II was shown by mass spectrometry to have a molecular weight of 282. When treatment of I with sodium nitrite and hydrochloric acid was followed by addition of benzoic anhydride, a neutral compound (III) was obtained which contained no nitrogen and had a saponification equivalent (SE) of 142 ± 1.

The ^{1}H NMR spectrum of compound I (60 MHz, $CDCl_3$) showed δ 1.08, s; δ 1.58, quintet (J = 7 Hz, relative intensities 1:4:6:4:1); δ 2.75, t. The relative signal areas were, from high to low field, 2:1:2.

2. A colorless liquid (I) gave no tests for halogen, nitrogen, sulfur, or metals. It was insoluble in water, dilute hydrochloric acid, dilute sodium hydroxide, and phosphoric acid, but soluble in cold concentrated sulfuric acid. It did not react with acetyl chloride or phenylhydrazine. It was boiled with dilute phosphoric acid, and an oil (II) separated when the solution was cooled. Compound II gave a precipitate with phenylhydrazine and with sodium bisulfite solution but did not react with acetyl chloride. When II was vigorously shaken with strong alkali, a compound (III) separated from the alkaline solution. This product (III) reacted with acetyl chloride but not with phenylhydrazine. Acidification of the alkaline solution gave IV, which had a neutralization equivalent of 136 ± 1. Strong oxidation of IV gave an acid (V) with a neutralization equivalent of 82 ± 1.

   The phosphoric acid solution, from which II was separated, was distilled. The distillate was saturated with potassium carbonate, and a compound (VI) was obtained. This compound reacted with sodium and acetyl chloride and gave a yellow precipitate with sodium hypoiodite. It did not react with Lucas reagent.

   The ^{1}H NMR spectrum of II (60 MHz, $CDCl_3$) showed δ 2.42, 3H, s; δ 7.18, 2H, d (J = 8 Hz); δ 7.66, 2H, d (J = 8 Hz); δ 9.81, 1H, s.

3. A colorless liquid (I) gave no tests for nitrogen, sulfur, or halogen. It was soluble in water and ether. It did not react with sodium, acetyl chloride, phenylhydrazine, or dilute permanganate solution. It did not decolorize bromine in carbon tetrachloride. When compound I was heated with an excess of hydrobromic acid, an oil (II) separated. This oil (II) contained bromine and readily gave a precipitate with alcoholic silver nitrate. It was insoluble in water, acids, and alkalies. After II was dried and purified, it was treated with magnesium in pure ether. A reaction occurred with the liberation of a gas (III). No Grignard reagent could be detected. Treatment of II with alcoholic potassium hydroxide liberated a gas (IV). Both III and IV decolorized bromine water and reduced permanganate solutions. A careful examination of the action of hydrobromic acid on compound I showed that II was the only organic compound produced and that no gases were evolved during this reaction.

   The mass spectrum of compound I showed peaks at m/z values (relative abundance) as follows: m/z 88 (65); m/z 89 (32); m/z 90 (3); and no peaks at higher mass. The ^{1}H NMR spectrum of I (60 MHz, $CDCl_3$) showed only a singlet at δ 3.69.

4. A compound (I) containing carbon, hydrogen, nitrogen, and oxygen was insoluble in dilute alkalies and acids. When heated for some time with hydrochloric acid, compound I yielded a solid acid (II) having a neutralization equivalent of 180 ± 1. If compound I was oxidized with potassium dichromate and sulfuric acid, it yielded a solid, nitrogen-containing acid (III) with a neutralization equivalent of 166 ± 1. Compound I reacted with benzaldehyde in the presence of alkalies to give a benzal derivative.

   If compound I was treated with tin (II) chloride and hydrogen chloride in dry ether and the resulting mixture was treated with water, a new compound (IV), whose molecular formula was C_8H_7N, resulted. Analysis showed IV to possess one active hydrogen atom. Compound IV was weakly basic and was resinified by acids.

The ^1H NMR spectrum of I (60 MHz, $CDCl_3$) showed signals at δ 4.25, 2H, s; δ 7.5–8.0, 3H, m; δ 8.25, 1H, d (distorted slightly). The spectrum of II (60 MHz, $CDCl_3$ containing DMSO-d_6) showed δ 4.02, 2H, s; δ 7.3–7.65, 3H, m; δ 8.0, 1H, d (distorted); δ 11.2, 1H, bs. The spectrum of III (60 MHz, $CDCl_3$ containing DMSO-d_6) showed δ 7.5–8.0, 4H, m; δ 12.15, 1H, s. The spectrum of IV (60 MHz, $CDCl_3$) showed δ 6.38, 1H, m; δ 6.76, 1H, t (distorted); δ 6.95–7.10, 4H, m; δ 7.4–7.65, 1H, m.

Problem Set 13

1. A colorless, oily liquid (I) having an agreeable odor and containing only carbon, hydrogen, and oxygen reacted with phenylhydrazine but not with acetyl chloride, and it readily decolorized potassium permanganate solution. When it was treated with a concentrated sodium hydroxide solution and the reaction mixture acidified, two products were obtained: an oxygen-containing compound (II) which reacted with acetyl chloride, and an acid (III) which at 200°C lost carbon dioxide to yield an oxygen-containing compound (C_4H_4O). The latter decolorized permanganate solutions but did not react with sodium or phenylhydrazine. When compound I was warmed with potassium cyanide, a compound (IV) was produced. Compound IV gave positive acetyl chloride and phenylhydrazine tests and on oxidation with periodic acid, was converted to the original compound (I) and to the acid (III).

   The ^1H NMR spectra of some of the above compounds showed the following signals.

Chemical shift	No. of protons	Multiplicity
Compound I in ($CDCl_3$)		
δ 6.63	1H	d of d (J = 5 Hz, 2 Hz)
δ 7.28	1H	d (J = 5 Hz)
δ 7.72	1H	m
δ 9.67	1H	s
Compound II (in $CDCl_3$)		
δ 2.83	1H	s (chemical shift concentration dependent)
δ 4.57	2H	s
δ 6.33	2H	m
δ 7.44	1H	m
Compound III (in DMSO-d_6)		
δ 6.56	1H	m
δ 7.20	1H	m
δ 7.65	1H	m
δ 12.18	1H	bs
Compound of formula C_4H_4O (in $CDCl_3$)		
δ 6.37	2H	t
δ 7.42	2H	t

2. A weakly acidic compound (I) contained nitrogen but no sulfur or halogens. It reacted with acetyl chloride but not with phenylhydrazine. When compound I was heated with dilute acids, a new compound (II) was obtained; it was isolated by distillation from the acid solution and saturation of the distillate with potassium car-

bonate. Compound II did not react with acetyl chloride or Tollens reagent but gave positive tests with phenylhydrazine and sodium bisulfite.

When compound I was warmed with phosphorus pentachloride and poured into water, a new compound (III) was obtained. Compound III still contained nitrogen but no halogen, was neutral, and was decomposed by alkalies. By the addition of benzoyl chloride to this alkaline solution an acid (IV) was obtained. It gave a neutralization equivalent of 220 ± 2. When compound IV was heated for some time with dilute acids and distilled, two products resulted. One (V) contained no nitrogen, was acidic, and gave a neutralization equivalent of 120 ± 2. The other proved to be identical with III. When III was treated with cold concentrated hydrochloric acid, a compound (VI) was obtained which contained nitrogen and chlorine and was soluble in water. It liberated a gas when treated with cold sodium nitrite solution.

The following ^{1}H NMR spectra were determined.

Chemical shift	No. of protons	Multiplicity
Compound I (in $CDCl_3$)		
δ 1.77	4H	m
δ 2.40	4H	m
δ 9.12	1H	bs
Compound III (in $CDCl_3$)		
δ 1.80	4H	m
δ 2.38	2H	m
δ 3.32	2H	m
δ 7.60	1H	bs

3. A compound (I) soluble in water but not in ether was decomposed by heat into a compound (II), which was insoluble in all tests, and a basic compound (III). When II and III were heated together, I was reformed. Both I and II gave a precipitate immediately when treated with silver nitrate. Compound III did not give a benzenesulfonamide but yielded a nitroso derivative.

The following ^{1}H NMR spectra were (60 MHz) determined.

Chemical shift	No. of protons	Multiplicity
Compound I (D_2O; 80 mg/0.5 mL)		
δ 3.70	9H	s
δ 7.4	5H	m
Compound II ($CDCl_3$ solvent)		
δ 2.20	—	s
Compound III ($CDCl_3$ solvent)		
δ 2.85	6H	s
δ 6.4–6.7	3H	m
δ 7.10	2H	m

4. A neutral compound (I) gave positive tests for chlorine, bromine, and iodine. Alcoholic silver nitrate gave a white precipitate which was readily soluble in ammonia. Phenylhydrazine produced a precipitate, but acetyl chloride failed to react. When compound I was shaken with cold dilute alkali for some time, it dissolved. Acidification of the alkaline solution produced a compound (II) which gave positive tests for bromine and iodine and possessed a neutralization equivalent of 369 ± 3. When compound I was boiled with dilute alkali and then acidified, a compound (III) precipitated which gave a positive test for iodine. Compound III possessed a neutralization equivalent of 306 ± 3 and reacted with both phenylhydrazine and acetyl chloride. Treatment of I or II with sodium hypochlorite and acidification yielded an acid containing iodine and having a neutralization equivalent of 146 ± 1.

5. A neutral compound (A) contained bromine but no nitrogen or other halogens. It did not react with hot alcoholic silver nitrate solution, acetyl chloride, phenylhydrazine, or bromine in carbon tetrachloride. It dissolved in boiling sodium hydroxide solution, but the distillate from this alkaline solution contained no organic compounds. Acidification of the alkaline solution with phosphoric acid caused the precipitation of a compound (B) containing bromine and having a neutralization equivalent of 200 ± 2. Steam distillation of the acid solution gave a distillate that was repeatedly extracted with chloroform. Removal of the chloroform left a colorless liquid (C) which was purified by distillation. Compound C contained no bromine and was soluble in sodium bicarbonate solution. It had a neutralization equivalent of 102 ± 1. After removal of C, the solution remaining in the steam distillation flask was made distinctly alkaline, benzoyl chloride added, and the mixture shaken vigorously. A new compound (D) separated from the alkaline solution. Compound D contained no bromine and was neutral. It had a saponification equivalent of 135 ± 1.

   The ^{1}H NMR spectra (60 MHz) of compounds B and C showed the following:

Chemical shift	**No. of protons**	**Multiplicity**
Compound B (in DMSO-d_6)		
δ 7.3	1H	bs (very broad)
δ 7.71	2H	d (distorted)
δ 7.90	2H	d (distorted)
Compound C (in CCl_4)		
δ 0.93	3H	t (distorted)
δ 1.2–1.8	4H	m
δ 2.31	2H	t
δ 11.7	1H	s

Problem Set 14

1. A compound (I) containing only carbon, hydrogen, and oxygen had a neutralization equivalent of 179 ± 1. When it was heated with aqueous sodium hydroxide and the reaction mixture was acidified with sulfuric acid, a solid (II) separated which was soluble in alkalies and had a neutralization equivalent of 138 ± 1. Compound II decolorized bromine water and gave a color with ferric chloride. Distillation of the filtrate from II yielded an acid (III) with a neutralization equivalent of 60 ± 1.

   The ^{1}H NMR spectrum of compound II (60 MHz, DMSO-d_6 plus $CDCl_3$) showed δ 6.7–7.0, 2H, m; δ 7.35, 1H, t of d ($J = 9$, 3 Hz); δ 7.75, 1H, d of d ($J = 9$, 3 Hz); δ 11.55, 2H, s.

2. A compound (I) containing nitrogen was insoluble in water and dilute alkalies but soluble in dilute hydrochloric acid. Treatment with benzenesulfonyl chloride and alkali gave a clear solution. Acidification of this solution produced a precipitate which dissolved in an excess of the acid. The original compound did not react with phenylhydrazine but was decomposed by boiling with hot sodium hydroxide solution. An oil (II) was separated from the alkaline solution (III). The oil still contained nitrogen and was soluble in dilute hydrochloric acid. Compound II reacted with acetyl chloride and sodium. When compound II was treated with benzenesulfonyl chloride and

alkali, an oil remained that proved to be soluble in hydrochloric acid. Compound II was dissolved in ether and the solution saturated with hydrogen chloride; a solid compound (IV) separated. Compound IV was soluble in water and had a neutralization equivalent of 187 ± 1.

Acidification of the alkaline solution (III) produced a precipitate (V) which dissolved in an excess of acid. Addition of sodium nitrite to a solution of V chilled in ice-water gave a clear solution without the evolution of nitrogen. Addition of this solution to a solution of sodium 2-naphthoxide gave a red solution. Compound V gave a neutralization equivalent of 137 ± 1.

Methylation (twice) of V followed by reduction gave II. The ^{1}H NMR spectrum of V (60 MHz, acetone) showed δ 6.55, 3H, bs; δ 6.76, 2H, d ($J = 8$ Hz); δ 7.83, 2H, d ($J = 8$ Hz).

3. A solid (I) gave tests for nitrogen, sulfur, and bromine. It was insoluble in ether but dissolved in water to give an acid solution. It gave a neutralization equivalent of 221 ± 1. Addition of cold alkali caused an oil (II) to separate which contained bromine and nitrogen but no sulfur. Compound II reacted with benzenesulfonyl chloride and alkali to give a clear solution from which acid (III) precipitated which contained bromine, nitrogen, and sulfur. Compound II did not give a precipitate with silver nitrate but decolorized both bromine water and dilute permanganate solution. Treatment of I with nitrous acid in the cold gave a solution without the evolution of any gas. This solution was poured into copper cyanide solution, and a compound (IV) separated. Compound IV still contained bromine and nitrogen but was insoluble in dilute acids and alkalies. When IV was boiled with dilute sulfuric acid for some time, it was converted to V. This compound no longer contained nitrogen but did contain bromine. It (V) was insoluble in water and dilute hydrochloric acid but soluble in sodium bicarbonate. A neutralization equivalent of 200 ± 2 was obtained. Compound V did not react with silver nitrate or permanganate.

The ^{1}H NMR spectrum of II (60 MHz, $CDCl_3$) showed δ 3.53, 2H, bs; δ 6.57, 2H, d ($J = 9$ Hz); δ 7.21, 2H, d ($J = 9$ Hz).

4. A pale yellow crystalline compound (I), giving tests for nitrogen and bromine, was insoluble in water, dilute acids, and alkalies. It did not react with phenylhydrazine, acetyl chloride, or cold dilute permanganate. Cold alcoholic silver nitrate did not react, but when the solution was boiled for some time, a precipitate of silver bromide formed. When compound I was heated with zinc and ammonium chloride solution and the mixture filtered, the filtrate was found to reduce Tollens reagent. Vigorous oxidation of I produced a new compound (II), still containing bromine and nitrogen, which was insoluble in hydrochloric acid but soluble in sodium bicarbonate solution. It had a neutralization equivalent of 145 ± 1.

Treatment of compound I with tin and hydrochloric acid followed by alkali gave a new compound (III) which contained nitrogen and bromine and was soluble in dilute hydrochloric acid. Compound III gave a precipitate with bromine water, reacted with acetyl chloride, and gave a clear solution when treated with benzenesulfonyl chloride and alkali. Vigorous oxidation of III produced a white crystalline acid (IV). Compound IV gave no tests for bromine or nitrogen and had a neutralization equivalent of 82 ± 1.

Compound IV, when heated, lost the elements of water. The ^{1}H NMR spectrum of III (60 MHz, $CDCl_3$) showed δ 3.96, 2H, bs; δ 6.45, 1H, d ($J = 8$ Hz); δ 7.25–7.75, 4H, m; δ 8.0–8.2, 1H, m.

5. A light yellow neutral solid (I) contained chlorine but not nitrogen. It did not react with hot alcoholic silver nitrate, acetyl chloride, or bromine in carbon tetrachloride.

It gave a precipitate with phenylhydrazine. Compound I was not attacked by cold alkalies, but when it was heated for some time with concentrated sodium hydroxide, a clear solution resulted. The distillate from this alkaline solution contained no organic compounds. Acidification of the alkaline solution with phosphoric acid gave a precipitate (II), which was removed by filtration. No organic compounds could be obtained from the filtrate by distillation or evaporation to dryness.

Compound II contained chlorine, had a neutralization equivalent of 297 ± 2, and reacted with acetic anhydride to produce a compound (III) having a neutralization equivalent of 340 ± 3. Compound III did not react with bromine water, bromine in carbon tetrachloride, or phenylhydrazine.

Vigorous oxidation of I with alkaline permanganate gave a very good yield of a product (IV) which contained chlorine and possessed a neutralization equivalent of 156 ± 1. No other oxidation product could be found.

Vigorous oxidation of II or III with potassium dichromate and sulfuric acid also produced IV but in very poor yield.

The ^{1}H NMR spectrum of IV (60 MHz, $CDCl_3$ plus DMSO-d_6) showed δ 7.37, 2H, d (J = 9 Hz); δ 7.45, 1H, bs; δ 7.81, 2H, d (J = 9 Hz). Treatment of the solution with deuterium oxide caused the δ 7.45 signal to disappear, leaving the two doublets.

Problem Set 15

1. An acid (I) was found to have a neutralization equivalent of 151 ± 2. Treatment in the cold with acetyl chloride converted it to a new acid (II) with a neutralization equivalent of 193 ± 2. Gentle oxidation of I with cold potassium permanganate solution transformed it to an acid (III) having a neutralization equivalent of 149 ± 2. Compound III yielded a derivative with phenylhydrazine. Vigorous oxidation of I, II, or III yielded an acid (IV) having a neutralization equivalent of 122 ± 1.

The ^{1}H NMR spectrum of compound I (60 MHz, acetone) showed δ 5.22, 1H, s; δ 6.93–7.88, 7H, m. The IR spectrum of compound I (in a mineral oil mull) showed, among other bands, a strong band at 1706 cm^{-1} and a broad band at 3333–2400 cm^{-1}.

2. A compound (I) had a neutralization equivalent of 223. Vigorous oxidation converted it to a new acid (II) having a neutralization equivalent of 167. Treatment of I with zinc and hydrochloric acid gave an acid (III) with a neutralization equivalent of 179. Compounds I, II, and III were found to contain nitrogen.

The ^{1}H NMR spectrum of compound II (60 MHz, $CDCl_3$ plus DMSO-d_6) showed δ 7.72, 1H, t (J = 8 Hz); δ 8.2–8.55, 2H, m; δ 8.71, 1H, t (J = 2 Hz); δ 12.98, 1H, s.

3. A sulfur-containing compound was soluble in strong alkalies but not in sodium bicarbonate solutions. Vigorous oxidation gave a sulfur-containing acid having a neutralization equivalent of 102 ± 1. When this compound was treated with superheated steam, a sulfur-free acid was formed.

The ^{1}H NMR spectrum of 100 mg of the sulfur-free acid (in 0.5 mL of acetone-d_6) showed δ 7.00–7.75, 5H, m; δ 8.00, 1H, s. The δ 8.00 signal position had a chemical shift that was concentration dependent. The ^{1}H NMR spectrum of the original compound (60 MHz, $CDCl_3$) showed δ 2.29, 3H, s; δ 3.37, 1H, s; δ 7.02, 4H, s (somewhat broadened at the base).

4. A compound (I) containing chlorine gave positive tests with alcoholic silver nitrate, sodium hypoiodite solution, and acetyl chloride. Hydrolysis with sodium bicarbonate yielded a chlorine-free compound (II) which gave a positive iodoform test and was oxidized by periodic acid to a single compound (III). Compound III also gave a positive iodoform test.

5. A compound (A), containing only carbon, hydrogen, and oxygen, was found to react with acetyl chloride but not with phenylhydrazine. Oxidation with periodic acid converted it to a new compound (B) which reduced Tollens but not Fehling's reagent.[10] Treatment of B with potassium cyanide in aqueous ethanol converted it to a new compound (C) which gave positive tests with acetyl chloride and phenylhydrazine. Oxidation of C with Fehling's solution or nitric acid converted it to a yellow compound (D) which yielded a derivative with *o*-phenylenediamine. When D was treated with hydrogen peroxide, it yielded an acid (E) having a neutralization equivalent of 135 ± 1. Catalytic hydrogenation of C or D produced the original compound (A).

The ^{1}H NMR spectrum of compound B (in $CDCl_3$) showed δ 2.42, 3H, s; δ 7.18, 2H, d (J = 10 Hz); δ 7.56, 2H, d (J = 10 Hz); δ 9.81, 1H, s.

6. A solid, neutral substance (A) contained nitrogen. It was recovered from attempted hydrolysis with dilute acids and bases. It gave negative tests with acetyl chloride, bromine in carbon tetrachlorde, sodium hypoiodite solution, and Tollens reagent. It reacted slowly with dinitrophenylhydrazine and, after treatment with zinc and ammonium chloride, reduced Tollens reagent.

When A was treated with hydroxylamine hydrochloride in pyridine, it was slowly converted to a new compound B. Substance B was treated with phosphorus pentachloride, which changed it to C. Boiling with acid converted C to D, a nitrogen-containing acid of neutralization equivalent 168 ± 2, and the salt of a base E. The hydrochloride of E had a neutralization equivalent (by titration with alkali) of 195 ± 2.

Acid D was treated with thionyl chloride, and the product was added to aqueous ammonia. The neutral substance so obtained was treated with bromine and sodium hydroxide solution. The product (F) was an acid soluble substance which, after treatment with hydrochloric acid and sodium nitrite, gave a color with 2-naphthol. Treatment of F with tin and hydrochloric acid produced G, a substance readily attacked by oxidizing agents.

Base E, obtained in the reaction with phosphorus pentachloride, reacted with sodium nitrite and dilute sulfuric acid. The resulting solution gave a color with 2-naphthol. Boiling the aqueous solution produced a nitrogen-free substance, H, which dissolved in aqueous alkali but not in aqueous sodium bicarbonate. Oxidation of either E or H produced an acid (neutralization equivalent 83) which was readily converted to an anhydride. Substance H did not undergo coupling with benzenediazonium solutions.

The ^{1}H NMR spectrum of compound F (60 MHz, $CDCl_3$ plus DMSO-d_6) showed δ 5.68, 2H, bs; δ 6.58, 2H, d (J = 10 Hz); δ 7.88, 2H, d (J = 10 Hz). The ^{1}H NMR spectrum of compound H (60 MHz, $CDCl_3$) showed δ 2.32, 3H, s; δ 5.12, 1H, bs; δ 7.1–7.55, 4H, m; δ 7.6–7.9, 1H, m; δ 8.0–8.2, 1H, m. The δ 5.12 signal moved to lower field upon addition of more compound H to the NMR tube.

[10] See Tollens, *Ber.,* 1950 (1881), and Daniels, Rush, and Bauer, *J. Chem. Educ., 37,* 205 (1960).

CHAPTER 10

Separation of Mixtures

The identification of the components of a mixture involves, first, a separation into individual components and, second, the characterization of each of the latter according to the procedures outlined in Chapter 7. It is very rarely possible to identify the constituents of a mixture without separation. The separation of the compounds in a mixture should be as nearly quantitative as possible in order to give some idea of the actual percentage of each component. It is far more important, however, to carry out the separation in such a manner that each compound is obtained in a pure form, because this renders the individual identifications much easier.

The method of separation chosen should be such that the compounds are obtained as they existed in the original mixture. Derivatives of the original compounds are not very useful unless they may be readily reconverted into the original compounds. This criterion of separation is necessary because the identification of a compound rests ultimately on agreement between physical constants of the original and of a derivative with similar data obtained from the literature.

The history of a mixture will frequently furnish sufficient information to indicate the group to which the mixture belongs and hence the general mode of separation to be used.

In recent years the field of analytical separation has been extensively developed and widely applied by organic chemists. It is thus necessary to be aware of the nature of the many techniques available in order to be able to choose the one which is most appropriate for the mixture in hand. There are a number of separation problems which frequently occur for the organic chemist. In one common situation the mixture is comprised of a number of components, all of reasonable purity and all of substantial proportion. Another type of separation is the isolation of a single component from large amounts of unreacted starting material or from undesired side products; these side products frequently are simply intractable tars or polymeric materials. We should try to place each new separation problem into one of these two categories in order to be able to select the most efficient separation approach.

Before selecting a separation procedure, the preliminary tests outlined in Section 10.1 should be carried out. As these tests are performed, one should constantly be concerned with the following:

1. Will the sample survive the separation procedure? That is, are the components of the mixture stable under the conditions of the procedure?
2. Is this the easiest and most efficient way to carry out the separation?

Thermal stability is always of concern. Stability of the sample under the conditions of the separation procedure may not be known until the separation is attempted. Compounds which are thermally unstable to the heat required for distillation at atmospheric pressure should be distilled at reduced pressure. Extractions and column chromatography do not involve heat and thus may be appropriate for samples which cannot be distilled. However, some samples may decompose because of chemical reactions with acid or base in extractions or with chromatographic packing or support in column chromatography. A TLC test (Chapter 3, pp. 59–63) is a fast and useful check for sample durability under chromatographic conditions.

10.1 PRELIMINARY EXAMINATIONS OF MIXTURES

1. The physical state is noted. Take advantage of existing separations. If a solid is suspended in a liquid, the solid is removed by filtration and is examined separately. If two immiscible liquids are present, they are separated and examined individually.

2. The solubility or insolubility of the mixture in water is determined. The compound can then be broadly classified according to Figure 5.1 (p. 96) and Table 5.1 (p. 96).

3. With liquid mixtures, 2 mL of the solution is evaporated to dryness on a watch glass or porcelain crucible cover and the presence or absence of a residue noted. The ignition test is applied to the residue or to 0.1 g of the liquid. For a solid, the ignition test is applied directly to a 0.1 g sample. Conclusions can be made from the results of the ignition test as described in Chapter 3 (p. 36).

4. In liquid mixtures the presence or absence of water is detected by (a) determining the miscibility of the solution with ether; (b) an anhydrous copper sulfate test; or (c) distillation test for water.

Copper sulfate will dissolve in water, turning the solution blue. Thus merely adding a small sample of anhydrous copper sulfate to a liquid may detect the presence of water by this color change. The distillation test is the most reliable and is carried out in the following manner. Five milliliters of the liquid mixture and 5 mL of anhydrous toluene are placed in a small distilling flask which is placed in a simple distillation apparatus. The mixture is heated gently with a flame until distillation occurs, and 2 mL of the distillate is collected. About 5–10 mL of anhydrous toluene is added to the distillate. The presence of two layers or distinct drops suspended in the toluene indicates water. If the solution is only cloudy, traces of water are indicated.

5. If water is absent, the presence or absence of a volatile solvent in a liquid mixture is determined by placing 1.0 mL of the mixture in a distilling flask in a simple distillation apparatus. The distilling flask is placed in a beaker of cold water which is heated to boiling. Any liquid distilling under these conditions is classified as a volatile solvent. The distillate, which may be a mixture of readily volatile compounds, and the residue in the flask are examined separately.

It frequently happens that distillation of a water soluble mixture yields a volatile solvent and a water insoluble residue. The separation of such a mixture is therefore carried out by removing all the volatile solvent. The residue is then treated as a water insoluble mixture.

If the residue after distillation is a water soluble liquid, it is best not to remove the solvent at this stage because the separation is usually not quantitative.

If, however, the residue after distillation is a water insoluble solid and the removal of the solvent seems quantitative, then it is desirable to remove all the volatile solvent and to examine the distillate and the residue separately.

It is to be noted that if water is present, no such separation should be attempted.

6. The reaction of an aqueous solution or suspension of the mixture to litmus and to phenolphthalein is determined. If the mixture is distinctly acidic, 1 mL (of a known exact weight) of the solution in 2.5 mL of water or ethanol should be titrated with a standardized 0.1 M sodium hydroxide solution in order to determine whether considerable amounts of free acid are present or whether the acidity is due to traces of acids formed by hydrolysis of esters. The titration must be performed in an ice-cold solution, and the first pink color of phenolphthalein taken as an end point. An IR spectrum of a mixture can be used to reveal the presence of any carboxylic acid groups.

7. Two milliliters of the mixture is acidified with 5% hydrochloric acid, and the solution is cooled. The evolution of a gas or the formation of a precipitate is noted. To this solution is added 5% sodium hydroxide solution until the solution is basic and the result is noted.

8. Two milliliters of the mixture is made distinctly basic with 5% sodium hydroxide solution. The separation of an oil or solid, the liberation of ammonia, and any color changes are noted. The solution should be heated just to boiling and then cooled. The odor is compared with that of the original mixture. The presence of esters is often indicated by a change in odor. Next, 5% hydrochloric acid is added until the solution is acidic, and the result is noted.

9. In the case of water insoluble mixtures, an elemental analysis (Chapter 4, pp. 90–93) should be performed. If water or a large amount of a volatile solvent is present in a water insoluble mixture, the elemental analysis of the mixture is omitted. If the water soluble mixture is composed of solids, an elemental analysis is performed.

10. If water is absent, the effect of the following classification reagents is cautiously determined: (a) metallic sodium (Experiment 5, p. 215); (b) acetyl chloride (Experiment 6, p. 217).

11. The action of the following classification reagents should be determined on an aqueous solution or suspension of the original mixture: (a) bromine water (Experiment 44, p. 298); (b) potassium permanganate solution (Experiment 36, p. 279); (c) ferric chloride solution (Experiment 43, p. 296); (d) alcoholic silver nitrate solution (Experiment 33, p. 270); (e) Fuchsin–aldehyde reagent (Experiment 15, p. 235); (f) 2,4-dinitrophenylhydrazine (Experiment 11, p. 229).

At this stage of the examination the results of the foregoing tests are summarized and as much information as possible is deduced from the behavior of the mixture. The preliminary study will show the group in which the mixture should be classified and will, therefore, indicate which of the following procedures should be used in its separation.

10.2 DISTILLATION AND SUBLIMATION

An introduction to simple distillation has been given on pages 44–50 (Chapter 3); in these earlier treatments simple sample purification and boiling point measurement were discussed. Other more sophisticated distillation techniques are also available for sample purification. Several of these methods will be discussed in the following sections.

Sublimation is a technique in which a solid is heated and vaporized, without passing through the liquid phase. The gas is then condensed and collected as a solid.

10.2.1 Distillation

For amounts from a few milligrams to 5 g, with appropriate variation of the sizes of the distillation flask and the receiver bulb, the Kugelrohr distillation apparatus (Figure 10.1) can be used. A converted coffee pot can be used as the heat source. The reciprocating motor moves the glassware back and forth so that the compound does not bump over. A vacuum pump is used as the source of the vacuum. The great advantage of this apparatus is its ability to apply good vacuums, sometimes as low as 0.1 mm Hg (torr), especially when the glassware is composed of only one piece rather than of a number of fitted pieces with many possibilities for leaks. An ice bath or dry ice–acetone bath is used to condense the material into the center receiver. The Kuglerohr apparatus is best for only purifying a product, therefore leaving impurities behind.

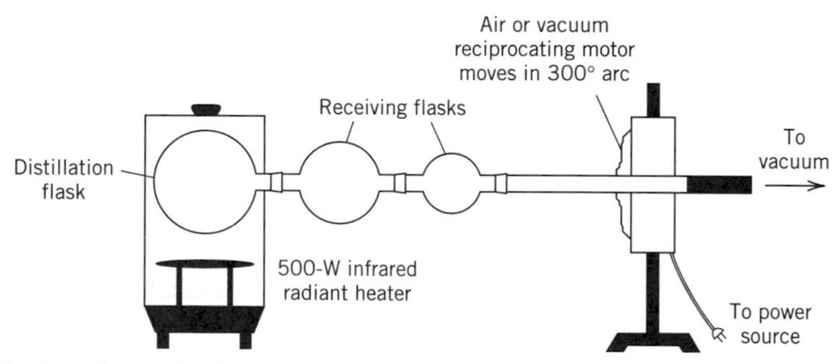

Figure 10.1 Kugelrohr distillation apparatus.

For larger amounts up to 50 mL of material, a short-path distillation apparatus (Figure 10.2) is used. The short-path process allows distillation of materials such as low-melting solids for which long exposure to elevated temperatures could be detrimental. This apparatus contains an inlet tube in the distillation flask allowing a stream of air to bleed into the pot to prevent bumping. Instead of an air inlet, the neck could be stoppered and a stirbar used.

A newer distillation apparatus that can be used with amounts of 0.5–2.0 mL is a Hickman–Hinkle distillation apparatus (Figure 10.3). An air condenser is used for condensation of the liquid. The side arm is capped while the distillation is in progress and uncapped when the product is removed via a pipette.

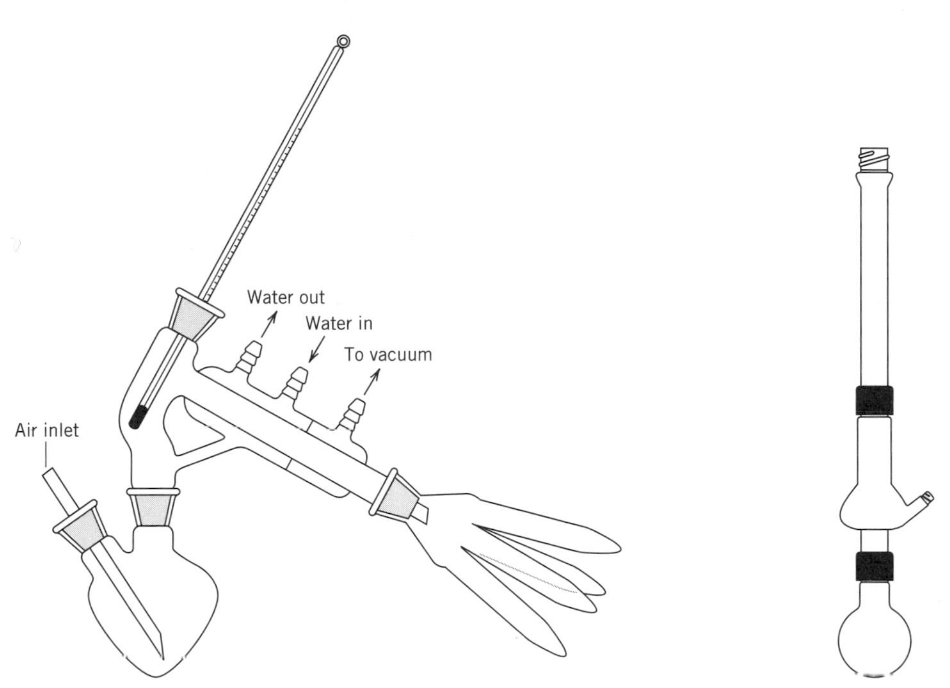

Figure 10.2 Short-path distillation apparatus.

Figure 10.3 Hickman–Hickle still head with a round-bottom flask and air condenser.

In order to improve the efficiency of a distillation, a column (Figure 10.4) can be placed between the vessel to be heated and the condenser tube. Frequently, the column is a condenser filled with glass beads or steel wool to provide increased surface area and/or increase cooling surfaces. Alternatively, the column may contain coils or glass indentations such as those in Vigreux columns (Figure 10.4a) which are available in some commercial condensers. In any case, this vertical column is not surrounded by a jacket of water as typically found in the traditional condenser.

A fractional distillation apparatus (Figure 10.5) uses the vertical column, in addition to the condenser. With this distillation apparatus, compounds with a difference in their boiling points of 5–10°C or more can be efficiently separated.

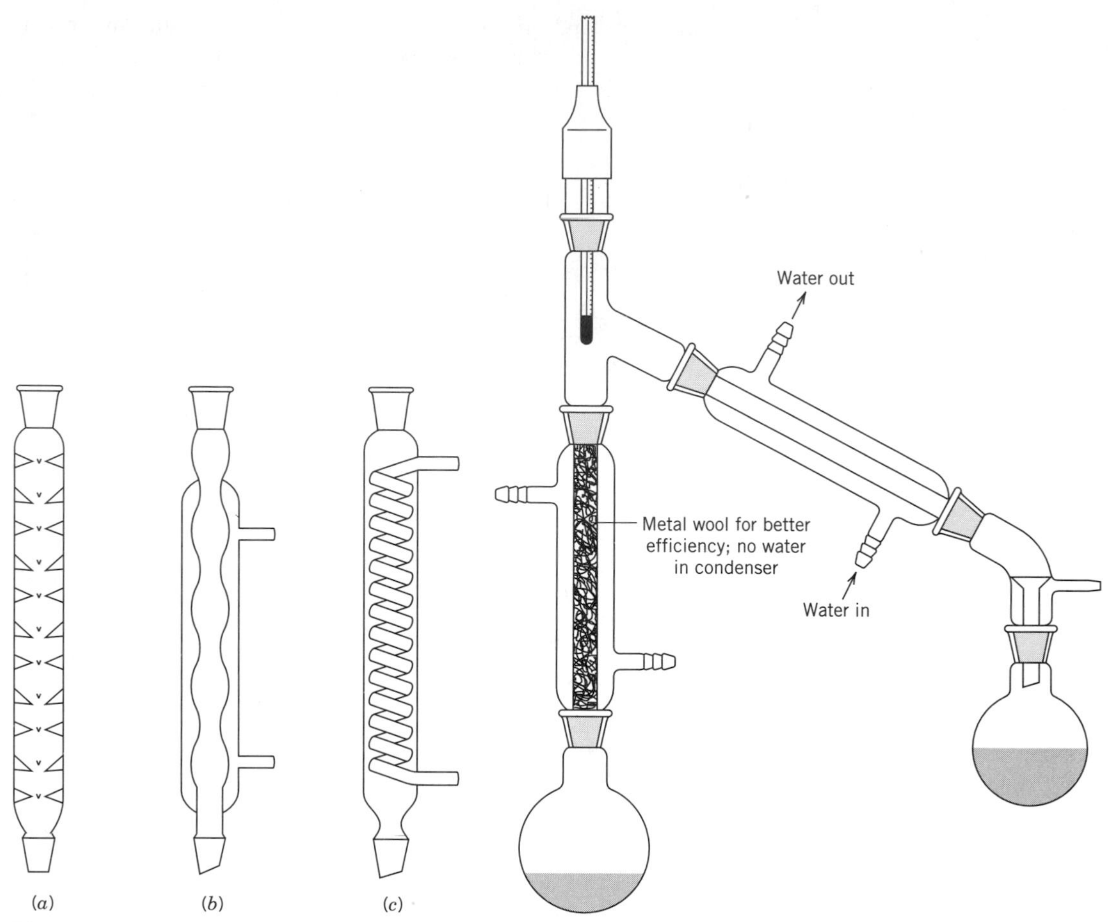

Figure 10.4 Distillation columns and condensers: *(a)* Vigreux column, *(b)* Allihn condenser, *(c)* coiled condenser.

Figure 10.5 A fractional distillation apparatus.

The spinning band apparatus, shown in a microscale form in Figure 10.6, allows a very efficient distillation because of the large number of theoretical plates provided for the distillate.

Frequently the result of a more efficient distillation apparatus is a longer time for the distilling compound to remain on the vertical column. To avoid heat loss, the column should be externally insulated with glass wool, cotton, or aluminum foil. In some cases the user must also be prepared to vary the amount of heat applied in different places during the distillation. Two commonly used techniques are heating tape and a heat gun. The use of heating tape provides uniform heating. A heat gun is used to heat localized places and is very similar to a hand-held hair dryer.

A Hickman flask (Figure 10.7) is another microscale approach to distilling small amounts of compound. The distilling flask, Vigreux column, and condenser are all one piece of glass in this particular apparatus. Normally a stopper is placed in the lower side tube of the Hickman flask. If a vacuum distillation is to be done, a small pipet is placed through the side tube to create a small stream of bubbles going through the compound. The distilled fractions are collected using test tubes attached to a cow adaptor.

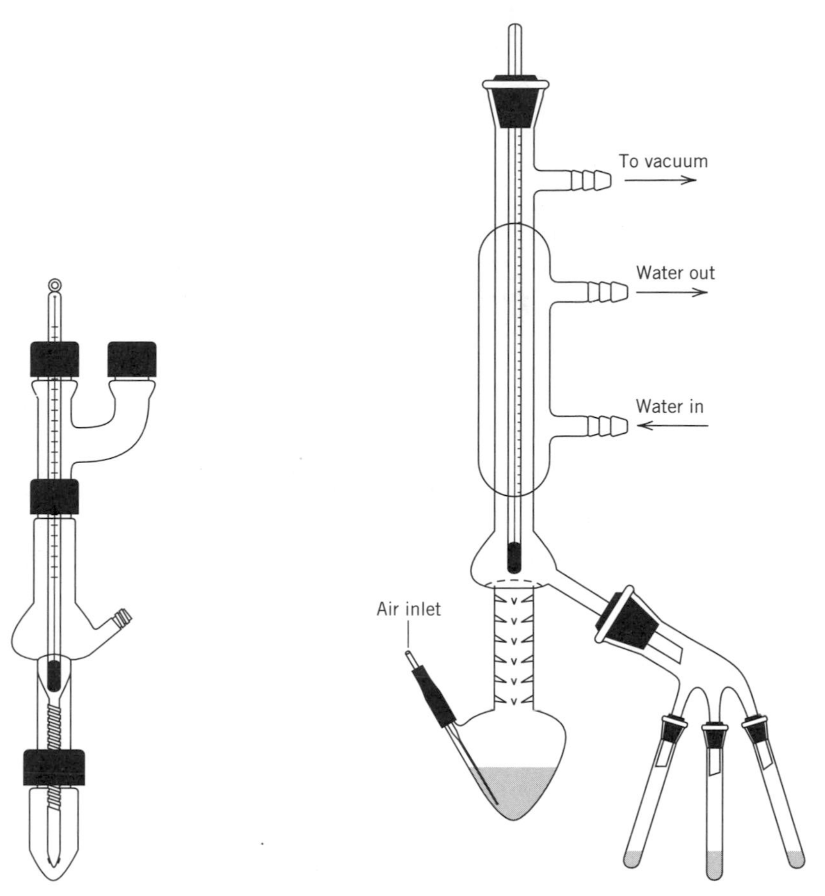

Figure 10.6 Microscale spinning band column.

Figure 10.7 Hickman flask.

In order to distill liquids and solids of low volatility which might be somewhat heat sensitive, a vacuum distillation apparatus (Figure 10.8) can be used. In this apparatus, the Vigreux column and the condenser are one piece of glass. The stopcocks are used to regulate pressure in the apparatus. An air inlet is optional. A laboratory aspirator can provide a vacuum of 15 mm Hg (torr) and a good vacuum pump can yield a range of 0.01–15 mm Hg (torr) for vacuum distillations.

An important aspect in distillation is the method of heating the distilling pot. For volatile liquids, a steam bath is used. Baths containing oils or other involatile, inert substances (see Liquid Media for Heating Baths, Appendix I, p. 484) can be used; such hot liquids provide a very even method of heat application and can be used to higher temperatures (ca. 250–400°C). Heating mantles attached to Variacs can be used for heat application; use of such mantles allows one to avoid the messy oils used for external heating.

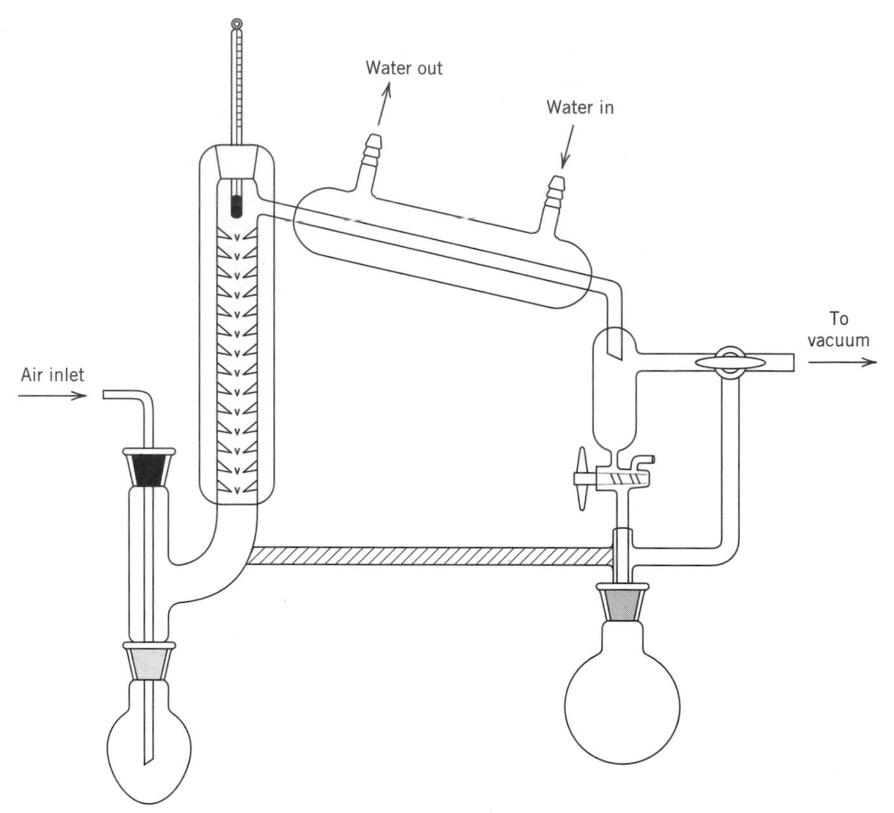

Figure 10.8 Vacuum distillation apparatus.

Sand baths (Figure 10.9), in which sand is placed in a small heating mantle (manufactured by Thermowell) and aluminum blocks (Figure 10.10), which have concave spaces for flasks, are popular as heating devices, particularly in microscale applications. Flaked graphite can be substituted for the sand. The heating mantles for the sand baths are controlled with a Variac. The aluminum blocks are heated by placing them on a hot-plate stirrer.

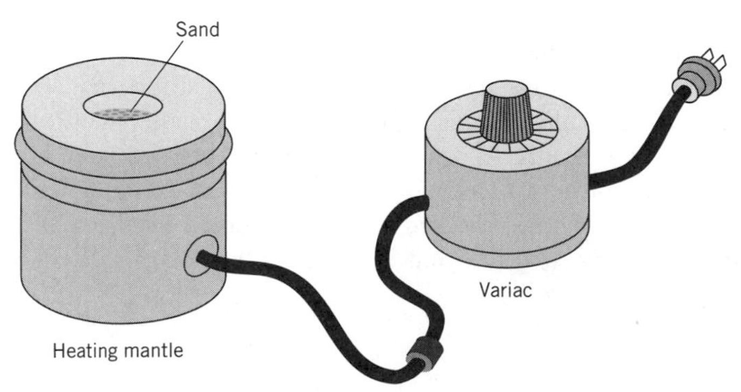

Figure 10.9 Sand bath with Variac.

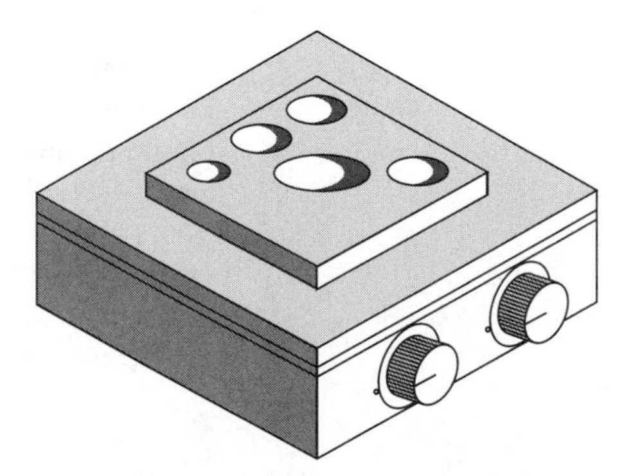

Figure 10.10 Aluminum block on hot-plate stirrer.

Frequently, an organic chemist is concerned with removal of volatile solvents from a solution during the work up of a reaction. A rotary evaporator ("rotovap," Figure 10.11) is useful for solvent removal. The distilling flask is spun to prevent bumping of the solution. Next, the vacuum is turned on, and then the distilling flask is heated by a hot water bath. The solvent is distilled into the receiver, leaving the product in the distilling flask.

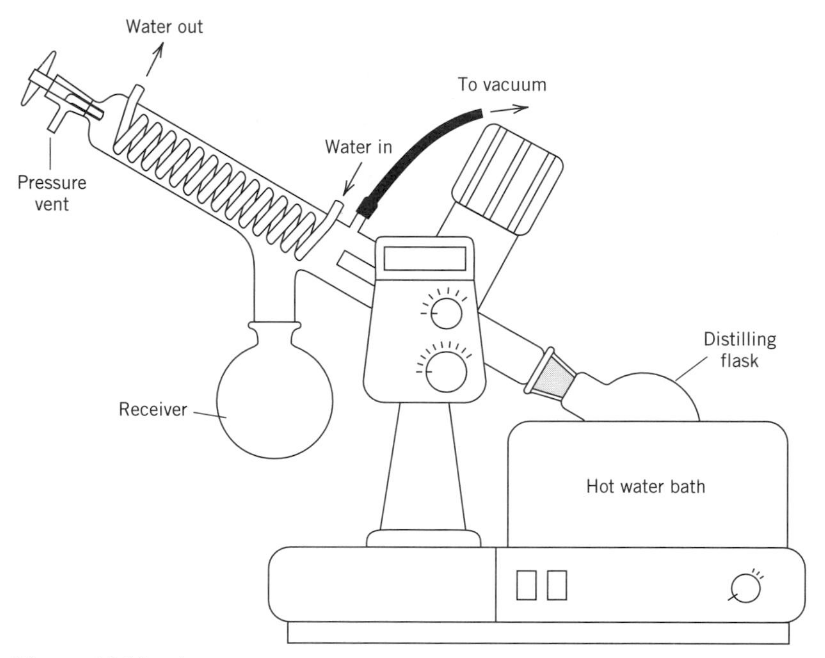

Figure 10.11 A rotary evaporator.

10.2.2 Steam Distillation

Steam distillation is a technique whereby a compound of relatively low volatility can be purified by co-distilling it with water. This distillation occurs because both of the liquid components contribute to the vapor pressure and thus the distillation can be carried out at a temperature slightly less than 100°C at 760 mm Hg (torr). The distillation is actually carried out by simply forcing steam through a vessel containing the mixture and collecting the distillate with a water-cooled condenser (Figure 10.12). An alternate way to the steam inlet is to heat the flask with a bunsen burner and add water, via an addition funnel, to maintain a constant volume in the flask.

A difference in polarity sufficient to permit separation by steam distillation is generally provided by a second functional group in the molecule. Polyfunctional compounds are normally more polar and thus have a higher boiling point than monofunctional compounds. Thus, monohydroxy alcohols can be separated from dihydroxy and polyhydroxy alcohols by this scheme. Similarly, simple acids, amines, and many other volatile compounds can be separated from the corresponding di- and polyfunctional compounds. Moreover, the additional group or groups need not be the same as the original. Amino acids, hydroxy acids, nitro acids, keto acids, keto alcohols, and cyano ketones are rarely volatile with steam. In fact, it is a general rule that the presence in a molecule of two or

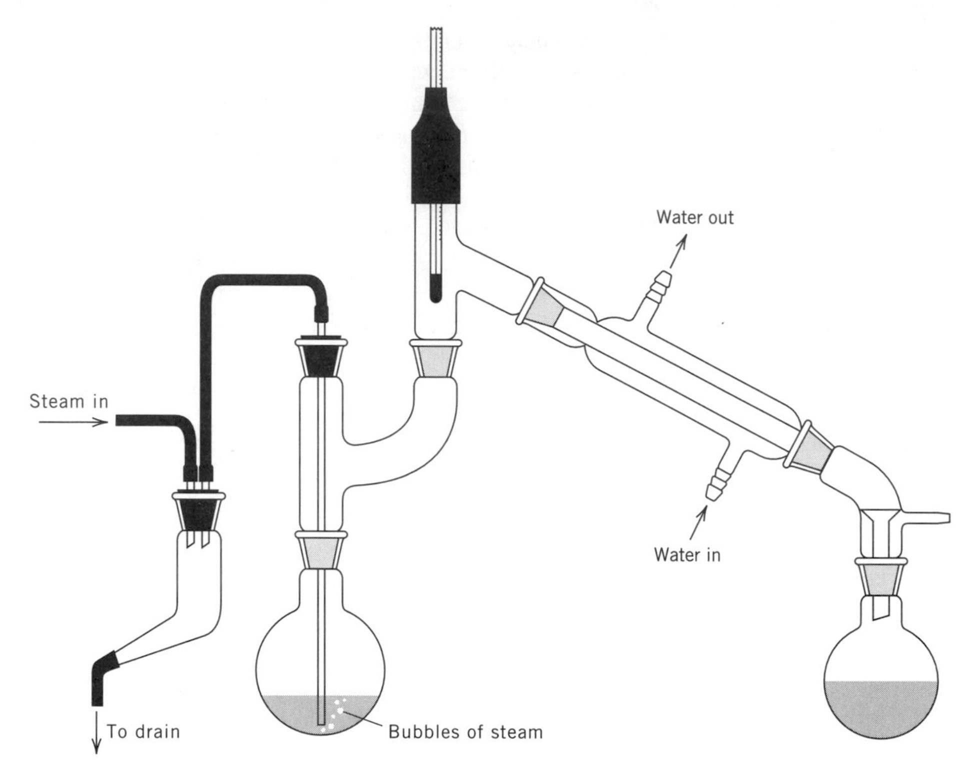

Figure 10.12 Steam distillation apparatus (macroscale).

more functional (polar) groups will render a compound nonvolatile. Table 10.1 shows which types of compounds are volatile with steam and which are not. Acetic acid and oxalic acid, ethyl alcohol and ethylene glycol, benzoic acid and 1,2-benzenedicarboxylic acid are mixtures that illustrate the point. In each pair the first named can be removed by steam distillation, whereas the other remains behind.

A very interesting group of exceptions to the multiple function rule is found in the aromatic series. 2-Hydroxybenzaldehyde, 2-nitrophenol, and many other *ortho*-disubstituted benzene derivatives are volatile with steam. The explanation for this apparently anomalous loss of polar character is found in the observation that *all these exceptional compounds are capable of intermolecular hydrogen-bonded forms.* These forms tend not to associate with thc water and are thus relatively volatile. The hydrogen-bonded structures of 2-hydroxybenzaldehyde and 2-nitrophenol are shown below.

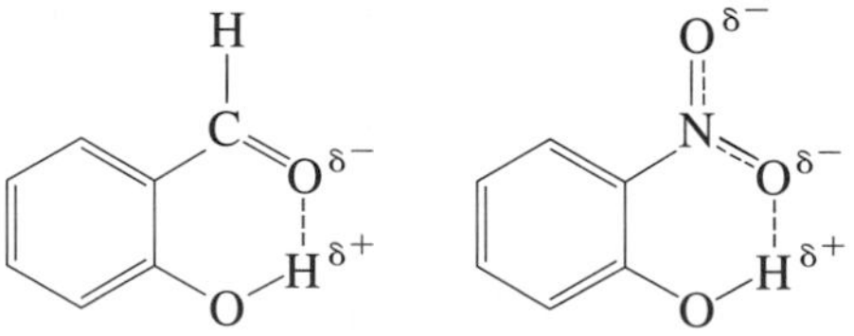

Another valuable use of steam distillation is the separation of reaction products from solvents such as *N,N*-dimethylformamide (DMF) and dimethyl sulfoxide (DMSO). DMF and DMSO are good solvents for carrying out many reactions, but their high boiling points (as well as other properties) make their removal from the reaction mixture a very difficult process when conducted by other procedures. Neither DMSO or DMF is volatile in a steam distillation. Thus in many cases we may merely dilute a reaction mixture with water and remove the products or the unreacted starting materials by steam distillation.

Table 10.1 Solubility and Steam Distillation

Solubility	Types of compounds	Volatility	Volatility with steam
Soluble in water and ether (S_A, S_B, S_1)	Low molecular weight alcohols, aldehydes, ketones, acids, esters, amines, nitriles, acid chlorides	Readily distill. Many compounds boil below 100°C	Volatile with steam
Soluble in water but insoluble in ether (S_2)	Polyhydroxy alcohols, diamines, carbohydrates, amine salts, metal salts, polybasic acids; hydroxy aldehydes, ketones, and acids; amino acids	Low volatility. With certain exceptions these compounds cannot be distilled at atmospheric pressure	Not volatile with steam
Insoluble in water but soluble in NaOH and $NaHCO_3$ (A_1)	High molecular weight acids; negatively substituted phenols	Low volatility	Usually not volatile, but there are some exceptions
Insoluble in water and $NaHCO_3$ but soluble in NaOH (A_2)	Phenols, sulfonamides of primary amines, primary and secondary nitro compounds; imides, thiophenols	High boiling points; many cannot be distilled	Usually not volatile
Insoluble in water but soluble in dilute HCl (B)	Amines containing not more than one aryl group attached to nitrogen; hydrazines	High boiling points	Many are volatile with steam
Insoluble in water, dilute NaOH, and HCl, but contain elements other than carbon, hydrogen, oxygen, and the halogens (MN)	Nitro compounds (tert), amides, negatively substituted amines; sulfonamides of secondary amines; azo and azoxy compounds; alkyl or aryl cyanides, nitrites, nitrates, sulfates, phosphates	High boiling points; many cannot be distilled	Some are volatile with steam
Insoluble in water, dilute NaOH, and HCl, but soluble in H_2SO_4 (N)	Alcohols, aldehydes, ketones, esters, unsaturated compounds	High boiling compounds	Usually volatile with steam
Insoluble in water, dilute NaOH, dilute HCl, and H_2SO_4 (I)	Aromatic and aliphatic hydrocarbons and their halogen derivatives	Volatile	Volatile with steam

10.2.3 Sublimation

Occasionally compounds may be purified by sublimation. In a sublimation apparatus (Figure 10.13), a cold finger is placed inside another container containing a side arm. The outer container may consist of a special type of glassware or may be as simple as a test tube or Erlenmeyer flask with a side arm. Ice, dry ice, or a dry ice–acetone slurry is placed inside the cold finger. The material to be sublimed is placed on the bottom of the inside of the outer container. Frequently, the apparatus is attached to vacuum. Heat is applied

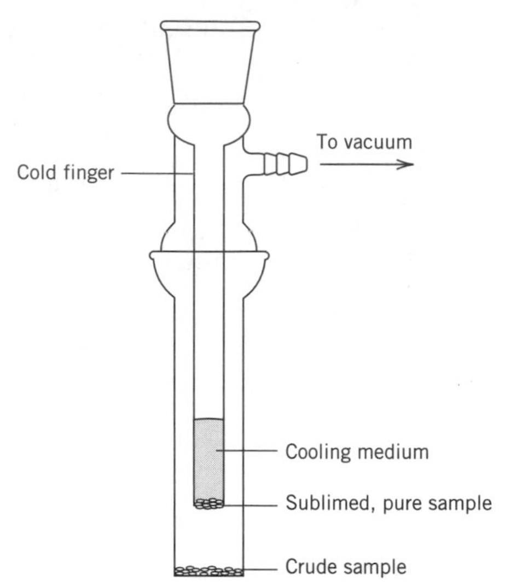

Figure 10.13 Sublimation apparatus.

externally, usually in the form of an oil or sand bath. Successful sublimation of material from the crude mixture will result in the formation of crystals on the bottom outside of the cold finger. It may be necessary to interrupt the sublimation periodically and scrape the solid off the surface of the cold finger.

Before attempting sublimation on a compound, look up the boiling point to see if the compound sublimes instead of boiling. The abbreviation "sub" indicates that the compound will sublime. Caffeine and camphor are examples of such compounds.

10.3 EXTRACTIONS: SEPARATIONS BASED UPON SALT FORMATION

The basic principle of this technique can be made clear by reference to simple examples. The separation of aniline from toluene is effected by extraction with dilute hydrochloric acid. The aniline goes into the aqueous layer as its salt, aniline hydrochloride. Whereas aniline is very soluble in tolucnc and virtually insoluble in water, its hydrochloride salt, because of its polar nature, is soluble in water and insoluble in toluene.

$$C_6H_5\ddot{N}H_2 + HCl \longrightarrow C_6H_5NH_3^+Cl^-$$

water soluble
toluene insoluble

Similarly, phenol is removed from toluene by shaking the mixture with a dilute sodium hydroxide solution. The phenol is transformed into its anionic form, sodium phen-oxide, whose highly polar character makes it insoluble in toluene and soluble in water.

$$C_6H_5OH + NaOH \longrightarrow C_6H_5O^-Na^+$$

water soluble
toluene insoluble

Benzaldehyde may be separated by a similar scheme. In this case the mixture is shaken with an aqueous solution of sodium bisulfite which converts the aldehyde into its bisulfite derivative. This is a typical salt and therefore insoluble in toluene but soluble in water.

$$C_6H_5CHO + NaHSO_3 \longrightarrow C_6H_5CH(O^-Na^+)SO_3H$$

water soluble
toluene insoluble

In each of these examples the original acid, base, or aldehyde is easily recovered by decomposition of the salt by familiar methods.

If the compounds to be separated are water soluble to any considerable degree, extraction methods usually have little value. Steam distillation, however, can generally be used instead. For example, a mixture of acetic acid and cyclohexanone can be separated by adding enough alkali to transform the acid to sodium acetate and steam-distilling the mixture. The ketone will be removed in the distillate while the salt, being nonvolatile, remains behind. Acidification with phosphoric acid regenerates the organic acid, which can now be distilled with steam.

Diethylamine can be separated from 1-butanol in a similar manner. Phosphoric acid is added in sufficient amount to neutralize the amine. Steam distillation will now remove the alcohol, and the amine can be recovered by adding dilute sodium hydroxide solution to the residue and repeating the steam distillation.

$$(CH_3CH_2)_2NH + CH_3CH_2CH_2CH_2OH \xrightarrow{H_3PO_4} (CH_3CH_2)_2NH_2^+\ H_2PO_4^- + CH_3CH_2CH_2CH_2OH$$

volatile with steam

$$(CH_3CH_2)_2NH_2^+\ H_2PO_4^- \xrightarrow{\text{dilute NaOH}} (CH_3CH_2)_2NH$$

Another useful method for establishing a marked difference in the polar character of the components is illustrated by the separation of mixtures of primary amines from tertiary amines. Acetylation or benzoylation converts the primary amine to a *neutral* amide. Extraction with dilute hydrochloric acid will then remove the tertiary amine and leave the amide behind. The amide can be reconverted to the original amine by hydrolysis.

A very similar principle is involved in the Hinsberg method (Chapter 7, p. 242) of separating and characterizing primary and secondary amines. The sulfonamide from the primary amine forms a salt with alkali and thus can be removed by extraction with dilute sodium hydroxide solution.

A general method can be developed for separating acidic compounds differing in acidity. Strong acids form salts when treated with sodium bicarbonate and can be extracted with this reagent. Thus, if a mixture of 2-methylphenol and benzoic acid is shaken with a dilute solution of sodium bicarbonate, the carboxylic acid passes into the water layer as sodium benzoate, leaving the less acidic 2-methylphenol (*o*-cresol) behind.

$$C_6H_5COOH + o\text{-}CH_3C_6H_4OH \xrightarrow{NaHCO_3} C_6H_5COO^-Na^+ + o\text{-}CH_3C_6H_4OH$$

water soluble

If a mixture contains more than two compounds, combinations of the following methods frequently lead to satisfactory separations. The necessary condition for successful separation is that *the components be such that a wide polarity difference exists or can be induced between any two of them.*

In practice, mixtures fall into two categories, depending on whether they are soluble in water. These two types will be considered separately.

10.3.1 Extraction of Water Insoluble Mixtures

After the removal of any volatile solvent one of the following procedures may be used for the separation of the compounds of a water insoluble mixture. For these procedures to work effectively, no component of the mixture can be soluble in water.

These procedures are used after the water solubility test on page 97. These procedures, as described below, assume that all possible fractions are obtained. Of course, if the mixture does not contain certain types of compounds, then those fractions will not be obtained.

Procedure A: Water Insoluble Mixtures (Figure 10.14). Five grams of the mixture is mixed with 15 mL of diethyl ether, and any insoluble compounds (Residue 1) are separated on a filter and washed with two 5-mL portions of diethyl ether. The ether washings are added to the original ether solution (Organic layer 1), and this layer is extracted with three 5-mL portions of 5% hydrochloric acid solution. If a solid amine hydrochloride separates during this extraction, water should be added until the amine hydrochloride is dissolved. The acidic aqueous layers are combined (Aqueous layer 1). The ether layer (Organic layer 2) is placed aside.

Aqueous layer 1 is rendered alkaline with 5% sodium hydroxide solution, and the resulting mixture is extracted with three 5-mL portions of diethyl ether. The ether layer (Organic layer 3) is dried with anhydrous sodium sulfate, and the ether is distilled off. The residue is composed of bases (Basic residue 1).

The remaining basic aqueous layer (Aqueous layer 2) is carefully neutralized with acetic acid. If a solid separates it may be removed by filtration, but it is better to extract the solution four or five times with 5-mL portions of ether in order to recover any amphoteric compounds. The ether is evaporated to leave the amphoteric compounds (Amphoteric residue 1).

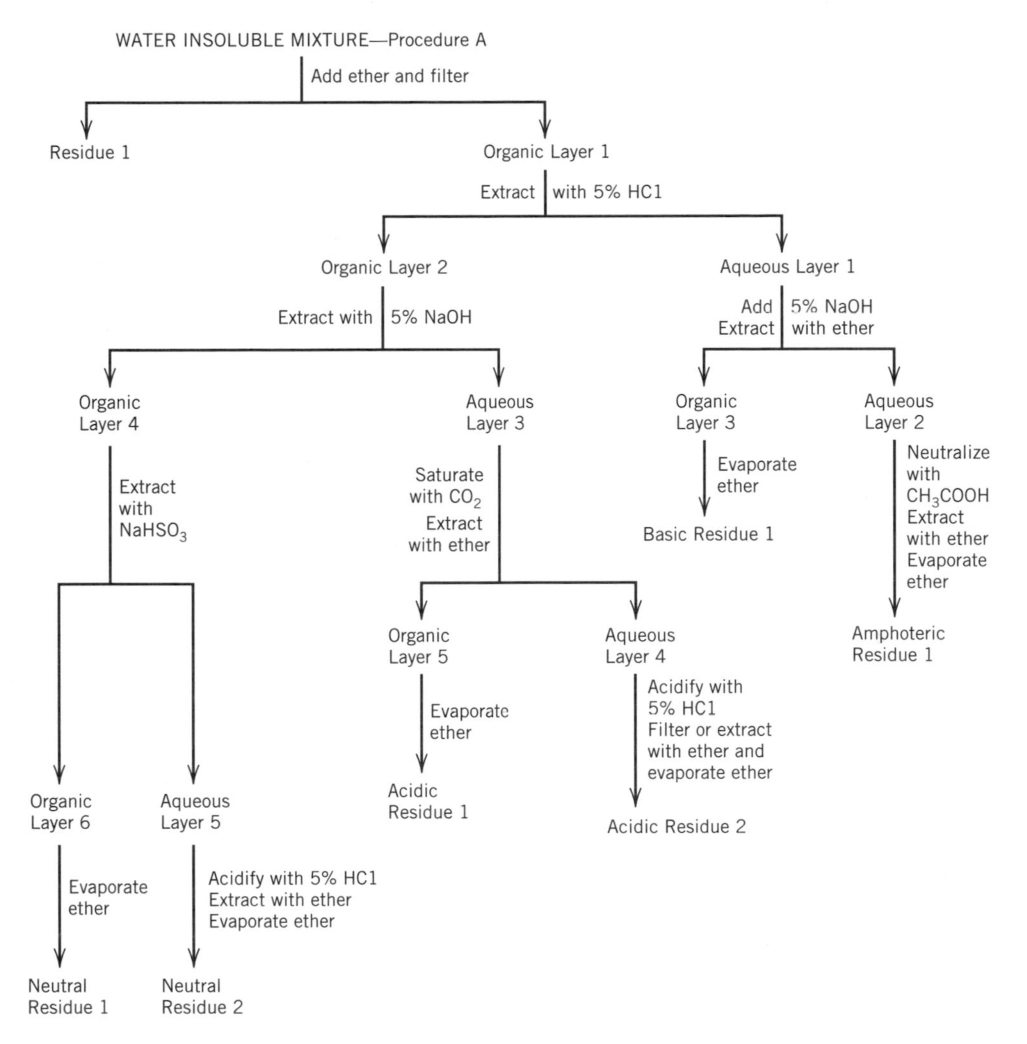

Figure 10.14 Separation of a water insoluble mixture as described in Procedure A.

The ether layer (Organic layer 2) is extracted with three 5-mL portions of 5% sodium hydroxide solution. If an emulsion is formed, more water and a few crystals of sodium chloride may be added to cause the separation of the two layers. The basic aqueous layers (Aqueous layer 3) are combined. The ether layer (Organic layer 4) is set aside.

The combined basic aqueous layers (Aqueous layer 3) are saturated with carbon dioxide (dry ice) by slowly adding pulverized dry ice, with caution, until the aqueous layer is no longer basic (pH paper). Any weak acids are extracted with three 5-mL portions of diethyl ether. The combined ether layers (Organic layer 5) are concentrated to yield any weak acids (Acidic residue 1).

The remaining aqueous layer (Aqueous layer 4) is acidified with 5% hydrochloric acid solution. The stronger organic acids (Acidic residue 2) are isolated by filtration or by extraction with three 5-mL portions of ether, followed by evaporation of the ether.

The ether layer (Organic layer 4) is extracted with four 5-mL portions of 40% aqueous sodium bisulfite solution. The ether layer (Organic layer 6) is placed aside. Seven milliliters of 5% hydrochloric acid solution is added to the combined bisulfite aqueous layers (Aqueous layer 5) and then the solution is extracted with three 5-mL portions of ether. The ether is removed by distillation to leave behind a residue (Neutral residue 2).

Ether is removed from the ether layer (Organic layer 6) to leave behind a residue (Neutral residue 1).

Frequently, the extent of the procedure given above in Procedure A is not needed. A more abbreviated version is described in Procedure B.

Procedure B: Water Insoluble Mixtures (Figure 10.15). Five grams of the mixture is mixed with 15 mL of diethyl ether and then the resulting solution is extracted with three 5-mL portions of 5% hydrochloric acid solution.

The combined acidic layers (Aqueous layer 1) are made alkaline with 5% sodium hydroxide solution, and the resulting mixture is extracted with three 5-mL portions of diethyl ether. The ether layer is dried with sodium sulfate and the ether distilled off. The residue is composed of bases (Basic residue 1).

The ether layer (Organic layer 1) is extracted with three 5-mL portions of 5% sodium hydroxide solution. The remaining ether layer (Organic layer 2) is set aside. The basic aqueous layers are combined to form Aqueous layer 2. This fraction is acidified with 5% hydrochloric acid solution, and then the acid is isolated either by filtration or by extraction with ether with three 5-mL portions, followed by evaporation of the ether (Acidic residue 1).

The ether layer (Organic layer 2) is extracted with four 5-mL portions of 40% sodium bisulfite solution. Seven milliliters of 5% hydrochloric acid solution is added to the combined bisulfite aqueous layers (Aqueous layer 3) and then the solution is extracted with three 5-mL portions of ether. The ether is removed by distillation to leave behind a residue (Neutral residue 2).

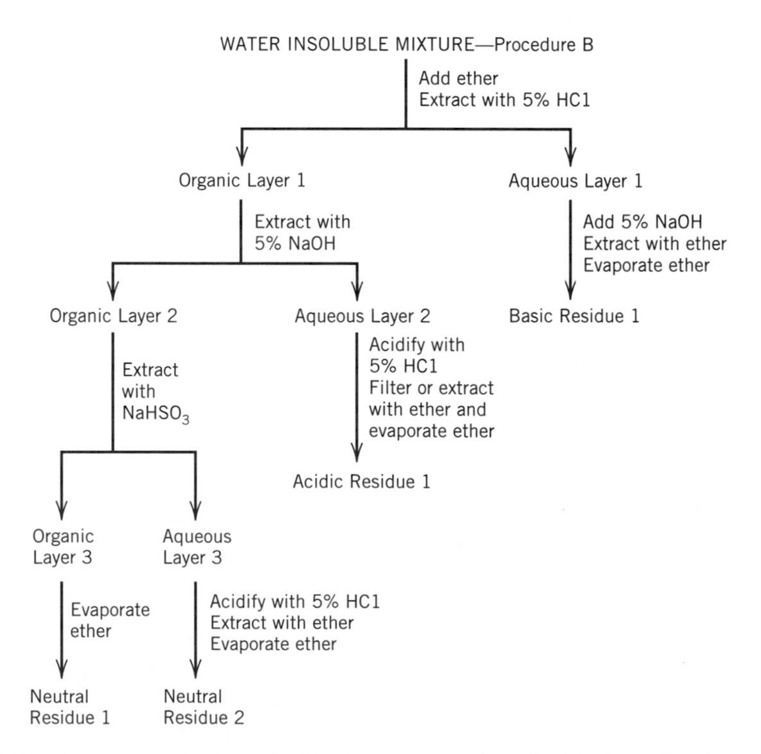

Figure 10.15 Separation of a water insoluble mixture as described in Procedure B.

Ether is removed from the ether layer (Organic layer 3) to leave behind a residue (Neutral residue 1).

Cleaning Up. For Procedures A and B, place the ether in the organic solvents container. Neutralize any aqueous filtrate before placing it into the aqueous solution container.

Questions

1. Show the equations for the formation of each salt at each step using Procedure B starting from a mixture of chlorobenzene, *N,N*-dimethylaniline, 2-naphthol, and 2-methoxybenzaldehyde.
2. A mixture of benzaldehyde, 4-butylaniline, 4-ethylphenol, and 4-propylbenzoic acid is subjected to the extraction procedure outlined in Procedure A (Figure 10.14). Where is each compound isolated and what are the reactions?
3. Although glucose is water soluble, its place of isolation in Figure 10.14 is predictable. Where would it be?

10.3.2 Extraction of Water Soluble Mixtures

If all components of the mixture are water soluble, steam distillation is the best method for the separation of the components. However, it may prove to be unsatisfactory if the mixture is not chosen carefully. Too often the components of the mixture, when improperly chosen, undergo reaction with each other or with boiling aqueous acid or alkali during steam distillation. Thus extraction, which does not involve heating of the mixture, would be preferable.

Procedure C: Water Soluble Mixtures (Figure 10.16). Five milliliters of the mixture is placed in a 125-mL round-bottom flask arranged for steam distillation (Figure

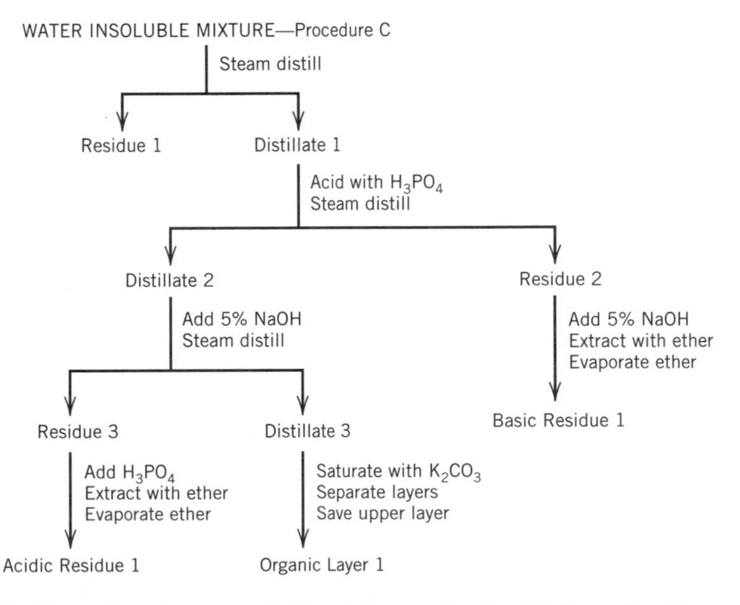

Figure 10.16 Separation of a water soluble mixture as described in Procedure C.

10.12). Five to six milliliters of the distilled solution (Distillate 1) is collected from the steam distillation. The solution (Residue 1) that remains in the distilling flask is placed in an evaporating dish and the water is evaporated by means of a steam bath. The last traces of water may be removed using a rotary evaporator (Figure 10.11).

The distilled solution (Distillate 1) is acidified with phosphoric acid and steam-distilled. About 5 mL of distilled compound (Distillate 2) is collected. The residue (Residue 2) left in the distilling flask is madc alkaline with 5% sodium hydroxide solution. The resulting mixture is extracted with three 5-mL portions of diethyl ether. Ether is removed from these extractions to yield the base (Basic residue 1).

The distilled solution (Distillate 2) is made just slightly alkaline with 5% sodium hydroxide solution and the mixture is steam-distilled. Five milliliters of the distilled solution (Distillate 3) is collected.

The solution (Residue 3) that remains in the flask is made acidic with phosphoric acid and then extracted with three 5-mL portions of diethyl ether. The ether layers are combined and concentrated to yield the acid (Acidic residue 1).

The distilled solution (Distillate 3) is saturated with potassium carbonate. The layers are separated and the upper layer is saved (Organic layer 1); and the classification tests are applied to this layer.

Cleaning Up. For Procedure C, place the ether in the organic solvents container. Neutralize any aqueous filtrate before placing it into the aqueous solution container.

Question

4. Show the equations for the formation of each salt at each step using Procedure C starting from a mixture of lactic acid, piperidine, acetic acid, and 2-propanol.

Hydrolysis of esters would occur using Procedure C. If the presence of an ester is indicated by odor or by change of color in the preliminary tests, then the procedure may be modified to use a milder base as described in Procedure D.

Procedure D: Water Soluble Mixture Containing Esters (Figure 10.17). Five milliliters is saturated with potassium carbonate to yield an upper organic layer (Organic layer 1) and an aqueous layer (Aqueous layer 1). The organic layer (Organic layer 1) is neutralized with 5% sulfuric acid solution using methyl orange as an indicator, and the resulting mixture is steam-distilled. The solution remaining in the flask (Residue 1) contains the amine sulfates.

The distillate (Distillate 1) is saturated with potassium carbonate. The upper layer (Organic layer 2), which contains the neutral compounds, is separated from the aqueous layer (Aqueous layer 2).

The original aqueous layer (Aqueous layer 1) is acidified with phosphoric acid and filtered. The precipitate (Solid 1) contains acidic compounds. The filtrate (Filtrate 1) is steam-distilled. The distilled solution (Distillate 2) contains acidic compounds.

The residue (Residue 2) is evaporated to dryness and extracted with hot ethanol to yield an alcohol solution of acids (Solution 1).

If the mixture contains no acidic constituents, the separation may start at the same point as the treatment of the first organic layer (Organic layer 1). In such a case the first residue from steam distillation (Residue 1) contains acidic compounds in addition to the amine sulfates. It may be separated by the first method given.

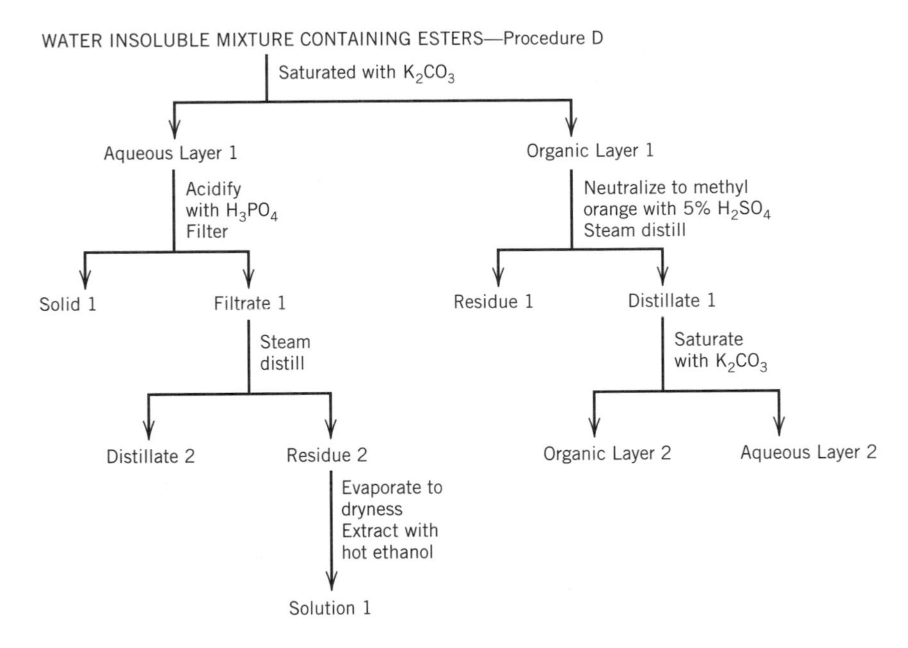

Figure 10.17 Separation of a water soluble mixture containing esters as described in Procedure D.

Cleaning Up. The aqueous filtrate is neutralized and placed in the aqueous solution container.

This modified procedure may also be used with mixtures containing acids and alcohols; it provides no opportunity for esterification to take place.

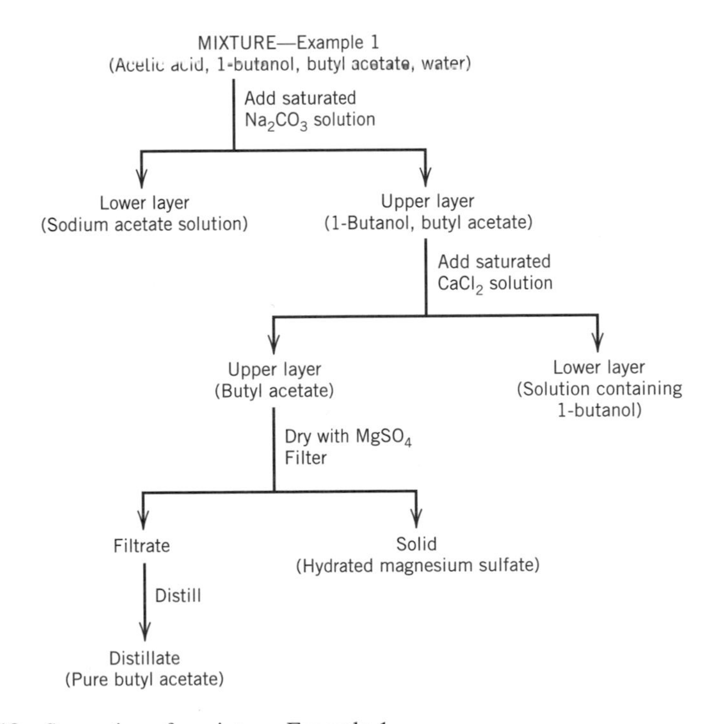

Figure 10.18 Separation of a mixture, Example 1.

The next three examples (Figures 10.18, 10.19, and 10.20) illustrate the separation of various types of mixtures.

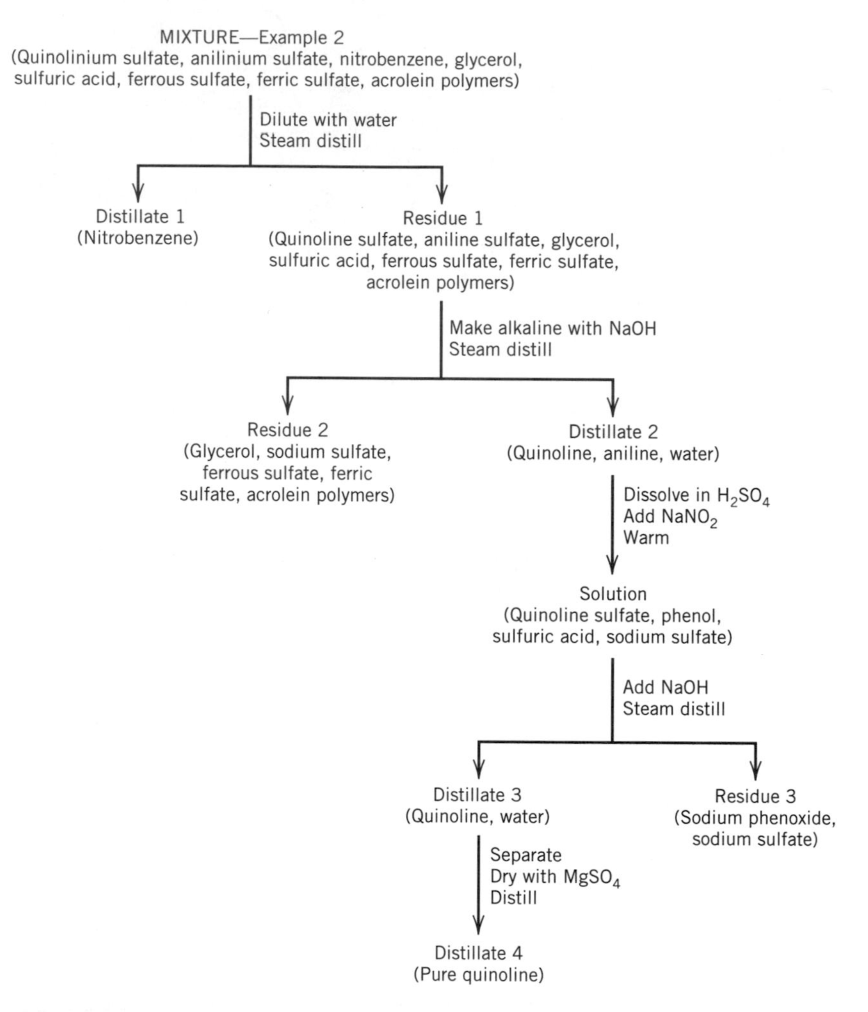

Figure 10.19 Separation of a mixture, Example 2.

Questions

5. Diagram a satisfactory procedure for the separation of the following mixtures and the identification of each component.

a. Water, 2-propanol, anilinium sulfate, 2-methylpropanoic acid.

b. Chloroform, aniline, *N,N*-diethylaniline, benzoic acid, naphthalene.

c. Toluene, diphenylamine, quinoline, nitrobenzene.

d. 2-Hydroxytoluene, 2-hydroxybenzoic acid, *N*-methylaniline, 4-aminotoluene, styrene.

e. Carbon tetrachloride, 2-hydroxybenzaldehyde, benzaldehyde, triphenylmethanol, benzyl alcohol.

f. Diethyl ether, 3-pentanone, diethylamine, acetic acid.

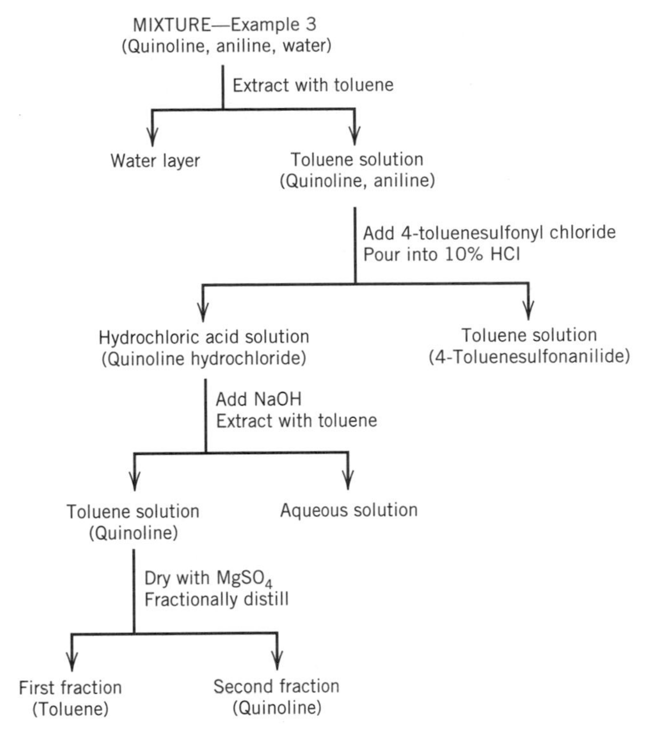

Figure 10.20 Separation of a mixture, Example 3.

g. Cyclohexane, 4-methylaniline, toluene, 3-hydroxytoluene, 2-chloro-4-aminobenzoic acid.

6. An instructor mixed the following compounds in the order named and gave out the mixture as an unknown: 2-methyl-2-propanol (2.47 g), benzyl alcohol (3.6 g), benzaldehyde (3.53 g), acetyl chloride (5.2 g), acetophenone (4.0 g), *N,N*-dimethylaniline (28.2 g). What might the student expect to find? Diagram a possible separation of the actual components after the mixture has stood for a week (assume complete reactions).

10.4 CHROMATOGRAPHY

Chromatography is the separation of the components of a mixture by the selective distribution of the components between a mobile phase and a stationary phase. The linguistic origin of the word chromatography is based on color (from the Greek word *chroma* meaning color and *graphy* meaning written). Early chromatography was carried out on paper using colored derivatives of naturally occurring compounds.

The mobile phase is a liquid or a gas and carries the compounds along a column. The stationary phase may be composed of various types of materials, for example, silica gel in column chromatography. The ability to separate various components of a mixture of organic compounds is based on selective and preferential adsorption of these components in the mobile phase by the stationary phase.

An illustration of the separation of a mixture in column chromatography is seen in Figure 10.21. The most polar compound, signified by dark circles, is more strongly adsorbed than the least polar compound, signified by light circles. Thus the least polar compound is eluted first from the column.

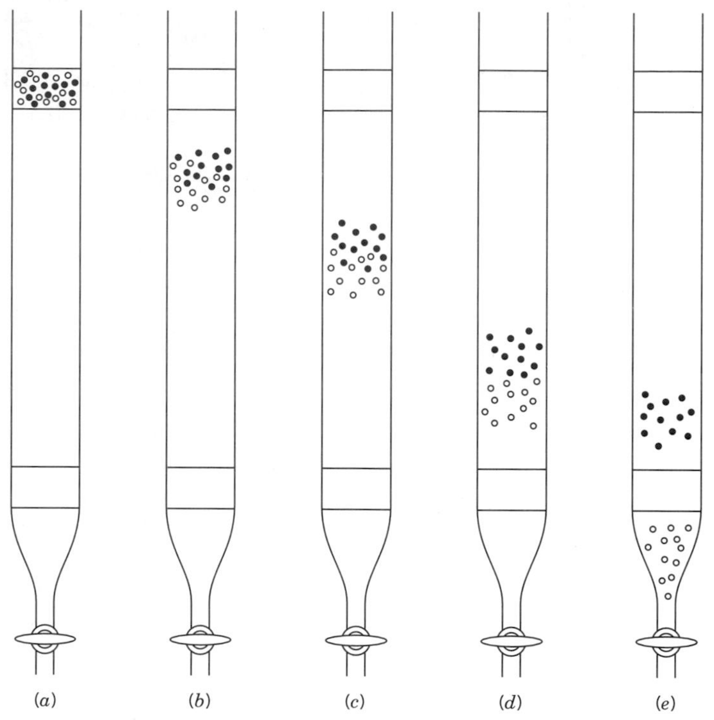

Figure 10.21 Illustration of separating a mixture in chromatography: *(a)* initial placement of the mixture; *(b)* components of mixture begin to elute down; *(c)*, *(d)* components are separating; *(e)* least polar component (light circles) is eluting first.

Organic chemists are interested in two major classes of chromatography: gas chromatography (GC) and liquid chromatography (LC). Gas chromatography is useful for relatively volatile and thermally stable organic compounds. This method involves a gaseous mobile phase, which is usually helium, or less frequently, nitrogen. The stationary phase is either a liquid adsorbed on a solid support, an organic compound bonded to a solid support, a solid, or a nonvolatile liquid.

Liquid chromatography (LC) uses a liquid mobile phase, which is usually a common organic solvent. The stationary phase may consist of a liquid adsorbed onto a solid support, an organic species spread over a solid support, a solid, or a resin. Examples of liquid chromatography are high-performance liquid chromatography (HPLC) and thin-layer chromatography (TLC).

In choosing to use either or both GC and LC, the following should be considered before making a decision.

Gas chromatography

1. The sample should be volatile and reasonably stable to heat. Specifically, the compound must be stable enough to survive the conditions necessary to convert it to the gas phase.
2. Simple gas chromatographs are inexpensive, easy to operate, and give results rapidly.

Liquid chromatography

1. With the simple gravity flow columns, the separation of the components is time consuming. Rapid analyses are carried out with HPLC and flash chromatography.
2. A small percentage of organic compounds may react with the stationary phase of some columns. Proper choice of conditions, in order to prevent undesirable side reactions, allows virtually any organic compound to be analyzed by LC.
3. Flash and other types of column chromatography are fairly cheap; HPLC has a much higher initial cost because of the high-quality pumps and column packings that are necessary.

10.4.1 Column Chromatography

Column chromatography is directly applicable to preparative-scale separations and purifications because, in principle, we can simply choose the size of the column and its contents to fit the amount of the sample to be fractionated. This approach usually may take a time commitment of many hours with the classical gravity flow columns. Flash chromatography is a type of column chromatography in which air pressure is applied to the top of the column to push the solvent through the column at a much faster rate. With this method, mixtures can be separated in a very short period of time. *It is imperative that a knowledge of thin-layer chromatography characteristics of a sample be known before column chromatography is employed.*

Types of Columns. A variety of chromatography columns (Figure 10.22) are available commercially. A small piece of Tygon tubing with a pinch clamp is attached to the bottom of the plain chromatographic column (Figure 10.22*a*) to regulate flow rate. Some columns come with a filter disk (fritted glass) above the stopcock (Figure 10.22*c*). Teflon

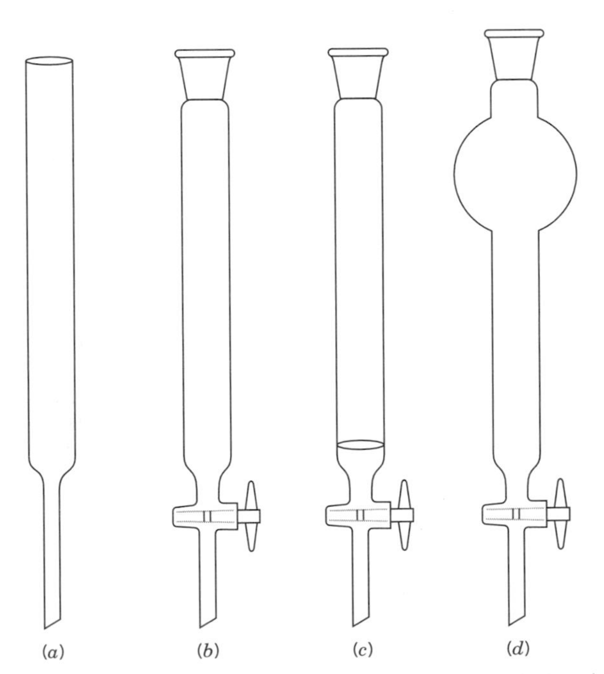

Figure 10.22 Chromatography columns: *(a)* plain; *(b)* with a ground glass joint at top; *(c)* with a fritted disk; *(d)* with a solvent reservoir.

stopcocks are recommended over glass stopcocks to avoid contamination by the grease from the glass stopcock. Burets and Pasteur pipets are used as columns when chromatographing microscale amounts of material.

A common alternative to the reservoir already built into the column is a separatory funnel. The separatory funnel would only be used with gravity columns. The separatory funnel, filled with solvent and closed at the stopcock and stoppered at the top, is hung such that the lower tip is under the surface of the solvent in the column. If the stopcock of the *stoppered* separatory funnel is opened, solvent will drip out of the funnel into the column whenever the solvent in the column drops below the tip of the funnel. This will supply a constant solvent head on the column until the contents of the separatory funnel are depleted.

(a) Slurry Packed Columns. The first type of column chromatography to be described is the slurry packed column, in which solvent is used to pack the column with adsorbent.

Column construction. Figure 10.23 illustrates a completed packed column. *LC (liquid column chromatography) columns are always packed the day they are to be used.* Push a small glass wool plug down to the bottom of the column. Place the column in a ring stand in a vertical position. Place the sand on top of the glass wool plug to a depth of 2–3 cm. Tap the column gently to level the sand. TLC is used to determine the choice of elution solvent(s). The column is now filled halfway with the proper elution solvent or solvent mixture.

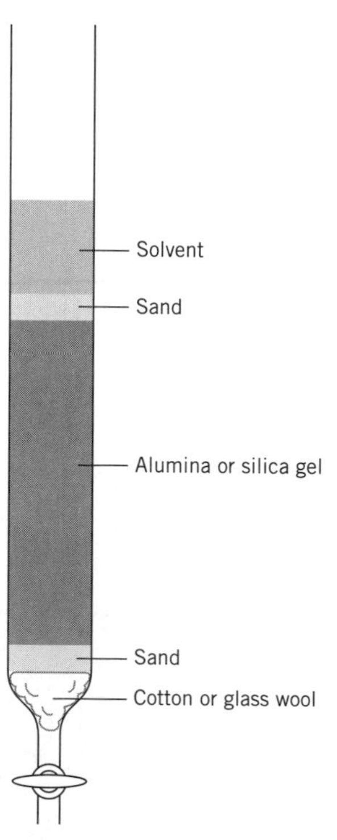

Figure 10.23 Packed chromatography column.

Thirty grams of adsorbent is weighed out per gram of mixture. The most commonly used adsorbents are silica gel and alumina (Al_2O_3). Silica gel is more commonly used and is especially favorable for sensitive organic compounds.

Prepare a slurry of the adsorbent and the solvent and *gently* pour on top of the solvent. Tap the column gently and drain the solvent slowly as the slurry is added so that the adsorbent is evenly packed. The solvent may be reused. *It is crucial that the adsorbent is never allowed to dry out; drying creates air bubbles in the adsorbent and results in poor separation due to solvent channeling.* Once all of the adsorbent has been added to the column and it has been leveled by gentle tapping, then slowly place 2–3 cm of sand on the top of the sand. Level the sand. Drain the solvent until it is halfway down the top sand layer.

Use of the packed chromatography column. Weigh the dried mixture to be analyzed. Dissolve this mixture in a minimum amount of solvent; use heat only if necessary. Using a pipet, transfer the solution onto the surface of the sand as gently as possible.

In those cases where it is very difficult to dissolve the sample in the solvent to be used for the column, the sample can be dissolved in a minimum amount of polar but volatile solvent and dispersed on a quantity of the adsorbent. The mass of adsorbent upon which the sample has been dispersed should be approximately five times that of the sample. The volatile solvent is then removed by placing this mixture in a rotary evaporator at a low temperature. When dried, the dispersion is evenly distributed onto the top of the partially packed column; this packed column is then covered with 2–3 cm of sand. This method is superior to pipetting a solution of the mixture in a very polar solvent directly onto the top layer of sand.

Carefully fill the rest of the column with the solvent, including the reservoir. Liquid should be allowed to drip from the bottom of the column at a rate of about one drop per second. Allow solvent to pass through the column until all pertinent sample bands are eluted.

During the course of the chromatographic elution, the solvent composition can be varied through a range of increasing polarities. This is recommended only if several components of a mixture need to be separated and isolated. For instance, the following solvent combinations of increasing polarity may be used on one particular column.

100% hexane, 0% ethyl acetate
80% hexane, 20% ethyl acetate
60% hexane, 40% ethyl acetate
40% hexane, 60% ethyl acetate
20% hexane, 80% ethyl acetate
0% hexane, 100% ethyl acetate

In the cases in which the solvent polarity is changing, the column should be treated with 10–20 mL of solvent per gram of adsorbent before changing polarity; this normally amounts to two to four fractions. Changing solvent polarity too abruptly may result in poor resolution between components of the mixture.

Collect the eluted liquid, in regular volume increments, into test tubes or Erlenmeyer flasks. Elution of colored sample may be monitored visually. Colorless samples must be monitored by TLC or GC, usually with fluorescent-sensitive sheets and looking at the plates under an ultraviolet light. Like fractions are combined. Carefully evaporate off the solvent by using a rotary evaporator with as little heat as possible.

The total mass of all compounds eluted should be monitored and compared to the mass of the mixture originally placed on the column. A good mass balance will not be obtained in those cases where substantial amounts of intractable tars are held behind near the top of the column.

The purity of the fractions is determined by TLC or GC. The identity of the components can be determined by NMR, IR, or mass spectrometry.

Band resolution problems such as streaking and unevenness can occur with column chromatography. Possible causes include impure solvents, channeled columns resulting from uneven adsorbent packing, contaminated adsorbent, or decomposition of the sample on the column. If the column has been packed too tightly, the elution flow may not reach the desired flow of one drop per second.

(b) Dry Column Chromatography. Dry column chromatography[1–4] provides several improvements over traditional column chromatography. These include the high degree of component resolution and that it is faster than traditional column chromatography. Another important characteristic is the near-quantitative applicability of TLC results to the dry column analysis.

The tremendous potential of dry column chromatography is shown by the separation of isomeric compounds A and B on silica gel. This is quite impressive when one realizes that compounds A and B have only very slight structural differences.

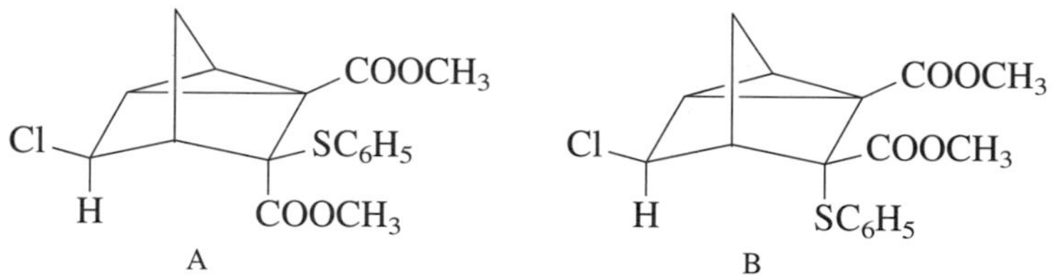

In dry column chromatography, the column is prepared dry without the use of solvent and developed just once with a single volume of solvent, in contrast to the slurry method and multiple volumes of solvent in traditional column chromatography.

Adsorbent. Careful control of the moisture content of the adsorbent is crucial to the dry column as well as other types of chromatography. Commercial and well-defined grades of adsorbent must be deactivated to match the TLC conditions. In the dry column, silica gel should contain 15% water and alumina should contain 3–6% water.

Initially the percent of water in the adsorbent is checked. Activity of adsorbents has been measured by extensively detailed procedures.[5] This allows deduction of the activity of the adsorbent using Tables 10.2 and 10.3. Deactivated adsorbent is then prepared by adding the appropriate quantity of water to the adsorbent and rotating this mixture in a rotary evaporator for about 3 hr without heat. The percent of water present is then checked again.

[1] B. Loev and M. M. Goodman, *Chem. Ind.,* 2026 (1967).

[2] B. Loev and M. M. Goodman, *Progress in Separation and Purification,* Vol. 3, edited by E. S. Perry and C. J. Van Oss (Wiley, New York, 1970), pp. 73–95.

[3] J. M. Bohen, M. M. Joullie, F. A. Kaplan, and B. Loev, *J. Chem. Educ., 50,* 367 (1973).

[4] B. Loev, P. E. Bender, and R. Smith, *Synthesis,* 362 (1973).

[5] H. Brockman and H. Schodder, *Ber., 74b,* 73 (1941)

Table 10.2 Activity of Alumina with 4-Aminoazobenzene

R_f of dye	Brockmann activity grade	Percent of water
0.00	I	0
0.12	II	3
0.24	III	6
0.46	IV	8
0.54	V	10

Table 10.3 Activity of Silica Gel with 4-*N,N*-Dimethylaminoazobenzene or 1,4-Di-*p*-toluidinoanthraquinone

R_f of dye	Brockmann activity grade	Percent of water
0.15	I	0
0.22	II	3
0.33	—	6
0.44	—	9
0.55	—	12
0.65	III	15

A fluorescent column adsorbent is extremely useful for monitoring the development of bands of colorless compounds. Band progress can be monitored by observing the column under a hand-held UV lamp and noting those bands on the column which have the fluorescence *blocked out.*

Column preparation for a glass column. A glass column cannot be used with a fluorescent column adsorbent since glass blocks the UV light. The size of the column chosen depends directly on the preliminary TLC results; the more difficult the separation, the larger the column. Compounds which are reasonably mobile with an R_f separation of 0.3 are usually involved in typical separations. The weight of the mixture and the column width are factors in choosing the correct height for the adsorbent (Figure 10.24). For instance, a 6.0-g sample with an average separation would need a 2-in.-thick column with a height of 12 in. of alumina. More efficiency is gained by using samples of 50–75% of the column capacity; for example, 3.0–4.5 g of the average mixture described above would be more efficiently separated than would 6.0 g.

These amounts correspond to a requirement of roughly 70 g of adsorbent per gram of mixture for average separations and of about 300 g of adsorbent per gram of mixture for difficult separations. The decreased capacity of grams of mixture per column height for silica gel compared to alumina is due largely to the fact that silica gel is about one-half as dense as alumina.

A standard glass column may be used (Figure 10.22). If a fritted disk is not present in the column, then place a small piece of glass wool in the bottom of the column.

The stopcock in the column should be left open during packing in order to minimize the trapping of air in the adsorbent. *Slowly* pour dry adsorbent into the column while gently tapping the column, preferably with something less dense than glass such as a cork ring.

Column preparation for a nylon column. For this separation, the nylon tubing must be transparent to UV light.

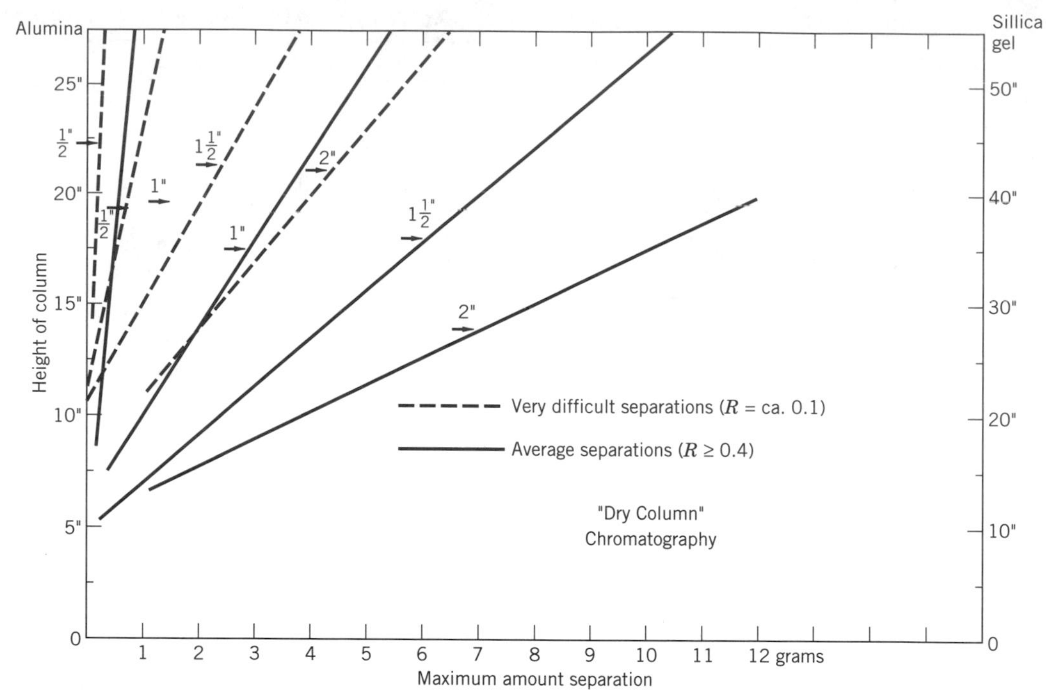

Figure 10.24 Graph for determining diameter and height of a dry column necessary to separate specified weights of mixtures. [The experimental procedure on which this graph is based is described in B. Loev and M. M. Goodman, *Chem. Ind.,* 2026 (1967); reproduced with permission.]

Cut a piece of the nylon tube which is longer than the length needed for the column (Figure 10.24). One end is sealed with heat using a hot flat iron or a hand sealer, or the end is folded a few times and stapled shut. A small piece of glass wool is placed at the bottom of the tube. Two or three small holes are made at the bottom to prevent air pocket formation during the packing of the column. About one-third of the adsorbent is added rapidly and compacted by tapping the bottom end of the tube two or three times on the bench top from a height of about 6 in. This packing procedure is done two more times until all of the adsorbent has been added. A properly packed column is quite sturdy and can be supported by a single clamp.

Use of the dry column. Mixtures are placed on the dry column in a minimum amount of the eluting solvent or on the adsorbent as described above for traditional column chromatography.

The solvent used for the column should ideally be one solvent and not a solvent mixture. The best solvent for the separation is determined by TLC. Approximate solvent amounts are shown in Table 10.4. Solvent should be added such that a constant head of

Table 10.4 Volume of Solvent for Development of Various Sizes of Dry Chromatographic Columns

Column size (in.)	Solvent volume (mL)
20×0.5	20
20×1.0	90
20×1.5	300
20×2.0	500

3–5 cm is maintained on top of the adsorbent; this can be done by placing a stoppered separatory funnel containing the solvent just above the top layer of sand. The quantity of the solvent required is enough to wet the entire column but not to drip out the bottom. *As soon as the solvent has reached the bottom, solvent addition should be halted; development is complete.*

With colored compounds the division between individual compounds is obvious. If the compounds are colorless and a fluorescent mixture has been used for an adsorbent, these divisions should be marked under UV light. The nylon column and its contents are then sliced cleanly and the separated units of adsorbent are extracted with methanol or diethyl ether to remove the desired components.

If a glass column is used, the column is inverted and the end of the column is attached to an air line. With a small amount of air pressure passing through the open stopcock, the contents of the column are carefully laid out in a pan in order to avoid mixing the separated components. The adsorbent is sliced into separate pieces and these pieces are extracted with methanol or diethyl ether.

Using the dry column technique, columns as long as 6 ft have been packed and mixtures of mass of up to 50 g have been separated.

(c) Flash Chromatography. Traditional column chromatography (LC) typically takes many hours. Flash chromatography allows separations of samples weighing 0.01–10.0 g in less than 15 min elution time[6] and produces quantitative applicability of TLC results. In flash chromatography, the column is pressurized with a flow control adaptor attached to the top of the column to pressure the column (Figure 10.25). The flow control adaptor is attached to an air line at a medium pressure. The stopcock on this adaptor is adjusted so that the air flow is not too strong.

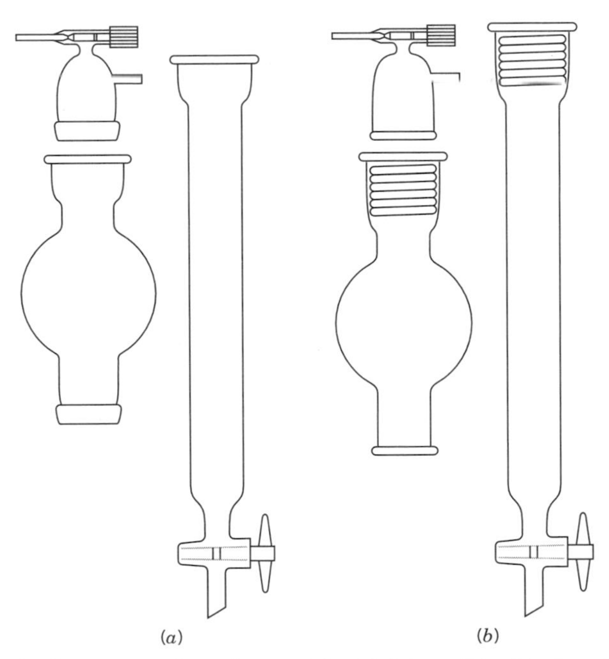

Figure 10.25 Flash chromatography columns: *(a)* with joints, *(b)* with thread connectors. Both types are available with or without the fritted disk.

[6]W. Still, M. Kahn, and A. Mitra, *J. Org. Chem.*, *43* (14), 2923 (1978).

Solvent system. A low-viscosity solvent system must be used for flash chromatography. The chosen solvent system should produce an R_f value centered around 0.35 on the TLC for the components to be effectively separated.

Column construction. After selection of a column of the correct size (Table 10.5), it is packed using either the slurry method (pp. 464–465) or the dry column method (pp. 467–468). The amount of silica gel used should be approximately 50 times the weight of the mixture to be separated.

Table 10.5 Column Diameter Needed for Flash Chromatography[a]

Column diameter (mm)	Vol of eluant[b] (mL)	Sample: typical loading (mg) $\Delta R_f \geq 0.2$	Sample: typical loading (mg) $\Delta R_f \geq 0.1$	Typical fraction size (mL)
10	100	100	40	5
20	200	400	160	10
30	400	900	360	20
40	600	1600	600	30
50	1000	2500	1000	50

[a]The experimental procedure on which this table is based is described in W. C. Still, M. Kahn, and A. Mitra, *J. Org. Chem.* 2923 (1978); reproduced with permission.

[b]Typical volume of eluant required for packing and elution.

If the dry column method is used, the column is filled with solvent above the dry adsorbent and the flow controller is attached. The inert gas line is attached to the inlet tube below the stopcock. The gas is barely turned on and the stopcock is slowly closed until a flow rate of 2 in./min is achieved. The solvent is replenished as needed until the adsorbent is saturated with the solvent and no air bubbles are present.

Use of the flash chromatography column. Once the column is prepared, the sample added (p. 465), and the column filled with solvent, the flow controller is attached to the top of the column. A flow rate of 2 in./min is maintained until all components have been eluted, replenishing the solvent as needed until the components are eluted. The total amount of solvent used for the separation is roughly 7–10 times, in mL, of the silica gel, in grams. The column is not allowed to run dry at any time.

The size of the fractions is one-third to one-sixth, in mL, of the amount of the silica gel, in grams. If the column is packed properly, one fraction should be collected every 10–20 sec.

A simple modification of flash chromatography uses a balloon as a reservoir of pressurized gas attached with glass tubing to a one-hole stopper which is placed on the top of the column.[7]

(d) Microscale Column Chromatography.[8] Microscale column chromatography uses either a Pasteur pipet or a 50-mL titration buret (Figure 10.26) for chromatographing mixtures.

[7]W. J. Thompson and B. A. Hansen, *J. Chem. Educ., 61* (7), 645 (1984).

[8]D. W. Mayo, R. M. Pike, and P. K. Trumper, *Microscale Organic Laboratory with Multistep and Multiscale Syntheses,* 3rd ed. (Wiley, New York, 1994), pp. 99–100.

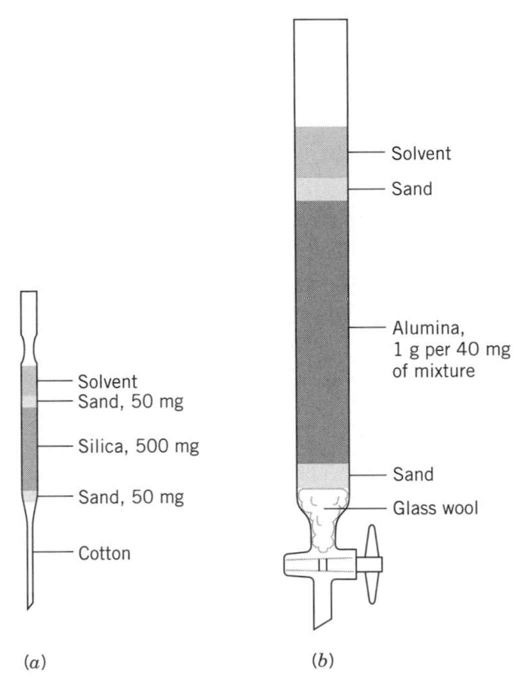

Figure 10.26 Microscale chromatography columns: *(a)* from a Pasteur pipet; *(b)* from a buret.

Pasteur pipet chromatography column. The Pasteur pipet is used to separate mixtures of 10–100 mg and is packed dry. The cotton is placed in the Pasteur buret, followed by 50 mg of sand. Only 500 mg of dry adsorbent is needed in the Pasteur pipet. The adsorbent is topped with another 50 mg of sand. The column is moistened with the solvent prior to use. No stopcock or clamp is attached to the bottom of the Pasteur pipet.

Buret chromatography column. A 50-mL titration buret, shortened to 10 cm above the stopcock, is used to separate mixtures of 50–200 mg and is packed as a slurry column. One centimeter of sand is used before and after the adsorbent. One gram of adsorbent is used for each 40 mg of mixture to be chromatographed.

Using the microscale chromatography columns. The microscale columns are used the same day they are prepared and are used in the same manner as the larger scale columns. Once moistened, they should not be allowed to run dry.

10.4.2 High-Performance Liquid Chromatography (HPLC)

High-performance liquid chromatography (HPLC), sometimes called high-pressure liquid chromatography, attacks problems associated with traditional column chromatography with great success.[9]

In traditional LC chromatography the stationary phase consists of a particle size which is large relative to that used in HPLC. In order to speed the diffusion of sample into the stationary phase, *very fine particles* of stationary phase are used in HPLC. These

[9]Consult Chapter 11 for analytical chemistry books which explain this method in more detail.

particles are on the order of a few micrometers in size. The small size of such fine particles produces a new problem, such as potentially slow flow rates; pressures well above atmospheric pressure are necessary to push the mobile phase through tightly packed columns of fine particles. Pumps delivering thousands of pounds per square inch push mobile phases through columns of very fine particles. Detectors that differentiate samples by refractive index, by UV absorption, and by fluorescence are commonly used.

High-performance liquid chromatography offers thc advantage of high speed, reusable columns, automatic and continuous solvent addition, reproducible programmed gradients of solvents, and automatic and continuous monitoring of the eluted samples. Less than 1 mg of sample is commonly analyzed. Preparative-scale instruments separate between several milligrams and a few grams of sample. The main disadvantage of preparative-scale separations is that large amounts of solvents are used.

Solvent system. In the schematic (Figure 10.27), the apparatus is equipped with one or more glass or steel solvent reservoirs. The solvent may be heated, stirred, or the solvent reservoirs may contain attachments inlets for inert gases for solvent degassing. The presence of dissolved gases results in poorer resolution of the peaks.

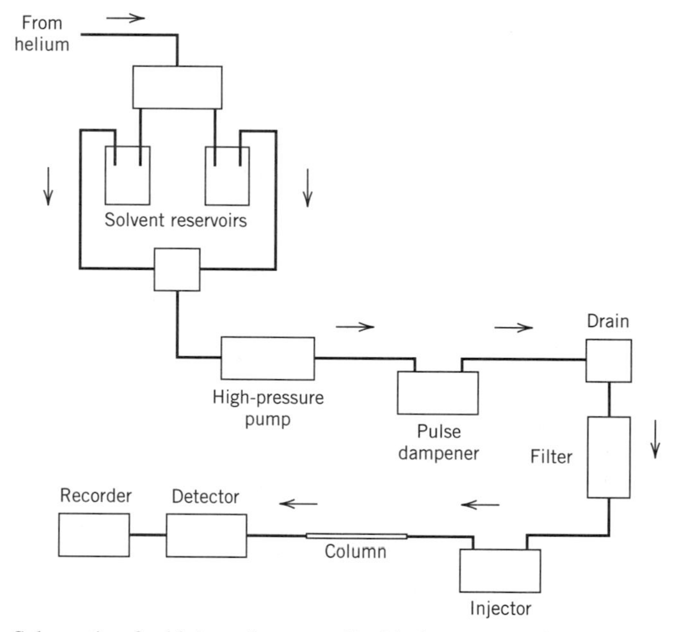

Figure 10.27 Schematic of a high-performance liquid chromatography (HPLC) instrument.

Pumps. The pump for the HPLC instrument must be pulseless and able to generate reproducible flow rates. The pump must be able to drive the mobile phase through long, narrow columns which are packed with very fine particles, and to generate pressures up to 6000 psi and a solvent flow of 0.1–10 mL/min.

Sample introduction. The sample is introduced into the system via a syringe. A sampling loop is the most common method of sample introduction. The sample is injected into a small loop. As a lever is moved, the loop is closed off from the outside and the solvent is allowed to pass through the loop, thus introducing the sample into the system.

Columns. Columns for HPLC are usually constructed from stainless steel tubing. Fittings and plugs must be inert and should not detract from the homogeneity of the flow. Analytical columns range in length from 10 to 150 cm long. The inside diameters vary from 1 to 20 mm. Columns of less than 8 mm are difficult for a novice to pack. The dimensions of a preparative LC column are typically 30 cm by 5 cm.

Silica is the most common type of packing material. The silica may be coated with a thin organic film, which is chemically bonded to the surface of the silica. By using a reverse-phase type of adsorbent, compounds may be separated on the basis of nonpolar–nonpolar interactions rather than polar–polar interactions.

Detectors. The most commonly used detector for HPLC is an ultraviolet spectrophotometer or a refractometer. Other detectors such as infrared and electrochemical have also been used. The spectrophotometers are considered to be much more versatile and can detect a wider range of compounds. By using a refractometer, changes in the solvent refractive index can be detected.

Analysis of chromatograms. The retention times and percent composition of each component are determined from the chromatogram. The retention times are calculated by dividing the distance by the chart speed (Figure 10.28). The distance is measured from the zero time mark, when the sample is injected, to the top of the peak. The zero time mark can be made by moving the pen on the recorder up and down before the sample is injected into the column. The chart speed is determined by looking at the recorder. The units of the chart speed may be in cm/min or mm/min. Care must be taken so that the units cancel out properly, leaving behind only the unit of time.

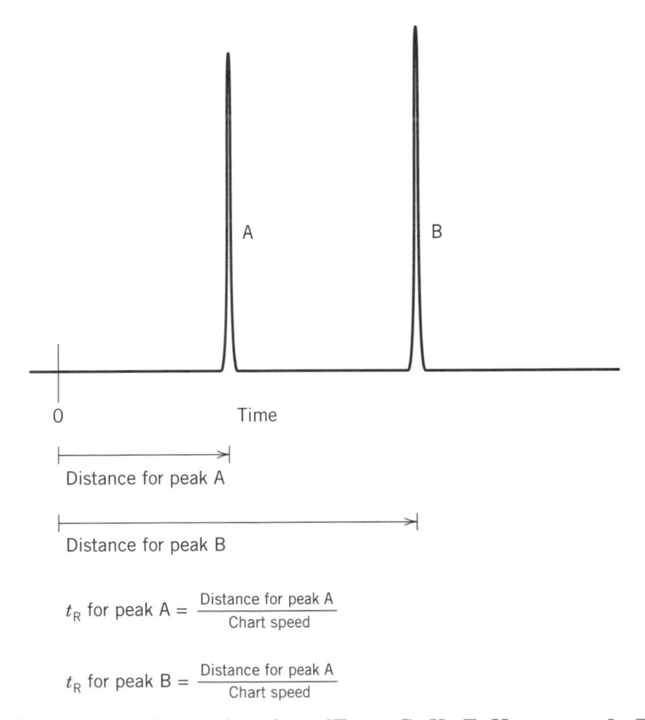

Figure 10.28 Calculation of retention time. [From C. K. F. Hermann, *J. Chem. Educ., 73,* 852–853 (1996). Copyright © 1996 by Division of Chemical Education, American Chemical Society, used with permission.]

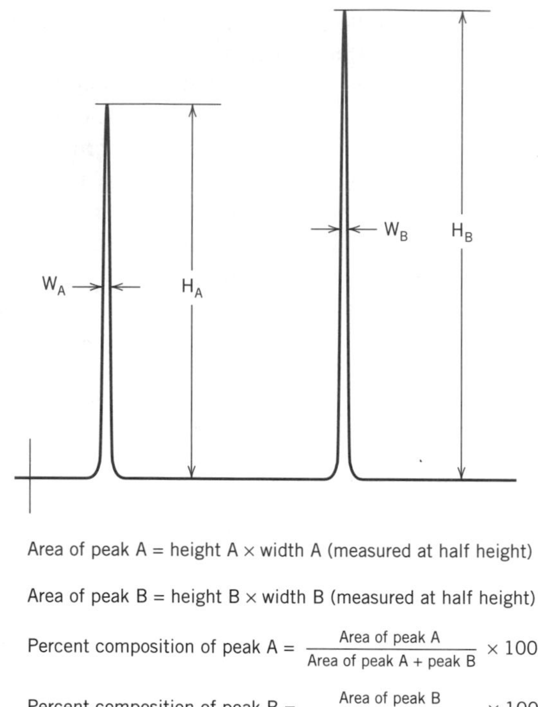

Figure 10.29 Calculation of peak area and molar percent composition. [From C. K. F. Hermann, *J. Chem. Educ., 73,* 852–853 (1996). Copyright © 1996 by Division of Chemical Education, American Chemical Society; used with permission.]

The area of each peak is roughly calculated by multiplying the height of the peak by the width of the peak, which is measured at half the height (Figure 10.29). The molar percent of a component is determined by dividing the area of the peak by the total area of all of the peaks combined and multiplying this number by 100.

10.4.3 Gas Chromatography

Gas chromatography has been introduced as an analytical and purity probe in Chapter 3 (pp. 63–70). If the sample to be analyzed can survive the conditions of the GC instrument, gas chromatography (GC) is one of the simplest, quickest, and most useful analytical methods available. A schematic (Figure 10.30) shows an overview of the gas chromatograph. The parts of the gas chromatograph have been summarized in Chapter 3 (pp. 64–65).

Sample Introduction. The amount of sample which can be introduced and analyzed for the gas chromatograph ranges from submilligram amounts to as much as 200 mg per injection. For preparative gas chromatography, 20–80 μL of sample can be separated into pure components. Too much sample will cause the column to be overloaded and the peaks will often be merged. Your instructor can help in the determination of the correct amount of sample to inject.

The plunger of the syringe must have some pressure on it at all times during injections. If during injection moderate pressure on the plunger has not been maintained, the

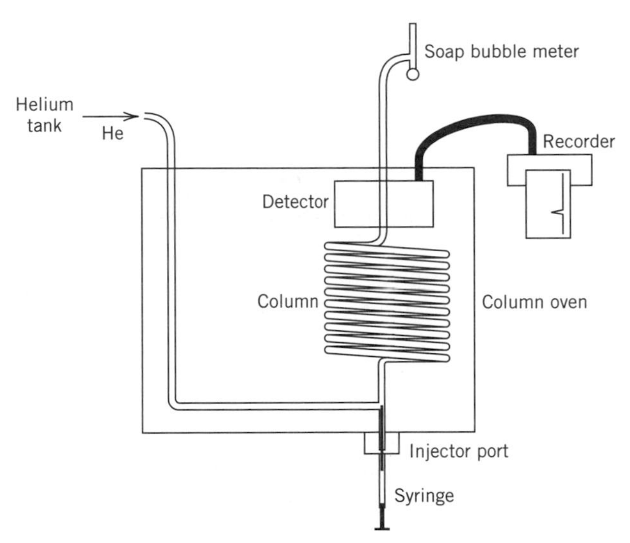

Figure 10.30 Schematic of a gas chromatograph.

plunger can be blown out of the barrel of the syringe due to the high pressure from the vaporization of the sample.

During injection, the needle of the syringe punctures a rubber or silicon septum located at the head of the inlet system. The sample is *quickly* injected into the gas chromatograph with a microsyringe; a prolonged injection of the sample results in poorer resolution of the peaks. The temperature of the injection port must be high enough so that all of the solution is instantly vaporized.

Columns. For preparative gas chromatography, packed columns are used. Packed columns usually have a diameter of one-eighth to one-fourth inch and a length of 4–12 feet and are constructed from glass, stainless steel, copper, or aluminum.

Carrier Gas. The sample is pushed through the column by a stream of an inert carrier gas, such as helium. The flow rate is commonly measured with a soap bubble flow meter. This flow meter may be as simple as a buret containing a few soap bubbles. The pressure of the carrier gas entering the gas chromatograph is usually 10–50 psi, with a resulting flow rate of 10–150 mL/min.

Column Oven. The column resides inside a thermostated oven, which is controlled by computer. The optimum column temperature required is dependent upon the boiling points of the components of the sample and the resolution needed. The column temperature is usually set 10–20°C above the highest boiling point. If the oven temperature is too low, the components may not elute off the column. With too high a column temperature, the components may elute off the column too quickly with poor resolution between the peaks.

The desire to expedite preparative work should not be allowed to result in the use of overly large samples. As one reaches the point where the sample sizes are so large that the column is overloaded, peaks observed on the recorder will begin to have more of a bell-shaped or shark-fin appearance, with possible overlap of the peaks.

Detectors. The most commonly used detector for preparative gas chromatography is the thermal conductivity detector. The thermal conductivity detector monitors changes in the thermal conductivity of the gas stream as different components pass through. This type of detector does not destroy the individual components.

Analysis of Chromatograms. The retention times of the components and the percent composition are calculated as described above for HPLC (pp. 473–474).

Collection Devices. In order to carry out a preparative-scale collection from a gas chromatograph, a proper collection device must be at hand. A number of commercial, automatic devices are available; some of these are built into the chromatograph or are available as separate units.

Most practicing organic chemists find it convenient and sufficient to use a collection device such as one of those shown in Figure 10.31. It should be pointed out that care must be exercised in the choice of a coolant; some compounds, when collected at their condensation temperatures, which are far too low, will form an aerosol which cannot be easily condensed in the collection trap. Centrifuge tubes are handy for collection when small samples require spinning down after collection.

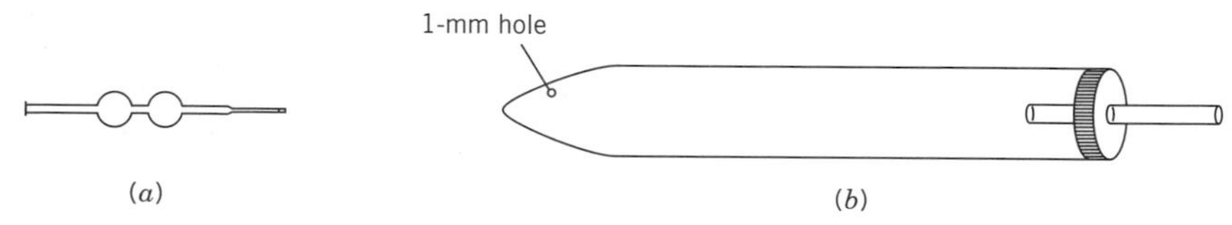

Figure 10.31 Collection devices for GC: *(a)* capillary size; *(b)* centrifuge tube. Both can be placed in an ice bath for volatile compounds.

Samples collected directly from GC are usually very pure and can be analyzed directly; for example, the sample can be washed with $CDCl_3$ solvent directly into a NMR tube. For many compounds, it is *necessary* that they be collected from the gas chromatograph and analyzed quickly so that the physical constants and other characteristics determined from these samples are accurate. The samples collected can be checked for purity by reinjecting a small sample into the GC.

Questions

7. For each of the following mixtures suggest a separation technique and explain a basis for your choice.

a. 1-octanol, 2-octanol, and 3-methyl-2-heptanol

b. aniline, benzamide, and 4-methylphenol (*p*-cresol)

c. ethyl benzoate, propyl benzoate, and butyl benzoate

8. It is common to convert carboxylic acids to esters before conducting separation analysis. Explain why this is important when using GC and LC.

CHAPTER 11

Chemical Literature

The chemical literature is important to the student of organic qualitative analysis. First, a student must be able to find the melting and boiling points of organic compounds and the melting points of their derivatives. The chemical literature also can be of use for procedures to supplement those found in this book. It greatly simplifies a laboratory situation when the melting point (and perhaps other characteristics) as well as a procedure for the preparation of a derivative of a *specific* compound are known *before* the derivative preparation is begun.

This chapter is not meant to be comprehensive. We provide additional references to the chemical literature at the end of this chapter. It is our desire here simply to provide adequate supplemental information such that the identification of unknowns can be carried out efficiently. We have somewhat arbitrarily broken the literature down into the following categories.

Handbooks (Section 11.1)
Compendia (Section 11.2)
Spectral Collections (Section 11.3)
Journals (Section 11.4)
Abstracts and Indexes (Section 11.5)
Monographs (Section 11.6)

The student in a qualitative analysis course likely is heavily focused on the first three sections: handbooks, compendia, and spectral collections. As organic qualitative analysis often serves as a lead into undergraduate and graduate research, it is very reasonable to find that the next step in the use of the literature involves journals, abstracts, and monographs, and these are described in three sections below.

11.1 HANDBOOKS

Certainly one of the best known handbooks is the *Handbook of Chemistry and Physics* (published by CRC press in Boca Raton, FL). Organic chemists routinely use the table entitled "Physical Constants for Organic Compounds" for properties of organic compounds used in reactions. This table includes molecular weights, solubilities, and physical appearance, as well as the melting and/or boiling points of most common organic compounds. The rest of the handbook also contains tables including those listing pK_a values for acids, properties of inorganic compounds, etc. This handbook is often called the "rubber" handbook (CRC is the former Chemical Rubber Co.) and it comes out in a new edition every year. Each year the data are revised. For example, it serves as a good source for the most up-to-date values of the abundance of naturally occurring isotopes of the elements. Other handbooks include Lange's *Handbook of Chemistry* (McGraw-Hill Book Company, New York) and the more limited Dean's *Handbook of Organic Compounds* (McGraw-Hill Book Company, New York, 1990). The latter includes tables describing thermodynamic data, spectral peaks, pK_a values, dipole moments, and bond lengths.

A very useful and inexpensive reference is the *Aldrich Catalog of Fine Chemicals,* published annually by Aldrich Chemical Co., Milwaukee, Wisconsin. This catalog has alphabetical entries for over 40,000 organic compounds including melting and/or boiling points, other physical properties, safety information, and references to Beilstein and other literature. There is also a molecular formula index.

One of the most useful books published is the *Merck Index* (the 12th edition was published by Merck and Co., Rahway, NJ, in 1996). The physical, chemical, and physiological properties of a large number of compounds are listed, and although this book is very useful for pharmaceutical studies it is certainly not limited to that. Older editions serve as useful brief supplements to organic handbooks, and the more recent editions include increasingly longer lists of compounds with more complex structures, a natural outcome of the modern drug industry. Moreover, the more recent editions provide more detailed and expanded information on the hazards associated with many of these compounds.

Other handbooks that contain information about compounds of a more limited scope are the *Handbook of Naturally Occurring Compounds* (3 vols., Academic Press, New York, 1972), the *Handbook of Organometallic Compounds* (W. A. Benjamin, New York, 1968), the *Handbook of Organometallic Compounds* (Van Nostrand, Princeton, NJ, 1968), and thc *Dictionary of Organometallic Compounds* (3 vols., Chapman & Hall, London, beginning 1984). More general works are Landolt-Bornstein's *Zahlenwerte und Funktionen aus Physik, Chemie, Astronomie, Geophysik und Technik* (3 vols., Springer, Berlin, beginning 1950) and Dreisbach's *Physical Properties of Chemical Compounds* (Series nos. 15, 22, 29, American Chemical Society, Washington, DC, 1955–1961).

A supplement to both this text and to the CRC "rubber" handbook is the *Handbook of Tables for Organic Compound Identification,* 3rd ed., edited by Z. Rappoport (CRC Press, Boca Raton, FL, 1967). This book provides melting points for many standard derivatives of many organic compounds and thus it is not a surprise to find that it is organized much like the derivative tables in Appendix III.

11.2 COMPENDIA

Compendia serve as a useful support for laboratory work and we shall review some of them here. One of the best known is Beilstein, an ambitious attempt to comprehensively cover the literature of organic chemistry. Every piece of information (synthetic methods and properties of organic compounds) is backed by a reference to the original literature and thus data may be checked. Moreover, information is not incorporated in Beilstein until it has been evaluated. Since more recent volumes allude to many compounds previously reported in earlier Beilstein volumes, corrections and updating are continually done. Beilstein up to (and including) 1959 was published in German. Beginning with 1960, it has been published in English.

Beilstein has a main work (*das Hauptwerk*) which covers the literature up to 1909. Thereafter time periods are covered by supplements (Erganzungswerks):

First Supplement (*erstes Erganzungswerk*)	1910–1919
Second Supplement (*zweites Erganzungswerk*)	1920–1929
Third Supplement (*drittes Erganzungswerk*)	1930–1949
Fourth Supplement (*viertes Erganzungswerk*)	1950–1959
Fifth Supplement (*funf Erganzungswerk*)	1960–1979

The book *Synthetic Organic Chemistry* by R. B. Wagner and H. D. Zook (published by Wiley in 1953) contains 39 chapters of 576 reaction types, limited to compounds with not more than two functional groups, for which yields, references, and physical constants are given. The methods described in Wagner and Zook are general; specific procedures may be obtained from the cited references.

Heilbron's Dictionary of Organic Compounds, edited by J. Buckingham (7 vol., Chapman & Hall, London, 1982), lists properties and derivatization procedures for over 150,000 compounds. References are cited to substantiate these listings. A similar *Dictionary of Steroids* (2 vols.) was published in 1991 by Chapman & Hall.

Rodd's Chemistry of Carbon Compounds, edited by S. Coffey, has been published (Elsevier Publishing, New York) beginning in 1964; this collection is not yet complete. An earlier edition of 10 volumes was published as the *Chemistry of Carbon Compounds* during 1951–1962. It is organized in the same fashion as an organic text, but it is much more extensively developed.

Reagents for Organic Synthesis, by L. F. Fieser and M. Fieser, is a series of volumes published by Wiley. Volume 1, published in 1967, is the cornerstone issue and is a 1457-page listing of reagents, catalysts, solvents, references, etc., used for standard organic procedures. Extensive amounts of practical information are included such as commercial suppliers of reagents, solvent drying procedures, and physical constants. Subsequent volumes are still being published, and these include both updates to Volume 1 entries and new and more elaborate reagents.

Wiley also publishes two well-known series that are useful in the organic laboratory: *Organic Synthesis* and *Organic Reactions. Organic Synthesis* is a collection of organic preparations with the distinct advantage of having been checked. Annual volumes began in 1921 and published cumulative volumes include Collective Volume I (annual volumes 1–9, 1932), Coll. Vol. II (10–19, 1943), Coll. Vol. III (20–29, 1953), Coll. Vol. IV (30–39, 1963), Coll. Vol. V (40–49, 1973), Coll. Vol. VI (50–59, 1988), Coll. Vol. VII (60–64, 1990), and Coll. Vol. VIII (65–69, 1993). Submitted procedures are sent to scientists at other research laboratories for independent testing of procedures and yields. All procedures are extensively supplemented by notes that describe the practical handling of solvents, reagents, and starting materials. Literature references are provided, and procedures for preparing starting materials are cited.

Organic Reactions is a continuing series that has been published by Wiley since 1942. Each chapter describes a given reaction in detail (e.g., the Clemmensen reduction) supported by literature references.

11.3 SPECTRAL COLLECTIONS

The literature provides a wealth of spectral information and it comes in two forms: books that list tables of data and bound volumes of illustrations that are direct reproductions of spectra. Both have their uses, but it is often advantageous for students to see the actual spectra. This is especially true for infrared spectra where peak shapes, as well as peak positions and intensities, are important characteristics for each compound.

When comparing spectra it is important to be aware of the details of how the spectra were obtained such as the type of spectrometer used (infrared spectra vary in the nature of the linear scale, and this greatly influences the shapes of the peaks—see Chapter 6) and the sampling techniques (solvent used, other sampling procedure, etc.). When comparing spectra for final identification, it is important to have both the unknown's spectrum and the reference spectrum for the known compound, hopefully under highly similar, if not identical, conditions. Also, it is of value to have a spectrum of any solvent or mulling compound used.

There are two well-known and long-used compilations of spectra: the Aldrich collection and the Sadtler collection. Aldrich Chemical Co. (Milwaukee, Wisconsin) publishes extensive collections of both infrared (IR) and nuclear magnetic resonance (NMR) spectra for many of the compounds sold by Aldrich Chemical Co. The IR spectra are available in both grating (in the older collections) and FT IR formats. The older ^{1}H NMR spectra were determined on instruments of 60 MHz field strength and the more recent FT NMR collections were determined at the higher field strength of 300 MHz. More recent NMR collections also provide ^{13}C NMR spectra for the same compound on the same page. The Aldrich NMR and IR collections both provide certain advantages: the spectra of a large number of compounds with similar structures in a given molecular class are on the same page (e.g., several spectra of various saturated alcohols will be found on one page). This allows the beginning student to obtain an overview of spectral information common to that structural class. It also provides a variety of related candidates for fingerprinting when completing a final structure assignment. A disadvantage of some of the Aldrich IR collections is the extensive use of Nujol mulls as a sampling technique, which blocks some of the IR bands occurring in the same regions as the C—H stretching or bending bands of Nujol oils. Also, the ^{1}H NMR spectra from Aldrich do not include the integrated areas for the signals.

The Sadtler collection also provides IR and NMR spectra of many organic compounds. The spectra are determined using samples from a variety of sources, which are listed. The NMR spectra have the added feature that the signals are assigned and the ^{1}H NMR spectra include the integrated areas for each signal.

Both the Sadtler and Aldrich and Sadtler collections are available in a computer format (e.g., on CD ROMs).

11.4 JOURNALS

Journals are where papers are published. A paper is frequently the initial formal report of laboratory research results. Sometimes these results may have already been presented in an oral paper or in a poster session at professional meetings such as those of the American Chemical Society. An extensive list of journals and their abbreviations is provided in Table 11.1.

There are a number of general formats for papers: they may be in complete paper form (thus usually including a historical background section, an experimental section, an extensive results and discussion section, and references) or in the briefer format of a preliminary report (thus often providing only a brief description of results, rather than an experimental section, and only a very brief recounting of the history behind the project). *The Journal of Organic Chemistry* includes examples of both of the foregoing formats. All of the papers described here are abstracted by *Chemical Abstracts* (Section 11.5 below).

The types of papers listed above are examples of primary literature. Many journals, for example, *Chemical Reviews,* include or totally comprise review papers. Review pa-

Table 11.1 Selected Journal Publications

Acc. Chem. Res. This journal, *Acc. Chem. Res.* (*Accounts of Chemical Research*), is a publication in which leading authors in certain areas of chemistry are invited to contribute a paper which is a review of an area in which the invited author is a leader. It usually is supported by a large number of references to primary publications.

Anal. Chem. This journal, *Anal. Chem.* (*Analytical Chemistry*), is a publication containing analytical results which very frequently contain data of use to organic chemists.

Angew. Chem. Int. Ed. (*Eng.*) This journal, *Angew. Chem. Int. Ed.* (*Eng.*) (*Angewandte Chemie International Edition, English*), although published in Germany, is normally in English. It contains review articles, papers, and brief communications in the area of organic chemistry.

Chem. Ber. This journal, *Chem. Ber.* (*Chemische Berichte*) is published in Germany, and articles appear in both German and English. At one time this journal was published under the name *Ber.* (*Berichte*).

J. Amer Chem. Soc. This journal (the *Journal of the American Chemical Society*) is published by the American Chemical Society and includes both complete papers and brief communications. It is intended for the chemistry community at large and thus the anticipation is that a paper is of interest to, for example, physical chemists, biochemists, and inorganic chemists as well as to organic chemists.

J. Chem. Soc. There are a variety of journals published under this general lead (*Journal of the Chemical Society*) for the British Chemical Society. *Perk. Trans.* (*Perkin Transactions*) *1 & 2* are places to publish papers, respectively, in organic and physical organic chemistry. *Chem. Commun.* is a British Chemical Society journal for brief communications in all areas of chemistry.

Science This journal is edited by members of the National Academy of Sciences. It includes papers of high significance that cross interdisciplinary boundaries. It also includes articles describing current events in the areas such as public policy in science.

Tetrahedron Letters This journal contains brief papers in areas of organic chemistry of timely interest. Chemists must submit papers in camera-ready form, and such papers appear as direct reproductions in the journal. The papers are brief in scope (normally four pages, double-spaced).

Zh. Org. Khim. This (*Zhurnal Organicheskoi Khimmi*) is the journal of organic chemistry published in Russia. As is the case for most major Russian journals, it is available in both Russian and English language form.

pers are descriptions of the contents of a large number of primary references and thus they normally survey and interrelate a much broader range of information than included in a single primary reference.

11.5 ABSTRACTS AND INDEXES

By far and away the best known abstracting service is *Chemical Abstracts* (*CA,* Cleveland, Ohio), which is published weekly. This publication endeavors to include all papers published within the broad domain of chemistry. Here an abstract means a brief, but succinct, synopsis of a paper published in a primary journal (Section 11.4). About 18,000 journals are covered from around the world. Moreover, patents obtained in eighteen countries, including Germany, Japan, England, the United Kingdom, as well as the United States, are abstracted. Textbooks and articles are listed but not abstracted. Fourteen of the eighty sections of *Chemical Abstracts* cover organic chemistry. The weekly issues of *Chemical Abstracts* contain more than 9500 abstracts.

A variety of indexes are available in *Chemical Abstracts*, including general subject indexes, patent indexes, key word indexes, author indexes, formula indexes, and chemical substance indexes. The general subject, patent, and author indexes have been published collectively since 1907, the formula indexes since 1920, and the chemical substance indexes since 1972. Collective indexes originally covered ten-year periods, but, because of great increases in the numbers of papers published, they more recently have been reduced to five-year periods. Beginning with the eighth collective index (1967), an *Index Guide* has been published. This guide is useful because *Chemical Abstracts* limits the number of types of entries for a given compound or other subject. This is especially pertinent when looking up organic compounds, as a given organic compound can have a variety of names and *Chemical Abstracts,* because of space constraints, must limit those entries. Thus if you know a compound only by a common or trivial name, especially one that is not often used, this index can be crucial. The 12th Collective Index (1993) is available in both print and CD ROM forms.

Chemical Abstracts began assigning registry numbers to compounds in 1965. More than ten million compounds have been assigned, and thousands more are added every week. Registry numbers are of course of great assistance in computer searches, since, for example, a given compound can often be known by more than one name.

Chemical Abstracts originated in a one-person operation of the American Chemical Society which began in 1907. In 1956 the ACS renamed the operation Chemical Abstracts Services. CAS has expanded its services to provide a computer format and to allow management of the large amount of information available in the literature. CAS provides a publication name *CASelects,* in a variety of subject areas each of which covers one of 250 subtopics of the chemical literature.

In 1983 CAS signed an agreement with Fachinformzentrum (FIZ) of Karlsruhe, Germany, to form a computer on-line service called STN International. Now more than 188 databases are available through STN International.

CAS has eight databases: CA, REGISTRY, CAOLD, CAPreviews, CIN, CHEMLIST, CASREACT, and MARPAT. CA is a file which allows access to all information published in *Chemical Abstracts* since 1967. REGISTRY is a companion file to CA. REGISTRY can be used for chemical identification and obtaining chemical substance files and it can be searched by a number of methods including structures, substructures, molecular formulas, and chemical names. CAOLD supplements the CA file by providing access to information about 700,000 papers published before 1967. CAPreviews provides a preview of entries that will be appearing in the CA file. The CINfile provides an on-line version of the entire publication. Chemicals regulated by the EPA under the Toxic Substances Control Act are described in the file CHEMLIST. CASREACT lists more than one million compounds appearing in the literature since 1985. MARPAT lists generic structures (called Markush structures) of compounds which are legally protected and have appeared in the literature since 1988.

Science Citation Index (*SCI*) is an index that lists all publications in a given year referring to a certain paper from an earlier year. This is especially useful when there is a key paper describing a procedure which will be used by a large number of chemists who publish in later years. For example, Wilkinson published his seminal paper on the use of his rhodium reagent to reduce balkiness in 1966. In subsequent years anyone who used this reagent to carry out reductions of alkenes will likely cite that paper. Thus a chemist can keep abreast of modern developments and modifications of the Wilkinson reduction approach. *SCI* is available in an on-line computer format.

11.6 MONOGRAPHS

A variety of organic qualitative analysis texts have been published. *Semimicro Qualitative Analysis,* 3rd ed., by N. D. Cheronis, J. B. Entrikin, and E. M. Hodnett, published by Wiley in 1965, describes many useful wet chemical tests. *Organic Structure Determination* by D. J. Pasto and C. R. Johnson (Prentice Hall, Upper Saddle River, NJ) provides chemical and spectroscopic methods of characterization of organic compounds. This book, originally published in 1969, was revised in 1979 and appeared in a greatly revised format under the title *Experiments and Techniques in Organic Chemistry* (third coauthor: M. J. Miller).

Spectroscopic information can be obtained from a variety of sources. The text *Spectrometric Identification of Organic Compounds,* the 5th edition (R. M. Silverstein, G. C. Bassler, T. C. Morrill) of which was published in 1991 by Wiley, describes the concepts of mass spectrometry, infrared (IR) spectrometry, nuclear magnetic resonance (NMR) spectrometry (both ^{1}H and ^{13}C NMR, as well as a chapter on 2D techniques). The main focus of the text is organic structure determination. The topics begin with the assumption that they are new to the reader, and they are then developed to the intermediate level. Moreover, the text includes many structure determination problems and ample tables to solve these problems.

General References

The ACS Style Guide: A Manual for Authors and Editors, edited by J. S. Dodd (American Chemical Society, Washington, DC, 1986).

IUPAC Nomenclature of Organic Chemistry, Secs. A–F, H (Pergamon: Elmsford, NY, 1979).

March, J., *Reactions, Mechanisms, and Structures,* 4th ed. (Wiley-Interscience, New York, 1992), Appendix A.

APPENDIX I

Handy Tables for the Organic Laboratory

Composition and Properties of Common Acids and Bases

	sp gr	Wt %	Moles per Liter	Grams/100 mL
Hydrochloric acid, conc	1.19	37	12.0	44.0
Constant-boiling (252 mL conc acid + 200 mL water, bp 110°)	1.10	22.2	6.1	24.4
10% (100 mL conc acid + 321 mL water)	1.05	10	2.9	10.5
5% (50 mL conc acid + 380. mL water)	1.03	5	1.4	5.2
1 N (41.5 mL conc acid diluted to 500 mL)	1.02	3.6	1	3.7
Hydrobromic acid, constant-boiling (bp 126°C)	1.49	47.5	8.8	70.8
Hydriodic acid, constant-boiling (bp 127°C)	1.7	57	7.6	97
Sulfuric acid, conc	1.84	96	18	177
10% (25 mL conc acid + 398 mL water)	1.07	10	1.1	10.7
1 N (13.9 mL conc acid diluted to 500 mL)	1.03	4.7	0.5	4.8
Nitric acid, conc	1.42	71	16	101
Sodium hydroxide, 10% solution	1.11	10	2.8	11.1
Ammonium hydroxide, conc	0.90	28.4	15	25.6
Phosphoric acid, conc (syrupy)	1.7	85	14.7	144

Composition of Common Buffer Solutions

pH	Components
0.1	1 N Hydrochloric acid
1.1	0.1 N Hydrochloric acid
2.2	15.0 g Tartaric acid per liter (0.1 M solution)
3.9	40.8 g Potassium acid phthalate per liter
5.0	14.0 g KH phthalate + 2.7 g $NaHCO_3$ per liter (heat to expel carbon dioxide, then cool)
6.0	23.2 g KH_2PO_4 + 4.3 g Na_2HPO_4 (anhyd) per liter
7.0	9.1 g KH_2PO_4 + 18.9 g Na_2HPO_4 per liter
8.0	11.8 g Boric acid + 9.1 g Borax ($Na_2B_4O_7 \cdot 10H_2O$) per liter
9.0	6.2 g Boric acid + 38.1 g Borax per liter
10.0	6.5 g $NaHCO_3$ + 13.2 g Na_2CO_3 per liter
11.0	11.4 g Na_2HPO_4 + 19.7 g Na_3PO_4 per liter
12.0	24.6 g Na_3PO_4 per liter (0.15 M solution)
13.0	4.1 g Sodium hydroxide pellets per liter (0.1 M)
14.0	41.3 g Sodium hydroxide pellets per liter (1 M)

Pressure–Temperature Nomograph for Vacuum Distillations

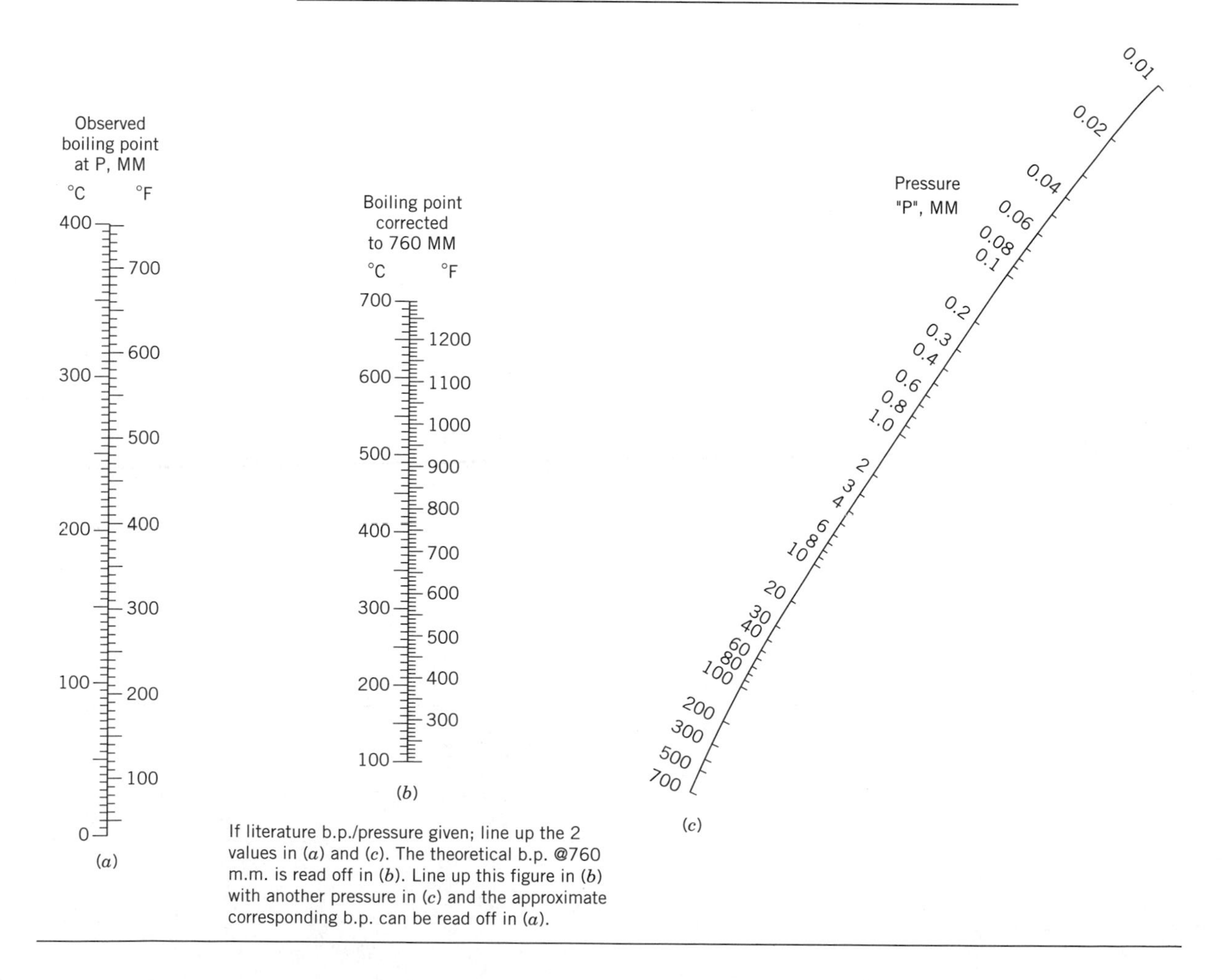

If literature b.p./pressure given; line up the 2 values in (*a*) and (*c*). The theoretical b.p. @760 m.m. is read off in (*b*). Line up this figure in (*b*) with another pressure in (*c*) and the approximate corresponding b.p. can be read off in (*a*).

Elution Solvents for Chromatography

Alumina adsorbent		Silica adsorbent
Fluoroalkanes	Increasing	Fluoroalkanes
Pentane	polarity[a]	Cyclohexane
Isooctane	↓	Heptane
Petroleum ether (light)		Pentane
Hexane		Carbon tetrachloride
Cyclohexane		Carbon disulfide
Cyclopentane		Chlorobenzene
Carbon tetrachloride		Ethylbenzene
Carbon disulfide		Toluene
Xylene		
Diisopropyl ether		Benzene
Toluene		2-Chloropropane
1-Chloropropane		Chloroform
Chlorobenzene		Nitrobenzene
		Diisopropyl ether
Benzene		Diethyl ether
Ethyl bromide		Ethyl acetate
Diethyl ether		2-Butanol
Diethyl sulfide		
Chloroform		Ethanol
Methylene chloride		Water
Tetrahydrofuran		Acetone
1,2-Dichloroethane		Acetic acid
Ethyl methyl ketone		Methanol
1-Nitropropane		Pyruvic acid
Acetone		
1,4-Dioxane		
Ethyl acetate		
Methyl acetate		
1-Pentanol		
Dimethyl sulfoxide		
Aniline		
Diethylamine		
Nitromethane		
Acetonitrile		
Pyridine		
Butyl cellosolve		
2-Propanol		
1-Propanol		
Ethanol		
Methanol		
Ethylene glycol		
Acetic acid		

[a]The ability of a solvent to elute chromatographically depends on the compound being eluted as well as the adsorbent; thus, exceptions to these orders will be found. In general, however, as one goes down these lists, one finds solvents of increasing "polarity" which, in principle, more readily elute the more polar compounds.

Salt–Ice Mixtures for Cooling Baths[a]

Substance	Initial temperature (°C)	g salt/100 g H_2O[b]	Final temperature (°C)
Na_2CO_3	−1 (ice)	20	−2.0
NH_4NO_3	20	106	−4.0
$NaC_2H_3O_2$	10.7	85	−4.7
NH_4Cl	13.3	30	−5.1
$NaNO_3$	13.2	75	−5.3
$Na_2S_2O_3 \cdot 5H_2O$	10.7	110	−8.0
$CaCl_2 \cdot 6H_2O$	−1 (ice)	41	−9.0
KCl	0 (ice)	30	−10.9
KI	10.8	140	−11.7
NH_4NO_3	13.6	60	−13.6
NH_4Cl	−1 (ice)	25	−15.4
NH_4NO_3	−1 (ice)	45	−16.8
NH_4SCN	13.2	133	−18.0
NaCl	−1 (ice)	33	−21.3
$CaCl_2 \cdot 6H_2O$	0 (ice)	81	−21.5
H_2SO_4 (66.2%)	0 (ice)	23	−25
NaBr	0 (ice)	66	−28
H_2SO_4 (66.2%)	0 (ice)	40	−30
C_2H_5OH(4°)	0 (ice)	105	−30
$MgCl_2$	0 (ice)	85	−34
H_2SO_4 (66.2%)	0 (ice)	91	−37
$CaCl_2 \cdot 6H_2O$	0 (ice)	123	−40.3
$CaCl_2 \cdot 6H_2O$	0 (ice)	143	−55

[a]Addition of the substance listed in the first column can, but does not always, lower the bath temperature to that listed in the last column. The final temperature may not be quite that low owing to insufficiently crushed ice, and so forth.

[b]H_2O means liquid water except where ice is listed parenthetically.

Liquid Media for Heating Baths[a]

Medium	mp (°C)	bp (°C)	Range (°C)	Flash point (°C)	Comments
H_2O (l)	0	100	0–80	None	Ideal in limited range
Ethylene glycol	−12	197	−10–180	115	Cheap; flammable; difficult to remove from apparatus
20% H_3PO_3, 80% H_3PO_4[b]	<20	—	20–250	None	Water soluble; nonflammable; corrosive; steam evolved at high temperature
Triethylene glycol	−5	287	0–250	156	Water soluble; stable
Glycerol	18	290	−20–260	160	Supercools; water soluble; viscous
Paraffin	~50	—	60–300	199	Flammable
Dibutyl phthalate	—	340	150–320	—	Viscous at low temperature
Wood's metal (50% Bi, 25% Pb, 12.5% Sn, 12.5% Cd)[b]	70	—	70–350	—	Oxidizes if used at >250° for long period of time
Tetracresyl silicate	<−48	~440	20–400	—	Noncorrosive; fire-resistant; expensive

[a]Range of useful temperatures for a bath which is open to the atmosphere; all of these baths except water should be used only in a hood.
[b]Wt %.

Solvents for Extraction of Aqueous Solutions[a]

Solvent	bp (°C)	Flammability[b]	Toxicity[b]	Comments
Benzene	80.1	3	3	Prone to emulsion;[c] good for alkaloids and phenols from buffered solutions
2-Butanol	99.5	1	3	High boiling; good for highly polar water-soluble materials from buffered solution
Carbon tetrachloride	76.5	0	4	Easily dried; good for nonpolar materials
Chloroform	61.7	0	4	May form emulsion;[c] easily dried
Diethyl ether	34.5	4	2	Absorbs large amounts of water; good general solvent
Diisopropyl ether	69	3	2	May form explosive peroxides on long storage; good for acids from phosphate-buffered solutions
Ethyl acetate	77.1	3	1	Absorbs large amount of water; good for polar materials
Methylenechloride	40	0	1	May form emulsions;[c] easily dried
Pentane	36.1	4	1	Hydrocarbons easily dried;
Hexane	69	4	1	poor solvents for polar
Heptane	98.4	3	1	compounds

[a]Data in this table are taken mostly from A. J. Gordon and R. A. Ford, *The Chemists Companion* (Wiley-Interscience, New York, 1972).
[b]4 means most toxic or flammable, 4 > 3 > 2 > 1; 0 = nonflammable.
[c]*Emulsions* may form during the extraction of aqueous solutions by organic solvents, making good separation very difficult, if not impossible. Their formation is especially liable to occur if the solution is alkaline; addition of dilute sulfuric acid (if permissible) may break up such an emulsion. The following are general methods for breaking up emulsions: saturation of the aqueous phase with a salt (NaCl, Na_2SO_4, etc.); addition of several drops of alcohol or ether (especially when $CHCl_3$ is the organic layer); centrifugation of the mixture, one of the most successful techniques.

Drying Agents of Moderate Strength for Organic Solvents

Drying agent	Capacity[a]	Speed[b]	Comments
$CaSO_4$	$\frac{1}{2}H_2O$	Very fast (1)	Sold commercially as Drierite with or without a color indicator; very efficient. When dry, the indicator ($CoCl_2$) is blue, but turns pink as it takes on H_2O (capacity $CoCl_2 \cdot 6H_2O$); useful in temperature range −50 to +86°C. Some organic solvents leach out, or change the color of $CoCl_2$ (acetone, alcohols, pyridine, etc.).
$CaCl_2$	$6H_2O$	Very fast (2)	Not very efficient; use only for hydrocarbons and alkyl halides (forms solvates, complexes, or reacts with many nitrogen and oxygen compounds).
$MgSO_4$	$7H_2O$	Fast (4)	Excellent general agent; very inert but may be slightly acidic (avoid with very acidic-sensitive compounds). May be soluble in some organic solvents.
Molecular sieve 4A	High	Fast (30)	Very efficient; predrying with a more common agent recommended. Sieve 3A also excellent.
Na_2SO_4	$10H_2O$	Slow (290)	Very mild, inefficient, slow, inexpensive, high capacity; good for gross predrying, but do not warm the solution.
K_2CO_3	$2H_2O$	Fast	Good for esters, nitriles, ketones, and especially alcohols; do not use with acidic compounds.
NaOH or KOH	Very high	Fast	Powerful, but used only with inert solutions in which agent is insoluble; especially good for amines.
H_2SO_4	Very high	Very fast	Very efficient, but use limited to saturated or aromatic hydrocarbons or halides (will remove olefins and other "basic" compounds).
Alumina (Al_2O_3) or silica gel (SiO_2)	Very high	Very fast	Especially good for hydrocarbons. Should be finely divided; can be reactivated after use by heating (300°C for SiO_2, 500°C for Al_2O_3).

[a]Moles of water per mole of agent (maximum).
[b]Relative rating. For first five entries, number in parentheses is relative drying speed for benzene—low number, rapid drying; order may change for slow agents with change in solvent. [B. Pearson and J. Ollerenshaw, *Chem. Ind.*, 370 (1966).]

More Powerful Dehydrating Agents for Organic Liquids

Agent[a]	Products formed with H_2O	Comments
Na[b]	NaOH, H_2	Excellent for saturated hydrocarbons and ethers; do *not* use with any halogenated compounds.
CaH	$Ca(OH)_2$, H_2	One of the best agents; slower than $LiAlH_4$ but just as efficient and safer. Use for hydrocarbons, ethers, amines, esters, C_4 and higher alcohols (not C_1, C_2, C_3 alcohols). Do *not* use for aldehydes and active carbonyl compounds.
$LiAlH_4$[c]	LiOH, $Al(OH)_3$, H_2	Use only with inert solvents [hydrocarbons, aryl (not alkyl) halides, ethers]; reacts with any acidic hydrogen and most functional groups (halo, carbonyl, nitro, etc.). Use caution; excess may be destroyed by slow addition of ethyl acetate.
BaO or CaO	$Ba(OH)_2$ or $Ca(OH)_2$	Slow but efficient; good mainly for alcohols and amines, but should not be used with compounds sensitive to strong base.
P_2O_5	HPO_3, H_3PO_4, $H_4P_2O_7$	Very fast and efficient; very acidic. Predrying recommended. Use only with inert compounds (especially hydrocarbons, ethers, halides, acids, and anhydrides).

[a]The best dehydrating agents are those that react rapidly and irreversibly with water (and not with the solvent or solutes); they are also the most dangerous and should be used only after gross predrying with a less vigorous drying agent (see previous table). These agents are almost always used only to dry a solvent prior to and/or during distillation. Although $MgClO_4$ is one of the most efficient drying agents, it is *not* recommended because it can cause explosions if mishandled. [See D. R. Burfield, K.-H. Lee, and R. H. Smithers, *J. Org. Chem.,* **42,** 3060, (1977), for studies of desiccant efficiency.]

APPENDIX II

Equipment and Chemicals for the Laboratory

The following representative list of laboratory supplies suitable for a course in identification has been included in response to many inquiries. Variation from one laboratory to another in the equipment available may require much of the equipment here to be modified or omitted.

The suggested supplementary microscale kit is included for those colleges and universities which already incorporate microscale equipment into their courses.

APPARATUS

Individual Desk Equipment

It has been found convenient to assign each student a standard organic laboratory desk equipped with the usual apparatus required for the first year's work in organic preparations. This is supplemented by a special kit containing apparatus for carrying out classifications tests and preparation of derivatives on a small scale.

Suggested Locker Equipment

1 Beaker, 50 mL
1 Beaker, 100 mL
1 Beaker, 250 mL
1 Beaker, 400 mL
1 Beaker, 800 mL
2 Brushes, test tube and regular
1 Burner, Bunsen
1 Burner, micro
1 Burner tip, wing top
1 Clamp, screw
1 Clamp, test tube
3 Cork rings, assorted

1 Cylinder, graduated, 10 mL
1 Cylinder, graduated, 100 mL
1 Flask, distilling, 10 mL
1 Flask, distilling, 25 mL
1 Flask, distilling, 50 mL
1 Flask, distilling, 100 mL
2 Flasks, Erlenmeyer, 25 mL
2 Flasks, Erlenmeyer, 50 mL
2 Flasks, Erlenmeyer, 125 mL
2 Flasks, Erlenmeyer, 250 mL
1 Flask, Erlenmeyer, 500 mL
1 Flask, filter, 125 mL
1 Flask, filter, 250 mL
2 Funnels, Buchner, 5.5 and 7.0 cm
1 Funnel, glass
2 Funnels, Hirsch
1 Funnel, powder
1 Goggles, pair
2 Rings, 4 and 2 in.
Several 3-ft sections of tubing
2 Spatulas, one flat and one curved
1 Striker (burner lighter)
6 Test tubes, 13 × 100 mm
6 Test tubes, 16 × 150 mm
2 Test tube racks, to fit above test tubes
1 Thermometer, −5 to 360°C
1 Thermometer adaptor
1 Tongs, crucible
2 Watch glasses, small and large

Suggested Supplementary Kit

1 Condenser
1 Crucible, porcelain with cover
1 Desiccator, small
2 Distillation adapters, needed for a simple distillation
1 Flask, filter, 50 mL
1 Funnel, separatory, 125 mL
2 magnetic stirring bars, small and large
2 NMR tubes
1 Pipet, graduated, 1 mL
1 Pipet, graduated, 10 mL
5 Pipets, disposable
2 Test tubes, side-arm, 15 × 125 mm
2 Test tubes, side-arm, 20 × 150 mm

Suggested Supplementary Microscale Kit

Aluminum block or sand bath in heating mantle with variac
Claisen adapter
Condenser, air
Condenser, jacketed
Drying tube
GC collection tube
Conical vials, 0.1 mL
Hickman–Hinkle still
Needles
Recrystallization tubes, Craig, 2 and 3 mL
Round-bottom flask, 5 and 10 mL
Spin vane
Syringe
Teflon band (used with certain distillation equipment)
Thermometer
Thermometer adapter
Vials, 0.1, 3.0, and 5.0 mL
Vial, reaction

General Laboratory Equipment

Boiling stones
Clamps, assorted
1 Desiccator, large, vacuum, with vacuum pump
Distilled water
Filter paper, assorted
Glass tubing and rods, assorted
Gloves, disposable
Grease
Hot plate stirrers, one per student

Ice machine
Melting point apparatus, one for each five students
Melting point tubes
Mortars and pestles
1 Oven, drying
Pipets, automatic, assorted sizes—stored in a holder
Pipets, Beral (box)
Pipet tips to fit automatic pipets
1 Rack, drying
Ring stands, assorted
Thiele tubes

Special Laboratory Equipment

The following equipment should be kept free from corrosive fumes.

Balances, quantitative, one for each five students, minimum requirements, 1 mg to 100 g range
1 Gas chromatograph mass spectrometer (GC–MS)
1 Gravitometer
1 Infrared (IR) spectrophotometer or Fourier transform IR spectrophotometer
1 Nuclear magnetic resonance (NMR) spectrometer or Fourier transform NMR spectrometer
1 Polarimeter with sodium lamp
1 Refractometer

Items Obtained on Temporary Loan from Instructor or Stockroom

Cylinder of dry nitrogen gas
Development chambers for TLC, such as wide-mouth jars
IR cells, salt plates, mulling oils, and calibration window
NMR solvents (CCl_4, $DCCl_3$, acetone-d_6, DMSO-d_6, D_2O)
Polarimeter tubes, 1 and 2 dm long
Syringes and needles for GC–MS
Thin-layer chromatography plates
UV lamp

Waste Containers Needed in the Laboratory

These containers should be conveniently located in a hood. Directions for the types of compounds to be placed in each container are discussed in the textbook.

Aqueous solution container
Aromatic organic solvent container
Halogenated organic waste container
Hazardous solid waste container
Hazardous waste container
Hazardous waste container for heavy metals

Nonhazardous solid waste container
Organic nonhazardous solid waste container
Organic solvent container

CHEMICALS NEEDED IN THE LABORATORY

The following compounds are useful for carrying out solubility and classification tests and for preparing the derivatives. It has been found convenient to provide a set of bottles of about 100 mL capacity, using small-mouth bottles for liquids and wide-mouth bottles for solids. Each reagent bottle should have a test tube taped to the side of the bottle which contains a disposable pipet or a spatula for the students to use with the reagent; providing this for students avoids waste of the chemical or contamination with use. For a class of 20 students, 20–50 g of the reagent may be placed in the shelf bottles. The actual amounts needed per student will naturally vary with the nature of the unknowns, the intelligence with which the classification tests are selected, and the manipulative skill of the student. The bottles should be grouped according to the information on the Material Data Safety Sheets (MSDS), and these sheets should be available in the laboratory in a three-ring binder.

Organic Compounds

Acetaldehyde
Acetic anhydride
Acetone
Acetonitrile
Acetyl chloride
Alizarin
Aniline
Azoxybenzene
Benzene
Benzenesulfonyl chloride
Benzoyl chloride
Benzylamine
Benzyl bromide
Benzyl chloride
4-Bromophenacyl bromide
4-Bromophenylhydrazine
Carbon tetrachloride
Chloroacetic acid
Chloroform
Cyclohexane
Diethylene glycol
Diethylene glycol dimethyl ether (diglyme)
Diethyl ether
1,2-Dinitrobenzene
3,5-Dinitrobenzoic acid
3,5-Dinitrobenzoyl chloride
2,4-Dinitrofluorobenzene
2,4-Dinitrophenylhydrazine
Dioxane
95% Ethanol
100% Ethanol
Ethyl acetate
Ethyl bromide
Fluorescein
Fuchsin (*p*-rosaniline hydrochloride)
Glycerol
Hexane
Ligroin
Methanol
Methone (dimedon)
2-Methoxyethanol
Methylene chloride
Methyl iodide
Methyl 4-toluenesulfonate
2-Naphthol
1-Naphthyl isocyanate
Ninhydrin (1,2,3-indanetrione monohydrate)
4-Nitrobenzaldehyde
4-Nitrobenzoyl chloride
4-Nitrobenzyl chloride
4-Nitrophenylhydrazine
3-Nitrophthalic anhydride
5-Nitrosalicylaldehyde
Oxalic acid
Petroleum ether
Phenol

Phenylhydrazine
Phenyl isocyanate
Phenyl isothiocyanate
Phthalic anhydride
Picric acid
Piperidine
2-Propanol
Propylene glycol
Pyridine
Sucrose
Thioglycolic acid (mercaptoacetic acid)
Thiourea
Toluene
4-Toluenesulfonyl chloride
4-Toluidine (4-methylaniline or 4-amino toluene)
Triethanolamine
Triethylamine
Xanthydrol

Inorganic Compounds

Aluminum chloride
Ammonium carbonate
Ammonium chloride
Ammonium hexanitratocerate
Ammonium polysulfide
Borax
Boron trifluoride–acetic acid complex
Bromine
Calcium sulfate
Carbon disulfide
Ceric ammonium nitrate
Chromic anhydride
Chloroplatinic acid
Chlorosulfonic acid
Copper sulfate
Copper wire
Ferric chloride (anhydrous)
Ferric ammonium sulfate
Ferrous ammonium sulfate
85% Hydrazine
3% Hydrogen peroxide
Hydroxylamine hydrochloride
Iodine
Iron nails
Lead acetate
Lead dioxide
Magnesium
Magnesium sulfate
Mercuric bromide
Mercuric chloride
Mercuric iodide
Mercuric nitrate
Nickel(II) chloride
Nickel chloride hexahydrate
Paraperiodic acid
Phenylhydrazine hydrochloride
Phosphorus pentachloride
Potassium acid phthalate
Potassium bromide
Potassium carbonate
Potassium cyanide
Potassium hydroxide
Potassium iodide
Potassium permanganate
Potassium persulfate
Potassium thiocyanate
Semicarbazide hydrochloride
Silver nitrate
Sodium
Sodium acetate
Sodium acetate trihydrate
Sodium bicarbonate
Sodium bisulfite
Sodium carbonate
Sodium chloride
Sodium citrate
Sodium dichromate
Sodium hydroxide
5.25% Sodium hypochlorite (household bleach)
Sodium iodide
Sodium–lead alloy
Sodium nitrite
Sodium nitroprusside (disodium pentacyano-nitrosoferrate III)
Sodium potassium tartrate

Sodium sulfate
Sodium thiosulfate (hydrated)
Tetrabutyl ammonium bromide
Thionyl chloride
Tin
Zinc dust
Zinc chloride
Zirconium chloride or nitrate

Acids and Bases

These acids and bases should be kept in 250 mL bottles that are easily refilled from the larger bottles.

Acetic acid
Ammonium hydroxide
Hydrochloric acid
Hydroiodic acid (57%)
Nitric acid
Fuming nitric acid
Phosphoric acid
Sulfuric acid
20% Fuming sulfuric acid

Solutions

Below is a listing of the various solutions needed in this textbook. The directions for the preparation of these solutions are given in Chapters 4, 7, and 8 and in textbooks on qualitative and quantitative inorganic chemistry. These solutions should be stored in 100 mL reagent bottles. The bottles that contain the sodium hydroxide solutions should have a rubber stopper, since the glass stoppers frequently get stuck on these bottles.

1% Ethanolic alizarin solution
2% Ammonia solution
5% Ammonium hydroxide solution
10% Ammonium hydroxide solution
10% Ammonium polysulfide solution
Benedict's solution
Bogen or Grammercy indicator–hydroxylamine hydrochloride reagent
1% Borax solution
5% Bromine in carbon tetrachloride
Bromine water
Ceric ammonium nitrate reagent
25% Chloroplatinic acid solution
1 M Copper sulfate solution
1.7% 1,2-Dinitrobenzene in 2-methoxyethanol
2,4-Dinitrophenylhydrazine reagent
50% Ethanol
60% Ethanol
Fehling's solution
1% Ferric chloride in chloroform
5% Alcoholic ferric chloride solution
5% Ferric chloride solution
Ferrous sulfate reagent
1% Fluorescein solution
0.25 M Hydrochloric acid solution
0.5 M Hydrochloric acid solution
1 M Hydrochloric acid solution
2 M Hydrochloric acid solution
1% Hydrochloric acid solution
2% Hydrochloric acid solution
5% Hydrochloric acid solution
10% Hydrochloric acid solution
25% Hydrochloric acid solution
0.5 M Hydroxylamine hydrochloride in 95% ethanol
1 M Hydroxylamine hydrochloride in propylene glycol
Saturated hydroxylamine hydrochloride in methanol
Indicator solution
Iodine in methylene chloride solution
Iodine–potassium iodide solution
Jones reagent
1% Lead acetate solution
Lucas reagent
5% Mercury(II) chloride solution
Mercuric nitrate solution
Nickel chloride hexahydrate in carbon disulfide
Nickel chloride hexahydrate in 5-nitrosalicylaldehyde

2 M Nitric acid solution
5% Nitric acid solution
20% Nitric acid solution
1.5% 4-Nitrobenzaldehyde in 2-methoxyethanol
Periodic acid reagent
10% Phosphoric acid solution
Picric acid in chloroform
10% Picric acid in ethanol
20% Picric acid in ethanol
Saturated picric acid in 95% ethanol
Potassium bromide and bromine solution
Alcoholic potassium hydroxide reagent
1 M Potassium hydroxide in diethylene glycol
1 M Potassium hydroxide solution
2 M Potassium hydroxide in methanol
1% Potassium permanganate solution
2% Potassium permanganate solution
Saturated potassium permanganate solution
Schiff's reagent
0.1 M Silver nitrate solution
2% Ethanolic silver nitrate solution
5% Silver nitrate solution
Satd sodium bicarbonate solution
10% Sodium bisulfite solution
40% Sodium bisulfite solution
Satd sodium bisulfite solution
2% Sodium carbonate solution
5% Sodium carbonate solution
10% Sodium carbonate solution
Satd sodium chloride solution
Alcoholic sodium hydroxide solution
0.1 M Sodium hydroxide solution (standardized)
1 M Sodium hydroxide solution
2 M Sodium hydroxide solution
3 M Sodium hydroxide solution
6 M Sodium hydroxide solution
0.1% Sodium hydroxide solution
1% Sodium hydroxide solution
2% Sodium hydroxide solution
5% Sodium hydroxide solution
5% Ethanolic sodium hydroxide solution
10% Sodium hydroxide solution
20% Sodium hydroxide solution
25% Sodium hydroxide solution
33% Sodium hydroxide solution
40% Sodium hydroxide solution
Sodium iodide in acetone
20% Sodium nitrite solution
2% Sodium nitroprusside solution
3 M Sulfuric acid solution
5% Sulfuric acid solution
10% Sulfuric acid solution
20% Sulfuric acid solution
25% Sulfuric acid solution
75% Sulfuric acid solution
0.4% Zirconium chloride or nitrate solution

Indicators

Bogen Universal Indicator
Grammercy Universal Indicator
Phenolphthalein
Thymol blue
Zirconium–alizarin test paper

Other Items

Blue litmus paper
Celite
Cheesecloth
Congo red paper
Cotton
Cottonseed oil

Glass wool
Norite (charcoal)
Paraffin wax
Phenolphthalein paper
pH paper (1–14)
Red litmus paper
Starch-potassium iodide paper

UNKNOWNS

Compounds for use as unknowns should be carefully selected for purity. Examples may be chosen from the tables in Appendix III, but not necessarily limited to the compounds listed there. The unknowns should not exceed 3 g of a solid and 5 mL of a liquid. Mixtures should also contain no more than 3 g of each solid and 5 mL of each liquid. If microscale techniques are to be used for the preparation of derivatives, then these amounts can be cut down considerably.

As the experience and technique of the student improve toward the end of the quarter or semester, much smaller amounts of the unknowns may be given and the solubility, classification tests, and preparation of derivatives carried out on a much smaller scale (about one-tenth the amounts specified in the experiments and procedures).

APPENDIX III

Tables of Derivatives

The tables on the following pages contain common organic compounds arranged according to classes. The compounds in each table are listed in the order of boiling points or melting points. Only compounds which have two or more solid derivatives are listed. This table of derivatives is not intended to be complete. One reason for this is the desirability of following the specific directions given in other literature for the conversion of a compound not listed on the table to its derivative. The use of such procedures will increase the student's chances of success since the directions given in Chapter 8 are compromise directions designed to apply to as many compounds of a particular class as possible. There are obviously many cases in which these generalizations are unsatisfactory, and detailed directions which have been found to work for the particular compound must be utilized.

In describing compounds in the literature it is customary to give a *range* of a number of degrees for the boiling point or melting point. In order to keep the following tables from being too cumbersome, only the highest point of the boiling or melting point range is listed; this value is rounded to the nearest whole degree. Specific gravities are given for a temperature of 20°C referred to water at 4°C unless otherwise indicated. Refractive indices are given at 20° for the sodium D line.

Students should bear in mind that the value obtained for a melting point depends somewhat on the observer and on the method used in the determination. Thus, it often happens that the literature gives several different values for the same compound. In such cases the highest value has generally been chosen for the tables that follow. If the literature melting point values vary by a great amount, the different values are included.

Acid Anhydrides (Liquids)

Name of compound	bp (°C)	Derivative mp (°C) Acid	Amide	Anilide	4-Toluidide
Trifluoroethanoic anhydride (trifluoroacetic anhydride)	39		75	88	
Ethanoic anhydride (acetic anhydride)	140	16	82	115	153
Propanoic anhydride (propionic anhydride)	168		81	106	126
2-Methylpropanoic anhydride (isobutyric anhydride)	182		128	105	108
2,2-Dimethylpropanoic anhydride (pivalic anhydride; trimethylacetic anhydride)	190	35	155	132	120
Butanoic anhydride (butyric anhydride)	198		116	96	75
Z-Butenedioic anhydride (maleic anhydride; *cis*-2-butenedioic anhydride)	198	130	181 (mono) 266 (di)	175 (mono) 187 (di)	142
Methyl *Z*-butenedioic anhydride (citraconic anhydride; methylmaleic anhydride; methyl *cis*-2-butenedioic anhydride)	214	92d	187d	175 (di) 153 (mono)	170 (mono)
3-Methylbutanoic anhydride (isovaleric anhydride)	215		137	110	107
Dichloroethanoic anhydride (dichloroacetic anhydride)	216d		98	118	153
Pentanoic anhydride (valeric anhydride)	218		106	63	74
2-Ethylbutanoic anhydride	229		112	127	116
E-2-Butenoic anhydride (crotonic anhydride; *trans*-2-butenoic anhydride)	248	72	161	118	132
(±)-Methylbutanedioic anhydride [(±)-methylsuccinic anhydride]	248	115	225	123; 159	164
Hexanoic anhydride (caproic anhydride)	257		100	95	75
Heptanoic anhydride	258		96	71	81
2-Methylpentanedioic anhydride (α-methylglutaric anhydride)	275			176	175
Cyclohexanecarboxylic anhydride (hexahydrobenzoic anhydride)	283	30	186		
Octanoic anhydride (caprylic anhydride)	285	16	110	57	70
4-Methyl-1,2-benzenedicarboxylic acid anhydride (4-methylphthalic anhydride)	295	152	188 (di)		
Z-1,3-Cyclohexanedicarboxylic acid anhydride (*cis*-hexahydroisophthalic anhydride)	304	189		299	

Acid Anhydrides (Solids)

Name of compound	mp (°C)	Derivative mp (°C)			
		Acid	Amide	Anilide	4-Toluidide
Z-9-Octadecenoic anhydride (oleic anhydride; *cis*-9-octadecenoic anhydride)	22		76	41	43
Decanoic anhydride (capric anhydride)	24	31	108	70	78
3-Ethyl-3-methylpentanedioic anhydride (β-ethyl-β-methylglutaric anhydride)	25	87		105	
Cyclohexanecarboxylic anhydride (hexahydrobenzoic anhydride)	25	30	186		
($\pm$)-Methylbutanedioic anhydride [($\pm$)-methylsuccinic anhydride]	37	115	225	123; 159	164
Undecanoic anhydride (hendecanoic anhydride)	37	30	103	71	80
2-Methylbenzoic anhydride (*o*-toluic anhydride)	39	105	143	125	144
3-Methylpentanedioic anhydride (β-methylglutaric anhydride)	41	87		200 (di) 121 (mono)	135
Bromoethanoic anhydride (bromoacetic anhydride)	42	50	91	131	
Dodecanoic anhydride (lauric anhydride)	42	44	110	78	87
Benzoic anhydride	42	122	130	163	158
E-2,3-Dimethylbutanedioic anhydride (*trans*-α,β-dimethylsuccinic anhydride)	43	208	238 (di) 167 (mono)		
Iodoethanoic anhydride (iodoacetic anhydride)	46	83	95	144	
Chloroethanoic anhydride (chloroacetic anhydride)	46	63	121	134	162
Tridecanoic anhydride	50	44	100	80	88
($\pm$)-Phenylbutanedioic anhydride [($\pm$)-phenylsuccinic anhydride]	54	168	210 (di) 159 (α) 145 (β)	222 (di) 175 (α) 171 (β)	175 (α) 169 (β)
Tetradecanoic anhydride (myristic anhydride)	54	54	107	84	93
Pentanedioic anhydride (glutaric anhydride)	56	97	176	224	218
Z-Butenedioic anhydride (maleic anhydride)	56	130	181 (mono) 266 (di)	175 (mono) 187 (di)	142
Hexadecanoic anhydride (palmitic anhydride)	64	63	107	90	98
Octanedioic anhydride (suberic anhydride)	65	144	127 (mono) 217 (di)	128 (mono) 186 (di)	219 (di)
2-Bromo-2-methylpropanoic anhydride (α-bromoisobutyric anhydride)	65	49	148	83	92
Heptadecanoic anhydride (margaric anhydride)	67	61	108		

Acid Anhydrides (Solids) (continued)

Name of compound	mp (°C)	Derivative mp (°C)			
		Acid	Amide	Anilide	4-Toluidide
Methylenebutanedioic anhydride (itaconic anhydride; methylenesuccinic anhydride)	68	165d	192	190	
Decanedioic anhydride (sebacic anhydride)	68	133	210 (di) 170 (mono)	201 (di) 122 (mono)	201
Octadecanoic anhydride (stearic anhydride)	70	70	109	95	102
3-Methylbenzoic anhydride (*m*-toluic anhydride)	71	113	94	126	118
Phenylethanoic anhydride (phenylacetic anhydride)	72	77	156	118	136
2-Bromobenzoic anhydride	76	150	156		
2-Chlorobenzoic anhydride	79	142	142	118	131
cis-α-Methylglutaconic anhydride	85	118		148	
cis-β-Methylglutaconic anhydride	86	149		143	
Z-2,3-Dimethylbutanedioic anhydride (*cis*-α,β-dimethylsuccinic anhydride)	87	129	149 (mono) 244 (di)	222	
4-Methyl-1,2-benzenedicarboxylic anhydride (4-methylphthalic anhydride)	92	152	188 (di)		
3-Chlorobenzoic anhydride	95	158	134	122	
4-Methylbenzoic anhydride (*p*-toluic anhydride)	95	180	160	145	160
Diphenylethanoic anhydride (diphenylacetic anhydride)	98	148	168	180	172
4-Methoxybenzoic anhydride (*p*-anisic anhydride)	99	186	167	171	186
3-Phenylpentanedioic anhydride (β-phenylglutaric anhydride)	105	140		171	154
4-Ethoxybenzoic anhydride	108	198	202	170	
3,5-Dinitrobenzoic anhydride	109	205	183	234	
4-Nitro-1,2-benzenedicarboxylic anhydride (4-nitrophthalic anhydride)	119	165	200d	192	172
Butanedioic anhydride (succinic anhydride)	120	188	157 (mono) 260 (di); 242	148 (mono) 230 (di)	180 (mono) 255 (di)
3-Pyridinecarboxylic anhydride (nicotinic anhydride)	123	238	128	85	150
1,2-Benzenedicarboxylic anhydride (phthalic anhydride)	132	184; 206	149 (mono) 220 (di)	170 (mono) 255 (di)	160 (mono) 201 (di)
3-Iodobenzoic anhydride	134	187	187		
2-Naphthoic anhydride	135	185	193	172	192
2-Nitrobenzoic anhydride	135	146	176	158	
E-3-Phenyl-2-propenoic anhydride (*trans*-cinnamic anhydride)	136	133	148	153	168
E-(±)-Cyclohexanecarboxylic acid anhydride [*trans*-(±)-hexahydrophthalic anhydride]	140	221	196		

Acid Anhydrides (Solids) (continued)

Name of compound	mp (°C)	Derivative mp (°C)			
		Acid	Amide	Anilide	4-Toluidide
2-Carboxyphenylethanoic anhydride (homophthalic anhydride; 2-carboxyphenyl-acetic anhydride)	141	181	230 (2-) 185 (α)	231	
1-Naphthoic anhydride	146	162	205	163	
3-Bromobenzoic anhydride	149	155	155		
2,4-Dinitrobenzoic anhydride	160	183	203		
3-Nitrobenzoic anhydride	160	140	143	154	162
3-Nitro-1,2-benzenedicarboxylic anhydride (3-nitrophthalic anhydride)	162	218	201d (di)	234 (di)	226 (di)
1,2-Naphthalic anhydride	169	175d	265d		
4-Nitrobenzoic anhydride	189	241	201	211	204
4-Chlorobenzoic anhydride	194	240	179	194	
β-phenylglutaconic anhydride	206	155	138	174	184
2,2′-Biphenyldicarboxylic acid anhydride (2,2′-diphenic anhydride)	217	229	191 (mono) 212 (di)	176 (mono) 230 (di)	
4-Bromobenzoic anhydride	218	251	189	197	
(+)-1,2,2-Trimethylcyclopentane-1,3-dicarboxylic acid anhydride (D-camphoric anhydride)	221	187	177 (mono)	210 (mono) 193 (di); 226 (di)	214 (α) 196 (β)
4-Iodobenzoic anhydride	228	270	218		
1,8-Naphthalenedicarboxylic anhydride	274	274		282	

Acyl Halides (Liquids)

Name of compound	bp (°C)	Derivative mp (°C)			
		Acid	Amide	Anilide	4-Toluidide
Ethanoyl fluoride (acetyl fluoride)	21		82	114	147
Propanoyl fluoride (propionyl fluoride)	46		81	106	126
Fluoroethanoyl fluoride (fluoroacetyl fluoride)	51	32	108		
Ethanoyl chloride (acetyl chloride)	55	16	82	115	153
Ethanedioic acid chloride (oxalyl chloride)	64	101	419d (di) 219 (mono)	246 (di) 148 (mono)	268 (di) 169 (mono)
Ethanedioic acid dibromide (oxalyl dibromide)	64	101	419d (di) 219 (mono)	254 (di) 148 (mono)	268 (di) 169 (mono)
Butanoyl fluoride (butyryl chloride)	67		115	96	75
Trichloroethanoyl fluoride (trichloroacetyl fluoride)	68	58	141	97	113
Fluoroethanoyl chloride (fluoroacetyl chloride)	73	32	108		
Chloroethanoyl fluoride (chloroacetyl fluoride)	75	63	120	137	162
Propenoic chloride (acrylyl chloride)	76		85	105	141
Propanoyl chloride (propionyl chloride)	80		81	106	126
Ethanoyl bromide (acetyl bromide)	81		82	115	147
2-Methylpropanoyl chloride (isobutyryl chloride)	92		129	105	108
2-Methylpropenoyl chloride (methacryl chloride)	95	16	102	87	
3-Butenoyl chloride (vinylacetyl chloride)	98		73	58	
Butanoyl chloride (butyryl chloride)	102		116	96	75
Propanoyl bromide (propionyl bromide)	103		81	106	126
2,2-Dimethylpropanoyl chloride (pivalyl chloride; trimethylacetyl chloride)	106	35	155	132	120
Dichloroethanoyl chloride (dichloroacetyl chloride)	108		98	118	153
Ethanoyl iodide (acetyl iodide)	108		82	114	147
Chloroethanoyl chloride (chloroacetyl chloride)	110	63	120	137	162
(±)-2-Chloropropanoyl chloride [(±)-α-chloropropionyl chloride]	111		80	92	124
Methoxyethanoyl chloride (methoxyacetyl chloride)	113		97	58	
3-Methylbutanoyl chloride (isovaleryl chloride)	115		135	110	107
(±)-2-Methylbutanoyl chloride [(±)-α-ethylmethylacetyl chloride]	116		112	110	93

Acyl Halides (Liquids) (continued)

Name of compound	bp (°C)	Derivative mp (°C)			
		Acid	Amide	Anilide	4-Toluidide
Trichloroethanoyl chloride (trichloroacetyl chloride)	118	58	141	97	113
Acetylglycyl chloride	118	206	137		
E-2-Butenoyl chloride (*trans*-crotonyl chloride)	126	72	161	118	132
Propanoyl iodide (propionyl chloride)	127		81	106	126
Pentanoyl chloride (valeryl chloride)	127		106	63	74
Chloroethanoyl bromide (chloroacetyl bromide)	127	63	120	137	162
Bromoethanoyl bromide (bromoacetyl bromide)	127		91	131	
Butanoyl bromide (butyryl bromide)	128		115	96	75
Bromoethanoyl chloride (bromoacetyl chloride)	135	50	91	131	
2-Ethylbutanoyl chloride	139		112	127	116
3-Methylbutanoyl bromide (isovaleryl bromide)	140		135		107
Trichloroethanoyl bromide (trichloroacetyl bromide)	143	58	163	97	113
3-Chloropropanoyl chloride (β-chloropropionyl chloride)	144	42		119	121
4-Methylpentanoyl chloride (isocaproyl chloride)	147		121	112	63
Butanoyl iodide (butyryl iodide)	148		115	96	75
Bromoethanoyl bromide (bromoacetyl bromide)	150	50	91	131	
($\pm$)-2-Bromobutanoyl chloride [($\pm$)-α-bromobutyryl chloride]	152		112	98	92
2-Bromopropanoyl bromide (α-bromopropionyl bromide)	153		123	99	
Hexanoyl chloride (caproyl chloride)	153		100	95	75
($\pm$)-2-Bromopropanoyl bromide [($\pm$)-α-bromopropionyl bromide]	155	26	123	110	
Benzoyl fluoride	159		128	160	
E-2-Butenedioic acid chloride (fumaroyl chloride)	160	287	267	314	
2-Bromo-2-methylpropanoyl bromide (α-bromoisobutyryl bromide)	164	49	148	83	93
Furoyl chloride	174	134	143	124	107
Heptanoyl chloride (enanthoyl chloride)	176		96	71	81

Acyl Halides (Liquids) (continued)

Name of compound	bp (°C)	Derivative mp (°C) Acid	Amide	Anilide	4-Toluidide
Hexanoyl bromide (caproyl bromide)	176		100	95	75
Cyclohexane carboxylic acid chloride (hexahydrobenzoyl chloride)	184	30	185	144	
3-Fluorobenzoyl chloride	189; 204	124	130		
Butanedioic acid dichloride (succinyl dichloride)	190d	188	260	230	255
Heptanoyl chloride	193		96	65	81
4-Fluorobenzoyl chloride	193	183	155		
(±)-2-Bromo-3-methylbutanoyl bromide [(±)-α-bromoisovaleryl bromide]	194	44	133	116	124
Octanoyl chloride (capryloyl chloride)	196		110	57	70
Benzoyl chloride	197	122	130	163	158
Diethylpropanedioic acid chloride (diethylmalonyl chloride)	197	125	224		
2-Fluorobenzoyl chloride	206	127	116		
Phenylethanoyl chloride (phenylacetyl chloride)	210	76	157	118	136
2-Methylbenzoyl chloride (*o*-toluyl chloride)	212	105	142	125	144
Nonanoyl chloride (pelargonyl chloride)	215		99	57	84
Pentanedioic acid dichloride (glutaryl dichloride)	218	98	176	224	218
3-Methylbenzoyl chloride (*m*-toluyl chloride)	218	111	96	126	118
Benzoyl bromide	219	122	130	163	158
4-Chlorobenzoyl chloride	222	242	179	194	
3-Phenylpropanoyl chloride (hydrocinnamyl chloride; β-phenylpropionyl chloride)	225d	48	105; 82	98	135
3-Chlorobenzoyl chloride	225	158	134	122	
1,2-Benzenedicarboxylic acid difluoride (phthaloyl difluoride)	226	206	220	253	201
Phenoxyethanoyl chloride (phenoxyacetyl chloride)	226	99	102	101	
4-Methylbenzoyl chloride (*p*-toluyl chloride)	226	180	160	145	165
Decanoyl chloride (capryl chloride)	232	31	108	70	78
2-Chlorobenzoyl chloride	238	142	142	118	131
3-Bromobenzoyl chloride	243	155	155	136	
2-Bromobenzoyl chloride	245	150	156	141	

Acyl Halides (Liquids) (continued)

Name of compound	bp (°C)	Derivative mp (°C)			
		Acid	Amide	Anilide	4-Toluidide
4-Bromobenzoyl chloride	247	253	190	197	
2-Methoxybenzoyl chloride (*o*-anisoyl chloride)	254	100	129	62; 131; 78	
E-3-Phenyl-2-propenoyl chloride (*trans*-cinnamoyl chloride)	258	133	148	151	168
4-Methoxybenzoyl chloride (*p*-anisyl chloride)	263	184	163	169	186
1,2-Benzenedicarboxylic acid chloride (phthaloyl chloride)	276	206	220	254 (di) 170 (mono)	201 (di) 155 (mono)
1,3-Benzenedicarboxylic acid chloride (isophthaloyl chloride)	276	347	280		
3-Nitrobenzoyl chloride	278	140	143	154	162
1-Chloronaphthalene (α-naphthyl chloride)	298	161	202		
2-Chloronaphthalene (β-naphthyl chloride)	306	185	192		

Acyl Halides (Solids)

Name of compound	mp (°C)	Derivative mp (°C) Acid	Amide	Anilide	4-Toluidide
4-Chlorobenzoyl chloride	16	242	179	194	
Butanedioic acid chloride (succinyl chloride)	20	188	260 (di) 157 (mono)	230 (di) 148 (mono)	255 (di) 179 (mono)
1-Naphthoyl chloride	20	161	202	163	
2-Nitrobenzoyl chloride	20	146	174	155	
2-Hydroxybenzoyl chloride (salicyloyl chloride)	20	158	142	136	156
Octadecanoyl chloride (stearyl chloride)	23	71	109	33; 95	102
4-Methoxybenzoyl chloride (*p*-anisyl chloride)	26	184	163	169	186
3-Nitrobenzoyl chloride	35	140	143	153	
E-3-Phenyl-2-propenoyl chloride (*trans*-cinnamoyl chloride)	36	133	148	153	168
2-Iodobenzoyl chloride	40	162	184		
4-Bromobenzoyl bromide	42	253	189	197	
2-Naphthoyl chloride	43	185	192	171	191
3-Nitrobenzoyl bromide	43	140	143	154	162
1,3-Benzenedicarboxylic acid dichloride (isophthaloyl dichloride)	44	347	280	250	
2,4-Dinitrobenzoyl chloride	46	183	203		
4-Nitrophenylethanoyl chloride (*p*-nitrophenylacetyl chloride)	48	153	198	198	
E-3-Phenyl-2-propenoyl bromide (*trans*-cinnamoyl bromide)	48	133	148	153	168
2-Naphthoxyethanoyl chloride (β-naphthoxyacetyl chloride)	54	156	147	145	
Diphenylethanoyl chloride (diphenylacetyl chloride)	57	148	168	180	173
4-Iodobenzoyl bromide	60	270	217	210	
3,5-Dinitrobenzoyl bromide	60	205	183	234	
Z-Butenedioic acid chloride (maleyl chloride)	60	139	266 (di) 181 (mono)	187	142
4-Nitrobenzoyl bromide	64	241	201	211	203
E-3-(2-Nitrophenyl)-2-propenoyl bromide (*trans*-2-nitrocinnamoyl chloride)	65	240	185		
3,5-Dinitrobenzoyl chloride	74	205	183	234	
4-Nitrobenzoyl chloride	75	241	201	211	204
3-Nitronaphthaloyl chloride	77	218	201	234	226
1,2-Benzenedicarboxylic acid dibromide (phthaloyl dibromide)	80	206	220	253	201
4-Iodobenzoyl chloride	83	270	218	210	
1,4-Benzenedicarboxylic acid dichloride (terephthaloyl dichloride)	84		250	337	
1-Hydroxy-2-naphthoyl chloride	85	195	202	154	
2,2′-Biphenyldicarboxylic acid dichloride (2,2′-diphenic acid dichloride)	94	229	212		

Acyl Halides (Solids) (continued)

Name of compound	mp (°C)	Derivative mp (°C)			
		Acid	Amide	Anilide	4-Toluidide
3-Hydroxy-2-naphthoyl chloride	95	222	217	243	221
2,2-Diphenylpropanoyl chloride (α,α-diphenylpropionyl chloride)	96	174	149		
2,5-Dihydroxybenzoyl chloride (gentisyl chloride)	98	200	218		
Phenanthrene-2-carboxylic acid chloride	101	260	243	218	
Phenanthrene-9-carboxylic acid chloride	102	252	232	218	
4-Phenylbenzoyl chloride (biphenyl-4-carbonyl chloride)	115	228	223		
Phenanthrene-3-carboxylic acid chloride	117	269	233	217	
E-3-(Nitrophenyl)-2-propenyl chloride (*trans*-nitrocinnamoyl chloride	124	286	217		
9,10-Anthraquinone-2-carboxylic acid chloride	147	290	280	260	

Alcohols (Liquids)

		Derivative mp (°C)				
Name of compound	bp (°C)	Phenyl-urethane	1-Naphthyl-urethane	4-Nitro-benzoate	3,5-Dinitro-benzoate	Hydrogen 3-nitro-phthalate
Methanol (methyl alcohol)	66	47	124	96	109	153
Ethanol (ethyl alcohol)	78	52	79	57	94	158
2-Propanol (isopropyl alcohol)	83	88; 76	106	111	123	154
2-Methyl-2-propanol (*tert*-butyl alcohol)	83	136	101	116	142	
2-Propen-1-ol (allyl alcohol)	97	70	109	29	50	124
1-Propanol	97	57	80	35	75	146
2-Butanol (*sec*-butyl alcohol)	100	65	98	26	76	131
2-Methyl-2-butanol (*tert*-pentyl alcohol)	102	42	72	85	118	
2-Methyl-1-propanol (isobutyl alcohol)	108	86	104	69	88	183
2,2-Dimethyl-1-propanol (neopentyl alcohol)	113	144	100			
(±)-3-Methyl-2-butanol	114	68	112		76	127
3-Pentanol	116	49	71; 95	17	101	121
1-Butanol	118	63	71	70; 36	64	147
(±)-2-Pentanol (*sec*-amyl alcohol; *sec*-pentyl alcohol)	120		76	17	62	103
3,3-Dimethyl-2-butanol	120	78			107	
2,3-Dimethyl-2-butanol	120	66	101		111	
3-Methyl-3-pentanol	123	44	84	69	97; 63	
2-Methyl-2-pentanol	123	239			72	
2-Methoxyethanol (methyl cellosolve; ethylene glycol monomethyl ether)	125		113	51		129
2-Methyl-3-pentanol	128	50			85	151
2-Methyl-1-butanol	129	31	82		70	158
2-Chloroethanol (ethylene chlorohydrin)	131	51	101	56	95	98
(±)-4-Methyl-2-pentanol	132	143	88	26	65	166
3-Methyl-1-butanol (isoamyl alcohol; isopentyl alcohol)	132	57	68	21	62	166
3-Methyl-2-butanol	134		72		44	
2-Ethoxyethanol (ethyl cellosolve; ethylene glycol monomethyl ether)	135		68		75	121 (anhyd) 94 (monohyd)
3-Hexanol	136				77	127
2,2-Dimethyl-1-butanol	137	66	81			51

Alcohols (Liquids) (continued)

		Derivative mp (°C)				
Name of compound	bp (°C)	Phenyl-urethane	1-Naphthyl-urethane	4-Nitro-benzoate	3,5-Dinitro-benzoate	Hydrogen 3-nitro-phthalate
1-Pentanol (pentyl alcohol; amyl alcohol)	138	46	68	11	46	136
(±)-2-Hexanol	139		61	40	39	
2,4-Dimethyl-3-pentanol	140	95	99	155; 40	38	151
Cyclopentanol	141	133	118	62	115	
2,3-Dimethyl-1-butanol	145	29			52	
2-Methyl-1-pentanol	148		76		51	145
2-Ethyl-1-butanol	149		61		52	147
2-Bromoethanol	150	76	86		85	172
2,2,2-Trichloroethanol	151	87	120	71	142	
4-Methyl-1-pentanol	153	48			72	140
3-Methyl-1-pentanol	154		40; 58		38	
(±)-4-Heptanol	156		80	35	64	
1-Hexanol	159	42	62		61	124
Cyclohexanol	160	82	129	50	113	160
(±)-2-Heptanol (*sec*-heptyl alcohol)	160		54		49	
3-Chloro-1-propanol (trimethylene chlorohydrin)	161d	38	76		77	
(±)-*cis*-2-Methylcyclo-hexanol	165	94	155	56	99	
4-Hydroxy-4-methyl-2-pentanone	166			48	55	
(±)-*trans*-2-Methylcyclo-hexanol	167	105	155	65	115	
Furfuryl alcohol	172	45	130	76	81	
(±)-4-Methyl-1-hexanol	174		50			149
(±)-*cis*-3-Methylcyclo-hexanol	174	88	129	65	92	
(±)-*cis*-4-Methylcyclo-hexanol	174	119	160	94	134	
(±)-*trans*-4-Methylcyclo-hexanol	174	125	160	67	140	
(±)-*trans*-3-Methylcyclo-hexanol	175	94	122	58	98	
2-Butoxyethanol	176	62				121
1-Heptanol	176	68	62		48	127
1,3-Dichloro-2-propanol	176	73	115		129	
Tetrahydrofurfuryl alcohol	178	61	90	48	84	
(±)-2-Octanol	179	114	64	28	32	
Cyclohexylmethanol	182		110		96	
2,3-Dichloro-1-propanol	183	73	93	38		
2-Ethyl-1-hexanol	184	34	61			108
1,2-Propanediol (propylene glycol)	187	153		127	147	

Alcohols (Liquids) (continued)

Name of compound	bp (°C)	Derivative mp (°C)				
		Phenyl-urethane	1-Naphthyl-urethane	4-Nitro-benzoate	3,5-Dinitro-benzoate	Hydrogen 3-nitro-phthalate
3,5,5-Trimethyl-1-hexanol	193				62	150
2-(2-Methoxyethoxy)-ethanol	194			92		92
1-Octanol	195	74	67		62	128
1,2-Ethanediol (ethylene glycol)	197	157	176	141	169	
(±)-2-Nonanol	198		56		43	
3,7-Dimethyl-1,6-octadien-3-ol (linalool)	199	66	53	70	135	
1-Phenylethanol	203	94	106	43	95	
Benzyl alcohol	205	78	134	86	113	183
(±)-1,3-Butanediol [(±)-1,3-butylene glycol]	208	123	184			
(±)-2-Decanol	211		69		44	
1-Nonanol	215	69	65	66	52	125
(−)-5-Methyl-2-(1-methylethyl)-cyclohexanol [(−)-menthol]	216	112	128	62	153	
1,3-Propanediol (trimethylene glycol)	216	137	164	119	164; 178	
1,7,7-Trimethyl-*exo*-hydroxybicyclo [2.2.1]heptane (isoborneol)	216			129	138	130
3-Methylbenzyl alcohol (α-hydroxy-3-xylene)	217		116	89	111	
2-Methylbenzyl alcohol (α-hydroxy-*o*-xylene)	219	79		101		
1-Phenyl-1-propanol	219		102	60		
2,3-Dibromo-1-propanol	219d		84		60	
1,3-Dibromo-2-propanol	219d	81		78		
2-Phenylethyl alcohol	219	79	119	62	108	123
α-Terpineol	221	113	152	139; 97	79	
1-Tetradecanol (Myristic alcohol)	221			51	67	123
E-3,7-Dimethylocta-2,6-dien-1-ol	229			35	63	117
1-Decanol	231	60	73	30	63	123
1,4-Butanediol (tetra-methylene glycol)	235	183	199	175		
3-Phenyl-1-propanol (hydrocinnamyl alcohol)	237	48		47	92	117

Alcohols (Liquids) (continued)

		Derivative mp (°C)				
Name of compound	bp (°C)	Phenyl-urethane	1-Naphthyl-urethane	4-Nitro-benzoate	3,5-Dinitro-benzoate	Hydrogen 3-nitro-phthalate
1,5-Pentanediol (pentamethylene glycol)	239	176	147	105		
1-Undecanol (1-hendecanol)	243	62	73	100; 30	55	123
Di-2-hydroxyethyl ether (diethylene glycol)	244		149; 122	151; 100 (di)	150	
2-Phenoxyethanol	245			63	105; 74	113
E-3-Phenyl-2-propen-1-ol (*trans*-cinnamyl alcohol)	257	91	114	78	121	
1-Dodecanol (lauryl alcohol)	259	74	80	45	60	124
4-Methoxybenzyl alcohol (*p*-anisyl alcohol)	259	93		94		
1,2,3-Propanetriol (glycerol; 1,2,3-trihydroxypropane)	290d	180	192	188	76	
Benzhydrol	297	140	136		141	
1-Heptadecanol	310		89	54	122	122
E-Octa-9-decen-1-ol (elaidyl alcohol)	333	57	71			

Alcohols (Solids)

Name of compound	mp (°C)	Derivative mp (°C) Phenyl-urethane	1-Naphthyl-urethane	4-Nitro-benzoate	3,5-Dinitro-benzoate	Hydrogen 3-nitro-phthalate
1-Undecanol	16	62	73	30; 100	55	123
1-Phenyl-1-butanol	16		99	58		
1,2,3-Propanetriol (glycerol; 1,2,3-trihydroxypropane)	18	180	192	188	76	
2,2,2-Trichloroethanol	19	87	120	71	142	
1,4-Butanediol (tetramethylene glycol)	19	183	198	175		
1-Phenylethanol	20	92	106	43	95	
(±)-*E*-2-Methylcyclo-hexanol	21	105	155	65	115	
1-Dodecanol (lauryl alcohol)	25	74	80	45	60	124
Cyclohexanol	25	82	129	52	113	160
2-Methyl-2-propanol (*tert*-butyl alcohol)	25	136	101	116	142	
4-Methoxybenzyl alcohol (*p*-anisyl alcohol)	25	93		94		
2,4-Hexadien-1-ol	31	79			85	
1-Tridecanol	31			37		124
E-3-Phenyl-2-propen-1-ol (*trans*-cinnamyl alcohol)	33	91	114	78	121	
α-Terpineol	35				78	
E-Octa-9-decen-1-ol (elaidyl alcohol)	37	57	71			
1-Tetradecanol (myristyl alcohol)	39	74	82	51	67	124
2-Methylbenzyl alcohol (α-hydroxy-2-xylene)	39	79		101		
(−)-5-Methyl-2-(1-methylethyl)-cyclohexanol [(−)-menthol]	44	112	128	62	153	
1-Pentadecanol	44	72	72	46		123
(±)-α-Propylbenzyl alcohol	49		99	58		
1-Hexadecanol (cetyl alcohol)	50	73	82	58	66	122
2,2-Dimethyl-1-propanol (neopentyl alcohol)	53	144	100			
1-Heptadecanol	54		89	54	122	122
2-Butyne-1,4-diol	55	132			191	
4-Methylbenzyl alcohol (α-hydroxy-4-xylene)	60	79			118	

Alcohols (Solids) (continued)

		Derivative mp (°C)				
Name of compound	**mp (°C)**	**Phenyl-urethane**	**1-Naphthyl-urethane**	**4-Nitro-benzoate**	**3,5-Dinitro-benzoate**	**Hydrogen 3-nitro-phthalate**
1-Octadecanol (stearyl alcohol)	60	80	89	64	66	119
Diphenylmethanol (benzhydrol)	69	140	139	132	142	
10-Nonadecanol (myricyl alcohol)	85	96	54			
(±)-Benzoin	137	165	140	123		
(−)-Cholesterol	148	168	160; 176	193		
(+)-1,7,7-Trimethyl-*endo*-3-hydroxy-bicyclo[2.2.1]heptane [(+)-borneol]	212			153		154
(−)-1,7,7-Trimethyl-*endo*-3-hydroxy-bicyclo[2.2.1]-heptane [(−)-borneol]	212	138	127	137	154	

Aldehydes (Liquids)

Name of compound	bp (°C)	Derivative mp (°C)					
		Semi-car-bazone	2,4-Di-nitro-phenyl-hydra-zone	4-Nitro-phenyl-hydra-zone	Phenyl-hydra-zone	Oxime	Dimedon
Methanal (formaldehyde)	−21	169	167	182	145		191
Ethanal (acetaldehyde)	21	163	147; 169	129	63; 99	47	141
Propanal (propionaldehyde)	50	89; 154	155	125		40	156
Ethanedial (glyoxal)	50	270	328	311	180	178	228 (di) 186 (mono)
Propenal (acrolein)	52	171	165	151	52		192
2-Methylpropanal (isobutyraldehyde)	64	126	187	132			154
2-Methyl-2-propenal (α-methylacrolein)	74	198	206		74		
Butanal (butyraldehyde)	75	106; 96	125	95	95		142; 134
2,2-Dimethylpropanal (pivalaldehyde; trimethylacetaldehyde)	75	190	210	119		41	
Methoxyethanal (methoxyacetaldehyde)	92		125	115			
3-Methylbutanal (isovaleraldehyde)	93	107; 132	123	111		49	155
2-Methylbutanal (α-Methylbutyraldehyde)	93	105	121				
Trichloroethanal (chloral)	98	90d	131	131		56	
2-Butenal (crotonaldehyde)	103	200	196	186	56	119	184
Pentanal (valeraldehyde)	104		107, 98	74		52	105
Ethoxyethanal (ethoxyacetaldehyde)	106		117	114			
5-Hydroxymethylfurfural	114	195; 166	184	185	141	78; 108	
2-Methylpentanal (α-methylvaleraldehyde)	116	102	103				
2-Ethylbutanal (α-ethylbutyraldehyde)	117	99	134; 95				102
4-Methylpentanal	121	127	99				
2-Pentenal	125	180		123			
Hexanal (caproaldehyde)	131	108	107	80		51	109
3-Methyl-2-butenal (3-methylcrotonaldehyde)	135	223	186				
2-Methyl-2-pentenal (3-ethyl-2-methylacrolein)	137	207	159		60	49	
5-Methylhexanal	144	117	117				
3-Furaldehyde	144	211			150		
Tetrahydrofurfural	145	166	134; 204				123
1-Cyclopentenylmethanal (1-cyclopentenylformaldehyde)	146	208		188			
2-Hexenal	150	176		139			
Heptanal (heptaldehyde)	156	109	108	73		57	135; 103
2-Furancarboxaldehyde (furfural)	162	203	230; 212; 185	54; 154	98	92; 76	162d

Aldehydes (Liquids) (continued)

		Derivative mp (°C)					
Name of compound	bp (°C)	Semi-car-bazone	2,4-Di-nitro-phenyl-hydra-zone	4-Nitro-phenyl-hydra-zone	Phenyl-hydra-zone	Oxime	Dimedon
1-Cyclohexanecarboxaldehyde (hexahydrobenzaldehyde)	162	176	172			91	
2-Ethylhexanal (α-ethylcaproaldehyde)	163	254d	121				
4-Cyclohexene-1-carboxaldehyde (1,2,3,6-tetrahydrobenzaldehyde)	165	154		163		76	
Butanedial (succinaldehyde)	170		280			172	
Octanal (caprylaldehyde)	171	101	106; 96	80		60	90
2-Ethylpentenal (2-ethyl-3-propylacrolein)	173	153	125				
3-Fluorobenzaldehyde	173			202	114	63	
4-Fluorobenzaldehyde	175			212	147	117; 86	
2-Fluorobenzaldehyde	175			205	90	63	
Benzaldehyde	179	222; 235	237	192; 236; 262	158	35; 130	195
5-Methylfurfural	187	211	212	130	148	112 (syn) 52 (anti)	
Pentanedial (glutaraldehyde)	189			169		178	
Nonanal (pelargonaldehyde)	190	100; 84	100			64	86
Phenylethanal (phenylacetaldehyde)	194	156; 163	121; 110	151	63; 102	103	165
2-Hydroxybenzaldehyde (salicylaldehyde)	197	231	252d	228	143	63	211
3-Methylbenzaldehyde (*m*-tolualdehyde)	199	213; 224	212; 194	157	84; 91	60	172
2-Methylbenzaldehyde (*o*-tolualdehyde)	200	208; 218	195	222	101; 111	49	167
4-Methylbenzaldehyde (*p*-tolualdehyde)	205	221; 234	239	201	114; 121	80; 110	
(+)-Citronellal	207	84	78				79
Decanal (capraldchyde)	209	102	104			69	92
3-Chlorobenzaldehyde	214	230	256	216	134	71	
2-Chlorobenzaldehyde	214	230; 146	209; 214	238; 249	86	76; 103	205d
Phenoxyethanal (phenoxyacetaldehyde)	215d	145			86	95	
4-Chlorobenzaldehyde	216	233	270; 254	239; 220	127	110 (α) 146 (β)	
2-Pyrrolecarboxaldehyde	219	184		183	139	164	
3-Phenylpropanal (hydrocinnamaldehyde)	224	127	149	123		97	
E-3,7-Dimethylocta-2,6-dien-1-al (geranial; citral)	229	164	116	195			
3-Methoxybenzaldehyde (*m*-anisaldehyde)	230	233d	218	171	76	40; 112	
2-Bromobenzaldehyde	230	214	203	240		102	
3-Bromobenzaldehyde	236	205	256	220	141	72	

Aldehydes (Liquids) (continued)

Name of compound	bp (°C)	Derivative mp (°C)					
		Semi-car-bazone	2,4-Di-nitro-phenyl-hydra-zone	4-Nitro-phenyl-hydra-zone	Phenyl-hydra-zone	Oxime	Dimedon
2-(1-Methylethyl)benzaldehyde (cumaldehyde; *p*-isopropylbenzaldehyde)	236	211	245	190	129	52; 111	171
Dodecanal (lauraldehyde)	238	106	106	90		78	
2-Methoxybenzaldehyde (*o*-anisaldehyde)	246	215d	253	205		92	188
4-Methoxybenzaldehyde (*p*-anisaldehyde)	248	210	254d	161	121	45	145 (α) 65 (α') 133 (β)
3,4-Dichlorobenzaldehyde	248			277		122	
4-Ethoxybenzaldehyde (salicylaldehyde ethyl ether)	249	219				59	
E-3-Phenyl-2-propenal (*trans*-cinnamaldehyde)	252	215	255d	195	168	139 (β) 65 (α)	219; 161
4-Ethoxybenzaldehyde	255	208				157 (syn) 118 (anti)	
3,4-Methylenedioxybenzaldehyde (piperonal)	263	237	266d	200	106	112; 146	178; 193
3,4-Dimethoxybenzaldehyde (veratraldehyde)	285	177	265		121	95	
1-Naphthaldehyde	292	221	254	234	80	98	
Diphenylethanal (diphenylacetaldehyde)	316	162				120 (α) 106 (β)	

Aldehydes (Solids)

Name of compound	mp (°C)	Derivative mp (°C)					
		Semi-car-bazone	2,4-Di-nitro-phenyl-hydra-zone	4-Nitro-phenyl-hydrazone	Phenyl-hydrazone	Oxime	Dimedon
2-Chlorobenzaldehyde	11	230; 146	209; 214	238; 249	86	76; 103	205
3-Chlorobenzaldehyde	18	230	256	216	135	71; 118	
2,3,5,6-Tetramethylbenzaldehyde	20	270d				125	
2-Ethoxybenzaldehyde (salicylaldehyde ethyl ether)	22	219				59	
2-Bromobenzaldehyde	22	214	230	240		102	
Tetradecanal (myristaldehyde)	24	107	108	95		84	
Pentadecanal	25	107	107	95		86	
Hexadecanal (palmitaldehyde)	34	109	108	97		88	
1-Naphthaldehyde	34	221	254	224; 234	80	90; 98	
Phenylethanal (phenylacetaldehyde)	34	156; 163	121; 110	151	63; 102	103	165
5-Hydroxymethylfurfural	36	195; 166	184	185	141	78; 108	
2-Iodobenzaldehyde	37	206			79	108	
3,4-Methylenedioxybenzaldehyde (piperonal)	37	237	266d	200	106	112; 146	178; 193
2-Methoxybenzaldehyde (*o*-anisaldehyde)	38	215d	253	205		92	188
Phenoxyethanal (phenoxyacetaldehyde)	38	145			86	95	
Octadecanal (stearaldehyde)	38	119	110	101		89	
2-Aminobenzaldehyde	40	247	250	220	221	135	
4-(*N*,*N*-Diethylamino)benzaldehyde	41	214d	206		103	93	
2-Nitrobenzaldehyde	44	256	265; 250	263	156	103; 154	
3,4-Dichlorobenzaldehyde	44			277		122	
3,3-Diphenyl-2-propenal (β-phenylcinnamaldehyde)	44	215	196		173		
2,4,5-Trimethylbenzaldehyde	44	243			127		
Dodecanal (lauraldehyde)	45	106	106	90		78	
4-Chlorobenzaldehyde	48	233	270; 254	239; 220	127	110 (α) 146 (β)	
2-Pyrrolecarboxaldehyde	50	184		183	139	164	
4-Chloro-2-hydroxybenzaldehyde	53	212		257		155	
Quinoline-4-carboxaldehyde	53			262		182	
2-(2-Furyl)propenal [2-(2-furyl)acrolein]	54	220			132	111	
2,3-Dimethoxybenzaldehyde (veratraldehyde)	54	231	264d		138	99	
3-Chloro-2-hydroxybenzaldehyde (3-chlorosalicylaldehyde)	56	243				168	
Isoquinoline-1-carboxaldehyde	56	197			172		
2-Hydroxy-5-methylbenzaldehyde (5-methylsalicylaldehyde)	56				149	105	
Trichloroethanal hydrate (chloral hydrate; trichloroacetaldehyde hydrate)	57	90	131			56	
4-Bromobenzaldehyde	57	229	128; 257	208	113	157 (syn) 111 (anti)	
3-Iodobenzaldehyde	57	226		212	155	62	
3-Nitrobenzaldehyde	58	246	293d	247	124	122	198

Aldehydes (Solids) (continued)

Name of compound	mp (°C)	Derivative mp (°C)					
		Semi-car-bazone	2,4-Di-nitro-phenyl-hydra-zone	4-Nitro-phenyl-hydrazone	Phenyl-hydrazone	Oxime	Dimedon
3,4-Dimethoxybenzaldehyde (veratraldehyde)	58	177	265		121	95	173
2,5-Dichlorobenzaldehyde	58				105	128	
2-Phenanthraldehyde	60	282				175	
4-Phenylbenzaldehyde	60	243	239		189	150	
3-Methoxy-1-naphthaldehyde	60	200		197		102	
2-Naphthaldehyde	61	245	270	230	206d; 218	156	
3,5-Dichlorobenzaldehyde	65				107	112	
5-Methoxy-1-naphthaldehyde	66	246		246		104	
Quinoline-2-carboxaldehyde	71			225; 250	204	188	
2,4-Dichlorobenzaldehyde	71		226	256		136	
4-Aminobenzaldehyde	72	153; 173			156	124	
4-(*N,N*-Dimethylamino)benzaldehyde	74	222	325; 237	182	148	185	
Quinoline-6-carboxaldehyde	76	239			185	191	
3,4,5-Trimethoxybenzaldehyde	78	220		202		84	
4-Iodobenzaldehyde	78	224	257	201	121		
3-Phenanthraldehyde	80	275				145	
4-Hydroxy-3-methoxybenzaldehyde (vanillin)	81	229; 240d	271d	228	105	122	198
2-Hydroxy-1-naphthaldehyde	82	240				157	
1,3-Benzenedicarboxaldehyde (isophthalaldehyde)	89				242	180	
3,4,5 Trichlorobenzaldehyde	91	254		342d	147		
2-Phenyl-2-oxoethanal (phenylglyoxal)	91	217d	296	310	152 (di)	129 (α) 168 (di)	
Quinoline-8-carboxaldehyde	95	239			176	115	
2,3-Diphenyl-2-propenal (α-phenylcinnamaldehyde)	95	195			126; 141	166	
3,5-Dichloro-2-hydroxybenzaldehyde (3,5-dichlorosalicylaldehyde)	96	227			153	196	
Benzaldehyde-2-carboxylic acid (*o*-formylbenzoic acid; phthalaldehyde acid)	99	202				120	
3-Hydroxy-2-naphthaldehyde	100	>270			248	207d	
5-Chloro-2-hydroxybenzaldehyde	100	287			152	128	
9-Phenanthraldehyde	101	223		265		157	
9-Anthraldehyde	105	219			207	187	
4-Nitrobenzaldehyde	106	221	320d	249	159	133	190 (anti) 184 (syn)
5-Bromo-2-hydroxybenzaldehyde	106	297	292			126	
3-Hydroxybenzaldehyde	108	199	260d	222	131; 147	90	
2,3-Dihydroxybenzaldehyde	108	226		167			
2-Chloro-5-hydroxybenzaldehyde	112	236		251		147	
2-Ethoxy-1-naphthaldehyde	115	215			91		
4-Hydroxybenzaldehyde	117	224; 280	280d; 260	266	178; 184	72; 112	190

Aldehydes (Solids) (continued)

		Derivative mp (°C)					
Name of compound	mp (°C)	Semi-car-bazone	2,4-Di-nitro-phenyl-hydra-zone	4-Nitro-phenyl-hydrazone	Phenyl-hydrazone	Oxime	Dimedon
3-Hydroxy-2,4,6-trichlorobenzaldehyde	117			272		172	
1,4-Benzenedicarboxaldehyde (terephthalaldehyde)	118	225		281	154; 278	200	
4-Chloro-3-hydroxybenzaldehyde	121	239		227		126	
2,4-Dihydroxybenzaldehyde (β-resorcylaldehyde)	136	260d	286d; 302	285	160	192	226
3-Chloro-4-hydroxybenzaldehyde	139	210				145	
2-Chloro-3-hydroxybenzaldehyde	140	237		245		149	
2,4-Dichloro-3-hydroxybenzaldehyde	141			278		188	
(±)-2,3-Dihydroxypropanal [(±)-glyceraldehyde]	142	160d	167		132	118	203
2-Chloro-4-hydroxybenzaldehyde	148	214		284d		194	
3,4-Dihydroxybenzaldehyde (protocatechualdehyde)	154	230d	275d		176	157	145
3,5-Dichloro-4-hydroxybenzaldehyde	156	237				185	
Diphenylhydroxyacetaldehyde (diphenylglycolaldehyde)	163	242				124	
Benzaldehyde-3-carboxylic acid (*m*-formylbenzoic acid)	175	265			164	188d	
2-Hydroxybenzaldehyde-3-carboxylic acid (3-formylsalicylic acid)	179				188	193	
Pentachlorobenzaldehyde	203				153	201	
3,4,5-Trihydroxybenzaldehyde (gallaldehyde)	212d			236		200	
4-Hydroxybenzaldehyde-3-carboxylic acid (5-formylsalicylic acid)	249				219	179	
Benzaldehyde-4-carboxylic acid (*p*-formylbenzoic acid)	256			226	210		

Amides (Liquids)

		Derivative mp (°C)					
			Acid			Amine	
Name of compound	bp (°C)	9-Acyl-amido-xanthene	Acid	4-Nitro-benzyl ester	4-Bromo-phen-acyl-ester	Acet-amide	Benz-amide
N,N-Dimethylmethanamide (*N,N*-dimethylformamide)	153	184		31	140		41
N,N-Diethylmethanamide (*N,N*-diethylformamide)	178			31	140		42
N-Methylmethanamide (*N*-methylformamide)	185			31	140	28	80
Methanamide (formamide)	195	184		31	140		
N-Ethylmethanamide (*N*-ethylformamide)	199			31	140		71
N-Formylpiperidine	222			31	140		48
N-Acetylpiperidine	226		166	78	86		48
N-Methylmethananilide (*N*-methylformanilide)	251			31	140	102	63
N-Ethylmethananilide (*N*-Ethylformanilide)	258			31	140	54	60
N-Propylmethananilide (*N*-propylformanilide)	267			31	140	47	
N-(2-Methylpropyl)methananilide (*N*-isobutylformanilide)	274			31	140		
N-(3-Methylbutyl)methananilide (*N*-isopentylformanilide)	286			31	140		

Amides (Solids)

		Derivative mp (°C)					
			Acid			Amine	
Name of compound	mp (°C)	9-Acylamidoxanthene	Acid	4-Nitrobenzyl ester	4-Bromophenacyl ester	Acetamide	Benzamide
Butananilide (butyranilide)	35				63	114	160
Z-9-Octadecenanilide (oleanilide)	41		16		46	114	160
N-Benzoylpiperidine (*N*-benzipiperidine)	48		122	89	119		48
Methananilide (formanilide)	50			31	140	114	160
N-Methyl-*N*-phenylethanamide (*N*-propylethananilide; *N*-propyl-*N*-phenylacetamide; *N*-propylacetanilide)	50		16	78	86	47	
Propanedioic acid monoamide (malonic acid monoamide; malonamic acid)	50		135	86 (di)			
N-Benzylhexanamide (*N*-benzylcaproamide)	53				72	60	105
N-Ethyl-*N*-phenylethanamide (*N*-ethylethananilide; *N*-ethyl-*N*-phenylacetamide; *N*-ethylacetanilide)	54		17	78	86	54	60
N-Benzyl-3-methylbutanamide (*N*-benzylisovaleramide)	54			68	137	60	105
N-Methyl-*N*-(2-methylphenyl)ethanamide *N*-methyl-*o*-acetotoluidide; *N*-acetyl-*N*-methyl-*o*-toluidine; *N*-methyl-*N*-(*o*-tolyl)acetamide]	56		17	78	86	56	66
Octananilide (caprylanilide)	57		16		67	114	160
Nonananilide (pelargonanilide)	57		12		69	114	160
N-Benzyl-*N*-phenylethanamide (*N*-benzylethananilide; *N*-benzyl-*N*-phenylacetamide; *N*-benzylacetanilide)	58		17	78	86	58	107
Methoxyethananilide (Methoxyacetanilide)	58					114	160
(±)-2-Hydroxypropananilide [(±)-lactanilide; (±)-β-hydroxypropionanilide]	59		18		113	114	160
N-Ethylbenzanilide	60		122	89	119	54	60
N-Benzylmethanamide (*N*-benzylformamide)	60			31	140	60	105
N-Benzylethanamide (*N*-benzylacetamide)	61		17	78	86	60	105
Pentananilide (valeranilide)	63				75	114	160
Z-13-Docosenanilide (erucanilide)	65		34		63	114	160
N-(3-Methylphenyl)ethanamide [*N*-acetyl-*m*-toluidine; *m*-acetotoluidide; *N*-(*m*-tolyl)acetamide]	66		17	78	86	65	125
N-Methyl-*N*-(3-methylphenyl)ethanamide [*N*-methyl-*m*-acetotoluidide; *N*-acetyl-*N*-methyl-*m*-toluidine; *N*-methyl-*N*-(*m*-tolyl)acetamide]	66		17	78	86	66	125
Ethyl *N*-phenylethanedioic acid monoamide (Ethyl oxanilate)	67					114	160
Benzindole	68		122	89	119	158	68
Heptanilide (Enanthanilide)	70				72	114	160
Decananilide (Capranilide)	70		31		60	114	160

Amides (Solids) (continued)

		Derivative mp (°C)					
			Acid			Amine	
Name of compound	mp (°C)	9-Acyl-amido-xanthene	Acid	4-Nitro-benzyl ester	4-Bromo-phen-acyl-ester	Acet-amide	Benz-amide
Undecananilide (Hendecananilide)	71		28; 16		68	114	160
N,N-Diphenylmethanamide (*N,N*-diphenylformamide)	73		8	31	140	101	180
3-Butenanilide (vinylacetanilide)	73					114	160
E-2-Methyl-2-butenamide (tiglamide)	76		65	64	68		
E-2-Methyl-2-butenanilide (tiglanilide)	77		65	64	68	114	160
4-Methylhexananilide	77					114	160
Dodecanilide (lauranilide)	78		44		76	114	160
(±)-2,3-Dimethylbutananilide	78					114	160
Pentadecananilide	78		52	40	77	114	160
E-13-Docosenanilide (brassidianilide)	79		60		94	114	160
(±)-2-Hydroxypropanamide [(±)-lactamide]	79		18		113		
N-(2-Ethoxyphenyl)ethanamide [*o*-acetophenetidide; *N*-acetyl-*o*-phenetidine; *N*-acetyl-2-ethoxyaniline; *N*-(*o*-ethoxyphenyl)acetamide]	79		17	78	86	79	104
Tridecananilide	80		43		75	114	160
Propanamide (propionamide)	81	214		31	63		
Ethanamide (acetamide)	82	245	17	78	86		
N-Methyl-*N*-(4-methylphenyl)ethanamide [*N*-methyl-*p*-acetotoluidide; *N*-acetyl-*N*-methyl-*p*-toluidine; *N*-methyl-*N*-(*p*-tolyl)acetamide]	83		17	78	86	83	
2-Bromo-2-methylpropananilide (α-bromoisobutyranilide)	83		49			114	160
Tetradecananilide (myristanilide)	84		54		81	114	160
2-Oxobutananilide (acetoacetanilide)	85					114	160
3-Pyridinecarboxylic acid anilide (Nicotinanilide)	85 (dihyd)		238			114	160
N-(4-Propylphenyl)ethanamide [*p*-propylacetanilide; *p*-propylethananilide; *N*-(*p*-propylphenyl)acetamide]	87		17	78	86	87	115
Propylanilide (propiolanilide)	87		18			114	160
N-(3-Bromophenyl)ethanamide [*m*-bromoacetanilide; *m*-bromoethananilide; *N*-(*m*-bromophenyl)acetamide]	87		17	78	86	87	120; 136
(±)-3-Methylpentanilide	87					114	160
Butyl oxamate	88		190 (anhyd) 101 (hyd)	204 (di)			
N-(2-Chlorophenyl)ethanamide [*o*-chloroacetanilide; *o*-chloroethananilide; *N*-(*o*-chlorophenyl)acetamide]	88		17	78	86	87	99

Amides (Solids) (continued)

		Derivative mp (°C)					
			Acid			Amine	
Name of compound	mp (°C)	9-Acyl-amido-xanthene	Acid	4-Nitro-benzyl ester	4-Bromo-phen-acyl-ester	Acet-amide	Benz-amide
(+)-13-(2-Cyclopentenyl)tridecananilide [(+)-chaulmoogranilide]	89		69			114	160
Hexadecananilide (palmitanilide)	90		63	43	86	114	160
Bromoethanamide (bromoacetamide)	91		50	88			
Z-N-Phenylbutenedioic acid imide (*N*-phenylmaleimide)	91		130	91 (di)	170; 190	114	160
Didecananilide (1-eicosananilide; arachidanilide)	92		77		89	114	160
(±)-2-Chloropropananilide [(±)-*a*-chloropropionanilide]	92					114	160
2,2-Dimethylbutanilide	92					114	160
Z-Butenedioic acid imide (maleimide)	93		130	91 (di)	170; 190		
N-(2-Nitrophenyl)ethanamide [*o*-nitro-acetanilide; *o*-nitroethananilide; *N*-(*o*-nitrophenyl)acetamide]	94		17	78	86	94	110
2-Ethylpentananilide	94					114	160
N-Methyl-*N*-(1-napththyl)ethanamide [*N*-methyl-*N*-(α-naphthyl)acetamide]	94		17	78	86	95	121
E-9-Octadecenamide (elaidamide)	94		51		65		
E-13-Docosenamide (transbrassidamide)	94		60		94		
Hexanilide (caproanilide)	95				72	114	160
Butananilide (butyranilide)	95			35	63	114	160
N-Benzylhexadecanamide (*N*-benzylpalmitamide)	95		63	43	86	60	105
Nonanedioic acid monoamide (azelaic acid monoamide)	95		106	44 (di)	131 (di)		
(±)-2-Methylpentanilide	95					114	160
Octadecananilide (stearanilide)	95		70		92	114	160
Heptanamide	96	155			72		
Semicarbazide	96					67 (mono) 138 (di)	172 (mono) 241 (di)
3-Methylbenzamide (*m*-toluamide)	97		113	87	108		
Trichloroethananilide (trichloroacetanilide)	97		58	80		114	160
N-(3-Ethoxyphenyl)ethanamide [*m*-acetophenetidide; *N*-acetyl-*m*-phenetidine; *N*-acetyl-3-ethoxyaniline; *N*-(*m*-ethoxyphenyl)acetamide]	97		17	78	86	37	103
N-Benzyloctadecanamide (*N*-benzylstearamide)	97		70		92	60	105
Hydroxyethananilide (glycolanilide; hydroxyacetanilide)	97		80	107	138	114	160
3-Phenylpropananilide (β-phenylpropionanilide; hydrocinnamanilide)	98		48	36	104	114	160
2-Hydroxy-2-methylpropanamide (α-hydroxyisobutyramide)	98		79	81	98		

Amides (Solids) (continued)

Name of compound	mp (°C)	Derivative mp (°C)					
			Acid			Amine	
		9-Acyl-amido-xanthene	Acid	4-Nitro-benzyl ester	4-Bromo-phen-acyl-ester	Acet-amide	Benz-amide
2-Methylhexananilide	98					114	160
Phenoxyethananilide (phenoxyacetanilide)	99		100		149	114	160
Decanamide (capramide)	99	148	31		67		
N-(4-Methyl-2-nitrophenyl)ethanamide [4-methyl-2-nitroacetanilide; 4-methyl-2-nitroethananilide; *N*-(4-methyl-2-nitrophenyl)acetamide]	99		17	78	86	99	148
Nonanamide (pelargonamide)	99	148	12		69		
N-(3-Bromophenyl)ethanamide [*m*-bromo-acetanilide; *m*-bromoethananilide; *N*-(*m*-bromophenyl)acetamide]	99		17	78	86	99	116
2-Bromopropananilide (α-bromopropionanilide)	99; 110		26			114	160
Tridecanamide	100		43		75		
Hexanamide (caproamide)	101	160			72		
N,*N*-Diphenylethanamide (*N*,*N*-diphenylacetamide)	101		17	78	86	101	180
N,*N*′-Diacetyl-1,3-diaminopropane (*N*,*N*′-diacetyltrimethylenediamine)	101		17	78	86	126 (mono) 101 (di)	140 (mono) 147 (di)
Phenoxyethanamide (phenoxyacetamide)	101		100		149		
Methylurea	102	230				28	80
N-Methyl-*N*-phenylethanamide (*N*-methyl-ethananilide; *N*-methyl-*N*-phenyl-acetamide; *N*-methylacetanilide)	102		17	78	86	102	63
4-Oxopentananilide (levulinanilide)	102		35	61	84	114	160
Pentadecanamide	102		52	40	77		
Z-2-Butenanilide (isocrotonanilide)	102		15		81	114	160
Z-2-Butenamide (isocrotonamide)	102		15		81		
N-Benzoyl-3-ethoxyaniline [*m*-benzophenetidide; *N*-benzoyl-*m*-phenetidine; *N*-(*m*-ethoxyphenyl)benzamide]	103			89	119	97	103
N,*N*-Dicyclohexylethanamide (*N*,*N*-dicyclohexylacetamide)	103		17	78	86	103	153
Undecanamide (hendecanamide)	103		29; 16		68		
3,4-Dimethylbenzanilide	104		166			114	160
N-Benzoyl-2-ethoxyaniline [*N*-benzoyl-*o*-phenetidine; *o*-benzophenetidide; *N*-(*o*-ethoxyphenyl)benzamide]	104		122	89	119	97	103
2-Oxopropananilide (pyruvanilide)	104		14			114	160
N-Cyclohexylethanamide (*N*-cyclohexylacetamide)	104		17	78	86	104	149
N-(2-Propylphenyl)ethanamide [*o*-propylacetanilide; *o*-propylethananilide; *N*-(*o*-propylphenyl) acetamide]	105		17	78	86	105	119

Amides (Solids) (continued)

		Derivative mp (°C)					
			Acid			Amine	
Name of compound	mp (°C)	9-Acyl-amido-xanthene	Acid	4-Nitro-benzyl ester	4-Bromo-phen-acyl-ester	Acet-amide	Benz-amide
2-Methylpropananilide (isobutyranilide)	105				77	114	160
N-(4-Butylphenyl)ethanamide [*p*-butyl-acetanilide; *p*-butylethananilide; *N*-(*p*-butylphenyl)acetamide]	105		17	78	86	105	126
Propenanilide (acrylanilide)	105		13			114	160
3-Phenylpropanamide (hydrocinnamamide; *β*-phenylpropionamide)	105	189	48	36	104		
Propananilide (propionanilide)	106			31	64	114	160
Pentanamide (valeramide)	106	167			75		
Hexadecanamide (palmitamide)	106	142	63	42	86		
N-Benzylbenzamide	106		122	89	119	60	105
N,N-Dimethyurea (*sym*-dimethylurea)	106					28	80
N-Benzylbenzanilide	107		122	89	119	58	107
Tetradecanamide (myristamide)	107		54		81		
Heptadecanamide (margaramide)	108		61	49	82		
4-Oxopentanamide (levulinamide)	108		35	61	84		
Nonanedioic acid (azelaic acid monoanilide)	108		107	144 (di)	131 (di)	114	160
N-(2-Iodophenyl)ethanamide [*o*-iodo-acetanilide; *o*-iodoethananilide; N-(*o*-iodophenyl)acetamide]	109		17	78	86	109	139
Octadecanamide (stearamide)	109	141	70		92		
Heptanedioic acid monoanilide (pimelic acid monoanilide)	109		105		137 (di)	114	160
Phenylpropynamide (phenylpropiolamide)	109		137	83			
Didecanamide (arachidamide)	109		77		89		
3-Methylbutananilide (isovaleranilide)	110				68	114	160
(±)-2-Methylbutananilide	110				55	114	160
Octanamide (caprylamide)	110	148	16		67		
N-(2-Ethylphenyl)ethanamide [*o*-ethyl-acctanilide; *o*-ethylethananilide; *N*-(*o*-ethylphenyl)acetamide]	111		17	78	86	111	147
4-Methylpentananilide (isocaproanilide)	112				77	114	160
N-(2-Methylphenyl)ethanamide [*N*-acetyl-*o*-toluidine; *o*-aceto-toluidide; *N*-(*o*-tolyl)acetamide]	112		17	78	86	111	146
N-Phenylethanamide (ethananilide; acetanilide; *N*-phenylacetamide)	114		17	78	86	114	160
N-Benzyl-*E*-2-butenamide (*N*-benzylcrotonamide)	114		72	67	96	60	105
Ethyl ethanedioic acid monoamide (ethyl oxamate)	114		189 (anhyd) 101 (dihyd)	204 (di)			71
Dihydroethanoic acid monoanilide (dihydroacetic acid monoanilide)	115		109			114	160
Butanamide (butyramide)	116	187			63		

Amides (Solids) (continued)

Name of compound	mp (°C)	Derivative mp (°C)					
			Acid			Amine	
		9-Acyl-amido-xanthene	Acid	4-Nitro-benzyl ester	4-Bromo-phen-acyl-ester	Acet-amide	Benz-amide
2-Bromo-3-methylbutananilide (α-bromoisovaleranilide)	116		44			114	160
2-Chlorobenzanilide	118		142	106	106	114	160
Z-2-Butenanilide (crotonanilide)	118		72	67	96	114	160
Phenylethananilide (phenylacetanilide; *N*,2-diphenylethanamide; *N*,2-diphenylacetamide)	118		77	65	89	114	160
Dichloroethananilide (dichloroacetanilide)	118				99	114	160
N-Ethyl-*N*-(4-nitrophenyl)ethanamide [*N*-ethyl-*p*-nitroacetanilide; *N*-ethyl-*p*-nitroethananilide; *N*-ethyl-*N*-(*p*-nitrophenyl)acetamide]	118		17	78	86	119	98
1,3,5-Benzenetricarboxylic acid trianilide (trimesic acid trianilide)	120		380		197 (tri)	114	160
N-(3-Iodophenyl)ethanamide [*m*-iodo-acetanilide; *m*-iodoethananilide; *N*-(*m*-iodophenyl)acetamide]	120		17			119	157
Hydroxyethanamide (glycolamide; hydroxyacetamide)	120		80	107	138		
Chloroethanamide (chloroacetamide)	120	209	52; 56; 61		104		
4-Methylpentanamide (isocaproamide)	121	160			77		
2-Chlorophenoxyethananilide (*o*-chlorophenoxyacetanilide)	121		146			114	160
Decanedioic acid monoanilide (sebacic acid monoanilide)	122		33	74 (di)	147 (di)	114	160
Cyanoethanamide (cyanoacetamide)	123	223	66				
Furanilide	124		134	134	139	114	160
3-Chlorobenzanilide	125		158	107	116	114	160
4-Chlorophenoxyethananilide (*p*-chlorophenoxyacetanilide)	125		158		136	114	160
N-Benzoyl-3-methylaniline (*N*-benzoyl-*m*-toluidine; *m*-benzotoluidine; *N*-benzoyl-*m*-aminotoluene)	125		122	89	119	65	125
2-Methylbenzanilide (*o*-toluanilide)	125		108	91	57	114	160
1-Acetyl-2-phenylhydrazine (acetyl-β-phenylhydrazine)	126		17	78	86	128; 107 (di)	168; 177 (di)
3-Methylbenzanilide (*m*-toluanilide)	126		113	87	108	114	160
Butanedioic acid imide (succinimide)	126	247	185	88 (di)	211 (di)		
Z-2-Methyl-2-butenanilide (angelanilide)	126		46			114	160
2,4-Dihydroxybenzanilide (β-resorcylanilide)	127		217	189		114	160
2-Ethylbutanilide	127					114	160
Octanedioic acid monoamide (suberic acid monoamide)	127		144	85 (di)	144 (di)		

Amides (Solids) (continued)

			Derivative mp (°C)				
			Acid			Amine	
Name of compound	mp (°C)	9-Acyl-amido-xanthene	Acid	4-Nitro-benzyl ester	4-Bromo-phen-acyl-ester	Acet-amide	Benz-amide
N-(4-Methoxyphenyl)ethanamide [*p*-methoxyacetanilide; *p*-methoxyethan-anilide; *N*-(*p*-methoxyphenyl)acetamide]	127		17	78	86	130	157
Phenylpropynanilide (phenylpropiolanilide)	128		137	83		114	160
Octanedioic acid monoanilide (suberic acid monoanilide)	129		144	85 (di)	144 (di)	114	160
N-(2-Bromo-4-nitrophenyl)ethanamide [2-bromo-4-nitroacetanilide; 2-bromo-4-nitroethananilide; *N*-(2-bromo-4-nitrophenyl)acetamide]	129		17	78	86	129	160
2-Methylpropanamide (isobutyramide)	129	211			77		
2-Methoxybenzamide	129		101		113		
Benzamide	130	224	122	89	119		
Hexanedioic acid monoamide (adipic acid monoamide)	130		154	106	154		
2-Aminobenzanilide (anthranilanilide)	131		147			114	160
Bromoethananilide (bromoacetanilide)	131		50	88		114	160
2-Methoxybenzanilide	131		101		113	114	160
3-Pyridinecarboxylic acid anilide (nicotinanilide)	132		238			114	160
3,3-Dimethylbutanilide	132					114	160
N-(2,5-Dichlorophenyl)ethanamide [2,5-dichloroacetanilide; 2,5-dichloroethananilide; *N*-(2,5-dichlorophenyl)acetamide]	132		17	78	86	132	120
Propanedioic acid monoamide (malonic acid monoamide)	132		135	86 (di)		114	160
N-(2,4-Dimethylphenyl)ethanamide [2,4-dimethylacetanilide; 2,4-dimethylethananilide; *N* (2,4-dimcthylphenyl)acetamide]	133		17	78	86	133	190
2,2-Dimethylpropananilide (pivalanilide; trimethylacetanilide)	133		36		76	114	160
4-Chlorophenoxyethanamide (*p*-chlorophenoxyacetamide)	133		158		136		
N-(2-Napththyl)ethanamide [*N*-(β-naphthyl)acetamide]	134		17	78	86	132	162
3-Chlorobenzamide	134		158	107	116		
Chloroethananilide (chloroacetanilide)	134		61; 56; 52		104	114	160
2-Hydroxy-2-phenylethanamide [(±)-mandelamide]	134		118	124			
N-(4-Ethoxyphenyl)ethanamide [*N*-acetyl-4-ethoxyaniline; *p*-acetophenetidide; phenacetin; *N*-acetyl-*p*-phenetidine; *N*-(*p*-ethoxyphenyl)acetamide]	135		17	78	86	137	173

Amides (Solids) (continued)

		Derivative mp (°C)					
			Acid			Amine	
Name of compound	mp (°C)	9-Acyl-amido-xanthene	Acid	4-Nitro-benzyl ester	4-Bromo-phen-acyl-ester	Acet-amide	Benz-amide
N-(2,3-Dimethylphenyl)ethanamide [2,3-dimethylacetanilide; 2,3-dimethylethananilide; *N*-(2,3-dimethylphenyl)acetamide]	135		17	78	86	135	189
Acetyl-2-hydroxybenzanilide (acetylsalicylanilide)	136		135	91		114	160
3-Bromobenzanilide	136		155	105	120	114	160
2-Hydroxybenzanilide (salicylanilide)	136		158	98	140	114	160
2-Hydroxy-2-methylpropananilide (α-hydroxyisobutyranilide)	136		79	81	98	114	160
3-Methylbutanamide (isovaleramide)	136	183			68		
N,N′-Diacetyl-1,4-diaminobutane (*N,N′*-diacetyltetramethylenediamine)	137		17	78 (di)	86 (di)	137	177
N-Benzoyl-2-iodoaniline (benzo-*o*-iodoanilide)	139		122	89	119	109	139
N-(2,5-Dimethylphenyl)ethanamide [2,5-dimethylacetanilide; 2,5-dimethylethananilide; *N*-(2,5-dimethylphenyl)acetamide]	139		17	78	86	139	140
3-Aminobenzanilide	140		174			114	160
2-Iodobenzanilide	141		162	111	143	114	160
2-Bromobenzanilide	141		150	110	102	114	160
Trichloroethanamide (trichloroacetamide)	141		58	80			
3-Methylphenylurea (*m*-tolylurea)	142					65	125
2-Chlorobenzamide	142		142	106	106		
2-Hydroxybenzamide (salicylamide)	142		158	98	140		
Furamide	143	210	134	134	139		
3-Nitrobenzamide	143		140	141	132		
2-Methylbenzamide (*o*-toluamide)	143	200	108	91	57		
Iodoethananilide (iodoacetanilide)	144		83		114	160	
N-(3,5-Dimethylphenyl)ethanamide [3,5-dimethylacetanilide; 3,5-dimethylethananilide; *N*-(3,5-dimethylphenyl)acetamide]	144		17	78	86	144	136
N-Benzoyl-2-methylaniline (*o*-benzotoluidide; *N*-benzoyl-*o*-toluidine; *N*-benzoyl-*o*-aminotoluene)	144		122	89	119	111	146
N-(2,4-Dichlorophenyl)ethanamide [2,4-dichloroacetanilide; 2,4-dichloroethananilide; *N*-(2,4-dichlorophenyl)acetamide]	145		17	78	86	145	177
Cyclohexanecarboxanilide (hexahydrobenzanilide)	146		31			114	160
Diethylpropanedioic acid monoamide (diethylmalonic acid monoamide)	146		125	91 (di)			
Phenylurea	147	225				114	160

Amides (Solids) (continued)

		Derivative mp (°C)					
			Acid			Amine	
Name of compound	mp (°C)	9-Acyl-amido-xanthene	Acid	4-Nitro-benzyl ester	4-Bromo-phen-acyl-ester	Acet-amide	Benz-amide
N,N′-Dibenzoyl-1,3-diaminopropane (*N,N′*-dibenzoyltrimethylenediamine)	147		122	89	119	126 (mono) 101 (di)	140 (mono) 147 (di)
N-Benzoyl-2-ethylaniline (benzo-*o*-ethylanilide)	147		122	89	119	111	147
4-Methylbenzanilide (*p*-toluanilide)	148		182	105	153	114	160
N-(2,4,6-Trichlorophenyl)ethanamide [2,4,6-trichloroacetanilide; 2,4,6-trichloroethananilide; *N*-(2,4,6-trichlorophenyl)acetamide]	148		17	78	86	148	172
E-3-Phenyl-2-propenamide (*trans*-cinnamide)	148		133	117	146		
Butanedioic acid monoanilide (succinic acid monoanilide)	149		185	88 (di)	211 (di)	114	160
N-Cyclohexylbenzamide	149		122	89	119	104	149
Benzylurea	149					60	105
1,2-Benzenedicarboxylic acid monoamide (phthalic acid monoamide)	149		206; 230	156 (di)	153 (di)		
2-Benzoylpropananilide (*β*-benzoylpropionanilide)	150		116			114	160
Ethylpropanedioic acid monoanilide (ethylmalonic acid monoanilide)	150		111	75		114	160
N-(4-Fluorophenyl)ethanamide [*p*-fluoroacetanilide; *p*-fluoroethananilide; *N*-(*p*-fluorophenyl)acetamide]	152		17	78	86	152	185
(±)-2-Hydroxy-2-phenylethananilide [(±)-mandelanilide; (±)-2-hydroxy-2-phenylacetanilide; 2-hydroxy-*N*,2-diphenyl-acetamide; 2-hydroxy-*N*,2-diphenyl-ethanamide]	152		118	124		114	160
Hexanedioic acid monoanilide (adipic acid monoanilide)	153		154	106	155	114	160
3-Phenyl-2-propenanilide (cinnamanilide)	153		133	118	146	114	160
N-(4-Methylphenyl)ethanamide [*N*-acetyl-*p*-toluidine; *p*-aceto-toluidide; *N*-(*p*-tolyl)acetamide]	153		17	78	86	147	158
N-Methyl-*N*-(4-nitrophenyl)ethanamide [*N*-methyl-*p*-nitroacetanilide; *N*-methyl-*p*-nitroethananilide; *N*-methyl-*N*-(*p*-nitrophenyl)acetamide]	153		17	78	86	153	112
3,4-Dimethoxybenzanilide (veratranilide)	154		181			114	160
3-Nitrobenzanilide	154		140	141	132	114	160
2,2-Diphenyl-2-hydroxyethanamide (benzilamide; 2,2-diphenyl-2-hydroxyacetamide)	155		150	100	152		
2-Bromobenzamide	155		150	110	102		

Amides (Solids) (continued)

		Derivative mp (°C)					
			Acid			Amine	
Name of compound	mp (°C)	9-Acyl-amido-xanthene	Acid	4-Nitro-benzyl ester	4-Bromo-phen-acyl-ester	Acet-amide	Benz-amide
Dibenzylethananilide (dibenzylacetanilide)	155		89			114	160
2-Nitrobenzanilide	155		146	112	107	114	160
3-Bromobenzamide	155		155	105	120		
N-(3-Nitrophenyl)ethanamide [*m*-nitroacetanilide; *m*-nitroethananilide; *N*-(*m*-nitrophenyl)acetamide]	155		17	78	86	155 (mono) 76 (di)	155 (mono) 150 (di)
N-Phenylbutanedioic acid imide (*N*-phenylsuccinimide)	156		185	88 (di)	211 (di)	114	160
Heptanedioic acid dianilide (pimelic acid dianilide)	156		105		137 (di)	114	160
Phenylethanamide (phenylacetamide)	157	196	77	65	89		
3-Hydroxybenzanilide	157		200	108	176	114	160
Butanedioic acid monoamide (succinic acid monoamide)	157		185	88 (di)	211 (di)		
2,2-Dimethylpropanamide (pivalamide; trimethylacetamide)	157		36		76		
Hydroxybutanedioic acid diamide (malamide; hydroxysuccinamide)	158		101	87 (mono) 125 (di)	179 (di)		
N-Benzoyl-4-methylaniline (*N*-benzoyl-*p*-toluidine; *p*-benzotoluidide; *N*-benzoyl-*p*-aminotoluene)	158		122	89	119	147	158
N-Acetylindole	158		17	78	86	158	68
N-(1-Naphthyl)ethanamide [*N*-(α-naphthyl)acetamide]	159		17	78	86	159	160
N-(2-Naphthyl)ethananilide [*N*-(β-naphthyl)acetanilide; *N*-(β-naphthyl)-*N*-phenylethanamide; *N*-(β-naphthyl)-*N*-phenylacetamide]	160		135			114	160
4-Methylbenzamide (*p*-toluamide)	160	225	180	105	153		
E-2-Butenamide (crotonamide)	161		72	67	96		
N-(1-Naphthyl)benzamide	161		122	89	119	159	160
N-(2-Naphthyl)benzamide	162		122	89	119	132	162
N-(4-Aminophenyl)ethanamide [*p*-aminoacetanilide; *p*-aminoethananilide; *N*-(*p*-aminophenyl)acetamide]	162		17	78	86	163 (mono) 304 (di)	128 (mono) 300 (di)
4-Hydroxybenzamide	162		215	182	192		
Benzanilide	163		122	89	119	114	160
1-Naphthanilide	164		162		136	114	160
2-Hydroxy-3,3,3-trichloropropananilide (trichlorolactanilide)	164		124			114	160
2-Benzoylbenzamide	165		128; 91 (hyd)	100			
3,4-Dihydroxybenzanilide (protocatechuanilide)	166		200			114	160
4-Methoxybenzamide (*p*-anisamide)	167		186	132	152		

Amides (Solids) (continued)

		Derivative mp (°C)					
			Acid			Amine	
Name of compound	mp (°C)	9-Acyl-amido-xanthene	Acid	4-Nitro-benzyl ester	4-Bromo-phen-acyl-ester	Acet-amide	Benz-amide
Acetyl-2-aminobenzanilide (acetylanthranilanilide)	167		185			114	160
N-(4-Bromophenyl)ethanamide [*p*-bromo-acetanilide; *p*-bromoethananilide; *N*-(*p*-bromophenyl)acetamide]	168		17	78	86	168	204
1-Benzoyl-2-phenylhydrazine (α-benzoyl-β-phenylhydrazine)	168		122	89	119	128	168
N-(4-Hydroxyphenyl)ethanamide [*p*-hydroxyacetanilide; *p*-hydroxyethananilide; *N*-(*p*-hydroxyphenyl)acetamide]	169		17	78	86	168 (mono) 150 (di)	217 (mono) 234 (di)
Propanedioic acid diamide (malonic acid diamide; malondiamide)	170	270	135	86 (di)			
3-Hydroxybenzamide	170		200	108	177		
Decanedioic acid monoamide (sebacic acid monoamide)	170		33	74 (di)	147 (di)		
Z-1,2,3-Propenetricarboxylic acid anilide (*cis*-aconitranilide)	170		125		186 (tri)	114	160
(±)-Phenylbutanedioic acid monoanilide [(±)-phenylsuccinic acid monoanilide]	170		168; 84 (anhyd)			114	160
4-Methoxybenzanilide (*p*-anisanilide)	171		186	132	152	114	160
4-Ethoxybenzanilide	172		198			114	160
2-Naphthylanilide	172		185			114	160
N,N'-Diacetyl-1,2-diaminoethane (*N,N'*-diacetylethylenediamine)	172		17	78	86	172 (di)	244 (di)
N-Benzoyl-2,4,6-trichloroaniline (benzo-2,4,6-trichloroanilide)	172		122	89	119	148	172
Nonanedioic acid diamide (azelaic acid diamide)	172		106	44 (di)	131 (di)		
Z-Butenedioic acid monoamide (maleic acid monoamide; maleamic acid)	173		138	91	190		
N-(4-Ethoxyphenyl)urea (*p*-phenethylurea)	173					137	173
N-Benzoyl-4-ethoxyaniline [*N*-benzoyl-*p*-phenetidine; *p*-benzophenetidide; *N*-(*p*-ethoxyphenyl)benzamide]	173		122	89	119	137	173
Heptanedioic acid diamide (pimelic acid diamide)	175		105		137 (di)		
Benzylanilide	175		150	100	152	114	160
(±)-Phenylbutanedioic acid monoanilide [(±)-phenylsuccinic acid monoanilide]	175		168; 84 (anhyd)			114	160
Z-Methylbutenedioic acid dianilide (citraconic acid dianilide; methylmaleic acid dianilide)	176		93; 22	71		114	160
Pentanedioic acid diamide (glutaric acid diamide)	176		98	69	140		
2-Nitrobenzamide	176		146	112	107		

Amides (Solids) (continued)

		Derivative mp (°C)					
			Acid			Amine	
Name of compound	**mp (°C)**	**9-Acyl-amido-xanthene**	**Acid**	**4-Nitro-benzyl ester**	**4-Bromo-phen-acyl-ester**	**Acet-amide**	**Benz-amide**
2-Carboxyphenyl-2-oxoethanoic acid monoanilide (phthalonic acid monoanilide; 2-carboxyphenyl-2-oxoacetic acid monoanilide)	176		146			114	160
E-Methylbutenedioic acid diamide (mesaconic acid diamide)	177		205	134 (di)			
1,2,2-Trimethylcyclopentane-1,3-dicarboxylic acid monoamide (D-camphoric acid monoamide)	177		188	65			
N,N'-Dibenzoyl-1,4-diaminobutane (*N,N'*-dibenzoyltetramethylenediamine)	177		122	89	119	137 (di)	177 (di)
N-(2,6-Dimethylphenyl)ethanamide [2,6-dimethylacetanilide; 2,6-dimethylethananilide; *N*-(2,6-dimethylphenyl)acetamide]	177		17	78	86	177	168
4-Cyanobenzanilide	179		219	189		114	160
N-Benzyl-1,2-benzenedicarboxylic acid amide (*N*-benzoylphthalamide)	179		206; 230	156 (di)	153 (di)	60	105
N-(4-Chlorophenyl)ethanamide [*p*-chloroacetanilide; *p*-chloroethananilide; *N*-(*p*-chlorophenyl)acetamide]	179		17	78	86	179	192
D-2,3-Dihydroxybutanedioic acid monoanilide (D-tartaric acid monoanilide)	180		171	163 (di)	204 (di)	114	160
1-Acetyl-2-methylurea (α-acetyl-β-methylurea)	180		17	78	86	28	80
Diphenylethananilide (diphenylacetanilide)	180		148			114	160
Z-Butenedioic acid diamide (maleic acid diamide)	181		137	91	170; 190		
4-Methylphenylurea (*p*-tolylurea)	182					147	158
N,N-Dimethylurea (*asym*-dimethylurea)	182	225; 250					41
N-Benzoylaminoethanamide (hippuramide; *N*-benzoylaminoacetamide)	183		187	136	151		
3,5-Dinitrobenzamide	183		205	157	159		
2-Iodobenzamide	184		162	111	143		
N-(4-Iodophenyl)ethanamide [*p*-iodo-acetanilide; *p*-iodoethananilide; *N*-(*p*-iodophenyl)acetamide]	184		17	78	86	184	222
N-Benzoyl-4-fluoroaniline (benzo-*p*-fluoroanilide)	185		122	89	119	152	185
N,N'-Diacetyl-2-diaminobenzene (*N,N'*-diacetyl-1,2-phenylenediamine)	185		17 (di)	78 (di)	86	185	301
3-(2-Nitrophenyl)-2-propenamide (2-nitrocinnamamide)	185		240	132	141		

Amides (Solids) (continued)

		Derivative mp (°C)					
			Acid			Amine	
Name of compound	mp (°C)	9-Acyl-amido-xanthene	Acid	4-Nitro-benzyl ester	4-Bromo-phen-acyl-ester	Acet-amide	Benz-amide
E-2-Methyl-2-butenedioic acid dianilide (Mesaconic acid dianilide)	186		204	134 (di)		114	160
Z-2-Chloro-2-butenedioic acid anilide (chlorofumaranilide)	186		192	139		114	160 (di)
3-Iodobenzamide	186		187	121	128		
Octanedioic acid dianilide (suberic acid dianilide)	187		144	85 (di)	144 (di)	114	160
Z-2-Butenedioic acid dianilide (maleic acid dianilide)	187		137	91 (di)	170; 190	114	160
Z-2-Butenedioic acid monoanilide (maleic acid monoanilide)	187		137	91 (di)	170; 190	114	160
Z-2-Methyl-2-butenedioic acid diamide (citraconic acid diamide)	187		93	71 (di)			
N-Benzoylaminoethanoic acid (hippuric acid; *N*-benzoylglycine; *N*-benzoylaminoacetic acid)	188		122	89	119	206	188
N,N-Diphenylurea (*asym*-diphenylurea)	189	180				101	180
1,2,3-Propanetricarboxylic acid dianilide (aconitic acid dianilide)	189		195		186 (tri)	114	160
4-Bromobenzamide	190		253	180			
(2*R*,3*S*)-2,3-Dihydroxybutanedioic acid diamide (*meso*-tartaric acid diamide)	190		140	93			
2-Methylenebutanedioic acid dianilide (itaconic acid dianilide; methylenesuccinic acid dianilide)	190		165	91	117	114 (di)	160 (di)
N,N′-Diacetyl-3-diaminobenzene (*N,N′*-diacetyl-3-phenylenediamine)	191		17	78	86	89 (mono) 191 (di)	125 (mono) 240 (di)
2-Methylenebutanedioic acid diamide (itaconic acid diamide; methylenesuccinic acid diamide)	192		165	91 (di)	117 (di)		
(2*R*,3*S*,4*R*,5*S*)-2,3,4,5-Tetrahydroxy-hexanedioic acid monoamide (mucic acid monoamide; galacratic acid monoamide)	192		214; 255	310	225		
4-Nitro-1,2-benzenedicarboxylic acid anilide (4-nitrophthalanilide)	192		165			114	160
2-Methylphenylurea (*o*-tolylurea)	192	228				111	146
N-Benzoyl-4-chloroaniline (benzo-*p*-chloroaniline)	192		122	89	119	179	192
D-1,2,2-Trimethylcyclopentane-1,3-dicarboxylic acid diamide (D-camphoric acid diamide)	193		188	66			
4-Chlorobenzanilide	194		243	130	126	114	160
2-Benzoylbenzanilide	195		128; 91 (hyd)	101		114	160

Amides (Solids) (continued)

		Derivative mp (°C)					
			Acid			Amine	
Name of compound	mp (°C)	9-Acyl-amido-xanthene	Acid	4-Nitro-benzyl ester	4-Bromo-phen-acyl-ester	Acet-amide	Benz-amide
3-(3-Nitrophenyl)-2-propenamide (3-nitrocinnamamide)	196		199	174	178		
(2*S*,3*S*)-2,3-Dihydroxybutanedioic acid diamide (D-tartaric acid diamide)	196		171	163 (di)	204 (di)		
Hydroxybutanedioic acid dianilide (L-malanilide; hydroxysuccinanilide)	197		101	87 (mono) 125 (di)	179 (di)	114	160
4-Hydroxybenzanilide	198		215	182	192	114	160
N-[4-(Nitrophenyl)phenyl]ethanamide {*p*-nitrophenylacetanilide; *p*-nitrophenylethananilide; *N*-[*p*-(nitrophenyl)phenyl]acetamide}	198		153		207	215	203
Cyanoethananilide (cyanoacetanilide)	199		66			114	160
2-Hydroxy-1,2,3-propanetricarboxylic acid trianilide (citric acid trianilide)	199		100; 153 (anhyd)	102 (tri)	148 (tri)	114	160
E-1,2,3-Propenetricarboxylic acid dianilide (aconitic acid dianilide)	200		125	186 (tri)	114	160	
2-Methylbutanedioic acid dianilide (methylsuccinic acid dianilide)	200		115			114	160
4-Nitrobenzamide	200	232	241	168	137		
3-Nitro-1,2-benzenedicarboxylic acid diamide (3-nitrophthalic acid diamide)	201		218	189			
Decanedioic acid dianilide (sebacic acid dianilide)	201		33	74 (di)	147 (di)	114	160
N-(2-Methyl-4-nitrophenyl)ethanamide [2-methyl-4-nitroacetanilide; 2-methyl-4-nitroethananilide; *N*-(2-methyl-4-nitrophenyl)acetamide]	202		17	78	86	202	
1-Naphthoamide	202		162		136		
2,4-Dinitrobenzamide	203		183	142	158		
D-1,2,2-Trimethylcyclopentane-1,3-dicarboxylic acid monoanilide (D-camphoric acid monoanilide)	204		188	66		114	160
N-Benzoyl-4-bromoaniline (benzo-4-bromoanilide)	204		122	89	119	168	204
N-Phenyl-1,2-benzenedicarboxylic acid imide (*N*-phenylphthalimide)	205		206; 230	156	153	114	160
3,4,5-Trihydroxybenzanilide (gallanilide)	207		254; 222	141	134	114	160
2-Carboxyphenyl-2-oxoethanoic acid dianilide (phthalonic acid dianilide; 2-carboxyphenyl-2-oxoacetic acid dianilide)	208		146			114	160
N-Benzoylaminoethananilide (*N*-benzoyl-aminoacetanilide; hippuranilide)	208		187	136	151	114	160
Decanedioic acid diamide (sebacic acid diamide)	210		133	74 (di)	147 (di)		

Amides (Solids) (continued)

		Derivative mp (°C)					
			Acid			Amine	
Name of compound	mp (°C)	9-Acyl-amido-xanthene	Acid	4-Nitro-benzyl ester	4-Bromo-phen-acyl-ester	Acet-amide	Benz-amide
4-Iodobenzanilide	210		270	141	146	114	160
4-Nitrobenzanilide	211		241	168	137	114	160
3,4-Dihydroxybenzamide (protocatechuamide)	212		200	188			
2,2′-Biphenyldicarboxylic acid diamide (2,2′-diphenic acid diamide)	212		233	187 (di)			
N-(4-Nitrophenyl)ethanamide [*p*-nitroacetanilide; *p*-nitroethananilide; *N*-(*p*-nitrophenyl)acetamide]	214		17	78	86	215	203
Ethylpropanedioic acid diamide (ethylmalonic acid diamide)	214		111	75			
2-Hydroxy-1,2,3-propanetricarboxylic acid triamide (citric acid triamide)	215		153	102 (tri)	148 (tri)		
3-Nitro-1,2-benzenedicarboxylic acid imide (3-nitrophthalimide)	216		218	189			
3-(4-Nitrophenyl)-2-propenamide (4-nitrocinnamamide)	217		285	186	191		
Octanedioic acid diamide (suberic acid diamide)	217		144	85	144		
4-Iodobenzamide	217		270	141	146		
Benzylpropanedioic acid dianilide (benzylmalonic acid dianilide)	217		121	120		114	160
Acetylurea	218		17	78	86		
sym-Di-(3-methylphenyl)urea (*sym*-di-*m*-tolylurea)	218					65	125
1,2-Benzenedicarboxylic acid diamide (phthalic acid diamide)	220		206; 230	156	153		
(2*R*,3*S*,4*R*,5*S*)-2,3,4,5-Tetrahydroxyhexanedioic acid diamide (mucic acid diamide)	220		214	310	225		
2,4-Dihydroxybenzamidc (*β*-resorcylamide)	222		213	189			
(±)-Phenylbutanedioic acid dianilide [(±)-phenylsuccinic acid dianilide]	222		168; 84 (anhyd)			114	160
4-Cyanobenzamide	223		219	189			
Diethylpropanedioic acid diamide (diethylmalonic acid diamide)	224		125	91 (di)			
Pentanedioic acid dianilide (glutaric acid dianilide)	224		98	69 (di)	140 (di)	114	160
5-Nitro-2-hydroxybenzanilide (5-nitrosalicylanilide)	224		230			114	160
Benzylpropanedioic acid diamide (benzylmalonic acid diamide)	225		121	120 (di)			
D-1,2,2-Trimethylcyclopentane-1,3-dicarboxylic acid dianilide (D-camphoric acid dianilide)	226		188	66		114	160

Amides (Solids) (continued)

Name of compound	mp (°C)	Derivative mp (°C)					
			Acid			Amine	
		9-Acyl-amido-xanthene	Acid	4-Nitro-benzyl ester	4-Bromo-phen-acyl-ester	Acet-amide	Benz-amide
(+)-2,3-Dihydroxybutanedioic acid monoamide [(±)-tartaric acid monoamide]	226		204	148			
Hexanedioic acid diamide (adipic acid diamide)	229		154	106	155		
Propanedioic acid dianilide (malonic acid dianilide)	230		135	86 (di)		114	160
Butanedioic acid dianilide (succinic acid dianilide)	230		185	88 (di)	211 (di)	114	160
2-Sulfobenzoic acid imide (saccharin)	230	199	206				
2,2′-Biphenyldicarboxylic acid anilide (2,2′-diphenanilide)	230		233	187 (di)		114	160
3,5-Dinitrobenzanilide	234		205	157	159	114	160
3-Nitro-1,2-benzenedicarboxylic acid dianilide (3-nitrophthalic acid dianilide)	234		218	189		114	160
(±)-2,3-Dihydroxybutanedioic acid monoanilide [(±)-tartaric acid monoanilide]	236		206			114	160
1,4-Benzenedicarboxylic acid monoanilide (terephthalic acid monoanilide)	237		300	264 (di)	225 (di)	114	160
N-Benzylpropenamide (*N*-benzylacrylamide)	237					60	105
1,2-Benzenedicarboxylic acid imide (phthalimide)	238	177	206; 230	89	119	232	204
N,N-Diphenylurea (*sym*-diphenylurea; carbanilide)	240					114	160
Hexanedioic acid dianilide (adipic acid dianilide)	241		154	106	155	114	160
N,N′-Dibenzoyl-1,2-diaminoethane (*N,N′*-dibenzoylethylenediamine)	244		122	89	119	172 (di)	244 (di)
3,4,5-Trihydroxybenzamide (gallamide)	245		254; 240	141	134		
3-Hydroxy-2-naphthanilide	249		223			114	160
N,N-Di-(2-methylphenyl)urea (*sym*-di-*o*-tolylurea)	250					111	146
1,2,3-Propanetricarboxylic acid trianilide (carballylic acid trianilide)	252		166			114	160
1,2-Benzenedicarboxylic acid dianilide (phthalic acid dianilide)	255		206; 230	156	153	114	160
N,N-Diphenylethanedioic acid dianilide (oxanilide; oxalic acid dianilide; ethanedioic acid dianilide)	257		189 (anhyd) 101 (dihyd)	204 (di)		114	160
Butanedioic acid diamide (succinic acid diamide)	260	275	185; 125	88 (di)	211 (di)		
(2*S*,3*S*)-2,3-Dihydroxybutanedioic acid dianilide (D-tartaric acid dianilide)	264		171	163	204	114	160

Amides (Solids) (continued)

		Derivative mp (°C)					
			Acid			Amine	
Name of compound	mp (°C)	9-Acyl-amido-xanthene	Acid	4-Nitro-benzyl ester	4-Bromo-phen-acyl-ester	Acet-amide	Benz-amide
3-Pyridinecarboxylic acid anilide (nicotinanilide)	265 (anhyd)		238			114	160
E-2-Butenedioic acid diamide (fumaric acid diamide)	266		295; 200	151			
N,N-Di-(4-methylphenyl)urea (*sym*-di-*o*-tolylurea)	268					147	158
1,3-Benzenedicarboxylic acid monoamide (isophthalic acid monoamide)	280		348	203	179		
1,3-Benzenedicarboxylic acid diamide (isophthalic acid diamide)	280		348	203	179		
(2*R*,3*S*,4*R*,5*S*)-2,3,4,5-Tetrahydroxy-hexanedioic acid monoanilide (mucic acid monoanilide)	310		214; 255	310	225	114	160
N,N'-Diacetyl-1,4-diaminobenzene (*N,N'*-diacetyl-1,4-phenylenediamine)	310		17	78	86	163 (mono) 304 (di)	128 (mono) 300 (di)
E-2-Butenedioic acid dianilide (fumaric acid dianilide)	314		286; 295; 200			114	160
N,N'-Diacetyl-4,4'-diaminobiphenyl (*N,N'*-diacetylbenzidine)	317		17	78	86	199 (mono) 317 (di)	205 (mono) 352 (di)
1,4-Benzenedicarboxylic acid dianilide (terephthalic acid dianilide)	337		300	264 (di)	225 (di)		
N,N'-Dibenzoyl-4,4'-diaminobiphenyl (*N,N'*-dibenzoylbenzidine)	352		122	89	119	199 (mono) 317 (di)	205 (mono) 351 (di)
1,3,5-Benzenetricarboxylic acid triamide (trimesic acid triamide)	365		380		197		
Ethanedioic acid diamide (oxalic acid diamide)	419		190 (anhyd) 101 (dihyd)	204			

Amines—Primary and Secondary (Liquids)

		Derivative mp (°C)					
Name of compound	**bp (°C)**	**Acetamide**	**Benzamide**	**Benzenesulfonamide**	**4-Toluenesulfonamide**	**Phenylthiourea**	**Amine hydrochloride**
Methylamine	−6	28	80	30	77	113	
Dimethylamine	7		42	47	79	135	171
Ethylamine	19		71	58	63	106; 135	110
2-Aminopropane (isopropylamine)	33		100	26	51	101	
2-Methyl-2-aminopropane (*tert*-butylamine)	46	102	134		114	120	280
Propylamine	49		84	36	52	63	
Cyclopropylamine	50		99	120			100
Diethylamine	56		42	42	60	34	230
3-Aminopropene (allylamine)	58			39	64	98	
2-Aminobutane (*sec*-butylamine)	63		76	70	55	101	
2-Methylpropylamine (isobutylamine)	69		57	53	78	82	
Butylamine	77		42		65	65	195
3-Methylbutylamine (isopentylamine)	96				65	102	
Piperidine	106		48	94	103	101	
Dipropylamine	110			51		69	
1,2-Diaminoethane (ethylenediamine)	116	172	249	168	160; 360	102	330
(±)-2-Methylpiperidine	119		45		55		207
1,2-Diaminopropane	120	139 (di)	192 (di)		103		
2,6-Dimethylpiperidine	128		111	50			
Hexylamine	130		40	96		77	
Morpholine	130		75	119	147	136	175
Cyclohexylamine	134	104	149	89	87	148	207
1,3-Diaminopropane (trimethylenediamine)	136	126 (di); 101	147 (di)	96	148		
Piperazine	139	144 (di)	196 (di)	292 (di)	173 (mono)		83 (di)
Di-(2-methylpropyl)amine (diisobutylamine)	139	86		57		113	
1,4-Diaminobutane (tetramethylenediamine)	159	137 (di)	177 (di)		224 (di)	168	315 (di)
2-Fluoroaniline	176	80	113				
1,5-Diaminopropane (pentamethylenediamine; cadaverine)	180		135 (di)	119		148	
Aniline	184	114	163	112	103	54; 154	198
Benzylamine	185	65	106	88	116; 185	156	258
(±)-1-Phenylethylamine	187	57	120				158
4-Fluoroaniline	188	152	185				
1,2,3-Triaminopropane	190	202 (tri)	218 (tri)				
N-Methylaniline	194	103	63	79	95	87	123
2-Phenylethylamine	198	114; 51	116	69	64	135	
2-Methylaniline (*o*-toluidine; *o*-aminotoluene)	200	112	146	124	110; 186	136	215
4-Methylaniline (*p*-toluidine; *p*-aminotoluene)	200	154	158	120	118	141	243
Nonylamine	201	35	49				
3-Methylaniline (*m*-toluidine; *m*-aminotoluene)	203	66	125	95	114; 172	94	228
N-Ethylaniline	205	55	60		88	89	176

Amines—Primary and Secondary (Liquids) (continued)

		Derivative mp (°C)					
Name of compound	bp (°C)	Acetamide	Benzamide	Benzenesulfonamide	4-Toluenesulfonamide	Phenylthiourea	Amine hydrochloride
1,6-Diaminohexane (hexamethylenediamine)	205	127 (di)	158 (di)	154 (di)			
3-Methylbenzylamine	207	240	150				208
2-Chloroaniline	208	88	99	129	105; 193	156	235
N,2-Dimethylaniline [*N*-methyl-*o*-toluidine; *o*-(*N*-methylamino)toluene]	208	56	66		120		
2-Methylbenzylamine	208	69	88				
4-Methylbenzylamine	208	108	137				
1-Phenylpropylamine	208		116	81			190
2-Amino-6-methylpyridine	209	90	90				155
N,4-Dimethylaniline [*N*-methyl-*p*-toluidine; *p*-(*N*-methyl)aminotoluene]	210	83	53	67	60	89	120
L-2-(1-Methylethyl)-5-methylcyclohexylamine (L-menthylamine)	212	145	157			135	
2,5-Dimethylaniline	215	139	140	138	119; 233	148	
2-Ethylaniline	216	112	147				
4-Ethylaniline	216	94	151		104	104	
2,4-Dimethylaniline	217	133	192	130	181	133; 152	
N-Ethyl-4-methylaniline [*N*-ethyl-*p*-toluidine; *p*-(*N*-ethylamino)toluene]	217		40	66	71		
2,6-Dimethylaniline	218	177	168		212	204	
2-Methoxyaniline	218	85	60		127		
N-Ethyl-2-methylaniline [*N*-ethyl-*o*-toluidine; *o*-(*N*-ethylamino)toluene]	218		72	62	75		
2-(*N,N*-Dimethylamino)aniline	219	72	51				
3,5-Dimethylaniline	220	144	145	136		153	
2,3-Dimethylaniline	222	136	189				254
N-Propylaniline	222	47		54		104	
2-Chloro-4-methylaniline (4-amino-3-chlorotoluene)	223	118	138	110	103		
E-9-Aminobicyclo[4.4.0]decane (*trans*-9-aminodecalin)	223	183	149				
2-Propylaniline	224	105	119				173
2-Amino-3-methylpyridine	224	64	220				
2-Methoxyaniline (*o*-anisidine)	225	88	60; 84	89	127	136	
4-(1-Methylethyl)aniline (*p*-isopropylaniline; *p*-cumidine)	225	102	162				
4-Propylaniline	225	94	115				204
3,4-Dimethylaniline	226	99		118	154		
α-Methyl-α-phenylhydrazine	227	92	153	132			
4-(1,1-Dimethylethyl)aniline (*p*-*tert*-butylaniline)	228	173	140				274
Z-9-Aminobicyclo[4.4.0]decane (*cis*-9-aminodecalin)	228	127	147				
2-Bromoaniline	229	99	116			146	
2-Ethoxyaniline (*o*-phenetidine)	229	79	104	102	164	137	
2-Aminoindane	230	127	155				241

Amines—Primary and Secondary (Liquids) (continued)

		Derivative mp (°C)					
Name of compound	bp (°C)	Acetamide	Benzamide	Benzenesulfonamide	4-Toluenesulfonamide	Phenylthiourea	Amine hydrochloride
2,4,6-Trimethylaniline	232	216	204	137	167	193	
1,2,3,4-Tetrahydroisoquinoline	233	46	129	154			
2-Aminothiophenol	234	135 (di)	154 (di)				
4-(2-Methylpropyl)aniline (*p*-isobutylaniline)	235	127			137		
4-Aminoindane	236	126	136				
3-Chloroaniline	236	79	120	121	138; 210	124	
2-Bromo-4-methylaniline (4-amino-3-bromotoluene)	240	118	149			154	221
1-Aminoundecane (1-aminohendecane)	240	48	60				190
N-Butylaniline	241		56		56		
4-Chloro-2-methylaniline (2-amino-5-chlorotoluene)	241	140	172; 142	125	145		
2-Amino-4-(1-methylethyl)toluene (5-isopropyl-2-methylaniline; 2-amino-*p*-cumene; *p*-cymidine)	242	71	102				207
Phenylhydrazine	243	128; 107 (di)	168; 177 (di)	148	151	172	
3-Chloro-2-methylaniline (2-amino-6-chlorotoluene)	245	159	173				
2-Chloro-6-methoxyaniline	246	123	135				
3-Ethoxyaniline (*m*-phenetidine)	248	97	103		157	138	
1,3-Di(aminomethyl)benzene (α,α'-diamino-*m*-xylene)	248	135	172				
1,2,3,4-Tetrahydroquinoline	250		76	67			181
2-Bromoaniline	250	99	116		90	146; 161	
3-Methoxyaniline (*m*-anisidine)	251	81			68		168
3-Bromoaniline	251	88	136; 120			143; 97	
2,5-Dichloroaniline	251	132	120			166	
2-Aminoacetophenone	252	77	98		148		168d
4-Ethoxyaniline (*p*-phenetidine)	254	137	173	143	106	136	
3-Bromo-2-methylaniline (2-amino-6-bromotoluene)	255	163	177				
Dicyclohexylamine	255	103	153		119		
3-Bromo-4-methylaniline (4-amino-2-bromotoluene)	257	118	132				
4-(*N*-Methylamino)aniline [*N*-methyl-*p*-toluidine; *p*-(*N*-methylamino)aniline]	258	63	165				
Methyl 2-aminobenzoate (methyl anthranilate)	260	101	100	107			
4-(*N,N*-Diethylamino)aniline	261	104	172		112		
4-Butylaniline	261	105	126				
4-(*N,N*-Dimethylamino)aniline	262	133	228				
Ethyl 2-aminobenzoate (ethyl anthranilate)	266	61	98	93	112		
Di-(2-hydroxyethyl)amine (diethanolamine)	270			130	99		

Amines—Primary and Secondary (Liquids) (continued)

		Derivative mp (°C)					
Name of compound	bp (°C)	Acetamide	Benzamide	Benzenesulfonamide	4-Toluenesulfonamide	Phenylthiourea	Amine hydrochloride
3-(*N,N*-Dimethylamino)aniline	272	87; 69	164				
1-Amino-*N*-methylnaphthalene (*N*-methyl-α-naphthylamine)	293	95	121		164		
Ethyl 3-aminobenzoate	294	110	148; 114				
N-Benzylaniline	298	58	107	119	149	103	
2-Aminobiphenyl	299	121	86; 102				
Dibenzylamine	300		112	68	159		256
Diphenylamine	302	103	180	124	144	152	
1,1-Diphenylmethylamine (benzhydrylamine; α-amino-diphenylmethane)	304	147	172				
N-Benzylaniline	306	58	107	119	140		
2-Amino-*N*-methylnaphthalene (*N*-methyl-β-naphthylamine)	317	51	84	107	78		183
2,4′-Diaminobiphenyl	363	202	278				

Amines—Primary and Secondary (Solids)

		Derivative mp (°C)					
Name of compound	mp (°C)	Acetamide	Benzamide	Benzenesulfonamide	4-Toluenesulfonamide	Phenylthiourea	Amine hydrochloride
Ethyl 2-aminobenzoate (ethyl anthranilate)	13	61	98	92	112		
2,5-Dimethylaniline	15	142	140	138	119; 233	148	
3-Bromoaniline	18	88	136; 120			97; 143	
1,2,3,4-Tetrahydroquinoline	20		76	67			181
Dicyclohexylamine	20	103	153		119		
2-Aminoacetophenone	20	76	98		148		
Phenylhydrazine	23	128; 107 (di)	168; 177 (di)	148	151	172	
4-Aminostyrene	24	142	161				
Methyl 2-aminobenzoate (methyl anthranilate)	24	101	100	107			
4-Amino-2-bromotoluene (3-bromo-4-methylaniline)	26	118	132				
2-Bromo-4-methylaniline (4-amino-3-bromotoluene)	26	118	149			154	
2-Amino-3-methylpyridine	26	64	220				
2-Aminothiophenol	26	135 (di)	154 (di)				
1,4-Diaminobutane (putrescine; tetramethylenediamine)	27	137 (di)	177 (di)		224 (di)		315 (di)
4-Chloro-2-methylaniline (2-amino-5-chlorotoluene)	29	140	172; 142	125	145		
2-Bromoaniline	32	99	116		90	146; 161	
1-Amino-2-methylnaphthalene (2-methyl-α-naphthylamine)	32	188	180				
3-Iodoaniline	33	119	157		128		
2-Aminodibenzyl	33	117	166				198
3-Iodoaniline	33	119	157		128		
N,N′-Diphenylhydrazine (*asym*-diphenylhydrazine)	34	184	192				
1,4-Di(aminomethyl)benzene (α,α′-diamino-1,4-dimethylbenzene)	35	194 (tetra)	193 (di)				
4-(*N*-Methylamino)aniline	36	63	165				
N-Benzylaniline	37	58	107	119	149	103	
5-Aminoindane	38	106	137				
2-Amino-5,6,7,8-tetrahydronaphthalene (2-aminotetralin)	38	107	167				
2-Amino-5-bromonaphthalene (5-bromo-β-naphthylamine)	38	165	109				
2,2-Diphenylethylamine	38	88	145				257
2-Iodo-4-methylaniline (4-amino-3-iodotoluene)	40	133	161				188
2-Chloro-4,6-dimethylaniline	40	206	148				
4-(*N,N*-Dimethylamino)aniline	41	133	228				
2-Amino-6-methylpyridine	41	90	90				155
1,6-Diaminohexane (hexamethylenediamine)	42	127 (di)	155 (di)	154 (di)			
4-Amino-3-methylbiphenyl	43	165	189				
2,4′-Diaminobiphenyl	45	202 (di)	278 (di)				

Amines—Primary and Secondary (Solids) (continued)

		Derivative mp (°C)					
Name of compound	mp (°C)	Acet-amide	Benz-amide	Ben-zene-sulfon-amide	4-Tolu-ene-sulfon-amide	Phenyl-thio-urea	Amine hydro-chlo-ride
4-Methylaniline (*p*-toluidine; *p*-aminotoluene)	45	154	158	120	118	141	240
4-Aminobenzyl cyanide	46	97 (mono) 153 (di)	177				
4-Aminothiophenol	46	144	180				
1-(2-Aminophenyl)-1-propanone (2-aminopropiophenone)	47	71	130				
3,4-Dimethylaniline	49	99	118	118	154		256
Heptadecylamine	49	62	91				
2-Aminobiphenyl	49	121	86; 102				
2-Bromo-4,6-dimethylaniline	50	197	186				
2,5-Dichloroaniline	50	133	120			166	192
1-Aminooxindole	50	187	189				
1-Aminonaphthalene (α-naphthylamine)	50	160	161	167	157; 147	165	286
2-Amino-1-methylnaphthalene (1-methyl-β-naphthylamine)	51	189	222				
1-Amino-3-methylnaphthalene (3-methyl-α-naphthylamine)	52	176	189				
1-Amino-4-methylnaphthalene (4-methyl-α-naphthylamine)	52	167	239				234
Indole	52	158	68	254			
2-Aminodiphenylmethane	52	135	116				137
4-(*N*,*N*-Dimethylamino)aniline	53	131	228				
4-Aminobiphenyl (*p*-phenylaniline)	53	175; 120 (di)	233		255; 160		
Diphenylamine	54	103	180; 107	124	142	152	
2-Amino-5-bromobiphenyl	57	130	162				
4-Methoxyaniline (*p*-anisidine)	58	130	157	96	114	157; 171; 144	
1-Amino-7-methylnaphthalene (7-methyl-α-naphthylamine)	59	183	204				
4-Bromo-2-methylaniline (2-amino-5-bromotoluene)	59	157	115				
2-Amino-1-chloronaphthalene (1-chloro-β-naphthylamine)	59	147	98	131			
2-Aminoazobenzene	59	126	122				
2-Aminopyridine	60	71	169 (di); 87			216	
2-Iodoaniline	61	110	139				154
N-Phenyl-1-aminonaphthalene (*N*-phenyl-α-naphthylamine)	62	115	152				
4-Iodoaniline	62	184	222			153	
1-Amino-3-chloronaphthalene (3-chloro-α-naphthylamine)	62	197	162				219
2-Amino-4,4′-dimethylbiphenyl	63	119	96				
1,3-Diaminobenzene (1,3-phenylenediamine)	63	191 (di) 89 (mono)	240 (di) 125 (mono)	194	172		
2,4-Dichloroaniline	63	146	117	128	126		

Amines—Primary and Secondary (Solids) (continued)

Name of compound	mp (°C)	Derivative mp (°C) Acetamide	Benzamide	Benzenesulfonamide	4-Toluenesulfonamide	Phenylthiourea	Amine hydrochloride
3-Aminopyridine	64	133 (mono) 88 (di)	119				
2-Amino-5-methylnaphthalene (5-methyl-β-naphthylamine)	64	124	156				
2,5-Diaminotoluene	64	220 (di)	307	147 (2-mono)	150 (2-mono)		
9-Aminofluorene	64	262	261				255
4-Methylphenylhydrazine (p-tolylhydrazine)	65	130	146; 70 (mono)				
4-Aminobenzyl alcohol	65	188	150				217
N-Methyl-3-nitroaniline	66	95	156	83			
4-Bromoaniline	66	168	204	134	101; 141	148	
2-Amino-5-methylbenzophenone	66	159	118				
1,8-Diaminonaphthalene	66	312 (di)			207 (di)		
1-Amino-8-methylnaphthalene (8-methyl-α-naphthylamine)	68	184	196				
2,4,5-Trimethylaniline (pseudocumine)	68	162	167	136			
4-Iodoaniline	68	184	222			153	
1,2-Bis(2-aminophenyl)ethane (2,2′-diaminobenzyl)	68	249 (di)	255 (di)				
2-Amino-4-methylnaphthalene (4-methyl-β-naphthylamine)	68	173	195				
N-Methyl-3-nitroaniline	68	95	105; 155	83			
1-Amino-3-bromonaphthalene (3-bromo-α-naphthylamine)	70	174	166				247
8-Aminoquinoline	70	103	98		156		209
2-Nitroaniline	71	94	98; 110	104	115; 142	142	
3,4-Diaminotriphenylmethane	72	226 (di)	243 (di)				
4-Chloroaniline	72	179	193	122	96; 121	152	
4-Aminophenylurethane	74	202; 181	230				242
3-Nitro-2,4,6-trimethylaniline (3-nitromesidine)	75	191	169	163			
4-Aminodiphenylamine	75	158	203				
2,4-Dimethyl-6-nitroaniline	76	176	185				
2-Bromo-1,4-diaminobenzene	76	200 (di)	235 (di)				
4-Amino-2-nitrotoluene (4-methyl-3-nitroaniline)	78	148	172	160	164	145; 171	
2,4,6-Trichloroaniline	78	206	174	154			
1-Amino-5-methylnaphthalene (5-methyl-α-naphthylamine)	78	195	174				
1-Amino-N-(4-methylphenyl)naphthalene [N-(p-tolyl)-α-naphthylamine]	79	124	140	83			
Di-(4-methylphenyl)amine (di-p-tolylamine)	79	88	125				
2,4-Dibromoaniline	79	146	134		134; 171		
2-Aminodiphenylamine	80	121 (mono)	136 (mono)				
2,4-Diaminophenol	80	222 (di) 182 (tri)	253 (di) 231				240 (di)

Amines—Primary and Secondary (Solids) (continued)

		Derivative mp (°C)					
Name of compound	mp (°C)	Acetamide	Benzamide	Benzenesulfonamide	4-Toluenesulfonamide	Phenylthiourea	Amine hydrochloride
2,2′-Diaminobiphenyl	81	89 (mono) 161 (di)	159 (mono) 190 (di)				
3-Aminoacenaphthene	81	193	209				
2-Aminobenzyl alcohol	82	114 (mono)	199 (mono)				108
4-Chloro-3-methylaniline (5-amino-2-chlorotoluene)	83	91	119				
2,5-Dimethoxyaniline	83	91	85		80		
5-Chloro-2-methoxyaniline	84	104	78				
4-Aminotriphenylmethane	84	168	198				
1-Amino-3-iodonaphthalene (3-iodo-α-naphthylamine)	84	207	174				
N-Methyl-4-hydroxyaniline [*p*-(*N*-methylamino)phenol]	85	240; 97; 43 (mono)	175 (mono)		135 (mono)		
2-Aminophenanthrene	85	225	216				
3,4-Dimethoxyaniline (4-aminoveratrol)	86	133	177				
4-Aminobenzonitrile	86	205	170				
N-Methyl-2-hydroxyaniline [*o*-(*N*-methylamino)phenol]	87	150	150				
4-Aminoacenaphthene	87	176	196				
3-Aminophenanthrene	88	201	213				
4-Amino-3-methyl-1-phenylpyrazole	88	95	181				
3-Aminoethananilide (*m*-aminoacetanilide)	89	191			241		
3,4-Diaminotoluene (4-methyl-1,2-phenylenediamine)	90	210 (di); 95; 131 (mono)	264 (di)	179 (di)	140 (mono)		
2-Nitrophenylhydrazine	90	141 (di); 58	166				
2-Amino-3-nitrotoluene (2-methyl-6-nitroaniline)	91	157	167				
1-Chloro-2,4-diaminobenzene (4-chloro-1,3-phenylenediamine)	91	170 (mono) 243 (di)	178		215 (di)		
1-Amino-2,6-dimethylnaphthalene (2,6-dimethyl-α-naphthylamine)	91	211	220				
4-Iodo-2-methylaniline (2-amino-5-iodotoluene)	92	170; 162	184				
Ethyl 4-aminobenzoate (benzocaine)	92	110	148				
2-Amino-3-nitrotoluene (2-methyl-6-nitroaniline)	92	158	167				
3-Bromo-2-hydroxy-5-methylaniline (6-amino-2-bromo-4-methylphenol; 5-amino-3-bromo-4-hydroxytoluene)	93	129 (mono) 169 (di)	185 (mono) 166 (di)	157 (mono) 230 (di)			
4-Chloro-2,6-dibromoaniline	93	226	194				
3-Nitrophenylhydrazine	93	150 (di)	153 (di)				
7-Aminoquinoline	94	167	189				
2-Aminodibenzofuran	94	178; 83 (di)	201				

Amines—Primary and Secondary (Solids) (continued)

		Derivative mp (°C)					
Name of compound	mp (°C)	Acetamide	Benzamide	Benzenesulfonamide	4-Toluenesulfonamide	Phenylthiourea	Amine hydrochloride
6-Chloro-2,4-dibromoaniline	95	227	192				
1,2-Diaminonaphthalene	95	234	291				
N-Ethyl-4-nitroaniline	96	119	98		107		
5-Amino-2-methylpyridine	96	126	111				218 (di)
2,4-Diiodoaniline	96	141; 171	181				
N-Methyl-2-hydroxyaniline [*p*-(*N*-methylamino)phenol]	96	64 (di)	160 (mono)				
1-Amino-8-nitronaphthalene (8-nitro-α-naphthylamine)	97	191		194			
2-Methyl-3-nitroaniline (2-amino-6-nitrotoluene)	97	158	168		122		
1-Amino-8-naphthol (8-hydroxy-α-naphthylamine)	97	181 (mono) 118 (di)	193 (mono) 206 (di)		189 (mono)		
3-Aminobenzyl alcohol	97	107 (mono)	115 (mono) 114 (di)				
2-Amino-4-methylpyridine	98	114; 103	114; 183 (di)	103			
2-Aminophenacyl alcohol	98	141 (mono)	167 (di)				
1,2-Diaminonaphthalene	98	234 (di)	291 (di)	215 (mono)			
5-Bromo-2-methoxyaniline	98	160	108				
2,4-Diaminotoluene (4-methyl-1,3-phenylenediamine)	99	224 (di)	224 (di)	178; 138 (mono) 192 (di)	192 (di) 160 (mono)		
3-Aminoacetophenone	99	129			130		
1,2-Diaminobenzene (1,2-phenylenediamine)	102	186 (di)	301 (di)	186	202; 260 (di)		
2-Amino-*N*-(4-methylphenyl)naphthalene [*N*-(*p*-tolyl)-β-naphthylamine]	103	85	139				
3,4-Diaminobiphenyl	103	211; 155 (mono) 163 (di)	186; 221 (mono) 248 (di)				
Piperazine	104	134; 52 (mono) 144 (di)	191; 75 (mono) 196 (di)	282 (di)	173 (mono)		
2-Amino-8-nitronaphthalene (8-nitro-β-naphthylamine)	104	196	162				
3-Amino-4,4′-dimethylbiphenyl	105	157	161				230
1,3-Diamino-4,6-dimethylbenzene	105	165 (mono) 295 (di)	259 (di)				
2-Bromo-4-nitroaniline	105	129	160				
Triphenylmethylamine	105	208	162				
4-Aminophenanthrene	105	190	224				
2-Aminobenzophenone	106	72; 89	80				
3-Amino-6-phenylpyridine	106	149	201				
3-Amino-4-methylpyridine	106	84	81				180
4-Aminoacetophenone	106	167	205	128	203		

Amines—Primary and Secondary (Solids) (continued)

		Derivative mp (°C)					
Name of compound	mp (°C)	Acetamide	Benzamide	Benzenesulfonamide	4-Toluenesulfonamide	Phenylthiourea	Amine hydrochloride
N-Methyl-9-aminophenanthrene (9-phenanthrylmethylamine)	107	185	167				
2-Amino-4-nitrotoluene (2-methyl-5-nitroaniline)	107	151	186	172			
N-Phenyl-2-aminonaphthalene (*N*-phenyl-β-naphthylamine)	108	93	148; 136; 111				
2-(4-Aminophenyl)ethyl alcohol	108	105	60 (mono) 136 (di)				171
5-Aminoacenaphthene	108	238 (mono) 122 (di)	210; 199				
2-Chloro-4-nitroaniline	108	139	161		164		
5-Aminoquinoline	110	178				204	
2,5-Diaminopyridine	110	290 (di)	230 (di)				
2-Aminobenzamide	111	177	215				
1,4-Diamino-2-iodobenzene	111	211	254				
4-Bromo-3-nitroaniline	111	104	138				
4-Bromo-2-nitroaniline	111	104	137				
2-Aminonaphthalene (β-naphthylamine)	112	134	162	102	133	129	260
4-Amino-3-methylbenzophenone	112	175	158				
4-Ethoxy-2-nitroaniline	113	104		72	94		
5-Bromo-2-hydroxy-3-methylaniline (2-amino-4-bromo-6-methylphenol; 3-amino-5-bromo-2-hydroxytoluene)	113	119 (mono) 200 (di)	195 (mono)				
1-Aminoethyl phenyl ketone (β-aminopropiophenone)	114	91	105				187
3-Nitroaniline	114	155	157 (mono) 76 (di)	136 150 (di)	139	160	
6-Aminoquinoline	114	138 (mono) 75 (di)	169	193			
Z-2,5-Dimethylpiperazine	114		152 (di)		147 (di)		
3-Aminocamphor	115	121	141				
3-Amino-2-phenylquinoline	116	124 (mono) 173 (di)	180				
5-Bromo-2-hydroxy-4-methylaniline (2-amino-4-bromo-5-methylphenol; 4-amino-2-bromo-5-hydroxytoluene)	116	199 (mono) 188 (di)	223 (mono)				
4-Methyl-2-nitroaniline (4-amino-3-nitrotoluene)	117	99	148	102	166; 146		171
4-Chloro-2-nitroaniline	117	104			110		
E-2,5-Dimethylpiperazine	118		229 (di)		225 (di)		
2-Methoxy-5-nitroaniline	118	176	161		128		
2-Amino-5′, 4-dimethylazobenzene	119	157	135				
2,4,6-Tribromoaniline	119 (di)	127 232 (mono)	198				
1-Amino-5-nitronaphthalene (5-nitro-α-naphthylamine)	119	220		183			

Amines—Primary and Secondary (Solids) (continued)

Name of compound	mp (°C)	Derivative mp (°C)					
		Acetamide	Benzamide	Benzenesulfonamide	4-Toluenesulfonamide	Phenylthiourea	Amine hydrochloride
5-Hydroxy-2,4,6-tribromoaniline (3-amino-2,4,6-tribromophenol)	119	136 (tri)			147		
1,4-Diaminonaphthalene	120	303 (di)	280 (di) 186 (mono)		188 (mono)		
2,2′-Diamino-4,4′-dimethylbiphenyl	120	189 (di)	170 (di)				
2,6-Diaminopyridine	121	203 (di)	176 (di)				
Z-4,4′-Diaminostilbene	121	172 (di)	253 (di)				
3-Hydroxyaniline (*m*-aminophenol)	123	101 (di) 148 (mono)	153; 204 174 (mono)		157 (mono)	156	229
3,4,5-Tribromoaniline	123	256	210				
2,4-Dimethyl-5-nitroaniline	123	159	200	149	192		
4-Aminobenzophenone	124	153	152				
2-Amino-5-nitrobiphenyl	125	133			169		
2-Amino-1-nitronaphthalene (1-nitro-β-naphthylamine)	126	124	168	156	160		
4-Aminoazobenzene	126	146	211				
Hydrazobenzene	127	159 (mono) 105 (di)	126 (mono) 162 (di)				
4,4′-Diaminobiphenyl (benzidine)	127	317 (di) 199 (mono)	352 (di) 205 (mono)	232 (di)	243 (di)		
4-Amino-2-hydroxytoluene (3-hydroxy-4-methylaniline; 5-amino-2-methylphenol)	127	119 (mono)	164 (di)				
1-Amino-6-bromonaphthalene (6-bromo-α-naphthylamine)	128	192	218				
2-Methylaniline (*o*-toluidine; *o*-aminotoluene)	129	314	265				
4-Methoxy-2-nitroaniline	129	117	140				
4,4′-Diamino-3,3′-dimethylbiphenyl	129	103 (mono) 315 (di) 211 (tetra)	198 (mono) 265 (di)				
2,4-Diaminodiphenylamine	130	188 (mono)	213 (di)				
2-Aminocoumarin	130	202	173				
2-Methyl-4-nitroaniline (2-amino-5-nitrotoluene)	130	202	178	158	174		
2-Amino-4-chloropyridine	131	116	120 (mono) 165 (di)				
2-Aminobenzothiazole	132	186	186				
3,5-Dimethyl-2-hydroxyaniline (2-amino-4,6-dimethylphenol)	135	96 (mono)	154 (di)				
2-Hydroxy-5-methylaniline (2-amino-4-methylphenol; 3-amino-4-hydroxytoluene)	135	160 (mono) 145 (di)	191 (mono) 190 (di)				
2-Amino-3-methylnaphthalene (3-methyl-β-naphthylamine)	135	182	190				
(±)-2,2′-Diamino-6,6′-dimethylbiphenyl	136	205 (di)	182 (di)		163 (di)		
1,4-Diamino-2-nitrobenzene	137	162, 189 (mono) 186 (di)	236				

Amines—Primary and Secondary (Solids) (continued)

Name of compound	mp (°C)	Derivative mp (°C)					
		Acetamide	Benzamide	Benzenesulfonamide	4-Toluenesulfonamide	Phenylthiourea	Amine hydrochloride
1-Aminophenanthrene	138	208	199				
4,4′-Diamino-3,3′-dimethoxybiphenyl (*o*-bianisidine)	138	242 (di)	236 (di)				
2-(4-Aminophenyl)quinoline	138	189 (mono) 154 (di)	234				
2-Methoxy-4-nitroaniline	140	154	150	181	175		
1-(4-aminophenyl)-1-propanone (4-aminopropiophenone)	140	161	190				
5-Amino-8-hydroxyquinoline	143	222 (mono) 207 (di)	205 (di)				
2-Amino-5-nitronaphthalene (5-nitro-β-naphthylamine)	144	186	182				
1-Amino-2-nitronaphthalene (2-nitro-α-naphthylamine)	144	199	175				
2-Aminobenzoic acid (anthranilic acid)	144	185	182	214	217		
2-Amino-4-hydroxytoluene (3-amino-4-methylphenol; 5-hydroxy-2-methylaniline)	144	178 (mono) 128 (di)		183 (mono)			
2,5-Dimethyl-4-nitroaniline	145	169		162	185		
4-Bromo-4′-bromobiphenyl	145	247			174		
1,4-Diaminobenzene (1,4-phenylenediamine)	147	304 (di) 163 (mono)	300, 338 (di) 128 (mono)	247 (di)	266 (di)		
4-Nitroaniline	148	216	203	139	191		
4,4′-Diamino-2,2′-dimethylazoxybenzene	148	281 (di)	290				
4-Bromo-3-hydroxyaniline (2-bromo-5-aminophenol)	150	212			136		
N-Methyl-4-nitroaniline	152	153	112	121			
4-Amino-2-chlorophenol (3-chloro-4-hydroxyaniline)	153	144 (mono) 124 (di)			117		
3-Aminobenzopyrazole	154	178 (di)	182 (di)				
(−)-2,2′-Diamino-6,6-dimethylbiphenyl	156	205 (di)	172 (di)				
3,3′-Diaminoazobenzene	156	272 (di)	286 (di)				
4-Nitrophenylhydrazine	157	205	193				
4-Aminopyridine	158	150	202				
4,4′-Diamino-3,3′-dimethyldiphenylmethane	159	224 (di) 119 (tetra)	215 (di)				
2-Amino-8-nitroquinoline	159	211	166				
2-Amino-4′-nitrobiphenyl	159	199			163		
3-Amino-2-methylquinoline	160	165	161				
1-Amino-4-hydroxy-3-nitronaphthalene (4-hydroxy-3-nitro-α-naphthylamine)	160	250	330				
1,3-Diamino-4-nitrobenzene	161	200 (mono) 246 (di)	222 (di)		169 (di)		
4-Amino-2-hydroxytoluene (3-hydroxy-4-methylaniline; 5-amino-2-methylphenol)	161	225 (mono) 133 (di)			112 (mono)		

Amines—Primary and Secondary (Solids) (continued)

		Derivative mp (°C)					
Name of compound	mp (°C)	Acetamide	Benzamide	Benzenesulfonamide	4-Toluenesulfonamide	Phenylthiourea	Amine hydrochloride
2-Hydroxy-4-methylaniline (4-amino-3-hydroxytoluene; 2-amino-5-methylphenol)	162	171 (mono)	169 (mono) 162 (di)				
2-Amino-3,5-dimethylphenol (2,4-dimethyl-6-hydroxyaniline)	163	187 (mono) 88 (di)	211 (mono) 149 (di)				
4-Aminophenacyl alcohol	165	177, 130 (mono) 162 (di)	188 (mono)				
4-Amino-2-bromophenol (3-bromo-4-hydroxyaniline)	165	157 (mono)	185 (mono) 192 (di)				
4-Aminobenzenesulfonamide (sulfanilamide)	165	219 (mono) 254 (di)	284 (mono) 268 (di)	211			
2,7-Diaminonaphthalene	166	261 (di)	267 (di)				
4-Amino-3-nitrobiphenyl	167	132	143				
3,4-Diaminophenol	168	207 (di)	203 (di) 225 (tri)				
4-Amino-2-phenylquinoline	168	117 (di)	182				
2-Amino-4,6-dinitrophenol (picramic acid; 3,5-dinitro-2-hydroxyaniline)	169	201; 193	229; 300		191		
6-Aminocoumarin	170	217	173	159			
2,6-Dinitro-4-methylaniline (4-amino-3,5-dinitrotoluene)	172	195	186				
2-Hydroxyaniline (*o*-aminophenol)	174	201 (mono) 124 (di)	185; 165	141	139	146	
3-Aminobenzoic acid	174	250	248				
4,5-Dimethyl-2-hydroxyaniline (2-amino-4,5-dimethylphenol)	175	191 (mono) 157 (di)	196 (mono) 153 (di)				
4-Amino-2-methylphenol (5-amino-2-hydroxytoluene; 4-hydroxy-3-methylaniline)	175	103 (di) 179 (mono)	194 (di)		110 (mono)		
2-Amino-4-hydroxy-3-(1-methylethyl) toluene [6-aminothymol; 3-amino-4-methyl-2-(1-methylethyl)phenol; 5-hydroxy-6-isopropyl-2-methylaniline]	179	74 (mono) 91 (tri)	179 (mono) 167 (di)				
4-Hydroxy-2-methylaniline (2-amino-5-hydroxytoluene; 4-amino-3-methylphenol)	179	130 (mono)	92 (mono)				215
1-Amino-3-hydroxynaphthalene (3-hydroxy-α-naphthylamine)	185	179 (mono)	309 (di)		137 (mono)		
4-Hydroxyaniline (*p*-aminophenol)	186	150 (di) 168 (mono)	234 (di) 217 (mono)	125	168 (di) 254; 142 (mono)	150	
6,6′-Diamino-3,3′-dimethyltriphenylmethane	186	217 (di)	196 (di)				
4-Aminobenzoic acid	187	252	278	212	223		
2,4-Dinitroaniline	188	121	202; 220		219		
2,4,6-Trinitroaniline (picramide)	190	230	196	211			
1-Amino-6-naphthol (6-hydroxy-α-naphthylamine)	190	218 (mono) 187 (di)	152 (mono) 223 (di)				
2-Amino-1,5-dinitronaphthalene (1,5-dinitro-β-naphthylamine)	191	201			182		

Amines—Primary and Secondary (Solids) (continued)

Name of compound	mp (°C)	Derivative mp (°C)					
		Acetamide	Benzamide	Benzenesulfonamide	4-Toluenesulfonamide	Phenylthiourea	Amine hydrochloride
(±)-2,2′-Diamino-1,1′-dinaphthyl	193	236 (di)	235 (di)				
1-Amino-4-nitronaphthalene (4-nitro-α-naphthylamine)	195	190	224	158; 173	185		
1,2-Diamino-4-nitrobenzene	198	205 (mono) 195 (di)	235 (di)				
2,4-Dinitrophenylhydrazine	200	198	207				
4-Amino-4′-nitrobiphenyl	200	264; 240		174			
2-Amino-5-nitrobenzaldehyde	200	161			182		
2-Amino-7-hydroxynaphthalene (7-hydroxy-β-naphthylamine)	201	232 (mono) 156 (di)	246 (mono) 181 (di)				
4,4′-Diamino-1,1′-dinaphthyl	202	363 (di)	320 (di)				
3-Amino-5-phenylacridine	204	156	246				
1-Amino-2-methylanthraquinone	205	177 (mono) 206 (di)			218		
1-Amino-7-naphthol (7-hydroxy-α-naphthylamine)	207	165 (mono)	209 (mono) 208 (di)				
1-Amino-8-naphthol (8-hydroxy-α-naphthylamine)	207	165	208				
4,4′-Diamino-2,5,2′,5′-tetramethyltriphenylmethane	210	217 (di)	250 (di)				
5-Bromo-4-hydroxy-2-methylaniline (2-amino-4-bromo-5-hydroxytoluene; 4-amino-2-bromo-5-methylphenol)	215	172 (di)	229 (di)				
2-Hydroxy-6-nitroaniline (2-amino-3-nitrophenol)	216	172 (mono)			136 (mono)		
1-Amino-5-chloroanthraquinone	219	219	218				
2-Amino-10-hydroxyphenanthrene	221	182 (di)	225 (di)				
5-Amino-4-nitroacenaphthene	222	252	233				
5-Chloro-4-hydroxy-2-methylaniline (2-amino-4-chloro-5-hydroxytoluene; 4-amino-2-chloro-5-methylphenol)	225	162 (di)	220 (di)				
2-Amino-1,8-dinitronaphthalene (1,8-dinitro-β-naphthylamine)	226	238			221		
E-4,4′-Diphenylethene (*trans*-4,4′-diaminostilbene)	231	353 (di)	352 (di)				
2-Amino-3-naphthol (3-hydroxy-β-naphthylamine)	234	188 (di)	235 (mono)				
1-Amino-2,4-dinitronaphthalene (2,4-dinitro-α-naphthylamine)	242	259	252		166		
1-Amino-3-bromoanthraquinone	243	214			227		
Dibenzopyrrole (carbazole)	246	69	98		137		
1-Aminoanthraquinone	252	218	255		229		
3-Aminodibenzopyrrole (3-aminocarbazole)	254	217 (mono) 200 (di) 175 (tri)	250 (mono)				
1,8-Diaminoanthraquinone	262	284 (di)	324 (di)				

Amines—Primary and Secondary (Solids) (continued)

		Derivative mp (°C)					
Name of compound	mp (°C)	Acet-amide	Benz-amide	Ben-zene-sulfon-amide	4-Tolu-ene-sulfon-amide	Phenyl-thio-urea	Amine hydro-chlo-ride
1,4-Diaminoanthraquinone	268	271 (di)	284 (di) 280 (mono)				
1,1′-Diamino-2,2′-dinaphthyl	281	230 (di)	278 (di)				
1,7-Diaminoanthraquinone	290	283 (di)	325 (di)				
1,6-Diaminoanthraquinone	292	295 (di)	275 (di)				
1-Amino-5-nitroanthraquinone	293	275	237				
2-Aminoanthraquinone	308	258 (di) 262 (mono)	228	271	304		
2-Amino-3-bromoanthraquinone	308	259; 217	279				

Amines—Tertiary (Liquids)

		mp (°C) of Derivative Formed Using				Derivative mp (°C)
Name of compound	bp (°C)	Chloroplatinic acid	Methyl 4-toluenesulfonate	Methyl iodide	Picric acid	Amine hydrochloride
Trimethylamine	3	242		230	216	278
N-Methylpyrrolidine	80	233			221	
Triethylamine	89			280	173	
1,2-Dimethylpyrrolidine	96	223			235	
1,3-Dimethylpyrrolidine	97	59			181; 115	
Pyridine	116	241; 262	139	118	167	
1,2,5-Trimethylpyrrolidine	116			310	163	
2-Dimethylaminodiethyl ether	121			165	121	
1-Methylpyrazole	127	198		190	148	
N-Ethylpiperidine	128	202			168	
2-Methylpyridine (α-picoline)	129	216; 195	150	230	169	
1,3-Dimethylpyrazole	136			256	138	
2-Methylpyrazine	137			130	133	
2,6-Dimethylpyridine (2,6-lutidine)	143	208		238	168	
4-Methylpyridine (γ-picoline)	143	231		152	167	
3-Methylpyridine (β-picoline)	144	202		92	150	
3-Chloropyridine	149	168			135	
2-Ethylpyridine	149	167			189	
Tripropylamine	156			208	117	
2,4-Dimethylpyridine (2,4-lutidine)	159	216		113	183	
2,5-Dimethylpyridine (2,5-lutidine)	160	194			169	
2-(Diethylamino)ethanol	161			249	79	
Tropidine	163	217		300	285	
2,3-Dimethylpyridine (2,3-lutidine)	164	195			188	
3,4-Dimethylpyridine (3,4-lutidine)	164	205			163	
3-Ethylpyridine	164	209			130	
4-Ethylpyridine	165	213			168	
Tropane	167	230			281	
2,4,5-Triethylpyridine	168	205			131	
3-Bromopyridine	170	175	156	165		
1,3,5-Trimethylpyrazole	170	191			147	
3,5-Dimethylpyridine (3,5-lutidine)	171	255			245	
2,4,6-Trimethylpyridine (γ-collidine)	172	223			156	
2,3,6-Trimethylpyridine	178	252			146	
2,3,5-Trimethylpyridine	184	228			183	
N,N,2-Trimethylaniline (*N,N*-dimethyl-*o*-toluidine)	185	193		210	122	156
N,N-Dimethylbenzylamine	185	192		179	93	
2,6-Dimethyl-4-ethylpyridine	186	210			120	
2,4-Diethylpyridine	188	171			100	
Methyl 2-pyridyl ketone	192	220		161	131	
2,3,4-Trimethylpyridine	193	259			164	
N,N-Dimethylaniline	193	173	161	228	163	90
3-Ethyl-4-methylpyridine	196	205; 234			150	
3,5-Dimethyl-2-ethylpyridine	198	189			152	

Amines—Tertiary (Liquids) (continued)

		mp (°C) of Derivative Formed Using				Derivative mp (°C)
Name of compound	bp (°C)	Chloro-plati-nic acid	Methyl 4-tolu-enesul-fonate	Methyl iodide	Picric acid	Amine hydro-chloride
N-Ethyl-*N*-methylaniline	201			125	134	114
N,*N*,2,5-Tetramethylaniline	204	196			158	
N,*N*,2,4-Tetramethylaniline	205	219			124	
2-Chloro-*N*,*N*-diethylaniline	207			152	132	
N,*N*-Diethyl-2-methylaniline (*N*,*N*-diethyl-*o*-toluidine)	210			224	180	
N,*N*,4-Trimethylaniline (*N*,*N*-dimethyl-*p*-toluidine)	210		85	220	130	
3,4-Diethylpyridine	211	221			139	
Tributylamine	212			186	107	
N,*N*,3-Trimethylaniline (*N*,*N*-dimethyl-*m*-toluidine)	212			177	131	
Methyl 4-pyridyl ketone	214	205			130	
N,*N*-Diethylaniline	218			104	142	
N,*N*-Diethyl-4-methylaniline (*N*,*N*-diethyl-*p*-toluidine)	229			184	110	157
2,3,4,5-Tetramethylpyridine	234	210			172	
Quinoline	239	227	126	72 (hyd) 133 (anhyd)	203	
Isoquinoline	243	263	163	159	223	
(±)-Nicotine	243	280		219	218	
2-(*N*,*N*-Dimethylamino)benzaldehyde	244	206		164		
N,*N*-Dipropylaniline	245			156	261	
2-Ethylquinoline	246	188		180	148	
2-Methylquinoline (quinaldine)	247	228	161; 134	195	195	
L-Nicotine	248	275			218	
8-Methylquinoline	248			193	203	
1-Ethylisoquinoline	250	200			210	
2,4-Dimethyl-5,6,7,8-tetrahydroquinoline	250			157	144	195
2,8-Dimethylquinoline	252			221	180	
N-Methyl-2-pyridone	255	141			145	
3-Ethylisoquinoline	257	180			172	
6-Methylquinoline	258		154	219	234	
3-Methylquinoline	259	249		221	187	
3-Chloroquinoline	260			276	182	
5-Methylquinoline	260			105	213	
6-Chloroquinoline	262		143	248		
4-Methylquinoline (lepidine)	263	230		174	212	
4-Chloroquinoline	263	278			212	
2,6,8-Trimethylquinoline	265	207			189	207
7-Chloroquinoline	268	253		250		
2,4-Dimethylquinoline	269	229		265	194	
2-Phenylpyridine	269	204			175	
6,8-Dimethylquinoline	269	235			289	
N,*N*-Dibutylaniline	271		180		125	

Amines—Tertiary (Liquids) (continued)

		mp (°C) of Derivative Formed Using				Derivative mp (°C)
Name of compound	bp (°C)	Chloro-plati-nic acid	Methyl 4-tolu-enesul-fonate	Methyl iodide	Picric acid	Amine hydro-chloride
4-Ethylquinoline	274	204		149	180	
3,5-Dimethyl-1-phenylpyrazole	275	186		190	103	
5,8-Dimethylquinoline	275	234			198	
6-Bromoquinoline	278			278	217	
2,6-Dimethylquinoline	280	238			237	
2,4,7-Trimethylquinoline	281	272		322	232	
4,7-Dimethylquinoline	283	227			224	
3,4-Dimethyl-1-phenylpyrazole	285	180			123	
8-Chloroquinoline	288	235		165		
8-Bromoquinoline	304	230		281		
6-Methoxyquinoline	305			236	305	
N-Benzyl-*N*-methylaniline	306			164	103; 127	
N-Benzyl-*N*-ethylaniline	312			161	117	

Amines—Tertiary (Solids)

		mp (°C) of Derivative Formed Using				Derivative mp (°C)
Name of compound	mp (°C)	Chloroplatinic acid	Methyl 4-toluenesulfonate	Methyl iodide	Picric acid	Amine hydrochloride
6-Bromoquinoline	19			278	217	
Isoquinoline	24		163	159	223	
2,8-Dimethylquinoline	27			221	180	
4-Chloroquinoline	31	278			212	
7-Chloroquinoline	32	253		250		
8-Iodoquinoline	36	251		200		
1,3,5-Trimethylpyrazole	37	191			147	
2,3-Dimethyl-5,6,7,8-tetrahydroquinoline	38			117	169	
7-Methylquinoline	39	224			237	
2,4,6-Trimethylquinoline	40 (hyd)			247; 225	201	272
6-Chloroquinoline	41		143	248		
2,4,8-Trimethylquinoline	42			229	193	
4-Chloro-2-methylquinoline	43			212	178	
5-Chloroquinoline	45	255		231; 172		
2,6,8-Trimethylquinoline	46	207			189	207
8-Methoxyquinoline	50			160	143	
4,8-Dimethylquinoline	55	226			217	
2,6-Dimethylquinoline	60		175	244	191	
N,N-Dimethyl-3-nitroaniline	60			205	119	
2,4,6-Trimethylquinoline	65 (anhyd)			247; 225	201	272
1,2-Dimethylbenzimidazole	65 (anhyd)			254	238	
2,3-Dimethylquinoline	69	230		218	231	
N,N-Dibenzylaniline	70			135	132	
3,4-Dimethylquinoline	74			191	215	290
8-Hydroxyaniline	76			143	204	
Tribenzylamine	91			184	190	227
4,4′-Bis(dimethylamino)diphenylmethane	91			214 (di)	185 (mono) 178 (di)	
2,3,4-Trimethylquinoline	92	215		260	216	274
2,2′-Dipyridylamine	95	160			228	
1-Phenylisoquinoline	96	242			165	235
4-Iodoquinoline	97	185		251		
5-Iodoquinoline	100	263		245		
2,2′-Bisquinolylmethane	103			205	239, 210 (di)	
Acridine	111			224	208	
1,2-Dimethylbenzimidazole	112 (anhyd)			254	238	
3,5-Dibromopyridine	112		219	274		
6,8′-Biquinolyl	148			126 (mono)	268	
2,7′-Biquinolyl	160			263	240	
7,7′-Biquinolyl	172			310 (mono)	300	
4,4′-Bis(dimethylamino)benzophenone (Michler's ketone)	174			105	156	
2,3′-Biquinolyl	176	278		286		
5-Hydroxyquinoline	224	230		224		240

Amines—Tertiary (Solids) (continued)

		mp (°C) of Derivative Formed Using				Derivative mp (°C)
Name of compound	mp (°C)	Chloro-platinic acid	Methyl 4-toluenesulfonate	Methyl iodide	Picric acid	Amine hydro-chloride
4-Hydroxy-2-methylquinoline	232 (anhyd)	215		201 (anhyd)	200	
7-Hydroxyquinoline	235			251	245	
Hexamethylenetetramine	280		205	190	179	

Amino Acids

Name of compound	Decomposition point (°C)	Derivative mp (°C)					
		4-Toluenesulfonamide	Phenylureido acid	Acetamide	Benzamide	3,5-Dinitrobenzamide	2,4-Dinitrophenyl derivative
2-Aminophenylacetic acid	119			158	179		
N-Phenylglycine	127		195	124; 194	63		
3-(4-Aminophenyl)propanoic acid	132			143 (anhyd)	195		
(4-aminohydrocinnamic acid)				124 (hyd)			
(+)- or (−)-Ornithine	140		190		240 (mono)		
					189 (di)		
2-Aminobenzoic acid	144	217	181	185	182	278	
(anthranilic acid)							
E-3-(2-Aminophenyl)-2-propenoic acid	158			250 (mono)	193		
(*trans*-2-aminocinnamic acid)				158 (di)			
3-Aminobenzoic acid	174		270	250	248	270	
E-3-(4-Aminophenyl)-2-propenoic acid	176			260	274		
(*trans*-4-aminocinnamic acid)							
E-3-(3-Aminophenyl)-2-propenoic acid	183			237	229		
(*trans*-3-aminocinnamic acid)							
4-Aminobenzoic acid	188	223	300	252	278	290	
(+)- or (−)-Glutamic acid	198	117		187	138; 157		
4-Aminophenylacetic acid	200			170	206		
3-Aminopropanoic acid (*β*-alanine)	200	117	174		120; 165	202	146
(±)-Proline	203		170			217	181
(+)- or (−)-Arginine	207				298 (mono)	150	
					235 (di)		
(+)- or (−)-Glutamic acid	211	131		199	138	217	
Sarcosine	212	102	102	135	104	154	
β-Hydroxyvaline	218		182		153 (mono)		
(+)- or (−)-Lysine	224		184		235 (mono)	169	171
					149 (di)		
(+)- or (−)-Proline	224	133	170		156		138
(+)- or (−)-Asparagine	227	175	164		189	196	181
(±)-Glutamic acid	227	213			156		149
(+)- or (−)-Serine	228	213			171 (mono)	95	174
					124 (di)		
(±)-*β*-Hydroxynorvaline	230		156		170		
(±)-3-Amino-3-phenylpropanoic acid	231			161	196		
[(±)-*β*-aminohydrocinnamic acid]							
Glycine	232	150	163; 197	206	187	179	204
3-Aminosalicylic acid	235			215	189		
(±)-Threonine	235; 251	147			174 (di)	145	178
					176 (mono);		
					150		
(±)-Isoserine	246		184		151		
(±)-Serine	246	213	169		150; 171		201
4-Hydroxyphenylglycine	248			203 (mono)	117		
				175 (di)			
(+)- or (−)-Threonine	253				148		145

Amino Acids (continued)

Name of compound	Decomposition point (°C)	Derivative mp (°C): 4-Toluenesulfonamide	Phenylureido acid	Acetamide	Benzamide	3,5-Dinitrobenzamide	2,4-Dinitrophenyl derivative
(±)-α-Aminophenylacetic acid	256			199	175		
(+)- or (−)-Cystine	260	205	117; 160		181 (di)	180	109
Glycylglycine	260	178	176		208	210	
(+)- or (−)-Aspartic acid	272	140	162		185		187
(±)-Methionine	272	105		114	151		117
(±)-Phenylalanine	273	135	182		188	93	186
(+)- or (−)-Hydroxyproline	274	153	175		100 (mono) 92 (di)		
(±)-Tryptophan	275	176			188	240	
(±)-Aspartic Acid	280				165		196
(+)- or (−)-Methionine	283			99	150	95 (hyd) 150 (anhyd)	
5-Aminosalicylic acid	283			218 (mono) 184 (di)	252		
(+)- or (−)-Isoleucine	285	132	121		117		113
(+)- or (−)-Histidine	288; 253; 277	204			230 (mono); 249	189	233
(+)- or (−)-Tryptophan	290; 252	176; 104	166		183	233	221
(±)-Isoleucine	292	141	121		118		175
(±)-Alanine	295	139	190; 174		166	177	
(+)- or (−)-Alanine	297	139	168; 190		151		
(±)-Valine	298	110	164		132	158	184
(2*R*,3*S*)-2,3-Diaminosuccinic acid	306			235		212 (di, hyd)	
(+)-Norvaline	307			137	64		
(±)-2-Aminobutanoic acid	307		170		147	194	143
(+)- or (−)-Valine	315	149	147		127	158; 181	132
(±)-Tyrosine	318	226		148 (*N*-); 172 (di)	197	254	
(±) or (−)-Phenylalanine	320; 283	165	181		146	93	189
(±)-Leucine	332; 293		165	157	141	187	
(+)- or (−)-Leucine	337	124	115		107; 118	187	94
(+)- or (−)-Tyrosine	344	119 (di) 188 (mono)	104	172	211 (di) 166 (mono)		180

Carbohydrates

Name of compound	Decomposition point (°C)	Specific rotation in water at 20°C	Time required for osazone formation (min)	Derivative mp (°C)			
				Phenylosazone	4-Nitrophenylhydrazone	4-Bromophenylhydrazone	Acetate
Turanose (hydrate)	65	+27.3 $\longrightarrow$ +75.8		220			141 (hepta)
1,3-Dihydroxyacetone	71 (monomer); 80 (dimer)			132	160		47
Melibiose (monohydrate)	85	+129.5		178			177 (β) 147 (α)
Gentiobiose (hydrated)	86	+9.5		164; 179			193 (β) 189 (α)
L-Ribose	87	+20.3 $\longrightarrow$ +20.7		166		165	
α-D-Glucose (hydrated)	90	+47.7	4–5	205	88; 196	166	132 (β) 112 (α)
L-Rhamnose (hydrated)	94	−8.6 $\longrightarrow$ +8.2		182; 222	191		
D-Ribose	95	−21.5		166		170	
Glycollic aldehyde	97			179			158 (mono)
β-Maltose (monohydrate)	101	+129.0		206			161 (β) 125 (α)
Levulose (D-fructose)	104	−92.0	2	210	176		109 (β) 70 (α)
α-L-Rhamnose (hydrated)	105	+9.4	9	190; 222	191		99
Lactic aldehyde	105			154	129		
L-Lyxose	106	+13.5		163	172	162	
D-Lyxose	107	+ 5.5 $\longrightarrow$ −14.0		164	172	162	
D-Galactose (hydrated)	120	+81.7	15–19	201	154; 197	168	142 (β) 95 (α)
β-L-Rhamnose (anhydrous)	126	+9.1	9	222; 190		191	99
β-D-Allose	128	+0.58 $\longrightarrow$ +14.41		178		145	
β-L-Allose	129	−1.9		165		145	
D-Talose	130	+30.0 $\longrightarrow$ +20.6		201		205	
D-Threose	132	+29 $\longrightarrow$ −19.6		164			114 (tri)
D-Mannose	132	+14.1	0.5	210	195	210	115 (β) 74 (α)
α-D-Mannoheptose	135	+85.05 $\longrightarrow$ +68.64		200		208	106 (hexa) 140
D-Xylose	145	+18.7	7	164	155	128	141; 126 (β) 59 (α)

Carbohydrates (continued)

				Derivative mp (°C)			
Name of compound	Decomposition point (°C)	Specific rotation in water at 20°C	Time required for osazone formation (min)	Phenylosazone	4-Nitrophenylhydrazone	4-Bromophenylhydrazone	Acetate
L-Fucose	145	−152.6		178	211	184	
		⟶ −75.9					
α-D-Glucose (anhydrous)	146	+52.8	4–5	210	88; 196	166	132 (β)
							112 (α)
β-D-Glucose	150	+18.7		210			132 (β)
		⟶ +52.7					
Turanose	157	+27.3		220			141 (hepta)
(anhydrous)		⟶ +75.8					
β-L-Arabinose	160	+104.0	9–10	166	186	155	86 (β)
							96 (α)
L-Sorbose	164	−43.0	4	162			97
D,L-Arabinose	164			169		160	
Maltose (anhydrous)	165	+129.0		206			161 (β)
							125 (α)
α-D-Galactose	170	+81.7	15–19	201	154; 197	168	142 (β)
(anhydrous)							95 (α)
D-Glucoheptolose	171			210			112 (hexa)
Sucrose	185	+66.5	30	205			89; 70
L-Ascorbic acid	190	+49.0			262 (di)	170 (di)	
α-D-Glucoheptose	193	−20		195			135 (β)
							164 (α)
Gentiobiose	195	+9.5		164; 179			193 (β)
(anhydrous)							189 (α)
Lactose (hydrate)	203	+52.5			200		100 (β)
							152 (α)
6-(β-D-Xyloside)-	210	+24.1		220			216 (β)
D-glucose		⟶ −3.3					
(primeverose)							
Cellobiose	225	+35.0			200		192 (β)
							230 (α)
Lactose	233; 252	+52.5			200		100 (β)
(anhydrous)							152 (α)

Carboxylic Acids (Liquids)

		Derivative mp (°C)						
Name of compound	bp (°C)	4-Toluidide	Anilide	4-Nitrobenzyl ester	4-Bromophenacyl ester	Amide	*S*-Benzylthiuronium salts	Phenylhydrazide
Trifluoroethanoic acid (trifluoroacetic acid)	72		91		74			
Thioethanoic acid (thioacetic acid)	93	131	76			115		
Methanoic acid (formic acid)	101	53	50	31	140		151	143
Ethanoic acid (acetic acid)	118	153	114	78	86	82	136	129
Propenoic acid (acrylic acid)	141	141	105			85		
Propanoic acid (propionic acid)	141	126	106	31	63	81	152	157
Propynoic acid (propiolic acid)	144d		87			62		
2-Methylpropanoic acid (isobutyric acid)	155	110	105		77	129	149	140
2-Methylpropenoic acid (methacrylic acid)	161		87			102		
Butanoic acid (butyric acid)	164	75	97	35	63	116	149	102
2,2-Dimethypropanoic acid (pivalic acid; trimethylacetic acid)	164	120	133		76	157		
2-Oxopropanoic acid (pyruvic acid)	165d	109; 130	104			125; 145		
Z-2-Butenoic acid (*cis*-crotonic acid; isocrotonic acid)	169	132	102		82	102		
3-Butenoic acid (vinylacetic acid)	169		58		60	73		
(±)-2-Methylbutanoic acid (ethylmethylacetic acid)	177	93	112		55	112		
3-Methylbutanoic acid (isovaleric acid)	177	109	110		68	137	159	
1-Heptyne-1-carboxylic acid (amylpropiolic acid)	180–220d	68				91		
3,3-Dimethylbutanoic acid (*tert*-butylacetic acid)	184	134	132			132		
Chloroethanoic acid (chloroacetic acid)	185	120	134		105			
2-Chloropropanoic acid (α-chloropropionic acid)	186	124	92			80		
Pentanoic acid (valeric acid)	186	74	63		75	106	156	109
2,2-Dimethylbutanoic acid (dimethylethylacetic acid)	190	83	92			103		
(±)-2,3-Dimethylbutanoic acid (isopropylmethylacetic acid)	192	113	78			132		
Dichloroethanoic acid (dichloroacetic acid)	194	153	125		99	98	178	
2-Ethylbutanoic acid (diethylacetic acid)	195	116	128	66		112		
(±)-2-Methylpentanoic acid	196	81	95			80		
(±)-3-Methylpentanoic acid	197	75	88			125		
Z-2-Methyl-2-butenoic acid (tiglic acid)	198	76	77	64	68			
4-Methylpentanoic acid (isocaproic acid; isobutylacetic acid)	200	63	112		79	121		144
Methoxyethanoic acid (glycolic acid methyl ester; methoxyacetic acid)	204		58			97		
Hexanoic acid (caproic acid)	205	75	95		72	101	159	98
2-Bromopropanoic acid (α-bromopropionic acid)	205	125	100			123		
Ethoxyethanoic acid (glycolic acid ethyl ester; ethoxyacetic acid)	207	32	95		105	82		
5-Methylhexanoic acid	207		75			103		
Bromoethanoic acid (bromoacetic acid)	208	91	131	88		91		
2-Ethylpentanoic acid (α-ethylpropylacetic acid)	209	129	94			105		
2-Methylhexanoic acid (butylmethylacetic acid)	210	85	98			73		
2-Bromobutanoic acid (α-bromobutyric acid)	217d	92	98	49		112		
4-Methylhexanoic acid	218		77			98		

Carboxylic Acids (Liquids) (continued)

		Derivative mp (°C)						
Name of compound	bp (°C)	4-Toluidide	Anilide	4-Nitrobenzyl ester	4-Bromophenacyl ester	Amide	S-Benzylthiuronium salts	Phenylhydrazide
Heptanoic acid	224	81	71		72	97		103
2-Ethylhexanoic acid	228	107	89			103		
Cyclohexanecarboxylic acid (hexahydrobenzoic acid)	233		146			186		
Octanoic acid (caprylic acid)	239	70	57		67	110	157	106
4-Oxopentanoic acid (levulinic acid; β-acetylpropionic acid)	250d	109	102	61	84	108		
Nonanoic acid (pelargonic acid)	254	84	57		69	101		
Decanoic acid (capric acid)	270	78	70		67	108		
2-Methyl-3-phenylpropanoic acid [(±)-α-methylhydrocinnamic acid; (±)-α-benzyl-propionic acid]	272	130; 116				108		
10-Undecenoic acid (undecylenic acid)	275	68	67			87	149	97
4-Oxohexanoic acid (4-acetylbutanoic acid; 4-acetobutyric acid)	275	123				114		
Undecanoic acid (undecylic acid; hendecanoic acid)	280	80	71		68	103		110
2-Phenylpropanoic acid (β-phenylpropionic acid)	280		92	36				
Z-9-Octadecenoic acid (oleic acid)	286	42	41		46	76		

Carboxylic Acids (Solids)

		Derivative mp (°C)						
Name of compound	**mp (°C)**	**4-Toluidide**	**Anilide**	**4-Nitrobenzyl ester**	**4-Bromophenacyl ester**	**Amide**	***S*-Benzylthiuronium salts**	**Phenylhydrazide**
Z-9-Octadecenoic acid (oleic acid)	16	43	41		46	76		
2-Methylpropenic acid (methacrylic acid)	16		87			106		
Ethanoic acid (acetic acid)	16	153	114	78	86	82	136	
Octanoic acid (caprylic acid)	16	70	57		67	110	157	
(±)-2-Hydroxypropanoic acid [(±)-lactic acid]	18	107	59		113	79	153	
Propynoic acid (propiolic acid)	18		87			62		
E-2-Methyl-2-pentanoic acid (*trans*-β-ethyl-α-methylacrylic acid)	24				91; 46	80		
10-Undecenoic acid	25	68	67			87	149	97
(±)-2-Bromopropanoic acid [(±)-α-Bromopropionic acid]	26	125	100			123		
Undecanoic acid (undecylic acid; hendecanoic acid)	29	80	71		68	103		
Cyclohexanecarboxylic acid (hexahydrobenzoic acid)	31		146			186		
Decanoic acid (capric acid)	32	78	70		67	108		105
Z-13-Docosenoic acid (erucic acid)	34	78	55		63	84		
4-Oxopentanoic acid (levulinic acid; β-acetylpropionic acid)	35	109	102	61	84	108		
2,2-Dimethylpropanoic acid (pivalic acid; trimethylacetic acid)	35	120	133		76	157		
2-Methyl-3-phenylpropanoic acid [(±)-α-methylhydrocinnamic acid; (±)-α-benzylpropionic acid]	37	130; 116				108		
Dodecanoic acid (lauric acid)	44	87	78		76	100	141	
Tridecanoic acid (tridecylic acid)	44	88	80		75	100		
(±)-2-Bromo-3-methylbutanoic acid [(±)-α-bromoisovaleric acid]	44	124	116			133		
E-2-Methyl-2-butenoic acid (angelic acid)	46		126			128		
3-Phenylpropanoic acid (hydrocinnamic acid; 3-phenylpropionic acid)	49	135	98	36	104	82; 105		
2-Bromo-2-methylpropanoic acid (α-bromoisobutanoic acid)	49	93	83			148		
Bromoethanoic acid (bromoacetic acid)	50	91	131	88		91		
E-9-Octadecenoic acid (elaidic acid; *trans*-oleic acid)	51				65	94		
4-Phenylbutanoic acid (4-phenylbutyric acid)	52				58	84		
Pentadecanoic acid (pentadecylic acid)	52		78	40	77	103		
Tetradecanoic acid (myristic acid)	57	93	84		81	103	139	
Trichloroethanoic acid (trichloroacetic acid)	58	113	97	80		141	148	123
E-13-Docosenoic acid (brassidic acid)	59		78		94	94		

Carboxylic Acids (Solids) (continued)

		Derivative mp (°C)						
Name of compound	mp (°C)	4-Toluidide	Anilide	4-Nitrobenzyl ester	4-Bromophenacyl ester	Amide	*S*-Benzylthiuronium salts	Phenylhydrazide
5-Phenylpentanoic acid	60		90			109		
Heptadecanoic acid (margaric acid)	61			49	83	108		
Z-3-Chloro-2-butenoic acid (β-chloroisocrotonic acid)	61		108			110		
Hexadecanoic acid (palmitic acid)	63	98	91	43	86	107	141	111
Chloroethanoic acid (chloroacetic acid)	63	162	137		105	121	160	111
Z-2-Methyl-2-butenoic acid (tiglic acid)	64	71	77	64	68	76		
Cyanoethanoic acid (cyanoacetic acid)	66		199			123		
D-13-(2-Cyclopentenyl)tridecanoic acid (chaulmoogric acid)	68	100	89			106		
3-Methyl-2-butenoic acid (β,β-dimethylacrylic acid)	68				104	107		
Octadecanoic acid (stearic acid)	71	102	96		92	109	143	115
E-2-Butenoic acid (*trans*-crotonic acid)	72	132	118	67	96	160	172	
Phenylethanoic acid (phenylacetic acid)	76	136	118	65	89	157	165	175
Didecanoic acid (eicosanic acid; arachidic acid)	77	96	92		89	109		
2-Hydroxy-2-methylpropanoic acid (α-hydroxyisobutyric acid)	79	133	136	81	98	98		
Hydroxyethanoic acid (glycolic acid; hydroxyacetic acid)	80	143	97	107	142	120	146	
Docosanoic acid (behenic acid)	82		102			111		
2-Benzoylpropanoic acid (α-benzoylpropionic acid)	83		138			146		
Iodoethanoic acid (iodoacetic acid)	84		143			95		
Dibenzylethanoic acid (dibenzylacetic acid)	89	175	155			129		
2-Benzoylbenzoic acid	90		195	100		165		
Z-Methylbutenedioic acid (citraconic acid; methylmaleic acid)	93	171 (mono)	176 (di) 153 (mono)	71 (di)		187d (di)		
E-3-Chloro-2-butenoic acid (β-chlorocrotonic acid)	94		124			101		
2-Chlorophenylethanoic acid (2-chlorophenylacetic acid)	95	170	139			175		
1,3-Pentanedicarboxylic acid (glutaric acid)	98	218 (di)	224 (di)	69 (di)	137 (di)	176 (di)	161	
Phenoxyethanoic acid (phenoxyacetic acid)	100		101		149	101		
3-Carboxy-3-hydroxypentanedioic acid (monohydrate) [2-hydroxy-1,2,3-propanetricarboxylic acid; citric acid (monohydrate)]	100	189 (tri)	199 (tri)	102 (tri)	148 (tri)	215 (tri)		
2-Methoxybenzoic acid (*o*-anisic acid)	101		131; 78	113	113	129		
2-Hydroxybutanedioic acid (L-malic acid; hydroxysuccinic acid)	101	207 (di)	197 (di)	124 (di) 87 (mono)	179 (di)	157 (di) 102 (mono)	124	

Carboxylic Acids (Solids) (continued)

		Derivative mp (°C)							
Name of compound	mp (°C)	4-Toluidide	Anilide	4-Nitrobenzyl ester	4-Bromophenacyl ester	Amide	*S*-Benzylthiuronium salts	Phenylhydrazide
Ethanedioic acid (dihydrate) [oxalic acid (dihydrate)]	101	268 (di) 169 (mono)	254 (di) 149 (mono)	204 (di)	252d	419d (di) 219 (mono)	198	
2-Butylpropanedioic acid (butylmalonic acid)	101		193 (di)			200 (di)		
3-Phenylpropanoic acid (benzylacetic acid)	104		108			113		
Heptanedioic acid (pimelic acid)	105	206 (di)	156 (di) 109 (mono)		137 (di)	175 (di)		
4-Chlorophenylethanoic acid (*p*-chlorophenylacetic acid)	106	190	165			175		
Nonanedioic acid (azelaic acid)	106	202 (di)	187 (di) 108 (mono)	44 (di)	131 (di)	175 (di) 95 (mono)		
2-Phenylpropenoic acid (2-phenylacrylic acid; atropic acid)	107		134			122		
2-Methylbenzoic acid (*o*-toluic acid)	108	144	125	91	57	143	146	
D-1,2,2,3-Trimethylcyclo-hexene-6-carboxylic acid (D-campholic acid)	109		91			80; 90		
E-2-Chloro-3-phenyl-2-propenoic acid (*cis*-α-chloroallocinnamic acid; *cis*-2-chlorocinnamic acid)	111		139			134		
2-Ethylpropanedioic acid (ethylmalonic acid)	111		150	75		214 (di)		
3-Methylbenzoic acid (*m*-toluic acid)	113	118	126	87	108	97	140	
2-Bromopropanedioic acid (bromomalonic acid)	113d	217 (di)				181 (di)		
2-Methylbutanedioic acid (methylsuccinic acid; pyrotartaric acid)	115	164	200 (di) 159 (mono)			165 (mono) 225 (di)		
2-Phenoxypropanoic acid (2-phenoxypropionic acid)	116	115	119			133		
3-Benzoylpropanoic acid (β-benzoylpropionic acid)	116		150			125; 146		
2-Hydroxy-2-phenylethanoic acid (mandelic acid; 2-hydroxy-2-phenylacetic acid)	119	172	152	124	113	134	166	
2-Chloro-2,2-diphenylethanoic acid (2-chloro-2,2-diphenylacetic acid)	119		88			115		
1-Cyclopentenylcarboxylic acid	121	122	126					
2-Benzylpropanedioic acid (benzylmalonic acid)	121d		217 (di)	120 (di)		225 (di)		
Benzoic acid	122	158	163	89	119	130	167	168
(±)-*trans*-Camphenic acid [(±)-*trans*-camphene-camphoric acid]	123		165 (di)			232 (di)		
3,3,3-Trichloro-2-hydroxypropanoic acid (3,3,3-trichlorolactic acid)	124		164			96		123
Trichloroethanoic acid (trichloroacetic acid)	124		164			145		

Carboxylic Acids (Solids) (continued)

		Derivative mp (°C)						
Name of compound	mp (°C)	4-Toluidide	Anilide	4-Nitrobenzyl ester	4-Bromophenacyl ester	Amide	*S*-Benzylthiuronium salts	Phenylhydrazide
2-Hydroxy-3-nitrobenzoic acid	125	165				145		
2,2-Diethylpropanedioic acid (diethylmalonic acid)	125			91 (di)		224 (di) 146 (mono)		
2,4-Dimethylbenzoic acid	127		141			181		
2-Benzoylbenzoic acid	128		195	100		165		
Z-Butenedioic acid (maleic acid)	132	142 (di)	187 (di) 198 (mono)	91 (di)	170; 190	153, 266 (di) 281 (mono) 181	163	
2,5-Dimethylbenzoic acid	132		140			186		
Z-2-Chloro-3-phenyl-2-propenoic acid (*cis*-β-chloroallocinnamic acid)	132	142	135			76		
Decanedioic acid (sebacic acid)	133	201 (di)	202 (di) 123 (mono)	72 (di)	147 (di)	210 (di) 170 (mono)	155	194
E-3-Phenyl-2-propenoic acid (*trans*-cinnamic acid)	133	168	153; 109	116	146	148	183	
2-Chloropropanedioic acid (chloromalonic acid)	133		118 (di)			170 (di)		
(±)-2-Hydroxy-2-phenylethanoic acid [(±)-mandelic acid; 2-hydroxy-2-phenylacetic acid]	134	172	152	124	113	134		
Furoic acid (pyromucic acid)	134	171; 108	124	134	139	143	211	
Propanedioic acid (malonic acid)	135d	253 (di) 156 (mono)	230 (di) 132 (mono)	86 (di)		170 (di) 110, 50 (mono)	147	194
1-Naphthaleneethanoic acid (α-naphthylacetic acid)	135		160		112	181		
2-Acetylbenzoic acid (aspirin; acetylsalicylic acid)	135		136	90		138	144	
1,5,5-Trimethylcyclopentene-2-carboxylic acid (β-campholytic acid)	135	114	104			130		
2,4-Hexadienoic acid (sorbic acid)	135		153		129	168		
3-Phenylpropynoic acid (phenylpropiolic acid)	137	142	126	83		102		
(±)-*cis*-Camphenic acid [(±)-*cis*-camphenecamphoric acid]	137		212 (di)			225 (di)		
E-2-Chloro-3-phenyl-2-propenoic acid (*trans*-α-chlorocinnamic acid)	138	116	118			122		
2-Methylpropanedioic acid (methylmalonic acid)	138d	228; 214 (di) 145d (mono)	182	75		217		
2-Pyridinecarboxylic acid (2-picolinic acid)	138	104	76			107		
5-Chloro-2-nitrobenzoic acid	139		164			154		

Carboxylic Acids (Solids) (continued)

Name of compound	mp (°C)	Derivative mp (°C)						
		4-Toluidide	Anilide	4-Nitrobenzyl ester	4-Bromophenacyl ester	Amide	*S*-Benzylthiuronium salts	Phenylhydrazide
(*R*,*S*)-2,3-Dihydroxybutanedioic acid (*meso*-tartaric acid)	140		194 (mono)	93		190 (di)		
3-Nitrobenzoic acid	141	162	154	141	137	143	163	
2-Chloro-4-nitrobenzoic acid	142		168			172		
4-Chloro-2-nitrobenzoic acid	142		131			172		
Z-3-Chloro-3-phenyl-2-propenoic acid (*trans*-β-chlorocinnamic acid)	142	125	128			118		
2-Chlorobenzoic acid	142	131	118	106	107	142; 202		
Octanedioic acid (suberic acid)	144	219 (di)	187 (di) 129 (mono)	85 (di)	144 (di)	216 (di) 127 (mono)		
2-Aminobenzoic acid (anthranilic acid)	144	151	131	205	172	109	149	
2,4,5-Trimethoxybenzoic acid (asaronic acid)	144		155			185		
2-Chlorophenoxyethanoic acid (*o*-chlorophenoxyacetic acid)	146		121			149		
2-Nitrobenzoic acid	146	203	155	112	107	176	159	
2-Carboxyphenyl-2-oxoethanoic acid [phthalonic acid (anhydrous); 2-carboxyphenyl-2-oxoacetic acid]	146		176 (mono) 208 (di)			179 (α) 155 (β)		
Diphenylethanoic acid (diphenylacetic acid)	148	173	180			168		
Oxodiethanoic acid (diglycolic acid; oxodiacetic acid)	148	148 (mono)	152 (di) 118 (mono)			135 (mono)		
N-Phenylethanedioic acid monoamide (*N*-Phenyloxalic acid monoamide; oxanilic acid)	149		154 (di)			228		
2-Bromobenzoic acid	150		141	110	102	155	171	
2-Hydroxy-2,2-diphenylethanoic acid (benzilic acid; 2-hydroxy-2,2-diphenylacetic acid)	150	190	175	100	152	155		
4-Nitrophenylethanoic acid (*p*-nitrophenylacetic acid)	153	210	198; 212		207	198		
3-Carboxy-3-hydroxypentanedioic acid (anhydrous) [citric acid (anhydrous); 2-hydroxy-1,2,3-propanetricarboxylic acid]	153	189 (tri)	192 (tri)	102 (tri)	148 (tri)	215d (tri)		
2-Hydroxy-5-methylbenzoic acid (5-methylsalicylic acid)	153			147	142	178	185	
Hexanedioic acid (adipic acid)	154	241	241 (di) 153 (mono)	106	155	220 (di) 130 (mono)	163	209
3-Bromobenzoic acid	155		146; 136	105	126	155	168	
2-Hydroxybenzoic acid (salicylic acid)	158	156	136	98	140	142	148	
4-Chlorophenoxyethanoic acid (*p*-chlorophenoxyacetic acid)	158		125		136	133		
3-Chlorobenzoic acid	158		124	107	117	134	155	
2-Iodobenzoic acid	162		142	111	143; 110	110; 184		

Carboxylic Acids (Solids) (continued)

		Derivative mp (°C)						
Name of compound	**mp (°C)**	**4-Toluidide**	**Anilide**	**4-Nitrobenzyl ester**	**4-Bromophenacyl ester**	**Amide**	***S*-Benzylthiuronium salts**	**Phenylhydrazide**
1-Naphthoic acid	162		164		135	205		
2,4-Dichlorobenzoic acid	164	168	165; 153			194		
4-Nitro-1,2-benzenedicarboxylic acid (4-nitrophthalic acid)	165	172	192			200d		
2-Methylenebutanedioic acid (itaconic acid; methylenesuccinic acid)	165		190; 152 (mono)	91	117	192 (di)		
5-Bromo-2-hydroxybenzoic acid (5-bromosalicylic acid)	165		222			232		
3,4-Dimethylbenzoic acid	166		108			130		
1,2,3-Propanetricarboxylic acid (tricarballylic acid)	166		252 (tri)		138 (tri)	207d (tri)		
2-(±)-Phenylbutanedioic acid [(±)-phenylsuccinic acid]	167	175 (mono)	222 (di) 175 (mono)			211 (di) 159, 145 (mono)		
2,3-Dihydroxybutanedioic acid (*D*- or *L*-tartaric acid)	169		180; 194, 180 (mono) 264 (di)	163 (di)	216, 204 (di)	196 (di) 172 (mono)		240
2-Hydroxy-3-methylbenzoic acid (3-methylsalicylic acid)	169	164		99		112	204	
3,5-Dinitro-2-hydroxybenzoic acid (3,5-dinitrosalicylic acid)	173		181			197		
3-Aminobenzoic acid	174		140	201	190	111		
2-Hydroxy-4-methylbenzoic acid	177			175			165	
8-Bromo-1-naphthoic acid	178		151			180		
4-Methylbenzoic acid (*p*-toluic acid)	180	165	148	105	153	160	190	
6-Bromo-3-nitrobenzoic acid	180		166			198		
3,4-Dimethoxybenzoic acid [veratric acid (anhydrous)]	181		166; 154		124	164		
N-Benzoyl-2-aminobenzoic acid (*N*-benzoylanthranilic acid; 2-benzaminobenzoic acid)	181		279			219		
4-Chloro-3-nitrobenzoic acid	182		131			156		
2,4-Dinitrobenzoic acid	183			142	158	204		
2-Naphthoic acid	185	192	173		211	195		
N-Acetyl-2-aminobenzoic acid (acetylanthranilic acid)	185		167			177		
2-Carboxyphenylethanoic acid (homophthalic acid; 2-carboxyphenylacetic acid)	185		232			228		
4-Methoxybenzoic acid (4-anisic acid)	186	186	171	132	152	167	185	
4-Aminobenzoic acid	186				200	114		
N-Benzoylaminoethanoic acid (hippuric acid; *N*-benzoylglycine; *N*-benzoylaminoacetic acid)	187		208	136	151	183		
3-Iodobenzoic acid	187			121	128	186		

Carboxylic Acids (Solids) (continued)

Name of compound	mp (°C)	Derivative mp (°C): 4-Toluidide	Anilide	4-Nitrobenzyl ester	4-Bromophenacyl ester	Amide	S-Benzylthiuronium salts	Phenylhydrazide
Coumarin-3-carboxylic acid	187		250			236		
Butanedioic acid (succinic acid)	188	260 (di) 180 (mono)	230 (di) 149 (mono)	88 (di)	211 (di)	260 (di) 157 (mono)	154	210
1,2,2-Trimethylcyclopentane-1,3-dicarboxylic acid (D-camphoric acid)	188	196; 214	210, 196 (mono) 226 (di)	67		192 (di) 183 (mono)		
Ethanedioic acid (anhydrous) [oxalic acid (anhydrous)]	188	268 (di) 169 (mono)	246 (di) 148 (mono)			419 (di) 219 (mono)		
1,2,3,4-Butanetetracarboxylic acid	189		187			181 (di) 310 (tetra)		
2-Chlorobutenedioic acid (chlorofumaric acid)	192		186	139				
Coumarilic acid (coumarone-2-carboxylic acid)	193		159			159		
2,2-Dimethylpropanedioic acid (dimethylmalonic acid)	193			84		269 (di)		
1-Hydroxy-2-naphthoic acid	195		154			202		
E-1,2,3-Propenetricarboxylic acid (aconitic acid)	195		189 (di) 170 (mono)	76	186 (tri)	250d (tri)		
4-Ethoxybenzoic acid	198		172	110		202		
E-3-(3-Nitrophenyl)-2-propenoic acid (*trans* 3 nitrocinnamic acid)	200			174	178	196		
3,4-Dihydroxybenzoic acid (protocatechuic acid)	200		167	188		212		
3-Hydroxybenzoic acid	201	163	157	108	176	170		
2-Methyl-*E*-butenedioic acid (mesaconic acid; methylfumaric acid)	204	212 (di) 196 (mono)	186 (di) 202, 163 (mono)	134 (di)		176 (di) 222 (mono)		
(±)-2,3-Dihydroxybutanedioic acid [(±)-tartaric acid]	204		236 (di)	147 (di)		226 (di)		
4-Bromo-3-nitrobenzoic acid	204		156			156		
3,5-Dinitrobenzoic acid	205		234	157	159	183		
1,2-Benzenedicarboxylic acid (phthalic acid)	208d	201 (di) 165, 150 (mono)	255 (di) 170 (mono)	155 (di)	153 (di)	220 (di) 149 (mono)	158	
E-3-(2-Hydroxyphenyl)-2-propenoic acid (*o*-coumaric acid; *trans*-2-hydroxycinnamic acid)	208			152		209d		
Ethanedioic acid monoamide (oxamic acid, oxalic acid monoamide)	210		149			419		
E-3-(2-Chlorophenyl)-2-propenoic acid (*trans*-2-chlorocinnamic acid)	212		176			168		
2,4-Dihydroxybenzoic acid (β-resorcylic acid)	213		127	189		222		

Carboxylic Acids (Solids) (continued)

		Derivative mp (°C)						
Name of compound	mp (°C)	4-Toluidide	Anilide	4-Nitrobenzyl ester	4-Bromophenacyl ester	Amide	*S*-Benzylthiuronium salts	Phenylhydrazide
2,3,4,5-Tetrahydroxyhexanedioic acid (mucic acid; galactaric acid)	214d			310	225	220 (di) 192 (mono)		
4-Hydroxybenzoic acid	215	204	202	198; 182	191	162	145	
3-Nitro-1,2-benzenedicarboxylic acid (3-nitrophthalic acid)	218	226 (di)	234 (di)	190	166	201 (di)		
4-Cyanobenzoic acid	219		179	189		223		
3-Hydroxy-2-naphthoic acid	223	223	249			218		
4-Hydroxy-2-naphthoic acid	226	206				218		
Biphenyl-2,2′-dicarboxylic acid (diphenic acid)	229		230 (di) 176 (mono)	187 (di)		212 (di) 193 (mono)		
2-Hydroxy-5-nitrobenzoic acid (5-nitrosalicylic acid)	230		224			225		
4-Chloro-1-hydroxy-2-naphthoic acid	234	144	181					
5-Bromo-2-hydroxy-3-methylbenzoic acid (5-bromo-2-hydroxy-3-toluic acid)	236		125			78		
1,2,3,4-Butanetetracarboxylic acid	237		168 (di)			169 (di)		
7-Bromo-1-naphthoic acid	237		202			247		
3-Pyridinecarboxylic acid (nicotinic acid)	238	150	85; 132; 265			128		
E-3-(2-Nitrophenyl)-2-propenic acid (*trans*-2-nitrocinnamic acid)	240			132	142	185		
4-Nitrobenzoic acid	241	204; 192	217	169	137	201	182	
4-Chlorobenzoic acid	243		194	130	126	179; 170		
7-Chloro-1-naphthoic acid	243		185			237		
2-Chloroquinoline-4-carboxylic acid (3-chlorocinchonic acid)	244		202			335; 278		
4-Bromobenzoic acid	253		197	139; 180		190		
3,4,5-Trihydroxybenzoic acid (gallic acid)	254d; 240d		207	141	134	245; 189		
3,4-Pyridinedicarboxylic acid (cinchomeronic acid)	260d		206 (di)			170, 200 (mono) 165 (di)		
4-Iodobenzoic acid	270		210	141	147	217		
5-Chloro-2-naphthoic acid	270		203			187		
1-Chloroanthraquinone-2-carboxylic acid	272		249			317		
2-Amino-3-phenylpropanoic acid (β-phenylalanine)	273			222		140		
E-3-(4-Nitrophenyl)-2-propenoic acid (*trans*-4-nitrocinnamic acid)	287			187	191	204; 217		
9,10-Anthraquinone-2-carboxylic acid	292		260			280		
9,10-Anthraquinone-1-carboxylic acid	294		289			280		
1,4-Benzenedicarboxylic acid (terephthalic acid)	300		337	263	225			

Carboxylic Acids (Solids) (continued)

		Derivative mp (°C)						
Name of compound	mp (°C)	4-Toluidide	Anilide	4-Nitrobenzyl ester	4-Bromophenacyl ester	Amide	*S*-Benzylthiuronium salts	Phenylhydrazide
E-Butenedioic acid (fumaric acid)	302; 287		314 (di) 235 (mono)	151	256d	267d (di) 302	195	
1,3-Benzenedicarboxylic acid (isophthalic acid)	348; 300		250	215; 203	186	280	216	
1,3,5-Benzenetricarboxylic acid (trimesic acid)	350; 380		120d (tri)		197 (tri)	365d (tri)		

Esters (Liquids)

			Derivative mp (°C)					
Name of compound	bp (°C)	sp gr	Acid	Amide	4-Toluidide	3,5-Dinitrobenzoate	*N*-Benzylamide	Hydrazide
Methyl methanoate (methyl formate)	32	0.998^{0}_{4} 0.974			53	108	60	54
Ethyl methanoate (ethyl formate)	54	0.938; 0.922			53	93	60	54
Methyl ethanoate (methyl acetate)	57	0.958; 0.927	17	82	153	108	61	77
1-Methylethyl methanoate (isopropyl formate)	71	0.883; 0.873		53	123		60	54
Ethenyl ethanoate (vinyl acetate; ethenyl acetate)	72	0.9317^{20}_{4}	17	82	153		61	77
Ethyl ethanoate (ethyl acetate)	77	0.924; 0.901	17	82	153	93	61	77
Methyl propanoate (methyl propionate)	80	0.937; 0.915		81	126	108	44	40
Propyl methanoate (propyl formate)	81	0.918; 0.904			53	74	60	54
1,1-Dimethylethyl methanoate (*tert*-butyl formate)	83				53	142	60	54
2-Propenyl methanoate (allyl formate; 2-propenyl formate)	84	0.932^{17}_{4}; 0.946			53	50	60	54
Methyl propenoate (methyl acrylate)	85	0.977^{0}_{4}; 0.961^{19}_{4}		85	141	108		
1-Methylethyl ethanoate (isopropyl acetate)	91	0.917; 0.872	17	82	153	123	61	77
Methyl 2-methylpropanoate (methyl isobutyrate)	92	0.912^{0}_{4}; 0.888		129	109	108	87	104
1-Methylpropyl methanoate (*sec*-butyl formate)	97	0.882			53	76	60	54
1,1-Dimethylethyl ethanoate (*tert*-butyl acetate)	98	0.867	17	82	153	142	61	77
2-Methylpropyl methanoate (isobutyl formate)	98	0.905; 0.876			53	87	60	54
Ethyl difluoroethanoate (ethyl difluoroacetate)	99			52		93		
Ethyl propanoate (ethyl propionate)	99	0.912; 0.889		81	126	93	44	40
Methyl 2-methylpropenoate (methyl methacrylate)	101	0.936	17	106		108		
Ethyl propenoate (ethyl acrylate)	101	0.925^{0}_{4}; 0.914^{15}_{4}; 0.909		85	141	93		
Methyl 2,2-dimethylpropanoate (methyl pivalate; methyl trimethylacetate)	101	0.891^{0}_{4}	36	157	120	108		
Propyl ethanoate (propyl acetate)	101	0.909; 0.883	17	82	153	74	61	77
Methyl butanoate (methyl butyrate)	102	0.919; 0.898		116	75	108	38	44

Esters (Liquids) (continued)

Name of compound	bp (°C)	sp gr	Derivative mp (°C) Acid	Amide	4-Toluidide	3,5-Dinitrobenzoate	*N*-Benzylamide	Hydrazide
2-Propenyl ethanoate (allyl acetate; 2-propenyl acetate)	104	0.938; 0.928	17	82	153	50	61	77
Trimethyl orthomethanoate (trimethyl orthoformate)	105	0.974_4^{23}; 0.967			53	108		
Butyl methanoate (butyl formate)	107	0.911; 0.888			53	64	60	54
Methyl *Z*-2-butenoate (methyl isocrotonate)	108			102	132	108		
Ethyl 2-methylpropanoate (ethyl isobutyrate)	111	0.890; 0.869		129	109	93	87	104
Chloromethyl ethanoate (chloromethyl acetate)	111	1.195_4^{14}; 1.094_4^{15}	17	82	153		61	77
1-Methylethyl propanoate (isopropyl propionate)	111	0.893_4^{0}		81	126	123	44	40
1-Methylpropyl ethanoate (*sec*-butyl acetate)	112	0.870	17	82	153	76	61	77
Methyl 3-methylbutanoate (methyl isovalerate)	117	0.901; 0.881		137	107	108	54	68
2-Methylpropyl ethanoate (isobutyl acetate)	118	0.892; 0.871	17	82	153	87	61	77
Ethyl 2,2-dimethylpropanoate (ethyl pivalate; ethyl trimethylacetate)	118	0.855	35	157	120	93		
Ethyl 2-methylpropenoate (ethyl methacrylate)	118	0.911	16	106		93		
Methyl *E*-2-butenoate (methyl crotonate)	119	0.981_4^{4}; 0.946	72	118	132	108	114	
1-Methylethyl 2-methylpropanoate (isopropyl isobutyrate)	120	0.847_4^{21}		129	109	123	87	104
Ethyl butanoate (ethyl butyrate)	122	0.900; 0.879		116	75	93	38	44
Propyl propanoate (propyl propionate)	123	0.902; 0.881		81	126	74	44	40
2,2-Dimethylpropyl ethanoate (*tert*-amyl acetate; dimethylethylcarbinyl acetate; 2,2-dimethylpropyl acetate)	124	0.874_4^{19}	17	82	153	118	61	77
2-Propenyl propanoate (allyl propionate)	124	0.914		81	126	50	44	40
3-Methylbutyl methanoate (isoamyl formate; isopentyl formate; 3-methylbutyl formate)	124	0.894; 0.882			53	61	60	54
Ethyl *Z*-2-butenoate (ethyl isocrotonate)	126	0.918	15	102	132	93		
Butyl ethanoate (butyl acetate)	126	0.902; 0.881	17	82	153	64	61	77

Esters (Liquids) (continued)

Name of compound	bp (°C)	sp gr	Derivative mp (°C) Acid	Amide	4-Toluidide	3,5-Dinitrobenzoate	*N*-Benzylamide	Hydrazide
1,1-Dimethylethyl 2-methylpropanoate (*tert*-butyl isobutyrate)	127			129	109	142	87	104
1-Methylethyl butanoate (isopropyl butyrate)	128	0.879^{0}_{4}; 0.865^{13}_{4}		116	75	123	38	44
Methyl pentanoate (methyl valerate)	130	0.910; 0.885		106	74	108	42	
Methyl methoxyethanoate (methyl methoxyacetate)	130	1.051		97		108		
Methyl chloroethanoate (methyl chloroacetate)	132	1.238	61; 53	121	162	108		
Ethyl methoxyethanoate (ethyl methoxyacetate)	132	1.012^{15}_{4}		97		93		
Pentyl methanoate (amyl formate)	132	0.902; 0.885			53	46	60	54
1-Ethylpropyl ethanoate (diethylcarbinyl acetate; 1-ethylpropyl acetate)	133	1.401	17	82	153	101	61	77
1-Methylbutyl ethanoate (methylpropylcarbinyl acetate; 1-methylbutyl acetate)	133	0.869^{18}_{4}	17	82	153	62	61	77
2-Propenyl 2-methylpropanoate (allyl isobutyrate)	134			129	109	50	87	104
Ethyl 3-methylbutanoate (ethyl isovalerate)	135	0.865		137	107	93	54	68
Propyl 2-methylpropanoate (propyl isobutyrate)	135	0.884^{0}_{4}; 0.864		129	109	74	87	104
Methyl 2-hydroxy-2-methylpropanoate (methyl α-hydroxy isobutyrate)	137		79		133	108		
Methyl 2-oxopropanoate (methyl pyruvate)	138	1.154	14	125; 145	109; 130	108		
2-Methylpropyl propanoate (isobutyl propionate)	138	0.888		81	126	87	44	40
Ethyl *E*-2-butenoate (ethyl crotonate)	138	0.9175^{20}_{4}	72	160	132	93	114	
2-Propenyl butanoate (allyl butyrate)	142	0.902		116	75	50	38	44
3-Methylbutyl ethanoate (isoamyl acetate; isopentyl acetate)	142	0.884; 0.867	17	82	153	61	61	77
1-Methylethyl 3-methylbutanoate (isopropyl isovalerate)	142	0.854^{17}_{4}		137	107	123	54	68
Propyl butanoate (propyl butyrate)	144	0.893; 0.872		116	75	74	38	44

Esters (Liquids) (continued)

Name of compound	bp (°C)	sp gr	Derivative mp (°C) Acid	Amide	4-Toluidide	3,5-Dinitrobenzoate	N-Benzylamide	Hydrazide
2-Methoxyethyl ethanoate (ethylene glycol monomethyl ether acetate; 2-methoxyethyl acetate)	144	1.0067^{20}_{20}; 1.088	17	82	153		61	77
Methyl bromoethanoate (methyl bromoacetate)	144	1.657^{19}_{4}	50	91		108		
2-Chloroethyl ethanoate (2-chloroethyl acetate)	145	1.178	17	82	153		61	77
Methyl (±)-2-hydroxypropanoate [methyl (±)-lactate]	145	1.090^{19}_{4}	17	79	107	108		
Ethyl chloroethanoate (ethyl chloroacetate)	145	1.158; 1.150	61; 52	121	162	93		
1,1-Dimethylethyl butanoate (*tert*-butyl butyrate)	146			116	75	142	38	44
Triethyl orthomethanoate (triethyl orthoformate)	146	0.897			53	93		
Ethyl pentanoate (ethyl valerate)	146	0.8765^{20}_{4}		106	74	93	42	
Ethyl 2-chloropropanoate (ethyl α-chloropropionate)	146	1.087		80	124	93		
Butyl propanoate (butyl propionate)	147	0.895; 0.875		81	126	64	44	40
Benzyl chloroethanoate (benzyl chloroacetate)	147	1.222^{4}_{4}	61; 52	121	162	113		
Methyl ethoxyethanoate (methyl ethoxyacetate)	148	1.011^{15}_{4}		82	32	108		
2-Methylpropyl 2-methylpropanoate (isobutyl isobutyrate)	149	0.875		129	109	87	87	104
Pentyl ethanoate (amyl acetate)	149	0.896; 0.875	17	82	153	46	61	77
Ethyl 2-hydroxy-2-methylpropanoate (ethyl α-hydroxyisobutyrate)	150		79		133	93		
Methyl hydroxyethanoate (methyl glycolate; methyl hydroxyacetate)	151	1.168^{18}_{4}	80	120	143	108	104	
Methyl hexanoate (methyl caproate)	151	0.904; 0.885		101	75	123	53	
1-Methylethyl pentanoate (isopropyl valerate)	153	0.858		106	74	61	42	
Ethyl (±)-2-hydroxypropanoate [ethyl (±)-lactate]	155	1.031^{19}_{4}	18	79	107	93		
Ethyl 2-oxopropanoate (ethyl pyruvate)	155	1.080^{14}_{4}; 1.055	14	125; 145	109; 130	93		
Propyl 3-methylbutanoate (propyl isovalerate)	156	0.864^{17}_{8}		137	107	74	54	68

Esters (Liquids) (continued)

			Derivative mp (°C)					
Name of compound	bp (°C)	sp gr	Acid	Amide	4-Toluidide	3,5-Dinitrobenzoate	*N*-Benzylamide	Hydrazide
Hexyl methanoate (hexyl formate)	156	0.898^{0}_{0}; 0.879			53	58	60	54
2-Methylpropyl butanoate (isobutyl butyrate)	157	0.888; 0.862		116	75	87	38	44
2-Ethoxyethyl ethanoate (ethylene glycol monoethyl ether acetate; 2-ethoxyethyl acetate)	158	0.9749^{20}_{20}; 0.970	17	82	153	73	61	77
Ethyl dichloroethanoate (ethyl dichloroacetate)	158	1.282		98	153	93		
Ethyl bromoethanoate (ethyl bromoacetate)	159	1.506	50	91		93		
Ethyl hydroxyethanoate (ethyl glycolate; ethyl hydroxyacetate)	160	1.0869^{15}_{4}; 1.082	80	120	143	93	104	
3-Methylbutyl propanoate (isoamyl propionate)	160	0.888; 0.870; 0.859		81	126	61	44	40
Ethyl 2-bromopropanoate (ethyl α-bromopropionate)	162	1.329; 1.524	26	123	125	93		
Cyclohexyl methanoate (cyclohexyl formate)	163	1.010; 0.994			53	113	60	54
2-Bromoethyl ethanoate (2-bromoethyl acetate)	163	1.524	17	82	153		61	77
Propyl pentanoate (propyl valerate)	167	0.8888^{0}_{4}; 0.870		106	74	74	42	
Butyl butanoate (butyl butyrate)	167	0.888; 0.869		116	75	64	38	44
Ethyl hexanoate (ethyl caproate)	168	0.889; 0.871		101	75	93	53	
Ethyl trichloroethanoate (ethyl trichloroacetate)	168	1.369^{15}_{4}; 1.380	58	141	113	93		
(±)-1-Methylethyl 2-hydroxypropanoate [isopropyl (±)-lactate]	168	0.998	18	79	107	123		
Pentyl propanoate (amyl propionate)	169	0.876; 0.881		81	126	46	44	40
3-Methylbutyl 2-methylpropanoate (isoamyl isobutyrate; isopentyl isobutyrate)	169	0.876^{0}_{4}		129	109	61	87	104
2-Methylpropyl 3-methylbutanoate (isobutyl isovalerate)	171	0.853		137	107	87	54	68
Methyl heptanoate (methyl enanthate)	174	0.898; 0.880		96	81	108		
1,2-Ethanediol dimethanoate (ethylene glycol diformate; 1,2-ethanediol formate)	174	1.193^{0}_{4}; 1.229			53	46	60	54
1-Methylpropyl pentanoate (*sec*-butyl valerate)	174	0.860^{20}_{4}		106	74	76	42	

Esters (Liquids) (continued)

Name of compound	bp (°C)	sp gr	Derivative mp (°C) Acid	Amide	4-Toluidide	3,5-Dinitrobenzoate	N-Benzylamide	Hydrazide
Cyclohexyl ethanoate (cyclohexyl acetate)	175	0.972_4^{19}	17	82	153	113	61	77
Butyl chloroethanoate (butyl chloroacetate)	175	1.081_4^{15}	61; 52	121	62	64		
Furfuryl ethanoate (furfuryl acetate)	177	1.1176_{20}^{20}	17	82	153	81	61	77
Heptyl methanoate (heptyl formate)	178	0.878			53	47	60	54
Hexyl ethanoate (hexyl acetate)	178	0.873	17	82	153	58	61	77
3-Methylbutyl butanoate (isoamyl butyrate; isopentyl butyrate)	179	0.882; 0.864		116	75	61	38	44
Ethyl 3-bromopropanoate (ethyl β-bromopropionate)	179	1.425	63	111		93		
2-Methylpropyl pentanoate (isobutyl valerate)	179	0.8625		106	74	87	42	
2-Hydroxyethyl methanoate (ethylene glycol monoformate; 2-hydroxyethyl formate)	180	1.199_4^{15}			53	46	60	54
Methyl 2-furoate (methyl pyromucate)	181	1.1786_4^{21}	134	143	171	108		80
Dimethyl propanedioate (dimethyl malonate, methyl malonate)	182	1.160_4^{15}; 1.119	135	106 (mono)	156 (mono) 170 (di)	108 253 (di)	142	154
Methyl cyclohexanecarboxylate (Methyl hexahydrobenzoate)	183	0.995_4^{15}; 0.990	31	186		108		
Ethyl hexadecanoate (ethyl palmitate)	185		63	107	98	93	95	111
Diethyl ethanedioate (diethyl oxalate; ethyl oxalate)	186	1.082	190 (anhyd) 101 (dihyd)	219 (mono) 419 (di)	169 (mono) 268 (di)	93	223	243
Pentyl butanoate (amyl butyrate)	186	0.8832_0^0; 0.866		116	75	46	38	44
Ethyl 2-methyl-3-oxobutanoate (ethyl methylacetoacetate)	187	1.024_4^{15}; 1.012		73		93		
Butyl pentanoate (butyl valerate)	187	0.868		106	74	64	42	
Propyl hexanoate (propyl caproate)	187	0.8844_0^0; 0.867		101	75	74	53	
Butyl 2-hydroxypropanoate (butyl lactate)	188	0.984_{20}^{20}	18	79	107	64		
2-Hydroxyethyl ethanoate (ethylene glycol monoacetate; 2-hydroxyethyl acetate)	189		17	82	153	46	61	77
Ethyl heptanoate (ethyl enanthate)	189	0.888; 0.869		97	81	93		
Hexyl propanoate (hexyl propionate)	190	0.870		81	126	58	44	40
3-Methylbutyl 3-methylbutanoate (isoamyl isovalerate; isopentyl isovalerate)	190	0.870		137	107	61	54	68

Esters (Liquids) (continued)

Name of compound	bp (°C)	sp gr	Derivative mp (°C) Acid	Amide	4-Toluidide	3,5-Dinitrobenzoate	*N*-Benzylamide	Hydrazide
Methyl 2-ethyl-3-oxobutanoate (methyl ethylacetoacetate)	190	0.989		96		108		
1,2-Ethanediol diethanoate (ethylene glycol diacetate; 1,2-ethanediol diacetate)	190	1.128; 1.104	17	82	153	46	61	77
Heptyl ethanoate (heptyl acetate)	192	0.8891; 0.865	17	82	153	47	61	77
Cyclohexyl propanoate (cyclohexyl propionate)	193	0.9718_4^0		81	126	113	44	40
Di-(1-methylethyl) ethanedioate (diisopropyl oxalate; isopropyl oxalate)	194	1.010_4^{18}; 0.995	190 (anhyd) 101 (dihyd)	219 (mono) 419 (di)	169 (mono) 268 (di)	123	223	243
1-Methylheptyl ethanoate (*sec*-octyl acetate)	195	0.861_4^{19}	17	82	153	32	61	77
Methyl octanoate (methyl caprylate)	195	0.894; 0.878	16	57	70	108		
2-Tetrahydrofurfuryl ethanoate (α-tetrahydrofurfuryl acetate)	195	1.0624_4^{25}	17	82	153	84	61	77
Dimethyl butenedioate (dimethyl succinate; methyl succinate)	196	1.120_4^{18}	185	157 (mono) 260d (di)	180 (mono) 256 (di)	108	206	168
Methyl 4-oxopentanoate (methyl levulinate)	196	1.068; 1.049	35	108	109	108		
Ethyl cyclohexanecarboxylate (ethyl hexahydrobenzoate)	196	0.967_4^{15}; 0.962	31	186		93		
Diethyl methylpropanedioate (diethyl methylmalonate; ethyl methylmalonate)	196	1.019_4^{15}	138	217	145 (mono) 228, 214 (di)	93		
Phenyl ethanoate (phenyl acetate)	197	1.081_4^{15}	17	82	153	146	61	77
Ethyl furoate (ethyl pyromucate)	197	1.117	134	143	171	93		80
Ethyl 2-ethyl-3-oxobutanoate (ethyl ethylacetoacetate)	198	0.972; 0.986		96		93		
Octyl methanoate (octyl formate)	199	0.8744			53	62	60	54
2-Ethyl-1-hexyl ethanoate (2-ethyl-1-hexyl acetate)	199	0.8733_{20}^{20}	17	82	153		61	77
Methyl benzoate	199	1.103; 1.089	122	130	158	108	106	112
Diethyl propanedioate (diethyl malonate; ethyl malonate)	199	1.077; 1.055	135	110 (mono) 170 (di)	156 (mono) 253 (di)	93	142	154
Methyl cyanoethanoate (methyl cyanoacetate)	200	1.0962_4^{25}	66	120		108	124	
Methyl *E*-2-methylbutenedioate (dimethyl mesaconate; methyl mesaconate)	203	1.0914; 1.120	205	222 (α) 174 (β) (mono) 177 (di)	196 (mono) 212 (di)	108		

Esters (Liquids) (continued)

Name of compound	bp (°C)	sp gr	Derivative mp (°C) Acid	Amide	4-Toluidide	3,5-Dinitrobenzoate	N-Benzylamide	Hydrazide
Benzyl methanoate (benzyl formate)	203	1.081_4^{23}			53	133	60	54
Ethenyl benzoate (vinyl benzoate)	203	1.065	122	130	158		106	112
Cyclohexyl 2-methylpropanoate (cyclohexyl isobutyrate)	204	0.949_4^0		129	109	113	87	104
Diethyl *Z*-butenedioate (diethyl maleate; ethyl maleate)	204	1.145_4^{15}	137	173 153 (mono) 181 (di)	142 (di)	108	150	
Ethyl 4-oxopentanoate (ethyl levulinate)	206	1.016	35	108	109	93		
Pentyl pentanoate (amyl valerate)	207	0.881_4^0		106	74	46	42	
2-Tetrahydrofurfuryl propanoate (α-tetrahydrofurfuryl propionate)	207			81	126	84	44	40
Butyl hexanoate (butyl caproate)	208	0.8653		101	75	64	53	
Hexyl butanoate (hexyl butyrate)	208	0.8652		116	75	58	38	44
2-Methylphenyl ethanoate (*o*-tolyl acetate; *o*-cresyl acetate)	208	1.048_4^{19}	17	82	153	135	61	77
Propyl heptanoate (propyl ethanthate)	208	0.866		97	81	74		
Ethyl octanoate (ethyl caprylate)	208	0.887	16	110	70	93		
Dimethyl methylenebutanedioate (dimethyl itaconate; methyl itaconate; dimethyl methylene-succinate; methyl methylenesuccinate)	208		165	192		108		
2-Methylpropyl heptanoate (isobutyl enanthate)	209	0.859		97	81	87		
1-Methylethyl 4-oxopentanoate (isopropyl levulinate)	209	0.987	34	108	109	123		
1,3-Propanediol diethanoate (trimethylene glycol diacetate; 1,3-diacetoxypropane; 1,3-propanediol diacetate)	210	1.070_4^{19}	17	82	153	178	61	77
Octyl ethanoate (octyl acetate)	210	0.885_4^0	17	82	153	62	61	77
Heptyl propanoate (heptyl propionate)	210	0.868		81	126	47	44	40
Dimethyl *Z*-methylbutenedioate (dimethyl citraconate; dimethyl *cis*-methylmaleate)	211	1.115	93	101		108		177
Propyl 2-furoate (propyl pyromucate)	211	1.0745_4^{26}	134	143	171	74		80
1,2-Ethanediol dipropanoate (ethylene glycol dipropionate)	211	1.045_4^{25}; 1.054_{15}^{15}		81	126	46	44	40

Esters (Liquids) (continued)

Name of compound	bp (°C)	sp gr	Derivative mp (°C) Acid	Amide	4-Toluidide	3,5-Dinitrobenzoate	*N*-Benzylamide	Hydrazide
Phenyl propanoate (phenyl propionate)	211	1.054_4^{15}		81	126	146	44	40
3-Methylphenyl ethanoate (*m*-tolyl acetate; *m*-cresyl acetate)	212	1.049	17	82	153	165	61	77
Cyclohexyl butanoate (cyclohexyl butyrate)	212	0.957_4^0		116	75	113	38	44
Ethyl benzoate	212	1.066; 1.047	122	130	158	93	106	112
4-Methylphenyl ethanoate (*p*-tolyl acetate; *p*-cresyl acetate)	213	1.050_4^{23}	17	82	153	189	61	77
Dipropyl ethanedioate (dipropyl oxalate; propyl oxalate)	214	1.038; 1.017	190 (anhyd) 101 (dihyd)	219 (mono) 419d (di)	169 (mono) 268 (di)	74	223	243
Methyl nonanoate (methyl pelargonate)	214	0.892		99	84	108		
Dimethyl pentanedioate (dimethyl glutarate; methyl glutarate)	214	1.0874	98	175 (di)	218 (di)	108	170	176
Ethyl 2,4-dioxopentanoate (ethyl acetopyruvate; ethyl aceto-2-oxopropanoate)	215	1.125	101	132		93		
Methyl 2-methylbenzoate (methyl *o*-toluate)	215	1.073_4^{15}; 1.061	108	143	144	108		124
2,6-Dimethylphenyl ethanoate (2,6-dimethylphenyl acetate)	216		17	82	153	159	61	77
Benzyl ethanoate (benzyl acetate)	217	1.062_4^{15}; 1.055	17	82	153	113	61	77
Diethyl butanedioate (diethyl succinate; ethyl succinate)	217	1.049_4^{15}; 1.0398	185	157 (mono) 260d (di)	180 (mono) 260 (di)	93	206	168
Diethyl *E*-butenedioate (diethyl fumarate; ethyl fumarate)	218	1.047_4^{25}; 1.052	287; 235; 200	270, 302 (mono) 266d (di)	234 (mono) 314 (di)		205	
2-Acetoxy-2′-ethoxydiethyl ether (diethylene glycol monoethyl ether acetate)	218	1.0114_{20}^{20}	17	82	153			
1-Methylethyl benzoate (isopropyl benzoate)	219	1.010_4^{24}	122	130	158	123	106	112
Methyl phenylethanoate (methyl phenylacetate)	220	1.044_4^{16}; 1.068	77	156	136	108	122	
L-1,5-Dimethyl-1-ethenyl-4-hexenyl ethanoate (L-3,7-dimethyl-1,6-octadien-3-ol acetate; L-linalyl acetate; L-1,5-dimethyl-1-ethenyl-4-hexenyl acetate)	220	0.895	17	82	153		61	77
Methyl 3-methylbenzoate (methyl *m*-toluate)	221	1.066_4^{15}	113	97	118	108	75	97
Propyl 4-oxopentanoate (propyl levulinate)	221	0.9895_4^{20}	35	108	109	74		

Esters (Liquids) (continued)

Name of compound	bp (°C)	sp gr	Derivative mp (°C) Acid	Amide	4-Toluidide	3,5-Dinitrobenzoate	*N*-Benzylamide	Hydrazide
Methyl 4-methylbenzoate (methyl *p*-toluate)	223		180	160	165	108	133	117
Methyl 2-hydroxybenzoate (methyl salicylate)	224	1.184_4^{20}	158	142	156	108	136	
Diethyl *Z*-butenedioate (diethyl maleate; ethyl maleate)	225	1.074_4^{15}; 1.066	137	173, 153 (mono) 181 (di)	142 (di)	93	150	
Diethyl (±)-2-hydroxypropanedioate [diethyl (±)-tartronate; ethyl (±)-tartronate; diethyl hydroxymalonate; ethyl hydroxymalonate]	225	1.152_4^{15}	158	198 (di) 196		93		
1-Methylpropyl 4-oxopentanoate (*sec*-butyl levulinate)	226	0.967	35	108	109	76		
Heptyl butanoate (heptyl butyrate)	226	0.8637		116	75	47	38	44
2,4-Dimethylphenyl ethanoate (2,4-dimethylphenyl acetate)	226	1.030_5^{15}	17	82	153	166	61	77
Pentyl hexanoate (amyl caproate)	226	0.863		101	75	46	53	
Butyl heptanoate (butyl enanthate)	226	0.864		97	81	64		
Hexyl pentanoate (hexyl valerate)	226	0.863		106	74	58	42	
Propyl octanoate (propyl caprylate)	226	0.8659	16	110	57	74		
Methyl 3-(2-furyl)propenoate [methyl *β*-(2-furyl)acrylate]	227			169		108		
Ethyl 2-methylbenzoate (ethyl *o*-toluate)	227	1.033_4^{21}	108	143	144	93		124
Ethyl nonanoate (ethyl pelargonate)	227	0.866_4^{17}		99	84	93		
L-Menthyl ethanoate (L-menthyl acetate)	227	0.9185_4^{20}	17	82	153	153	61	77
Octyl propanoate (octyl propionate)	228	0.866		81	126	62	44	40
Diethyl methylenebutanedioate (diethyl itaconate; ethyl itaconate; diethyl methylenesuccinate; ethyl methylenesuccinate)	228	1.047	165	192 (di)		93		
Ethyl phenylethanoate (ethyl phenylacetate)	229	1.031	77	156	136	93	122	
Diethyl *E*-methylbutenedioate (diethyl mesaconate; ethyl mesaconate; diethyl methylfumarate; ethyl methylfumarate)	229	1.0453	205	222 (α) 174 (β) (mono) 177 (di)	196 (α) (mono) 212 (di)	93		

Esters (Liquids) (continued)

Name of compound	bp (°C)	sp gr	Derivative mp (°C)					
			Acid	Amide	4-Toluidide	3,5-Dinitrobenzoate	*N*-Benzylamide	Hydrazide
Di-(2-methylpropyl) ethanedioate (diisobutyl oxalate; isobutyl oxalate)	229	1.002^{14}_{4}; 0.974	190 (anhyd) 101 (dihyd)	219 (mono) 419d (di)	169 (mono) 268 (di)	87	223	243
2-Propenyl benzoate (allyl benzoate)	230	1.067^{4}_{4}; 1.052	122	130	158	50	106	112
2-Methylpropyl 4-oxopentanoate (isobutyl levulinate)	231	0.9677	35	108	109	87		
Propyl benzoate	231	1.025^{15}_{4}	122	130	158	74	106	112
Diethyl *Z*-methylbutenedioate (diethyl citraconate; ethyl citraconate; diethyl methylmaleate)	231	1.049	93	101		93		177
Methyl 3-chlorobenzoate	231		158	134		108		158
Methyl decanoate (methyl caprate)	232	0.876^{18}_{4}	32	108	78	108		
2-Phenylethyl ethanoate (β-phenylethyl acetate)	232	1.057^{22}_{4}	17	82	153	108	61	77
Ethyl 3-(2-furyl)propenoate [ethyl β-(2-furyl)acrylate]	232	1.089^{15}_{4}	141	169		93		
Diethyl pentanedioate (diethyl glutarate; ethyl glutarate)	234	1.022; 1.424	98	176 (di)	218 (di)	93	170	176
Ethyl 3-methylbenzoate (ethyl *m*-toluate)	234	1.0265^{21}_{4}	113	97	118	93	75	97
Ethyl 2-hydroxybenzoate (ethyl salicylate)	234	1.147^{4}_{4}; 1.125	158	142	156	93	136	
Methyl 2-chlorobenzoate	234		144	142; 202	131	108		110
Ethyl 4-methylbenzoate (ethyl *p*-toluate)	235	1.027^{18}_{4}	180	160	165	93	133	117
Diethyl 2-bromopropanedioate (diethyl bromomalonate; ethyl bromomalonate)	235	1.426^{15}_{15}	113	181 (di)	217 (di)	93		
3,5-Dimethylphenyl ethanoate (3,5-dimethylphenyl acetate)	235		17	82	153	182	61	77
2,4,6-Trimethylphenyl ethanoate (mesityl acetate; 2,4,6-trimethylphenyl acetate)	236		17	82	153		61	77
1,2-Ethanediol dibutanoate (ethylene glycol dibutyrate)	237	1.0005		116	75	46	38	44
2,5-Dimethylphenyl ethanoate (2,5-dimethylphenyl acetate)	237	1.0264^{15}_{4}	17	82	153	137	61	77
Butyl 4-oxopentanoate (butyl levulinate)	238	0.9735	35	108	109	64		
Methyl 3-phenylpropanoate (methyl hydrocinnamate; methyl β-phenylpropionate)	239	1.0455^{0}_{4}	48	105; 82	135	108	85	

Esters (Liquids) (continued)

			Derivative mp (°C)					
Name of compound	bp (°C)	sp gr	Acid	Amide	4-Toluidide	3,5-Dinitrobenzoate	*N*-Benzylamide	Hydrazide
Diethyl 2-butylpropanedioate (diethyl butylmalonate; ethyl butylmalonate)	240	1.425		200 (di)		93		
Benzyl butanoate (benzyl butyrate)	240	1.033^{16}_{4}		116	75	113	38	44
2-Methoxyphenyl ethanoate (guaiacol acetate; *o*-methoxyphenyl acetate)	240	1.129^{25}_{4}	17	82	153	141	61	77
1-Methylethyl 2-hydroxybenzoate (isopropyl salicylate)	242	1.095^{19}_{4}; 1.073	158	142	156	123	136	
Dimethyl L-hydroxybutanedioate (dimethyl L-malate; methyl L-malate; dimethyl hydroxysuccinate; methyl hydroxysuccinate)	242	1.233	101	157 (di)	207 (di)	108	157	178
2-Methylpropyl benzoate (isobutyl benzoate)	242	1.002^{15}_{4}	122	130	158	87	106	112
Methyl 2-bromobenzoate	244		150	155		108		
Octyl butanoate (octyl butyrate)	244	0.863		116	75	62	38	44
Ethyl decanoate (ethyl caprate)	245	0.868^{18}_{4}	32	108	78	93		
Diethyl hexanedioate (diethyl adipate; ethyl adipate)	245	1.009	154	130 (mono) 220 (di)	241	93	189	171
Methyl phenoxyethanoate (methyl phenoxyacetate)	245	1.150^{17}_{4}	99	102		108		
5-Methyl-2-(1-methylethyl)-phenyl ethanoate (thymyl acetate; 2-isopropyl-5-methylphenyl acetate)	245	1.009	17	82	153	103	61	77
Butyl octanoate (butyl caprylate)	245	0.8628	16	110	70	64		
Heptyl pentanoate (heptyl valerate)	245	0.8622		106	74	47	42	
Pentyl heptanoate (pentyl enanthate)	245	0.8623		97	81	46		
Hexyl decanoate (hexyl caproate)	245	0.8622	32	101	75	58		
Dibutyl ethanedioate (dibutyl oxalate; butyl oxalate)	246	1.010; 0.987	190 (anhyd) 101 (dihyd)	219 (mono) 419d (di)	169 (mono) 268 (di)	64	223	243
2-Acetoxy-2′-butyldiethyl ether (diethylene glycol monobutyl ether acetate)	246	0.9871^{20}_{20}	17	82	153			
2,4,5-Trimethylphenyl ethanoate (pseudomenyl acetate; 2,4,5-trimethylphenyl acetate)	246		17	82		153	61	77
2-Methylpropyl phenylethanoate (isobutyl phenylacetate)	247	0.999^{18}_{4}	77	156	136	87	122	
Ethyl 3-phenylpropanoate (ethyl hydrocinnamate; ethyl *β*-phenylpropionate)	247	1.0147	49	105; 82	135	93	85	

Esters (Liquids) (continued)

Name of compound	bp (°C)	sp gr	Derivative mp (°C) Acid	Amide	4-Toluidide	3,5-Dinitrobenzoate	*N*-Benzylamide	Hydrazide
Dipropyl butenedioate (dipropyl succinate; propyl succinate)	248	1.016_4^4	185	157 (mono) 260d (di)	180 (mono) 256 (di)	74	206	168
Methyl 2-methoxybenzoate	248	1.157_4^{19}	101	129		108		
Methyl undecanylenate (methyl hendecanoate)	248	0.889_4^{15}	25	87		108		
3-Methylbutyl 4-oxopentanoate (isobutyl levulinate)	249	0.9614	35	108	109	62		
Butyl benzoate	250	1.000; 1.005	122	130	158	64	106	112
Propyl 2-hydroxybenzoate (propyl salicylate)	251	1.098_4^{15}	158	142	156	74	136	
2,2′-Diacetoxy diethyl ether (diethylene glycol diacetate)	251	1.108_{15}^{15}		82	153	149		
Ethyl phenoxyethanoate (ethyl phenoxyacetate)	251	1.104_4^{17}	99	102		93		
Diethyl L-hydroxybutanedioate (diethyl L-malate; ethyl L-malate; diethyl hydroxysuccinate; ethyl hydroxysuccinate)	253	1.129	101	157 (di)	207 (di)	108	157	178
Pentyl 4-oxopentanoate (amyl levulinate)	253	0.9614	35	108	109	46		
Ethyl (±)-2-hydroxy-2-phenylethanoate [ethyl (±)-mandelate; ethyl (±)-2-hydroxy-2-phenylacetate]	254		120	134	172	93		
Butyl phenylethanoate (butyl phenylacetate)	254	0.996_4^{18}	77	156	136	64	122	
2-Methoxyethyl benzoate	255	1.0891_4^{25}	122	130	158		106	112
Methyl 4-methoxybenzoate (methyl *p*-anisate)	255		185	167	186	108	132	
Diethyl heptanedioate (diethyl pimelate; ethyl pimelate)	255	0.9945_4^{20}; 0.9929	105	175 (di)	206 (di)	93	154	182
Ethyl benzoylmethanoate (ethyl benzoylformate)	257	1.222_4^{25}	66	91		93		
1,2,3-Propanetriol triethanoate (glyceryl triacetate; 1,2,3-propanetriol triacetate; triacetin)	258	1.161_4^{15}	17	82	153		61	77
Pentyl octanoate (amyl caprylate)	260	0.8613	16	110	70	46		
2-Ethoxyethyl benzoate	261	1.058_{25}^{25}	122	130	158	78	106	112
Hexyl heptanoate (hexyl enanthate)	261	0.8611		97	81	58		
Heptyl hexanoate (heptyl caproate)	261	0.8611		101	75	47	53	
Ethyl 2-methoxybenzoate	261	1.112	101	129		93		

Esters (Liquids) (continued)

Name of compound	bp (°C)	sp gr	Derivative mp (°C) Acid	Amide	4-Toluidide	3,5-Dinitrobenzoate	N-Benzylamide	Hydrazide
Octyl pentanoate (octyl valerate)	261	0.8615		106	74	62	42	
Methyl *E*-3-phenyl-2-propenoate (methyl cinnamate)	261		133	148	168	108	225	
2-Methylpropyl 2-hydroxybenzoate (isobutyl salicylate)	262	1.0639	158	142	156	87	136	
3-Methylbutyl benzoate (isoamyl benzoate; isopentyl benzoate)	262	1.004; 0.986	122	130	158	61	106	112
Ethyl 4-bromobenzoate	262					92		164
D-Dimethyl 1,2,2-trimethylcyclopentane-1,3-dicarboxylate (dimethyl D-camphor; methyl D-camphor)	263	1.0747	188	176, 183 (mono) 193 (di)	214 (α) 196 (β)	108		
Ethyl undecylenate (ethyl hendecylenate)	264	0.8827_{15}^{15}		87		93		
Di-(2-methylpropyl) butanedioate (diisobutyl succinate; isobutyl succinate)	265	0.974	185	157 (mono) 260d (di)	180 (mono) 260 (di)	87	206	185
Di-(3-methylbutyl) ethanedioate (diisoamyl oxalate; isoamyl oxalate)	268	0.968_4^{11}; 0.961	190 (anhyd) 101 (dihyd)	219 (mono) 419d (di)	169 (mono) 268 (di)	61	223	243
Dimethyl octanedioate (dimethyl suberate; methyl suberate)	268	1.0198	144	127 (mono) 217 (di)	219 (di)	108		
Methyl dodecanoate (methyl laurate)	268	0.869_4^{19}	44	100	87	108	83	105
Ethyl 4-methoxybenzoate (ethyl *p*-anisate)	269	1.119_4^4; 1.1038	185	167	186	93	132	
Ethyl dodecanoate (ethyl laurate)	269	0.867_4^{13}	44	100	87	93	83	105
Ethyl *E*-3-phenyl-2-propenoate (ethyl cinnamate)	271	1.050	133	148	168	93	225	
Butyl 2-hydroxybenzoate (butyl salicylate)	272	1.074_4^{19}	158	142	156	64	136	
Dibutyl butanedioate (dibutyl succinate; butyl succinate)	274	0.976	185	157 (mono) 260d (di)	180 (mono) 260 (di)	64	206	168
Di-(1-Methylethyl) D-2,3-dihydroxybutanedioate (diisopropyl D-tartarate; isopropyl D-tartarate)	275			172 (mono) 196d (di)		123		
Di-(1-Methylethyl) (±)-2,3-dihydroxybutanedioate [diisopropyl (±)-tartarate; isopropyl (±)-tartarate]	275	1.1274	171	226		123	199	
Ethyl 4-ethoxybenzoate	275	1.076_4^{21}	170	202		93		
Octyl hexanoate (octyl caproate)	275	0.8603		101	75	62	53	
Ethyl 2-nitrobenzoate	275		146	176		93		

Esters (Liquids) (continued)

			Derivative mp (°C)					
Name of compound	bp (°C)	sp gr	Acid	Amide	4-Toluidide	3,5-Dinitrobenzoate	*N*-Benzylamide	Hydrazide
Heptyl heptanoate (heptyl enanthate)	277	0.8604		96	81	47		
Hexyl octanoate (heptyl caproate)	277	0.8603	16	110	70	58		
3-Methylbutyl 2-hydroxybenzoate (isoamyl salicylate; isopentyl salicylate)	278	1.065_4^{15}; 1.0535	158	143	156	61	136	
2,4-Dihydroxybenzoic acid diethanoate (resorcinol diacetate; 2,4-dihydroxybenzoic acid diacetate)	278	1.180_4^{19}	17	82	153	201 (di)	61	77
Diethyl D-2,3-dihydroxybutanedioate (diethyl D-tartarate; ethyl D-tartarate)	280	1.2028		172 (mono) 196d (di)		93		
Diethyl octanedioate (diethyl suberate; ethyl suberate)	282	0.9822_4^{20}; 1.9807	144	127 (mono) 217 (di)	219 (di)	93		
2,4-Dihydroxybenzoic acid monoethanoate (resorcinol monoacetate; 2,4-dihydroxybenzoic acid monoacetate)	283		17	82	153	201 (di)	61	77
Dimethyl 1,2-benzenedicarboxylate (dimethyl phthalate; methyl phthalate)	283	1.196_4^{19}; 1.191	206	149 (mono) 220 (di)	150, 165 (mono) 201 (di)	108		
Ethyl 3,4-methylenedioxybenzoate (ethyl piperonylate)	286			169		93		
Diethyl 1,3-benzenedicarboxylate (diethyl isophthalate; ethyl isophthalate)	286	1.121	348	280 (mono) 280 (di)		93		
Dipropyl (±)-2,3-dihydroxybutanedioate [dipropyl (±)-tartarate; propyl (±)-tartarate]	286		171	226		74	199	
D-Diethyl 1,2,2-trimethylcyclopentane-1,3-dicarboxylate (diethyl D-camphor; ethyl D-camphor)	286	1.0298	188	176, 183 (mono) 193 (di)	214 (α) 196 (β)	93		
Dimethyl decanedioate (dimethyl sebacate;) methyl sebacate	286		133	170 (mono) 210 (di)	201 (di)	108	167	
1,2,3-Propanetriol tripropanoate (glyceryl tripropionate)	289	1.083_4^{19}	16	81	126		44	40
Heptyl octanoate (heptyl caprylate)	291	0.859		110	70	47		

Esters (Liquids) (continued)

			Derivative mp (°C)					
Name of compound	**bp (°C)**	**sp gr**	**Acid**	**Amide**	**4-Toluidide**	**3,5-Dinitrobenzoate**	***N*-Benzylamide**	**Hydrazide**
Octyl heptanoate (octyl enanthate)	291	0.860		97	81	62		
Diethyl nonanedioate (diethyl azelate; ethyl azelate)	291	0.9766_{0}^{15}; 0.973	107	95 (mono) 175 (di)	202 (di)	93		
Triethyl 2-hydroxy-1,2,3-tripropanecarboxylate (triethyl citrate)	294	1.137	153 (anhyd) 100 (hyd)	215 (tri)	189 (tri)	93	170	
Ethyl 3-nitrobenzoate	296		141	190		93	101	
Dipropyl D-2,3-dihydroxybutanedioate (dipropyl D-tartarate; propyl D-tartarate)	297	1.139		172 (mono) 196d (di)		74		
Di-(3-methylbutyl) butanedioate (diisoamyl succinate; isoamyl succinate)	297	0.961_{4}^{13}	185	157 (mono) 260d (di)	180 (mono) 260 (di)	61	206	168
Diethyl 1,2-benzenedicarboxylate (diethyl phthalate; ethyl phthalate)	298	1.117	206	149 (mono) 220 (di)	150, 165 (mono) 201 (di)	93		
Diethyl 2-benzylpropanoate (diethyl benzylmalonate; ethyl benzylmalonate)	300	1.077_{4}^{15}	117	225		93		
Methyl 2-aminobenzoate (methyl anthranilate)	300		146	109	151	93		
2-Tetrahydrofurfuryl benzoate (α-tetrahydrofurfuryl benzoate)	302	1.137_{0}^{20}	122	130	158	84	106	112
Diethyl 1,4-benzenedicarboxylate (diethyl terephthalate; ethyl terephthalate)	302	1.065_{4}^{19}		>225		93		
Di-(1-methylethyl)-1,2-benzenedicarboxylate (diisopropyl phthalate; isopropyl phthalate)	302	1.065_{4}^{19}	206	149 (mono) 220 (di)	150, 165 (mono) 201 (di)	123		
Ethyl 2-naphthoate	304			195	192	93		
Ethyl tetradecanoate (ethyl myristate)	306	0.865_{4}^{19}; 0.857_{4}^{25}	58	103	93	93	90	
Octyl octanoate (octyl caprylate)	307	0.859		110	70	62		
Diethyl decanedioate (diethyl sebacate; ethyl sebacate)	307	0.965_{4}^{16}	133	170 (mono) 210 (di)	201 (di)	93	167	
2-Methylphenyl benzoate (*o*-tolyl benzoate)	307	1.114_{4}^{19}	122	130	158	138	106	112
Ethyl 1-naphthoate	309	1.127_{15}^{15}	162	205		93		
1,2,3-Propanetriol tributanoate (glyceryl tributyrate)	318	1.033_{4}^{17}		116	75		38	44

Esters (Liquids) (continued)

Name of compound	bp (°C)	sp gr	Derivative mp (°C) Acid	Amide	4-Toluidide	3,5-Dinitrobenzoate	N-Benzylamide	Hydrazide
Dibutyl (±)-2,3-dihydroxybutanedioate [dibutyl (±)-tartarate; butyl (±)-tartarate]	320	1.0879_4^{18}	206 (anhyd) 204 (hyd)	226 (di)		64		
Benzyl 2-hydroxybenzoate (benzyl salicylate)	320		158	142	156	113	136	
Benzyl benzoate	323	1.114_4^{18}	122	130	158	113	106	112
Methyl tetradecanoate (methyl myristate)	323; 295	0.873_4^{19}	58	103	93	108	90	
Dibutyl 1,2-benzenedicarboxylate (dibutyl phthalate; butyl phthalate)	340	1.050_4^{19}	206	149 (mono) 220 (di)	150, 165 (mono) 201 (di)	64		
Dibutyl decanedioate (dibutyl sebacate; butyl sebacate)	345	0.9329_4^{15}	133	170 (mono) 210 (di)	201 (di)	64	167	
Di-(3-methylbutyl)-1,2-benzenedicarboxylate (diisoamyl phthalate; isoamyl phthalate)	349	1.024_4^{17}	206	149 (mono) 220 (di)	150, 165 (mono) 201 (di)	61		

Esters (Solids)

Name of compound	mp (°C)	Derivative mp (°C)					
		Acid	Amide	4-Toluidide	3,5-Dinitrobenzoate	*N*-Benzylamide	Hydrazide
Methyl dodecanoate (methyl laurate)	5	44	100	87	108	83	105
Ethyl 4-methoxybenzoate (ethyl *p*-anisate)	7	185	167	186	93	132	
Ethyl tetradecanoate (ethyl myristate)	11	58	103	93	93	90	
Diethyl D-2,3-dihydroxybutanedioate (diethyl D-tartarate; ethyl D-tartarate)	17		172 (mono) 196d (di)		93		
Dimethyl butanedioate (dimethyl succinate; methyl succinate)	19	185	157 (mono) 260 (di)	180 (mono) 260 (di)	108	206	168
Methyl tetradecanoate (methyl myristate)	19	58	103	93	108	90	
Ethyl 3,4-methylenedioxybenzoate (ethyl piperonylate)	19	229	169		93		
Diethyl D-2,3-dihydroxybutanedioate (diethyl D-tartarate; ethyl D-tartarate)	19		172 (mono) 196 (di)		93		
Phenyl propanoate (phenyl propionate)	20		81	126	146	44	40
Ethyl heptadecanoate (ethyl margarate)	21		108		93		
Methyl 3-chlorobenzoate	21	158	134		108		158
Benzyl benzoate	21	122	130	158	113	106	112
Dibutyl D-2,3-dihydroxybutanedioate (dibutyl D-tartarate; butyl D-tartarate)	22		172 (mono) 196 (di)		64		
3,4-Dimethylphenyl ethanoate (3,4-dimethyldiphenyl acetate)	22	17	82	153	182	61	77
3-Methylbutyl octadecanoate (isoamyl stearate; isopentyl stearate)	23	70	109	102	61	97	
Hexadecyl ethanoate (cetyl acetate; hexadecyl acetate)	24; 19	17	82	153	66	61	77
Ethyl hexadecanoate (ethyl palmitate)	24; 19	63	107	98	93	95	111
Methyl 2-aminobenzoate (methyl anthranilate)	24	146	109	151	93		
Dipropyl (±)-2,3-dihydroxybutanedioate [dipropyl (±)-tartarate; propyl (±)-tartarate]	25		226		74		
Methyl 3-(2-furyl)propenoate [methyl β-(2-furyl)acrylate]	27		169		108		
Dimethyl decanedioate (dimethyl sebacate; methyl sebacate)	28	133	170 (mono) 210 (di)	201 (di)	108	167	
Butyl octadecanoate (butyl stearate)	28	70	109	102	64	97	
2-Methylpropyl octadecanoate (isobutyl stearate)	29; 23	70	109	102	87	97	
Methyl heptadecanoate (methyl margarate)	29		108		108		
Methyl hexadecanoate (methyl palmitate)	30	63	107	98	108	95	111
Pentyl octadecanoate (amyl stearate)	30	70	109	102	136	97	
Ethyl 2-nitrobenzoate	30	147	176		93		
Octadecyl ethanoate (octadecyl acetate)	32	17	82	153	66	61	77
Ethyl 2-naphthoate	32		195	192	93		
Methyl 3-bromobenzoate	32	155	155		108		
Methyl 4-methylbenzoate (methyl *p*-toluate)	33	180	160	165	108	133	117

Esters (Solids) (continued)

Name of compound	mp (°C)	Derivative mp (°C)					
		Acid	Amide	4-Toluidide	3,5-Dinitrobenzoate	*N*-Benzylamide	Hydrazide
5-Methyl-2-(1-methylethyl)phenyl benzoate (thymyl benzoate; 2-isopropyl-5-methylphenyl benzoate)	33	122	130	158	103	106	112
Di-(2-ethoxyethyl) 1,2-benzenedicarboxylate [di-(β-ethoxyethyl) phthalate]	33	208	149 (mono) 220 (di)	150, 165 (mono) 201 (di)	75	179	
Ethyl octadecanoate (ethyl stearate)	34	70	109	102	93	97	
Di-(1-methylethyl) (±)-2,3-dihydroxybutanedioate [diisopropyl (±)-tartarate; isopropyl (±)-tartarate]	34		226		123		
Ethyl furoate (ethyl pyromucate)	34	134	143	171	93		80
Diethyl 4-nitro-1,2-benzenedicarboxylate (diethyl 4-nitrophthalate; ethyl 4-nitrophthalate)	34	165	200	172 (mono)	93		
Ethyl 2,2-diphenyl-2-hydroxyethanoate (ethyl benzilate; ethyl 2,2-diphenyl-2-hydroxyacetate)	34		155	190	93		
2,4,5-Trimethylphenyl ethanoate (pseudocumenyl acetate; 2,4,5-trimethylphenyl acetate)	35	17	82	153		61	77
Methyl 3-phenyl-2-propenoate (methyl cinnamate)	36	133	148	168	108	225	
Ethyl (±)-2-hydroxy-2-phenylethanoate [ethyl (±)-mandelate; ethyl (±)-2-hydroxy-2-phenylacetate]	37	120	134	172	93		
Dimethyl methylenebutanedioate (dimethyl itaconate; methyl itaconate; dimethyl methylenesuccinate; methyl methylenesuccinate)	38	165	192 (di)		108		
Dimethyl decanedioate (dimethyl sebacate; methyl sebacate)	38	133	170 (mono) 210 (di)	201 (di)	108	167	
Methyl octadecanoate (methyl stearate)	39	70	109	102	108	97	
Benzyl 3-phenyl-2-propenoate (benzyl cinnamate)	39	133	147	168	113	225	
Methyl dibenzylethanoate (methyl dibenzylacetate)	41		129	175	108		
Phenyl 2-hydroxybenzoate (phenyl salicylate; salol)	42	158	142	156	146	136	
Butanedioic acid monobenzyl ester (succinic acid monobenzyl ester)	42	185	157 (mono) 260 (di)	180 (mono) 260 (di)	113	168	206
Dibenzyl 1,2-benzenedicarboxylate (dibenzyl phthalate; benzyl phthalate)	43	208	149 (mono) 220 (di)	150, 165 (mono) 201 (di)	113	179	
Diethyl 1,4-benzenedicarboxylate (diethyl terephthalate; ethyl terephthalate)	44	380	>225		93	266	

Esters (Solids) (continued)

Name of compound	mp (°C)	Derivative mp (°C) Acid	Amide	4-Toluidide	3,5-Dinitrobenzoate	*N*-Benzylamide	Hydrazide
3-Phenyl-2-propenyl 3-phenyl-2-propenoate (cinnamyl cinnamate)	44	133	148	168	121	225	
Ethyl 3-(2-nitrophenyl)-2-propenoate (ethyl 2-nitrocinnamate)	44	240	185		93		
Methyl 3-(2-chlorophenyl)-2-propenoate (methyl 2-chlorocinnamate)	44		168		108		
Diethyl 3-nitro-1,2-benzenedicarboxylate (diethyl 3-nitrophthalate; ethyl 3-nitrophthalate)	46	219	201 (di)	226 (di)	93		
Ethyl 3-nitrobenzoate	47	141	190		93	101	
Dicyclohexyl ethanedioate (dicyclohexyl oxalate; cyclohexyl oxalate)	47	190 (anhyd) 101 (dihyd)	219 (mono) 419d (di)	168 (mono) 268 (di)	113	223	243
2-Phenylethyl 3-phenyl-2-propenoate (2-phenylethyl cinnamate)	48	133	148	168	108	225	
1-Naphthyl ethanoate (α-naphthyl acetate)	49	17	82	153	217	61	77
Methyl 4-methoxybenzoate (methyl *p*-anisate)	49	184	167	186	108	132	
Phenacyl ethanoate (benzoylcarbinyl acetate; phenacyl acetate)	49	17	82	153		61	77
Dibenzyl D-2,3-dihydroxybutanedioate (dibenzyl D-tartarate; benzyl D-tartarate)	50		172 (mono) 196 (di)		113		
Dibenzyl butanedioate (dibenzyl succinate; benzyl succinate)	52	185	157 (mono) 260 (di)	180 (mono) 260 (di)	113	206	168
Methyl 3,4-methylenedioxybenzoate (methyl piperonylate)	52		169		108		
Hexadecyl hexadecanoate (cetyl palmitate)	52	63	107	98	66	95	111
Furfuryl diethanoate (furfuryl diacetate)	52	17	82	153		61	77
1,2-Ethanediol di(dodecanoate) (ethylene glycol dilaurate)	52	44	100	87	169	83	105
Phenyl octadecanoate (phenyl stearate)	52	70	109	102	146	97	
Methyl (±)-2-hydroxy-2-phenylethanoate [methyl (±)-mandelate; methyl (±)-2-hydroxy-2-phenylacetate]	53	120	134	172	108		
Dimethyl 2-hydroxypropanedioate (dimethyl tartronate; methyl tartronate; dimethyl hydroxymalonate; methyl hydroxymalonate)	53		198 (di)		108		
Dimethyl ethanedioate (dimethyl oxalate; methyl oxalate)	54	190 (anhyd) 101 (dihyd)	219 (mono) 419d (di)	168 (mono) 268 (di)	108	223	243
3-Methylphenyl benzoate (*m*-tolyl benzoate)	55	122	130	158	165	106	112
Diethyl *R,S*-2,3-dihydroxybutanedioate (diethyl *meso*-tartarate; ethyl *meso*-tartarate)	55	140	190 (di)		93	205	

Esters (Solids) (continued)

Name of compound	mp (°C)	Derivative mp (°C)					
		Acid	Amide	4-Toluidide	3,5-Dinitrobenzoate	*N*-Benzylamide	Hydrazide
Ethyl 4-nitrobenzoate	56	239	201	204	93	142	
1-Naphthyl benzoate	56	122	130	158	217	106	112
Hexadecyl octadecanoate (cetyl stearate)	57	70	109	102	66	97	
Di-(2-methylpropyl) (±)-2,3-dihydroxybutanedioate [diisobutyl (±)-tartarate; isobutyl (±)-tartarate]	58		226		87		
Ethyl diphenylethanoate (ethyl diphenylacetate)	58		168	173	93		
Ethyl 2-benzoylbenzoate	58		165		93		
Methyl diphenylethanoate (methyl diphenylacetate)	60		168	173	108		
Methyl 2-(4-methylphenyl)benzoate [methyl 2-(*p*-tolyl)benzoate]	61		176		108		
Dimethyl D-2,3-dihydroxybutanedioate (dimethyl D-tartarate; methyl D-tartarate)	62		172 (mono) 196 (di)		108		
1,2-Ethanediol di(tetradecanoate) (ethylene glycol dimyristate)	63	58	103	93	169	90	
Dimethyl 4-nitro-1,2-benzenedicarboxylate (dimethyl 4-nitrophthalate; methyl 4-nitrophthalate)	66	165	200	172 (mono)	108		
Dicyclohexyl 1,2-benzenedicarboxylate (dicyclohexyl phthalate; cyclohexyl phthalate)	66	208	149 (mono) 220 (di)	150, 165 (mono) 201 (di)	113	179	
Ethyl *N*-phenylethanedioic acid monoamide (ethyl oxalinate; ethyl *N*-phenyl oxalic acid monoamide)	67		228		93		
Dimethyl 1,3-benzenedicarboxylate (dimethyl isophthalate; methyl isophthalate)	68	347	280 (mono) 280 (di)		108		220
Ethyl 2-(4-methylphenyl)benzoate [ethyl 2-(*p*-tolyl)benzoate]	69		176		93		
Dimethyl 3-nitro-1,2-benzenedicarboxylate (dimethyl 3-nitrophthalate; methyl 3-nitrophthalate)	69	219	201 (di)	226 (di)	108		
Methyl 3-hydroxybenzoate	70	201	170	163	108	142	
Phenyl benzoate	71	122	130	158	146	106	112
1,2-Ethanediol di(hexadecanoate) (ethylene glycol dipalmitate)	71	63	107	98	169	95	111
2-Naphthyl ethanoate (α-naphthyl acetate)	71	17	82	153	210	61	77
1,2,3-Propanetriol tri(octadecanoate) (glyceryl tristearate)	71	70	109	102		97	
4-Methylphenyl benzoate (*p*-tolyl benzoate)	71	122	130	158	189	106	112
Phenyl 3-phenyl-2-propenoate (phenyl cinnamate)	72	133	148	168	146	225	

Esters (Solids) (continued)

		Derivative mp (°C)					
Name of compound	mp (°C)	Acid	Amide	4-Toluidide	3,5-Dinitrobenzoate	*N*-Benzylamide	Hydrazide
1,2-Ethanediol dibenzoate (ethylene glycol dibenzoate)	73	122	130	158	169 (di)	106	112
Methyl *E*-3-(2-nitrophenyl)-2-propenoate (methyl 2-nitrocinnamate)	73	240	185		108		
Di(2-methylpropyl) D-2,3-dihydroxy butenedioate (diisobutyl D-tartarate; isobutyl D-tartarate)	74		172 (mono) 196 (di)		87		
Ethyl 3-hydroxybenzoate	74	201	170	163	93	142	
Diphenyl 1,2-benzenedicarboxylate (diphenyl phthalate; phenyl phthalate)	75	208	149 (mono) 220 (di)	150, 165 (mono) 201 (di)	146	179	
Methyl 3-hydroxy-2-naphthoate	75		228	223	108		
Methyl 2,2-diphenyl-2-hydroxy-ethanoate (methyl benzilate; methyl 2,2-diphenyl-2-hydroxyacetate)	75		155	190	108		
1,2,3-Propanetriol tribenzoate (glyceryl tribenzoate)	76	122	130	158		106	112
1,2-Ethanediol di(octadecanoate) (ethylene glycol distearate)	76	70	109	102	169 (di)	97	
Methyl 2-naphthoate	77		193	192	108		
Methyl 2-phenylbutanoate (methyl α-phenylbutyrate)	78		87		108		
Methyl 3-nitrobenzoate	78	141	143	162	108	101	
Trimethyl 2-hydroxy-1,2,3-propane-tricarboxylate (trimethyl citrate; methyl citrate)	79	100	215 (tri)	189 (tri)	108	170	
Ethyl *E*-3-(3-nitrophenyl)-2-propenoate (ethyl 3-nitrocinnamate)	79	205	196		93		
Methyl 2-benzylbenzoate	80		165		108		
Dibenzyl 1,2-ethanedioate (dibenzyl oxalate; benzyl oxalate)	80	190 (anhyd) 101 (dihyd)	219 (mono) 419d (di)	169 (mono) 268 (di)	113	223	243
Methyl 4-bromobenzoate	81	252	190		108		163
1-Acetoxybenzyl phenyl ketone (benzoin acetate)	83		82	153			
1,2-Diacetoxybenzene (catechol diacetate)	84	16	130	158	152 (di)		
Ethyl 3-hydroxy-2-naphthoate	85		218	223	93		
Dimethyl (±)-2,3-dihydroxybutanedioate [dimethyl (±)-tartarate; methyl (±)-tartarate]	90		226		108		
Di(2-methylphenyl) 1,2-ethanedioate (di-*o*-tolyl oxalate; *o*-tolyl oxalate)	91	190 (anhyd) 101 (dihyd)	219 (mono) 419d (di)	169 (mono) 268 (di)	138	223	243
Ethyl 3,5-dinitrobenzoate	94	207	183		93		
2-Naphthyl 2-hydroxybenzoate (β-naphthyl salicylate)	96	158	142	156	210	136	
Methyl 4-nitrobenzoate	96	239	201	204	108	142	
Propyl 4-hydroxybenzoate	96	213	162	204	123		
1,2,4-Triacetoxybenzene (hydroquinone triacetate)	97	17	82	153		61	77

Esters (Solids) (continued)

Name of compound	mp (°C)	Derivative mp (°C)					
		Acid	Amide	4-Toluidide	3,5-Dinitrobenzoate	*N*-Benzylamide	Hydrazide
Ethyl 3,5-dinitro-2-hydroxybenzoate (ethyl 3,5-dinitrosalicylate)	99		181		93		
Dimethyl *E*-butenedioate (dimethyl fumarate; methyl fumarate)	102	286	270; 302 (mono) 266 (di)		108	205	
Ethyl 5-nitro-2-hydroxybenzoate (ethyl 5-nitrosalicylate)	102		225		93		
Di-(3-methylphenyl) 1,2-ethanedioate (di-*m*-tolyl oxalate; *m*-tolyl oxalate)	105	190 (anhyd) 101 (dihyd)	219 (mono) 419d (di)	169 (mono) 268 (di)	165	223	243
1,3,5-Triacetoxybenzene (phloroglucinol triacetate)	106	16	82	153	162 (tri)	61	77
Diphenyl hexanedioate (diphenyl adipate; phenyl adipate)	106	152	130 (mono) 220 (di)	241	146	189	171
2-Naphthyl benzoate	107	122	130	158	210	106	112
Methyl 3,5-dinitrobenzoate	108	207	183		108		
Dimethyl *R,S*-2,3-dihydroxybutanedioate (dimethyl *meso*-tartarate; methyl *meso*-tartarate)	111	140	190 (di)		108	205	
Ethanedioic acid monoamide monoethyl ester (ethyl oxamate; oxalic acid monoamide monoethyl ester)	115		419d		93		
Ethyl 4-hydroxybenzoate	116	213	162	204	93		
2,4-Dihydroxyphenyl dibenzoate (resorcinol dibenzoate)	117	122	130	158	201	106	112
Ethyl 3-nitro-2-hydroxybenzoate (ethyl 3-nitrosalicylate)	118		145		93		
Methyl 5-nitro-2-hydroxybenzoate (methyl 5-nitrosalicylate)	119		225		108		
Diphenyl butanedioate (diphenyl succinate; phenyl succinate)	121	185	157 (mono) 260 (di)	180 (mono) 260 (di)	149	206	168
Di-(4-methylphenyl) butanedioate (di-*p*-tolyl succinate; *p*-tolyl succinate)	121	185	157 (mono) 260 (di)	180 (mono) 260 (di)	189	206	168
Hydroquinone diethanoate (hydroquinone diacetate)	124	16	82	153	317	61	77
Methyl *E*-3-(3-nitrophenyl)-2-propenoate (methyl 3-nitrocinnamate)	124	205	196		108		
Methyl 3,5-dinitro-2-hydroxybenzoate (methyl 3,5-dinitrosalicylate)	127		181		108		
Methyl 4-hydroxybenzoate	131	213	162	204	108		
Methyl 3-nitro-2-hydroxybenzoate (methyl 3-nitrosalicylate)	132		145		108		
Triethyl 1,3,5-benzenetricarboxylate (triethyl trimesate)	133	380	365 (tri)		93		
Ethyl *E*-3-(4-nitrophenyl)-2-propenoate (ethyl 4-nitrocinnamate)	141	287	217		93		
Dimethyl 1,4-benzenedicarboxylate (dimethyl terephthalate; methyl terephthalate)	141	300	>225 (di)		108	266	

Esters (Solids) (continued)

		Derivative mp (°C)					
Name of compound	mp (°C)	Acid	Amide	4-Toluidide	3,5-Dinitrobenzoate	*N*-Benzylamide	Hydrazide
Trimethyl 1,3,5-benzenetricarboxylate (trimethyl trimesate)	144	380	365 (tri)		108		
Di(4-methylphenyl) 1,2-ethanedioate (di-*p*-tolyl oxalate; *p*-tolyl oxalate)	149	190 (anhyd) 101 (dihyd)	219 (mono) 419d (di)	169 (mono) 268 (di)	189	223	243
Methyl *E*-3-(4-nitrophenyl)-2-propenoate (methyl 4-nitrocinnamate)	161	287	217		108		
Diethyl 2,3,4,5-tetrahydroxyhexanedioate (diethyl mucate; ethyl mucate)	164		192 (mono) 220 (di)		93		
Dimethyl 2,3,4,5-tetrahydroxyhexanedioate (dimethyl mucate; methyl mucate)	167		192 (mono) 220 (di)		108		
Hydroquinone benzoate	204	122	130	158	317	106	112

Ethers—Aromatic (Liquids)

Name of compound	bp (°C)	sp gr	Derivative mp (°C)			
			Picrate	Sulfonamide	Nitro derivative	Bromo derivative
Methoxybenzene (anisole)	158	0.988	81 (4-)	113	87, 95 (2,4-)	61 (2,4-)
2-Methoxytoluene (*o*-cresol methyl ether; *o*-methylanisole; methyl *o*-tolyl ether)	171	0.966^{0}_{4}; 0.985	119	137(5-)	69 (3,5-) 64 (5-)	
Ethoxybenzene (phenetole)	172	0.979^{4}_{4}; 0.967	92	150 (4-)	59 (4-)	
4-Methoxytoluene (*p*-cresyl methyl ether; *p*-methylanisole; methyl *p*-tolyl ether)	176	0.987^{0}_{4}; 0.970	89	182 (3-)	122 (3,5-)	
3-Methoxytoluene (*m*-cresyl methyl ether; *m*-methylanisole; methyl *m*-tolyl ether)	177	0.985^{4}_{4}; 0.972	114	130 (6-)	92 (2,4,6-) 55 (2-)	
4-Ethoxytoluene (*p*-cresyl ethyl ether; ethyl *p*-tolyl ether)	192	0.949	111	138		
3-Ethoxytoluene (*m*-cresyl ethyl ether; ethyl *m*-tolyl ether)	192	0.949	115	111		
2-Ethoxytoluene (*o*-cresyl ethyl ether; ethyl *o*-tolyl ether)	192	0.953	118 (di)	149	51	
2-Chloro-1-methoxybenzene (*o*-chloroanisole)	195	1.191		130 (4-)	95 (4-)	
4-Chloro-1-methoxybenzene (*p*-chloroanisole)	200	1.185^{13}_{4}		151 (2-)	98 (2-)	
2-Methoxyphenol (guaiacol; catechol monomethyl ether)	205	1.129	87			116 (4,5,6-)
1,2-Dimethoxybenzene (veratole)	207	1.086^{15}_{4}	57	136 (4-)	95 (4-) 132 (4,5-)	93 (4,5-)
2-Chlorophenyl ethyl ether (*o*-chlorophenetole)	208			133	82	
Butyl phenyl ether	210	0.950	112	104		
4-Chloro-1-ethoxybenzene (*p*-chlorophenetole)	212	1.123^{20}_{20}		134	61	54 (2,6-)
1,3-Dimethoxybenzene (resorcinol dimethyl ether)	217	1.080^{0}_{4}; 1.055^{25}_{25}	58	167 (4-)	72 (2,4-) 157 (4,6-) 124 (2,4,6-)	140 (4,6-)
2-Bromo-1-methoxybenzene (*o*-bromoanisole)	218			140 (4-)	106 (4-)	
4-Bromo-1-methoxybenzene (*p*-bromoanisole)	223	1.494^{9}_{4}		148 (2-)	88 (2-)	
2-Bromophenyl ethyl ether (*o*-bromophenetole)	224			135 (4-)	98 (4-)	
1,2-Methylenedioxy-4-(2-propenyl)benzene (safrole)	233	1.096^{18}_{4}	105			108 (tri) 170 (penta) 51 (1,3,5-)
4-Bromophenyl ethyl ether (*p*-bromophenetole)	233			145	47	
1-Methoxy-4-propenylbenzene (anethole)	235	0.989^{28}_{4}	70			67 (di) 108 (tri)
1,3-Diethoxybenzene (resorcinol diethyl ether)	235		58; 109	184		69 (tri)
1,2,3-Trimethoxybenzene	241		81	123 (2,3,4-)	106 (5-)	74 (4,5,6-)

Ethers—Aromatic (Liquids) (continued)

			Derivative mp (°C)			
Name of compound	bp (°C)	sp gr	Picrate	Sulfon-amide	Nitro derivative	Bromo derivative
1,2-Dimethoxy-4-(2-propenyl)benzene (eugenol methyl ether)	244	1.055_4^{15}; 1.034	115			78 (tri)
Ethyl 2-iodophenyl ether (*o*-iodophenetole)	246	1.800	84		110 (tri)	
1,2-Methylenedioxy-4-propylbenzene (isosafrole)	248	1.125_4^{14}	75		53 (di) 110 (tri)	
Diphenyl ether	259	1.073	110	159 (4,4′-)	144 (4,4′) 197 (2,2′,4,4′-)	58 (4,4′-)
1,2-Dimethoxy-4-propenylbenzene (isoeugenol methyl ether)	264	1.0528	45		101 (di)	
Methyl 1-naphthyl ether (α-methoxynaphthalene)	271	1.096_4^{14}	113; 131	157 (4-)	80 (2-) 85(4-) 128 (2,4,5-)	46, 68 (mono) 55 (2,4-)
Ethyl 1-naphthyl ether (α-ethoxynaphthalene)	280	1.074; 1.060	100; 119	165 (4-)	84 (2-) 117 (4-) 149 (2,4,5-)	48 (4-)
Ethyl 2-naphthyl ether (β-ethoxynaphthalene)	282	1.064	44; 101	163		66 (1-) 94 (1,6-)
Benzyl ether	300	1.043	78			108 (di)
2-Naphthyl pentyl ether (amyl β-naphthyl ether)	328		67	159	135 (di)	58 (di)

Ethers—Aromatic (Solids)

Name of compound	mp (°C)	Derivative mp (°C)			
		Picrate	Sulfon-amide	Nitro derivative	Bromo derivative
4-Chloro-1-ethoxybenzene (*p*-chlorophenetole)	21		134	54 (2,6-)	
1,2-Dimethoxybenzene (veratole)	22	57	136 (4-)	95(4-) 132 (4,5-)	93 (4,5-)
1-Methoxy-4-propenylbenzene (anethole)	22	70			67 (di) 108 (tri)
2-Naphthyl pentyl ether (amyl *β*-naphthyl ether)	25	67	159	135 (di)	58 (di)
Phenyl ether	28	110	159 (4,4′)	144 (4,4′-) 197 (2,2′,4,4′-)	58
2-Methoxyphenol (guaiacol; catechol monomethyl ether)	28	87			116 (4,5,6-)
Ethyl 2-naphthyl ether (*β*-ethoxynaphthalene)	37	44; 101	163		66 (1-) 94 (1,6-)
1,2-Diethoxybenzene (catechol diethyl ether)	43	71	163 (3,4-)	122 (tri)	
1,2,3-Trimethoxybenzene (pyrogallol trimethyl ether)	47	81	124 (2,3,4-)	106 (5-)	74 (4,5,6-)
1,4-Dimethoxybenzene (hydroquinone dimethyl ether)	57	119; 48	148 (2-)	72 (2-) 177 (2,3-) 202 (2,5-)	142 (di)
Methyl 2-naphthyl ether (*β*-methoxynaphthalene)	72	117	151 (8-)	128 (1-) 215 (1,6,8-)	63 (mono) 84 (1-) 78 (3-) 108 (6-)
Biphenylene oxide (dibenzofuran)	87	94		182 (3-) 245 (di)	
4-Methoxybiphenyl (4-biphenyl methyl ether; *p*-phenylanisole)	90		92 (3-) 138 (3,5-) 171 (3,4′-)	79 (3-) 144 (4′-) 134 (3,4′-) 87 (3,5-)	
1,2-Diphenoxyethane (ethylene glycol diphenyl ether)	98		229 (4,4′-)	215 (2′,4′-)	135 (4,4′-)

Halides—Alkyl, Cycloalkyl, and Aralkyl (Liquids)

		Derivative mp (°C)					
Name of compound	bp (°C)	Anilide	*N*-Naphthylamide	Alkylmercuric halide	Alkyl 2-naphthyl ether	Alkyl 2-naphthyl ether picrate	*S*-Alkylthiuronium picrate
Chlorides:							
Choroethane (ethyl chloride)	12	104	126	192	37	104	188
2-Chloropropane (isopropyl chloride)	36	104		97	41	95	196; 148
3-Chloropropene (allyl chloride)	45	114			16	99	155
1-Chloropropane (propyl chloride)	46	92	121	147	39	81	181
2-Chloro-2-methylpropane (*tert*-butyl chloride)	54	128	147	123			161
2-Chlorobutane (*sec*-butyl chloride)	68	108	129	39	34	86	190; 166
1-Chloro-2-methylpropane (isobutyl chloride)	69	110	126		33	85	174
1-Chlorobutane (butyl chloride)	78	63	112	128	33	67	180
1-Chloro-2,2-dimethylpropane (neopentyl chloride)	85	131		118			
2-Chloro-2-methylbutane (*tert*-pentyl chloride; *tert*-amyl chloride)	86	92	138				
(±)-2-Chloropentane	97	96	103				
3-Chloropentane	97	127	118				
1-Chloro-3-methylbutane (isopentyl chloride; isoamyl chloride)	100	108	111	86	28	94	179
1-Chloropentane (pentyl chloride; amyl chloride)	107	96	112	110	25	67	154
2-Chloro-2-methylpentane	113	74	118				
1-Chloro-3,3-dimethylbutane	115	139		133			
1-Chlorohexane (hexyl chloride)	134	69	106	125			157
Cyclohexyl chloride	143	146	188	164	116		
1-Chloroheptane (heptyl chloride)	160	57	95	120			142
Benzyl chloride	179	117	166		99	123	188
1-Chlorooctane (octyl chloride)	184	57	91	115; 151			134
1-Chloro-2-phenylethane (*β*-phenylethyl chloride)	190	97			70	84	139
1-Chlorohexadecane (cetyl chloride)	286			102			155
Bromides:							
Bromomethane (methyl bromide)	5	114	160	160; 172	72	118	224
Bromoethane (ethyl bromide)	38	104	126	198	37	104	188
2-Bromopropane (isopropyl bromide)	60	104		94	41	92	196; 148
3-Bromopropene (allyl bromide)	71	114			16	99	155
1-Bromopropane (propyl bromide)	71	92	121	138	40	81	181
2-Bromo-2-methylpropane (*tert*-butyl bromide)	72	128	147				151
2-Bromobutane (*sec*-butyl bromide)	91	108	129	39	34	86	190; 166
1-Bromo-2-methylpropane (isobutyl bromide)	91	109	126	56	33	84	174
1-Bromobutane (butyl bromide)	101	63	112	136	33	67	180
2-Bromo-2-methylbutane (*tert*-pentyl bromide; *tert*-amyl bromide)	108	92	138				

Halides—Alkyl, Cycloalkyl, and Aralkyl (Liquids) (continued)

		Derivative mp (°C)					
Name of compound	bp (°C)	Anilide	*N*-Naphthylamide	Alkylmercuric halide	Alkyl 2-naphthyl ether	Alkyl 2-naphthyl ether picrate	*S*-Alkylthiuronium picrate
1-Bromo-3-methylbutane (isopentyl bromide; isoamyl bromide)	118	110	111	80	28	94	179
(±)-2-Bromopentane	118	93	104				
1-Bromopentane (pentyl bromide; amyl bromide)	129	96	112	127	25	67	154
1-Bromohexane (hexyl bromide)	157	69	106	119; 127			157
Cyclohexyl bromide	165	146	188	153	116		
1-Bromoheptane (heptyl bromide)	174	57	95	118			142
Benzyl bromide	198	117	166	119	99	123	188
1-Bromooctane (octyl bromide)	204	57	91	109			134
1-Bromo-2-phenylethane (2-phenylethyl bromide)	218	97		169	70	84	139
1-Bromononane	220			109			131
β-Bromostyrene	221	115	217	91			
Iodides:							
Iodomethane (methyl iodide)	43	114	160	152	72	118	224
Iodoethane (ethyl iodide)	73	104	126	186	37	104	188
2-Iodopropane (isopropyl iodide)	89	103		125	41	95	196; 148
1-Iodopropane (propyl iodide)	102	92	121	113	40	81	181
2-Iodo-2-methylpropane (*tert*-butyl iodide)	103	128	147				151
3-Iodopropene (allyl iodide)	103	114	121	112	16	99	155
2-Iodobutane (*sec*-butyl iodide)	120	108	129		34	86	190; 166
1-Iodo-2-methylpropane (isobutyl iodide)	120	109	125	72	33	84	174
2-Iodo-2-methylbutane (*tert*-pentyl iodide; *tert*-amyl iodide)	128	92	138				
1-Iodobutane (butyl iodide)	130	63	112	117	33	67	180
1-Iodo-3-methylbutane (isopentyl iodide; isoamyl iodide)	148	108	111	122	28	94	179
1-Iodopentane (pentyl iodide; amyl iodide)	156	96	112	110	25	67	154
Cyclohexyl iodide	179	146	188		116		
1-Iodohexane (hexyl iodide)	180	69	106	110			157
1-Iodoheptane (heptyl iodide)	204	57	95	103			142

Halides—Aromatic (Liquids)

		Nitration product		Sulfonamide		Oxidation	
Name of compound	bp (°C)	Position	mp (°C)	Position	mp (°C)	Name of product	mp (°C)
Fluorobenzene	85	4	27	4	125		
2-Fluorotoluene	114			5	105	2-Fluorobenzoic acid	127
3-Fluorotoluene	116			6	174	3-Fluorobenzoic acid	124
4-Fluorotoluene	117			2	141	4-Fluorobenzoic acid	182
Chlorobenzene	132	2,4	52	4	144		
Bromobenzene	157	2,4	75	4	166		
2-Chlorotoluene	159	3,5	63	5	128	2-Chlorobenzoic acid	141
3-Chlorotoluene	162	4,6	91	6	185	3-Chlorobenzoic acid	158
4-Chlorotoluene	162	2	38	2	143	4-Chlorobenzoic acid	242
		2,6	76				
1,3-Dichlorobenzene	173	4,6	103	4	182		
				2,4	180		
1,2-Dichlorobenzene	179	4,5	110	4	140		
				3,4	135		
2-Bromotoluene	182	3,5	82	5	146	2-Bromobenzoic acid	150
3-Bromotoluene	184	4,6	103	6	168	3-Bromobenzoic acid	155
4-Bromotoluene	185	2	47	2	165	4-Bromobenzoic acid	251
2-Chloro-1,4-dimethylbenzene	185	5	77	5	155		
		5,6	101				
1-Chloro-2,4-dimethylbenzene	192	6	42	6	192	4-Chloro-3-methylbenzoic acid	210
						4-Chlorobenzene-1,3-dicarboxylic acid (4-chloroisophthalic acid)	295
1-Chloro-3,4-dimethylbenzene	195	5	63	5	207		
2,6-Dichlorotoluene	199	3	50	3	204	2,6-Dichlorobenzoic acid	139
		3,5	121				
2,5-Dichlorotoluene	199	4	51			2,5-Dichlorobenzoic acid	154
		4,6	101				
2,4-Dichlorotoluene	200	3,5	104	5	176	2,4-Dichlorobenzoic acid	164
3,5-Dichlorotoluene	201	2	62			3,5-Dichlorobenzoic acid	188
		2,6	100				
3-Iodotoluene	204	4,6	108			3-Iodobenzoic acid	187
2-Chloro-1,3,5-trimethylbenzene	206	4,6	178		166	2-Chlorobenzene-1,3,5-tricarboxylic acid	285 (anhyd) 278 (hyd)
2,3-Dichlorotoluene	207	4	51			2,3-Dichlorobenzoic acid	163
		4,6	72				
3,4-Dichlorotoluene	209	6	64	6	190	3,4-Dichlorobenzoic acid	208
		2,6	92				
2-Iodobenzene	211	6	103			2-Iodobenzoic acid	162
1,3-Dibromobenzene	219	4	62	4	190		
		4,6	117				
1,2-Dibromobenzene	224	4,5	114	4	176		
1-Chloronaphthalene	259	4,5	180	4	186		
1-Bromonaphthalene	281	4	85	4	193		
3-Chlorobiphenyl	285	4,4′	203			3-Chlorobenzoic acid	158

Halides—Aromatic (Solids)

		Nitration product		Sulfonamide		Oxidation	
Name of compound	mp (°C)	Position	mp (°C)	Position	mp (°C)	Name of product	mp (°C)
4-Bromotoluene	28	2	47	6	165	4-Bromobenzoic acid	251
2,4,6-Trichlorotoluene	38	3	54			2,4,6-Trichlorobenzoic acid	161
		3,5	180				
2,3,4-Trichlorotoluene	41	5	60			2,3,4-Trichlorobenzoic acid	187
		6	60				
		5,6	141				
3,4,5-Trichlorotoluene	45	2	82			3,4,5-Trichlorobenzoic acid	203
		2,6	164				
2,3,5-Trichlorotoluene	46	4	59			2,3,5-Trichlorobenzoic acid	162
		6	59				
		4,6	150				
1,6-Dichloronaphthalene	48	4	119	4	216		
1,2,3-Trichlorobenzene	53	4	56	4	230		
		4,6	93				
1,4-Dichlorobenzene	53	2	56	2	186		
2-Chloronaphthalene	61	1,8	175		126		
				8	232		
1,3,5-Trichlorobenzene	63	2	68	2	212		
1,7-Dichloronaphthalene	64		139	4	226		
1,4-Dichloronaphthalene	68	8	92	6	244	3,6-Dichlorobenzene-1,2-dicarboxylic acid (3,6-dichlorophthalic acid)	194
2,4,5-Trichlorotoluene	82	3	90			2,4,5-Trichlorobenzoic acid	168
		3,6	227				
1,4-Dibromobenzene	89	2	84	2	195		
		2,5	84				
1,5-Dichloronaphthalene	107	8	142	3	204		
2,7-Dichloronaphthalene	115	mono	142	3	218		
1,3,5-Tribromobenzene	120	2,4	192	2	222		

Hydrocarbons—Aromatic (Liquids)

Name of compound	bp (°C)	sp gr	Nitration product: Position	Nitration product: mp (°C)	Derivative mp (°C): Aroyl-benzoic acid	Derivative mp (°C): Picrate
Benzene	80	0.874	1,3	90	128	84
			1,3,5	122		
Toluene	111	0.881_4^4; 0.867	2,4	71	137	88
Ethylbenzene	136	0.876_4^{19}; 0.867	2,4,6	37	128	96
1,4-Dimethylbenzene (*p*-xylene)	138	0.866_4^4; 0.861	2,3,5	139	132; 148	90
1,3-Dimethylbenzene (*m*-xylene)	139	0.871_4^{12};	2,4	83	126	91
		0.864	2,4,6	183	142	
1,2-Dimethylbenzene (*o*-xylene)	144	0.890_4^4; 0.880	4,5	71; 118	178; 167	88
1-Methylethylbenzene (isopropylbenzene; cumene)	153	0.875_4^4; 0.862	2,4,6	109	134	
Propylbenzene	159	0.861			125	103
1,3,5-Trimethylbenzene (mesitylene)	164	0.869_4^{10};	2,4	86	212	97
		0.865	2,4,6	235		
1,2,4-Trimethylbenzene (pseudocumene)	169	0.895; 0.876	3,5,6	185	149	97
1-Methyl-4-(1-methylethyl)benzene (*p*-cymene; *p*-isopropyltoluene)	177	0.857	2,6	54	124	
			2,3,6	118		
1,3-Diethylbenzene	182	0.860	2,4,6	62	114	
1,2,3,5-Tetramethylbenzene (isodurene)	198	0.890	4,6	157; 181	213	
1,2,3,4-Tetramethylbenzene (prehnitene)	205	0.905	5,6	176		95
1,2,3,4-Tetrahydronaphthalene(tetralin)	207	0.971	5,7	95	155	
1,3,5-Triethylbenzene	218	0.863; 0.857_4^{20}	2,4,6	112	129	
2-Methylnaphthalene	241		1	81	190	116
1-Methylnaphthalene	245	1.001_4^{19};	4	71	169	142
		1.020	4,5	143	68	

Hydrocarbons—Aromatic (Solids)

Name of compound	mp (°C)	Nitration product		Derivative mp (°C)	
		Position	mp (°C)	Aroyl-benzoic acid	Picrate
2-Methylnaphthalene	38	1	81	190	116
Pentamethylbenzene	54	6	154		131
Biphenyl	71	4,4′	237	225	
		2,2′,4,4′	150		
1,2,4,5-Tetramethylbenzene (durene)	80	3,6	207	264	95
Naphthalene	80	1	61	173	150
Acenaphthalene	96	5	101	200	162
Fluorene	117	2,7	199	228	87
		2	156		
Hexamethylbenzene	162		176		170
Dibiphenylenethylene (bifluroenylidene)	195		171		178
1,2-Benzphenanthrene (chrysene)	254			214	273

Ketones (Liquids)

Name of compound	bp (°C)	Derivative mp (°C)				
		Semi-carba-zone	2,4-Di-nitro-phenyl-hydra-zone	4-Nitro-phenyl-hydra-zone	Phenyl-hydra-zone	Oxime
Propanone (acetone)	56	190	128	152	42	59
2-Butanone (ethyl methyl ketone)	80	146	117	129		
3-Butyn-2-one (ethynyl methyl ketone)	86		181	143		
2,3-Butanedione (biacetyl)	88	235 (mono) 278 (di)	315 (di)	230 (mono)	134 (mono) 245 (di)	74 (mono) 246, 234 (di)
3-Methyl-2-butanone (isopropyl methyl ketone)	94	114	120	109		
2-Pentanone (methyl propyl ketone)	102	112	144	117		58
3-Pentanone (diethyl ketone)	102	139	156	144		69
3,3-Dimethyl-2-butanone (pinacolone; *tert*-butyl methyl ketone)	106	158	125	139		79
1-Methoxy-2-propanone (methoxymethyl methyl ketone)	115		163	111		
3-Benzoylpropanoic acid	116	181	191			
3-Methyl-2-pentanone (*sec*-butyl methyl ketone)	118	95	71			
4-Methyl-2-pentanone (isobutyl methyl ketone)	119	135	95	79		58
1-Chloro-2-propanone (chloroacetone)	119	164d	125	83		
2,4-Dimethyl-3-pentanone (diisopropyl ketone)	125	160; 149	98; 107			34
3-Hexanone (ethyl propyl ketone)	125	113	130			
2-Hexanone (butyl methyl ketone)	129	122	110	88		49
4-Methyl-3-penten-2-one (mesityl oxide)	130	164 (α) 134 (β)	203	134	142	49 (β)
Cyclopentanone	131	210	146	154	55	57
5-Hexen-2-one (allyl acetone)	132	102	108			
2,2,4-Trimethyl-3-pentanone (*tert*-butyl isopropyl ketone)	135	132				144
1-Bromo-2-propanone (bromoacetone)	136	135				36
Methyl 2-oxopropanoate (methyl pyruvate)	136	208	187			
4-Methyl-3-hexanone (*sec*-butyl ethyl ketone)	136	137	78			
4-Methyl-2-hexanone (ethyl isobutyl ketone)	136	152	75			
2-Methyl-3-hexanone (isopropyl propyl ketone)	136	119	97			
3-Methyl-1-penten-4-one	138	201				76
2,4-Pentanedione (acetylacetone)	139	107	122 (mono) 209 (di)	209		149 (di)
3-Hydroxy-3-methyl-2-butanone	140	165				87
D-3-Methylcyclopentanone	143	185				92 (α) 69 (β)
5-Methyl-2-hexanone (isopentyl methyl ketone)	144	143	95			
4-Heptanone (dipropyl ketone)	145	133	75			
1-Hydroxy-2-propanone (hydroxyacetone; acetol)	146	196	129	173	103	71
3-Hydroxy-2-butanone (acetoin)	148	185; 202	318		243d	
2-Heptanone (methyl pentyl ketone; amyl methyl ketone)	151	127	89	73	207	
Cyclohexanone	155	167	162	147	82	91
Ethyl 2-oxopropanoate (ethyl pyruvate)	155	206	155			
2-Methyl-3-cyclopentenone	161	220				127

Ketones (Liquids) (continued)

Name of compound	bp (°C)	Derivative mp (°C)				
		Semi-carba-zone	2,4-Di-nitro-phenyl-hydra-zone	4-Nitro-phenyl-hydra-zone	Phenyl-hydra-zone	Oxime
2-Oxopropanoic acid (pyruvic acid)	165	222	218	220		
2-Methylcyclohexanone	166	197	137	132		43
4-Hydroxy-4-methyl-2-pentanone (diacetone alcohol)	166		159; 203	209		58
2,6-Dimethyl-4-heptanone (diisobutyl ketone)	168	126	92; 66			210
3-Methylcyclohexanone	170	180; 191	155	119	94	43
2,2-Dimethylcyclohexanone	170	201	142			
Methyl 3-oxobutanoate (methyl acetoacetate)	170	152	119			
6-Methyl-2-heptanone (isohexyl methyl ketone)	171	154	77			
4-Methylcyclohexanone	171	203	134	129	110	39
(±)-2,5-Dimethylcyclohexanone	173	122 (α) 173 (β)				111
2-Octanone (hexyl methyl ketone)	173	123	58	93		
2-Acetylfuran (2-furyl methyl ketone)	173	150	220	186	86	104
D-2,5-Dimethylcyclohexanone	174	177				98
Acetoxyacetone	175	145		144	60	
cis-2,4-Dimethylcyclohexanone	176	200				99
2,2,6-Trimethylcyclohexanone	179	209	141			
Cyclohexyl methyl ketone	180	177	140	154		60
Ethyl 3-oxobutanoate (ethyl acetoacetate)	181	133	93	218		
Cycloheptanone	182	163	148	137		23
cis-3,5-Dimethylcyclohexanone	183	203				74
5-Nonanone (dibutyl ketone)	187	90	41			
2,5-Dimethylcyclohexen-3-one	190	165				93; 169
2,5-Hexanedione (acetonylacetone)	194	185 (mono) 224 (di)	257 (di)	212 (di); 115	120 (di)	137 (di)
2-Nonanone (methyl octyl ketone)	194	120	56			
Fenchone	194	184	140			167
Methyl 4-oxopentanoate (methyl levulinate)	196	144	142	136	105; 96	
Cyclooctanone	196	167	163			
4-Fluoroacetophenone	196	219	235			80
(±)-2-Ethyl-5-methylcyclohexanone	197	181				80
1-Acetyl-4-methylcyclohexanone	197	159 (α) 175 (β)				59
2,6-Dimethyl-2,5-heptadiene-4-one (phorone; diisopropylideneacetone)	198	186; 221	118			48
2-Propylcyclohexanone	199	133				68
1-Acetylcyclohexene	200	220				59
Acetophenone (methyl phenyl ketone)	205	203	250	185	105	60
Ethyl 4-oxopentanoate (ethyl levulinate)	206	150	102	157	104	
L-Menthone	209	189	146		53	59
1,5-Dimethylcyclohexen-3-one	209	180			78	
2-Decanone (methyl octyl ketone)	209	124	74			
2-Acetylthiophene (methyl 2-thienyl ketone)	214	191		181	96	81
1,5,5-Trimethylcyclohexen-3-one (isophorone)	215	191; 200	130		68	80
1-Phenyl-2-propanone (benzyl methyl ketone)	216	200	156	145	87	70
2-Methylacetophenone (methyl *o*-tolyl ketone)	216	210	159			61

Ketones (Liquids) (continued)

Name of compound	bp (°C)	Derivative mp (°C)				
		Semi-carba-zone	2,4-Di-nitro-phenyl-hydra-zone	4-Nitro-phenyl-hydra-zone	Phenyl-hydra-zone	Oxime
Propyl 2-pyridyl ketone	218				82	48
2-Hydroxyacetophenone (2-acetylphenol)	218	210	213		110	118
1-Phenyl-1-propanone (ethyl phenyl ketone; propiophenone)	220	174	191	147	147	54
Methyl 3-pyridyl ketone	220				137	113
3-Methylacetophenone (methyl *m*-tolyl ketone)	220	203	207			57
2-Methyl-1-phenyl-1-propanone (isopropyl phenyl ketone; isobutyrophenone)	222	181	163		73	61; 94
2,2-Dimethyl-1-phenyl-1-propanone (*tert*-butyl phenyl ketone; pivalophenone)	224	150	195			167
1-Phenyl-2-butanone (benzyl ethyl ketone)	226	135; 146	140			
4-Methylacetophenone (methyl *p*-tolyl ketone)	226	205	260	198	97	88
3-Chloroacetophenone	228	232		176	176	88
2-Undecanone (2-hendecanone; methyl nonyl ketone)	228	123	63	91		45
2-Chloroacetophenone	229	160; 179	206	215		113
Carvone	230	142; 163	193	175	110	72 (α) 57 (β); 94
1-Phenyl-1-butanone (phenyl propyl ketone; butyrophenone)	230	191	190		200	50
2,4-Dimethylacetophenone	235	187				64
4-Phenyl-2-butanone (methyl β-phenylethyl ketone)	235	142	128			87
3-Methyl-1-phenyl-1-butanone (isobutyl phenyl ketone; isovalerophenone)	236	210	240			76
4-Chloroacetophenone	236	204; 160; 146	231	239	114	95
3,5-Dimethylacetophenone	237			180		114
3-Methoxyacetophenone	240	196	189			
1-Phenyl-1-pentanone (butyl phenyl ketone; valerophenone)	242	166	166	162	162	52
2-Methoxyacetophenone (*o*-acetylanisole)	245	183			114	83; 97
3-Acetylpropanoic acid (levulinic acid)	246	187	206	175	108	46
2,4,5-Trimethylacetophenone	247	204				86
1-Phenyl-1-pentanone (butyl phenyl ketone; valerophenone)	248	166	166	162	162	52
5-Isopropyl-2-methylacetophenone (2-acetyl-*p*-cymene)	250	147	142			92
Diethyl 3-oxopentanedioate (ethyl acetonedicarboxylate)	250	94	86			
3,4-Dimethylacetophenone	251	234				85
Propyl 3-pyridyl ketone	252	170			182	
2-Aminoacetophenone	252	290			108	109
α-Tetralone	257	226	257	231		
4-Methoxyacetophenone	258	198	220; 232	196	142	87

Ketones (Liquids) (continued)

		Derivative mp (°C)				
Name of compound	bp (°C)	Semi-carba-zone	2,4-Di-nitro-phenyl-hydra-zone	4-Nitro-phenyl-hydra-zone	Phenyl-hydra-zone	Oxime
Benzylidenepropanone (benzalacetone)	262	186	227	166	157	116
2-Tridecanone (methyl undecyl ketone)	263	126	69	102		57
1-Phenyl-1-hexanone (pentyl phenyl ketone; amyl phenyl ketone)	265	133	168			
1-Phenyl-1-heptanone (enanthophenone; hexyl phenyl ketone)	283	119		128		55
3-Phenylcyclohexanone	288	167				129
Methyl 1-naphthyl ketone (α-acetylnaphthalene)	302	289; 233	255		149	140
3-Methylphenyl phenyl ketone (phenyl *m*-tolyl ketone)	314		221			101
Phenyl 2-pyridyl ketone (2-benzoylpyridine)	317		199		136	150; 165
1,3-Diphenyl-2-propanone (dibenzyl ketone)	330	146; 126	100		129	125
1,3-Diphenyl-2-propen-1-one (dypnone; α-methylstyryl phenyl ketone)	345	151				134 (syn) 78 (anti)

Ketones (Solids)

Name of compound	mp (°C)	Derivative mp (°C) Semicarbazone	2,4-Dinitrophenylhydrazone	4-Nitrophenylhydrazone	Phenylhydrazone	Oxime
2-Aminoacetophenone	20	290d			108	109
Acetophenone (methyl phenyl ketone)	20	203	250; 237	185	105	60
2-Chloroacetophenone	20	201	231	239	114	95
1-Phenyl-1-hexanone (pentyl phenyl ketone; amyl phenyl ketone)	25	133	168			
1-Phenyl-2-propanone (benzyl methyl ketone; phenylacetone)	27	199	156	145	87	70
2-Nitroacetophenone	27	210	154			117
2,6-Dimethyl-2,5-heptadiene-4-one (phorone)	28	186; 221	118			48
4-Methylacetophenone	28	205	260	198	97	88
2-Hydroxyacetophenone (*o*-acetylphenol)	28	210	212		110	118
2,6-Dimethyl-2,5-heptadiene-4-one (phorone; diisopropylideneacetone)	28	186; 221	112			48
2-Tridecanone (methyl undecyl ketone)	28	126	69	102		57
4-Cyclohexylcyclohexanone	31	216	137			105
2-Acetylfuran (2-furyl methyl ketone)	33	150	220	186	86	104
3-Acetylpropanoic acid (levulinic acid)	33	187	206	175	108	46
Methyl 1-naphthyl ketone (α-acetylnaphthalene)	34	235; 289	255		149	140
1,3-Diphenyl-2-propanone (dibenzyl ketone; 1,3-diphenylacetone)	35	146; 126	100		129	125
2,2,6,6-Tetramethyl-4-piperidone	35	220				153
1-(4-Chlorophenyl)-1-propanone (*p* chloropropiophenone)	36	177	222			63
4-Methoxyacetophenone	38	198	220; 232	196	142	87
Furfuralacetone	39		241		132	
Propiopiperone (3,4-methylenedioxypropiophenone)	39	188			97	104
2-Hydroxybenzophenone	39; 153				155	143
4,4-Dimethylcyclohexanone	41	204			107	
Benzylidenepropane	41	186	227	166		
4-Phenyl-3-buten-2-one (benzalacetone)	41	187	227	166	157	116
1-Indanone (α-hydrindone)	42	239	258	235	135	146
2,4-Dichloroacetophenone	42	208				152
1-(3-Aminophenyl)-1-propanone (*m*-aminopropiophenone)	42	197				113
4-Phenyl-3-buten-2-one (methyl styryl ketone)	42	198; 142	227	167	159	116; 87
2-Isopropyl-5-methyl-1,4-benzoquinone (thymoquinone)	45	202 (mono) 237 (di)	180[a] (mono) 162		93[a]	162 (mono)
1,3-Dichloropropanone (1,3-dichloroacetone)	45	120	133			
1-(4-Bromophenyl)-1-propanone (*p*-bromopropiophenone)	46	171				91
1-Phenyl-1-dodecanone (phenyl undecyl ketone; laurophenone)	47	95				64
1-(2-Aminophenyl)-1-propanone (*o*-aminopropiophenone)	47	190				89
Benzophenone (diphenyl ketone)	49	167	239	159	138	144
2,4,6-Heptanetrione (diacetylacetone)	49	203 (mono)			142 (di)	69 (di)

Ketones (Solids) (continued)

Name of compound	mp (°C)	Derivative mp (°C)				
		Semi-carba-zone	2,4-Di-nitro-phenyl-hydra-zone	4-Nitro-phenyl-hydra-zone	Phenyl-hydra-zone	Oxime
5-Phenoxy-2-pentanone (methyl 3-phenoxypropyl ketone)	50	110	110			
9-Heptadecanone (dioctyl ketone)	50			54		112
α-Bromoacetophenone (phenacyl bromide)	51	146	220			97
4-Bromoacetophenone	51	208	237	248	126	129
3,4-Dimethoxyacetophenone	51	218	207	227	131	140
Methyl 2-naphthyl ketone (β-acetylnaphthalene)	54	237	262		177	149
1-Chloro-4-phenyl-2-butanone (chloromethyl 2-phenylethyl ketone)	54	156	147			89
2-Benzoylthiophene (phenyl 2-thienyl ketone)	56	197				93
2-Nonadecanone (heptadecyl methyl ketone)	56	126				77
2-Chloro-1,4-benzoquinone	57	185d				184; 148d
2-Indanone	59	218		232		155
α-Chloroacetophenone (phenacyl chloride)	59	156	215			89
1-Phenyl-2-propanone (deoxybenzoin; benzyl phenyl ketone)	60	148	204	163	116	98
1-(2-Naphthyl)-1-propanone (ethyl β-naphthyl ketone)	60	202				133
4-Methylbenzophenone (phenyl *p*-tolyl ketone)	60	122	202		109	154; 136; 115
1-Phenyl-1,3-butadione (benzoylacetone; methyl phenacyl ketone)	61		151	101	153	
1,1-Diphenyl-2-propanone (1,1-diphenylacetone)	61	170			131	165
1,3-Diphenyl-2-propen-1-one (benzalacetophenone; phenyl styryl ketone; chalone)	62	168; 180	248; 208		120	75; 140
4-Methoxybenzophenone (*p*-anisyl phenyl ketone)	63		180; 228	199	132; 90	116; 138
2-Phenylcyclohexanone	63	90	139			169
Benzoylformic acid (phenylglyoxylic acid)	66		197			127(α) 115(β)
2,2′-Dimethylbenzophenone (di-*o*-tolyl ketone)	67		190			105
2-Methyl-1,4-benzoquinone (*p*-toluquinone)	68	179 (mono) 240 (di)	128[a] 269 (di)		130[a]	135 (mono) 220d (di)
3,5-Dibromoacetophenone	68	268			110	
6-Phenyl-5-hexen-2,4-dione (cinnamalacetone)	68	186	223		180	153
12-Tricosanone (laurone; diundecyl ketone)	69	179				40
1,3-Dihydroxy-2-propanone (dihydroxyacetone)	72		278	160		84
3-Acetylphenanthrene	72	230				194
4-(4-Methoxyphenyl)-3-buten-2-one (4-methoxybenzalacetone; anisalacetone)	73		229			120
1,3-Diphenyl-1-propanone (benzylacetophenone)	73	144				87
9-Acetylphenanthrene	74	201				155
1-Naphthyl phenyl ketone	76	385	247			161
2-Benzoylfuryl methyl ketone	76	207			154	
2-Naphthoxy-2-propanone (β-naphthoxyacetone)	78	203			154	123
4-Phenylcyclohexanone	78	212; 229				110
4-Chlorobenzophenone	78		185		106	163; 106

Ketones (Solids) (continued)

Name of compound	mp (°C)	Derivative mp (°C)				
		Semi-carbazone	2,4-Dinitrophenylhydrazone	4-Nitrophenylhydrazone	Phenylhydrazone	Oxime
1,4-Cyclohexanedione	79	222 (mono) 231 (di)	240			188
1,3-Diphenyl-1,3-propadione (dibenzoylmethane; benzoylacetophenone; phenacyl phenyl ketone)	81	205			105 (mono)	165 (mono)
3-Nitroacetophenone	81	259	232	135	135	132
4-Bromobenzophenone	82	350	230		126	117; 169
Fluorene (diphenylene ketone; 9-oxofluorene)	83	234	284	269	152	196
α-Hydroxyacetophenone (phenacyl ketone)	86	146			112	70
4,4′-Dimethylbenzophenone (di-*p*-tolyl ketone)	95	144	229; 219		100	163
Benzil (dibenzoyl)	95	182 (mono) 244 (di)	189 (mono) 189 (di) 314 (di)	193 (mono) 290 (di) 192 (di)	134 (mono) 225 (di)	140 (mono) 237 (di)
3-Hydroxyacetophenone (*m*-acetylphenol)	96	191	261			
2,3-Dihydroxyacetophenone (3-acetylcatechol)	98	167				97
3-Aminoacetophenone	99	196				194; 148
3-Benzoylpropenoic acid (3-benzoylacrylic acid)	99	190			197	168d
2-Acetyl-1-hydroxynaphthalene	102	250			137	169
1,5-Diphenyl-4-penten-1,3-dione (cinnamacetophenone)	102		222	135		
2-Methyl-1,4-naphthoquinone	106	178 (4-) 247 (di)	299d (mono)			160 (mono) 168 (di)
Benzoylnitromethane	106				106	96
4-Aminoacetophenone	106	250	267			148
α,4-Dibromoacetophenone (4-bromophenacyl bromide)	108		218			115
4-Hydroxyacetophenone (*p*-acetylphenol)	109	199	210; 225; 261		151	145
Piperonalacetone	111	217 (α) 168 (β)			163	186
1,5-Diphenyl-1,4-pentadien-3-one (dibenzalacetone; dibenzylidenepropanone)	112	190	180	173	153	144
1,4-Benzoquinone (quinone)	116	166, 178 (mono) 243 (di)	186 (mono) 231[a] (di) 240		152[a]	144 (mono) 240 (di)
3-Benzoylpropanoic acid	116	181	191			
α-Hydroxyacetophenone (phenacyl alcohol; benzoyl carbinol)	118	147			112	70
1,2-Naphthoquinone	120; 147	184 (mono)	162	251, 236 (mono)	138[a] (mono)	110, 164 (mono) 169 (di)
Acenaphthenone (1-oxoacenaphthene)	121				90	175; 184, 222 (di)
4-Phenylacetophenone (*p*-acetylbiphenyl; methyl *p*-xenyl ketone)	121		242			187
1,4-Naphthoquinone (α-naphthoquinone)	125	247 (mono)	278[a] (mono) 198	279 (mono)	206[a] (mono)	198 (mono) 207 (di)

Ketones (Solids) (continued)

		Derivative mp (°C)				
Name of compound	mp (°C)	Semi-carba-zone	2,4-Di-nitro-phenyl-hydra-zone	4-Nitro-phenyl-hydra-zone	Phenyl-hydra-zone	Oxime
2,5-Dimethyl-1,4-benzoquinone (*p*-xylo-*p*-quinone)	125				124; 155	168 (mono) 272, 254 (di)
1,5-Di(4-methoxyphenyl)-3-pentanone (dianisalacetone)	129		83			148
4-(4-Hydroxy-3-methoxyphenyl)-3-buten-2-one (vanillidineacetone; 4-hydroxy-3-methoxystyryl methyl ketone)	130		230		128	
2,6-Dibromo-1,4-benzoquinone	131	225 (mono)				170 (mono)
4,4′-Dimethoxydibenzoyl (anisil; 4,4′-dimethoxybenzil)	133	255 (di)				133 (mono) 195, 217 (di)
Benzoylphenylmethanol (benzoin)	135	206d	245		159 (α) 106 (β)	152 (α) 99 (β)
3,4,5-Tibromoacetophenone	135	265			134	
Furoin	135		217		81	161 (α) 102 (β)
4-Hydroxybenzophenone (*p*-benzoylphenol)	135	194	244		144	152
2-Acetylphenanthrene	143	260			188	
1,9-Diphenyl-1,8-nonadien-3,5,7-trione (dicinnamalacetone)	144		196		166	
1,2-Naphthoquinone	146	184 (mono)		235	138	162 (2-) 109 (1-) 169 (di)
1,2-Dibenzoylethane	147		265 (di)		116 (mono) 179 (di)	204
2,4-Dihydroxyacetophenone (resacetophenone)	147	220	208; 218; 244		159	203
3,5-Dihydroxyacetophenone	148	206		237		
4,4′-Dichlorobenzophenone	148		240			135
Furil	165		215	199	184	100 (di)
Quinhydrone	171				152	161
2,3,4-Trihydroxyacetophenone (Gallacetophenone)	173	225	199			163
4,4′-Di-(dimethylamino)benzophenone (Michler's ketone)	174		273		175	233
Diphenylene ketone oxide (xanthone)	174				152	161
(±)-Camphor	178	237; 248	164; 177	217	233	118
(+)-Camphor	179	238	177		233	118
3,4,5-Trihydroxyacetophenone	188	217		260		
7-Isopropyl-1-methyl-9,10-phenanthraquinone (retenequinone)	197	200		223	160	131
Camphorquinone	199	236 (α) 147	36 (mono) 190[a] (di)	239	170,[a] 190	170, 153, 115 (mono) 201, 248, 140 194 (di)

Ketones (Solids) (continued)

Name of compound	mp (°C)	Derivative mp (°C)				
		Semi-carba-zone	2,4-Di-nitro-phenyl-hydra-zone	4-Nitro-phenyl-hydra-zone	Phenyl-hydra-zone	Oxime
2,5-Dihydroxyacetophenone (quinacetophenone)	202			216		150
9,10-Phenanthrenequinone	208	220d (mono)	313d (mono) 162	245 (mono)	165[a] (mono)	158 (mono) 202 (di)
Ninhydrin (triketohydrindene hydrate)	243				208	201
Acenaphthenequinone	261	193 (mono) 271 (di)	222 (di)	247 (mono)	179 (mono) 219 (di)	230 (mono)
3-Bromo-9,10-phenanthraquinone	268	242 (mono)			177 (mono)	198 (mono) 212 (di)
Aceanthrenequinone (3,4-benzacenaphthenequinone)	270				203 (mono)	251 (mono)
9,10-Anthraquinone	286		183[a]		183	224 (mono)

[a]These melting points are those of the addition compounds of the quinones and substituted hydrazines. The products are not true hydrazones.

Nitriles (Liquids)

		Acid derivative mp (°C)			Amine derivative mp (°C)			Derivative mp (°C)
Name of compound	bp (°C)	Acid	Amide	Anilide	Benzamide	Benzenesulfonamide	Phenylthiourea	α-(Imidioylthio)-acetic acid hydrochloride
Propynenitrile (cyanoacetylene)	42	18	62	87				
Propenenitrile (acrylonitrile; vinyl cyanide)	78		85	105				
Fluoroethanenitrile (fluoroacetonitrile; fluoromethyl cyanide)	80	33	77					
Ethanenitrile (acetonitrile; methyl cyanide)	82		82	114	71	58	106; 135	115
Trichloroethanenitrile (trichloroacetonitrile; trichloromethyl cyanide)	86	58	141	97				
Propanenitrile (propionitrile; ethyl cyanide)	97		79	106	84	36	63	128
2,2-Dimethylpropanenitrile (trimethylacetonitrile; tert-butyl cyanide)	106	35	154	129				
2-Methylpropanenitrile (isobutyronitrile; isopropyl cyanide)	108		129	110	57	53	82	137
2-Chloro-2-methylpropanenitrile (2-chloroisobutyronitrile; α-chloroisopropyl cyanide)	116	31	70					
Butanenitrile (butyronitrile; propyl cyanide)	117		116	96	42	65	136	
E-2-Butenenitrile (*trans*-crotononitrile)	119	72	158	115		110		
3-Butenenitrile (vinylacetonitrile; allyl cyanide)	119		73	58		39	127; 57	
Methoxyethanenitrile (methoxyacetonitrile; methoxymethyl cyanide)	120		96	58				
2-Hydroxy-2-methylpropanenitrile (2-hydroxyisobutyronitrile; acetone cyanohydrin)	120	79	98	136				
2-Methylbutanenitrile (2-methylbutyronitrile; α-methylpropyl cyanide)	126		112	110				
Chloroethanenitrile (chloroacetonitrile; chloromethyl cyanide)	127	63	120	137				
3-Methylbutanenitrile (isovaleronitrile; isobutyl cyanide)	130		136	113			102	133
Z-2-Chloro-2-butenenitrile (2-chlorocrotononitrile)	136	99	212					
2,4-Pentadienonitrile	138	72	124					

Nitriles (Liquids) (continued)

		Acid derivative mp (°C)			Amine derivative mp (°C)			Derivative mp (°C)
Name of compound	bp (°C)	Acid	Amide	Anilide	Benzamide	Benzenesulfonamide	Phenylthiourea	α-(Imidioylthio)-acetic acid hydrochloride
2-Methyl-2-butenenitrile (2-methylcrotononitrile)	138	64	76	77				
2-Bromo-2-methylpropanenitrile (2-bromoisobutyronitrile; α-bromoisopropyl cyanide)	140	49	148	83				
2,3-Dicyanothiophene (thiophene-2,3-dicarbonitrile)	140	274	228 (di)					
Pentanenitrile (valeronitrile; butyl cyanide)	141		106	63			69	138
3,3-Dimethylpropenenitrile (3,3-dimethylacrylonitrile; β,β-dimethylvinyl cyanide)	142	70	108; 66	127			106	
2-Chlorobutanenitrile (2-chlorobutyronitrile; α-chloropropyl cyanide)	143		76	75				
2-Furancarbonitrile (α-furonitrile; 2-furyl cyanide)	147	134	142	124				
2-Methyl-3-oxobutanenitrile (2-methylacetoacetonitrile)	147	73	140					
Cyclobutanecarbonitrile (cyclobutyl cyanide)	150		155	113				
4-Methylpentanenitrile (isocapronitrile; isopentyl cyanide)	155		121	112				128
2-Methylhexanenitrile (2-methylcapronitrile; α-methylpentyl cyanide)	165		72	98				
Hexanenitrile (capronitrile; pentyl cyanide)	165		100	95	40	96	77	136
Z-Chlorobutenedinitrile (chlorofumaronitrile)	172	192		186 (di)				
3-Chlorobutanenitrile (3-chlorobutyronitrile; β-chloropropyl cyanide)	176	44		90				
3-Chloropropanenitrile (3-chloropropionitrile; β-chloroethyl cyanide)	178	41		119				
3-Cyanothiophene (thiophene-3-carbonitrile)	179; 205	138	180					
5-Methylhexanenitrile (5-methylcapronitrile; isohexyl cyanide)	180		104	75				
2-Hydroxyethanenitrile (glycolonitrile; α-hydroxyacetonitrile formaldehyde cyanohydrin)	183	80	120	97				

Nitriles (Liquids) (continued)

Name of compound	bp (°C)	Acid derivative mp (°C): Acid	Acid derivative mp (°C): Amide	Acid derivative mp (°C): Anilide	Amine derivative mp (°C): Benzamide	Amine derivative mp (°C): Benzenesulfonamide	Amine derivative mp (°C): Phenylthiourea	Derivative mp (°C): α-(Imidioylthio)-acetic acid hydrochloride
(±)-2-Hydroxypropanenitrile [(±)-lactonitrile; α-hydroxypropionitrile; acetaldehyde cyanohydrin]	184		79	59				
Heptanenitrile (hexyl cyanide)	187		95	71			75	133
Benzonitrile (phenyl cyanide)	190	122	129	162	105	88	156	124
2-Cyanothiophene (thiophene-2-carbonitrile)	192	130; 192	180	140				
2-Octynenitrile	196		91	44				
4-Chlorobutanenitrile (4-chlorobutyronitrile; β-chloropropyl cyanide)	197	16	89	70				
2-Methylbenzonitrile (*o*-tolunitrile; *o*-tolyl cyanide)	205	104	142	125	88			
2,3,3-Trimethyl-1-cyclopentene-1-carbonitrile (β-campholytonitrile)	205; 225	135	130	104				
Octanenitrile (caprylnitrile) heptyl cyanide)	206; 199	16	110	57				134
1,1-Dicyanopropane (ethyl malononitrile)	206	112	216 (di)					
1,1-Dicyanobutane (propylmalonitrile)	210	96	184 (di)	198 (di)				
3-Methylbenzonitrile (*m*-tolunitrile; *m*-tolyl cyanide)	212	113	96	126				168
Cyclohexylethanenitrile (cyclohexylacetonitrile; cyclohexylmethyl cyanide)	215	33	172		81			
Dibenzylethanenitrile (dibenzylacetonitrile; dibenzylmethyl cyanide)	215	89	129	155				
2-Cyanopyridine (picolinonitrile; 2-pyridinecarbonitrile)	215		107	76				
4-Methylbenzonitrile (*p*-tolunitrile; *p*-tolyl cyanide)	218		165	148	137			181
1,1-Dicyano-3-methylbutane (isobutylmalonitrile)	222	108	196 (di)					
Nonanenitrile (octyl cyanide)	224	15	99	57	49			
Z-2-Phenyl-2-butenenitrile (2-phenyl-2-crotononitrile)	226	136	99					
Phenylethanenitrile (phenylacetonitrile; benzyl cyanide)	234	77	157	118	116	69	135	146

Nitriles (Liquids) (continued)

		Acid derivative mp (°C)			Amine derivative mp (°C)			Derivative mp (°C)
Name of compound	bp (°C)	Acid	Amide	Anilide	Benzamide	Benzenesulfonamide	Phenylthiourea	α-(Imidioylthio)acetic acid hydrochloride
Phenoxyethanenitrile (phenoxyacetonitrile; phenoxymethyl cyanide)	240	99	102					
3-Methylphenylethanenitrile (*m*-tolylacetonitrile; *m*-xylyl cyanide; *m*-tolylmethyl cyanide)	241	61	141					
4-Methylphenylethanenitrile (*p*-tolylacetonitrile; *p*-xylyl cyanide; *p*-tolylmethyl cyanide)	243	94	185		96			
4-(1-Methylethyl)benzonitrile (*p*-isopropylbenzonitrile; *p*-isopropylphenyl cyanide)	244	118	153; 133					
2-Methylphenylethanenitrile (*o*-tolylacetonitrile; *o*-xylyl cyanide; *o*-tolylmethyl cyanide)	244	89	161					
Decanenitrile (nonyl cyanide)	245	31	108	70				
3-Methyl-2-phenylbutanenitrile (3-methyl-2-phenylbutyronitrile; β-methyl-α-phenylpropyl cyanide)	249	62	112	133				
2-Chlorophenylethanenitrile (*o*-chlorobenzyl cyanide; *o*-chlorophenylacetonitrile)	251		175	139				
1,2-Dicyanopropane (methyl succinonitrile)	254	115	225 (di)	200 (di)	154			
1-Undecanenitrile (1-hendecanonitrile; decyl cyanide)	254	28	99	71				
2-Phenylpentanenitrile (2-phenylvaleronitrile; α-phenylbutyl cyanide)	255	58	85					
3-Phenyl-2-propenenitrile (cinnamonitrile)	256		153; 109	147				
2-Methoxybenzonitrile (*o*-methoxyphenylcyanide)	256		129	131				
3-Phenylpropanenitrile (3-phenylpropionitrile; β-phenylethyl cyanide)	261	49	105	98	58			
N-Cyano-*N*-methylaniline (*N*-methylanilinonitrile)	266	100	163					
4-Chlorophenylethanenitrile (*p*-chlorobenzyl cyanide; *p*-chlorophenylacetonitrile)	267		175	165				
3-(2-Chlorophenyl)propanenitrile [3-(*o*-chlorophenyl)propionitrile; β-(*o*-chlorophenyl)ethyl cyanide]	268	97	119					

Nitriles (Liquids) (continued)

		Acid derivative mp (°C)			Amine derivative mp (°C)			Derivative mp (°C)
Name of compound	bp (°C)	Acid	Amide	Anilide	Benzamide	Benzenesulfonamide	Phenylthiourea	α-(Imidioylthio)-acetic acid hydrochloride
1,3-Dicyano-2-methylpropane	271	79	176 (di)					
Dodecanenitrile (undecyl cyanide; lauronitrile)	277	44	110	78				
Pentanedinitrile (glutaronitrile; 1,3-dicyanopropane; trimethylene cyanide)	286	97	175	224	135 (di)	119	148	
Hexanedinitrile (adipononitrile; 1,4-dicyanobutane; tetramethylene cyanide)	295	153	220	239	158 (di)	154 (di)		
1-Naphthonitrile (1-cyanonaphthalene)	302				148			
Pentanedecanenitrile	322		102	78				

Nitriles (Solids)

Name of compound	mp (°C)	Acid derivative mp (°C)			Amine derivative mp (°C)	
		Acid	Amide	Anilide	Benzamide	Benzenesulfonamide
2-Cyanodiphenylmethane	19	117	163			
Tetradecanenitrile	20	54	103	82		
3-Phenyl-2-propenenitrile (cinnamonitrile)	20	133	153; 109	147		
3,3,3-Trichloropropenenitrile (trichloroacrylonitrile; trichlorovinyl cyanide)	20	76	98; 87	98		
(±) -2-Hydroxy-2-phenylethanenitrile [(±)-mandelonitrile; benzaldehyde cyanohydrin; 2-hydroxy-2-phenylacetonitrile; *α*-hydroxybenzyl cyanide]	22	119	134	152	149	
Pentanedecanenitrile	23	52	102	78		
2-Methoxybenzonitrile (*o*-methoxyphenyl cyanide)	25	101	129	131		
2-Chlorophenylethanenitrile (*o*-chlorophenylacetonitrile; *o*-chlorobenzyl cyanide)	25	95	175	139		
1,1-Dicyanoethane (methylmalonitrile)	26	138	217 (di)	182 (di)		
2-Cyanopyridine (picolinonitrile; 2-pyridinecarbonitrile)	26	137	107	76		
4-Methylbenzonitrile (*p*-tolunitrile; *p*-tolyl cyanide)	29	180	165	148	137	
D-2-Hydroxy-2-phenylethanenitrile (D-mandelonitrile; D-2-hydroxy-2-phenylacetonitrile; D-*α*-hydroxybenzyl cyanide)	29	133	123			
2-Bromophenylethanenitrile (*o*-bromophenylacetonitrile; *o*-bromobenzyl cyanide)	29	84	148			
4-Chlorophenylethanenitrile (*p*-chlorophenylacetonitrile; *p*-chlorobenzyl cyanide)	30	105	175	165		
Propanedinitrile (dicyanomethane; malonitrile; methylene cyanide)	30	135	170		140	96
Hexadecanenitrile	31	63	107	90		
Z-Butenedinitrile (maleonitrile; *cis*-1,2-dicyanoethylene; *cis*-1,2-dicyanoethene)	31	130	181	187	179 (di)	155 (di)
2,2-Dicyanopropane (dimethylmalonitrile)	32	193	269 (di)	204	152 (di)	
2,2-Dimethylpropanenitrile (*tert*-butylacetonitrile; neopentyl cyanide)	33		132	131		
1-Naphthylethanenitrile (*α*-naphthylacetonitrile; *α*-naphthylmethyl cyanide)	33	131	154; 180			
4,4-Dicyanoheptane (dipropylmalononitrile)	34 (anhyd)	161	214 (di)	168 (di)	154 (di)	
Heptadecanenitrile	34	61	106		91	
2-Cyanopropanoic acid (2-cyanopropionic acid)	35	135; 120	206 (di)	182 (di); 214		
4-Fluorobenzonitrile (*p*-fluorophenyl cyanide)	35	183	155			
Indole-3-ethanenitrile (indole-3-acetonitrile)	36	165; 199	151	150	138	
1-Naphthonitrile (*α*-cyanonaphthalene)	37	162	202			148
3-Bromobenzonitrile (*m*-bromophenyl cyanide)	38	155	155		136	
2-(Phenylamino)butanenitrile [2-(*N*-anilino)-butanenitrile; 2-(*N*-anilino)butyronitrile]	39	141	123	92		

Nitriles (Solids) (continued)

Name of compound	mp (°C)	Acid derivative mp (°C)			Amine derivative mp (°C)	
		Acid	Amide	Anilide	Benzamide	Benzenesulfonamide
E-3-(2-Chlorophenyl)-2-propenenitrile (*trans-o*-chlorocinnamonitrile)	40	212	168	176		
3-Chlorobenzonitrile (*m*-chlorophenyl cyanide)	41	158	134	125	214	
2-Chlorobenzonitrile (*o*-chlorophenyl cyanide)	43	141	142	118	117	
4-Chloro-2-hydroxy-2-phenylethanenitrile (4-chloromandelonitrile; 4-chloro-2-hydroxy-2-phenylacetonitrile; *p*-chloro-α-hydroxybenzyl cyanide)	43	122; 112	123			
Octadecanenitrile	43	70	109	95		
Nonadecanenitrile	43	69	110	96		
4-Cyanobutanoic acid (4-cyanobutyric acid)	45	98	183	222 (di)	105	
3,3-Dicyanopentane (diethylmalonitrile)	45	125	146 (mono) 224 (di)			
R,S-2,3-Dimethylbutanedinitrile (*meso*-2,3-dimethylsuccinonitrile)	46	209; 198	313 (di)	235 (di)		
4-Aminophenylethanenitrile (*p*-aminophenylacetonitrile; 4-aminobenzyl cyanide)	46	199	162			
3-Bromo-4-methylbenzonitrile (3-bromo-4-tolunitrile; 3-bromo-4-tolyl cyanide)	47	140	137			
4-Bromophenylethanenitrile (*p*-bromophenylacetonitrile; *p*-bromobenzyl cyanide)	47	114	194			
2-(Phenylamino)ethanenitrile [*N*-anilinoacetonitrile; 2-(phenylamino)acetonitrile]	48	128	136	113		
3,3-Diphenylpropenenitrile (3,3-diphenylacrylonitrile; β-phenylcinnamonitrile; β,β-diphenylvinyl cyanide)	49	167		131		
3-Bromo-2-hydroxybenzonitrile (3-bromo-2-hydroxyphenyl cyanide)	50	184	165			
4,4-Dicyanoheptane (dipropylmalonitrile)	50 (hyd)	161	214 (di)	169 (di)	154 (di)	
3-Cyanopropanoic acid (3-cyanopropionic acid)	50	235	157 (mono) 269 (di)	149 (mono) 230 (di)	80	
Eicosanonitrile	50	77	109	92		
3-Cyanopyridine (nicotinonitrile)	50	232	122	85; 132	132	
2-(Phenylamino)pentanenitrile [2-(*N*-anilino) valeronitrile; 2-(*N*-anilino)butyl cyanide]	51	148	99			
E-2,3-Diphenylpropenenitrile (*trans*-2,3-diphenylacrylonitrile; *trans*-α,β-diphenylvinyl cyanide)	51	172	127	141		
2-Aminobenzonitrile (anthranilonitrile; *o*-aminophenyl cyanide)	51	147	111	131	167	
5-Cyanothiazole	53	218	186			
2-Bromobenzonitrile (*o*-bromophenyl cyanide)	53	150	156			
3-Aminobenzonitrile (*m*-aminophenyl cyanide)	54	174	79	114		
2-Iodobenzonitrile (*o*-iodophenyl cyanide)	55	162	184		154	
2,4,6-Trimethylbenzonitrile (2,4,6-trimethylphenyl cyanide)	55	155	188		154	

Nitriles (Solids) (continued)

Name of compound	mp (°C)	Acid derivative mp (°C)			Amine derivative mp (°C)	
		Acid	Amide	Anilide	Benz-amide	Ben-zene-sulfon-amide
2-Amino-2-phenylethanenitrile (α-aminobenzyl cyanide; 2-amino-2-phenylacetonitrile)	55	256	132		225 (di)	
Butanedinitrile (succinonitrile; 1,2-dicyanoethane; ethylene cyanide)	57	188	260	230	177	
2-Iodo-4-methylbenzonitrile (2-iodo-4-tolunitrile; 2-iodo-4-tolyl cyanide)	58	206	167			
(±)-2,3-Dimethylbutanedinitrile [(±)-2,3-dimethylsuccinonitrile; (±)-1,2-dicyano-1,2-dimethylethane]	59	135	149 (mono) 244 (di)	222 (di)		
4-Methoxybenzonitrile (*p*-methoxyphenyl cyanide)	62	186	163	169		
2-Chloro-4-methylbenzonitrile (2-chloro-4-tolunitrile; 2-chloro-4-tolyl cyanide)	62	156	182			
2,4-Dichlorobenzonitrile (2,4-dichlorophenyl cyanide)	62	164	194			
4-Methoxy-3-phenyl-2-propenenitrile (*p*-methoxycinnamonitrile)	64	170	186			
Bromopropanedinitrile (bromomalonitrile; bromomethylene cyanide)	66	113	181			
2-Naphthonitrile (β-cyanonaphthalene)	66	184	195			
2-Cyano-2-ethylbutanoic acid (diethylcyanoacetic acid)	66; 57	125	146 (mono) 224 (di)			
Cyanoethanoic acid (cyanoacetic acid)	66	136	170; 123	227; 198	120	
2-Chloro-3-methylbenzonitrile (2-chloro-3-tolunitrile; 2-chloro-3-tolyl cyanide)	67	132	124			
4-Chloro-2-methylbenzonitrile (4-chloro-2-tolunitrile; 4-chloro-2-tolyl cyanide)	67	172	183			
2-Cyanobenzyl cyanide (phenylmalono-nitrile; phenylmethylene cyanide)	69	153	233			
5-Bromo-2-methylbenzonitrile (5-bromo-2-tolunitrile; 5-bromo-2-tolyl cyanide)	70	187	170			
4-Hydroxyphenylethanenitrile (*p*-hydroxyphenyl-acetonitrile; *p*-hydroxybenzyl cyanide)	70	150	175			
2-Methyl-6-nitrobenzonitrile (6-nitro-2-tolunitrile; 6-nitro-2-tolyl cyanide)	70	184	163			
3,4-Dichlorobenzonitrile (3,4-dichlorophenyl cyanide)	72	209	133			
2-Aminophenylethanenitrile (*o*-aminophenyl-acetonitrile; *o*-aminobenzyl cyanide)	72	119	93			
2,2-Diphenylpentanedinitrile (2,2-diphenyl-glutaronitrile; α,α-diphenyltrimethylene cyanide)	72	197; 183	144 (mono)			
1,2,2,3-Tetramethylcyclopentene-1-carbo-nitrile (campholic acid)	73	106	80	91	98	
2,2-Diphenylethanenitrile (2,2-diphenyl-acetonitrile; α-phenylbenzyl cyanide)	75	148	168	180	145	

Nitriles (Solids) (continued)

		Acid derivative mp (°C)			Amine derivative mp (°C)	
Name of compound	mp (°C)	Acid	Amide	Anilide	Benz-amide	Ben-zene-sulfon-amide
4-Cyano-*N,N*-dimethylaniline	76	243	206	183		
Dicyanodimethylamine [bis(cyanomethyl)amine]	77	247; 225	143 (di)	141 (di)	166 (tri)	
4-Cyanobenzyl chloride (α-chloro-*p*-tolunitrile; α-chloro-*p*-tolyl cyanide)	80	203	173			
3-Methyl-6-nitrobenzonitrile (6-nitro-3-tolunitrile; 6-nitro-3-tolyl cyanide)	80	219	151			
3-Cyanobenzaldehyde (*m*-formylbenzonitrile)	81	175	190			
Benzoylethanenitrile (benzoylacetonitrile; benzoylmethyl cyanide)	81	104	113	108		
4-Cyanopyridine	83	315	156			
6-Chloro-2-methylbenzonitrile (6-chloro-2-tolunitrile; 6-chloro-2-tolyl cyanide)	83	102	167			
2,3,4,5-Tetrachlorobenzonitrile (2,3,4,5-tetrachlorophenyl cyanide)	84	195	208	197		
8-Cyanoquinoline	84	187	173			
3-Methyl-2-nitrobenzonitrile (2-nitro-3-tolunitrile; 2-nitro-3-tolyl cyanide)	84	223	192			
4-Aminobenzonitrile (*p*-cyanoaniline; *p*-aminophenyl cyanide)	86	188	183			
Z-2,3-Diphenylpropenenitrile (*cis*-2,3-diphenylacrylonitrile; *cis*-α,β-diphenylvinyl cyanide)	86	138	168	179		
4-Cyanobiphenyl	86	228	223			
2-Naphthylethanenitrile (β-naphthylacetonitrile; β-naphthylmethyl cyanide)	86	142	200			
1-Cyano-2-phenylpropenenitrile (1-cyano-2-phenylacrylonitrile; benzalmalononitrile; 1-cyano-2-phenylvinyl cyanide)	87	196	190 (di)			
5-Bromo-2,4-dimethylbenzonitrile (5-bromo-2,4-dimethylphenyl cyanide)	90	181	198			
2,6-Dimethylbenzonitrile (2,6-dimethylphenyl cyanide)	91	116	139; 125			
2,4-Dibromobenzonitrile (2,4-dibromophenyl cyanide)	92	174	198			
2-(2-Nitrophenyl)propenenitrile [β-(2-nitrophenyl)-acrylonitrile; β-(2-nitrophenyl)vinyl cyanide]	92	147 (*cis*) 240 (*trans*)	185			
2-(Phenylamino)propanenitrile [2-(*N*-anilino)-propionitrile; 2-(*N*-anilino)ethyl cyanide]	92	162	144	127		
2-Methyl-2-(phenylamino)propanenitrile [2-(*N*-anilo)-isobutyronitrile; 2-(*N*-anilino)isopropyl cyanide]	94	185	136	155		
3-Methyl-4-nitrobenzonitrile (4-nitro-3-tolunitrile; 4-nitro-3-tolyl cyanide)	94	134	177			
2-Cyanoquinoline	94	157	133			
E-Butenedinitrile (fumaronitrile; *trans*-1,2-dicyanoethene)	96	300	266	314		

Nitriles (Solids) (continued)

Name of compound	mp (°C)	Acid derivative mp (°C) Acid	Amide	Anilide	Amine derivative mp (°C) Benzamide	Benzenesulfonamide
4-Chlorobenzonitrile (*p*-chlorophenyl cyanide)	96	240	179	194	140	
4-Cyanopentanenitrile (4-cyanovaleric acid; 2-methylglutaromononitrile)	96	79	176 (di)			
3,5-Dibromobenzonitrile (3,5-dibromophenyl cyanide)	97	220	187			
9-Phenanthrylethanenitrile (9-phenanthrylacetonitrile)	97	224	252			
4-Chloro-2-nitrobenzonitrile (4-chloro-2-nitrophenyl cyanide)	98	142	172			
2-Hydroxybenzonitrile (*o*-cyanophenol; *o*-hydroxyphenyl cyanide)	98	159	133	135		
4-Cyanotriphenylmethane	100	165		196		
4-Methyl-3-nitrobenzonitrile (3-nitro-4-tolunitrile; 3-nitro-4-tolyl cyanide)	101	165	153			
4-Chloro-3-nitrobenzonitrile (4-chloro-3-nitrophenyl cyanide)	101	182	156	131		
2-Cyano-3-phenylpropanoic acid (2-cyano-3-phenylpropionic acid)	102; 75	121	225			
4-Cyanoquinoline	102	254	181			
2,3,3-Triphenylpropanenitrile (2,3,3-triphenylpropionitrile; α,β,β-triphenylethyl cyanide)	102	223	213			
3-Cyanophenanthrene	102	269	234	217		
4-Bromo-2,5-dimethylbenzonitrile (4-bromo-2,5-dimethylphenyl cyanide)	104	172	210			
3-Methyl-5-nitrobenzonitrile (5-nitro-3-tolunitrile; 5-nitro-3-tolyl cyanide)	105	174	165			
2,4-Dinitrobenzonitrile (2,4-dinitrophenyl cyanide)	105	183	204			
2-Methyl-4-nitrobenzonitrile (4-nitro-2-tolunitrile; 4-nitro-2-tolyl cyanide)	105	179	174			
6-Chloro-3-nitrobenzonitrile (6-chloro-3-nitrophenyl cyanide)	106	165	178			
5-Bromo-2-methyl-3-nitrobenzonitrile (5-bromo-3-nitro-2-tolunitrile; 5-bromo-3-nitro-2-tolyl cyanide)	107	226	235			
4-Methyl-2-nitrobenzonitrile (2-nitro-4-tolunitrile; 2-nitro-4-tolyl cyanide)	107	190	166			
3-Cyanoquinoline	108	275	199			
2-Cyanophenanthrene	109	260	243	218		
2-Methyl-3-nitrobenzonitrile (3-nitro-2-tolunitrile; 3-nitro-2-tolyl cyanide)	110	152	158			
2-Nitrobenzonitrile (*o*-nitrophenyl cyanide)	110	146	176	155		
4-Chloro-1-naphthonitrile (1-chloro-4-cyanonaphthalene)	110	224; 210	236			
9-Cyanophenanthrene	111	257	233	218	167	
5-Cyanoacenaphthene (acenaphthene-5-carbonitrile)	111	219	198		184	149

Nitriles (Solids) (continued)

		Acid derivative mp (°C)			Amine derivative mp (°C)	
Name of compound	mp (°C)	Acid	Amide	Anilide	Benzamide	Benzenesulfonamide
4-Bromobenzonitrile (*p*-bromophenyl cyanide)	112	251	190		143	
4-Hydroxybenzonitrile (*p*-cyanophenol; *p*-hydroxyphenyl cyanide)	113	214	162 (hyd)	197		
2,4,5-Trimethoxybenzonitrile (2,4,5-trimethoxyphenyl cyanide)	114	144	185	155		
4-Nitrophenylethanenitrile (*p*-nitrophenylacetonitrile; *p*-nitrobenzyl cyanide)	116	153	198	198		
6-Bromo-3-nitrobenzonitrile (6-bromo-3-nitrophenyl cyanide)	117	180	198	166		
3-Nitrobenzonitrile (*m*-nitrophenyl cyanide)	118	140	143	154		
2-Hydroxyphenylethanenitrile (*o*-hydroxyphenylacetonitrile; *o*-hydroxybenzyl cyanide)	119	149	118			
4-Bromo-3-nitrobenzonitrile (4-bromo-3-nitrophenyl cyanide)	120	204	156			
4-Cyanoazobenzene	121	241	225			
2,6-Dicyanopyridine (dipicolonitrile)	123; 113	252; 228	302 (di)			
Dibromopropanedinitrile (dibromomalononitrile; dibromodicyanomethane; dibromomethylene cyanide)	124	147	206 (di)			
2-Cyanohexanoic acid	126	101	200 (di)	193 (di)		
1-Cyanoanthracene	126	245	260			
2,2,3-Triphenylpropanenitrile (2,2,3-triphenylpropionitrile; α,α,β-triphenylethyl cyanide)	126	162; 132	111			
1-Cyanophenanthrene	128	233	284	245		
5-Bromo-4-methyl-3-nitrobenzonitrile (5-bromo-3-nitro-4-tolunitrile; 5-bromo-3-nitro-4-tolyl cyanide)	130	206	171			
2,5-Dichlorobenzonitrile (2,5-Dichlorophenyl cyanide)	130	154	155			
2,3-Diphenylbutanenitrile (2,3-diphenylbutyronitrile; α,β-diphenylpropyl cyanide)	130	134; 186	174; 193			
5-Bromo-4-methyl-2-nitrobenzonitrile (5-bromo-2-nitro-4-tolunitrile; 5-bromo-2-nitro-4-tolyl cyanide)	132	203	191			
2-Hydroxy-3-nitrobenzonitrile (2-cyano-6-nitrophenol; 2-hydroxy-3-nitrophenyl cyanide)	133	148	155			
4-Nitro-1-naphthonitrile (1-cyano-4-nitronaphthalene)	133	221	218			
4-Acetamidobenzonitrile	133	185	177	168		
6-Cyanoquinoline	135	292	174			
3,5-Dichloro-2-hydroxybenzonitrile (3,5-dichloro-2-hydroxyphenyl cyanide)	139	223	209			
4-Cyano-2-phenylquinoline (2-phenyl-4-quinolinonitrile)	140	218	196	198		

Nitriles (Solids) (continued)

Name of compound	mp (°C)	Acid derivative mp (°C): Acid	Acid derivative mp (°C): Amide	Acid derivative mp (°C): Anilide	Amine derivative mp (°C): Benzamide	Amine derivative mp (°C): Benzenesulfonamide
3,3,3-Triphenylpropanenitrile (3,3,3-triphenylpropionitrile; β,β,β-triphenylethyl cyanide)	140	177	198			
1,2-Dicyanobenzene (phthalonitrile)	141	206	220 (di)	253 (di)	184 (di)	
8-Nitro-2-naphthonitrile (2-cyano-8-nitronaphthalene)	143	295; 288	218			
5-Chloro-2-cyanonaphthalene (5-chloro-2-cyanonaphthonitrile)	144	270	187	202		
5-Chloro-1-naphthonitrile (5-chloro-1-cyanonaphthalene)	145	245	239			
5-Bromo-1-naphthonitrile (1-bromo-5-cyanonaphthalene)	147	261	241			
4-Nitrobenzonitrile (*p*-nitrophenyl cyanide)	149	241	201	211		
2-Cyano-5-iodonaphthalene (5-iodo-2-naphthonitrile)	149	264	196	203		
2-Cyano-2-propylpentanamide (dipropylcyanoacetamide; 2-cyano-2-propylvaleramide)	153	161	214 (di)	169 (di)		
3-Chloro-4-hydroxybenzonitrile (3-chloro-4-hydroxyphenyl cyanide)	155	170	182			
2,6-Dibromobenzonitrile (2,6-dibromophenyl cyanide)	155	157	209			
5-Chloro-2,4-dinitrobenzonitrile (5-chloro-2,4-dinitrophenyl cyanide)	156	183	212	226		
4-Benzamidobenzonitrile (*N*-benzoylanthranilonitrile)	156	181	219	279		
5-Bromo-2-hydroxybenzonitrile (5-bromo-2-hydroxyphenyl cyanide)	159	165	232	222		
(±)-2,3-Diphenylbutanedinitrile [(±)-2,3-diphenylsuccinonitrile; (±)-α,β-diphenylethylene cyanide]	160	183 (hyd)		173 (mono)		
2-Hydroxy-4-nitrobenzonitrile (2-cyano-5-nitrophenol; 2-hydroxy-4-nitrophenyl cyanide)	161	235	194			
1,3-Dicyanobenzene (isophthalonitrile)	162	347	280 (di)			
(±)-4-Cyano-3,4-diphenylbutanoic acid [(±)-4-cyano-3,4-diphenylbutyric acid; (±)-2,3-diphenylglutaromononitrile]	163	210	205 (mono)	202 (mono)		
2,3-Diphenyl-3-phenyl-2-propenenitrile (2,3-diphenylcinnamonitrile)	167	213	223			
5-Chloro-2-hydroxybenzonitrile (4-chloro-2-cyanophenol; 5-chloro-2-hydroxyphenyl cyanide)	167	172	226			
2-Cyano-2'-biphenylcarboxylic acid (2,2'-diphenic acid mononitrile; 2-carboxy-2'-cyanobiphenyl)	172	234	193 (mono) 212 (di)	183 (mono)		
5-Nitro-2-naphthonitrile (2-cyano-5-nitronaphthalene)	173	295	263			
2,3-Dicyanopyridine	176	229	169 (mono) 165 (di)			

Nitriles (Solids) (continued)

		Acid derivative mp (°C)			Amine derivative mp (°C)	
Name of compound	mp (°C)	Acid	Amide	Anilide	Benz-amide	Ben-zene-sulfon-amide
2-Cyano-3-phenyl-2-propanoic acid (2-cyanocinnamic acid)	183	196	190 (di)			
1,2-Dicyanonaphthalene	190	175	265 (di)			
2-Cyanobenzoic acid	192	206	219 (di)	253 (di)	184 (di)	
2-Hydroxy-5-nitrobenzonitrile (2-cyano-4-nitrophenol; 2-hydroxy-5-nitrophenyl cyanide)	196	230	225			
5-Nitro-1-naphthonitrile (1-cyano-5-nitronaphthalene)	205	242	236			
3-Cyanobenzoic acid	217	345	280			
4-Cyanobenzoic acid	219	300		337 (di)		
1,4-Dicyanobenzene (terephthalonitrile)	222	300		337 (di)		
1,8-Dicyanonaphthalene	232	260	252 (di)			
1-Cyano-9,10-anthraquinone	247	294	280	289		

Nitro Compounds (Liquids)

Name of compound	bp (°C)	Acyl derivative mp (°C) of amine		
		Acet-amide	Benz-amide	Benzene-sulfon-amide
Nitroethene (nitroethylene)	99		71	58
Nitromethane	101	28	80	30
Nitroethane	114		71	58
1-Nitropropane	132		84	36
2-Nitrobutane	140		76	70
2-Methyl-1-nitropropane	141		58	53
2-Methyl-1-nitropropene (1-nitroisobutene)	158		57	53
1-Nitrohexane	193		40	96
Nitrocyclohexane	206	104	147	
Nitrobenzene	211	114	160	112
2-Nitrotoluene	224	111	146	124
1-Ethyl-2-nitrobenzene	224	112	147	
1,3-Dimethyl-2-nitrobenzene	226	177	168	
Phenylnitromethane	226	60	105	88
3-Nitrotoluene	233	65	125	95
1,4-Dimethyl-2-nitrobenzene	234	139	140	138
6-Chloro-2-nitrotoluene	238	159; 136	173	
1-Ethyl-4-nitrobenzene	241	94	151	
2,4-Dimethyl-1-nitrobenzene	246	133	192	130
2,3-Dimethyl-1-nitrobenzene (1,2-dimethyl-3-nitrobenzene)	250	135	189	
2,4-Dichloro-1-nitrobenzene	258	146	115	128
1-Isopropyl-4-methyl-2-nitrobenzene	264	71	102	
2-Methoxy-1-nitrobenzene (*o*-nitroanisole)	265	88	84; 60	89
1-(2-Methylpropyl)-4-nitrobenzene (*o*-*tert*-butylnitrobenzene)	267	170	136	
2-Ethoxy-1-nitrobenzene (*o*-nitrophenetole)	268	79	104	102

Nitro Compounds (Solids)

Name of compound	mp (°C)	Acyl derivative mp (°C) of amine: Acet-amide	Benz-amide	Benzene-sulfon-amide
3-Nitrotoluene	16		125	95
4-Fluoro-1-nitrobenzene	27	152	185	
2-Chloro-1-nitrobenzene	32	87	99	129
2,4-Dichloro-1-nitrobenzene	33	146	115	128
3-Ethoxy-1-nitrobenzene (*m*-nitrophenetole)	34	97	103	
2-Nitrobiphenyl	37	121	102	
6-Chloro-2-nitrotoluene	37	159; 136	173	
3-Iodo-1-nitrobenzene	38	119	157	
5-Nitroindane	40	106	137	
2-Bromo-1-nitrobenzene	43	99	116	
4-Nitroindane	44	126	136	
2-Nitro-1,3,5-trimethylbenzene (nitromesitylene)	44	217	204	
3-Chloro-1-nitrobenzene	44	78	120	121
2-Iodo-1-nitrobenzene	49	109	139	
1-Chloro-2,4-dinitrobenzene	52	142 (di)	178 (di)	
4-Methoxy-1-nitrobenzene (*p*-nitroanisole)	54	130	154	95
4-Nitrotoluene	54	147	158	120
2,5-Dichloro-1-nitrobenzene	54	132	120	
3-Bromo-1-nitrobenzene	56	87	120; 136	
1-Methyl-2-nitronaphthalene	59	189	222	
4-Ethoxy-1-nitrobenzene (*p*-nitrophenetole)	60	137	173	143
1-Nitronaphthalene	61	159	160	167
3,4-Dinitrotoluene	61	132 (mono) 210 (di)	194 (mono) 264 (di)	179 (di)
1-Methyl-8-nitronaphthalene	64	184	196	
2,4-Dinitrotoluene	72	224 (di)	224 (di)	138 (mono) 191 (di)
1-Methyl-4-nitronaphthalene	72	167	239	
5-Methyl-2-nitronaphthalene	72	124	156	
2-Nitronaphthalene	78	132	162	136; 102
2-Methyl-1-nitronaphthalene	81	188	180	
4-Nitrophenanthrene	81	190	224	
5-Methyl-1-nitronaphthalene	83	195	174	
4-Chloro-1-nitrobenzene	83	179	192	122
1,3-Dinitrobenzene	90	89 (mono) 191 (di)	125 (mono) 240 (di)	194
1,5-Dimethyl-2,4-dinitrobenzene	93	165 (mono) 295 (di)	259 (di)	
2,4′-Dinitrobiphenyl	93	202 (di)	278 (di)	
4-Chloro-1-methoxy-2-nitrobenzene (4-chloro-2-nitroanisole)	98	104	78	
2-Nitrophenanthrene	99	225	216	

Nitro Compounds (Solids) (continued)

Name of compound	mp (°C)	Acyl derivative mp (°C) of amine: Acetamide	Benzamide	Benzenesulfonamide
1,2-Dinitronaphthalene	103	234 (di)	291 (di)	215 (mono)
5-Nitroacenaphthene	106	238	210; 199	
4-Nitrobiphenyl	114	171	230	
9-Nitrophenanthrene	117	208	199	
1,2-Dinitrobenzene	118	185 (di)	301 (di)	186
1-Nitro-2,4,6-tribromobenzene	125	232	198	
4-Bromo-1-nitrobenzene	126	168	204	134
2,2′-Dinitrobiphenyl	128	90 (mono) 161 (di)	160 (mono) 191 (di)	
1,4-Dinitronaphthalene	132	304 (di)	280 (di)	
3-Nitroacenaphthene	152	193	210	
1,6-Dinitronaphthalene	166	263 (di)	265 (di)	
3-Nitrophenanthrene	171	201	213	
1,4-Dinitrobenzene	173	163 (mono) 304 (di)	128 (mono) 300 (di)	247 (di)
4-Iodo-1-nitrobenzene	173	184	222	
9-Nitrofluorene	182	262	261	
2-Nitroanthraquinone	185	262	228	
1-Nitroanthraquinone	230	218	255	
2,7-Dinitronaphthalene	234	261 (di)	267 (di)	
4,4′-Dinitrobiphenyl	240	199 (mono) 317 (di)	205 (mono) 352 (di)	
1,6-Dinitroanthraquinone	257	295 (di)	275 (di)	
E-4,4′-Dinitro-1,2-diphenylethene (*trans*-4,4′-dinitrostilbene)	288	353 (di)	352 (di)	
1,7-Dinitroanthraquinone	295	283 (di)	325 (di)	
1,8-Dinitroanthraquinone	312	284 (di)	324 (di)	

Phenols (Liquids)

		Derivative mp (°C)							
Name of compound	bp (°C)	Phenyl-urethane	1-Naph-thyl-urethane	4-Nitro-ben-zoate	3,5-Di-nitro-ben-zoate	Acetate	Ben-zoate	Aryl-oxy-acetic acid	Bromo deriv-ative
2-Chlorophenol	176	121	120	115	143			145	49 (mono) 76 (di)
Phenol	183	126	133	127	146		69	99	95 (tri)
2-Methylphenol (*o*-cresol; *o*-hydroxytoluene)	192	143	142	94	138			152	56 (di)
2-Bromophenol	195		129					143	95 (tri)
2-Chloro-4-methylphenol (3-chloro-4-hydroxytoluene)	196							72	108
2-Hydroxybenzaldehyde (salicylaldehyde)	197	133		128		39		132	
4-Methylphenol (*p*-cresol; *p*-hydroxytoluene)	202	115	146	98	189		71	136	49 (di) 199 (tetra)
3-Methylphenol (*m*-cresol; *m*-hydroxytoluene)	203	125	128	90	165		55	103	84 (tri)
2-Methoxyphenol (guaiacol)	205	136	118	93	142		58	121	116 (tri)
2-Ethylphenol	207	144		57	108		39	141	
2,4-Dichlorophenol	210						96	140	140 (mono)
2,4-Dimethylphenol	212	112	135	105	165		38	142	179 (tri)
3-Chlorophenol	214		158	99	156		71	110	
2-Hydroxyacetophenone (*o*-acetylphenol)	215					89	88		
3-Ethylphenol	217	138		68			52	77	
4-Chlorophenol	217	149	166	171	186		89	156	34 (mono) 90 (di)
2-(2-Propenyl)phenol (2-allylphenol)	220	116						150	50 (mono)
2-Chloro-4,6-dimethylphenol	223	130		95					
Methyl 2-hydroxybenzoate (methyl salicylate)	224	117		128		52	92		
2-Propylphenol	226	111			96			100	
2-(1-Methylpropyl)phenol (2-*sec*-butylphenol)	228	86						110	
4-Propylphenol	232	129			123		38		
Ethyl 2-hydroxybenzoate (ethyl salicylate)	234	100		108			87		
3-Bromophenol	236		108				86	108	
2-Butylphenol	237				97	106		105	
2-Methyl-5-(1-methylethyl)-phenol (carvacrol; 5-isopropyl-2-methylphenol)	238	138	116	51	83			151	46 (mono)
2,4-Dibromophenol	239			184		36	98	153	95 (tri) 96 (mono)
3-Methoxyphenol (resorcinol monomethyl ether)	244		129					118	104 (tri)

Phenols (Liquids) (continued)

		Derivative mp (°C)							
Name of compound	bp (°C)	Phenyl-urethane	1-Naph-thyl-urethane	4-Nitro-ben-zoate	3,5-Di-nitro-ben-zoate	Acetate	Ben-zoate	Aryl-oxy-acetic acid	Bromo deriv-ative
4-Butylphenol	248	115		68	92		27; 127	81	
2-Methoxy-4-(2-propenyl)-phenol (eugenol; 4-allyl-2-methoxyphenol)	255	96	122	81	131	30	70	81 (hyd) 100	118 (tetra)
2-Methoxy-4-(1-propenyl) phenol (isoeugenol)	267	118 (*cis*) 152 (*trans*)	150	109	158	80	68 (*cis*) 106 (*trans*)	94; 116	94 (di)

Phenols (Solids)

		Derivative mp (°C)							
Name of compound	mp (°C)	Phenyl-urethane	1-Naph-thyl-urethane	4-Nitro-ben-zoate	3,5-Di-nitro-ben-zoate	Acetate	Ben-zoate	Aryl-oxy-acetic acid	Bromo deriv-ative
4-Propylphenol	22	129			123		38		
4-Butylphenol	22	115		68	92		27; 127	81	
4-Pentylphenol (*p*-amylphenol)	23						52	90	
3-Propylphenol	26				75; 118			97	
2,4-Dimethylphenol	28	112	135	105	165		38	142	179 (tri)
2-Hydroxyacetophenone (*o*-acetylphenol)	28					89	88		
2-Methylphenol (*o*-cresol; 2-hydroxytoluene)	31	142	142	94	138			152	56 (di)
2-Methoxyphenol (guaiacol)	32		118	93	142		58	119	116 (tri)
3-Bromophenol	33		108				86	108	
3-Chlorophenol	33		158	99	156		71	110	
2-Bromo-4-chlorophenol	34						100	140	
4-Methylphenol (*p*-cresol; *p*-hydroxytoluene)	36	115	146	98	189		71	136	108, 199 (tetra) 49 (di)
4-Methyl-2-nitrophenol (4-hydroxy-3-nitrotoluene)	36				192		101		
2,4-Diethylphenol	38	171						68	
2,4-Dibromophenol	40			184		36	98	153	95 (tri) 96 (mono)
3-Ethoxyphenol (resorcinol monoethyl ether)	40			184			97		
3-Iodophenol	40	138		133	183	38	73	115	
3-Methyl-2-nitrophenol (3-hydroxy-2-nitrotoluene)	41					59	79		
Phenol	42	126	133	127	146		69	99	95 (tri)
Phenyl 2-hydroxybenzoate (phenyl salicylate)	42	112		111		99	81		
2-Iodophenol	43	122				101	34	135	
4-Chlorophenol	43	149	166	171	186		93	156	34 (mono) 90 (di)
2,4-Dichlorophenol	45				143		97	141	68 (mono)
2-Nitrophenol	45		113	141	142; 155	41	59	158	117 (di)
4-Ethylphenol	47	120	128	81	133		60	97	
3-Methyl-2,4,6-trichlorophenol	47					32	53		
4-Chloro-2-methylphenol (5-chloro-2-hydroxytoluene)	48						71	117	
2,6-Dimethylphenol	49		176		159		39	140	79 (mono)
5-Methyl-2-(1-methylethyl)phenol (thymol; 2-isopropyl-5-methylphenol)	52	107	160	70	103		33	149	55 (mono)
2,6-Dibromo-4-methylphenol (3,5-dibromo-4-hydroxytoluene)	54			141		67	95		

Phenols (Solids) (continued)

		Derivative mp (°C)							
Name of compound	mp (°C)	Phenyl-urethane	1-Naph-thyl-urethane	4-Nitro-ben-zoate	3,5-Di-nitro-ben-zoate	Acetate	Ben-zoate	Aryl-oxy-acetic acid	Bromo deriv-ative
4-Methoxyphenol (*p*-hydroxy-anisole; hydroquinone monomethyl ether)	56				166	32	87	112	
3-Methyl-6-nitrophenol (3-hydroxy-4-nitrotoluene)	56					48	77		
2,6-Dibromophenol	57						46	68	93 (mono)
4,6-Dibromo-2-methylphenol (3,5-dibromo-2-hydroxy-toluene)	57			137			62		
3-Hydroxy-6-nitrobiphenyl	58			135	171				
3,5-Dihydroxytoluene (orcinol; 5-methyl-resorcinol)	58 (hyd)	154	160	214	190	25 (di)	88 (di)	217	104 (tri)
4-Chloro-5-methyl-2-(1-methylethyl)phenol (4-chloro-2-isopropyl-5-methylphenol)	60				129		72		
5,6,7,8-Tetrahydro-2-naphthol	62			113			96		
3,4-Dimethylphenol	65	120	142		182	22	59	163	171 (tri)
2-Methyl-1-naphthol	65					82	95		
3,4-Dihydroxytoluene (4-methylcatechol)	65	166 (di)					58 (di)	58 (di)	
4-Bromophenol	66	140	169	180	191	22	102; 58	157	95, 171 (tri)
4-Chloro-2-methylphenol (2-chloro-5-hydroxytoluene)	66		154				86		
4-Chloro-3-methylphenol (4-chloro-3-hydroxytoluene)	66		154				86	178	
4-Hydroxy-3-nitrobiphenyl	66					86	111		
2,4,6-Trichlorophenol	67		188				70		
4-Methyl-2,3,5-trichloro-phenol (4-hydroxy-2,3,6-trichlorotoluene)	67					38	89		
2-Phenylphenol (2-hydroxybiphenyl)	68					63	76	107	
3,5-Dichlorophenol	68					38	55		189 (tri)
3,5-Dimethylphenol	68	151		109	195		24	111; 86 (hyd)	166 (tri)
2,3,5-Trichlorophenol	68						93	157	
2,4,5-Trichlorophenol	68						93	157	
2,4,6-Trichlorophenol	70			106	136		76	186	
2,4,6-Trimethylphenol (mesitol)	70	142					62	142	158 (di)
2,3,4,6-Tetrachlorophenol	70					66	108		
2,4,5-Trimethylphenol (pseudodocumenol)	71	110			179	35	63	132	35 (mono)

Phenols (Solids) (continued)

Name of compound	mp (°C)	Derivative mp (°C)							
		Phenylurethane	1-Naphthylurethane	4-Nitrobenzoate	3,5-Dinitrobenzoate	Acetate	Benzoate	Aryloxyacetic acid	Bromo derivative
1-Chloro-2-naphthol	72					43	100		
2,4-Diiodophenol	72					71	98		
5-Chloro-2-methylphenol (4-chloro-2-hydroxytoluene)	74						54		190 (tri)
Ethyl 3-hydroxybenzoate	74					35	58		
2,5-Dimethylphenol	75	166	173	87	137		61	118	178 (tri)
5,6,7,8-Tetrahydro-1-naphthol	75					75	46		
2,3-Dimethylphenol	75	193		104				187	
8-Hydroxyquinoline	76			175		174	120		
4-Bromo-2,6-dinitrophenol	78					111	154		
4-Chloro-2-iodophenol	78	128				57	88		
2,3,4,6-Tetramethylphenol (isodurenol)	79	179					72		
4-Hydroxy-3-methoxybenzaldehyde (vanillin)	81	117				102	78	189	160 (mono)
3,5-Dibromophenol	81					53	77		
3-Methyl-2,4,6-tribromophenol (3-hydroxy-2,4,6-tribromotoluene)	82					68	85		
4-(1,1,3,3-Tetramethylbutyl)phenol	84						82	103	
3-(Dimethylamino)phenol (3-hydroxydimethylaniline)	85					37	94		
3-Hydroxy-2-nitrobiphenyl	86					62	131		
2-Hydroxyphenylmethanol (saligenin; 2-hydroxybenzyl alcohol)	87						51, 85 (di)	120	
4,6-Dinitro-2-methylphenol (3,5-dinitro-2-hydroxytoluene)	87					96	135		
4-(Methylamino)phenol (4-hydroxy-*N*-methylaniline)	87					43	174	214	
2-Methyl-3,4,6-tribromophenol (6-hydroxy-2,3,5-tribromotoluene; 2-hydroxy-3,5,6-tribromotoluene)	91					77	85; 133		
4-(1,1-Dimethylpropyl)phenol (*p-tert*-amylphenol; *p-tert*-pentylphenol)	93	108					61		
4-Iodophenol	94	148				32	119	156	
1-Naphthol	94	178	152	143	217	49	56	194	105 (di)
2,4,6-Tribromophenol (bromol)	95		153	153	174	87	81	200	120 (tetra)
3-Bromo-4-hydroxybiphenyl	95					75	94		
2,3,5-Trimethylphenol	95	174				241	50		
2-Naphthyl 2-hydroxybenzoate (2-naphthyl salicylate; betol)	96	268				136			

Phenols (Solids) (continued)

Name of compound	mp (°C)	Derivative mp (°C)							
		Phenyl-urethane	1-Naph-thyl-urethane	4-Nitro-ben-zoate	3,5-Di-nitro-ben-zoate	Acetate	Ben-zoate	Aryl-oxy-acetic acid	Bromo deriv-ative
3-Hydroxyacetophenone (*m*-acetylphenol)	96					45	53		
5-Iodo-2-nitrophenol	96					95	122		
3-Nitrophenol	97	129	167	174	159	56	95	156	91 (di)
2,3-Dibromo-5,6-dimethylphenol	97					78	153		
4-(1,1-Dimethylethyl)phenol (*p-tert*-butylphenol)	100	149	110				83	87	50 (mono) 67 (di)
2-Acetyl-1-naphthol	102					108	128		
3,5-Diiodophenol	104					79	93		
3-Hydroxy-4-nitrobiphenyl	105			157	199				
1,2-Dihydroxybenzene (catechol; pyrocatechol)	105	169 (di)	175	159 (mono) 170 (di)	152 (di)	58 (mono) 65 (di)	181, 84 (di)	138	193 (tetra)
2-Hydroxypyridine	107			120			42		
3,5-Dihydroxytoluene (orcinol; 5-methyl-resorcinol)	107 (anhyd)	154	160	214	190	25 (di)	88 (di)	217	104 (tri)
1,2-Dihydroxynaph-thalene (1,2-naphthalenediol)	108					109 (di)	106 (di)	106 (di)	
1,2,3,4-Tetrahydroanthranol	108					109	142		
4-Methyl-5,6,7,8-tetra-hydro-2-naphthol	108			116			89		
3-Hydroxybenzaldehyde	108	160					49; 38	148	
3 Methyl 2,4,6 trinitrophenol (3-hydroxy-2,4,6-trinitro-toluene)	110					135	140		
1,3-Dihydroxybenzene (resorcinol)	110	164 (di)	206	182 (di)	201 (di)	58	136 (mono) 117 (di)	175; 195	112 (di) 112 (tri)
2,2′-Dihydroxybiphenyl (2,2′-biphenol)	110	145 (di)				95 (di)	101 (di)		188 (di)
2-Bromo-1,4-dihydroxy-benzene (bromohydroquinone)	110					72 (di)			186 (di)
4-Hydroxyacetophenone	110				138	54	134	177	
1-Methyl-2-naphthol	111					66	117		
1,3-Dihydroxy-2,4,6-tribromobenzene (2,4,6-tribromoresorcinol)	112					114 (mono) 108 (di)	120 (mono)		
4′,5-Dimethyl-2-hydroxyazo-benzene	113					91	95		
4,4′-Dihydroxy-2,2′-dimethylbiphenyl	114					75 (di)	127 (di)		
2-Bromo-4-nitrophenol	114					62; 86	132		
4-Nitrophenol	114	156	151	159	188	83	143	187	142 (di)
2,4-Dinitrophenol	114			139		72	132	148	118 (mono)
4-Hydroxy-3-methoxy-benzyl alcohol	115					51 (mono) 48 (di)	90 (mono) 121 (di)		

Phenols (Solids) (continued)

Name of compound	mp (°C)	Derivative mp (°C): Phenylurethane	1-Naphthylurethane	4-Nitrobenzoate	3,5-Dinitrobenzoate	Acetate	Benzoate	Aryloxyacetic acid	Bromo derivative
1-Hydroxyfluorene	115					131	129		
4-Chloro-3,5-dimethylphenol	115						68	141	
3,5-Dimethyl-2,4-dinitrophenol	116					148	156		
4-Hydroxy-2′-nitrobiphenyl	116					122	157	161	
4-Hydroxybenzaldehyde	117						91	198	181 (di)
1,3,5-Trihydroxybenzene (phloroglucinol)	117 (hyd)	191 (tri)		283	162 (tri)	104 (tri)	185 (tri) 174		151 (tri)
2,4′-Dihydroxydiphenylmethane	118					70 (di)	108 (di)		
2-Bromo-4,6-dinitrophenol	119					105	94		
3,4-Dihydroxy-acetophenone (4-acetylcatechol)	119					58 (mono) 91 (di)	118 (di)		
2-Chloro-5-nitrophenol	120					82	128		
9-Hydroxyanthracene	120					131	288		
2-Chloro-3-nitrophenol	121					51	94		
3-Aminophenol (*m*-hydroxyaniline)	122			143	179		153 (di)		
1,3-Dihydroxy-4-nitrobenzene (4-nitroresorcinol)	122					91 (di)	124, 189 (mono) 110 (di)		
2,4,6-Trinitrophenol (picric acid)	122			143		76			
2-Naphthol	123	156	157	169	210	72	107	95; 154	84 (mono)
2,3,4,5-Tetrabromophenol	123					111	133		226
3,4-Dimethyl-1-naphthol	123				224	91			
3,3′-Dihydroxybiphenyl (3,3′-biphenol)	124					83 (di)	92 (di)		
3-Iodo-4-nitrophenol	124					77	119		
2,5-Dihydroxytoluene (2-methylhydroquinone)	125					92 (mono) 49 (di)	120 (di)	153	84
(4-Hydroxyphenyl)methanol (4-hydroxybenzyl alcohol)	125					84 (mono) 75 (di)	89		
4-(Phenylmethyl)-1-naphthol (4-benzyl-1-naphthol)	126					88	103		
Pentamethylphenol	126	215				273	127		
4-Chloro-3-nitrophenol	127					84	97		
1,3-Dibromo-2,4-dihydroxynaphthalene (1,3-dibromo-2,4-naphthalenediol)	129					148 (mono) 125 (di)			186
3-Methyl-4-nitrophenol (5-hydroxy-2-nitrotoluene)	129					34	74		
4-Hydroxy-3-nitroazobenzene	129					121	132		
Methyl 4-hydroxybenzoate	131	135				85	135		
4-Cyclohexylphenol	132	146		137	168	35	119		

Phenols (Solids) (continued)

		Derivative mp (°C)							
Name of compound	mp (°C)	Phenyl-urethane	1-Naph-thyl-urethane	4-Nitro-ben-zoate	3,5-Di-nitro-ben-zoate	Acetate	Ben-zoate	Aryl-oxy-acetic acid	Bromo deriv-ative
1,2,3-Trihydroxybenzene (pyrogallol)	133	173 (tri)		230 (tri)	205 (tri)	173 (tri); 111 (di)	90 (tri); 108, 126 (di) 140 (mono)	198	158 (tri); 158 (di)
4,6-Dibromo-2-naphthol	135					128	129		
2,3-Dichloro-1,4-di-hydroxynaphthalene)	135; 156					240 (di)	252 (di)		
4-Hydroxybenzophenone (*p*-benzoylphenol)	135					81	115; 95		
1,4-Dihydroxy-2,6-dinitrobenzene (2,6-dinitrohydroquinone)	136					96 (mono) 136 (di)	151 (mono)		
1,4-Dimethyl-2-naphthol	136					78	125		
5-Iodo-3-nitrophenol	136					110	101		
1,6-Dihydroxynaphthalene (1,6-naphthalenediol)	138					73 (di)	104 (di)		
2,6-Dibromo-3,4,5-tri-hydroxybenzoic acid (Dibromogallic acid)	139					168 (di)	96 (di)		
2-Hydroxybenzamide (salicylamide)	139				224	138	143		
2-Hydroxy-2′-nitrobiphenyl	140			116	180		116		149 (di)
1,8-Dihydroxynaphthalene (1,8-naphthalenediol)	142		220			155 (di)	175 (di)		
4-Hydroxy-2-nitrobiphenyl	143					169	106		
3-Chloro-1-naphthol	143					69	119		
5-Methyl-2-(1-methylethyl)-1,4-dihydroxybenzene (thymoquinol; 2-isopropyl-5-methylhydroquinone)	143	233	148			75 (di)	142 (di)		
1,2,4-Trihydroxybenzene (hydroxyhydroquinone)	145					97 (tri)	120 (tri)		
1,3-Diphenyl-2,4′-dihydroxy-2-propen-1-one (2,4′-dihydroxychalcone)	145					95 (di)	120 (di)		
2,4-Dihydroxyacetophenone (4-acetylresorcinol; resacetophenone)	147					120, 74 (mono) 38 (di)	67, 107 (mono)		
3-Chloro-5-nitrophenol	147					84	78		
2-Chloro-4-hydroxy-benzaldehyde	148					52	97		
1,3-Dihydroxy-2,4-dinitrobenzene (2,4-dinitroresorcinol)	148					120	184	155	
4-Hydroxypropiophenone (*p*-propionylphenol)	148					62	108		

Phenols (Solids) (continued)

		Derivative mp (°C)							
Name of compound	mp (°C)	Phenyl-urethane	1-Naph-thyl-urethane	4-Nitro-ben-zoate	3,5-Di-nitro-ben-zoate	Acetate	Ben-zoate	Aryl-oxy-acetic acid	Bromo deriv-ative
9,10-Dihydroxyphenanthrene (9,10-phenanthrenediol)	148					170 (mono) 202 (di)	217 (di)		
4-Hydroxypyridine	149					150	81		
4-Hydroxyazobenzene	152					89	138		
2-Hydroxy-5-methyl-benzoic acid (5-methylsalicylic acid)	153					152	155	185	
6-Hydroxybiphenyl-2-carboxylic acid	154					89	121; 150		
2-Azophenol (2,2′-dihydroxy-azoxybenzene)	155					150 (di)	108 (di)		
E-3,5-Dihydroxy-1,2-diphenylethene (3,5-dihydroxystilbene; 3,5-stilbenediol)	156					101 (di)	151 (di		
9-Hydroxyphenanthrene (9-phenanthrol)	158					78	100		
4,4′-Dihydroxydiphenyl-methane	158					70 (di)	156 (di)		
2-Hydroxybenzoic acid (salicylic acid)	159			205		135	132	191	
2,4,6-Triiodophenol	159					181	156	137	
1,3-Diphenyl-2,2′-dihydroxy-2-propen-1-one (2,2′-dihydroxychalcone)	161						86 (di)	114 (di)	
4,4′-Dihydroxy-3,3′-dimethylbiphenyl	161						136 (di)	185	
3-Hydroxy-1,2-benzene-dicarboxylic acid (3-hydroxyphthalic acid)	161						116	148	
1,2,3,4-Tetrahydroxybenzene (apionol; phenetrol)	161						142 (tctra)	192 (tetra)	
2,3-Dihydroxynaphthalene	162						105	152	
2-Hydroxy-3-methyl-benzoic acid (3-methylsalicylic acid)	163						113		204
2,2′-Dihydroxy-6,6′-dimethylbiphenyl	164						87 (di)	136 (di)	
2-Methoxy-4-nitrophenol	165					158	188		
4-Hydroxybiphenyl (*p*-phenylphenol)	165	168				89	151	190	
2-Hydroxyphenanthrene (2-phenanthrol)	168					143	140		
1,5-Di-(2-hydroxyphenyl)-1,4-pentadiene-3-one	168					128 (di)	135		
3,3′-Dihydroxy-benzophenone	170					90	102		

Phenols (Solids) (continued)

		Derivative mp (°C)							
Name of compound	mp (°C)	Phenyl-urethane	1-Naph-thyl-urethane	4-Nitro-ben-zoate	3,5-Di-nitro-ben-zoate	Acetate	Ben-zoate	Aryl-oxy-acetic acid	Bromo deriv-ative
1,4-Dihydroxy-2-methyl-naphthalene (2-methyl-1,4-naphthohydroquinone)	170					113 (di)	181		
5-Nitro-1-naphthol	171					114	109		
1,4-Dihydroxybenzene (hydroquinone)	172	207, 224 (di)	247	258 (di)	317 (di)	124 (di)	204 (di)	250	186 (di)
2,3,4-Trihydroxy-acetophenone (gallacetophenone)	173					85 (tri)	118		
1,4-Dihydroxynaphthalene (1,4-naphthalenediol; 1,4-naphthohydroquinone)	176		220			130 (di)	169 (di)		
1,2-Dihydroxy-4-nitro-benzene (4-nitrocatechol)	176					98 (di)	156 (di)		
1,7-Dihydroxynaphthalene (1,7-naphthalenediol)	178	204		183		108 (di)	102 (di)		
9,10-Dihydroxyanthracene (9,10-anthradiol)	180					260 (di)	292 (di)		
4-Hydroxy-3-iodo-5-methoxybenzaldehyde (5-iodovanillin)	180					106	136		
3,3′-Dihydroxyazoxy-benzene (3-azoxyphenol)	183					102 (di)	75 (di)		
N-Benzylidene-4-aminophenol	183					92	144		
4-Aminophenol (*p*-hydroxyaniline)	184				179	168	234		
4,6-Diacetyl-1,3-dihydroxybenzene (4,6-diacetylresorcinol; resodiacetophenone)	185					120 (di)	215 (mono) 118 (di)		
7-Hydroxy-4-methylcoumarin	186	156		143		150	160		
α,2,4-Trihydroxy-acetophenone (2, 4-dihydroxyphenyl hydroxymethyl ketone)	189					129 (tri)	200 (mono)		
2,7-Dihydroxynaphthalene (2,7-naphthalenediol)	190					172 (mono) 136 (di)	199 (mono) 139 (di)	149	
Pentachlorophenol	190					150	165	196	
1,5-Diacetyl-2,3,4-trihydroxybenzene	191					109 (di)	189 (tri)		
1,2-Dihydroxynaphthalene (1,2-naphthalenediol)	192					130 (di)	169 (di)		
1,2-Dihydroxy-3,4,5,6-tetrabromobenzene (tetrabromocatechol)	193					216 (di)	198 (di)		
6-Hydroxyquinoline	193						38	231	

Phenols (Solids) (continued)

Name of compound	mp (°C)	Derivative mp (°C) Phenyl-urethane	1-Naph-thyl-urethane	4-Nitro-ben-zoate	3,5-Di-nitro-ben-zoate	Acetate	Ben-zoate	Aryl-oxy-acetic acid	Bromo deriv-ative
3-Methyl-2,4,5,6-tetrabromophenol	194						166	154	
1,5-Dichloro-4,8-di-hydroxynaphthalene	194						160 (mono) 154 (di)	158 (mono) 179 (di)	
Hexahydroxybenzene	200						203 (hexa)	313 (hexa)	
3-Hydroxybenzoic acid	200						131		206
Methyl 3,4,5-trihydroxy-benzoate (methyl gallate)	201						120 (tri)	139 (tri)	
2,5-Dihydroxyacetophenone (2-acetylhydroquinone)	202						68 (di)	113 (di)	
4-Hydroxy-4′-nitrobiphenyl	203						139	210	
3,5-Dimethoxy-4-hydroxy-benzoic acid	205						191	232	
2,3,5-Trihydroxy-acetophenone	207						107 (tri)	107 (tri)	
3,3′-Dihydroxyazobenzene (3-azophenol)	207						144 (di)	188 (di)	
2-Acetamidophenol (2-hydroxyacetanilide)	209						122	140	
4-Hydroxy-3-methoxy benzoic acid (vanillic acid)	210				141		110	178	
2,4-Dihydroxybenzoic acid	213						136 (di)	152	
4-Hydroxybenzoic acid	215						187	223	278
1,3-Dihydroxy-4,6-dinitrobenzene (4,6-dinitroresorcinol)	215				178 (di)		139 (di)	344 (di)	
4,4′-Dihydroxyazobenzene (4-azophenol)	216						119 (di)	212; 251 (di)	
2,5-Dimethyl-1,4-dihydroxy-benzene (hydrophlorone; 2,5-dimethylhydroquinone)	217					117 (mono) 135 (di)	163 (mono) 159 (di)		
1,3,5-Trihydroxybenzene (phloroglucinol)	218 (anhyd)	191		283	162 (tri)	105 (tri)	185, 174 (tri) 126 (di) 196 (mono)	151 (tri)	
2,2′-Dihydroxy-1,1′-binaphthyl	218					109 (di)	204 (mono) 160 (di)		
2,6-Dihydroxynaphthalene (2,6-naphthalenediol)	218					175 (di)	215 (di)		
4-Hydroxy-4′-nitro-azobenzene	220					147	195		
3-Hydroxy-2-naphthoic acid	222						184	204	
2,4,6-Trihydroxy-acetophenone (acetylphloroglucinol)	222						103 (tri)	168, 211 (mono) 118 (tri)	

Phenols (Solids) (continued)

		Derivative mp (°C)							
Name of compound	mp (°C)	Phenyl-urethane	1-Naph-thyl-urethane	4-Nitro-ben-zoate	3,5-Di-nitro-ben-zoate	Acetate	Ben-zoate	Aryl-oxy-acetic acid	Bromo deriv-ative
4,4′-Dihydroxyazoxybenzene	224						163 (di)	200, 212 (mono) 190 (di)	
4-Hydroxybenzanilide (4-benzamidophenol)	227						171	235	
2,6-Dihydroxyphenanthrene (2,6-phenanthrenediol)	234						123 (di)	253 (di)	
5-Hydroxy-1-naphthoic acid	235						202	241	
3,5-Dihydroxybenzoic acid	236						160 (di)	227	
1,4-Dihydroxy-2,3,5,6-tetrachlorobenzene (tetrachlorohydroquinone)	237					245 (di)	233 (di)		
3-Hydroxynaphthoic acid	248					170	223		
2,3-Dihydroxyquinoline (2,3-quinolinediol)	258					211 (mono)	287 (mono) 46 (di)		
1,5-Dihydroxynaphthalene (1,5-naphthalenediol)	265					161 (di)	242 (di)		
Phenolphthalein	265	135 (di)				143 (di)	169 (di)		
4,4′-Dihydroxyphenol (4,4′-biphenol)	275					161 (di)	241 (di)	274	

Sulfonamides

Name of compound	mp (°C)	Derivative mp (°C)			
		Sulfonic acid	Sulfonyl chloride	Sulfon-anilide	*N*-Xan-thylsul-fonamide
2,4,5-Trimethoxybenzenesulfonamide	76		130	170	
Methanesulfonamide	90	20		100	
2,6-Dimethylbenzenesulfonamide	96	98	39		
Hexadecane-1-sulfonamide (cetylsulfonamide)	97	54	54		
Benzylsulfonamide	105		93	102	188
4-Methylbenzenesulfonamide (*p*-toluenesulfonamide)	105 (dihyd)	105; 92	71	103	197
Pyridine-3-sulfonamide	111	357	144	145	
3-Methoxynaphthalene-2-sulfonamide	113		138	174	
2,6-Dimethylbenzenesulfonamide	113	98	39		
2-Phenylethane-1-sulfonamide	122	91	33	77	
4-Chloro-3-methylbenzenesulfonamide	128		65	92	
D-Camphor-10-sulfonamide	132	193	67	121; 88	
2-Fluoronaphthalene-6-sulfonamide	133	105 (hyd)	97	129	
3-Chloro-4-methylbenzenesulfonamide	134		38	96	
(±)-Camphor-8-sulfonamide	135	58	106		
Tetralin-6-sulfonamide	135		58	156	
3,5-Dimethylbenzenesulfonamide	135		94	129	
Indane-5-sulfonamide	136	92	47	129	
4-Methylbenzenesulfonamide (*p*-toluenesulfonamide)	139 (anhyd)	105; 92	71	103	197
2,4-Dimethylbenzenesulfonamide	139	62 (hyd)	34	110	188
7-Ethoxynaphthalene-2-sulfonamide	142		103	153	
2,4,6-Trimethylbenzenesulfonamide	142	78	57	109	203
D-Camphor-3-sulfonamide	143	77	88	124	
3,4-Dimethylbenzenesulfonamide	144	64	52		
4-Chlorobenzenesulfonamide	144	93; 69	53	104	
4-Methyl-3-nitrobenzenesulfonamide	145	92	36		
3-Bromocamphor-8-sulfonamide	145	196 (anhyd)	137		
2,5-Dimethylbenzenesulfonamide	148	48 (anhyd) 86 (hyd)	26		176
Naphthalene-1-sulfonamide	150	90	68	112; 152	
4,6-Dichloro-2,5-dimethylbenzenesulfonamide	150		81	175	
6-Methoxynaphthalene-1-sulfonamide	150		81	178	
8-Chloro-7-methoxynaphthalene-1-sulfonamide	153		137	196	
2-Carboxybenzenesulfonamide	154	69 (hyd) 134 (anhyd)	79; 40	195	
6-Ethoxynaphthalene-1-sulfonamide	154		118	195	
Benzenesulfonamide	156	66 (anhyd) 44 (monohyd)		112	206
2-Chloro-5-methylbenzenesulfonamide	156	56		230	
2-Methylbenzenesulfonamide (*o*-toluenesulfonamide)	156	57	68	136	183
3-Bromocamphor-10-sulfonamide	157	48	65		
2,4-Dinitrobenzenesulfonamide	157	108 (hyd) 130 (anhyd)	102		
2-Methyl-4-nitrobenzenesulfonamide	157	130 (anhyd)	106		
Benzophenone-3,3′-disulfonamide	157		138 (di)	178 (di)	197
4-Methoxynaphthalene-2-sulfonamide	157		75	145	
2-Ethoxynaphthalene-1-sulfonamide	158		116	187	

Sulfonamides (continued)

Name of compound	mp (°C)	Derivative mp (°C) Sulfonic acid	Sulfonyl chloride	Sulfon-anilide	*N*-Xan-thylsul-fonamide
2-Methoxynaphthalene-1-sulfonamide	159		121	197	
3-Nitrobenzylsulfonamide	159	74 (hyd)	100		
2-Nitrodiphenylamine-4-sulfonamide	162	220	157		
2-Ethoxybenzenesulfonamide	163		66	158	
4-Chloro-2-nitrobenzenesulfonamide	164	82	75	138	
3,6-Dichloro-2,5-dimethylbenzenesulfonamide	165		71	171	
4-Bromobenzenesulfonamide	166	103; 89	76	119	
3-Nitrobenzenesulfonamide	167	48	64	126	
3,4-Dimethoxybenzenesulfonamide	167	192	71		
4-Hydroxynaphthalene-1-sulfonamide	167	170		200	
Propane-1,3-disulfonamide	169	92 (di)	45 (di)	129 (di)	
4-Ethoxynaphthalene-1-sulfonamide	170		103	180	
3-Carboxybenzenesulfonamide	170	98 (hyd) 148 (anhyd)	20 (di)		
4-Methyl-2-nitrobenzenesulfonamide	170	141 (anhyd)	99		
2,4-Dimethyl-3-nitrobenzenesulfonamide	172	144 (anhyd)	96		
2,5-Dimethyl-3-nitrobenzenesulfonamide	173	128; 200	61	144	
4-Nitrodiphenylamine-2-sulfonamide	173		104	164	
3,4-Dibromobenzenesulfonamide	175	68 (anhyd)	34		
7-Chloronaphthalene-2-sulfonamide	176	118 (anhyd) 68 (tetrahyd)	87		
4-Methylnaphthalene-1-sulfonamide	177		81	158	
4-Nitrobenzenesulfonamide	180	95, 111	80	136, 171	
3-Chloro-2-methylbenzenesulfonamide	180	72	72		
2,4,5-Trimethylbenzenesulfonamide	181	112	62		
5-Chloro-4-methyl-2-nitrobenzenesulfonamide	181	128	99		
2,5-Dichlorobenzenesulfonamide	181	93	38	160	
2,4-Dichlorobenzenesulfonamide	182	86	55		
4-Iodobenzenesulfonamide	183		85	143	
4-Ethoxynaphthalene-2-sulfonamide	183		85	144	
5-Ethoxynaphthalene-1-sulfonamide	183		121	130	
6-Ethoxynaphthalene-2-sulfonamide	183		108	153	
Quinoline-8-sulfonamide	184	312	124		
5-Nitronaphthalene-2-sulfonamide	184	119	125		
4-Chloro-2,5-dimethylbenzenesulfonamide	185	100	50	155	
2-Carboxy-5-methylbenzenesulfonamide	185	190, 158 (anhyd)	59 (di)		
2,5-Dichlorobenzenesulfonamide	186	97	38	160	
2-Chloro-5-nitrobenzenesulfonamide	186	169 (hyd)	90		
2-Methyl-5-nitrobenzenesulfonamide	186	134 (dihyd)	47	148	
6-Chloro-3-nitrobenzenesulfonamide	186	169	90		
2,4-Dimethyl-5-nitrobenzenesulfonamide	187	132; 122	98		
2-Methyl-5-nitrobenzenesulfonamide	187	131	47	148	
4-Chloronaphthalene-1-sulfonamide	187	133	95	146	
8-Iodonaphthalene-1-sulfonamide	187		115	140	
2,4-Diaminobenzene-1,5-disulfonamide	187		275	236	
5-Methylnaphthalene-2-sulfonamide	189		122	250; 134	
6-Methoxynaphthalene-2-sulfonamide	189		93	120	

Sulfonamides (continued)

Name of compound	mp (°C)	Derivative mp (°C) Sulfonic acid	Sulfonyl chloride	Sulfon-anilide	N-Xan-thylsul-fonamide
Phenanthrene-3-sulfonamide	190	176 (anhyd) 121 (monohyd) 89 (dihyd)	111		
2,4-Dibromobenzenesulfonamide	190	110 (anhyd)	79		
2-Nitrobenzenesulfonamide	191	70	68	115	
8-Nitronaphthalene-1-sulfonamide	191	115 (trihyd)	165	179	
4-Methylbenzene-2,4-disulfonamide	191		56	189	
2,5-Dimethyl-6-nitrobenzenesulfonamide	192	145 (anhyd)	110	182	
4-Carboxy-3-nitrobenzenesulfonamide	192 (mono)	111 (di)	160 (di)		
2-Nitrobenzenesulfonamide	193	70; 85	69	115	
Phenanthrene-9-sulfonamide	194	174 (anhyd) 134 (hyd)	127		
2-Carboxybenzenesulfonamide	194	69 (hyd) 134 (anhyd)	79; 40	195	
2,5-Dibromobenzenesulfonamide	195	128 (anhyd)	71		
5-Methoxynaphthalene-1-sulfonamide	195		120	157	
7-Methylnaphthalene-1-sulfonamide	196		96	164	
5-Fluoronaphthalene-1-sulfonamide	197	106 (hyd)	123		
2,5-Dimethyl-4-nitrobenzenesulfonamide	198	140	75	131	
Acenaphthene-3-sulfonamide	199	89	114	286	
3,4-Dicarboxybenzenesulfonamide	200	140 (monohyd)	170		
3,4-Dichloro-2,5-dimethylbenzenesulfonamide	201		62	157	
5-Amino-2-hydroxybenzenesulfonamide	202	100 (anhyd)		159	
4-Nitrobenzylsulfonamide	204	71	90	220	
4-Iodonaphthalene-1-sulfonamide	206		124	136	
4-Fluoronaphthalene-1-sulfonamide	206	100 (hyd)	86	144	
2,6-Dimethyl-4-hydroxybenzene-1,3-disulfonamide	208		119	207	
Fluorene-2-sulfonamide	213	155 (hyd)	164		
1-Nitronaphthalene-2-sulfonamide	214	105	121	202	
5-Methylbenzene-1,3-disulfonamide	216		94	153	
Naphthalene-2-sulfonamide	217	91	79	132	
4-Acetamidobenzenesulfonamide	219		149	214	
7-Methoxynaphthalene-2-sulfonamide	220		83	121	
Acenaphthene-5-sulfonamide	223		111	178	
8-Nitronaphthalene-2-sulfonamide	223	136 (monohyd)	169	173	
3-Chlorobenzene-1,5-disulfonamide	224	100	106		
2-Methylbenzene-1,4-disulfonamide	224		98	178 (di)	
4-Carboxy-3-nitrobenzenesulfonamide	226 (di)	111 (di)	160 (di)		
4-Methoxynaphthalene-1-sulfonamide	226		99	148	
5-Chloronaphthalene-1-sulfonamide	226		95	138	
3,4-Diiodobenzenesulfonamide	227	125	82		
8-Nitronaphthalene-2-sulfonamide	228	136 (anhyd)	169	173	
2,4,6-Tribromobenzenesulfonamide	228	64	64	222	
Benzene-1,3-disulfonamide	229		63	150	170
Biphenyl-4-sulfonamide	230		115	125	
2,4-Diiodobenzenesulfonamide	230	162 (anhyd)	78		
2-Carboxybenzenesulfonamide	230	69 (hyd) 134 (anhyd)	79; 40	195	199

Sulfonamides (continued)

Name of compound	mp (°C)	Derivative mp (°C) Sulfonic acid	Sulfonyl chloride	Sulfon-anilide	*N*-Xan-thylsul-fonamide
25-Nitronaphthalene-1-sulfonamide	236		113	123	
4-Carboxybenzenesulfonamide	236 (di)	94 (hyd) 260 (anhyd)	57 (di)	252 (di)	
6-Hydroxynaphthalene-2-sulfonamide	237	167 (anhyd) 129 (hyd)		161	
4-Chloro-2-nitrobenzenesulfonamide	237	82	75	138	
4-Methylbenzene-1,2-disulfonamide	239		111	190	
4,5-Dimethylbenzene-1,3-disulfonamide	239		79	200	
4-Hydroxybenzene-1,3-disulfonamide	239	100 (di)	89 (di)	205 (di)	
4-Methoxybenzene-1,3-disulfonamide	240		86	209	
2,4,6-Trimethylbenzene-1,3-disulfonamide	244		125 (di)	151 (di)	
4,6-Dimethylbenzene-1,3-disulfonamide	249		130	196	
1-Chloronaphthalene-2-sulfonamide	250	133 (anhyd)	85	172	
Azobenzene-4,4′-disulfonamide	250	169 (anhyd)	222; 170 (di)		
Phenanthrene-2-sulfonamide	254	150	156	158	
Benzene-1,2-disulfonamide	254		143	241	
2-Amino-5-methylbenzene-1,3-disulfonamide	257	290 (di)	156 (di)	197 (di)	
2-Methylbenzene-1,3-disulfonamide	260		88	162	
Anthracene-2-sulfonamide	261		122	201	
Anthraquinone-2-sulfonamide	261		197	193	
Azoxybenzene-3,3′-disulfonamide	273	126	138		
Naphthalene-1,4-disulfonamide	273		166	179	
9,10-Dichloroanthracene-2-sulfonamide	279		221	248	
Benzene-1,4-disulfonamide	288		139	249	
2,5-Dimethyl-1,3-disulfonamide	295		81	174	
Naphthalene-1,6-disulfonamide	298	125 (anhyd)	129		
Biphenyl-4,4′-disulfonamide	300	72	203		
Naphthalene-1,5-disulfonamide	310	245 (di) (anhyd)	183 (di)	249 (di)	
2,5-Dimethyl-1,4-disulfonamide	310		164	223	
Benzene-1,3,5-trisulfonamide	315	100 (tri)	187 (tri)	237 (tri)	
Anthracene-1,5-disulfonamide	330		249 (di)	293 (di)	
Anthracene-1,8-disulfonamide	333		225	224	
Anthraquinone-1,8-disulfonamide	340	294	223	238	
Anthraquinone-1,5-disulfonamide	350	311 (anhyd)	270	270	

Sulfonic Acids

Name of compound	mp (°C)	Derivative mp (°C)				
		Sulfonyl chloride	Sulfon-amide	Sulfon-anilide	Benzyl thiuro-nium salt	4-Tolu-idine salt
Methanesulfonic acid	20		90	100		
Benzenesulfonic acid	44 (hyd)		156	112	148	205
3-Bromocamphor-10-sulfonic acid	48	65	157			
3-Nitrobenzenesulfonic acid	48	64	167	126	146	222
2,5-Dimethylbenzenesulfonic acid	48 (anhyd)	26	148		184	
Hexanedecane-1-sulfonic acid (cetylsulfonic acid)	54	54	97			
2-Methylbenzenesulfonic acid (2-toluenesulfonic acid)	57	68	156	136	170	204
(±)-Camphor-8-sulfonic acid	58	106	135			
Diphenylmethane-4,4′-disulfonic acid	59	124		178		
2,4-Dimethylbenzenesulfonic acid	62 (hyd)	34	139	110	146	
3,4-Dimethylbenzenesulfonic acid	64	52	144		208	
2,4,6-Tribromobenzenesulfonic acid	64	64	228	222		
Benzenesulfonic acid	66 (anhyd)		156	112	148	205
4-Chlorobenzenesulfonic acid	68	53	144	104	175	210
3,4-Dibromobenzenesulfonic acid	68	34	175			
7-Chloronaphthalene-2-sulfonic acid	68 (tetrahyd)	87	176			
2-Carboxybenzenesulfonic acid (*o*-sulfobenzoic acid)	69 (hyd)	79; 40	194; 154; 230	195	206	200
4-Nitrobenzylsulfonic acid	71	90	204	220		
3-Chloro-2-methylbenzenesulfonic acid	72	72	180			
Biphenyl-4,4′-disulfonic acid	72	203	300		171	330
3-Nitrobenzylsulfonic acid	74 (hyd)	100	159			
D-Camphor-3-sulfonic acid	77	88	143	124		197
2,4,6-Trimethylbenzenesulfonic acid	78	56	142	109		
4-Chloro-2-nitrobenzensulfonic acid	82	75	164; 237	138		
2-Nitrobenzenesulfonic acid	85	69	193	115		
2,4-Dichlorobenzenesulfonic acid	86	55	182			206
2,5-Dimethylbenzenesulfonic acid	86 (hyd)	26	148		184	
Phenanthrene-3-sulfonic acid	89 (dihyd)	111	190			222
Acenaphthene-3-sulfonic acid	89	114	199	286		
4-Bromobenzenesulfonic acid	90	76	166	119	170	216
Naphthalene-1-sulfonic acid	90	68	150	112; 152	137	181
2-Phenylethane-1-sulfonic acid	91	33	122	77		
Naphthalene-2-sulfonic acid	91	79	217	132	191	221
Indane-5-sulfonic acid	92	47	136	129		
4-Methyl-3-nitrobenzenesulfonic acid	92	36	144			131
4-Methylbenzenesulfonic acid (*p*-toluenesulfonic acid)	92	71	139 (anhyd) 105 (dihyd)	103	182	198
Propane-1,3-disulfonic acid	92d	45 (di)	169 (di)	129 (di)		
4-Chlorobenzenesulfonic acid	93	53	144	104	175	210
4-Carboxybenzenesulfonic acid (*p*-sulfobenzoic acid)	94 (hyd)	57 (di)	237 (di)	252 (di)	213	
4-Nitrobenzenesulfonic acid	95	80	180	171		180
2,5-Dichlorobenzenesulfonic acid	97	38	186	160	170	248
2,6-Dimethylbenzenesulfonic acid	98	39	113; 96			
3-Carboxybenzenesulfonic acid	98 (hyd)	20	170 (di)		163	226

Sulfonic Acids (continued)

Name of compound	mp (°C)	Derivative mp (°C) Sulfonyl chloride	Sulfon-amide	Sulfon-anilide	Benzyl thiuro-nium salt	4-Tolu-idine salt
4-Chloro-2,5-dimethylbenzenesulfonic acid	100	50	185	155		
5-Amino-2-hydroxybenzenesulfonic acid	100 (anhyd)		202	159		
4-Fluoronaphthalene-1-sulfonic acid	100	86	206	144		
3-Chlorobenzene-1,5-disulfonic acid	100d	106	224			
Benzene-1,3,5-trisulfonic acid	100d	187	312	237		
4-Bromobenzenesulfonic acid	103	76	166	119	170	216
Ethanedisulfonic acid	104	95 (di)		69	202	270
2-Fluoronaphthalene-6-sulfonic acid	105 (hyd)	97	133	129	191	221
4-Methylbenzenesulfonic acid (*p*-toluenesulfonic acid)	105	71	139 (anhyd) 105 (dihyd)	103	182	198
1-Nitronaphthalene-2-sulfonic acid	105	121	214	202		
5-Fluoronaphthalene-1-sulfonic acid	106 (hyd)	123	197			
2,4-Dinitrobenzenesulfonic acid	108 (hyd)	102	157			
2,4-Dibromobenzenesulfonic acid	110 (anhyd)	79	190			
4-Nitrobenzenesulfonic acid	111	80	180	136; 171		180
4-Carboxy-3-nitrobenzenesulfonic acid (2-nitro-4-sulfobenzoic acid)	111	160 (di)	226 (di) 192 (mono)			
2,4,5-Trimethylbenzenesulfonic acid	112	62	181			
8-Nitronaphthalene-1-sulfonic acid	115 (trihyd)	165	192	179		
8-Iodonaphthalene-1-sulfonic acid	115	140	187			
7-Chloronaphthalene-2-sulfonic acid	118 (anhyd)	87	176			
5-Nitronaphthalene-2-sulfonic acid	119	125	184			
Anthraquinone-1,7-disulfonic acid	120 (hyd)	232	238			
Phenanthrene-3-sulfonic acid	121 (monohyd)	111	190			222
Naphthalene-2-sulfonic acid	122	79	217	132	191	221
3,4-Diiodobenzenesulfonic acid	125	82	227			
Naphthalene-1,6-disulfonic acid	125 (anhyd)	129	298		81; 235	315
Azoxybenzene-3,3′-disulfonic acid	126	138	273 (di)			
2,5-Dimethyl-3-nitrobenzenesulfonic acid	128	61	173	144		136
5-Chloro-4-methyl-2-nitrobenzenesulfonic acid	128	99	181			
2,5-Dibromobenzenesulfonic acid	128 (anhyd)	71	195			
6-Hydroxynaphthalene-2-sulfonic acid	129 (hyd)		237	161	217	248
2-Methyl-4-nitrobenzenesulfonic acid	130	106	157			
2,4-Dinitrobenzeneusulfonic acid	130 (anhyd)	102	157			
2,4-Dimethyl-5-nitrobenzenesulfonic acid	132	98	187			
4-Chloronaphthalene-1-sulfonic acid	133	95	187	146		
1-Chloronaphthalene-2-sulfonic acid	133 (anhyd)	85	250	172		
2-Methyl-5-nitrobenzenesulfonic acid	134 (dihyd)	47	186	148		257
Phenanthrene-9-sulfonic acid	134 (hyd)	127	194			235
2-Carboxybenzenesulfonic acid (2-sulfobenzoic acid)	134 (anhyd)	79; 40	194; 154; 230	195	206	200
8-Nitronaphthalene-2-sulfonic acid	136 (hyd)	169	228	173		
3,4-Dicarboxybenzenesulfonic acid (phthalic acid-4-sulfonic acid)	140 (monohyd)	170	200			
2,5-Dimethyl-4-nitrobenzenesulfonic acid	140	75	198	131	144	
4-Methyl-2-nitrobenzenesulfonic acid	141 (anhyd)	99	170			
2,4-Dimethyl-3-nitrobenzenesulfonic acid	144 (anhyd)	96	172			

Sulfonic Acids (continued)

Name of compound	mp (°C)	Derivative mp (°C) Sulfonyl chloride	Sulfon-amide	Sulfon-anilide	Benzyl-thiuro-nium salt	4-Tolu-idine salt
2,5-Dimethyl-6-nitrobenzenesulfonic acid	145 (anhyd)	110	192	182		159
3-Carboxybenzenesulfonic acid	148 (anhyd)	20	170 (di)		163	226
Phenanthrene-2-sulfonic acid	150	156	254	158		291
Fluorene-2-sulfonic acid	155 (hyd)	164	213			
5-Carboxy-2-methylbenzenesulfonic acid	158 (anhyd)	59	185			
2,4-Diiodobenzenesulfonic acid	167 (anhyd)	78	230			
6-Hydroxynaphthalene-2-sulfonic acid	167 (anhyd)		237	161	217	248
2-Chloro-5-nitrobenzenesulfonic acid	169 (hyd)	90	186			
Azobenzene-4,4′-disulfonic acid	169 (anhyd)	222, 170 (di)	250 (di)			
4-Hydroxynaphthalene-1-sulfonic acid	170		167	200	103	196
Phenanthrene-9-sulfonic acid	174 (anhyd)	127	194			235
Phenanthrene-3-sulfonic acid	176 (anhyd)	111	190			222
2-Carboxy-5-methylbenzenesulfonic acid	190	59 (di)	185			
2,4-Dimethoxybenzenesulfonic acid	192	71	167			
D-Camphor-10-sulfonic acid	193	67	132	121; 88	210	
3-Bromocamphor-8-sulfonic acid	196	137	145			
2,5-Dimethyl-3-nitrobenzenesulfonic acid	200	61	173		144	136
Anthraquinone-1,6-disulfonic acid	217	198		227d		
Anthraquinone-1-sulfonic acid	218	218		216	191	
2-Nitrodiphenylamine-4-sulfonic acid	220d	157	162			
Naphthalene-1,5-disulfonic acid	245 (anhyd)	183 (di)	310, 340 (di)	249 (di)	257	332
4-Carboxybenzenesulfonic acid (4-sulfobenzoic acid)	260 (anhyd)	57 (di)	236 (di)	252 (di)	213	
2-Amino-5-methylbenzene-1,4-disulfonic acid	290d	156 (di)	257 (di)	197 (di)		
Anthraquinone-1,8-disulfonic acid	294	223	> 340	238		
Anthraquinone-1,5-disulfonic acid	311 (hyd)	270	246	270		
Quinoline-8-sulfonic acid	312	124	184			
Pyridine-3-sulfonic acid	357		111	145		

Sulfonyl Chlorides

Name of compound	mp (°C)	Derivative mp (°C)		
		Sulfonic acid	Sulfonamide	Sulfonanilide
2-Methylbenzenesulfonyl chloride (*o*-toluenesulfonyl chloride)	10	57	153	136
3-Methylbenzenesulfonyl chloride (*m*-toluenesulfonyl chloride)	12		108	96
Benzenesulfonyl chloride	14	66 (anhyd) 44 (monohyd)	156	110
3-Carboxybenzenesulfonyl chloride (*m*-sulfobenzoic acid)	20	98 (hyd) 148 (anhyd)	170 (di)	
2,5-Dimethylbenzenesulfonyl chloride	25	48 (anhyd) 86 (hyd)	148	
2-Phenylethane-1-sulfonyl chloride	33	91	122	77
2,4-Dimethylbenzenesulfonyl chloride	34	62 (hyd)	139	110
3,4-Dibromobenzenesulfonyl chloride	34	68 (anhyd)	175	
4-Methyl-3-nitrobenzenesulfonyl chloride	36	92	144	109
3-Chloro-4-methylbenzenesulfonyl chloride	38		134	96
2,5-Dichlorobenzenesulfonyl chloride	38	97	186	160
2,6-Dimethylbenzenesulfonyl chloride	39	98	96; 113	
2-Carboxybenzenesulfonyl chloride	40	69 (hyd) 134 (anhyd)	194 (anhyd) 154	195
Propane-1,3-disulfonyl chloride	45	92	169 (di)	129 (di)
2-Methyl-5-nitrobenzenesulfonyl chloride	47	134 (dihyd)	187	148
Indane-5-sulfonyl chloride	47	92	136	129
4-Chloro-2,5-dimethylbenzenesulfonyl chloride	50	100	185	155
3,4-Dimethylbenzenesulfonyl chloride	52	64	144	
4-Chlorobenzenesulfonyl chloride	53	69; 93	144	104
Hexane-1-sulfonyl chloride (cetyl sulfonyl chloride)	54	54	97	
2,4-Dichlorobenzenesulfonyl chloride	55	86	182	
4-Methylbenzene-1,3-disulfonyl chloride	56		191 (di)	189
2,4,6-Trimethylbenzenesulfonyl chloride	57	77	142	109
2-Chloro-5-methylbenzenesulfonyl chloride	56		156	230
4-Carboxybenzenesulfonyl chloride	57	94 (di) 260 (anhyd)	236 (di)	252 (di)
Tetralin-6-sulfonyl chloride	58		135	156
2-Carboxy-5-methylbenzenesulfonyl chloride	59	190; 158 (anhyd)	185	
2,4,5-Trimethylbenzenesulfonyl chloride	61	112	181	
2,5-Dimethyl-3-nitrobenzenesulfonyl chloride	61	200; 128	173	144
3,4-Dichloro-2,5-dimethylbenzenesulfonyl chloride	62		201	157
Benzene-1,3-disulfonyl chloride	63		229	150
3-Nitrobenzenesulfonyl chloride	64	48	167	126
2,4,6-Tribromobenzenesulfonyl chloride	64	64	228	222
4-Chloro-3-methylbenzenesulfonyl chloride	65		128	92
3-Bromocamphor-10-sulfonyl chloride	65	48	157	
2-Ethoxybenzenesulfonyl chloride	66		163	158
D-Camphor-10-sulfonyl chloride	67	193	132	121
Naphthalene-1-sulfonyl chloride	68	90	150	112; 152
2-Methylbenzenesulfonyl chloride (*o*-toluenesulfonyl chloride)	68	57	156	136
2-Nitrobenzenesulfonyl chloride	69	70; 85	193	115
4-Methylbenzenesulfonyl chloride (*p*-toluenesulfonyl chloride)	71	92; 105	138 (anhyd) 105 (dihyd)	103
2,4-Dimethoxybenzenesulfonyl chloride	71	192	167	

Sulfonyl Chlorides (continued)

Name of compound	mp (°C)	Derivative mp (°C)		
		Sulfonic acid	Sulfonamide	Sulfonanilide
3,6-Dichloro-2,5-dimethylbenzenesulfonyl chloride	71		165	171
2,5-Dibromobenzenesulfonyl chloride	71	128 (anhyd)	195	
3-Chloro-2-methylbenzenesulfonyl chloride	72	72	180	
2,5-Dimethyl-4-nitrobenzenesulfonyl chloride	75	140	198	131
4-Chloro-2-nitrobenzenesulfonyl chloride	75		164; 237	138
4-Methoxynaphthalene-2-sulfonyl chloride	75		157	145
4-Bromobenzenesulfonyl chloride	76	103; 90	166	119
2,4-Diiodobenzenesulfonyl chloride	78	167 (anhyd)	230	
4,5-Dimethoxybenzene-1,3-disulfonyl chloride	79		239	200
Napthalene-2-sulfonyl chloride	79	91; 122	217	132
2,4-Dibromobenzenesulfonyl chloride	79	110 (anhyd)	190	
1,2-Dimethylbenzene-3,5-disulfonyl chloride	79		239	200
2-Carboxybenzenesulfonyl chloride	79	69 (hyd) 134 (anhyd)	194 (anhyd) 154; 230	195
4-Nitrobenzenesulfonyl chloride	80	95; 111	180	171; 136
6-Methoxynaphthalene-1-sulfonyl chloride	81		150	178
2,5-Dimethylbenzene-1,3-disulfonyl chloride	81		295	174
4,6-Dichloro-2,5-dimethylbenzenesulfonyl chloride	81		150	175
1-Methylnaphthalene-4-sulfonyl chloride	81		177	158
3,4-Diiodobenzenesulfonyl chloride	82	125	227	
7-Methoxynaphthalene-2-sulfonyl chloride	83		220	121
4-Iodobenzenesulfonyl chloride	85		183	143
1-Chloronaphthalene-2-sulfonyl chloride	85	133 (anhyd)	250	172
4-Ethoxynaphthalene-2-sulfonyl chloride	85		183	144
4-Fluoronaphthalene-1-sulfonyl chloride	86	100 (hyd)	206	144
4-Methoxybenzene-1,3-disulfonyl chloride	86		240	209
7-Chloronaphthalene-2-sulfonyl chloride	87	118 (anhyd) 68 (tetrahyd)	176	
D-Camphor-3-sulfonyl chloride	88	77	143	124
2-Methylbenzene-1,3-disulfonyl chloride	88		260	162
4-Hydroxybenzene-1,3-disulfonyl chloride	89	100	239 (di)	205 (di)
2-Chloro-5-nitrobenzenesulfonyl chloride	90	169 (hyd)	186	
4-Nitrobenzylsulfonyl chloride	90	71	204	220
6-Methoxynaphthalene-2-sulfonyl chloride	93		189	120
Benzylsulfonyl chloride	93		105	102
3,5-Dimethylbenzenesulfonyl chloride	94		135	129
Ethanedisulfonyl chloride	95	104		69
3-Methylbenzene-1,5-disulfonyl chloride	95		216	153
4-Chloronaphthalene-1-sulfonyl chloride	95	133	187	146
5-Chloronaphthalene-1-sulfonyl chloride	95		226	138
7-Methylnaphthalene-1-sulfonyl chloride	96		196	164
2,4-Dimethyl-3-nitrobenzenesulfonyl chloride	96	144 (anhyd)	172	
2-Fluoronaphthalene-6-sulfonyl chloride	97 (anhyd)	105	133	129
2-Methylbenzene-1,4-disulfonyl chloride	98		224 (di)	178 (di)
2,4-Dimethyl-5-nitrobenzenesulfonyl chloride	98	132	187	
5-Chloro-4-methyl-2-nitrobenzenesulfonyl chloride	99	128	181	
4-Methyl-2-nitrobenzenesulfonyl chloride	99	141 (anhyd)	170	
4-Methoxynaphthalene-1-sulfonyl chloride	99		226	148
3-Nitrobenzylsulfonyl chloride	100	74	159	

Sulfonyl Chlorides (continued)

Name of compound	mp (°C)	Derivative mp (°C)		
		Sulfonic acid	Sulfonamide	Sulfonanilide
2,4-Dinitrobenzenesulfonyl chloride	102	108 (hyd) 130 (anhyd)	157	
4-Ethoxynaphthalene-1-sulfonyl chloride	103		170	180
7-Ethoxynaphthalene-2-sulfonyl chloride	103		142	153
4-Nitrodiphenylamine-2-sulfonyl chloride	104		173	164
3-Chlorobenzene-1,5-disulfonyl chloride	106	100	224	
(±)-Camphor-8-sulfonyl chloride	106	58	135	
2-Methyl-4-nitrobenzenesulfonyl chloride	106	130 (anhyd)	157	
6-Ethoxynaphthalene-2-sulfonyl chloride	108		183	153
2,5-Dimethyl-6-nitrobenzenesulfonyl chloride	110	145 (anhyd)	192	182
Phenanthrene-3-sulfonyl chloride	111	176 (anhyd) 121 (monohyd) 89 (dihyd)	190	
4-Methylbenzene-1,2-disulfonyl chloride	111		239	190
Acenaphthene-5-sulfonyl chloride	111		223	178
5-Nitronaphthalene-1-sulfonyl chloride	113		236	123
Acenaphthene-3-sulfonyl chloride	114	89	199	286
8-Iodonaphthalene-1-sulfonyl chloride	115		187	140
Biphenyl-4-sulfonyl chloride	115		230	125
2-Ethoxynaphthalene-1-sulfonyl chloride	116		158	187
6-Ethoxynaphthalene-1-sulfonyl chloride	118		154	195
2,6-Dimethyl-4-hydroxybenzene-1,3-disulfonyl chloride	119		208	207
5-Methoxynaphthalene-1-sulfonyl chloride	120		195	157
1-Iodonaphthalene-4-sulfonyl chloride	121		202	133
1-Nitronaphthalene-2-sulfonyl chloride	121	105	214	202
5-Ethoxynaphthalene-1-sulfonyl chloride	121		183	130
2-Methoxynaphthalene-1-sulfonyl chloride	121		159	197
5-Methylnaphthalene-2-sulfonyl chloride	122		189	250; 134
Anthracene-2-sulfonyl chloride	122		261	201
5-Fluoronaphthalene-1-sulfonyl chloride	123	106 (hyd)	197	
Quinoline-8-sulfonyl chloride	124	312	184	
Diphenylmethane-4,4′-disulfonyl chloride	124	59		178
4-Iodonaphthalene-1-sulfonyl chloride	124		206	136
5-Nitronaphthalene-2-sulfonyl chloride	125	119	184	
2,4,6-Trimethylbenzene-1,3-disulfonyl chloride	125		244 (di)	151 (di)
Phenanthrene-9-sulfonyl chloride	127	174 (anhyd) 134 (hyd)	194	
Naphthalene-1,6-disulfonyl chloride	129	125 (anhyd)	298	
2,4-Dimethylbenzene-1,5-disulfonyl chloride	129		249	196
2,4,5-Trimethoxybenzenesulfonyl chloride	130		76	170
8-Chloro-7-methoxynaphthalene-1-sulfonyl chloride	137		153	196
3-Bromocamphor-8-sulfonic chloride	137	196 (anhyd)	145	
Benzophenone-3,3′-disulfonyl chloride	138		157 (di)	178 (di)
Azoxybenzene-3,3′-disulfonyl chloride	138	126	273 (di)	
3-Methoxynaphthalene-2-sulfonyl chloride	138		113	174
Benzene-1,4-disulfonyl chloride	139		288	249
8-Iodonaphthalene-1-sulfonyl chloride	140	115	187	
Benzene-1,2-disulfonyl chloride	143		254	241
Pyridine-3-sulfonyl chloride	144	357	111	145

Sulfonyl Chlorides (continued)

Name of compound	mp (°C)	Derivative mp (°C)		
		Sulfonic acid	Sulfonamide	Sulfonanilide
4-Acetamidobenzenesulfonyl chloride	149		219	214
2-Amino-5-methylbenzene-1,3-disulfonyl chloride	156		257 (di)	192 (di)
Phenanthrene-2-sulfonyl chloride	156	150	254	158
2-Nitrodiphenylamine-4-sulfonyl chloride	157	220	162	
4-Carboxy-3-nitrobenzenesulfonyl chloride	160 (di)	111 (dihyd)	226 (di) 192 (mono)	
Fluorene-2-sulfonyl chloride	164	155 (anhyd)	213	
2,5-Dimethylbenzene-1,4-disulfonyl chloride	164		310	223
8-Nitronaphthalene-1-sulfonyl chloride	165	115 (trihyd)	191	179
Naphthalene-1,4-disulfonyl chloride	166		273	179
8-Nitronaphthalene-2-sulfonyl chloride	169	136 (hyd)	228	173
3,4-Dicarboxybenzenesulfonyl chloride	170	140 (monohyd)	200	
Azobenzene-4,4′-disulfonyl chloride	170	169 (anhyd)	250 (di)	
Naphthalene-1,5-disulfonyl chloride	183	245 (anhyd) (di)	310; 340 (di)	249 (di)
Benzene-1,3,5-trisulfonyl chloride	187	100 (tri)	315 (tri)	237 (tri)
Anthraquinone-2-sulfonyl chloride	197		261	193
Anthraquinone-1,6-disulfonyl chloride	198	217		228
Biphenyl-4,4′-disulfonyl chloride	203	72	300	
Anthraquinone-1-sulfonyl chloride	218	218		216
9,10-Dichloroanthracene-2-sulfonyl chloride	221		279	248
Azobenzene-4,4′-disulfonyl chloride	222	169 (anhyd)	250 (di)	
Anthraquinone-1,8-disulfonyl chloride	223	294 (di)	340	238
Anthracene-1,8-disulfonyl chloride	225		333	224
Anthraquinone-1,7-disulfonyl chloride	232	120 (hyd)		238
Anthracene-1,5-disulfonyl chloride	249		330 (di)	293 (di)
Anthraquinone-1,5-disulfonyl chloride	270	310 (hyd) (di)	246	270
2,4-Diaminobenzene-1,5-disulfonyl chloride	275		187	236
Anthraquinone-1,7-disulfonyl chloride	302	120 (hyd)		238

INDEX

Derivatization Procedures by Functional Groups

Compound	Derivative	Procedure Number	Page
Acid anhydride	Acid	2	308
	Amide	3b	309
	Anilide	4c	311
	4-Toluidide	4d	311
Acyl halide	Acid	2	308
	Amide	3b	309
	Anilide	4d	311
	4-Toluidide	4d	311
Alcohol	3,5-Dinitrobenzoate	10	317
	Hydrogen 3-nitrophthalate	11	318
	1-Naphthylurethane	8	315
	4-Nitrobenzoate	9	316
	Phenylurethane	8	315
Aldehyde	Dimedon derivative	16	323
	2,4-Dinitrophenylhydrazone	13	321
	4-Nitrophenylhydrazone	14	322
	Oxidation to an acid	17	324
	Oxime	15	323
	Phenylhydrazone	14	322
	Semicarbazone	12	321
Amide	Acetamide	19, then 20a	329, 333
	9-Acylamidoxanthene	18	328
	Benzamide	19, then 20b or 20c	329, 333 or 334
	4-Bromophenacyl ester	19, then 5	329, 312
	Hydrolysis to acid and amine	19	329
	4-Nitrobenzyl ester	19, then 5	329, 312
Amine—1° or 2°	Acetamide	20a	333
	Amine hydrochloride	26	338
	Benzamide	20b or 20c	333 or 334
	Benzenesulfonamide	21	335
	Phenylthiourea	22	336
	4-Toluenesulfonamide	21	335
Amine—3°	Amine hydrochloride	26	338
	Chloroplatinate	24	337
	Methyl iodide	23a	337
	Methyl 4-toluenesulfonate	23b	337
	Picrate	25	338
Amino acid	Acetamide	20d	335
	Benzamide	20d	335
	3,5-Dinitrobenzamide	29	344
	2,4-Dinitrophenyl derivative	30	344
	Phenylureido acid	28	343
	4-Toluenesulfonamide	27	343
Carbohydrate	Acetate	33	347
	4-Bromophenylhydrazone	32	347
	4-Nitrophenylhydrazone	32	347
	Phenylosazone	31	346

Derivatization Procedures by Functional Groups

Compound	Derivative	Procedure Number	Page
Carboxylic acid	Amide	3a	309
	Anilide	4a or 4b	310
	S-Benzylthiouronium salt	6	313
	4-Bromophenacyl ester	5	312
	Neutralization equivalent (NE)	1	307
	4-Nitrobenzyl ester	5	312
	Phenylhydrazide	7	313
	4-Toluidide	4a or 4b	310
Ester	Acid hydrazide	39	359
	Amide	34, then 3a	351, 309
	N-Benzylamide	38	358
	3,5-Dinitrobenzoate	37	358
	Saponification and hydrolysis	34	351
	Saponification equivalent (SE)	35	354
	4-Toluidide	36; 34, then 4a or 4b	356 351, 310
Ether—Alkyl	3,5-Dinitrobenzoate	40	359
Ether—Aromatic	Bromo derivative	44	364
	Nitro derivative	49	373
	Picrate	41	361
	Sulfonamide	42, then 43	362, 363
Halide—Alkyl	Alkylmercuric halide	45	366
	Alkyl β-naphthyl ether	47	367
	Alkyl β-naphthyl ether picrate	48	368
	S-Alkylthiouronium picrate	49	368
	Anilide	46	366
	N-Naphthylamide	46	366
Halide—Aromatic	Nitration	51	373
	Oxidation	50	370
	Sulfonamide	42, then 43	362, 363
Hydrocarbon—Aromatic	Aroylbenzoic acid	52	374
	Nitration	51	373
	Picrate	41	361
Ketone	2,4-Dinitrophenylhydrazone	13	321
	4-Nitrophenylhydrazone	14	322
	Oxime	15	323
	Phenylhydrazone	14	322
	Semicarbazone	12	321
Nitrile	Amide	53, then 3; 54	377, 309 378
	Anilide	53, then 4a or 4b	377, 310
	Benzamide	55, then 20b or 20c	378, 333 or 334
	Benzenesulfonamide	55, then 21	378, 335
	Hydrolysis of nitrile	53	377
	α-(Imidioylthio)acetic acid hydrochloride	56	379
	Phenylthiourea	55, then 22	378, 336
	Reduction of nitrile	55	378